Oxford Graduate Texts in Mathematics

Series Editors

R. Cohen   S. K. Donaldson   S. Hildebrandt
T. J. Lyons   M. J. Taylor

OXFORD GRADUATE TEXTS IN MATHEMATICS

Algebraic Geometry and Arithmetic Curves

Qing Liu

Professor
CNRS Laboratoire de Théorie des Nombres et d'Algorithmique Arithmétique
Université Bordeaux 1

Translated by

Reinie Erné

Institut de Recherche Mathématique de Rennes
Université Rennes 1

OXFORD
UNIVERSITY PRESS

OXFORD
UNIVERSITY PRESS

Great Clarendon Street, Oxford OX2 6DP

Oxford University Press is a department of the University of Oxford.
It furthers the University's objective of excellence in research, scholarship,
and education by publishing worldwide in

Oxford New York

Auckland   Cape Town Dar es Salaam   Hong Kong   Karachi
Kuala Lumpur Madrid Melbourne Mexico City   Nairobi
New Delhi Shanghai Taipei Toronto
With offices in
Argentina Austria Brazil Chile Czech Republic France Greece
Guatemala Hungary Italy Japan South Korea Poland Portugal
Singapore Switzerland Thailand Turkey Ukraine Vietnam

Oxford is a registered trade mark of Oxford University Press
in the UK and in certain other countries

Published in the United States
by Oxford University Press Inc., New York

ISBN 978-0-19-920249-2

Printed in the United Kingdom by
Lightning Source UK Ltd., Milton Keynes

To my mother

Preface

This book begins with an introduction to algebraic geometry in the language of schemes. Then, the general theory is illustrated through the study of arithmetic surfaces and the reduction of algebraic curves. The origin of this work is notes distributed to the participants of a course on arithmetic surfaces for graduate students. The aim of the course was to describe the foundation of the geometry of arithmetic surfaces as presented in [56] and [90], and the theory of stable reduction [26]. In spite of the importance of recent developments in these subjects and of their growing implications in number theory, unfortunately there does not exist any book in the literature that treats these subjects in a systematic manner, and at a level that is accessible to a student or to a mathematician who is not a specialist in the field. The aim of this book is therefore to gather together these results, now classical and indispensable in arithmetic geometry, in order to make them more easily accessible to a larger audience.

The first part of the book presents general aspects of the theory of schemes. It can be useful to a student of algebraic geometry, even if a thorough examination of the subjects treated in the second part is not required. Let us briefly present the contents of the first seven chapters that make up this first part. I believe that we cannot separate the learning of algebraic geometry from the study of commutative algebra. That is the reason why the book starts with a chapter on the tensor product, flatness, and formal completion. These notions will frequently recur throughout the book. In the second chapter, we begin with Hilbert's Nullstellensatz, in order to give an intuitive basis for the theory of schemes. Next, schemes and morphisms of schemes, as well as other basic notions, are defined. In Chapter 3, we study the fibered product of schemes and the fundamental concept of base change. We examine the behavior of algebraic varieties with respect to base change, before going on to proper morphisms and to projective morphisms. Chapter 4 treats local properties of schemes and of morphisms such as normality and smoothness. We conclude with an elementary proof of Zariski's Main Theorem. The global aspect of schemes is approached through the theory of coherent sheaves in Chapter 5. After studying coherent sheaves on projective schemes, we define the Čech cohomology of sheaves, and we look at some fundamental theorems such as Serre's finiteness theorem, the theorem of formal functions, and as an application, Zariski's connectedness principle. Chapter 6 studies particular coherent sheaves: the sheaf of differentials, and, in certain favorable cases (local complete intersections), the relative dualizing sheaf. At the end of that chapter, we present Grothendieck's duality theory. Chapter 7 starts with a rather general study of divisors, which is then restricted to the case of projective curves over a field. The theorem of Riemann–Roch, as well as Hurwitz's theorem, are proven with the help of duality theory. The chapter concludes with a detailed study of the Picard group of a not necessarily reduced projective curve over an algebraically closed field. The necessity of studying singular curves arises, among other things, from the fact that an arithmetic (hence regular) surface in general has fibers that are singular. These seven chapters can be used for a basic course on algebraic geometry.

The second part of the book is made up of three chapters. Chapter 8 begins with the study of blowing-ups. An intermediate section digresses towards commutative algebra by giving, often without proof, some principal results concerning Cohen–Macaulay, Nagata, and excellent rings. Next, we present the general aspects of fibered surfaces over a Dedekind ring and the theory of desingularization of surfaces. Chapter 9 studies intersection theory on an arithmetic surface, and its applications. In particular, we show the adjunction formula, the factorization theorem, Castelnuovo's criterion, and the existence of the minimal regular model. The last chapter treats the reduction theory of algebraic curves. After discussing general properties that essentially follow from the study of arithmetic surfaces, we treat the different types of reduction of elliptic curves in detail. The end of the chapter is devoted to stable curves and stable reduction. We describe the proof of the stable reduction theorem of Deligne–Mumford by Artin–Winters, and we give some concrete examples of computations of the stable reduction.

From the outset, the book was written with arithmetic geometry in mind. In particular, we almost never suppose that the base field is algebraically closed, nor of characteristic zero, nor even perfect. Likewise, for the arithmetic surfaces, in general we do not impose any hypothesis on the base (Dedekind) rings. In fact, it does not demand much effort to work in general conditions, and does not affect the presentation in an unreasonable way. The advantage is that it lets us acquire good reflexes right from the beginning.

As far as possible, the treatment is self-contained. The prerequisites for reading this book are therefore rather few. A good undergraduate student, and in any case a graduate student, possesses, in principle, the background necessary to begin reading the book. In addressing beginners, I have found it necessary to render concepts explicit with examples, and above all exercises. In this spirit, all sections end with a list of exercises. Some are simple applications of already proven results, others are statements of results which did not fit in the main text. All are sufficiently detailed to be solved with a minimum of effort. This book should therefore allow the reader to approach more specialized works such as [25] and [15] with more ease.

Acknowledgements

It is my great pleasure to thank Michel Matignon and Martin Taylor, who encouraged me to write up my lecture notes. Reinie Erné combined her linguistic and mathematical talents to translate this book from French to English. I thank her for her patience and generous help. I thank Philippe Cassou-Noguès, Reinie Erné, Arnaud Lacoume, Thierry Sageaux, Alain Thiéry, and especially Dino Lorenzini and Sylvain Maugeais for their careful reading of the manuscript. It is due to their vigilance that many errors were found and corrected. My thanks also go to Jean Fresnel, Dino Lorenzini, and Michel Matignon for mathematical discussions during the preparation of the book. I thank the Laboratoire de Mathématiques Pures de Bordeaux for providing me with such an agreeable environment for the greatest part of the writing of this book.

I cannot thank my friends and family enough for their constant encouragement and their understanding. I apologize for not being able to name them individually. Finally, special thanks to Isabelle, who supported me and who put up with me during the long period of writing. Without her sacrifices and the encouragement that she gave me in moments of doubt, this book would probably be far from being finished today.

Numbering style

The book is organized by chapter/section/subsection. Each section ends with a series of exercises. The statements and exercises are numbered within each section. References to results and definitions consist of the chapter number followed by the section number and the reference number within the section. The first one is omitted when the reference is to a result within the same chapter. Thus a reference to Proposition 2.7; 3.2.7; means, respectively, Section 2, Proposition 2.7 of the same chapter; and Chapter 3, Section 2, Proposition 2.7. On the contrary, we always refer to sections and subsections with the chapter number followed by the section number, and followed by the subsection number for subsections: e.g., Section 3.2 and Subsection 3.2.4.

Errata

Future errata will be listed at
http://www.math.u-bordeaux.fr/~liu/Book/errata.html

Q.L.
Bordeaux
June 2001

Preface to the paper-back edition

I am very much indebted to many people who have contributed comments and corrections since this book was first published in 2002. My hearty thanks to Robert Ash, Michael Brunnbauer, Oliver Dodane, Rémy Eupherte, Xander Faber, Anton Geraschenko, Yves Laszlo, Yogesh More, and especially to Lars Halvard Halle, Carlos Ivorra, Dino Lorenzini and René Schmidt.

The list of all changes made from the first edition is found on my web page
http://www.math.u-bordeaux.fr/ liu/Book/errata.html
This web page will also include the list of errata for the present edition.

Q.L.
Bordeaux
March 2006

Contents

1

Some topics in commutative algebra

Unless otherwise specified, all rings in this book will be supposed commutative and with unit.

In this chapter, we introduce some indispensable basic notions of commutative algebra such as the tensor product, localization, and flatness. Other, more elaborate notions will be dealt with later, as they are needed. We assume that the reader is familiar with linear algebra over a commutative ring, and with Noetherian rings and modules.

1.1 Tensor products

In the theory of schemes, the fibered product plays an important role (in particular the technique of base change). The corresponding notion in commutative algebra is the tensor product of modules over a ring.

1.1.1 Tensor product of modules

Definition 1.1. Let A be a commutative ring with unit. Let M, N be two A-modules. The *tensor product of M and N over A* is defined to be an A-module H, together with a bilinear map $\phi : M \times N \to H$ satisfying the following universal property:

> For every A-module L and every bilinear map $f : M \times N \to L$, there exists a unique homomorphism of A-modules $\tilde{f} : H \to L$ making the following diagram commutative:

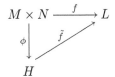

Proposition 1.2. *Let A be a ring, and let M, N be A-modules. The tensor product (H, ϕ) exists, and is unique up to isomorphism.*

Proof As the solution of a universal problem, the uniqueness is automatic, and its proof is standard. We give it here as an example. Let (H, ϕ) and (H', ϕ') be two solutions. By the universal property, ϕ and ϕ' factor respectively as $\phi = \tilde\phi \circ \phi'$ and $\phi' = \tilde\phi' \circ \phi$. It follows that $\phi = (\tilde\phi \circ \tilde\phi') \circ \phi$. As $\phi = \mathrm{Id} \circ \phi$, it follows from the uniqueness of the decomposition of ϕ that $(\tilde\phi \circ \tilde\phi') = \mathrm{Id}$. Thus we see that $\tilde\phi : H \to H'$ is an isomorphism.

Let us now show existence. Consider the free A-module $A^{(M \times N)}$ with basis $M \times N$. Let $\{e_{x,y}\}_{(x,y) \in M \times N}$ denote its canonical basis. Let L be the submodule of $A^{(M \times N)}$ generated by the elements having one of the following forms:

$$\begin{cases} e_{x_1+x_2,y} - e_{x_1,y} - e_{x_2,y} \\ e_{x,y_1+y_2} - e_{x,y_1} - e_{x,y_2} \\ e_{ax,y} - e_{x,ay}, \quad ae_{x,y} - e_{ax,y}, \quad a \in A. \end{cases}$$

Let $H = A^{(M \times N)}/L$, and $\phi : M \times N \to H$ be the map defined by $\phi(x, y) =$ the image of $e_{x,y}$ in H. One immediately verifies that the pair (H, ϕ) verifies the universal property mentioned above. □

Notation. We denote the tensor product of M and N by $(M \otimes_A N, \phi)$. In general, the map ϕ is omitted in the notation. For any $(x, y) \in M \times N$, we let $x \otimes y$ denote its image by ϕ. By the bilinearity of ϕ, we have $a(x \otimes y) = (ax) \otimes y = x \otimes (ay)$ for every $a \in A$.

Remark 1.3. By construction, $M \otimes_A N$ is generated as an A-module by its elements of the form $x \otimes y$. Thus every element of $M \otimes_A N$ can be written (though not in a unique manner) as a finite sum $\sum_i x_i \otimes y_i$, with $x_i \in M$ and $y_i \in N$. In general, an element of $M \otimes_A N$ cannot be written $x \otimes y$.

Example 1.4. Let $A = \mathbb{Z}$, $M = A/2A$, and $N = A/3A$. Then $M \otimes_A N = 0$. In fact, for every $(x, y) \in M \times N$, we have $x \otimes y = 3(x \otimes y) - 2(x \otimes y) = x \otimes (3y) - (2x) \otimes y = 0$.

Proposition 1.5. *Let A be a ring, and let M, N, M_i be A-modules. We have the following canonical isomorphisms of A-modules:*

(a) $M \otimes_A A \simeq M$;

(b) *(commutativity)* $M \otimes_A N \simeq N \otimes_A M$;

(c) *(associativity)* $(L \otimes_A M) \otimes_A N \simeq L \otimes_A (M \otimes_A N)$;

(d) *(distributivity)* $(\oplus_{i \in I} M_i) \otimes_A N \simeq \oplus_{i \in I}(M_i \otimes_A N)$.

Proof Everything follows from the universal property. Let us, for example, show (a) and (d).

(a) Let $\phi : M \times A \to M$ be the bilinear map defined by $(x, a) \mapsto ax$. For any bilinear map $f : M \times A \to L$, set $\tilde f : M \to L, x \mapsto f(x, 1)$. Then $f = \tilde f \circ \phi$, and $\tilde f$ is the unique linear map $M \to L$ having this property. Hence (M, ϕ) is the tensor product of M and A.

(d) Let $\phi : (\oplus_{i \in I} M_i) \times N \to \oplus_{i \in I}(M_i \otimes_A N)$ be the map defined by

$$\phi : \left(\sum_i x_i, y \right) \mapsto \sum_i (x_i \otimes y).$$

Let $f : (\oplus_{i \in I} M_i) \times N \to L$ be a bilinear map. For every $i \in I$, f induces a bilinear map $f_i : M_i \times N \to L$ which factors through $\widetilde{f}_i : M_i \otimes_A N \to L$. One verifies that f factors uniquely as $f = \widetilde{f} \circ \psi$, where $\psi : (\oplus_{i \in I} M_i) \times N \to (\oplus_{i \in I} M_i) \otimes N$ is the canonical map and $\widetilde{f} = \oplus_i \widetilde{f}_i$. Hence $\oplus_{i \in I}(M_i \otimes_A N)$ is the tensor product of $(\oplus_{i \in I} M_i)$ with N. $\qquad\square$

Corollary 1.6. *Let M be a free A-module with basis $\{e_i\}_{i \in I}$. Then every element of $M \otimes_A N$ can be written uniquely as a finite sum $\sum_i e_i \otimes y_i$, with $y_i \in N$. In particular, if A is a field and $\{e_i\}_{i \in I}$ (resp. $\{d_j\}_{j \in J}$) is a basis of M (resp. of N), then $\{e_i \otimes d_j\}_{(i,j) \in I \times J}$ is a basis of $M \otimes_A N$.*

Remark 1.7. The associativity of the tensor product allows us to define the tensor product $M_1 \otimes_A \cdots \otimes_A M_n$ of a finite number of A-modules. This tensor product has a universal property analogous to that of the tensor product of two modules, with the bilinear maps replaced by multilinear ones.

Definition 1.8. Let $u : M \to M'$, $v : N \to N'$ be linear maps of A-modules. By the universal property of the tensor product, there exists a unique A-linear map $u \otimes v : M \otimes_A N \to M' \otimes_A N'$ such that $(u \otimes v)(x \otimes y) = u(x) \otimes v(y)$. In fact, the map $g : M \times N \to M' \otimes_A N'$ defined by $g(x, y) = u(x) \otimes v(y)$ is clearly bilinear, and hence factors uniquely as $(u \otimes v) \circ \phi$, where ϕ is the canonical map $M \times N \to M \otimes N$. The map $u \otimes v$ is called the *tensor product of u and v*. The notation is justified by Exercise 1.2.

Let $\rho : A \to B$ be a ring homomorphism, and N a B-module. Then ρ induces, in a natural way, the structure of an A-module on N: for any $a \in A$ and $y \in N$, we set $a \cdot y = \rho(a)y$. We denote this A-module by $\rho_* N$, or simply by N.

Definition 1.9. Let M be an A-module. We can endow $M \otimes_A N$ with the structure of a B-module as follows. Let $b \in B$. Let $t_b : N \to N$ denote the multiplication by b, and for any $z \in M \otimes_A N$, set $b \cdot z := (\mathrm{Id}_M \otimes t_b)(z)$. One easily verifies that this defines the structure of a B-module. We denote the B-module $M \otimes_A B$ by $\rho^* M$. This is called the *extension of scalars of M by B*. By construction, we have $b(x \otimes y) = x \otimes (by)$ for every $b \in B$, $x \in M$, and $y \in N$.

Proposition 1.10. *Let $\rho : A \to B$ be a ring homomorphism, M an A-module, and let N, P be B-modules. Then there exists a canonical isomorphism of B-modules*

$$M \otimes_A (N \otimes_B P) \simeq (M \otimes_A N) \otimes_B P.$$

Proof Let us show that there exist A-linear maps

$$f : M \otimes_A (N \otimes_B P) \to (M \otimes_A N) \otimes_B P, \quad g : (M \otimes_A N) \otimes_B P \to M \otimes_A (N \otimes_B P)$$

such that for every $x \in M$, $y \in N$, and $z \in P$, we have $f(x \otimes (y \otimes z)) = (x \otimes y) \otimes z$ and $g((x \otimes y) \otimes z) = x \otimes (y \otimes z)$. This will imply that f is an isomorphism, with inverse g. The B-linearity of f follows from this identity.

Let us fix $x \in M$. Let $t_x : N \to M \otimes_A N$ denote the A-linear map defined by $t_x(y) = x \otimes y$. Consider the map $h : M \times (N \otimes_B P) \to (M \otimes_A N) \otimes_B P$ defined by $h(x, u) = (t_x \otimes \mathrm{Id}_P)(u)$. This map is A-bilinear, and hence induces an A-linear map f as desired. The construction of g is similar.   □

Taking $N = B$ in the proposition above, we obtain:

Corollary 1.11. *Let $\rho : A \to B$ be a ring homomorphism, let M be an A-module, and N a B-module. There exists a canonical isomorphism of B-modules*

$$(M \otimes_A B) \otimes_B N \simeq M \otimes_A N \qquad \text{(simplification by } B\text{)}.$$

1.1.2   Right-exactness of the tensor product

Let us recall that a *complex* of A-modules consists of a (finite or infinite) sequence of A-modules M_i, together with linear maps $f_i : M_i \to M_{i+1}$, such that $f_{i+1} \circ f_i = 0$. A complex is written more visually as

$$\cdots \to M_i \xrightarrow{f_i} M_{i+1} \xrightarrow{f_{i+1}} M_{i+2} \to \cdots .$$

The complex is called *exact* if $\mathrm{Ker}(f_{i+1}) = \mathrm{Im}(f_i)$ for all i. An exact complex is also called an *exact sequence*. For example, a sequence

$$0 \to M \xrightarrow{f} N \quad (\text{resp. } M \xrightarrow{f} N \to 0)$$

is exact if and only if f is injective (resp. surjective).

Let $f : N' \to N$ be a linear map of A-modules. For simplicity, for any A-module M, we denote the linear map $f \otimes \mathrm{Id}_M : N' \otimes_A M \to N \otimes_A M$ by f_M.

Proposition 1.12. *Let A be a ring, and let*

$$N' \xrightarrow{f} N \xrightarrow{g} N'' \to 0$$

be an exact sequence of A-modules. Then for any A-module M, the sequence

$$N' \otimes_A M \xrightarrow{f_M} N \otimes_A M \xrightarrow{g_M} N'' \otimes_A M \to 0$$

is exact.

Proof   The surjectivity of g_M follows from that of g (use Remark 1.3). It remains to show that $\mathrm{Ker}(g_M) = \mathrm{Im}\, f_M$; in other words, that the canonical homomorphism $\widetilde{g} : (N \otimes_A M)/(\mathrm{Im}\, f_M) \to N'' \otimes_A M$ is an isomorphism. Let $h : N'' \times M \to (N \otimes_A M)/(\mathrm{Im}\, f_M)$ be defined by $h(x, z) =$ the image of $y \otimes z$ in the quotient, where $y \in g^{-1}(x)$. The map h is well defined and moreover bilinear. It therefore induces a linear map $\widetilde{h} : N'' \otimes_A M \to (N \otimes_A M)/(\mathrm{Im}\, f_M)$, and it is easy to see that this is the inverse of $\widetilde{g}$.   □

Corollary 1.13. *Let A be a ring. Let N, M be A-modules, and $i : N' \to N$ a submodule of N. There exists a canonical isomorphism*

$$(N \otimes_A M)/(\operatorname{Im} i_M) \simeq (N/N') \otimes_A M.$$

In particular, if I is an ideal of A, then we have $M \otimes_A (A/I) \simeq M/(IM)$.

Proof It suffices to apply Proposition 1.12 to the exact sequence $N' \to N \to (N/N') \to 0$. If $N = A$ and $N' = I$, we also use Proposition 1.5(a). $\square$

1.1.3 Tensor product of algebras

An *A-algebra* is a commutative ring B with unit, endowed with a ring homomorphism $A \to B$. Let B and C be A-algebras. We can canonically endow $B \otimes_A C$ with the structure of an A-algebra, as follows. Let $g : B \times C \times B \times C \to B \otimes_A C$ be defined by $g(b, c, b', c') = (bb') \otimes (cc')$. This is a multilinear map, and hence factors through $\widetilde{g} : B \otimes_A C \otimes_A B \otimes_A C \to B \otimes_A C$ (see Remark 1.7). We can then define the product on $B \otimes_A C$ using the composition of $(B \otimes_A C) \times (B \otimes_A C) \to (B \otimes_A C) \otimes_A (B \otimes_A C)$ with $\widetilde{g}$. More precisely, we set $\sum_i (b_i \otimes c_i) \cdot \sum_j (b'_j \otimes c'_j) = \sum_{i,j} (b_i b'_j) \otimes (c_i c'_j)$. The point was to see that this is well defined.

We have homomorphisms of A-algebras $p_1 : B \to B \otimes_A C$, $p_2 : C \to B \otimes_A C$ defined by $p_1(b) = b \otimes 1$ and $p_2(c) = 1 \otimes c$. The following proposition can be verified immediately:

Proposition 1.14. *Let us keep the notation above. The triplet $(B \otimes_A C, p_1, p_2)$ satisfies the following universal property:*

> *For every A-algebra D, and for every pair of homomorphisms of A-algebras $q_1 : B \to D$, $q_2 : C \to D$, there exists a unique homomorphism of A-algebras $q : B \otimes_A C \to D$ such that $q_i = q \circ p_i$.*

Example 1.15. Let $A[T_1, \ldots, T_n]$ be the polynomial ring in n variables over A. Let B be an A-algebra. One easily verifies (either with the universal property given above or using Proposition 1.5(d)) that the homomorphism of B-algebras

$$A[T_1, \ldots, T_n] \otimes_A B \to B[T_1, \ldots, T_n]$$

defined by $(\sum_\nu a_\nu T^\nu) \otimes b \mapsto \sum_\nu (b a_\nu T^\nu)$ is an isomorphism. In particular, taking a polynomial ring for B, we obtain

$$A[T_1, \ldots, T_n] \otimes_A A[S_1, \ldots, S_m] \simeq A[T_1, \ldots, T_n, S_1, \ldots, S_m].$$

Exercises

1.1. Let $\{M_i\}_{i \in I}$, $\{N_j\}_{j \in J}$ be two families of modules over a ring A. Show that

$$(\oplus_{i \in I} M_i) \otimes_A (\oplus_{j \in J} N_j) \simeq \oplus_{(i,j) \in I \times J} (M_i \otimes_A N_j).$$

1.2. Show that there exists a unique A-linear map

$$f : \operatorname{Hom}_A(M, M') \otimes_A \operatorname{Hom}_A(N, N') \to \operatorname{Hom}_A(M \otimes_A N, M' \otimes_A N')$$

such that $f(u \otimes v) = u \otimes v$.

1.3. Let M, N be A-modules, and $i : M' \to M$, $j : N' \to N$ submodules of M and N, respectively. Then there exists a canonical isomorphism

$$(M/M') \otimes_A (N/N') \simeq (M \otimes_A N)/(\operatorname{Im} i_N + \operatorname{Im} j_M).$$

Show that $(\mathbb{Z}/n\mathbb{Z}) \otimes_{\mathbb{Z}} (\mathbb{Z}/m\mathbb{Z}) = \mathbb{Z}/l\mathbb{Z}$, where $l = \gcd(m, n)$.

1.4. Let M, N be A-modules, and let B, C be A-algebras.

  (a) If M and N are finitely generated over A, then so is $M \otimes_A N$.

  (b) If B and C are finitely generated over A, then so is $B \otimes_A C$.

  (c) Taking $A = \mathbb{Z}$, $M = B = \mathbb{Z}/2\mathbb{Z}$, and $N = C = \mathbb{Q}$, show that the converse of (a) and (b) is false.

1.5. Let $\rho : A \to B$ be a ring homomorphism, M an A-module, and N a B-module. Show that there exists a canonical isomorphism of A-modules

$$\operatorname{Hom}_A(M, \; \rho_* N) \simeq \operatorname{Hom}_B(\rho^* M, \; N).$$

1.6. Let $(N_i)_{i \in I}$ be a direct system of A-modules. Then for any A-module M, there exists a canonical isomorphism $\varinjlim(N_i \otimes_A M) \simeq (\varinjlim N_i) \otimes_A M$. (*Hint*: show that $\varinjlim(N_i \otimes_A M)$ verifies the universal property of the tensor product $\varinjlim(N_i) \otimes_A M$.)

1.7. Let B be an A-algebra, and let M, N be B-modules. Show that there exists a canonical surjective homomorphism $M \otimes_A N \to M \otimes_B N$.

1.2   Flatness

The flatness of a module over a ring is a property concerning extensions of scalars. In algebraic geometry, it assures a certain 'continuity' behavior. In this section, we study a few elementary aspects of flatness. We conclude the section with faithful flatness.

1.2.1   Left-exactness: flatness

Definition 2.1. An A-module M is called *flat* (over A) if for every injective homomorphism of A-modules $N \to N'$, $N \otimes_A M \to N' \otimes_A M$ is injective. An A-algebra B is called *flat* if B is flat over A for its A-module structure, and the canonical homomorphism $A \to B$ will be called a *flat homomorphism*.

It follows from Proposition 1.12 that if $0 \to N' \to N \to N'' \to 0$ is an exact sequence of A-modules, and M is flat over A, then

$$0 \to N' \otimes_A M \to N \otimes_A M \to N'' \otimes_A M \to 0$$

is an exact sequence.

Proposition 2.2. *Let A be a ring. We have the following properties:*

(a) *Every free A-module is flat.*

(b) *(Product) The tensor product of modules that are flat over A is flat over A.*

(c) *(Base change) Let B be an A-algebra. If M is flat over A, then $M \otimes_A B$ is flat over B.*

(d) *(Transitivity) Let B be a flat A-algebra. Then every B-module that is flat over B is flat over A.*

Proof These easily follow from the definition and the general properties of the tensor product (Proposition 1.5 and Corollary 1.11). □

Example 2.3. Let $A = \mathbb{Z}$, $n \geq 2$, and $M = A/nA$. Then M is not flat over A. In fact, if we tensor the canonical injection $nA \to A$ by M, the image of $nA \otimes_A M$ in $A \otimes_A M = M$ is equal to $nM = 0$, while $nA \otimes_A M \simeq A \otimes_A M = M \neq 0$.

Theorem 2.4. *Let M be an A-module. Then M is flat if and only if for every ideal I of A, the canonical homomorphism $I \otimes_A M \to IM$ is an isomorphism.*

Proof If M is flat over A, then for any ideal I of A, we can tensor the canonical injection $I \hookrightarrow A$ by M. This shows that $I \otimes_A M \to A \otimes_A M = M$ is injective. The image of this map is clearly IM, whence the isomorphism $I \otimes_A M \simeq IM$.

Conversely, let us suppose that we have this isomorphism for every ideal I, and let us show that M is flat. Let $N' \to N$ be an injective homomorphism of A-modules. We need to show that $N' \otimes M \to N \otimes M$ is injective.

Let us first suppose that N is free of finite rank, and let us show the injectivity by induction on the rank n of N. The case $n = 1$ follows from the hypothesis. Let us suppose that $n \geq 2$ and that the result holds for every free module of rank $< n$. The module N is a direct sum of two free submodules N_1 and N_2, different from N. Let $N_1' = N_1 \cap N'$, and let N_2' be the image of N' in $N_2 = N/N_1$. We then have the following commutative diagram

$$
\begin{array}{ccccc}
N_1' & \longrightarrow & N' & \longrightarrow & N_2' \\
\downarrow & & \downarrow & & \downarrow \\
N_1 & \longrightarrow & N & \longrightarrow & N_2
\end{array}
$$

whose horizontal lines are exact, and whose vertical arrows are injective. Tensoring by M gives the commutative diagram

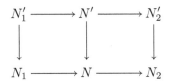

whose horizontal lines are still exact (Proposition 1.12). The map α is injective because N_1 is a direct factor of N (Proposition 1.5(d)), and β, γ are injective by the induction hypothesis applied to the N_i. It follows that $N' \otimes M \to N \otimes M$ is injective.

Let us now suppose N is free of arbitrary rank. Let N_0 be a direct factor of N of finite rank. By the above, the map $(N' \cap N_0) \otimes M \to N_0 \otimes M$ is injective. Hence so is $(N' \cap N_0) \otimes M \to N \otimes M$, since N_0 is a direct factor of N. Because for every $x \in N' \otimes M$ there exists an N_0 such that x is contained in the image of $(N' \cap N_0) \otimes M \to N' \otimes M$ (use Corollary 1.6), we see that $N' \otimes M \to N \otimes M$ is injective.

Let N now be an arbitrary A-module. There exist a free A-module L and a surjective homomorphism $p : L \to N$. Let us set $L' = p^{-1}(N')$. We have a commutative diagram

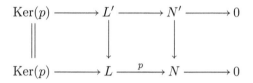

whose horizontal lines are exact, whence the commutative diagram

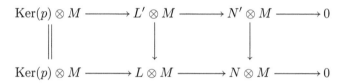

whose horizontal lines are also exact. As the middle vertical arrow is injective, this shows the injectivity of $N' \otimes M \to N \otimes M$.   □

Let A be an integral domain and M an A-module. An element $x \in M$ is called a *torsion* element if there exists a non-zero $a \in A$ such that $ax = 0$. We call M *torsion-free* (over A) if there is no non-zero torsion element in M. When A is a principal ideal domain, flatness can be expressed in a very simple way.

Corollary 2.5. *Let A be a principal ideal domain. An A-module M is flat if and only if it is torsion-free over A.*

Proof Let $I = aA \neq 0$ be an ideal of A. Let t_a (resp. u_a) denote multiplication by a in A (resp. in M). Then $t_a : A \to I$ is an isomorphism. We have the following commutative diagram:

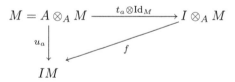

where f is the canonical homomorphism. Consequently, f is an isomorphism if and only if u_a is an isomorphism, which is equivalent to saying that $ax = 0$ for $x \in M$ implies that $x = 0$. Hence the corollary follows from Theorem 2.4.   □

Proposition 2.6. *Let A be a ring. Let $0 \to M' \to M \to M'' \to 0$ be an exact sequence of A-modules. Let us suppose that M'' is flat. Then the sequence $0 \to M' \otimes N \to M \otimes N \to M'' \otimes N \to 0$ is also exact for any A-module N.*

Proof It suffices to show the injectivity of $M' \otimes N \to M \otimes N$ (Proposition 2.12). Let us write N as the quotient of a free A-module L by a submodule K. We then have a commutative diagram of exact sequences:

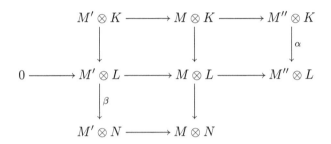

with α injective (M'' is flat) and β surjective. Some diagram chasing now immediately implies that $M' \otimes N \to M \otimes N$ is injective. □

1.2.2 Local nature of flatness

Let A be a ring. The intersection of all maximal ideals of A is denoted by $\mathrm{Rad}(A)$. If A is a *local ring*, i.e. a ring having only one maximal ideal $\mathfrak{m}$, then $\mathrm{Rad}(A) = \mathfrak{m}$. The following lemma is very useful:

Lemma 2.7 (Nakayama's lemma). *Let A be a ring, $I = \mathrm{Rad}(A)$ and M a finitely generated A-module such that $M = IM$. Then $M = 0$.*

Proof Let $\{x_1, \ldots, x_n\}$ be a system of generators of M. We may suppose n minimal. There exist $\alpha_i \in I$ such that $x_n = \sum \alpha_i x_i$. Hence $(1 - \alpha_n)x_n = \sum_{i<n} \alpha_i x_i$. As $1 - \alpha_n$ is invertible, and n is assumed to be minimal, it follows that $n = 1$ and $x_n = 0$. □

Remark 2.8. Let us note that following Corollary 1.13, the condition $M = IM$ is equivalent to $M \otimes_A A/I = 0$.

Corollary 2.9. *Let A be a ring, $I = \mathrm{Rad}(A)$, M a finitely generated A-module, and N a submodule of M. Suppose that $M \subseteq N + IM$. Then $M = N$.*

Proof In fact, M/N is a finitely generated A-module, and $(M/N) = I(M/N)$. Hence $M/N = 0$. □

Some properties of the localization of modules

Let A be a ring, S a *multiplicative subset of A* (i.e., a subset of A that is stable for multiplication, and which contains the unit element of A), and M an A-module. We define the *localization $S^{-1}M$ of M with respect to S* as follows. It is the set of elements written (formally) m/s, with $m \in M$ and $s \in S$, modulo the equivalence relation $m/s \sim m'/s' \iff$ there exists an $s'' \in S$ such that

$s''(s'm - sm') = 0$. We endow $S^{-1}M$ with the structure of a commutative group by setting

$$m/s + m'/s' = (s'm + sm')/(ss').$$

Note that if $0 \in S$, then $S^{-1}M = 0$. If $M = A$, we define a ring structure on $S^{-1}A$ by setting

$$(a/s) \cdot (a'/s') = (aa')/(ss').$$

The canonical homomorphism $\rho : A \to S^{-1}A$ defined by $\rho(a) = a/1$ makes $S^{-1}A$ into an A-algebra. Moreover, the correspondences $I \mapsto I(S^{-1}A)$, $J \mapsto \rho^{-1}(J)$ establish a bijection between those prime ideals of A that are disjoint from S and the prime ideals of $S^{-1}A$. Let $\phi : A \to B$ be a ring homomorphism. Then $\phi(S) \subseteq B^*$ (invertible elements of B) if and only if ϕ factors through some A-algebra homomorphism $S^{-1}A \to B$.

For any A-module M, we have the structure of an $S^{-1}A$-module on $S^{-1}M$ by setting

$$(a/s) \cdot (m/s') = (am)/(ss').$$

For $f \in A$, we let M_f denote the localization of M with respect to the multiplicative set $\{f^n \mid n \geq 1\} \cup \{1\}$. For a prime ideal $\mathfrak{p}$ of A, the localization of M with respect to the multiplicative set $A \setminus \mathfrak{p}$ is denoted $M_{\mathfrak{p}}$.

Lemma 2.10. *Let A be a ring, S a multiplicative subset of A, and M an A-module. There exists a canonical isomorphism of $S^{-1}A$-modules $M \otimes_A S^{-1} A \simeq S^{-1}M$.*

Proof We have a canonical homomorphism of A-modules $M \otimes_A S^{-1}A \to S^{-1}M$ defined by $m \otimes (a/s) \mapsto (am)/s$. This is an isomorphism because it has an inverse given by $m/s \mapsto m \otimes (1/s)$ (this map is well defined). Moreover, this homomorphism is clearly compatible with the structure of $S^{-1}A$-modules. □

Corollary 2.11. *For any multiplicative subset S of A, the canonical homomorphism $A \to S^{-1}A$ is flat.*

Proof Let N be a sub-A-module of M. It is clear, by construction, that $S^{-1}N$ is a submodule of $S^{-1}M$. Hence $N \otimes_A S^{-1}A \to M \otimes_A S^{-1}A$ is injective. □

Lemma 2.12. *Let M be an A-module. Then $M = 0$ if and only if $M_{\mathfrak{m}} = 0$ for every maximal ideal $\mathfrak{m}$ of A.*

Proof Let $x \in M$. Let us consider the ideal $I = \{a \in A \mid ax = 0\}$. If $I \neq A$, there exists a maximal ideal $\mathfrak{m}$ of A such that $I \subseteq \mathfrak{m}$. As $M_{\mathfrak{m}} = 0$, there exists an $s \in A \setminus \mathfrak{m}$ such that $sx = 0$. Hence $s \in I$, which contradicts the assumption that $I \subseteq \mathfrak{m}$. Consequently, $I = A$ and $1 \in I$, and hence $x = 0$. □

Proposition 2.13. *Let M be an A-module. The following properties are equivalent:*

(i) *M is flat over A.*

(ii) *$M_{\mathfrak{p}}$ is flat over $A_{\mathfrak{p}}$ for every prime ideal $\mathfrak{p}$ of A.*

(iii) *$M_{\mathfrak{m}}$ is flat over $A_{\mathfrak{m}}$ for every maximal ideal $\mathfrak{m}$ of A.*

Proof (i) $\implies$ (ii) follows from Lemma 2.10 and Proposition 2.2(c). As (ii) trivially implies (iii), it remains to show that (iii) implies (i). Let $N' \to N$ be an injective homomorphism. Let L denote the kernel of $N' \otimes_A M \to N \otimes_A M$. We have an exact sequence

$$0 \to L \to N' \otimes_A M \to N \otimes_A M. \tag{2.1}$$

For any maximal ideal $\mathfrak{m}$ of A, we have an exact sequence

$$0 \to L \otimes_A A_{\mathfrak{m}} \to (N' \otimes_A M) \otimes_A A_{\mathfrak{m}} \to (N \otimes_A M) \otimes_A A_{\mathfrak{m}}$$

(by virtue of Corollary 2.11). As $(N' \otimes_A M) \otimes_A A_{\mathfrak{m}} = (N'_{\mathfrak{m}}) \otimes_{A_{\mathfrak{m}}} M_{\mathfrak{m}}$, and $N'_{\mathfrak{m}} \to N_{\mathfrak{m}}$ is injective, we have $L_{\mathfrak{m}} = L \otimes_A A_{\mathfrak{m}} = 0$. It follows by Lemma 2.12 that $L = 0$. Hence M is flat over A. $\square$

Let us recall that a *Dedekind domain* is a Noetherian integral domain A whose localizations $A_{\mathfrak{p}}$ at the prime ideals $\mathfrak{p}$ are principal ideal domains (we will see the equivalence with the usual definition in Subsection 4.1.1). The following corollary is an immediate consequence of the proposition above and Corollary 2.5.

Corollary 2.14. *Let A be a Dedekind domain. An A-module is flat if and only if it is torsion-free over A. In particular, every injective ring homomorphism $A \to B$ with B an integral domain is flat.*

Corollary 2.15. *Let $\rho : A \to B$ be a ring homomorphism. The following properties are equivalent:*

(i) *$A \to B$ is flat.*

(ii) *For every prime ideal $\mathfrak{q}$ of B, $B_{\mathfrak{q}}$ is flat over $A_{\mathfrak{p}}$, where $\mathfrak{p} = \rho^{-1}(\mathfrak{q})$.*

(iii) *For every maximal ideal $\mathfrak{q}$ of B, $B_{\mathfrak{q}}$ is flat over $A_{\mathfrak{p}}$.*

Proof (i) implies (ii): it is easy to see that $B_{\mathfrak{q}}$ is a localization of (and hence is flat over) $B \otimes_A A_{\mathfrak{p}}$. Now, the latter is flat over $A_{\mathfrak{p}}$ since B is flat over A. Hence $B_{\mathfrak{q}}$ is flat over $A_{\mathfrak{p}}$ by the transitivity of flatness (Proposition 2.2(d)).

The proof of (iii) $\implies$ (i) is analogous to that of Proposition 2.13: one can replace M by B in exact sequence (2.1); then L is a B-module, and $L \otimes_B B_{\mathfrak{q}} = 0$ for every maximal ideal $\mathfrak{q}$ of B. Hence $L = 0$ and B is flat over A. $\square$

Let us conclude with the following theorem, which is a partial converse to Proposition 2.2(a). This converse is false in general, as is shown by the example $A = \mathbb{Z}_{p\mathbb{Z}}$ and $M = \mathrm{Frac}(A)$.

Theorem 2.16. *Let M be a finitely generated flat module over a local ring A. Then M is free over A.*

Proof Let $\mathfrak{m}$ be the maximal ideal of A, and $k = A/\mathfrak{m}$ the residue field of A. Then $M/\mathfrak{m}M = M \otimes_A k$ is a vector space over k. Let $\{x_1, \ldots, x_n\}$ be a set of elements of M whose images in $M/\mathfrak{m}M$ form a free set over k. Let us show that the set is free over A. Suppose that $\sum_i a_i x_i = 0$ with $a_i \in A$. Consider the linear map $f : A^n \to A$ defined by $f(b_1, \ldots, b_n) = \sum_i a_i b_i$. We have an exact sequence

$$\mathrm{Ker}(f) \to A^n \xrightarrow{f} A.$$

This gives an exact sequence when we tensor by M (here we use the flatness assumption on M):

$$\mathrm{Ker}(f) \otimes_A M \to M^n \xrightarrow{f_M} M,$$

where f_M is defined by $f_M(y_1, \ldots, y_n) = \sum_i a_i y_i$. Hence $(x_1, \ldots, x_n) \in \mathrm{Ker}(f) \otimes M$. There exist $r \in \mathbb{N}$ and elements $b_j = (b_{1j}, \ldots, b_{nj}) \in \mathrm{Ker}(f)$, $y_j \in M$, $1 \le j \le r$, such that $(x_1, \ldots, x_n) = \sum_j b_j \otimes y_j$. At least one of the coordinates b_{ij} is not an element of $\mathfrak{m}$. We can, for example, suppose that $b_{11} \notin \mathfrak{m}$ (hence it is invertible). From the equality $\sum_i b_{i1} a_i = 0$, we deduce that

$$a_1 + a_2 c_2 + \cdots + a_n c_n = 0, \quad \text{where } c_i = b_{i1} b_{11}^{-1}.$$

In particular, if $n = 1$, then $a_1 = 0$. If $n \ge 2$, we have

$$a_2(x_2 - c_2 x_1) + \cdots + a_n(x_n - c_n x_1) = 0.$$

A reasoning using induction on n shows that $\{x_i\}_i$ is free.

Now let $\{x_1, \ldots, x_n\}$ be a set of elements of M whose images in $M/\mathfrak{m}M$ form a basis over k. Let N be the submodule of M generated by the x_i. Then the canonical homomorphism $N \otimes k \to M \otimes k$ is surjective. Hence $(M/N) \otimes k = 0$. By Nakayama's lemma, $M/N = 0$. Hence the x_i form a basis of M.   □

1.2.3 Faithful flatness

Proposition 2.17. *Let M be a flat A-module. Then the following properties are equivalent:*

(i) *$M \ne \mathfrak{m}M$ for every maximal ideal $\mathfrak{m}$ of A.*

(ii) *Let N be an A-module. If $M \otimes_A N = 0$, then $N = 0$.*

(iii) *Let $f : N_1 \to N_2$ be a homomorphism of A-modules. If $f_M : N_1 \otimes M \to N_2 \otimes M$ is an isomorphism, then so is f.*

Proof (i) implies (ii): it suffices to show that for every finitely generated sub-module N_0 of N, we have $N_0 = 0$. As it is, for every submodule N_0 of N, we have $N_0 \otimes M \hookrightarrow N \otimes M$, so $N_0 \otimes M = 0$. We may assume that N is itself finitely generated. Moreover, by virtue of Lemma 2.12, we may assume that A is local with maximal ideal $\mathfrak{m}$. By tensoring with $k := A/\mathfrak{m}$, we obtain $M/\mathfrak{m}M \otimes_k N/\mathfrak{m}N = 0$. It follows that $N/\mathfrak{m}N = 0$ (Corollary 1.6). Hence $N = 0$, by Nakayama's lemma.

(ii) implies (iii): we have $\mathrm{Coker}(f_M) = \mathrm{Coker}(f) \otimes M$, and since M is flat, $\mathrm{Ker}(f_M) = \mathrm{Ker}(f) \otimes M$. Hence if f_M is an isomorphism, then so is f.

Finally, (iii) implies (i) since if $M = \mathfrak{m}M$, then the constant map $0 \to A/\mathfrak{m}$ is an isomorphism after tensoring with M, so $A/\mathfrak{m} = 0$. This is impossible.   □

Definition 2.18. Let M be a flat module over a ring A. We say that M is *faithfully flat over A* if it verifies one of the properties of the proposition above. Let $f : A \to B$ be a ring homomorphism. We say that B is faithfully flat over A if it is faithfully flat as an A-module. We will also say that f is faithfully flat.

Remark 2.19. One can immediately verify that Proposition 2.2 remains true when we replace 'flat' by 'faithfully flat' and take only non-zero modules.

Corollary 2.20. *Let $f : A \to B$ be a flat ring homomorphism. The following properties are equivalent:*

(i) *f is faithfully flat.*

(ii) *For every prime ideal $\mathfrak{p}$ of A, there exists a prime ideal $\mathfrak{q}$ of B such that $f^{-1}(\mathfrak{q}) = \mathfrak{p}$.*

(iii) *For every maximal ideal $\mathfrak{m}$ of A, there exists a maximal ideal $\mathfrak{n}$ of B such that $f^{-1}(\mathfrak{n}) = \mathfrak{m}$.*

Proof Let us show the only non-trivial implication, (i) $\Longrightarrow$ (ii). Let us first note that f is necessarily injective, because $\mathrm{Ker}(f) \otimes_A B \simeq \mathrm{Ker}(f)B = 0$. Since $A/\mathfrak{p} \to B/\mathfrak{p}B$ is faithfully flat (see Remark 2.19), we may assume that $\mathfrak{p} = 0$, and that A is an integral domain. Let $\rho : B \to B\otimes_A \mathrm{Frac}(A)$ be the canonical map. Let $\mathfrak{m}$ be a maximal ideal of $B \otimes_A \mathrm{Frac}(A)$. Then $\mathfrak{q} := \rho^{-1}(\mathfrak{m})$ is a prime ideal of B. Since $\rho \circ f : A \to B \otimes_A \mathrm{Frac}(A)$ factorizes into $A \to \mathrm{Frac}(A) \to B \otimes_A \mathrm{Frac}(A)$, and the inverse image of $\mathfrak{m}$ in $\mathrm{Frac}(A)$ is zero, we have $f^{-1}(\mathfrak{q}) = (\rho \circ f)^{-1}(\mathfrak{m}) = 0$. $\square$

Exercises

2.1. Let M be an A-module. We call the ideal $\{a \in A \mid aM = 0\}$ of A the *annihilator of M*, and we denote it by $\mathrm{Ann}(M)$. Let $I \subseteq \mathrm{Ann}(M)$ be an ideal.

(a) Show that M is endowed, in a natural way, with the structure of an A/I-module, and that $M \simeq M \otimes_A A/I$.

(b) Let N be another A-module such that $I \subseteq \mathrm{Ann}(N)$. Show that the canonical homomorphism $M \otimes_A N \to M \otimes_{A/I} N$ is an isomorphism.

2.2. Let $\rho : A \to B$ be a ring homomorphism, S a multiplicative subset of A, and $T = \rho(S)$. Show that T is a multiplicative subset of B, and that $T^{-1}B \simeq B \otimes_A S^{-1}A$ as A-algebras.

2.3. Show that Nakayama's lemma is false for modules M that are not finitely generated.

2.4. Let I be a finitely generated ideal of A. Show that the following properties are equivalent:

(i) A/I is flat over A;

(ii) $I = I^2$;

(iii) there exists an $e \in A$ such that $e^2 = e$ and $I = eA$.

2.5. Let A be an integral domain. Show that every A-module is flat if and only if A is a field.

2.6. Let B be a flat A-algebra.

(a) Show that for any *finite* family $\{I_\lambda\}_{\lambda \in \Lambda}$ of ideals of A, we have $\cap_{\lambda \in \Lambda}(I_\lambda B) = (\cap_{\lambda \in \Lambda} I_\lambda)B$.

(b) Let us suppose that B is faithfully flat. Show that $IB \cap A = I$ for any ideal I of A.

(c) Let k be a field, $A = k[t, s]$, and $C = A[z]/(tz - s)$. By considering the ideals tA and sA, show that C is not flat over A.

2.7. Give an example of a finitely generated flat module that is not free (over a suitable ring A).

2.8. Let A be a Noetherian ring, M a finitely generated A-module, and N an A-module. Let B be a flat A-algebra. Let us consider the canonical homomorphism

$$\rho : \operatorname{Hom}_A(M, N) \otimes_A B \to \operatorname{Hom}_B(M \otimes_A B, N \otimes_A B).$$

(a) Show that ρ is an isomorphism if M is free of finite rank.

(b) Let $0 \to K \to L \to M \to 0$ be an exact sequence of A-modules. Show that we have a canonical exact sequence

$$0 \to \operatorname{Hom}_A(M, N) \to \operatorname{Hom}_A(L, N) \to \operatorname{Hom}_A(K, N).$$

(c) By taking for L a free module of finite rank, show that ρ is injective. By applying the injectivity to K, show that ρ is an isomorphism.

2.9. Let A be an integral domain, and K its field of fractions. Let M be a finitely generated sub-A-module of K. Show that M is flat if and only if it is locally free of rank 1 (i.e., $M_{\mathfrak{p}}$ is free of rank 1 over $A_{\mathfrak{p}}$ for every prime ideal $\mathfrak{p}$ of A).

2.10. Let A be an integral domain, and B its integral closure in the field of fractions $\operatorname{Frac}(A)$. Suppose that B is a finitely generated A-module. Show that B is flat over A if and only if $B = A$. One can show that this result is true without the assumption of finiteness of B over A.

2.11. Let M be a flat A-module, and let $\rho : A \to B$ be a ring homomorphism. Show that the following properties are true.

(a) If A is an integral domain, then M is torsion-free.

(b) The canonical homomorphism $M \to M \otimes_A B$ is injective if and only if $\operatorname{Ker}(\rho)$ is contained in $\operatorname{Ann}(M)$.

2.12. Let B be a flat algebra over a Dedekind domain A. Let $f \in B$ be such that for every maximal ideal $\mathfrak{m}$ of A, the image of f in $B/\mathfrak{m}B$ is not a zero divisor. Show that B/fB is still flat over A. See also Lemma 4.3.16.

2.13. Let M be a faithfully flat A-module. Let $N' \to N \to N''$ be a sequence of A-modules. Show that it is exact if and only if the sequence $N' \otimes M \to N \otimes M \to N'' \otimes M$ is exact.

2.14. Let $A \to B$ be a ring homomorphism, and let J be an ideal of B such that B/J is flat over A. Show that for any ideal I of A, we have $(IB) \cap J = IJ$ (tensor the injection $I \to A$ by B/J).

2.15. Let $0 \to M' \to M \to M'' \to 0$ be an exact sequence of modules over a ring A. Let us suppose that M'' is flat. Show that M' is flat if and only if M is flat.

2.16. Let $(A, \mathfrak{m})$ be a Noetherian local ring, and

$$C^\bullet : 0 \to M' \to M \to M'' \to 0$$

a complex of finitely generated flat A-modules. Show that if there exists an ideal $I \subseteq \mathfrak{m}$ such that $C^\bullet \otimes_A A/I$ is exact, then $C^\bullet$ is exact.

2.17. Let M be a finitely generated flat A-module. Show that it is faithfully flat if and only if $\mathrm{Ann}(M) = 0$.

2.18. Let B be an A-algebra, and let E be a faithfully flat B-module. Show that E is flat over A if and only if B is flat over A.

2.19. Let $f : A \to B$ be faithfully flat ring homomorphism.

(a) Show that f is injective and that $I \to I \otimes_A B$ is injective for every ideal I of A.

(b) Let $N = \mathrm{Coker}(f)$ be the cokernel of f. Let I be an ideal of A. Using the commutative diagram

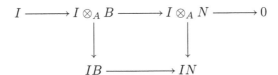

and Exercise 2.6(b), show that $I \otimes_A N \to IN$ is injective, and hence N is a flat A-module.

(c) Show that for any A-module M, the canonical map $M \to M \otimes_A B$ is injective.

1.3 Formal completion

1.3.1 Inverse limits and completions

Let us first recall some notions and properties of topological groups. An *(Abelian) topological group* is an Abelian group G endowed with the structure of a topological space for which the homomorphism $G \times G \to G$ defined by $(x, y) \mapsto x - y$ is continuous. Such a structure is entirely determined by giving a fundamental system $\mathcal{V}$ of neighborhoods of 0 such that every element V of $\mathcal{V}$ contains the difference $V_1 - V_2$ of two elements of $\mathcal{V}$.

A (descending) *filtration* $(G_n)_n$ (i.e., a descending chain of subgroups $(G_n)_n$ of G) defines a unique structure of topological group on G for which the G_n form a fundamental system of neighborhoods of 0. In this section we are essentially interested in topologies of this type. For this topology, G is *separated* (i.e., Hausdorff) if and only if $\cap_n G_n = \{0\}$. Two filtrations $(G_n)_n$, $(G'_n)_n$ define the same topology on G if and only if for every n, there exists an m such that $G'_m \subseteq G_n$, and vice versa.

Let us note that a group homomorphism $f : G \to H$ between two topological groups is continuous if and only if for every neighborhood V of 0 in H, $f^{-1}(V)$ is a neighborhood of 0 in G.

A *(commutative) topological ring* is a ring A endowed with a topology for which the maps $A \times A \to A$ defined by $(x, y) \mapsto x - y$ and $(x, y) \mapsto xy$ are continuous. A *topological A-module* M is an A-module whose additive structure is a topological group, and for which the map $(a, x) \mapsto ax$ from $A \times M$ to M is continuous. Let A be a commutative ring, and I an ideal. The ideals I^n of A define a topology on A called the *I-adic topology*. Let M be an A-module; then the filtration $(I^n M)_n$ defines a structure of topological A-module, called the *I-adic topology on M*.

Let G be a topological group defined by a filtration $(G_n)_n$. A sequence $(x_m)_m$ of elements of G is called a *Cauchy sequence* if for every n there exists an m_0 such that $x_m - x_{m_0} \in G_n$ for every $m \geq m_0$. The topological group G is *complete* if every Cauchy sequence has a limit in G. One way to construct complete groups is to construct inverse limits.

An *inverse system* (of sets) consists of a collection of sets $(A_n)_{n \geq 0}$ and maps $\pi_n : A_{n+1} \to A_n$ for every n. The *inverse limit of the* $(A_n)_n$ is the set

$$\varprojlim_n A_n := \{(a_n)_n \in \prod_n A_n \mid a_n = \pi_n(a_{n+1}) \text{ for all } n\}.$$

For every m, the projection onto the mth coordinate defines a canonical map $p_m : \varprojlim_n A_n \to A_m$.

Let B be a set, and let $f_n : B \to A_n$ be maps such that $f_n = \pi_n \circ f_{n+1}$ for every n. Then there exists a unique map $f : B \to \varprojlim A_n$ such that $f_m = p_m \circ f$. This universal property characterizes the inverse limit.

We define inverse limits of groups, of rings, and of modules in the same way, imposing that the transition maps π_n be compatible with the group, ring, or module structure. The inverse limits are then groups, rings, or modules.

A *homomorphism* of inverse systems $(A_n)_n \to (B_n)_n$ of Abelian groups consists of homomorphisms $A_n \to B_n$ that are compatible with the projection maps $A_{n+1} \to A_n$ and $B_{n+1} \to B_n$. This canonically induces a group homomorphism $\varprojlim A_n \to \varprojlim B_n$. A sequence of homomorphisms

$$0 \to (A_n)_n \to (B_n)_n \to (C_n)_n \to 0$$

of inverse systems of Abelian groups is called an *exact sequence* if $0 \to A_n \to B_n \to C_n \to 0$ is exact for all n.

Lemma 3.1. *Let* $0 \to (A_n)_n \to (B_n)_n \to (C_n)_n \to 0$ *be an exact sequence of inverse systems of Abelian groups. Then* $0 \to \varprojlim A_n \to \varprojlim B_n \to \varprojlim C_n$ *is exact. Moreover, the last homomorphism is surjective if there exists an* n_0 *such that* $A_{n+1} \to A_n$ *is surjective for every* $n \geq n_0$ *(see also Exercise 3.15).*

Proof Let us check the surjectivity of $\varprojlim B_n \to \varprojlim C_n$; the rest follows immediately from the definition. As the inverse limit of a system does not depend on its first terms, we may assume that $n_0 = 0$. Let $\pi_n : B_{n+1} \to B_n$ denote the transition homomorphism. Let $(c_n)_n \in \varprojlim C_n$. Let $b_n \in B_n$ be an arbitrary preimage of c_n. Identifying A_n with a submodule of B_n, we have $a_n := \pi_n(b_{n+1}) - b_n \in A_n$. Hence there exists an $a_{n+1} \in A_{n+1}$ such that $\pi_n(a_{n+1}) = a_n$. It follows that $\pi_n(b_{n+1} - a_{n+1}) = b_n$. Thus we can modify the b_n one after the other to obtain $\pi_n(b_{n+1}) = b_n$ for all n. Consequently, $(b_n)_n \in \varprojlim B_n$, and its image in $\varprojlim C_n$ is $(c_n)_n$.   □

Let G be a topological group defined by a filtration $(G_n)_n$. We then have a natural inverse system $(G/G_n)_n$. Consider $\hat{G} := \varprojlim(G/G_n)$. Let

$$\hat{G}_n := \{(a_m)_m \in \hat{G} \mid a_m = 0 \text{ for every } m \leq n\}.$$

This defines a filtration $(\hat{G}_n)_n$ on $\hat{G}$ and makes the latter into a topological group. We have a natural homomorphism $\pi : G \to \hat{G}$ defined by $\pi(g) = (g_n)_n$, where g_n is the canonical image of g in G/G_n. As $\pi^{-1}(\hat{G}_n) = G_n$, π is continuous.

Let G be an Abelian topological group. The *completion of* G is a complete separated Abelian topological group K together with a continuous homomorphism $\phi : G \to K$ such that every continuous homomorphism from G to a complete separated Abelian topological group factors uniquely through ϕ. We will in general omit the term 'separated' and use 'complete' to mean 'separated and complete'.

Proposition 3.2. *Let us keep the notation above. Then* $\pi : G \to \hat{G}$ *is the completion of* G. *Moreover,* π *induces an isomorphism* $G/G_n \simeq \hat{G}/\hat{G}_n$ *for every* n.

Proof It is clear that $\cap_n \hat{G}_n = 0$. Hence $\hat{G}$ is separated. Let $(g^m)_m$ be a Cauchy sequence in $\hat{G}$. For every $r \geq 0$, there exists an n_r such that for every $n \geq n_r$, we have $g^n - g^{n_r} \in \hat{G}_r$. If we write $g^n = (g_1^n, \ldots, g_k^n, \ldots)$, this implies that $g_i^n = g_i^{n_r}$ for every $i \leq r$ and every $n \geq n_r$. Let $\hat{g} = (g_1^{n_1}, g_2^{n_2}, \ldots)$. One immediately verifies that $\hat{g} \in \hat{G}$, and that this is the limit of the sequence $(g^m)_m$. Hence $\hat{G}$ is indeed complete.

Let $(g_n)_n \in \hat{G}$. If $g \in G$ is a preimage of g_m in G, then $\pi(g) - (g_n)_n \in \hat{G}_m$. Therefore $\pi(G)$ is dense in $\hat{G}$. The isomorphism $G/G_n \simeq \hat{G}/\hat{G}_n$ immediately follows from the construction.

Let us now show the factoring property. Let $f : G \to H$ be a continuous homomorphism to a complete Abelian topological group. Let $\alpha \in \hat{G}$.

Then there exists a sequence $(\alpha^m)_m$ of elements of G such that $\pi(\alpha^m)$ tends to α. As $\pi^{-1}(\hat{G}_n) = G_n$, this implies that $(\alpha^m)_m$, and therefore $(f(\alpha^m))_m$, is a Cauchy sequence. Let β be the limit of $(f(\alpha^m))_m$. Since H is separated, $\operatorname{Ker}\pi = \cap_{n\geq 0}G_n \subseteq \operatorname{Ker} f$, which implies that β does not depend on the choice of the sequence $(\alpha^m)_m$. We can therefore factor f as $\tilde{f} \circ \pi$ by setting $\tilde{f}(\alpha) = \beta$. The uniqueness of $\tilde{f}$ comes from the density of $\pi(G)$, and its continuity is easy to verify. □

Lemma 3.3. *Let M, N be Abelian groups endowed respectively with filtrations $(M_n)_n$, $(N_n)_n$. Let $\phi : M \to N$ be a homomorphism such that for some $n_0 \geq 0$, we have $N = \phi(M) + N_n$ and $N_n = \phi(M_n) + N_{n+1}$ for every $n \geq n_0$. Then the canonical homomorphism $\hat{M} \to \hat{N}$ is surjective.*

Proof As explained in the proof of Lemma 3.1, we may assume that $n_0 = 0$. The hypotheses imply that the canonical homomorphisms $\phi_n : M/M_n \to N/N_n$ and $\operatorname{Ker}\phi_n = \phi^{-1}(N_n)/M_n \to \operatorname{Ker}\phi_{n-1} = \phi^{-1}(N_{n-1})/M_{n-1}$ are surjective. The surjectivity of $\hat{M} \to \hat{N}$ then results from Lemma 3.1 applied to the exact sequence of inverse systems $0 \to (\operatorname{Ker}\phi_n)_n \to (M/M_n)_n \to (N/N_n)_n \to 0$. □

Let A be a ring endowed with the I-adic topology. We denote the completion of A by $\hat{A}$. We also call $\hat{A}$ the *formal completion of A for the I-adic topology*. As A is a ring, so is $\hat{A}$. Let M be an A-module. Then any submodule filtration (M_n) of M defines the structure of a topological A-module on M. We let $\hat{M}$ denote the completion of M for the I-adic topology (i.e., that defined by the filtration $(I^n M)$). This is an $\hat{A}$-module.

Example 3.4. Let p be a prime number. The completion of $\mathbb{Z}$ for the p-adic topology is denoted $\mathbb{Z}_p$, and is called the ring of *p-adic integers*.

Definition 3.5. Let A be a commutative ring with unit. The *ring of formal power series* in one variable $A[[T]]$ is defined in the following way. Let $A^{\mathbb{N}}$ be the group of sequences with coefficients in A. To simplify, we denote a sequence $(a_n)_{n\geq 0}$ by

$$a_0 + a_1 T + a_2 T^2 + \dots .$$

We endow $A^{\mathbb{N}}$ with a multiplicative law by setting

$$\left(\sum_{i\geq 0} a_i T^i\right)\left(\sum_{j\geq 0} b_j T^j\right) = \sum_{k\geq 0} c_k T^k,$$

where $c_k = \sum_{i+j=k} a_i b_j$. This ring clearly contains the polynomial ring $A[T]$. We define inductively the ring of formal power series

$$A[[T_1,\dots,T_r]] = A[[T_1,\dots,T_{r-1}]][[T_r]].$$

Example 3.6. Let $B = A[T_1,\dots,T_r]$ be a polynomial ring with coefficients in a commutative ring A. Let $\mathfrak{m}$ be the ideal of B generated by $T_1, T_2,\dots, T_r$. Then the $\mathfrak{m}$-adic completion $\hat{B}$ is isomorphic to $A[[T_1,\dots,T_r]]$, the latter being endowed with the $\mathfrak{n}$-adic topology for the ideal $\mathfrak{n}$ generated by the T_i.

In fact, the following facts are very easy to verify: $A[[T_1, \ldots, T_r]]$ is complete, and the canonical map $B \to A[[T_1, \ldots, T_r]]$ is continuous and induces an isomorphism $B/\mathfrak{m}^n \simeq A[[T_1, \ldots, T_r]]/\mathfrak{n}^n$ for every $n \geq 1$. This implies our assertion.

Proposition 3.7. *Let A be a Noetherian ring; then the ring of formal power series $A[[T_1, \ldots, T_r]]$ is also Noetherian.*

Proof By induction on r, it suffices to show that $A[[T]]$ is Noetherian. Let I be an ideal of $A[[T]]$. We have to show that I is finitely generated. For each $i \geq 0$, let $J_i = \{\alpha \in A \mid \alpha T^i \in I + T^{i+1} A[[T]]\}$. This defines an ascending sequence of ideals of A. Hence there exists a $d \geq 0$ such that $J_n = J_d$ for every $n \geq d$. Since $I/(I \cap T^d A[[T]])$ is a submodule of $A[[T]]/(T^d) = A[T]/(T^d)$ which is finitely generated over A, it is enough to show that $I \cap T^d A[[T]]$ is finitely generated over $A[[T]]$. Let $f_1, \ldots, f_m$ be generators of J_d, and let $F_j \in I$ be such that $F_j - f_j T^d \in T^{d+1} A[[T]]$. Then $F_j \in I \cap T^d A[[T]]$. We are going to show that $F_1, \ldots, F_m$ generate $I \cap T^d A[[T]]$. We can suppose that $I \subseteq T^d A[[T]]$.

Let $F = \sum_{i \geq 0} a_i T^i \in I$. Let $q = \min\{i \geq 0 \mid a_i \neq 0\} \geq d$. Then $a_q \in J_q = J_d$. Hence there exist $b_1, \ldots, b_m \in A$ such that $a_q = \sum_{1 \leq j \leq m} b_j f_j$. It follows that $F - \sum_{1 \leq j \leq m} (b_j T^{q-d}) F_j \in I \cap T^{q+1} A[[T]]$. By induction on $n \geq q$, we see that F can be written as

$$F = G_q + G_{q+1} + \cdots + G_n + H_n,$$

with $G_i \in T^{i-d}(F_1, \ldots, F_m)$ and $H_n \in I \cap T^{n+1} A[[T]]$. It is clear that the series $G_q + G_{q+1} + \ldots$ tends to an element of $(F_1, \ldots, F_m)$ and that H_n tends to 0. Hence $F \in (F_1, \ldots, F_m)$.   □

Corollary 3.8. *Let A be a Noetherian ring, and let I be an ideal of A. Then the formal completion of A for the I-adic topology is a Noetherian ring.*

Proof Let $t_1, \ldots, t_r$ be a system of generators of I. Let us consider the surjective homomorphism of A-algebras $\phi : B = A[T_1, \ldots, T_r] \to A$ defined by $\phi(T_i) = t_i$, and endow B with the $\mathfrak{m}$-adic topology, where $\mathfrak{m}$ is the ideal generated by the T_i. For any $n \geq 1$, we have $\phi(\mathfrak{m}^n) = I^n$. It follows from Lemma 3.3 that $A[[T_1, \ldots, T_r]] = \hat{B} \to \hat{A}$ is surjective. Hence $\hat{A}$ is Noetherian by the proposition above.   □

Let M, N be two I-adic A-modules. It is clear that the product topology on $M \oplus N = M \times N$ is also the I-adic topology. Consequently, $(M \oplus N)^\wedge = \hat{M} \oplus \hat{N}$. In particular, every isomorphism of A-modules $A^r \to L$ canonically induces an isomorphism of $\hat{A}$-modules $\hat{A}^r \to \hat{L}$. Let M be an I-adic A-module. There exists a canonical homomorphism $M \otimes_A \hat{A} \to \hat{M}$. It is in general neither injective nor surjective. However, we do have the following lemma:

Lemma 3.9. *Let M be a finitely generated A-module. Then $M \otimes_A \hat{A} \to \hat{M}$ is surjective.*

Proof Let $p : L \to M$ be a surjective homomorphism of A-modules, with L free of finite rank. We have $p(I^n L) = I^n M$. It follows from Lemma 3.3 that $\hat{L} \to \hat{M}$ is surjective. It results from the commutative diagram

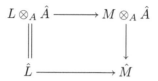

that $M \otimes_A \hat{A} \to \hat{M}$ is surjective.   □

1.3.2   The Artin–Rees lemma and applications

Let us first make a digression towards graded rings and modules. The topological notions do not intervene at first.

Let B be a commutative ring (with unit). A *grading on B* consists of a decomposition of B in subgroups

$$B = \oplus_{d \geq 0} B_d,$$

such that $B_d B_e \subseteq B_{d+e}$. The elements of B_d are called the *homogeneous elements of degree d*. If B is an algebra over a ring A, we also ask that the image of A in B be contained in B_0. We then call B a *graded algebra over A*. A polynomial ring $B = A[T_0, \ldots, T_r]$, for example, is naturally graded by taking for B_d the sub-A-module generated by the monomials of degree d.

A *graded B-module* is a B-module E which has a decomposition in subgroups $E = \oplus_{d \geq 0} E_d$ such that $B_d E_e \subseteq E_{d+e}$.

Let A be a commutative ring, and I an ideal of A. Let $\tilde{A}$ denote the graded ring $\tilde{A} = \oplus_{d \geq 0} I^d$. We can also see $\tilde{A}$ as the subring $\sum_{d \geq 0} I^d T^d \subseteq A[T]$. Let us note that if I is finitely generated, then $\tilde{A}$ is *a finitely generated A-algebra*; that is, it is the quotient of a polynomial ring $A[T_1, \ldots, T_r]$ by an ideal. In fact, let $t_1, \ldots, t_r$ be a system of generators of I. Then the map $A[T_1, \ldots, T_r] \to \tilde{A}$ defined by $T_i \mapsto t_i \in \tilde{A}_1$ is a surjective ring homomorphism. In particular, if A is Noetherian, then $\tilde{A}$ is Noetherian because $A[T]$ is Noetherian.

Let M be an A-module. An *I-filtration of M* is a filtration $(M_n)_n$ of M by submodules M_n such that $IM_n \subseteq M_{n+1}$. We call the filtration *stable* if there exists an n_0 such that $M_{n+1} = IM_n$ for every $n \geq n_0$.

Let $\widetilde{M} := \oplus_{n \geq 0} M_n$. This is a graded $\tilde{A}$-module.

Lemma 3.10. *Let A be a Noetherian ring, M a finitely generated A-module, and $(M_n)_n$ an I-filtration on M. The following properties are equivalent:*

(i) *The filtration $(M_n)_n$ is stable.*

(ii) *The module $\widetilde{M}$ is finitely generated over $\tilde{A}$.*

Proof Let $N_n = \oplus_{i \leq n} M_i$. This is a finitely generated A-module. Let

$$P_n = N_n \oplus (\oplus_{j \geq 1} I^j M_n) \subseteq \widetilde{M}$$

(where $I^j M_n$ is in the component M_{n+j}). Then the P_n are finitely generated $\widetilde{A}$-modules, and form an ascending sequence whose union equals all of $\widetilde{M}$. As $\widetilde{A}$ is a Noetherian ring, $\widetilde{M}$ is finitely generated if and only if there exists an n_0 such that $P_{n+1} = P_n$ for every $n \geq n_0$. Now, this equality is equivalent to $I M_n = M_{n+1}$.   □

Proposition 3.11. *Let A be a Noetherian ring, I an ideal of A, and M a finitely generated A-module endowed with a stable I-filtration $(M_n)_n$. Then for any submodule N of M, the I-filtration $(M_n \cap N)_n$ of N is also stable.*

Proof In fact, the filtration $(M_n \cap N)_n$ induces on $\widetilde{N}$ the structure of a sub-module of $\widetilde{M}$. The proposition results from Lemma 3.10.   □

Taking $M_n = I^n M$, we obtain the following corollary:

Corollary 3.12 (The Artin–Rees lemma). *Let A be a Noetherian ring, I an ideal of A, M a finitely generated A-module, and N a submodule of M. Then there exists an n_0 such that*

$$(I^{n+1}M) \cap N = I((I^n M) \cap N)$$

for every $n \geq n_0$.

An immediate application of this result is a theorem of Krull:

Corollary 3.13. *Let A be a Noetherian ring, I an ideal of A, and M a finitely generated A-module. Then $\cap_{n \geq 0}(I^n M)$ is the set of elements $x \in M$ for which there exists an $\alpha \in I$ such that $(1 + \alpha)x = 0$.*

Proof If $(1 + \alpha)x = 0$ for an $\alpha \in I$, then $x = -x\alpha \in IM$. We see by induction that $x \in I^n M$ for every n. Conversely, let us suppose that $x \in \cap_{n \geq 0}(I^n M)$. Let us consider the submodule $N := xA$ of M. Then by the Artin–Rees lemma, there exists an integer $n \geq 1$ such that $I^n M \cap N \subseteq IN$, and hence $x \in IN = xI$, whence the result.   □

Let us return to topological modules.

Corollary 3.14. *Let A be a Noetherian ring, I an ideal of A, and M a finitely generated A-module. Let $\hat{M}$ be the completion of M for the I-adic topology. Then the canonical homomorphism $M \otimes_A \hat{A} \to \hat{M}$ is an isomorphism. In particular, if A is complete for the I-adic topology, then so is M.*

Proof Let $0 \to N \to L \to M \to 0$ be an exact sequence of A-modules with L free of finite rank. As the filtrations $(I^n N)_n$, $((I^n L) \cap N)_n$ define the same topology on N by the Artin–Rees lemma, $0 \to \hat{N} \to \hat{L} \to \hat{M} \to 0$ is an exact sequence (Lemma 3.1). Let us consider the commutative diagram

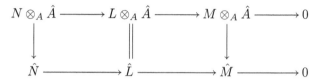

whose lines are exact. As the vertical arrows are surjective (Lemma 3.9), the map $M \otimes_A \hat{A} \to \hat{M}$ is bijective.   □

Theorem 3.15. *Let A be a Noetherian ring, and I an ideal of A. Then the completion $\hat{A}$ of A for the I-adic topology is a flat ring over A.*

Proof Let J be an ideal of A. By Corollary 3.14, we have $J \otimes_A \hat{A} \simeq \hat{J}$. Let us consider the canonical homomorphism $\hat{J} \to \hat{A}$. Let us show that it is injective, which will imply the flatness of $\hat{A}$ over A (Theorem 2.4).

Let $\alpha = (\alpha_n)_n$ be an element of the kernel of $\hat{J} \to \hat{A}$. Let us fix an $n \geq 1$. By virtue of the Artin–Rees lemma, there exists an $m \geq n$ such that $I^m \cap J \subseteq I^n J$. Let $\beta_m \in J$ be an element whose image in $J/I^m J$ equals α_m; then $\beta_m \in I^n J$. As α_n is the image of β_m in $J/I^n J$, it follows that $\alpha_n = 0$ and therefore $\alpha = 0$.   □

1.3.3   The case of Noetherian local rings

Theorem 3.16. *Let $(A, \mathfrak{m})$ be a Noetherian local ring, and $\hat{A}$ its $\mathfrak{m}$-adic completion. We have the following properties:*

(a) *For every $n \geq 1$, we have a canonical isomorphism $A/\mathfrak{m}^n \simeq \hat{A}/\mathfrak{m}^n \hat{A}$.*

(b) *$\hat{A}$ is a local ring with maximal ideal $\mathfrak{m}\hat{A}$, which is faithfully flat over A.*

(c) *Let $(B, \mathfrak{n})$ be a local ring such that $A \subseteq B \subseteq \hat{A}$ and $\mathfrak{m}B = \mathfrak{n}$. Then the $\mathfrak{n}$-adic completion $\hat{B}$ is isomorphic to $\hat{A}$.*

Proof (a) The topology on $A/\mathfrak{m}^n$ induced by the $\mathfrak{m}$-adic topology on A is discrete, giving the exact sequence $0 \to \widehat{\mathfrak{m}^n} \to \hat{A} \to A/\mathfrak{m}^n \to 0$, and moreover $\widehat{\mathfrak{m}^n} \simeq \mathfrak{m}^n \hat{A}$ (Corollary 3.14 and Theorem 3.15). This proves (a).

(b) Since $\hat{A}/\mathfrak{m}\hat{A} \simeq A/\mathfrak{m}$ is a field, $\mathfrak{m}\hat{A}$ is a maximal ideal. Let $\alpha \in \hat{A}$. We can write $\alpha = a + \varepsilon$ with $a \in A$ and $\varepsilon \in \mathfrak{m}\hat{A}$. If $\alpha \notin \mathfrak{m}\hat{A}$, then $a \notin \mathfrak{m}$, and hence $\alpha = a(1 - \delta)$ with $\delta \in \mathfrak{m}\hat{A}$. It follows that

$$a^{-1}(1 + \delta + \delta^2 + \dots)$$

is the inverse of α in $\hat{A}$. Hence this is a local ring. We already know that it is flat over A, and the faithful flatness follows from Corollary 2.20.

(c) Let $n \geq 1$. We have $\mathfrak{n}^n = \mathfrak{m}^n B$. Since the composition $A/\mathfrak{m}^n \to B/\mathfrak{m}^n B \to \hat{A}/\mathfrak{m}^n \hat{A}$ is an isomorphism, $B/\mathfrak{m}^n B \to \hat{A}/\mathfrak{m}^n \hat{A}$ is surjective. It remains to show that it is injective; that is, that $\mathfrak{m}^n \hat{A} \cap B = \mathfrak{m}^n B$. We have $B = A + \mathfrak{m}B = A + \mathfrak{m}^2 B = \dots = A + \mathfrak{m}^n B$, so every element $b \in B$ can be written $b = a + \varepsilon$ with $a \in A$, $\varepsilon \in \mathfrak{m}^n B \subseteq \mathfrak{m}^n \hat{A}$. If, moreover, $b \in \mathfrak{m}^n \hat{A}$, then $a \in \mathfrak{m}^n \hat{A} \cap A = \mathfrak{m}^n$, so $b \in \mathfrak{m}^n B$.   □

Exercises

3.1. Is the usual topology on $\mathbb{R}$ defined by a subgroup filtration?

3.2. Show that $\mathbb{Z}_p$ is a principal local ring, with maximal ideal $p\mathbb{Z}_p$.

3.3. Let A be the ring of germs of real analytic functions in $0 \in \mathbb{R}$. Let $x \in A$ be the identity map on $\mathbb{R}$. Show that the map $f \mapsto \sum_n (f^{(n)}(0)/n!)T^n$ from A to $\mathbb{R}[[T]]$ is injective, and induces an isomorphism from the xA-adic completion of A onto $\mathbb{R}[[T]]$. Show that this is false if we replace A by the C^∞-functions in 0.

3.4. Find examples where $M \otimes_A \hat{A} \to \hat{M}$ is not surjective, or is not injective.

3.5. Let $B = \oplus_{n \geq 0} B_n$ be a graded ring.

    (a) Show that B_0 is a subring of B, with unit.

    (b) Show that if B is Noetherian, then so is B_0, and B is a finitely generated algebra over B_0.

3.6. Let A be an integral domain that is not a field, and let K be its field of fractions. Show that the conclusion of Proposition 3.11 does not hold if we take $M = K$, $N = A$, and $I \neq 0$.

3.7. Let $(A, \mathfrak{m})$ be a Noetherian local ring. Show that $\cap_{n \geq 0} \mathfrak{m}^n = 0$. Give a counter-example with A not Noetherian.

3.8. Let A be a Noetherian ring, and I, J ideals of A. Let $\hat{A}$ be the I-adic completion of A and $(A/J)^\wedge$ the completion of A/J for the $(I+J)/J$-adic topology. Show that there is a canonical isomorphism $\hat{A}/J\hat{A} \simeq (A/J)^\wedge$.

3.9. (*nth root*) Let $n \geq 2$ be an integer. Let $D = \mathbb{Z}[1/n]$.

    (a) Consider the polynomial $S = (1+T)^n - 1 \in D[T]$. Show that $D[[S]] = D[[T]]$ and that there exists an $f(S) \in SD[[S]]$ such that $1 + S = (1 + f(S))^n$.

    (b) Let A be a complete ring for the I-adic topology, where I is an ideal of A. Suppose that n is invertible in A. Let $x \in I$. Show that there exists a unique continuous homomorphism $\phi : D[[S]] \to A$ such that $\phi(S) = x$. Conclude that there exists a $y \in I$ such that $1+x = (1+y)^n$.

    (c) Show, by giving an example, that statement (b) is false if n is not invertible in A.

3.10. Let k be a field of characteristic different from 2, and $A = k[x, y]/(y^2 - x^2(x + 1))$. Let $\hat{A}$ be the $\mathfrak{m}$-adic completion, where $\mathfrak{m} = (x, y)A$. Show that $\hat{A} \simeq k[[u, v]]/uv$.

3.11. Let A be a complete ring for the I-adic topology, where I is an ideal of A. Let $(M_n)_{n \geq 0}$ be A-modules such that $I^{n+1}M_n = 0$ and that there exist surjective homomorphisms $\pi_n : M_{n+1} \to M_n$ with $\mathrm{Ker}(\pi_n) = I^{n+1}M_{n+1}$.

Let $M = \varprojlim_n M_n$ and denote the (surjective) canonical homomorphisms by $u_n : M \to M_n$.

(a) Fix $d \geq 0$. Show that for any $n \geq d$, there is an exact sequence

$$0 \to I^{d+1}M_n \to M_n \to M_d \to 0.$$

(b) Let us suppose that M_0 is generated over A by a finite number of elements $e_{0,1}, \ldots, e_{0,m}$. Let $e_1, \ldots, e_m \in M$ be such that $u_0(e_i) = e_{0,i}$, and define $\phi_n : A^m \to M_n$ by $(a_1, \ldots, a_m) \mapsto \sum_i a_i u_n(e_i)$. Show that ϕ_n is surjective, and, using Lemma 3.1, that M is generated by the e_i.

(c) Let us moreover suppose that A is Noetherian and that M_0 is finitely generated over A. We are going to show that $\operatorname{Ker} u_n = I^{n+1}M$ (in other words, u_n induces an isomorphism $M/I^{n+1}M \simeq M_n$).

(1) Let $K_n = \operatorname{Ker} u_n$. Show that $K_n = I^{n+1}M + K_{n+1}$.

(2) Show that the canonical homomorphism $\varprojlim_n M/I^{n+1}M \to \varprojlim_n M_n$ is an isomorphism (use Corollary 3.14).

(3) Apply Lemma 3.1 to the exact sequence of inverse systems

$$0 \to (K_n/I^{n+1}M)_n \to (M/I^{n+1}M)_n \to (M_n)_n \to 0$$

and show that $K_n = I^{n+1}M$.

3.12. Let A be a Noetherian ring, I an ideal of A, and $\hat{A}$ the I-adic completion of A. Let $a \in A$. Show that if a is not a zero divisor in A, then it is not a zero divisor in $\hat{A}$. However, show, by giving an example, that A can be an integral domain without $\hat{A}$ being one.

3.13. Let A be a Noetherian ring, I an ideal of A, and M a finitely generated A-module. Suppose that A is complete for the I-adic topology. Show that every submodule of M is closed.

3.14. Let A be a Noetherian ring, I an ideal of A, and $\hat{A}$ the formal I-adic completion of A. Show that the topology on $\hat{A}$ induced by that on A is the $I\hat{A}$-adic topology.

3.15. (*Mittag–Leffler condition*) Let $(A_n, \pi_n)_n$ be an inverse system of sets. For any pair $m > n$, let us denote the map $\pi_n \circ \pi_{n+1} \circ \cdots \circ \pi_{m-1} : A_m \to A_n$ by $\pi_{m,n}$. We say that the system $(A_n, \pi_n)_n$ satisfies the *Mittag–Leffler condition* if for every n, the descending sequence $(\pi_{m,n}(A_m))_{m>n}$ is stationary.

(a) Let $A'_n = \cap_{m>n} \pi_{m,n}(A_m)$. Show that $(A'_n, \pi_{n-1}|_{A'_n})_n$ is an inverse system and that the canonical map $\varprojlim_n A'_n \to \varprojlim_n A_n$ is bijective.

(b) Let us suppose that $(A_n, \pi_n)_n$ satisfies the Mittag–Leffler condition and that $A_n \neq \emptyset$ for all n. Show that $A'_n \neq \emptyset$, $A'_{n+1} \to A'_n$ is surjective, and that $\varprojlim_n A'_n \neq \emptyset$. Deduce from this that $\varprojlim_n A_n \neq \emptyset$.

(c) Let $0 \to (A_n)_n \to (B_n)_n \overset{\rho}{\to} (C_n)_n \to 0$ be an exact sequence of inverse systems of Abelian groups such that $(A_n)_n$ satisfies the Mittag–Leffler condition. Let $(c_n)_n \in \varprojlim_n C_n$ and $X_n = \rho_n^{-1}(c_n)$, where $\rho_n : B_n \to C_n$ is the nth component of ρ. Show that $(X_n)_n$ is an inverse system of sets satisfying the Mittag–Leffler condition. Deduce from this that $\varprojlim_n B_n \to \varprojlim_n C_n$ is surjective. This generalizes Lemma 3.1.

2

General properties of schemes

In this chapter we introduce the notion of schemes. The first three sections are devoted to the definitions of schemes and morphisms of schemes, as well as to examples. After that, in Sections 2.4 and 2.5, we consider some elementary properties, more particularly topological properties (irreducible components, dimension), of schemes.

2.1   Spectrum of a ring

A (topological or differential) variety is made up of local charts which are open subsets of an $\mathbb{R}^n$. The construction of schemes is done in a similar way. The local charts are affine schemes. In this section, we define the underlying topological space of an affine scheme. To keep a certain intuition in algebraic geometry, we will study the particular case of algebraic sets.

2.1.1   Zariski topology

Let A be a (commutative) ring (with unit). We let $\operatorname{Spec} A$ denote the set of prime ideals of A. We call it the *spectrum of A*. By convention, the unit ideal is not a prime ideal. Thus $\operatorname{Spec}\{0\} = \emptyset$.

We will now endow $\operatorname{Spec} A$ with a topological structure. For any ideal I of A, let $V(I) := \{\mathfrak{p} \in \operatorname{Spec} A \mid I \subseteq \mathfrak{p}\}$. If $f \in A$, let $D(f) := \operatorname{Spec} A \setminus V(fA)$.

Proposition 1.1. *Let A be a ring. We have the following properties:*

(a) *For any pair of ideals I, J of A, we have $V(I) \cup V(J) = V(I \cap J)$.*

(b) *Let $(I_\lambda)_\lambda$ be a family of ideals of A. Then $\cap_\lambda V(I_\lambda) = V(\sum_\lambda I_\lambda)$.*

(c) *$V(A) = \emptyset$ and $V(0) = \operatorname{Spec} A$.*

In particular, there exists a unique topology on Spec A whose closed subsets are the sets of the form $V(I)$ for an ideal I of A. Moreover, the sets of the form $D(f)$, $f \in A$, constitute a base of open subsets on Spec A.

Proof This follows immediately. For the last assertion, we note that by definition, every open subset of Spec A is of the form Spec $A \setminus V(I)$ for some ideal I. This is equal to the union of the $D(f)$ where f runs through the elements of I. □

Definition 1.2. Let A be a ring. We call the topology defined by Proposition 1.1 the *Zariski topology* on Spec A. An open set of the form $D(f)$ is called a *principal open subset*, while its complement $V(f) := V(fA)$ is called a *principal closed subset*.

In the remainder of the book, the set Spec A will always be endowed with the Zariski topology.

Remark 1.3. Let $\mathfrak{p} \in$ Spec A. Then the singleton $\{\mathfrak{p}\}$ is closed for the Zariski topology if and only if $\mathfrak{p}$ is a maximal ideal of A. We will then say that $\mathfrak{p}$ is a *closed point of* Spec A. More generally, a point x of a topological space is said to be *closed* if the set $\{x\}$ is closed.

Example 1.4. Let k be a field, and let $\mathbb{A}_k^1 :=$ Spec $k[T]$ be the *affine line* over k. Then $\mathbb{A}_k^1$ consists of the 'generic' point ξ (we will come back to this in Section 2.4) corresponding to the prime ideal $\{0\}$, and of the points corresponding to the maximal ideals of $k[T]$. The proper closed subsets of $\mathbb{A}_k^1$ are finite sets. Indeed, such a subset is of the form $V(I)$, with $I \neq 0$. Let $P(T) \in k[T]$ be a generator of I, let $P_1(T), \ldots, P_r(T)$ be the irreducible factors of the polynomial $P(T)$. Then $V(I)$ is the set of prime ideals $\{P_1(T)k[T], \ldots, P_r(T)k[T]\}$.

The point ξ is not closed since 0 is not a maximal ideal of $k[T]$, while all of the other points are closed. Moreover, if $\{\xi\} \subset V(I)$, then $I \subseteq \{0\}$, and hence $V(I) = \mathbb{A}_k^1$. This means that the closure of $\{\xi\}$ is all of $\mathbb{A}_k^1$. The existence of a non-closed point implies, in particular, that the topological space $\mathbb{A}_k^1$ is not separated in the usual sense. However, we will see later on that $\mathbb{A}_k^1$ is separated as a scheme (Proposition 3.3.4).

Example 1.5. The arithmetic counterpart of the preceding example is Spec $\mathbb{Z}$. All of the statements above hold for Spec $\mathbb{Z}$. The points of Spec $\mathbb{Z}$ are the generic point $\xi = \{0\}$ and the closed points $p\mathbb{Z}$, with p a prime number.

Figure 1. The spectrum of $\mathbb{Z}$.

Let us note that in a space Spec A, two ideals I and J can define the same closed subset. Indeed, let $\sqrt{I}$ be the *radical* of I (this is the set of elements $a \in A$ such that $a^n \in I$ for some $n \geq 1$); then $V(I) = V(\sqrt{I})$.

Lemma 1.6. *Let A be a ring. Let I, J be two ideals of A. The following properties are true.*

(a) *The radical $\sqrt{I}$ equals the intersection of the ideals $\mathfrak{p} \in V(I)$.*

(b) *We have $V(I) \subseteq V(J)$ if and only if $J \subseteq \sqrt{I}$.*

Proof (a) It is clear that $\sqrt{I} \subseteq \cap_{\mathfrak{p} \in V(I)}\mathfrak{p}$. Let us show the other inclusion. By replacing A by A/I, we may assume that $I = 0$ and therefore $V(I) = \operatorname{Spec} A$. Let $f \in \cap_{\mathfrak{p} \in \operatorname{Spec} A}\mathfrak{p}$. We want to show that f is *nilpotent* (i.e., $f \in \sqrt{0}$). Let us suppose that the contrary is true. Then the localization A_f is a non-zero ring. Let $\mathfrak{p}'$ be a prime ideal of A_f. It induces a prime ideal $\mathfrak{p} \in \operatorname{Spec} A$ such that $f \notin \mathfrak{p}$ (see Subsection 1.2.2). Which contradicts the hypothesis on f.

(b) This follows immediately from (a).   □

Let $\varphi : A \to B$ be a ring homomorphism. Then we have a map of sets $\operatorname{Spec} \varphi : \operatorname{Spec} B \to \operatorname{Spec} A$ defined by $\mathfrak{p} \mapsto \varphi^{-1}(\mathfrak{p})$ for every $\mathfrak{p} \in \operatorname{Spec} B$.

Lemma 1.7. *Let $\varphi : A \to B$ be a ring homomorphism. Let $f = \operatorname{Spec} \varphi$ be the map associated to φ as above. The following properties are true.*

(a) *The map f is continuous.*

(b) *If φ is surjective, then f induces a homeomorphism from $\operatorname{Spec} B$ onto the closed subset $V(\operatorname{Ker} \varphi)$ of $\operatorname{Spec} A$.*

(c) *If φ is a localization morphism $A \to S^{-1}A$, then f is a homeomorphism from $\operatorname{Spec}(S^{-1}A)$ onto the subspace $\{\mathfrak{p} \in \operatorname{Spec} A \mid \mathfrak{p} \cap S = \emptyset\}$ of $\operatorname{Spec} A$.*

Proof (a) Let I be an ideal of A. We let IB denote the ideal of B generated by $\varphi(I)$. It is easy to see that $f^{-1}(V(I)) = V(IB)$. Hence f is continuous.

(b) If φ is surjective, f clearly establishes a continuous bijection from $\operatorname{Spec} B$ onto $V(\operatorname{Ker} \varphi)$. Moreover, it sends a closed set $V(J)$ onto a closed set $V(\varphi^{-1}(J))$. It follows that f is closed, and therefore a homeomorphism.

(c) The assertion concerning the image of f, as well as the injectivity of f, follows directly from the definition (see 1.2.2). Let $B = S^{-1}A$, and let J be an ideal of B. Then we easily see that $f(V(J)) = V(\varphi^{-1}(J)) \cap f(\operatorname{Spec} B)$, which shows that f is a closed map.   □

Example 1.8. Let us consider a more sophisticated example than previously. Let $\mathbb{A}^1_{\mathbb{Z}} := \operatorname{Spec} \mathbb{Z}[T]$. Let $f : \mathbb{A}^1_{\mathbb{Z}} \to \operatorname{Spec} \mathbb{Z}$ denote the continuous map induced by the canonical homomorphism $\mathbb{Z} \to \mathbb{Z}[T]$. Then we have a partition

$$\mathbb{A}^1_{\mathbb{Z}} = f^{-1}(\{0\}) \cup \left(\cup_{p \text{ prime}} f^{-1}(p\mathbb{Z}) \right).$$

Let us now study the parts of this partition. Let S be the multiplicative part $\mathbb{Z} \setminus \{0\}$ of $\mathbb{Z}[T]$. Then a prime ideal $\mathfrak{p} \in \mathbb{A}^1_{\mathbb{Z}}$ is contained in $f^{-1}(\{0\})$ if and only if $\mathfrak{p} \cap \mathbb{Z} = 0$, which is equivalent to $\mathfrak{p} \cap S = \emptyset$. As $S^{-1}\mathbb{Z}[T] = \mathbb{Q}[T]$, we therefore have a canonical homeomorphism between $f^{-1}(\{0\})$ and $\operatorname{Spec} \mathbb{Q}[T] = \mathbb{A}^1_{\mathbb{Q}}$, by Lemma 1.7(c). Let p be a prime number; then $\mathfrak{p} \in f^{-1}(p\mathbb{Z})$ if and only if $p \in \mathfrak{p}$. As $\mathbb{Z}[T]/(p) = \mathbb{F}_p[T]$, it follows from Lemma 1.7(b) that we have a homeomorphism between $f^{-1}(p\mathbb{Z})$ and $\operatorname{Spec} \mathbb{F}_p[T] = \mathbb{A}^1_{\mathbb{F}_p}$.

To summarize, we see that $\operatorname{Spec} \mathbb{Z}[T]$ can be seen as a family of affine lines, parameterized by the points of $\operatorname{Spec} \mathbb{Z}$, and over fields of different characteristics. In a way, we have brought the affine lines $\mathbb{A}^1_{\mathbb{Q}}$, $\mathbb{A}^1_{\mathbb{F}_p}$ together in a single space. We will come back to this in Section 3.1.

2.1.2   Algebraic sets

We fix a field k. Let A be a *finitely generated k-algebra* (i.e., A is an algebra which is the quotient of a polynomial ring over k). We will give a more concrete description of $\operatorname{Spec} A$, or more precisely of its closed points, via the Nullstellensatz (Corollary 1.15).

We say that a ring homomorphism $\varphi : A_0 \to A$ is *integral*, or that A is *integral over A_0*, if every element $a \in A$ is integral over A_0, that is to say that there exists a monic polynomial $\sum_i \alpha_i T^i \in A_0[T]$ such that $\sum_i \varphi(\alpha_i)a^i = 0$. Such an equality is called an *integral equation* for a over A_0. The set of elements of A that are integral over A_0 form a subring of A. We say that A is *finite over A_0* if it is a finitely generated A_0-module. The composition of two finite homomorphisms is clearly a finite homomorphism. It is easy to see that A is finite over A_0 if and only if it is integral and finitely generated over A_0.

Proposition 1.9 (Noether normalization lemma). *Let A be a non-zero finitely generated algebra over a field k. Then there exist an integer $d \geq 0$ and a finite injective homomorphism $k[T_1, \ldots, T_d] \hookrightarrow A$.*

Proof   By assumption, we have $A = k[X_1, \ldots, X_n]/I$. We will use induction on n. If $n = 0$, or if $I = 0$, there is nothing to prove. Let us suppose that $n \geq 1$ and $I \neq 0$. Let $P(X) = \sum_{\nu \in \mathbb{N}^n} \alpha_\nu X^\nu \in I \setminus \{0\}$. Let $m = (m_1, \ldots, m_{n-1}, 1) \in \mathbb{N}^n$. Let us set $S_i = X_i - X_n^{m_i}$ for $i \leq n-1$. Then $k[X_1, \ldots, X_n] = k[S_1, \ldots, S_{n-1}, X_n]$. So $P(X)$ is a polynomial in $S_1, \ldots, S_{n-1}$ and X_n. Let us show that by choosing m well, we can arrange for $P(X)$ to be monic in X_n, that is of the form

$$P(X) = \alpha X_n^e + Q_1(S)X_n^{e-1} + \cdots + Q_e(S),$$

with $\alpha \in k^*$, $e \geq 1$, and the $Q_j(S) \in k[S] := k[S_1, \ldots, S_{n-1}]$. Let ν_0 be the maximum (for the lexicographical order) of the indices ν for which $\alpha_\nu \neq 0$. Then it is easy to see that there exists an $m \in \mathbb{N}^n$, with $m_n = 1$, such that the scalar product $\langle m, \nu_0 \rangle$ is strictly superior to $\langle m, \nu \rangle$ for every $\nu \neq \nu_0$ with $\alpha_\nu \neq 0$. With such an m, $P(X)$ is indeed of the described form, with $e = \langle m, \nu_0 \rangle$ and $\alpha = \alpha_{\nu_0}$.

The polynomials $S_1, \ldots, S_{n-1} \in k[X]$ canonically induce an injective homomorphism $k[S]/(I \cap k[S]) \hookrightarrow A$. This homomorphism is finite by virtue of the way in which $P(X)$ has been written. We conclude by applying the induction hypothesis to $k[S]/(I \cap k[S])$. $\square$

Example 1.10. Let $A = k[X_1, X_2]/(X_1 X_2)$. Then the homomorphism $k[T] \to A$ defined by $T \mapsto X_1 + X_2$ is finite and injective.

The method of Proposition 1.9 in fact allows us to obtain a somewhat better result.

Proposition 1.11. *Let k be a field, and I a proper ideal of $k[X_1, \ldots, X_n]$. Then there exist a polynomial sub-k-algebra $k[S_1, \ldots, S_n]$ of $k[X_1, \ldots, X_n]$ and an integer $0 \le r \le n$ such that:*

(a) *$k[X_1, \ldots, X_n]$ is finite over $k[S_1, \ldots, S_n]$;*

(b) *$k[S_1, \ldots, S_n] \cap I = (S_1, \ldots, S_r)$ (this is the zero ideal if $r = 0$);*

(c) *$k[S_{r+1}, \ldots, S_n] \to k[X_1, \ldots, X_n]/I$ is finite injective.*

Proof Property (c) (which is Proposition 1.9) follows from (a) and (b). We will show (a) and (b) by induction on n. There is nothing to show if $n = 0$. Let us suppose $n \ge 1$ and $I \ne 0$ (otherwise we take $S_i = X_i$ and $r = 0$). As in the proof of Proposition 1.9, after, if necessary, applying a k-automorphism to $k[X_1, \ldots, X_n]$, there exists a non-zero $P \in I$ that is monic in X_1. By the induction hypothesis, we can find a sub-k-algebra $k[S_2, \ldots, S_n]$ of $k[X_2, \ldots, X_n]$ and an $r \ge 0$ such that $I \cap k[S_2, \ldots, S_n] = (S_2, \ldots, S_r)$, and that $k[X_2, \ldots, X_n]$ is finite over $k[S_2, \ldots, S_n]$. Let us set $S_1 = P$. It can then immediately be verified that $k[S_1, \ldots, S_n] \cap I = (S_1, \ldots, S_r)$, and that $k[X_1, \ldots, X_n]$ is finite over $k[S_1, \ldots, S_n]$. □

Corollary 1.12. *Let A be a finitely generated algebra over a field k. Let $\mathfrak{m}$ be a maximal ideal of A. Then $A/\mathfrak{m}$ is a finite algebraic extension of k.*

Proof As $A/\mathfrak{m}$ is a finitely generated k-algebra, there exists a finite injective homomorphism $A_0 \hookrightarrow A/\mathfrak{m}$, where $A_0 = k[T_1, \ldots, T_d]$ is a polynomial ring over k. Let us suppose that $d \ge 1$. We have $1/T_1 \in A/\mathfrak{m}$ since $A/\mathfrak{m}$ is a field. Hence $1/T_1$ is integral over A_0. By considering an integral equation for $1/T_1$ over A_0, we see that T_1 is invertible in A_0, which is impossible. Hence $A_0 = k$ and $A/\mathfrak{m}$ is finite over k. □

This corollary makes it possible to describe the maximal ideals of a finitely generated algebra over k concretely. If $(\alpha_1, \ldots, \alpha_n) \in k^n$, then the ideal of $k[T_1, \ldots, T_n]$ generated by the $T_i - \alpha_i$ is a maximal ideal since the quotient algebra is isomorphic to k. Conversely, we have the following corollary:

Corollary 1.13 (Weak Nullstellensatz). *Let k be an algebraically closed field. Then for any maximal ideal $\mathfrak{m}$ of $k[T_1, \ldots, T_n]$, there exists a unique point $(\alpha_1, \ldots, \alpha_n) \in k^n$ such that $\mathfrak{m}$ is the ideal generated by $T_1 - \alpha_1, \ldots, T_n - \alpha_n$.*

Proof By the preceding corollary, $k \to k[T_1, \ldots, T_n]/\mathfrak{m}$ is an isomorphism. Let $\alpha_i \in k$ be the image of T_i in $k[T_1, \ldots, T_n]/\mathfrak{m}$. Then $T_i - \alpha_i \in \mathfrak{m}$. It follows that $\mathfrak{m}$ contains the ideal $(T_1 - \alpha_1, \ldots, T_n - \alpha_n)$. Since the latter is maximal, there is equality. The uniqueness of the α_i is immediate. □

Definition 1.14. *Let k be an algebraically closed field. Let $P_1(T), \ldots, P_m(T)$ be polynomials in $k[T_1, \ldots, T_n]$. Let*

$$Z(P_1, \ldots, P_m) := \{(\alpha_1, \ldots, \alpha_n) \in k^n \mid P_j(\alpha_1, \ldots, \alpha_n) = 0, \ 1 \le j \le m\}.$$

We call such a set an *algebraic set*. It is therefore the set of solutions of a system of polynomial equations. Let us note that it is also the set of common zeros of the polynomials belonging to the ideal generated by the $P_j(T)$.

For example, if p is a positive integer, the set of $(a, b) \in k^2$ such that

$$a^p + b^p = 1$$

is an algebraic set.

Corollary 1.15. *Let k be an algebraically closed field. Let $A = k[T_1, \ldots, T_n]/I$ be a finitely generated algebra over k. Then there is a bijection between the closed points of $\operatorname{Spec} A$ and the algebraic set*

$$Z(I) := \{(\alpha_1, \ldots, \alpha_n) \in k^n \mid P(\alpha_1, \ldots, \alpha_n) = 0, \text{ for every } P(T) \in I\}.$$

Proof   We can identify the closed points of $\operatorname{Spec} A$ with the maximal ideals of $k[T_1, \ldots, T_n]$ containing I. Let $\mathfrak{m} = (T_1 - \alpha_1, \ldots, T_n - \alpha_n)$ be a maximal ideal of $k[T_1, \ldots, T_n]$. Let $P(T) \in k[T_1, \ldots, T_n]$. Using the Taylor expansion of $P(T)$ at $\alpha := (\alpha_1, \ldots, \alpha_n)$, we see that $P(T) \in \mathfrak{m}$ if and only if $P(\alpha) = 0$. It follows that $I \subseteq \mathfrak{m}$ if and only if $P(\alpha) = 0$ for every $P(T) \in I$.   □

Remark 1.16. The object of algebraic geometry is the study of solutions of systems of polynomial equations over a field k. Corollary 1.15 explains why such a study corresponds to that of the spectra of finitely generated algebras over k.

Example 1.17. Let k be an algebraically closed field of characteristic different from 2. The set of closed points of $\operatorname{Spec} k[X, Y]/(Y^2 - X^2(X + 1))$ corresponds to the algebraic set $\{(x, y) \in k^2 \mid y^2 - x^2(x + 1) = 0\}$ (see Figure 2).

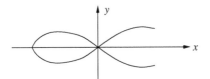

Figure 2. An algebraic set.

Lemma 1.18. *Let A be a non-zero finitely generated algebra over a field k. Then the intersection of the maximal ideals of A is equal to the nilradical $\sqrt{0}$ of A.*

Proof   Let f be an element of the intersection. We must show that f is nilpotent. Let us suppose that the contrary is true. Then the localization A_f (see 1.2.2) is a finitely generated k-algebra, because it is isomorphic to $A[T]/(Tf - 1)$. Let $\rho : A \to A_f$ be the canonical homomorphism. There exists a maximal ideal $\mathfrak{m} \subset A_f$ because $A_f \neq 0$. Then $A/\rho^{-1}(\mathfrak{m})$ is a sub-k-algebra of $A_f/\mathfrak{m}$. Since the latter is an algebraic extension of k, $A/\rho^{-1}(\mathfrak{m})$ is a field. Hence $\rho^{-1}(\mathfrak{m})$ is a maximal ideal of A that does not contain f, whence a contradiction.   □

Let us finish with another form of the Nullstellensatz:

Proposition 1.19. *Let k be an algebraically closed field. Let I be an ideal of $k[T_1, \ldots, T_n]$. If a polynomial $F \in k[T_1, \ldots, T_n]$ is such that $F(\alpha) = 0$ for every $\alpha \in Z(I)$, then $F \in \sqrt{I}$.*

Proof Let $A = k[T_1, \ldots, T_n]/I$ and let f denote the image of F in A. We must show that f is nilpotent. Every maximal ideal $\mathfrak{m}$ of A corresponds to a point $\alpha \in Z(I)$ (Corollary 1.15). Hence $F(\alpha) = 0$ and $f \in \mathfrak{m}$. By Lemma 1.18, f is indeed nilpotent.   □

Remark 1.20. This proposition says that we can recover the ideal I, up to its radical, from its set of zeros $Z(I)$.

Exercises

1.1. Let $A = k[[T]]$ be the ring of formal power series with coefficients in a field k. Determine $\operatorname{Spec} A$.

1.2. Let $\varphi : A \to B$ be a homomorphism of finitely generated algebras over a field. Show that the image of a closed point under $\operatorname{Spec} \varphi$ is a closed point.

1.3. Let $k = \mathbb{R}$ be the field of real numbers. Let $A = k[X, Y]/(X^2 + Y^2 + 1)$. We wish to describe $\operatorname{Spec} A$. Let x, y be the respective images of X, Y in A.

    (a) Let $\mathfrak{m}$ be a maximal ideal of A. Show that there exist $a, b, c, d \in k$ such that $x^2 + ax + b, y^2 + cy + d \in \mathfrak{m}$. Using the relation $x^2 + y^2 + 1 = 0$, show that $\mathfrak{m}$ contains an element $f = \alpha x + \beta y + \gamma$ with $(\alpha, \beta) \neq (0, 0)$. Deduce from this that $\mathfrak{m} = fA$.

    (b) Show that the map $(\alpha, \beta, \gamma) \mapsto (\alpha x + \beta y + \gamma)A$ establishes a bijection between the subset $\mathbb{P}(k^3) \setminus \{(0, 0, 1)\}$ of the projective space $\mathbb{P}(k^3)$ (see the notation before Lemma 3.43) and the set of maximal ideals of A.

    (c) Let $\mathfrak{p}$ be a non-maximal prime ideal of A. Show that the canonical homomorphism $k[X] \to A$ is finite and injective. Deduce from this that $\mathfrak{p} \cap k[X] = 0$. Let $g \in \mathfrak{p}$, and let $g^n + a_{n-1}g^{n-1} + \ldots + a_0 = 0$ be an integral equation with $a_i \in k[X]$. Show that $a_0 = 0$. Conclude that $\mathfrak{p} = 0$.

1.4. Let A be a ring.

    (a) Let $\mathfrak{p}$ be a minimal prime ideal of A. Show that $\mathfrak{p}A_{\mathfrak{p}}$ is nilpotent. Deduce from this that every element of $\mathfrak{p}$ is a zero divisor in A.

    (b) Show that if A is reduced, then any zero divisor in A is an element of a minimal prime ideal. Show with an example that this is false if A is not reduced (use Lemma 1.6). See also Corollary 7.1.3(a).

1.5. Let k be a field. Let $\mathfrak{m}$ be a maximal ideal of $k[T_1, \ldots, T_n]$.

    (a) Let $P_1(T_1)$ be a generator of $\mathfrak{m} \cap k[T_1]$ and $k_1 = k[T_1]/(P_1)$. Show that we have an exact sequence
$$0 \to P_1(T_1)k[T_1, \ldots, T_n] \to k[T_1, \ldots, T_n] \to k_1[T_2, \ldots, T_n] \to 0.$$

    (b) Show that there exist n polynomials $P_1(T_1), P_2(T_1, T_2), \ldots,$ $P_n(T_1, \ldots, T_n)$ such that $k[T_1, \ldots, T_i] \cap \mathfrak{m} = (P_1, \ldots, P_i)$ for all $i \leq n$. In particular, $\mathfrak{m}$ is generated by n elements.

1.6. We say that a topological space X is *quasi-compact* if from any open covering $\{U_i\}_i$ of X, we can extract a finite subcovering. Show that any closed subspace of a quasi-compact space is quasi-compact. Show that Spec A is quasi-compact.

1.7. Describe the closed subsets of Spec $\mathbb{Z}[T]$. Show that they are not all principal.

1.8. Let $\varphi : A \to B$ be an integral homomorphism.

(a) Show that Spec $\varphi :$ Spec $B \to$ Spec A maps a closed point to a closed point, and that any preimage of a closed point is a closed point.

(b) Let $\mathfrak{p} \in$ Spec A. Show that the canonical homomorphism $A_{\mathfrak{p}} \to B \otimes_A A_{\mathfrak{p}}$ is integral.

(c) Let $T = \varphi(A \setminus \mathfrak{p})$. Let us suppose that φ is injective. Show that T is a multiplicative subset of B, and that $B \otimes_A A_{\mathfrak{p}} = T^{-1}B \neq 0$. Deduce from this that Spec φ is surjective if φ is integral and injective.

1.9. Let A be a finitely generated algebra over a field k.

(a) Let us suppose that A is finite over k. Show that Spec A is a finite set, of cardinality bounded from above by the dimension $\dim_k A$ of A as a vector space. (Construct a strictly descending chain of ideals by taking intersections of maximal ideals.) Show that every prime ideal of A is maximal.

(b) Show that Spec $k[T_1, \ldots, T_d]$ is infinite if $d \geq 1$.

(c) Show that Spec A is finite if and only if A is finite over k.

2.2 Ringed topological spaces

One way of defining schemes is to consider them as ringed topological spaces which are locally affine schemes. Before introducing the notion of ringed topological space, we briefly recall the theory of sheaves. This is an essential tool in algebraic geometry. It makes it possible to collect local data in order to deduce global information.

2.2.1 Sheaves

We recall here some definitions and local properties of sheaves on a topological space. One can consult [40] for a detailed account of the theory of sheaves.

Definition 2.1. Let X be a topological space. A *presheaf $\mathcal{F}$ (of Abelian groups) on X* consists of the following data:

– an Abelian group $\mathcal{F}(U)$ for every open subset U of X, and

– a group homomorphism (*restriction map*) $\rho_{UV} : \mathcal{F}(U) \to \mathcal{F}(V)$ for every pair of open subsets $V \subseteq U$

which verify the following conditions:

(1) $\mathcal{F}(\emptyset) = 0$;

(2) $\rho_{UU} = \mathrm{Id}$;

(3) if we have three open subsets $W \subseteq V \subseteq U$, then $\rho_{UW} = \rho_{VW} \circ \rho_{UV}$.

An element $s \in \mathcal{F}(U)$ is called a *section of $\mathcal{F}$ over U*. We let $s|_V$ denote the element $\rho_{UV}(s) \in \mathcal{F}(V)$, and we call it the *restriction of s to V*.

Definition 2.2. We say that a presheaf $\mathcal{F}$ is a *sheaf* if we have the following properties:

(4) (*Uniqueness*) Let U be an open subset of X, $s \in \mathcal{F}(U)$, $\{U_i\}_i$ a covering of U by open subsets U_i. If $s|_{U_i} = 0$ for every i, then $s = 0$.

(5) (*Glueing local sections*) Let us keep the notation of (4). Let $s_i \in \mathcal{F}(U_i)$, $i \in I$, be sections such that $s_i|_{U_i \cap U_j} = s_j|_{U_i \cap U_j}$. Then there exists a section $s \in \mathcal{F}(U)$ such that $s|_{U_i} = s_i$ (this section s is unique by condition (4)).

We can define in the same way *sheaves of rings*, *sheaves of algebras* over a fixed ring, etc. There is a natural notion of *subsheaf $\mathcal{F}'$ of $\mathcal{F}$* : $\mathcal{F}'(U)$ is a subgroup of $\mathcal{F}(U)$, and the restriction ρ'_{UV} is induced by ρ_{UV}.

Example 2.3. Let X be a topological space. For any open subset U of X, let $\mathcal{C}(U) = \mathrm{C}^0(U, \mathbb{R})$ be the set of continuous functions from U to $\mathbb{R}$. The restrictions ρ_{UV} are the usual restrictions of functions. Then $\mathcal{C}$ is a sheaf on X. If we let $\mathcal{F}(U) = \mathbb{R}^U$ be the set of functions on U with values in $\mathbb{R}$, this defines a sheaf $\mathcal{F}$ of which $\mathcal{C}$ is a subsheaf.

Example 2.4. Let A be a non-trivial Abelian group. Let X be a topological space. Let $\mathcal{A}_X(U) = A$ and $\rho_{UV} = \mathrm{Id}_A$ if U and V are non-empty. This defines a presheaf on X. In general, $\mathcal{A}_X$ is not a sheaf. For example, if X is the disjoint union of two non-empty open sets, then condition (5) for sheaves is not verified. See also Exercise 2.1.

Remark 2.5. If U is an open subset of X, every presheaf $\mathcal{F}$ on X induces, in an obvious way, a presheaf $\mathcal{F}|_U$ on U by setting $\mathcal{F}|_U(V) = \mathcal{F}(V)$ for every open subset V of U. This is the *restriction of $\mathcal{F}$ to U*. If $\mathcal{F}$ is a sheaf, then so is $\mathcal{F}|_U$.

Remark 2.6. Let $\mathcal{B}$ be a base of open subsets on X (we mean that $\mathcal{B}$ is a set of open subsets of X, that any open subset of X is a union of open subsets in $\mathcal{B}$, and that $\mathcal{B}$ is stable by finite intersection). We can define *$\mathcal{B}$-presheaves* and *$\mathcal{B}$-sheaves* by replacing 'open subset U of X' by 'open set U belonging to $\mathcal{B}$' in the definition above. Then every $\mathcal{B}$-sheaf $\mathcal{F}_0$ extends in a unique way (up to isomorphism to be more precise, see Definition 2.10) to a sheaf $\mathcal{F}$ on X. Indeed, for any open subset U of X, we can write U as a union of open sets $U_i \in \mathcal{B}$, in which case $\mathcal{F}(U)$ is the set of elements $(s_i)_i \in \prod_i \mathcal{F}_0(U_i)$ such that $s_i|_{U_i \cap U_j} = s_j|_{U_i \cap U_j}$. In other words, a sheaf is completely determined by its sections over a base of open sets.

Let $\mathcal{U} = \{U_i\}_i$ be a family of open subsets of X. Let $U = \cup_i U_i$ and $U_{ij} = U_i \cap U_j$. For any presheaf $\mathcal{F}$ on X, we have a complex of Abelian groups $C^\bullet(\mathcal{U}, \mathcal{F})$:

$$0 \to \mathcal{F}(U) \xrightarrow{d_0} \prod_i \mathcal{F}(U_i) \xrightarrow{d_1} \prod_{i,j} \mathcal{F}(U_{ij}),$$

defined by $d_0 : s \mapsto (s|_{U_i})_i$ and $d_1 : (s_i)_i \mapsto (s_i|_{U_{ij}} - s_j|_{U_{ij}})_{i,j}$.

Lemma 2.7. *With the above notation, $\mathcal{F}$ is a sheaf if and only if $C^\bullet(\mathcal{U}, \mathcal{F})$ is exact for every family of open subsets $\mathcal{U}$ of X.*

Proof We note that the injectivity of d_0 is equivalent to condition (4) of Definition 2.2, while the equality $\operatorname{Ker} d_1 = \operatorname{Im} d_0$ is equivalent to condition (5). $\quad\square$

Definition 2.8. Let $\mathcal{F}$ be a presheaf on X, and let $x \in X$. The *stalk of $\mathcal{F}$ at x* is the group

$$\mathcal{F}_x = \varinjlim_{U \ni x} \mathcal{F}(U),$$

the direct limit being taken over the open neighborhoods U of x. If $\mathcal{F}$ is a presheaf of rings, then $\mathcal{F}_x$ is a ring. Let $s \in \mathcal{F}(U)$ be a section; for any $x \in U$, we denote the image of s in $\mathcal{F}_x$ by s_x. We call s_x the *germ of s at x*. The map $\mathcal{F}(U) \to \mathcal{F}_x$ defined by $s \mapsto s_x$ is clearly a group homomorphism.

Lemma 2.9. *Let $\mathcal{F}$ be a sheaf on X. Let $s, t \in \mathcal{F}(X)$ be sections such that $s_x = t_x$ for every $x \in X$. Then $s = t$.*

Proof We may assume that $t = 0$. For every $x \in X$, there exists an open neighborhood U_x of x such that $s|_{U_x} = 0$, because $s_x = 0$. As the U_x cover X when x varies, we indeed have $s = 0$. $\quad\square$

Definition 2.10. Let $\mathcal{F}, \mathcal{G}$ be two presheaves on X. A *morphism of presheaves* $\alpha : \mathcal{F} \to \mathcal{G}$ consists of, for every open subset U of X, a group homomorphism $\alpha(U) : \mathcal{F}(U) \to \mathcal{G}(U)$ which is compatible with the restrictions ρ_{UV}.

A morphism of presheaves α is called *injective* if for every open subset U of X, $\alpha(U)$ is injective (take care: a surjective morphism is not defined in the same way, see further on). We can, of course, compose two morphisms of presheaves. An *isomorphism* is an invertible morphism α. This amounts to saying that $\alpha(U)$ is an isomorphism for every open subset U of X.

Let $\alpha : \mathcal{F} \to \mathcal{G}$ be a morphism of presheaves on X. For any $x \in X$, α canonically induces a group homomorphism $\alpha_x : \mathcal{F}_x \to \mathcal{G}_x$ such that $(\alpha(U)(s))_x = \alpha_x(s_x)$ for any open subset U of X, $s \in \mathcal{F}(U)$, and $x \in U$. We say that α is *surjective* if α_x is surjective for every $x \in X$. By definition of the direct limit, this means that for every $t_x \in \mathcal{G}_x$, there exist an open neighborhood U of x and a section $s \in \mathcal{F}(U)$ such that $(\alpha(U)(s))_x = t_x$.

Example 2.11. Let $X = \mathbb{C} \setminus \{0\}$, and let $\mathcal{F}$ be the sheaf of holomorphic functions on X, $\mathcal{G}$ the sheaf of invertible holomorphic functions on X. Let us consider the homomorphism $\alpha : \mathcal{F} \to \mathcal{G}$ defined by $\alpha(U)(f) = \exp(f)$ for every open set U and for every $f \in \mathcal{F}(U)$. Then it is well known that α is surjective. However, $\alpha(X) : \mathcal{F}(X) \to \mathcal{G}(X)$ is not surjective. For example, the identity function is not the exponential of a holomorphic function on X.

Proposition 2.12. *Let* $\alpha : \mathcal{F} \to \mathcal{G}$ *be a morphism of sheaves on* X. *Then* α *is an isomorphism if and only if* α_x *is an isomorphism for every* $x \in X$.

Proof Suppose that α_x is an isomorphism for every $x \in X$. Let us show that α is an isomorphism (the converse is trivial). Let $s \in \mathcal{F}(U)$ be a section. If $\alpha(U)(s) = 0$, then for every $x \in U$, we have $\alpha_x(s_x) = (\alpha(U)(s))_x = 0$. It follows that $s_x = 0$ for every $x \in U$. Hence $s = 0$ (Lemma 2.9) and α is injective. Next, let $t \in \mathcal{G}(U)$. Then we can find a covering of U by open sets U_i and sections $s_i \in \mathcal{F}(U_i)$ such that $\alpha(U_i)(s_i) = t|_{U_i}$. As we have just seen that α is injective, s_i and s_j coincide on $U_i \cap U_j$. The s_i therefore glue to a section $s \in \mathcal{F}(U)$ (i.e., $s|_{U_i} = s_i$). By construction, $\alpha(U)(s)$ and t coincide on every U_i, and are therefore equal. This proves that $\alpha(U)$ is surjective. $\square$

A similar proof shows that $\mathcal{F} \to \mathcal{G}$ is injective if and only if $\mathcal{F}_x \to \mathcal{G}_x$ is injective for every $x \in X$.

Corollary 2.13. *Let* $\alpha : \mathcal{F} \to \mathcal{G}$ *be a morphism of sheaves. Then it is an isomorphism if and only if it is injective and surjective.*

There exists a canonical way to construct a sheaf from a presheaf while preserving the stalks.

Definition 2.14. Let $\mathcal{F}$ be a presheaf on X. We define the *sheaf associated to* $\mathcal{F}$ to be a sheaf $\mathcal{F}^\dagger$ together with a morphism of presheaves $\theta : \mathcal{F} \to \mathcal{F}^\dagger$ verifying the following universal property:

> For every morphism $\alpha : \mathcal{F} \to \mathcal{G}$, where $\mathcal{G}$ is a sheaf, there exists a unique morphism $\tilde{\alpha} : \mathcal{F}^\dagger \to \mathcal{G}$ such that $\alpha = \tilde{\alpha} \circ \theta$.

Proposition 2.15. *Let* $\mathcal{F}$ *be a presheaf on* X. *Then the sheaf* $\mathcal{F}^\dagger$ *associated to* $\mathcal{F}$ *exists and is unique up to isomorphism. Moreover,* $\theta_x : \mathcal{F}_x \to \mathcal{F}_x^\dagger$ *is an isomorphism for every* $x \in X$.

Proof The uniqueness of $\mathcal{F}^\dagger$ immediately results from the universal property. Let us show existence. Let U be an open subset of X. Let us consider the set $\mathcal{F}^\dagger(U)$ of functions $f : U \to \coprod_{x \in U} \mathcal{F}_x$ such that for every $x \in U$, there exist an open neighborhood V of x and a section $s \in \mathcal{F}(V)$ verifying $f(y) = s_y$ for every $y \in V$. We have an obvious natural homomorphism $\theta(U) : \mathcal{F}(U) \to \mathcal{F}^\dagger(U)$. We leave it to the reader to verify that $\mathcal{F}^\dagger$ is a sheaf and that the pair $(\mathcal{F}^\dagger, \theta)$ satisfies the conditions of the proposition. Another construction is given in Exercise 2.3. $\square$

Let us note that if $\mathcal{F}$ is already a sheaf, we have $\mathcal{F}^\dagger \simeq \mathcal{F}$ by virtue of the universal property.

Let us now look at some notions relating to exact sequences. Let $\mathcal{F}$ be a sheaf on X and $\mathcal{F}'$ a subsheaf of $\mathcal{F}$. Then $U \mapsto \mathcal{F}(U)/\mathcal{F}'(U)$ is a presheaf on X. The associated sheaf $\mathcal{F}/\mathcal{F}'$ is called the *quotient sheaf*. Let $\alpha : \mathcal{F} \to \mathcal{G}$ be a morphism of sheaves. Then $U \mapsto \mathrm{Ker}(\alpha(U))$ is a subsheaf of $\mathcal{F}$, denoted by $\mathrm{Ker}\,\alpha$. This sheaf is called the *kernel of* α. On the other hand, $U \mapsto \mathrm{Im}(\alpha(U))$ is in general only a presheaf, and the associated sheaf $\mathrm{Im}\,\alpha$ is called the *image of* α. The next lemma follows immediately.

2.2. Ringed topological spaces

Lemma 2.16. Let $\mathcal{F}$, $\mathcal{G}$ be sheaves on X, $\mathcal{F}'$ a subsheaf of $\mathcal{F}$, $\alpha : \mathcal{F} \to \mathcal{G}$ a morphism. We have $(\mathcal{F}/\mathcal{F}')_x = \mathcal{F}_x/\mathcal{F}'_x$, $(\operatorname{Im}\alpha)_x = \operatorname{Im}(\alpha_x)$, and $\operatorname{Im}\alpha$ is a subsheaf of $\mathcal{G}$.

Definition 2.17. We say that a sequence of sheaves $\mathcal{F} \xrightarrow{\alpha} \mathcal{G} \xrightarrow{\beta} \mathcal{H}$ is exact if $\operatorname{Im}\alpha = \operatorname{Ker}\beta$.

Proposition 2.18. A sequence of sheaves $\mathcal{F} \to \mathcal{G} \to \mathcal{H}$ is exact if and only if $\mathcal{F}_x \to \mathcal{G}_x \to \mathcal{H}_x$ is an exact sequence of groups for every $x \in X$.

Proof This is an immediate consequence of the lemma above and Proposition 2.12. □

Up to now, we have only considered sheaves on a single topological space. We can, however, transfer sheaves from one space to another by continuous functions. Let $f : X \to Y$ be a continuous map of topological spaces. Let $\mathcal{F}$ be a sheaf on X and $\mathcal{G}$ a sheaf on Y. Then $V \mapsto \mathcal{F}(f^{-1}(V))$ defines a sheaf $f_*\mathcal{F}$ on Y which we call the *direct image of $\mathcal{F}$*. On the other hand,

$$U \mapsto \varinjlim_{V \supseteq f(U)} \mathcal{G}(V)$$

defines a presheaf on X; the associated sheaf is called the *inverse image of $\mathcal{G}$* and is denoted by $f^{-1}\mathcal{G}$. The construction of $f^{-1}\mathcal{G}$ is a bit complex, but we will remember the property (easy to prove using Proposition 2.15)

$$(f^{-1}\mathcal{G})_x = \mathcal{G}_{f(x)}, \quad \text{for every } x \in X.$$

Let us note that if V is an open subset of Y and if we let $i : V \to Y$ denote the canonical injection, then $i^{-1}\mathcal{G} = \mathcal{G}|_V$.

2.2.2 Ringed topological spaces

Definition 2.19. A *ringed topological space (locally ringed in local rings)* consists of a topological space X endowed with a sheaf of rings $\mathcal{O}_X$ on X such that $\mathcal{O}_{X,x}$ is a local ring for every $x \in X$. We denote it $(X, \mathcal{O}_X)$. The sheaf $\mathcal{O}_X$ is called the *structure sheaf* of $(X, \mathcal{O}_X)$. When there is no confusion possible, we will omit $\mathcal{O}_X$ from the notation.

Let $\mathfrak{m}_x$ be the maximal ideal of $\mathcal{O}_{X,x}$; we call $\mathcal{O}_{X,x}/\mathfrak{m}_x$ the *residue field of X at x*, and we denote it $k(x)$.

Definition 2.20. A *morphism of ringed topological spaces*

$$(f, f^\#) : (X, \mathcal{O}_X) \to (Y, \mathcal{O}_Y)$$

consists of a continuous map $f : X \to Y$ and a morphism of sheaves of rings $f^\# : \mathcal{O}_Y \to f_*\mathcal{O}_X$ such that for every $x \in X$, the stalk $f_x^\# : \mathcal{O}_{Y,f(x)} \to \mathcal{O}_{X,x}$ is a local homomorphism (i.e., $f_x^{\#-1}(\mathfrak{m}_x) = \mathfrak{m}_{f(x)}$ or, equivalently, $f_x^\#(\mathfrak{m}_{f(x)}) \subseteq \mathfrak{m}_x$).

We define in an obvious manner the composition of two morphisms (Exercise 2.6). An *isomorphism* is an invertible morphism.

Example 2.21. Let $X = \mathbb{C}^n$. For any open subset U, we let $\mathcal{O}_X^h(U)$ be the set of holomorphic functions on U. Then $(X, \mathcal{O}_X^h)$ is a complex analytic variety. It is a ringed topological space. The property that we need to verify is that the stalks of $\mathcal{O}_X^h$ are indeed local rings.

Let $z \in \mathbb{C}^n$. Then $\mathcal{O}_{X,z}^h$ can be identified with the holomorphic functions defined on a neighborhood of z. Let $\mathfrak{m}_z$ be the set of those which vanish in z. This is a maximal ideal of $\mathcal{O}_{X,z}^h$ because $\mathcal{O}_{X,z}^h/\mathfrak{m}_z \simeq \mathbb{C}$. If a holomorphic function f does not vanish in z, then $1/f$ is still holomorphic in z. This shows that $\mathfrak{m}_z$ is the unique maximal ideal of $\mathcal{O}_{X,z}^h$ and hence the latter is a local ring.

Definition 2.22. We say that a morphism $(f, f^\#) : (X, \mathcal{O}_X) \to (Y, \mathcal{O}_Y)$ is an *open immersion* (resp. *closed immersion*) if f is a topological open immersion (resp. closed immersion) and if $f_x^\#$ is an isomorphism (resp. if $f_x^\#$ is surjective) for every $x \in X$.

Let V be an open subset of Y. Then $(V, \mathcal{O}_Y|_V)$ is clearly a ringed topological space. In all that follows, an open subset V will always be endowed with this structure induced by $\mathcal{O}_Y$. It is clear that $(f, f^\#)$ is an open immersion if and only if there exists an open subset V of Y such that $(f, f^\#)$ induces an isomorphism from $(X, \mathcal{O}_X)$ onto $(V, \mathcal{O}_Y|_V)$ (see Proposition 2.12).

In what follows, we will give a characterization of the closed immersions. Let $(X, \mathcal{O}_X)$ be a ringed topological space. Let $\mathcal{J}$ be a sheaf of ideals of $\mathcal{O}_X$ (i.e., $\mathcal{J}(U)$ is an ideal of $\mathcal{O}_X(U)$ for every open subset U). Let

$$V(\mathcal{J}) := \{x \in X \mid \mathcal{J}_x \neq \mathcal{O}_{X,x}\}.$$

Lemma 2.23. *Let X be a ringed topological space. Let $\mathcal{J}$ be a sheaf of ideals of $\mathcal{O}_X$. Let $j : V(\mathcal{J}) \to X$ denote the inclusion. Then $V(\mathcal{J})$ is a closed subset of X, $(V(\mathcal{J}), j^{-1}(\mathcal{O}_X/\mathcal{J}))$ is a ringed topological space, and we have a closed immersion $(j, j^\#)$ of this space into $(X, \mathcal{O}_X)$, where $j^\#$ is the canonical surjection*

$$\mathcal{O}_X \to \mathcal{O}_X/\mathcal{J} = j_*(j^{-1}(\mathcal{O}_X/\mathcal{J})).$$

Proof If $x \in X \setminus V(\mathcal{J})$, there exist an open neighborhood $U \ni x$ and $f \in \mathcal{J}(U)$ such that $f_x = 1$. It follows that $f|_V = 1$ in an open neighborhood $V \subseteq U$ of x. We then have $V \subseteq X \setminus V(\mathcal{J})$. The latter is therefore open.

For every $x \in V(\mathcal{J})$, $(j^{-1}(\mathcal{O}_X/\mathcal{J}))_x = (\mathcal{O}_X/\mathcal{J})_x = \mathcal{O}_{X,x}/\mathcal{J}_x$ is a local ring. The rest follows immediately. $\qquad\square$

In the rest of this book, given a sheaf of ideals $\mathcal{J}$, we will always endow $V(\mathcal{J})$ with the structure described in this lemma.

Proposition 2.24. *Let $f : Y \to X$ be a closed immersion of ringed topological spaces. Let Z be the ringed topological space $V(\mathcal{J})$ where $\mathcal{J} = \operatorname{Ker} f^\# \subseteq \mathcal{O}_X$. Then f factors into an isomorphism $Y \simeq Z$ followed by the canonical closed immersion $Z \to X$.*

Proof As $f(Y)$ is closed in X, we have

$$(f_*\mathcal{O}_Y)_x = \begin{cases} 0 & \text{if } x \notin f(Y) \\ \mathcal{O}_{Y,y} & \text{if } x = f(y). \end{cases}$$

Using the exact sequence $0 \to \mathcal{J} \to \mathcal{O}_X \to f_*\mathcal{O}_Y \to 0$ and Proposition 2.18, we deduce that $\mathcal{J}_x = \mathcal{O}_{X,x}$ if and only if $x \notin f(Y)$, whence the equality of sets $V(\mathcal{J}) = f(Y)$. Let $j : Z \to X$ be the canonical injection. Let $g : Y \to Z$ be the homeomorphism induced by f. Then $f = j \circ g$ as maps and $j_*\mathcal{O}_Z = \mathcal{O}_X/\mathcal{J} \simeq f_*\mathcal{O}_Y$. We show without difficulty that $f_*\mathcal{O}_Y = j_*g_*\mathcal{O}_Y$ (Exercise 2.6). It follows that (Exercise 2.13)

$$\mathcal{O}_Z = j^{-1}j_*\mathcal{O}_Z \simeq j^{-1}j_*g_*\mathcal{O}_Y = g_*\mathcal{O}_Y,$$

whence an isomorphism of ringed topological spaces $g : Y \simeq Z$. We leave it to the reader to verify that $f = j \circ g$ as morphisms of ringed topological spaces. $\square$

Exercises

2.1. Determine the sheaf associated to the presheaf $\mathcal{A}_X$ of Example 2.4. Such a sheaf is called a *constant sheaf*. Show that $\mathcal{A}_X$ is a sheaf if and only if every non-empty open subset of X is connected.

2.2. Let $\mathcal{F}$ be a sheaf on X. Let $s, t \in \mathcal{F}(X)$. Show that the set of $x \in X$ such that $s_x = t_x$ is open in X.

2.3. (*Sheaf associated to a presheaf*) Let us fix a topological space X. Let $\mathcal{F}$ be a presheaf on X. We keep the notation introduced just before Lemma 2.7.

 (a) Let $\mathcal{U} = \{U_i\}_i$ be an open covering of X. Let $\mathcal{F}_\mathcal{U}(X) = \text{Ker}\, d_1$. For any open subset W of X, we define in the same way a group $\mathcal{F}_\mathcal{U}(W)$ relative to the covering $\{W \cap U_i\}_i$ of W. Show that $\mathcal{F}_\mathcal{U}$ is a presheaf on X and that we have a morphism of presheaves $\mathcal{F} \to \mathcal{F}_\mathcal{U}$.

 (b) Let $\mathcal{V} = \{V_k\}_k$ be another open covering of X. We define a partial ordering: $\mathcal{V} \preceq \mathcal{U}$ if every V_k is contained in a U_i. Show that we have a canonical homomorphism $\mathcal{F}_\mathcal{U}(W) \to \mathcal{F}_\mathcal{V}(W)$ for every open subset W if $\mathcal{V} \preceq \mathcal{U}$.

 (c) Show that the ordering $\preceq$ makes the set of open coverings of X into a direct system. Let

$$\mathcal{F}'(W) = \varinjlim_\mathcal{V} \mathcal{F}_\mathcal{V}(W),$$

the direct limit being taken over the open coverings of W. Show that $\mathcal{F}'$ is the sheaf associated to $\mathcal{F}$ (see also Proposition 2.15).

2.4. Let $0 \to \mathcal{F}' \to \mathcal{F} \to \mathcal{F}''$ be an exact sequence of sheaves on X. Show that the sequence of Abelian groups $0 \to \mathcal{F}'(X) \to \mathcal{F}(X) \to \mathcal{F}''(X)$ is exact.

2.5. (*Supports of sheaves*) Let $\mathcal{F}$ be a sheaf on X. Let $\mathrm{Supp}\,\mathcal{F} = \{x \in X \mid \mathcal{F}_x \neq 0\}$. We want to show that in general, $\mathrm{Supp}\,\mathcal{F}$ is not a closed subset of X. Let us fix a sheaf $\mathcal{G}$ on X and a closed point $x_0 \in X$. Let us define a presheaf $\mathcal{F}$ by $\mathcal{F}(U) = \mathcal{G}(U)$ if $x_0 \notin U$ and $\mathcal{F}(U) = \{s \in \mathcal{G}(U) \mid s_{x_0} = 0\}$ otherwise. Show that $\mathcal{F}$ is a sheaf and that $\mathrm{Supp}\,\mathcal{F} = \mathrm{Supp}\,\mathcal{G} \setminus \{x_0\}$.

2.6. Let $f : X \to Y$, $g : Y \to Z$ be continuous maps, $\mathcal{F}$ a sheaf on X, and $\mathcal{H}$ a sheaf on Z. Show that

$$f^{-1}(g^{-1}\mathcal{H}) = (g \circ f)^{-1}\mathcal{H}, \quad g_*(f_*\mathcal{F}) = (g \circ f)_*\mathcal{F}.$$

2.7. Let $\mathcal{B}$ be a base of open subsets on a topological space X. Let $\mathcal{F}, \mathcal{G}$ be two sheaves on X. Suppose that for every $U \in \mathcal{B}$ there exists a homomorphism $\alpha(U) : \mathcal{F}(U) \to \mathcal{G}(U)$ which is compatible with restrictions. Show that this extends in a unique way to a homomorphism of sheaves $\alpha : \mathcal{F} \to \mathcal{G}$. Show that if $\alpha(U)$ is surjective (resp. injective) for every $U \in \mathcal{B}$, then α is surjective (resp. injective).

2.8. (*Glueing sheaves*) Let $\{U_i\}_i$ be an open covering of X. We suppose given sheaves $\mathcal{F}_i$ on U_i and isomorphisms $\varphi_{ij} : \mathcal{F}_i|_{U_i \cap U_j} \simeq \mathcal{F}_j|_{U_i \cap U_j}$ such that $\varphi_{ii} = \mathrm{Id}$ and $\varphi_{ik} = \varphi_{jk} \circ \varphi_{ij}$ on $U_i \cap U_j \cap U_k$. Show that there exists a unique sheaf $\mathcal{F}$ on X with isomorphisms $\psi_i : \mathcal{F}|_{U_i} \simeq \mathcal{F}_i$ such that $\psi_j = \varphi_{ij} \circ \psi_i$ on $U_i \cap U_j$. The sheaf $\mathcal{F}$ is called the *glueing of the* $\mathcal{F}_i$ *via the isomorphisms* φ_{ij}.

2.9. Let $\mathcal{F}$ be a sheaf on X whose support Y is a finite set of closed points. We say that $\mathcal{F}$ is a *skyscraper sheaf*. Show that for every open subset U of X, we have $\mathcal{F}(U) = \oplus_{x \in U} \mathcal{F}_x$. In particular, $\mathcal{F} = 0$ if and only if $\mathcal{F}(X) = 0$.

2.10. Let $(X, \mathcal{O}_X)$ be a ringed topological space. Let Y be a subset of X endowed with the induced topology. Let $i : Y \hookrightarrow X$ denote the canonical inclusion. Show that $(Y, i^{-1}\mathcal{O}_X)$ is a ringed topological space.

2.11. Let X, Y be two ringed topological spaces. We suppose given an open covering $\{U_i\}_i$ of X and morphisms $f_i : U_i \to Y$ such that $f_i|_{U_i \cap U_j} = f_j|_{U_i \cap U_j}$ (as morphisms) for every i, j. Show that there exists a unique morphism $f : X \to Y$ such that $f|_{U_i} = f_i$. This is the *glueing of the morphisms* f_i.

2.12. Let $f : X \to Y$ be a morphism of ringed topological spaces. Let V be an open subset of Y containing $f(X)$. Show that there exists a unique morphism $g : X \to V$ whose composition with the open immersion $V \to Y$ is f.

2.13. Let $f : X \to Y$ be a continuous map of topological spaces. Let $\mathcal{F}$ be a sheaf on X and $\mathcal{G}$ a sheaf on Y.

(a) Show that there exist canonical morphisms of sheaves

$$\mathcal{G} \to f_*f^{-1}\mathcal{G}, \quad f^{-1}f_*\mathcal{F} \to \mathcal{F}. \tag{2.1}$$

Moreover, if f is a closed immersion, then the first morphism is surjective. If f is an open immersion, then the second morphism is an isomorphism.

(b) Using the maps of (2.1), show that there exists a canonical bijection

$$\mathrm{Mor}_Y(\mathcal{G}, f_*\mathcal{F}) \simeq \mathrm{Mor}_X(f^{-1}\mathcal{G}, \mathcal{F}).$$

2.14. (*Quotient space*) Let G be a group acting on a ringed topological space $(X, \mathcal{O}_X)$ (i.e., there is a group homomorphism $G \to \mathrm{Aut}(X)$). A *quotient space* of X by G consists of a ringed topological space $(Y, \mathcal{O}_Y)$ and a morphism $p : X \to Y$ (called *quotient morphism*), verifying the following universal property:

> $p = p \circ \sigma$ for every $\sigma \in G$; and any morphism of ringed topological spaces $f : X \to Z$ verifying the same property factors in a unique way through p; that is, there exists a unique morphism $\tilde{f} : Y \to Z$ such that $f = \tilde{f} \circ p$.

The purpose of this exercise is to show that the quotient exists. Let $Y = X/G$ be the quotient set, and $p : X \to Y$ the canonical projection. We endow Y with the quotient topology, that is, $V \subseteq Y$ is open if and only if $p^{-1}(V)$ is open in X.

(a) Show that p is an open map.

(b) Show that G acts on the ringed topological space $p^{-1}(V)$. In particular, G acts on the ring $\mathcal{O}_X(p^{-1}(V))$.

(c) For any open subset V of Y, let $\mathcal{O}_Y(V) = \mathcal{O}_X(p^{-1}(V))^G$ be the set of elements which are invariant under G. Show that $\mathcal{O}_Y$ is a sheaf of rings on Y.

(d) Let $y \in Y$. We wish to show that $\mathcal{O}_{Y,y}$ is a local ring. Let us fix a point $x \in p^{-1}(y)$. Show that $\mathcal{O}_{Y,y} \subseteq \mathcal{O}_{X,x}$. Let $\mathfrak{m}_x$ denote the maximal ideal of $\mathcal{O}_{X,x}$ and let us set $\mathfrak{m}_y = \mathfrak{m}_x \cap \mathcal{O}_{Y,y}$. Show that $\mathfrak{m}_y$ is the unique maximal ideal of $\mathcal{O}_{Y,y}$.

(e) Show that $(Y, \mathcal{O}_Y)$ is a ringed topological space, that p induces a morphism of ringed topological spaces from $(X, \mathcal{O}_X)$ to $(Y, \mathcal{O}_Y)$, and that $p : X \to Y$ verifies the universal property of the quotient given above.

Remark. The quotient in the universal sense does not, in general, exist in the category of schemes. And even in the case when the quotient scheme exists, it does not always coincide with the quotient as a ringed topological space. See Exercise 3.22.

2.3 Schemes

In this section, we give the definitions of schemes and of morphisms of schemes. Examples of important schemes such as projective schemes, Noetherian schemes, and algebraic varieties are given in the last two subsections.

2.3.1   Definition of schemes and examples

As we have said above, a scheme is a ringed topological space which is locally
an affine scheme. Let us therefore begin by defining affine schemes. Let A be a
ring. Let us consider the space $X = \operatorname{Spec} A$ endowed with its Zariski topology
(Proposition 1.1). We will construct a sheaf of rings $\mathcal{O}_X$ on X.

Let $D(f)$ be a principal open subset of X. Let us set $\mathcal{O}_X(D(f)) = A_f$.
Let $D(g) \subseteq D(f)$, in which case $g \in \sqrt{fA}$ (Lemma 1.6(b)). Hence there exist
$m \geq 1$ and $b \in A$ such that $g^m = fb$. Thus f is invertible in A_g. This canonically induces a restriction homomorphism $A_f \to A_g$ which maps $af^{-n} \in A_f$ to
$(ab^n)g^{-nm} \in A_g$. If $D(f) = D(g)$, we easily verify that $A_f \to A_g$ is an isomorphism. Therefore $\mathcal{O}_X(D(f))$ does not depend on the choice of f. We have thus
constructed a $\mathcal{B}$-presheaf, where $\mathcal{B}$ is the base of open subsets on X made up of
the principal open sets (Remark 2.6).

Proposition 3.1. *Let A be a ring and $X = \operatorname{Spec} A$. We have the following
properties:*

- (a) *$\mathcal{O}_X$ is a $\mathcal{B}$-sheaf of rings. It therefore induces a sheaf of rings $\mathcal{O}_X$ on
  $\operatorname{Spec} A$, and we have $\mathcal{O}_X(X) = A$.*

- (b) *For any $\mathfrak{p} \in X$, the stalk $\mathcal{O}_{X,\mathfrak{p}}$ is canonically isomorphic to $A_{\mathfrak{p}}$. In
  particular, $(X, \mathcal{O}_X)$ is a ringed topological space.*

Proof (a) We will verify conditions (4), (5) of Definition 2.2 for the open set
$U = X$. The case of an arbitrary principal open set is shown in a similar way.
Let us write $U_i = D(f_i)$. We then have $V(\sum_i f_i A) = \cap_i V(f_i A) = \emptyset$, and hence
$A = \sum_i f_i A$. There therefore exists a finite subset $F \subseteq I$ such that $1 \in \sum_{i \in F} f_i A$.

Let $s \in A$ be such that $s|_{U_i} = 0$ for every i. Then there exists an $m \geq 1$
such that $f_i^m s = 0$ for every $i \in F$. Now $X = \cup_{i \in F} D(f_i) = \cup_{i \in F} D(f_i^m)$. It
follows that $1 \in \sum_{i \in F} f_i^m A$, and hence $s \in \sum_{i \in F} s f_i^m A = \{0\}$. This implies the
uniqueness condition (Definition 2.2).

Let $s_i \in \mathcal{O}_X(U_i)$ be sections which coincide on the intersections $U_i \cap U_j$. We
must find an $s \in A$ such that $s|_{U_i} = s_i$ for every i. There exists an $m \geq 1$ such
that $s_i = b_i f_i^{-m} \in A_{f_i}$ for every $i \in F$. Moreover, the condition $s_i|_{U_i \cap U_j} =
s_j|_{U_i \cap U_j}$ implies that there exists an $r \geq 1$ such that $(b_i f_j^m - b_j f_i^m)(f_i f_j)^r = 0$
for every $i, j \in F$. As above, we have a relation $1 = \sum_{j \in F} a_j f_j^{m+r}$ with $a_j \in A$.
Let us set $s = \sum_{j \in F} a_j b_j f_j^r \in A$. For every $i \in F$, we have

$$ f_i^{m+r} s = \sum_{j \in F} a_j b_j f_j^r f_i^{m+r} = \sum_{j \in F} a_j b_i f_i^r f_j^{m+r} = b_i f_i^r. $$

It follows that $s|_{U_i} = s_i$ for every $i \in F$. More generally, if $i \in I$, then
$(s|_{U_i})|_{U_i \cap U_j} = s_i|_{U_i \cap U_j}$ for every $j \in F$. As $U_i = \cup_{j \in F} U_i \cap U_j$, we have $s|_{U_i} = s_i$.

(b) A principal open set $D(f)$ contains $\mathfrak{p}$ if and only if $f \notin \mathfrak{p}$. It suffices to
show that the canonical homomorphism

$$ \varphi : \varinjlim_{f \notin \mathfrak{p}} A_f \to A_{\mathfrak{p}} $$

is an isomorphism. Every element α of $A_{\mathfrak{p}}$ can be written $\alpha = af^{-1}$ for some $f \notin \mathfrak{p}$. It follows that α is in the image of A_f. Hence φ is surjective. On the other hand, if $af^{-n} \in A_f$ ($f \notin \mathfrak{p}$) is mapped to 0 in $A_{\mathfrak{p}}$, there exists a $g \notin \mathfrak{p}$ such that $ga = 0$. It follows that $af^{-n} = 0$ in A_{gf}. Hence φ is injective. $\qquad\square$

Definition 3.2. We define an *affine scheme* to be a ringed topological space isomorphic to some $(\operatorname{Spec} A, \mathcal{O}_{\operatorname{Spec} A})$ constructed as above. By abuse of notation, the latter will often be denoted simply by $\operatorname{Spec} A$.

Before studying some examples, let us show a convenient lemma concerning $\mathcal{O}_{\operatorname{Spec} A}$.

Lemma 3.3. *Let A be an integral domain, with field of fractions K. Let ξ be the point of $X = \operatorname{Spec} A$ corresponding to the prime ideal 0. Then $\mathcal{O}_{X,\xi} = K$. Moreover, for every non-empty open subset U of X, we have $\xi \in U$, and the canonical homomorphism $\mathcal{O}_X(U) \to \mathcal{O}_{X,\xi}$ is injective. If $V \subseteq U$, then the restriction $\mathcal{O}_X(U) \to \mathcal{O}_X(V)$ is injective.*

Proof The equality $\mathcal{O}_{X,\xi} = K$ comes from Proposition 3.1(b). If $U = D(f)$, then $\mathcal{O}_X(U) = A_f \subset K$. In the general case, $U = \cup_i D(f_i)$. If $s \in \mathcal{O}_X(U)$ is mapped to 0 in K, then $s|_{D(f_i)} = 0$ for every i, so $s = 0$. Therefore $\mathcal{O}_X(U) \to K$ is injective. The injectivity of $\mathcal{O}_X(U) \to \mathcal{O}_X(V)$ follows immediately. $\qquad\square$

Thus we see that for an integral domain A, $\mathcal{O}_{\operatorname{Spec} A}$ is a subsheaf of the constant presheaf $K = \operatorname{Frac}(A)$ (which in this case is a sheaf).

Example 3.4. Let k be a field. Let us study the example of the affine line $X = \mathbb{A}_k^1$ (Example 1.4). Every non-empty subset U of X is of the form $U = D(P(T))$, where $P(T) \in k[T] \setminus \{0\}$. We have $\mathcal{O}_X(U) = k[T]_{P(T)} = k[T, 1/P(T)]$. This is the set of rational fractions whose denominator is only divisible by the irreducible factors of $P(T)$. If k is algebraically closed, we can consider a rational fraction as a function $k \to k \cup \{+\infty\}$. Then $\mathcal{O}_X(U)$ consists exactly of those without a pole in U.

Example 3.5. Let $X = \operatorname{Spec} \mathbb{Z}$. A proper open subset of X is always of the form $D(f)$, where f is an non-zero integer. We have $\mathcal{O}_X(D(f)) = \mathbb{Z}_f \subseteq \mathbb{Q}$. A rational number a/b (with $(a, b) = 1$) belongs to $\mathcal{O}_X(D(f))$ if and only if every prime number dividing b also divides f.

Example 3.6. Let $X = \operatorname{Spec} \mathbb{Z}[T]$. Let p be a prime number. Then (T, p) is a maximal ideal of $\mathbb{Z}[T]$. Let $U \subset X$ denote the complement of the corresponding closed point. We have $U = D(p) \cup D(T)$. Therefore $\mathcal{O}_X(U) \subseteq \mathcal{O}_X(D(T)) \cap \mathcal{O}_X(D(p)) = \mathbb{Z}[T, 1/T] \cap \mathbb{Z}[T, 1/p] = \mathbb{Z}[T]$. This implies that $\mathcal{O}_X(U) = \mathbb{Z}[T] = \mathcal{O}_X(X)$.

Lemma 3.7. *Let $X = \operatorname{Spec} A$ be an affine scheme, and let $g \in A$. Then the open subset $D(g)$, endowed with the structure of a ringed topological space induced by that of X, is an affine scheme isomorphic to $\operatorname{Spec} A_g$ (as a ringed topological space).*

Proof Let Y denote the affine scheme $\operatorname{Spec} A_g$. By Lemma 1.7(c), we canonically have a topological open immersion $i : Y \to X$ whose image is $D(g)$. Let $D(h)$ be a principal open subset of X contained in $D(g)$. Let $\overline{h}$ be the image of h in A_g. We canonically have $\mathcal{O}_X(D(h)) = A_h \simeq (A_g)_{\overline{h}} = \mathcal{O}_Y(D(\overline{h})) = i_* \mathcal{O}_Y(D(h))$. As the $D(h)$ form a base of open subsets on $D(g)$, this shows that i induces an isomorphism from $(Y, \mathcal{O}_Y)$ onto $(D(g), \mathcal{O}_X|_{D(g)})$ (Exercise 2.7). □

Definition 3.8. A *scheme* is a ringed topological space $(X, \mathcal{O}_X)$ admitting an open covering $\{U_i\}_i$ such that $(U_i, \mathcal{O}_X|_{U_i})$ is an affine scheme for every i. We will denote it simply by X when there is no confusion possible. Let U be an open subset of a scheme X. The elements of $\mathcal{O}_X(U)$ are called, somewhat improperly, *regular functions on U*. See, however, Example 3.4 and Proposition 4.4, which illustrate well the function aspect of the sheaf $\mathcal{O}_X$.

An affine scheme is a scheme. If a ringed topological space X admits an open covering $\{U_i\}_i$ such that $(U_i, \mathcal{O}_X|_{U_i})$ is a scheme for every i, then X is a scheme. Conversely, we have the following proposition:

Proposition 3.9. *Let X be a scheme. Then for any open subset U of X, the ringed topological space $(U, \mathcal{O}_X|_U)$ is also a scheme.*

Proof By definition, we have $X = \cup_i U_i$, where U_i is open and affine. It suffices to show that $U \cap U_i$ is a scheme. Now, $U \cap U_i$ is a union of principal open subsets of U_i, each of these principal open sets being an affine scheme by Lemma 3.7, so $U \cap U_i$ is indeed a scheme. □

Definition 3.10. Let X be a scheme. Let U be an open subset of X. We will say that $(U, \mathcal{O}_X|_U)$ (or more simply that U) is an *open subscheme* of X. We will say that U is an *affine open subset* if $(U, \mathcal{O}_X|_U)$ is an affine scheme.

In all that follows, an open subset U of X will always be endowed with the scheme structure $(U, \mathcal{O}_X|_U)$. Let us note that the affine open subsets of X form a base of open sets. We can see that in Example 3.6 above, the open subscheme U is not affine (by using the fact that $\mathcal{O}_X(U) = \mathcal{O}_X(X)$ and Proposition 3.25). We will encounter other non-affine schemes at the end of this section.

The notion of principal open subsets of an affine scheme can be generalized.

Definition 3.11. Let X be a scheme and $f \in \mathcal{O}_X(X)$. We denote by X_f the set of $x \in X$ such that $f_x \in \mathcal{O}_{X,x}^*$.

Let us consider the following condition:

$$\left\{ \begin{array}{l} X \text{ admits a covering by a finite number of affine} \\ \text{open subsets } \{U_i\}_i \text{ such that } U_i \cap U_j \text{ also admits} \\ \text{a finite covering by affine open subsets.} \end{array} \right. \qquad (3.2)$$

Affine schemes always verify this condition, as do Noetherian schemes (Proposition 3.46(a)).

Proposition 3.12. *Let X be a scheme and $f \in \mathcal{O}_X(X)$. Then X_f is an open subset of X. Moreover, if X verifies condition (3.2) above, then the restriction $\mathcal{O}_X(X) \to \mathcal{O}_X(X_f)$ induces an isomorphism $\mathcal{O}_X(X)_f \simeq \mathcal{O}_X(X_f)$.*

Proof If $x \in X_f$, there exist an open neighborhood $U \ni x$ and a $g \in \mathcal{O}_X(U)$ such that $f_x g_x = 1$. It follows that $(fg)|_V = 1$ in an open neighborhood $V \subseteq U$ of x. We then have $V \subseteq X_f$. The latter is therefore open. Moreover, by glueing the inverses $g \in \mathcal{O}_X(V)$ of f as the V vary, we obtain an element of $\mathcal{O}_X(X_f)$, the inverse of $f|_{X_f}$. Consequently, $\mathcal{O}_X(X) \to \mathcal{O}_X(X_f)$ induces a homomorphism $\alpha : \mathcal{O}_X(X)_f \to \mathcal{O}_X(X_f)$.

By hypothesis (3.2), X admits a finite covering $\mathcal{U} = \{U_i\}_i$ by affine open subsets. Hence X_f is the union of the $V_i := U_i \cap X_f = D(f|_{U_i})$. For simplicity, let us also denote by f its restriction to any open subset of X. We have $\mathcal{O}_X(U_i)_f = \mathcal{O}_X(V_i)$ by virtue of Lemma 3.7. With the notation of Lemma 2.7, we have an exact complex $C^\bullet(\mathcal{U}, \mathcal{O}_X)$ where we can replace $\prod$ by $\oplus$ since there are only a finite number of terms. Therefore $C^\bullet(\mathcal{U}, \mathcal{O}_X) \otimes_{\mathcal{O}_X(X)} \mathcal{O}_X(X)_f$ is still exact owing to the flatness of $\mathcal{O}_X(X) \to \mathcal{O}_X(X)_f$ (Corollary 1.2.11). This gives a commutative diagram:

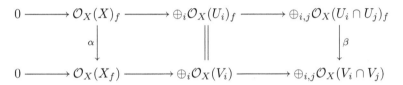

where the horizontal lines are exact. It follows that α is injective. Let us note that we have only used the hypothesis that X admits a covering by a finite number of affine open subsets. We can therefore apply the same reasoning to $U_i \cap U_j$, which implies that β is injective. Returning to the commutative diagram above, this shows that α is an isomorphism. □

2.3.2 Morphisms of schemes

Definition 3.13. A *morphism of schemes* $f : X \to Y$ is a morphism of ringed topological spaces. An *isomorphism of schemes* is an isomorphism of ringed topological spaces. An *open or closed immersion of schemes* is an open or closed immersion of ringed topological spaces (Definition 2.22).

The proposition that follows makes it possible to construct morphisms of schemes. Let $\varphi : A \to B$ be a ring homomorphism. We have defined a continuous map $\operatorname{Spec} \varphi : \operatorname{Spec} B \to \operatorname{Spec} A$ in Lemma 1.7. For simplicity, let us denote it by f_φ.

Proposition 3.14. *Let $\varphi : A \to B$ be a ring homomorphism. Then there exists a morphism of schemes $(f_\varphi, f_\varphi^\#) : \operatorname{Spec} B \to \operatorname{Spec} A$ such that $f_\varphi^\#(\operatorname{Spec} A) = \varphi$.*

Proof Let $X = \operatorname{Spec} B$ and $Y = \operatorname{Spec} A$. For every $g \in A$, we have $f_\varphi^{-1}(D(g)) = D(\varphi(g))$, and φ canonically induces a ring homomorphism $\mathcal{O}_Y(D(g)) = A_g \to B_{\varphi(g)} = (f_\varphi)_* \mathcal{O}_X(D(g))$. This homomorphism is compatible with the restrictions; we therefore have a morphism of sheaves $f_\varphi^\# : \mathcal{O}_Y \to f_{\varphi*}\mathcal{O}_X$ since the principal open sets form a base. Moreover, for every $x \in X$ corresponding to a prime ideal $\mathfrak{q}$ of B, the canonical homomorphism $A_{\varphi^{-1}(\mathfrak{q})} \to B_\mathfrak{q}$ induced by φ is a local homomorphism and coincides with $(f_\varphi^\#)_x$. Therefore $(f_\varphi, f_\varphi^\#)$ is indeed a morphism of ringed topological spaces. By construction, we have $f_\varphi^\#(Y) = \varphi$. □

Example 3.15. Let A be a ring and $g \in A$. Proposition 3.14 associates a morphism $f_\varphi : \operatorname{Spec} A_g \to \operatorname{Spec} A$ to the localization homomorphism $\varphi : A \to A_g$. We easily see that this morphism factors into the isomorphism $\operatorname{Spec} A_g \simeq D(g)$ of Lemma 3.7, followed by the open immersion $D(g) \to \operatorname{Spec} A$.

Example 3.16. Let X be a scheme, $x \in X$. Then we have a canonical morphism $\operatorname{Spec} \mathcal{O}_{X,x} \to X$ defined as follows. Let us take an affine open set $U \ni x$. The canonical homomorphism $\mathcal{O}_X(U) \to \mathcal{O}_{X,x}$ induces a morphism $\operatorname{Spec} \mathcal{O}_{X,x} \to U$. Composing this morphism with the open immersion $U \to X$, we obtain a morphism $\operatorname{Spec} \mathcal{O}_{X,x} \to X$. This morphism does not depend on the choice of U. See also Exercise 4.2.

We will now turn our attention to closed subschemes and closed immersions.

Lemma 3.17. *Let A be a ring. Let I be an ideal of A. Then the morphism of schemes*

$$i : \operatorname{Spec} A/I \to \operatorname{Spec} A,$$

induced by the canonical surjection $A \to A/I$, is a closed immersion of schemes whose image is $V(I)$. Moreover, for any principal open subset $D(g)$ of $\operatorname{Spec} A$, we have $(\operatorname{Ker} i^\#)(D(g)) = I \otimes_A A_g$.

Proof By Lemma 1.7(b), the map i is a closed immersion whose image is $V(I)$. It remains to show the properties concerning $i^\#$. For every principal open subset $D(g)$ of $\operatorname{Spec} A$, $i^\#(D(g))$ is, by construction, the canonical surjection $\mathcal{O}_{\operatorname{Spec} A}(D(g)) = A_g \to (A/I)_{\bar{g}} = i_*\mathcal{O}_{\operatorname{Spec} A/I}(D(g))$, where $\bar{g}$ is the image of g in A/I. This proves the surjectivity of $i^\#$ (Exercise 2.7) and that $(\operatorname{Ker} i^\#)(D(g)) = I_g = I \otimes_A A_g$. □

Example 3.18. Let X be a scheme and $x \in X$. Let $k(x)$ be the residue field of X at x (Definition 2.19). Then the canonical surjection $\mathcal{O}_{X,x} \to k(x)$ induces a closed immersion $\operatorname{Spec} k(x) \to \operatorname{Spec} \mathcal{O}_{X,x}$. Composing with the morphism $\operatorname{Spec} \mathcal{O}_{X,x} \to X$ of Example 3.16, we obtain a morphism $\operatorname{Spec} k(x) \to X$. Set-theoretically, this morphism maps the unique point of $\operatorname{Spec} k(x)$ to the point $x \in X$.

We have examined above the notion of open subschemes. That of closed subschemes is a bit more delicate.

Definition 3.19. A *closed subscheme* of X is a closed subset Z of X endowed with the structure $(Z, \mathcal{O}_Z)$ of a scheme and with a closed immersion $(j, j^\#) : (Z, \mathcal{O}_Z) \to (X, \mathcal{O}_X)$, where $j : Z \to X$ is the canonical injection.

In general, on a given closed subset Z, the structure of a closed subscheme of X is not unique (see, however, Proposition 4.2). It is therefore necessary always to specify the scheme structure that we want to put on Z.

Proposition 3.20. *Let $X = \operatorname{Spec} A$ be an affine scheme. Let $j : Z \to X$ be a closed immersion of schemes. Then Z is affine and there exists a unique ideal J of A such that j induces an isomorphism from Z onto $\operatorname{Spec} A/J$.*

Proof Let us first show that Z verifies condition (3.2) above. There exist open subsets U_p of X such that $\{j^{-1}(U_p)\}_p$ is an affine open covering of Z. Each U_p is a union of principal open subsets $\{U_{pk}\}_k$ of X, and $j^{-1}(U_{pk})$ is affine because it is a principal open subset of $j^{-1}(U_p)$. By adding principal open sets covering $X \setminus j(Z)$, we obtain (since X is quasi-compact, see Exercise 1.6) a covering of X by a finite number of principal open sets $\{V_l\}_l$ such that $j^{-1}(V_l)$ is affine for every l. As $j^{-1}(V_l) \cap j^{-1}(V_{l'})$ is a principal open subset of $j^{-1}(V_l)$, and therefore affine, we indeed have condition (3.2).

Let $\mathcal{J} = \operatorname{Ker} j^{\#}$ and $J = \mathcal{J}(X)$. We know that Z is isomorphic to $V(\mathcal{J})$ endowed with the sheaf $\mathcal{O}_X/\mathcal{J}$ (Proposition 2.24). Let $g \in A$. We let h denote the image of g in $\mathcal{O}_Z(Z)$. From the exact sequence $0 \to J \to A \to \mathcal{O}_Z(Z)$, we deduce an exact sequence by $\otimes_A A_g$:

$$0 \to J \otimes_A A_g \to A_g = \mathcal{O}_X(D(g)) \to \mathcal{O}_Z(Z)_h.$$

Now $\mathcal{O}_Z(Z)_h = \mathcal{O}_Z(Z_h) = j_* \mathcal{O}_Z(D(g))$ (Proposition 3.12); therefore we have $\mathcal{J}(D(g)) = J \otimes_A A_g$. Let $i : \operatorname{Spec} A/J \to X$ be the closed immersion defined in the preceding lemma. Then $(\operatorname{Ker} i^{\#})(D(g)) = \mathcal{J}(D(g))$ for every principal open subset $D(g)$ of X. It follows that $\operatorname{Ker} i^{\#} = \mathcal{J}$. Hence $Z \simeq \operatorname{Spec} A/J$. Finally, the uniqueness of J comes from the fact that if we have an isomorphism $A/J \to A/J'$ which is compatible with the canonical surjections $A \to A/J$ and $A \to A/J'$, then $J = J'$. $\qquad\square$

Definition 3.21. Let S be a scheme. An *S-scheme* or a *scheme over S* is a scheme X endowed with a morphism of schemes $\pi : X \to S$. The morphism π is then called the *structural morphism* and S the *base scheme*. When $S = \operatorname{Spec} A$, we will also say *scheme over A* or *A-scheme* instead of scheme over S, and A will be called *base ring*.

If $\pi : X \to S$, $\rho : Y \to S$ are S-schemes, a *morphism of S-schemes* $f : X \to Y$ is a morphism of schemes that is compatible with the structural morphisms (i.e., $\rho \circ f = \pi$).

Example 3.22. Let A be a ring. For any A-algebra B, $\operatorname{Spec} B$ is canonically an A-scheme (Proposition 3.14). We let $\mathbb{A}_A^n = \operatorname{Spec} A[T_1, \ldots, T_n]$. This is therefore an A-scheme, called the *affine space* (of relative dimension n) over A.

Morphisms into an affine scheme

Let us study morphisms to an affine scheme more closely. Let X, Y be two schemes. Let $\operatorname{Mor}(X, Y)$ denote the set of morphisms of schemes from X to Y. For two rings A, B, we let $\operatorname{Hom}_{\mathrm{rings}}(A, B)$ (or more simply $\operatorname{Hom}(A, B)$ if there is no ambiguity possible) denote the set of ring homomorphisms from A to B. We have a canonical map

$$\rho : \operatorname{Mor}(X, Y) \to \operatorname{Hom}_{\mathrm{rings}}(\mathcal{O}_Y(Y), \mathcal{O}_X(X)), \tag{3.3}$$

which to $(f, f^\#)$ associates $f^\#(Y) : \mathcal{O}_Y(Y) \to f_*\mathcal{O}_X(Y) = \mathcal{O}_X(X)$. This map is 'functorial' in X in the sense that for any morphism of schemes $g : Z \to X$, we have a commutative diagram

$$
\begin{array}{ccc}
\mathrm{Mor}(X,Y) & \longrightarrow & \mathrm{Hom}_{\mathrm{rings}}(\mathcal{O}_Y(Y), \mathcal{O}_X(X)) \\
\downarrow & & \downarrow \\
\mathrm{Mor}(Z,Y) & \longrightarrow & \mathrm{Hom}_{\mathrm{rings}}(\mathcal{O}_Y(Y), \mathcal{O}_Z(Z))
\end{array}
\tag{3.4}
$$

the vertical arrow on the left being the composition with g and that on the right the composition with $g^\#(X) : \mathcal{O}_X(X) \to \mathcal{O}_Z(Z)$.

Lemma 3.23. *Let X, Y be affine schemes. Then the map (3.3) defined above is bijective.*

Proof Let $f \in \mathrm{Mor}(X,Y)$ and $\varphi = \rho(f) = f^\#(Y)$. Proposition 3.14 gives a morphism of schemes $f_\varphi : X \to Y$. It suffices to show that $f = f_\varphi$. Let $B = \mathcal{O}_X(X)$ and $A = \mathcal{O}_Y(Y)$. For any $x \in X$ corresponding to a prime ideal $\mathfrak{q} \subset B$, we have a commutative diagram

$$
\begin{array}{ccc}
A & \xrightarrow{\;\varphi\;} & B \\
\downarrow & & \downarrow \\
A_\mathfrak{p} & \xrightarrow{\;f_x^\#\;} & B_\mathfrak{q}
\end{array}
$$

where $\mathfrak{p}$ is the prime ideal corresponding to $f(x)$, and the vertical homomorphisms are localization homomorphisms. By considering the invertible elements of $A_\mathfrak{p}$ and $B_\mathfrak{q}$, we see that $\varphi(A \setminus \mathfrak{p}) \subseteq B \setminus \mathfrak{q}$. Hence $\varphi^{-1}(\mathfrak{q}) \subseteq \mathfrak{p}$. The fact that $f_x^\#$ is a local homomorphism implies that there is equality. Therefore f and f_φ coincide set-theoretically, and $f_x^\# = f_{\varphi,x}^\#$ for every $x \in X$. It follows from Lemma 2.9 that $f^\# = f_\varphi^\#$. □

Remark 3.24. This lemma expresses the fact that the category of affine schemes is equivalent to that of (commutative) rings (with unit).

Proposition 3.25. *Let Y be an affine scheme. For any scheme X, the canonical map*

$$
\rho : \mathrm{Mor}(X,Y) \to \mathrm{Hom}_{\mathrm{rings}}(\mathcal{O}_Y(Y), \mathcal{O}_X(X))
$$

is bijective.

Proof We have $X = \cup_i U_i$ where the U_i are affine open subsets. Using the commutative diagram (3.4) above with each $U_i \to X$, we obtain a new one:

$$
\begin{array}{ccc}
\mathrm{Mor}(X,Y) & \xrightarrow{\;\rho\;} & \mathrm{Hom}(\mathcal{O}_Y(Y), \mathcal{O}_X(X)) \\
\alpha \downarrow & & \downarrow \beta \\
\prod_i \mathrm{Mor}(U_i,Y) & \xrightarrow{\;\gamma\;} & \prod_i \mathrm{Hom}(\mathcal{O}_Y(Y), \mathcal{O}_X(U_i))
\end{array}
$$

By virtue of Lemma 3.23, γ is bijective. As α is clearly injective, it follows that ρ is injective. Let $\varphi \in \mathrm{Hom}(\mathcal{O}_Y(Y), \mathcal{O}_X(X))$; then φ_i, the composition of φ

with $\mathcal{O}_X(X) \to \mathcal{O}_X(U_i)$, pulls back to a morphism $f_i \in \text{Mor}(U_i, Y)$. We have $f_i|_{U_i \cap U_j} = f_j|_{U_i \cap U_j}$. Indeed, for every affine open $V \subseteq U_i \cap U_j$, $f_i|_V$ and $f_j|_V$ have the same image in $\text{Hom}(\mathcal{O}_Y(Y), \mathcal{O}_X(V))$, and are therefore equal. It follows that the f_i glue to a morphism $f \in \text{Mor}(X, Y)$ (Exercise 2.11). We have $\rho(f) = \varphi$ by the injectivity of β, which proves that ρ is surjective. □

Corollary 3.26. *Let A be a ring. Then giving a scheme X over A is equivalent to giving a scheme X and the structure of a sheaf of A-algebras on $\mathcal{O}_X$.*

Example 3.27. Every scheme X is in a unique way a $\mathbb{Z}$-scheme, because every ring has one and only one $\mathbb{Z}$-algebra structure.

Let us now return to arbitrary base schemes.

Definition 3.28. Let $\pi : X \to S$ be an S-scheme. A *section of X* is a morphism of S-schemes $\sigma : S \to X$. This amounts to saying that $\pi \circ \sigma = \text{Id}_S$. The set of sections of X is denoted by $X(S)$ (and also by $X(A)$ if $S = \text{Spec } A$).

Example 3.29. Let X be a scheme over a field k. Then we can identify $X(k)$ with the set of points $x \in X$ such that $k(x) = k$. Indeed, let $\sigma \in X(k)$, and let x be the image of the point of $\text{Spec } k$. The homomorphism $\sigma_x^\#$ induces a field homomorphism $k(x) \to k$. As $k(x)$ is a k-algebra, this implies that $k(x) = k$. Conversely, if $x \in X$ verifies $k(x) = k$, there exists a unique section $\text{Spec } k \to X$ whose image is x (Example 3.18).

Definition 3.30. Let X be a scheme over a field k. With the identification above, we call the points of $X(k)$ *(k-)rational points of X*. The notion of rational points is fundamental in arithmetic geometry.

Remark 3.31. Let Y be an open or closed subscheme of X. For any point $y \in Y$, the residue fields of $\mathcal{O}_Y$ and of $\mathcal{O}_X$ at y are isomorphic. So if X is a scheme over a field k, then $Y(k) = X(k) \cap Y$.

Example 3.32. Let k be a field and $X = \text{Spec } k[T_1, \ldots, T_n]/I$ an affine scheme over k. Let Z be the set of $\alpha = (\alpha_1, \ldots, \alpha_n) \in k^n$ such that $P(\alpha) = 0$ for every $P(T) \in I$ (see Subsection 2.1.2 for the case when k is algebraically closed). Then we have a canonical bijection $\lambda : Z \to X(k)$.

Indeed, let $\alpha = (\alpha_1, \ldots, \alpha_n) \in Z$, and consider the maximal ideal $\mathfrak{m}_\alpha := (T_1 - \alpha_1, \ldots, T_n - \alpha_n)$. Then $I \subseteq \mathfrak{m}_\alpha$ (see the proof of Corollary 1.15). Therefore $\mathfrak{m}_\alpha$ corresponds to a point $x = \lambda(\alpha)$ of $X = V(I)$ and the residue field $k(x)$ is $k[T_1, \ldots, T_n]/\mathfrak{m}_\alpha = k$. Conversely, if $x \in X(k)$, we set α_i equal to the image of T_i in $k(x) = k$. We have $P(\alpha) = 0$ for every $P(T) \in I$ by a reasoning similar to that of Corollary 1.15. It is now clear that λ is a bijection.

Thus, solving a system of polynomial equations in k amounts to determining the set of rational points of a scheme of the type $\text{Spec } k[T_1, \ldots, T_n]/I$.

Let us finish with a lemma which is useful for constructing schemes.

Lemma 3.33 (Glueing schemes). *Let S be a scheme. Let us consider a family $\{X_i\}_i$ of schemes over S. We suppose given open subschemes X_{ij} of X_i and isomorphisms of S-schemes $f_{ij} : X_{ij} \xrightarrow{\sim} X_{ji}$ such that $f_{ii} = \text{Id}_{X_i}$, $f_{ij}(X_{ij} \cap X_{ik}) =$*

$X_{ji} \cap X_{jk}$, and $f_{ik} = f_{jk} \circ f_{ij}$ on $X_{ij} \cap X_{ik}$. Then there exists an S-scheme X, unique up to isomorphism, with open immersions (of S-schemes) $g_i : X_i \to X$ such that $g_i = g_j \circ f_{ij}$ on X_{ij}, and that $X = \cup_i g_i(X_i)$.

Proof We first construct a topological space X by taking the disjoint union $\coprod_i X_i$ modulo the equivalence relation $x \sim y$ if $x \in X_i$, $y \in X_j$, and $y = f_{ij}(x)$, and we endow X with the quotient topology. We then have topological open immersions $g_i : X_i \hookrightarrow X$ such that $g_i = g_j \circ f_{ij}$ on X_{ij}. Let us set $U_i = g_i(X_i)$ and $\mathcal{O}_{U_i} = g_{i*}\mathcal{O}_{X_i}$. It is then easy to verify that $\mathcal{O}_{U_i}|_{U_i \cap U_j} = \mathcal{O}_{U_j}|_{U_i \cap U_j}$. Let $\mathcal{O}_X$ be the sheaf on X such that $\mathcal{O}_X|_{U_i} = \mathcal{O}_{U_i}$. Then it is clear that $(X, \mathcal{O}_X)$ is a scheme and that the g_i induce isomorphisms $X_i \simeq U_i$. Finally, let $h_i : U_i \to S$ be the composition of g_i^{-1} and the structural morphism $X_i \to S$. Then $h_i|_{U_i \cap U_j} = h_j|_{U_i \cap U_j}$. By glueing, this gives a morphism $X \to S$, which is compatible with the S-scheme structure of the X_i, whence the existence of an X with the required properties. The uniqueness is clearly true.   □

The scheme X is called the *glueing of the X_i* via the isomorphisms f_{ij}, or *along the X_{ij}*.

Example 3.34. Let us fix a ring A and an integer $n \geq 0$. For every $0 \leq i \leq n$, let us set

$$X_i = \operatorname{Spec} A[\, T_i^{-1}T_j \,]_{0 \leq j \leq n}, \quad X_{ij} = D(T_i^{-1}T_j) \subseteq X_i.$$

We have $\mathcal{O}_{X_i}(X_{ij}) = A[\, T_i^{-1}T_j, T_j^{-1}T_i, T_i^{-1}T_k \,]_{0 \leq k \leq n} = \mathcal{O}_{X_j}(X_{ji})$. We therefore have a canonical isomorphism $X_{ij} \simeq X_{ji}$. The A-schemes X_i can be glued along the X_{ij}. We denote the resulting A-scheme by $\mathbb{P}_A^n$. This is the *projective space (of relative dimension n) over A*. Below, we will see a different construction of $\mathbb{P}_A^n$.

2.3.3   Projective schemes

In this subsection we will consider a very important example of schemes, that of projective schemes. In terms of polynomial equations, it concerns homogeneous equations.

Let us fix a ring A. Let $B = \oplus_{d \geq 0} B_d$ be a graded A-algebra (see 1.3.2). An ideal I of B is said to be *homogeneous* if it is generated by homogeneous elements. This is equivalent to $I = \oplus_{d \geq 0}(I \cap B_d)$. The quotient B/I then has a natural grading $(B/I)_d = B_d/(I \cap B_d)$. We let $\operatorname{Proj} B$ denote the set of homogeneous prime ideals of B which do not contain the ideal $B_+ := \oplus_{d > 0} B_d$. In what follows, we will endow this set with the structure of a scheme (Proposition 3.38).

For any homogeneous ideal I of B, we let $V_+(I)$ denote the set of $\mathfrak{p} \in \operatorname{Proj} B$ containing I. We trivially have the equalities

$$\cap_i V_+(I_i) = V_+(\textstyle\sum_i I_i), \quad V_+(I) \cup V_+(J) = V_+(I \cap J),$$
$$V_+(B) = \emptyset, \quad V_+(0) = \operatorname{Proj} B.$$

This makes it possible to endow $\operatorname{Proj} B$ with the topology whose closed sets are of the form $V_+(I)$. We call this topology the *Zariski topology on $\operatorname{Proj} B$*.

Let I be an arbitrary ideal of B. We can associate to it a homogeneous ideal $I^h = \oplus_{d \geq 0}(I \cap B_d)$. It follows from the definition that I is homogeneous if and only if $I = I^h$.

Lemma 3.35. *Let I, J be ideals of a graded ring B. Then the following properties are true.*

   (a) *If I is prime, then the associated homogeneous ideal I^h is prime.*

   (b) *Let us suppose I and J are homogeneous. Then $V_+(I) \subseteq V_+(J)$ if and only if $J \cap B_+ \subseteq \sqrt{I}$.*

   (c) *We have $\operatorname{Proj} B = \emptyset$ if and only if B_+ is nilpotent.*

Proof (a) Let us suppose that I is prime. Let $a, b \in B$ be such that $ab \in I^h$ and that $a, b \notin I^h$. We write the decomposition in homogeneous elements

$$a = \sum_{0 \leq i \leq n} a_i, \quad b = \sum_{0 \leq j \leq m} b_j, \quad a_d, b_d \in B_d.$$

We may assume that $a_n, b_m \notin I^h$. The decomposition of the product ab in homogeneous elements is $ab = a_n b_m + \{\text{elements of degree} < n + m\}$. Hence $a_n b_m$ is the homogeneous component of degree $n + m$ of ab. It follows that $a_n b_m \in I^h \subseteq I$. As I is supposed to be prime, this leads to a contradiction.

(b) Let us first suppose that $J \cap B_+ \subseteq \sqrt{I}$. For every $\mathfrak{p} \in V_+(I)$, we have $\mathfrak{p} \supseteq J \cap B_+ \supseteq JB_+$. As $\mathfrak{p}$ is prime and does not contain B_+, we have $\mathfrak{p} \supseteq J$. Therefore $\mathfrak{p} \in V_+(J)$. Conversely, let us suppose that $V_+(I) \subseteq V_+(J)$. For every prime ideal $\mathfrak{p} \in V(I)$, the homogeneous ideal $\mathfrak{p}^h$ is prime and contains I. If $\mathfrak{p}^h$ does not contain B_+, then $\mathfrak{p}^h \in V_+(I)$; hence $\mathfrak{p} \supseteq \mathfrak{p}^h \supseteq J \cap B_+$. Otherwise we still have $\mathfrak{p} \supseteq \mathfrak{p}^h \supseteq J \cap B_+$. Consequently $J \cap B_+ \subseteq \cap_{\mathfrak{p} \in V(I)} \mathfrak{p} = \sqrt{I}$ (Lemma 1.6).

(c) The property $\operatorname{Proj} B = \emptyset$ is equivalent to $V_+(0) \subseteq V_+(B_+)$. The latter is equivalent to $B_+ \subseteq \sqrt{0}$ by (b). □

Let $f \in B$ be a homogeneous element, and write $D_+(f) = \operatorname{Proj} B \setminus V_+(fB)$. The open sets of this form are called *principal open sets*. They constitute a base of open sets of $\operatorname{Proj} B$. In fact, we can restrict ourselves to the open sets $D_+(f)$ with $f \in B_+$. Indeed, $\emptyset = V_+(B_+) = \cap_i V_+(f_i)$ where the f_i are homogeneous elements that generate B_+, so $\operatorname{Proj} B = \cup_i D_+(f_i)$. It follows that for every homogeneous $g \in B$, we have $D_+(g) = \cup_i D_+(gf_i)$ with $gf_i \in B_+$.

If $f \in B$ is homogeneous, we let $B_{(f)}$ denote the subring of B_f made up of the elements of the form af^{-N}, $N \geq 0$, $\deg a = N \deg f$. These are the *elements of degree 0 of B_f*. For example, if $B = k[T_0, \ldots, T_n]$, then $B_{(T_i)} = k[\, T_i^{-1}T_j\,]_{0 \leq j \leq n}$.

Lemma 3.36. *Let $f \in B_+$ be a homogeneous element of degree r.*

   (a) *There exists a canonical homeomorphism $\theta : D_+(f) \to \operatorname{Spec} B_{(f)}$.*

   (b) *Let $D_+(g) \subseteq D_+(f)$ and $\alpha = g^r f^{-\deg g} \in B_{(f)}$. Then $\theta(D_+(g)) = D(\alpha)$.*

   (c) *We have a canonical homomorphism $B_{(f)} \to B_{(g)}$ which induces an isomorphism $(B_{(f)})_\alpha \simeq B_{(g)}$.*

(d) *Let I be a homogeneous ideal of B. Then the image under θ of $V_+(I) \cap D_+(f)$ is the closed set $V(I_{(f)})$, where $I_{(f)} := IB_f \cap B_{(f)}$.*

(e) *Let $\{h_1, \ldots, h_n\}$ be homogeneous elements generating I. Then for any $f \in B_1$, $I_{(f)}$ is generated by the $h_i/f^{\deg h_i}$.*

Proof (a)–(b) By construction, $\operatorname{Proj} B$ is a subset of $\operatorname{Spec} B$. Moreover, for any $f \in B$, we have $V(f) \cap \operatorname{Proj} B = \cap_i V_+(f_i)$ if $f = f_0 + f_1 + \ldots + f_d$ with $f_i \in B_i$. So $D(f) \cap \operatorname{Proj} B = \cup_i D_+(f_i)$. Hence the topology of $\operatorname{Proj} B$ is induced by that of $\operatorname{Spec} B$. Let $\theta : D_+(f) \to \operatorname{Spec} B_{(f)}$ denote the restriction of the canonical map $D(f) = \operatorname{Spec} B_f \to \operatorname{Spec} B_{(f)}$ to $D_+(f)$. It is continuous.

Let us show that θ is surjective. Let us note that B_f is a graded $B_{(f)}$-algebra, the homogeneous elements of degree n ($n \in \mathbb{Z}$) being of the form bf^{-N} with $b \in B$ and $\deg b = Nr + n$. Let $\mathfrak{q} \in \operatorname{Spec} B_{(f)}$. Using this grading it is easy to see that $\sqrt{\mathfrak{q}B_f}$ is a prime ideal of B_f (we note that if $a, b \in B_f$ are two homogeneous elements, of respective degrees n and m, such that $ab \in \mathfrak{q}B_f$, then $(a^r f^{-n})(b^r f^{-m}) \in \mathfrak{q}$). It is homogeneous because $\mathfrak{q}B_f$ is homogeneous. Let $\rho : B \to B_f$ be the localization homomorphism. This is a homomorphism of graded rings. It is then easy to deduce from this that $\mathfrak{p} := \rho^{-1}(\sqrt{\mathfrak{q}B_f})$ is a homogeneous prime ideal of B and that $\mathfrak{p} \in D_+(f)$ (see Lemma 3.40 for more detail). It remains to show that $\theta(\mathfrak{p}) = \mathfrak{q}$, which will imply the surjectivity of θ. This equality can easily be shown by once more using the grading of B_f.

Let $\mathfrak{p}, \mathfrak{p}' \in D_+(f)$ be such that $(\mathfrak{p}B_f) \cap B_{(f)} = (\mathfrak{p}'B_f) \cap B_{(f)}$. For every homogeneous $b \in \mathfrak{p}$, we have $b^r f^{-\deg b} \in \mathfrak{p}B_f \cap B_{(f)} \subset \mathfrak{p}'B_f$, and hence $\mathfrak{p} \subseteq \mathfrak{p}'$. By symmetry, we obtain $\mathfrak{p} = \mathfrak{p}'$. This shows that θ is injective.

To conclude, let us show that $\theta(D_+(g)) = D(\alpha)$. This will imply that θ is open and therefore a homeomorphism. Let $\mathfrak{p} \in D_+(f)$. If $g \in \mathfrak{p}$, then $\alpha \in \mathfrak{p}B_f \cap B_{(f)}$. Conversely, if $g \notin \mathfrak{p}$, it is just as easy to see that $\alpha \notin \mathfrak{p}B_f \cap B_{(f)}$. Hence $\theta(D_+(g)) = D(\alpha)$.

(c) By Lemma 3.35(b), there exist $n \geq 1$ and $b \in B$ such that $g^n = fb$. We can choose b homogeneous by replacing it, if necessary, by its component of degree $n \deg g - \deg f$. We therefore have a homomorphism $B_{(f)} \to B_{(g)}$ which maps af^{-N} to $(ab^N)g^{-nN}$. It is easy but somewhat tedious to verify that this map is well defined and that it induces an isomorphism from the localization $(B_{(f)})_\alpha$ onto $B_{(g)}$.

(d) This follows from the definition of θ and the construction of θ^{-1}. Finally, (e) is easily deduced from the definition.   □

Remark 3.37. Lemma 3.36 is much easier to prove if $\deg f = 1$.

Proposition 3.38. *Let A be a ring. Let B be a graded algebra over A. Then we can endow $\operatorname{Proj} B$ with a unique structure of an A-scheme such that for any homogeneous $f \in B_+$, the open set $D_+(f)$ is affine and isomorphic to $\operatorname{Spec} B_{(f)}$.*

Proof Let $X = \operatorname{Proj} B$ and let $\mathcal{B}$ be the base for X made up of the principal open sets $D_+(f)$ with $f \in B_+$. For any $D_+(f) \in \mathcal{B}$, let $\mathcal{O}_X(D_+(f)) = B_{(f)}$. Using Lemma 3.36(c), we see that $B_{(f)}$ is canonically isomorphic to $B_{(f')}$ if $D_+(f) = D_+(f')$ and that we have a canonical restriction homomorphism $\mathcal{O}_X(D_+(f)) \to \mathcal{O}_X(D_+(g))$ if $D_+(g) \subseteq D_+(f)$. Hence $\mathcal{O}_X$ is a $\mathcal{B}$-presheaf. Using the homeomorphism θ of Lemma 3.36, we see moreover that $\mathcal{O}_X$ is a $\mathcal{B}$-sheaf. We can then extend it to a sheaf $\mathcal{O}_X$ on X. The proposition is now clear, because $(D_+(f), \mathcal{O}_X|_{D_+(f)})$ is isomorphic, via θ, to the affine scheme $\operatorname{Spec} B_{(f)}$. Finally, $B_{(f)}$ is naturally an A-algebra (since, by hypothesis, the image of A in B is contained in B_0), which gives the structure of an A-scheme on X.   □

Example 3.39. Let A be a ring. We endow $B = A[T_0, \ldots, T_n]$ with the grading by the degree. Then $B_{(T_i)} = A[\, T_i^{-1}T_j \,]_{0 \le j \le n}$. We easily deduce from this an isomorphism from $\operatorname{Proj} B$ onto the A-scheme $\mathbb{P}_A^n$ of Example 3.34.

Lemma 3.40. Let $\varphi : C \to B$ be a homomorphism of graded rings (i.e., there exists an $r \ge 1$ such that for every $d \ge 0$, we have $\varphi(C_d) \subseteq B_{rd}$). Let M be the homogeneous ideal $\varphi(C_+)B$ of B. Then φ induces a morphism of A-schemes $f : \operatorname{Proj} B \setminus V_+(M) \to \operatorname{Proj} C$ such that for every homogeneous $h \in C_+$, we have $f^{-1}(D_+(h)) = D_+(\varphi(h))$, and that $f|_{D_+(h)}$ coincides with the morphism of affine schemes induced by the homomorphism $C_{(h)} \to B_{(\varphi(h))}$.

Proof Let $\mathfrak{p}$ be a homogeneous ideal contained in $\operatorname{Proj} B$; then $\varphi^{-1}(\mathfrak{p})$ is clearly a homogeneous prime ideal of C. Moreover, $\varphi^{-1}(\mathfrak{p})$ does not contain C_+ if and only if $\mathfrak{p}$ does not contain M, whence the map $f : \operatorname{Proj} B \setminus V_+(M) \to \operatorname{Proj} C$. The properties of f are elementary to verify.   □

A particularly interesting class of graded algebras is that of the *homogeneous algebras* over A. These are, by definition, the quotients of $A[T_0, \ldots, T_n]$ by a homogeneous ideal.

Lemma 3.41. Let $B = A[T_0, \ldots, T_n]/I$ be as above. Then $\operatorname{Proj} B$ is isomorphic to a closed subscheme of $\mathbb{P}_A^n$, with support (i.e., with underlying topological space) $V_+(I)$.

Proof Let $C = A[T_0, \ldots, T_n]$ and let $\varphi : C \to B$ denote the canonical surjection. Then $\varphi(C_+) = B_+$. The lemma now follows from the preceding lemma.   □

Definition 3.42. Let A be a ring. A *projective scheme over A* is an A-scheme that is isomorphic to a closed subscheme of $\mathbb{P}_A^n$ for some $n \ge 0$.

The lemma above asserts that for every homogeneous A-algebra B, $\operatorname{Proj} B$ is a projective scheme over A. We can show that the converse is also true (Proposition 5.1.30). See also Exercise 4.12.

We will conclude the subsection with an identification of the rational points of projective schemes over a field. Let k be a field and V a vector space of finite dimension over k. We have an equivalence relation on $V \setminus \{0\} : u \sim v$ if there exists a $\lambda \in k^*$ such that $u = \lambda v$. The quotient $\mathbb{P}(V) = (V \setminus \{0\})/k^*$ is a projective space in the sense of classical projective geometry. This space represents the set of lines in V passing through the origin.

Let us fix an isomorphism $V \simeq k^{n+1}$. If $(\alpha_0, \ldots, \alpha_n) \in k^{n+1} \setminus \{0\}$ is a representative of a point $\alpha \in \mathbb{P}(k^{n+1})$, then the α_i are called the *homogeneous coordinates* of α.

Lemma 3.43. Let k be a field. Let $\alpha \in \mathbb{P}(k^{n+1})$ be a point with homogeneous coordinates $(\alpha_0, \ldots, \alpha_n)$. Then the ideal $\rho(\alpha) \subseteq k[T_0, \ldots, T_n]$ generated by the

$\alpha_j T_i - \alpha_i T_j$, $0 \le i, j \le n$, is an element of $\mathbb{P}_k^n$. Moreover, $\rho : \mathbb{P}(k^{n+1}) \to \mathbb{P}_k^n$ is a bijection from $\mathbb{P}(k^{n+1})$ onto the set $\mathbb{P}_k^n(k)$ of rational points of $\mathbb{P}_k^n$.

Proof We may assume that $\alpha_0 \ne 0$. It is clear that $\rho(\alpha)$ is homogeneous. It is prime because $k[T_0, \ldots, T_n]/\rho(\alpha) \simeq k[T_0]$. Finally, $\rho(\alpha) \in \mathbb{P}_k^n$ because the ideal $\rho(\alpha)$ does not contain the (maximal) ideal $k[T_0, \ldots, T_n]_+ = (T_0, \ldots, T_n)$. Let us now show that $\rho(\alpha)$ is rational. We have $T_i - \alpha_0^{-1}\alpha_i T_0 \in \rho(\alpha)$ for every i. Hence $\rho(\alpha) \in D_+(T_0)$, and corresponds (via the identification of Lemma 3.36(a)) to the ideal of $k[\, T_0^{-1}T_i]_i$ generated by the $T_0^{-1}T_i - \alpha_0^{-1}\alpha_i$. Hence $k(\rho(\alpha)) = k$. Let $\beta \in \mathbb{P}(k^{n+1})$ be a point with homogeneous coordinates $(\beta_0, \ldots, \beta_n)$ such that $\rho(\beta) = \rho(\alpha)$. Then $\beta_0 \ne 0$ because otherwise $T_0 \in \rho(\beta)$. By considering the points $\rho(\alpha)$, $\rho(\beta)$ in $D_+(T_0) \simeq \mathbb{A}_k^n$, we obtain $\alpha_0^{-1}\alpha_i = \beta_0^{-1}\beta_i$ for every i. Therefore $\beta_i = (\alpha_0^{-1}\beta_0)\alpha_i$ for every i. It follows that $\beta = \alpha$ and that ρ is injective.

It remains to show the surjectivity of ρ. Let $x \in \mathbb{P}_k^n(k)$. We may assume, for example, that $x \in D_+(T_0)$. Let α_i be the image of $T_0^{-1}T_i \in \mathcal{O}(D_+(T_0))$ in $k = k(x)$. Consider the point $\alpha \in \mathbb{P}(k^{n+1})$ with homogeneous coordinates $(\alpha_0, \ldots, \alpha_n)$. Then we have $\rho(\alpha) = x$ by what we have just seen. □

We can now give the projective equivalent of Corollary 1.15. Let k be a field. Let $P_1(T), \ldots, P_m(T) \in k[T_0, \ldots, T_n]$ be homogeneous polynomials. Consider the system of polynomial equations

$$P_1(\alpha_0, \ldots, \alpha_n) = \cdots = P_m(\alpha_0, \ldots, \alpha_n) = 0, \quad (\alpha_0, \ldots, \alpha_n) \in k^{n+1}.$$

If $(\alpha_0, \ldots, \alpha_n)$ is a solution, then so is $(\lambda\alpha_0, \ldots, \lambda\alpha_n)$ for any $\lambda \in k^*$. We can therefore speak of solutions (called *homogeneous*) in $\mathbb{P}(k^{n+1})$. We denote by $Z_+(P_1, \ldots, P_m) \subseteq \mathbb{P}(k^{n+1})$ the set of these solutions. For example, and let $a, b \in \mathbb{Q}$ be given rational numbers, and let $P(X, Y, Z)$ be the homogeneous polynomial $Y^2 Z - X^3 - aXZ^2 - bZ^3$. Then $Z_+(P)$ corresponds to the set of $(x, y, z) \in (\mathbb{Q}^3 \setminus \{0\})/\mathbb{Q}^*$ such that $y^2 z = x^3 + axz^2 + bz^3$. These are the rational points of an elliptic curve if $4a^3 + 27b^2 \ne 0$.

Corollary 3.44. *Let k be a field and $P_1(T), \ldots, P_m(T)$ homogeneous polynomials in $k[T_0, \ldots, T_n]$. Then there exists a bijection between $Z_+(P_1, \ldots, P_m)$ and the set of k-rational points of the scheme $\operatorname{Proj} k[T_0, \ldots, T_n]/I$, where I is the ideal generated by the $P_j(T)$.*

Proof Let us write $B = k[T_0, \ldots, T_n]$ and $Z_+ = Z_+(P_1, \ldots, P_m)$. By virtue of Lemma 3.41, there exists a canonical bijection between the rational points of $\operatorname{Proj} B/I$ and $V_+(I) \cap \mathbb{P}_k^n(k)$. Let $\rho : \mathbb{P}(k^{n+1}) \to \mathbb{P}_k^n(k)$ be the bijection of Lemma 3.43. It suffices to show that $\rho(Z_+) = V_+(I) \cap \mathbb{P}_k^n(k)$.

Let us fix $0 \le i \le n$. Let $U_i = \rho^{-1}(D_+(T_i)) \subset \mathbb{P}(k^{n+1})$. This is also the set of points $(\alpha_0, \ldots, \alpha_n)$ such that $\alpha_i \ne 0$. It suffices to show that $\rho(Z_+ \cap U_i) = V_+(I) \cap D_+(T_i)(k)$. We have a commutative diagram of canonical bijections:

$$
\begin{array}{ccc}
U_i & \xrightarrow{\;\rho|_{U_i}\;} & D_+(T_i)(k) \\
\downarrow{\scriptstyle p} & & \downarrow{\scriptstyle \theta} \\
k^n & \xrightarrow{\;\;\lambda\;\;} & \operatorname{Spec} B_{(T_i)}(k)
\end{array}
$$

with p the projection $(\alpha_0, \dots, \alpha_n) \mapsto (\alpha_i^{-1}\alpha_0, \dots, \alpha_i^{-1}\alpha_{i-1}, \alpha_i^{-1}\alpha_{i+1}, \alpha_i^{-1}\alpha_n)$, θ given by Lemma 3.36, and λ given by Example 3.32. For any homogeneous polynomial P, we write

$$P_{(i)} = P(T_i^{-1}T_0, \dots, T_i^{-1}T_n) \in k[T_i^{-1}T_0, \dots, T_i^{-1}T_n] = B_{(T_i)}.$$

Then it is clear that $\lambda \circ p(Z_+ \cap U_i) = V(P_{1(i)}, \dots, P_{m(i)})(k)$ (use Example 3.32). On the other hand, by Lemma 3.36(d), we have

$$\theta(V_+(I) \cap D_+(T_i)(k)) = V(I_{(T_i)})(k) \subseteq \operatorname{Spec} B_{(T_i)}.$$

As $I_{(T_i)}$ is the ideal of $B_{(T_i)}$ generated by the $P_{1(i)}, \dots, P_{m(i)}$, this concludes the proof of the corollary. $\qquad\square$

2.3.4 Noetherian schemes, algebraic varieties

Definition 3.45. A scheme X is said to be *Noetherian* if it is a finite union of affine open X_i such that $\mathcal{O}_X(X_i)$ is a Noetherian ring for every i. We say that a scheme is *locally Noetherian* if every point has a Noetherian open neighborhood.

In practice, most of the schemes that we will come across are Noetherian or locally Noetherian, and most fundamental theorems in algebraic geometry are proven in this setting.

Proposition 3.46. *Let X be a Noetherian scheme.*

 (a) *Any open or closed subscheme of X is Noetherian. For any point $x \in X$, the local ring $\mathcal{O}_{X,x}$ is Noetherian.*

 (b) *For any affine open subset U of X, $\mathcal{O}_X(U)$ is Noetherian.*

Proof By assumption, X is a finite union of affine open X_i with $\mathcal{O}_X(X_i)$ Noetherian.

(a) Let Z be an open or closed subscheme of X. It suffices to show that $Z \cap X_i$ is Noetherian. We may therefore assume that $X = \operatorname{Spec} A$ is affine. If Z is open, then $Z = X \setminus V(I)$. Since the ideal I is finitely generated, Z is a finite union of principal open sets $D(f_j)$. Now every localization of A (in particular A_{f_j}) is Noetherian, and hence Z is Noetherian. If Z is closed, then the assertion follows from Proposition 3.20. The ring $\mathcal{O}_{X,x}$ is a localization of a Noetherian ring, and hence it is also Noetherian.

(b) As we have seen above, $U \cap X_i$ is a finite union of Noetherian affine open subsets (i.e., spectra of Noetherian rings) of X_i. We therefore reduce to the case when U is a finite union of Noetherian affine open sets U_j. Let I be an ideal of $A = \mathcal{O}_X(U)$. As $I\mathcal{O}_X(U_j)$ is finitely generated, there exists a finitely generated ideal $J \subseteq I$ such that $J\mathcal{O}_X(U_j) = I\mathcal{O}_X(U_j)$ for every j. For every point $x \in U$, we therefore have $I\mathcal{O}_{U,x} = J\mathcal{O}_{U,x}$, that is $I/J \otimes_A A_{\mathfrak{p}} = 0$ for every $\mathfrak{p} \in \operatorname{Spec} A$. Whence $I/J = 0$, by virtue of Lemma 1.2.12, and $I = J$ is finitely generated. $\quad\square$

Definition 3.47. Let k be a field. An *affine variety over k* is the affine scheme associated to a finitely generated algebra over k. An *algebraic variety over k* is a k-scheme X such that there exists a covering by a finite number of affine open subschemes X_i which are affine varieties over k. A *projective variety over k* is a projective scheme over k. Projective varieties are algebraic varieties (Example 3.34). By definition, a *morphism of algebraic varieties over k* is a morphism of k-schemes.

Remark 3.48. An algebraic variety is a Noetherian scheme. It is easy to see that an open or closed subscheme of an algebraic variety X is an algebraic variety (see also Proposition 3.2.2 and Exercise 3.2.2). We will then say *subvariety* instead of subscheme. On the other hand, $\operatorname{Spec} \mathcal{O}_{X,x}$ is not an algebraic variety, unless x is an isolated point in X (i.e., $\{x\}$ is an open subset of X).

Remark 3.49. Let X be an algebraic variety. Then the set of closed points of X is dense in X. Indeed, any affine open subset U of a scheme contains a closed point, and in the case of algebraic varieties, every closed point of an open subset is closed in X (Lemma 4.3).

Remark 3.50. Let X be an algebraic variety. Let X^0 be the topological subspace made up of the closed points of X. Let $i : X^0 \to X$ denote the canonical injection. Then $(X^0, i^{-1}\mathcal{O}_X)$ is a ringed topological space. In the study of algebraic varieties, we will often restrict ourselves to considering this ringed topological space. In general, the properties of this space are equivalent to those of the scheme X because X^0 is dense in X.

Exercises

3.1. Let X be an open subscheme of an affine scheme Y. Show that the canonical morphism $X \to Y$ corresponds by the map ρ of Proposition 3.25 to the restriction $\mathcal{O}_Y(Y) \to \mathcal{O}_X(X)$.

3.2. Let $U = \operatorname{Spec} B$ be an affine open subscheme of $X = \operatorname{Spec} A$. Show that the restriction $A \to B$ is a flat homomorphism.

3.3. Closure of a subset.

    (a) Let F be a subset of an affine scheme $\operatorname{Spec} A$. Show that the closure $\overline{F}$ of F in (the underlying topological space of) $\operatorname{Spec} A$ is equal to $V(I)$, where $I = \cap_{\mathfrak{p} \in F} \mathfrak{p}$.

    (b) Let $\varphi : A \to B$ be a ring homomorphism. Let $f : \operatorname{Spec} B \to \operatorname{Spec} A$ be the morphism of schemes associated to φ. Show that $\overline{\operatorname{Im} f} = V(\operatorname{Ker} \varphi)$. Study the situation where B is the localization of A at a prime ideal.

3.4. Let X be a scheme and $f \in \mathcal{O}_X(X)$. Show that $U \mapsto f|_U \mathcal{O}_X(U)$ for every affine open subset U defines a sheaf of ideals on X. We denote this sheaf by $f\mathcal{O}_X$. Show that $\operatorname{Supp} f\mathcal{O}_X$ is closed (see Exercise 2.5).

3.5. Let Y be a scheme that satisfies the conclusion of Proposition 3.25 for every affine scheme X. Show that Y is affine.

3.6. Generalize Example 3.32 to the case when k is an arbitrary ring.

3.7. Let X be a scheme over a field k. Let $\varphi : k[T_1, \ldots, T_n] \to \mathcal{O}_X(X)$ be a homomorphism of k-algebras and $f : X \to \mathbb{A}^n_k$ the morphism induced by φ. Show that for any rational point $x \in X(k)$, via the identification $\mathbb{A}^n_k(k) = k^n$, we have $f(x) = (f_1(x), \ldots, f_n(x))$, where $f_i = \varphi(T_i)$ and $f_i(x)$ is the image of f_i in $k(x) = k$.

3.8. Let X be a scheme over a ring A. Let $f_0, \ldots, f_n \in \mathcal{O}_X(X)$ be such that the $f_{i,x}$ generate the unit ideal of $\mathcal{O}_{X,x}$ for every $x \in X$. Show that X is the union of the X_{f_i} (see Definition 3.11), and that we have a morphism $f : X \to \mathbb{P}^n_A$ such that $f^{-1}(D_+(T_i)) = X_{f_i}$ and that $f|_{X_{f_i}}$ is induced by the homomorphism $A[T_i^{-1}T_j]_j \to \mathcal{O}_X(X_{f_i})$ given by $T_i^{-1}T_j \mapsto f_i^{-1}f_j$. If $A = k$ is a field and $x \in X(k)$, determine $f(x)$ as in the preceding exercise.

3.9. Let B be a graded ring. Let $f \in B$ be non-nilpotent, homogeneous of degree 0. Show that B_f possesses a natural grading, and that $D_+(f) \simeq \operatorname{Proj} B_f$.

3.10. Let A be a ring, $X = \mathbb{P}^n_A$. Show that $\mathcal{O}_X(X) = A$ (use the covering $\mathcal{U} = \{D_+(T_i)\}_i$ of X and the complex associated to $\mathcal{U}$ and $\mathcal{O}_X$ as in Lemma 2.7). Deduce from this that X is affine if and only if $n = 0$.

3.11. Let $B = \oplus_{d \geq 0} B_d$ be a graded ring. Let $e \geq 1$ be an integer.

    (a) Let us denote the graded ring $\oplus_{d \geq 0} B_{de}$ by C (so $C_d := B_{de}$). Show that $\operatorname{Proj} B \simeq \operatorname{Proj} C$.

    (b) Let us suppose that for every $d \geq 1$, B_{ed} is generated by B_e^d, and that B_e is finitely generated over a ring A. Show that $\operatorname{Proj} B$ is a projective scheme over A.

3.12. Let B be a Noetherian graded ring.

    (a) Show that for any homogeneous $f \in B_+$, $B_{(f)}$ is Noetherian.

    (b) Show that $\operatorname{Proj} B$ is Noetherian.

3.13. Let $B = A[X, Y, Z]$ be a polynomial ring. Let B_d be the sub-A-module of B generated by the elements of the form $X^a Y^b Z^c$ with $a + 2b + 3c = d$.

    (a) Show that the B_d induce a grading on B.

    (b) Determine B_6 and show that B_{6d} is generated by the elements of B_6^d. Deduce from this that $\operatorname{Proj} B$ is isomorphic to a closed subscheme of $\mathbb{P}^6_A$.

3.14. We say that a scheme X is *quasi-compact* if it is quasi-compact as a topological space (Exercise 1.6). Show that a scheme is quasi-compact if and only if it is a finite union of affine schemes.

3.15. Let X be a quasi-compact scheme, $A = \mathcal{O}_X(X)$. Let us consider the morphism $f : X \to \operatorname{Spec} A$ induced by the identity on A (Proposition 3.25). Show that $f(X)$ is dense in $\operatorname{Spec} A$.

3.16. Let X be a scheme. Show that X is locally Noetherian if and only if any affine open subscheme of X is Noetherian.

3.17. Let $f : X \to Y$ be a morphism of schemes. We say that f is *quasi-compact* if the inverse image of any affine open subset is quasi-compact.

(a) Show that every closed immersion is quasi-compact.

(b) Show that an open immersion $f : X \to Y$ is quasi-compact if Y is locally Noetherian.

Let us suppose in what follows that f is quasi-compact.

(c) Let $\mathcal{J} = \mathrm{Ker}(f^{\#} : \mathcal{O}_Y \to f_*\mathcal{O}_X)$. Show that the ringed topological subspace $Z := V(\mathcal{J})$ of Y (see Lemma 2.23) is a scheme (reduce to the case when Y is affine and draw inspiration from the proofs of Propositions 3.12 and 3.20).

(d) Let $j : Z \to Y$ be the closed immersion. Show that we have a morphism $g : X \to Z$ such that $f = j \circ g$ and that if f decomposes into a morphism $g' : X \to Z'$ followed by a closed immersion $j' : Z' \to Y$, then Z is a closed subscheme of Z'.

(e) Show that $f(X)$ is dense in Z (reduce to the case when Z is affine). We call Z the *scheme-theoretic closure* of $f(X)$ in Y. If X is reduced, show that Z is reduced.

3.18. Let X be an affine algebraic variety over a field k. Show that there exists a projective variety $\overline{X}$ over k such that X is isomorphic to a dense open subscheme of $\overline{X}$. See also Exercise 3.3.20(b).

3.19. Let K be a number field. Let $\mathcal{O}_K$ be the ring of integers of K (i.e., the set of elements of K which are integral over $\mathbb{Z}$). Using the finiteness theorem of the class group $\mathrm{cl}(K)$, show that every open subset of $\mathrm{Spec}\,\mathcal{O}_K$ is principal. Deduce from this that every open subscheme of $\mathrm{Spec}\,\mathcal{O}_K$ is affine. See also Exercise 4.1.9.

3.20. Let A be a ring, G a finite group of automorphisms of A, and A^G the subring of elements of A which are invariant under G. Let $p : \mathrm{Spec}\,A \to \mathrm{Spec}\,A^G$ denote the morphism induced by the inclusion $A^G \to A$.

(a) Show that G acts naturally on $\mathrm{Spec}\,A$. Show that $p(x_1) = p(x_2)$ if and only if there exists a $\sigma \in G$ such that $\sigma(x_1) = x_2$.

(b) Show that A is integral over A^G (let $a \in A$, and consider the polynomial $P(T) := \prod_{\sigma \in G}(T - \sigma a) \in A^G[T]$). Deduce from this that p is surjective (Exercise 1.8).

(c) Let $a \in A$. Let $P(T)$ be as above and let us write $P(T) = T^d + \sum_{i \le d-1} b_i T^i$, $b_i \in A^G$. Show that $p(D(a)) = \cup_i D(b_i)$. Show that p is open (in the topological sense).

(d) Show that for any $b \in A^G$, we have $p^{-1}(D(b)) = D(bA)$ and $(A^G)_b = (A_b)^G$. Let V be an open subset of $\mathrm{Spec}\,A^G$. Show that G acts on the scheme $p^{-1}(V)$ and that $\mathcal{O}_{\mathrm{Spec}\,A^G}(V) = \mathcal{O}_{\mathrm{Spec}\,A}(p^{-1}(V))^G$.

3.21. Let X be a scheme, and let G be a finite group acting on X (i.e., G is endowed with a group homomorphism $G \to \text{Aut}(X)$). We define the *quotient scheme* X/G by the universal property of Exercise 2.14, where we replace the ringed topological spaces by schemes. It does not always exists.

   (a) Let A be a ring, and G a finite group of automorphisms of A, which we identify with a group of automorphisms of the scheme $\text{Spec } A$. Show that the quotient scheme $(\text{Spec } A)/G$ exists and is isomorphic to $\text{Spec}(A^G)$. Show that this is also the quotient as a ringed topological space (Exercise 2.14).

   (b) Let U be an open subscheme of $\text{Spec } A$ that is stable under G. Preserving the notation of the preceding exercise, show that the quotient scheme U/G exists and is isomorphic to $p(U)$.

   (c) Let G be a finite group acting on a scheme X. We suppose that every point $x \in X$ has an affine open neighborhood that is stable under G (see Exercise 3.3.23 for examples of such X). Show that the quotient scheme X/G exists.

   Remark. See [72] for quotients of algebraic varieties by algebraic groups.

3.22. Let k be a field of characteristic 0. We let $G := \mathbb{Z}$ act on the polynomial ring $k[T]$ by $n : T \mapsto T + n$ if $n \in \mathbb{Z}$.

   (a) Show that G is a subgroup of the group of automorphisms (of k-algebras) of $k[T]$. We identify it with a group of automorphisms of $\mathbb{A}_k^1$. Show that the only open subschemes of $\mathbb{A}_k^1$ that are stable under G are $\emptyset$ and $\mathbb{A}_k^1$ itself.

   (b) Show that the quotient scheme $\mathbb{A}_k^1/G$ exists and is equal to $\text{Spec } k$.

   (c) Show that $\mathbb{A}_k^1/G$ is not the quotient as a ringed topological space (Exercise 2.14).

2.4 Reduced schemes and integral schemes

In this section we study some basic properties of schemes. The first subsection deals with reduced schemes. The link between the sheaf of regular functions $\mathcal{O}_X$ and functions in the usual sense will be established in the case of algebraic varieties. Next we introduce a topological notion, that of irreducible components. This makes it possible to conclude with integral schemes and rational functions.

2.4.1 Reduced schemes

Recall that a ring A is called *reduced* if 0 is the only nilpotent element of A.

Definition 4.1. Let X be a scheme, $x \in X$. We say that X is *reduced at x* if the ring $\mathcal{O}_{X,x}$ is reduced. We say that X is *reduced* if it is reduced at every point of X.

Proposition 4.2. *Let X be a scheme. The following properties are true.*

(a) *X is reduced if and only if $\mathcal{O}_X(U)$ is reduced for every open subset U of X.*

(b) *Let $\{X_i\}_i$ be a covering of X by affine open subsets X_i. If the $\mathcal{O}_X(X_i)$ are reduced, then X is a reduced scheme.*

(c) *There exists a unique reduced closed subscheme $i : X_{\mathrm{red}} \to X$ having the same underlying topological space as X. Moreover, if X is quasi-compact, then the kernel of $i^\#(X) : \mathcal{O}_X(X) \to \mathcal{O}_X(X_{\mathrm{red}})$ is the nilradical of $\mathcal{O}_X(X)$.*

(d) *Let Y be a reduced scheme. Then any morphism $f : Y \to X$ factors in a unique way into a morphism $Y \to X_{\mathrm{red}}$ followed by $i : X_{\mathrm{red}} \to X$.*

(e) *Let Z be a closed subset of X. Then there exists a unique structure of reduced closed subscheme on Z.*

Proof For any ring A, let $N(A)$ denote the nilradical of A. Let us define a sheaf of ideals $\mathcal{N} \subseteq \mathcal{O}_X$ by

$$\mathcal{N}(U) := \{s \in \mathcal{O}_X(U) \mid s_x \in N(\mathcal{O}_{X,x}), \; \forall x \in U\}.$$

Then X is reduced if and only if $\mathcal{N} = 0$. It is easy to verify that for any quasi-compact (e.g., affine) open subset U of X (Exercise 3.14), we have $\mathcal{N}(U) = N(\mathcal{O}_X(U))$. This immediately implies (a).

Let X_{red} be the ringed topological space $(X, \mathcal{O}_X/\mathcal{N})$. Let us verify that this defines the structure of a scheme; properties (c) and (d) will then follow immediately. For this, we may assume that $X = \mathrm{Spec}\, A$ is affine. Let $N = N(A)$. Let $i : \mathrm{Spec}\, A/N \to \mathrm{Spec}\, A$ be the closed immersion of schemes defined by the canonical surjection $A \to A/N$. For any principal open subset $D(g)$ of X, we have $\mathrm{Ker}\, i^\#(D(g)) = N \otimes_A A_g$ (Lemma 3.17). Now it is easy to see that $N \otimes_A A_g = N(A_g) = \mathcal{N}(D(g))$. Therefore $\mathrm{Ker}\, i^\# = \mathcal{N}$, which shows that $X_{\mathrm{red}} = \mathrm{Spec}\, A/N$. At the same time this shows (b), because if $\mathcal{O}_X(X_i)$ is reduced, then $\mathcal{N}|_{X_i} = 0$.

Let Z be a closed subset of X. Let us first show the uniqueness of the reduced closed subscheme structure $(Z, \mathcal{O}_Z)$ on Z. Let U be an affine open subset of X. Then $(Z \cap U, \mathcal{O}_{Z \cap U})$ is determined by an ideal I of $\mathcal{O}_X(U)$ (Proposition 3.20), and we have $\mathcal{O}_Z(Z \cap U) = \mathcal{O}_X(U)/I$, $Z \cap U = V(I)$. If $(Z, \mathcal{O}_Z)$ is reduced, then $I = \sqrt{I}$ by (a). The uniqueness then follows from Lemma 1.6(b).

Let $\{X_i\}_i$ be a covering of X by affine open subsets X_i. Let I_i be an ideal of $\mathcal{O}_X(X_i)$ such that $Z \cap X_i = V(I_i)$. We endow $Z \cap X_i$ with the reduced closed subscheme structure $\mathrm{Spec}(\mathcal{O}_X(X_i)/\sqrt{I_i})$. By the uniqueness of such a structure, we may glue the $\mathrm{Spec}(\mathcal{O}_X(X_i)/\sqrt{I_i})$ to endow Z with the structure of a reduced closed subscheme of X. $\qquad\square$

Let X be an algebraic variety over a field k. We will compare the regular functions on X (i.e., elements of $\mathcal{O}_X(X)$) to functions in the usual sense. Let X^0 denote the topological subspace made up of the closed points of X.

Lemma 4.3. *Let X be an algebraic variety over a field k, and let U be an open subset of X. Then $U^0 = U \cap X^0$.*

Proof This amounts to showing that $U^0 \subseteq X^0$. We may assume that U is affine. Let x be a closed point of U. It suffices to show that x is closed in every affine open subset V that contains it. The point $x \in V$ corresponds to a prime ideal $\mathfrak{p}$ of $A := \mathcal{O}_X(V)$ and we have $k \subseteq A/\mathfrak{p} \subseteq k(x)$. The fact that x is closed in U implies that $k(x)$, and therefore $A/\mathfrak{p}$, is a finite extension of k (Remark 1.3 and Corollary 1.12). It follows that $A/\mathfrak{p}$ is a field and therefore that $\mathfrak{p}$ is a maximal ideal, which proves that x is closed in V. □

Let us fix an algebraic closure $\bar{k}$ of k. Let $f \in \mathcal{O}_X(X)$. We can associate to it a function $\tilde{f} : X^0 \to \bar{k}$ defined by $\tilde{f}(x) =$ the image of f_x in $k(x)$ (as we have just seen, $k(x)$ is a subextension of $\bar{k}$). This clearly induces a group homomorphism from $\mathcal{O}_X(X)$ to the set $\mathrm{F}(X^0, \bar{k})$ of maps from X^0 to $\bar{k}$. Let $\mathcal{F}_X$ be the sheaf $U \mapsto \mathrm{F}(U^0, \bar{k})$. Then more generally we have a homomorphism of sheaves from $\mathcal{O}_X$ to $\mathcal{F}_X$. The following proposition explains, in a way, the meaning of 'regular functions'.

Proposition 4.4. *Let X be an algebraic variety over a field k. Then the homomorphism $\mathcal{O}_X \to \mathcal{F}_X$ is injective if and only if X is reduced.*

Proof Let us first assume that X is reduced. Let U be an open subset of X and $f \in \mathcal{O}_X(U)$ be such that $\tilde{f}(x) = 0$ for every $x \in U^0$. Then for any affine open subset V of U, and for any closed point x of V, we have $\tilde{f}(x) = 0$. So $f|_V$ is contained in the intersection of the maximal ideals of $\mathcal{O}_X(V)$. It follows by Lemma 1.18 that $f|_V$ is nilpotent. By Proposition 4.2(a), this implies that $f|_V = 0$, whence $f = 0$.

Conversely, if X is not reduced, there exist an affine open subset U of X and an $f \in \mathcal{O}_X(U) \setminus \{0\}$ which is nilpotent. It is then clear that $\tilde{f} = 0$. Hence $\mathcal{O}_X \to \mathcal{F}$ is not injective. In general, the kernel of $\mathcal{O}_X \to \mathcal{F}$ is the sheaf of nilpotent elements $\mathcal{N}$ as defined in the proof of Proposition 4.2. □

2.4.2 Irreducible components

Let us now move to a more topological property of schemes. Let us recall that a topological space X is called *irreducible* if the condition $X = X_1 \cup X_2$ with closed subsets X_i implies that $X_1 = X$ or $X_2 = X$. This amounts to saying that the intersection of two non-empty open subsets of X is non-empty. For example, a space made up of one point is irreducible; the closure of an irreducible subspace (e.g., of a point) is irreducible. The notion of irreducibility is not interesting for the classical (e.g., metric) topology, since an irreducible separated space is reduced to a point. But this notion is very important in algebraic geometry.

The set of irreducible subspaces of X admits maximal elements for the inclusion relation. Indeed, an increasing union of irreducible subsets stays irreducible, and we then apply Zorn's lemma. Such a maximal element is a closed subset since the closure of an irreducible subset is irreducible. We call these maximal elements the *irreducible components of X*. Their union is equal to X since for every $x \in X$, $\overline{\{x\}}$ is irreducible and therefore contained in a maximal element.

Proposition 4.5. *Let X be a topological space.*

(a) *If X is irreducible, then any non-empty open subset of X is dense in X and is irreducible.*

(b) *Let U be an open subset of X. Then the irreducible components of U are the $\{X_i \cap U\}_i$, where the X_i are the irreducible components of X which meet U.*

(c) *Suppose that X is a finite union of irreducible closed subsets Z_j. Then every irreducible component Z of X is equal to one of the Z_j. If, moreover, there is no inclusion relation between the Z_j, then the Z_j are exactly the irreducible components of X.*

Proof (a) Let V be a non-empty open subset of X. Let $\overline{V}$ be its closure in X. Then $X = (X \setminus V) \cup \overline{V}$. It follows that $X = \overline{V}$. Finally, V is irreducible because the intersection of two non-empty open subsets of V is non-empty since they are dense in X.

(b) Let X_i be an irreducible component of X that meets U. If $X_i \cap U$ is contained in an irreducible closed subset Z of U, then $X_i = \overline{X_i \cap U} \subseteq \overline{Z}$ by (a). Hence $X_i = \overline{Z}$ since the latter is irreducible. It follows that $Z \subseteq X_i \cap U$. Therefore $X_i \cap U$ is an irreducible component of U. Conversely, let Z be an irreducible component of U. Then there exists an irreducible component X_i of X such that $\overline{Z} \subseteq X_i$, and hence $Z \subseteq X_i \cap U$. There is therefore equality. This proves (b).

(c) We have $Z = \cup_j (Z_j \cap Z)$. It follows, by induction on the number of the Z_j, that Z is contained in (and therefore equal to) one of the Z_j. For every j, Z_j is contained in an irreducible component, which is a Z_k by what we have just seen. Hence the set of Z_j is equal to the set of irreducible components of X if there is no inclusion relation between the Z_j.   □

Definition 4.6. We say that a scheme X is *irreducible* if the underlying topological space of X is irreducible.

Proposition 4.7. *Let $X = \mathrm{Spec}\, A$. Let I be an ideal of A.*

(a) *The space $V(I)$ is irreducible if and only if $\sqrt{I}$ is prime.*

(b) *Let $\{\mathfrak{p}_i\}_i$ be the minimal prime ideals of A; then the $\{V(\mathfrak{p}_i)\}_i$ are the irreducible components of X.*

(c) *X is irreducible if and only if A admits a unique minimal prime ideal. In particular, X is irreducible if A is an integral domain.*

Proof (a) Let us first suppose that $\sqrt{I}$ is prime. If $V(I) = V(J_1) \cup V(J_2)$, then $\sqrt{I} = \sqrt{J_1 J_2} \supseteq J_1 J_2$ (Lemma 1.6(b)). It follows that J_1 (for example) is contained in $\sqrt{I}$, and hence $V(I) = V(J_1)$, which shows that $V(I)$ is irreducible. Conversely, if $\sqrt{I}$ is not prime, there exist $a, b \in A \setminus \sqrt{I}$ such that $ab \in \sqrt{I}$. It follows that $V(I)$ is the union of the closed sets $V(a) \cap V(I)$, $V(b) \cap V(I)$ and is not equal to either of them. Hence $V(I)$ is not irreducible.

(b) and (c) follow immediately from (a). □

Example 4.8. Let k be a field and $X = \operatorname{Spec} k[x, y, z]/(xz, z(z^2 - y^3)) \subset \mathbb{A}_k^3$. Then $X = X_1 \cup X_2$, where $X_1 = V(z)$ and $X_2 = V(x, z^2 - y^3)$. As $k[x, y, z]/(z)$ and $k[x, y, z]/(x, z^2 - y^3)$ are integral domains, X_1 and X_2 are irreducible (Proposition 4.7) and without an inclusion relation between them. They are therefore the irreducible components of X. See Figure 3.

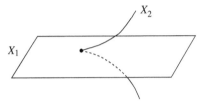

Figure 3. An affine scheme with two irreducible components X_1 and X_2.

Proposition 4.9. *Let X be a Noetherian scheme. Then X has only a finite number of irreducible components.*

Proof By covering X with a finite number of affine open subsets U_i and using Proposition 4.5(b), we see that X has a finite number of irreducible components as soon as this is the case for every U_i. We therefore reduce to the case when X is itself affine. Let $A = \mathcal{O}_X(X)$. Let us first show that any *radical ideal I of A* (i.e., $I = \sqrt{I}$) is a finite intersection of prime ideals. Let S be the set of radical ideals of A that do not verify this property. If S is non-empty, then it admits a maximal element I, since A is Noetherian. As I cannot be prime, there exist $a, b \notin I$ such that $ab \in I$. The radical ideals $\sqrt{I + aA}$ and $\sqrt{I + bA}$ do not belong to S because I is maximal; they are therefore finite intersections of prime ideals. But $I = \sqrt{I + aA} \cap \sqrt{I + bA}$ is then also a finite intersection of prime ideals, which is a contradiction. Thus $S = \emptyset$.

Applying this result to the ideal $\sqrt{0}$, we see that X is a finite union of closed irreducible subsets. The assertion then follows from Proposition 4.5(c). □

Definition 4.10. Let x, y be points of a topological space X. We say that y is a *specialization of x* or that *x specializes to y* if $y \in \overline{\{x\}}$, the closure of $\{x\}$. We say that $x \in X$ is a *generic point of X* if x is the unique point of X that specializes to x.

Lemma 4.11. *Let $X = \operatorname{Spec} A$. Let x, y be two points of X corresponding to prime ideals $\mathfrak{p}$ and $\mathfrak{q}$. Then*

(a) $\overline{\{x\}} = V(\mathfrak{p})$.

(b) *The point y is a specialization of x if and only if $\mathfrak{p} \subseteq \mathfrak{q}$.*

(c) *The point x is a generic point if and only if $\mathfrak{p}$ is a minimal prime ideal.*

Proof This is an easy exercise using Lemma 1.6. □

Proposition 4.12. *Let X be a scheme.*

(a) *For any generic point ξ of X, $\overline{\{\xi\}}$ is an irreducible component of X. This establishes a bijection between the irreducible components of X and the generic points of X.*

(b) *Let X be a scheme and $x \in X$. Then the irreducible components of $\operatorname{Spec} \mathcal{O}_{X,x}$ correspond bijectively to the irreducible components of X passing through x.*

Proof (a) Let Z be an irreducible component of X. Let U be an affine open subset of X such that $Z \cap U \neq \emptyset$. Then $Z \cap U$ is an irreducible component of U, by Proposition 4.5(b). Let ξ be the generic point of $Z \cap U$ (which exists by Lemma 4.11(c)). Then $Z = \overline{\{\xi\}}$. Conversely, let ξ be a generic point of X. Let Z be an irreducible component of X containing ξ. By what we have just seen, there exists a $\xi' \in X$ such that $Z = \overline{\{\xi'\}}$. It follows that ξ is a specialization of ξ', and hence $\xi' = \xi$ and $\overline{\{\xi\}} = Z$.

(b) By Proposition 4.5(b), we can replace X by an affine open subset containing x. The property that must be shown then follows from Lemma 4.11(c). □

Example 4.13. Let $\mathcal{O}_K$ be a non-trivial discrete valuation ring (i.e., $\mathcal{O}_K$ is a local principal ideal domain which is not a field); then $\operatorname{Spec} \mathcal{O}_K$ has exactly two points: the generic point η corresponds to the ideal $\{0\}$, and the closed point s corresponds to the maximal ideal of $\mathcal{O}_K$. The subset $\{\eta\}$ is open in $\operatorname{Spec} \mathcal{O}_K$.

Remark 4.14. Let X be a scheme, $x \in X$, and $y \in \overline{\{x\}}$ a specialization of x. Then we have a canonical ring homomorphism $\mathcal{O}_{X,y} \to \mathcal{O}_{X,x}$. Indeed, let U be an affine open set containing y; then $x \in U$ and $\mathcal{O}_{X,y} \to \mathcal{O}_{X,x}$ is a localization at a prime ideal (see Lemma 4.11(b)).

Remark 4.15. Let X be an irreducible scheme with generic point ξ. Let $\mathcal{P}$ be a property of schemes. Suppose that the set $X_{\mathcal{P}}$ of points of X satisfying $\mathcal{P}$ is open (which is true for many properties in algebraic geometry, see a first example in Exercise 4.9). If $\xi \in X_{\mathcal{P}}$, then $X_{\mathcal{P}}$ is a dense open set in X. Conversely, as soon as $X_{\mathcal{P}}$ is non-empty, it contains ξ. This generic point in a way occupies a privileged position in X. The (suitable) properties of X at ξ 'spread' to the 'sufficiently general' points of X.

For algebraic varieties, we often use the notion of 'general point', which is different from that of generic point. The latter is a very specific point of a scheme, while a general point is any closed point of a suitable dense open subset of X. The discussion above says that (suitable) properties of X at ξ are true for general points.

2.4.3　Integral schemes

Definition 4.16. We say that a scheme X is *integral at* $x \in X$ if $\mathcal{O}_{X,x}$ is an integral domain. This is equivalent to saying that X is reduced at x and that there is a single irreducible component of X passing through x. We say that X

is *integral* if it is reduced and irreducible. This implies that X is integral at all of its points.

For example, $\operatorname{Spec} A$ is integral if and only if A is an integral domain (Proposition 4.2(a) and (b), Lemma 4.11).

Proposition 4.17. *Let X be a scheme. Then X is integral if and only if $\mathcal{O}_X(U)$ is an integral domain for every open subset U of X.*

Proof Let us suppose that X is integral. As every open subscheme of X is integral (Proposition 4.5(a)), we can restrict ourselves to showing that $\mathcal{O}_X(X)$ is an integral domain. Let $f, g \in \mathcal{O}_X(X)$ be such that $fg = 0$. Let us fix a non-empty affine open subset W of X. Let us consider the closed subsets $V(f|_W)$ and $V(g|_W)$ of W; their union is equal to W. Since the latter is irreducible, we have, for example, $W = V(f|_W)$, so $f|_W = 0$ since $\mathcal{O}_X(W)$ is reduced (Proposition 4.2(a)). Let $U \neq \emptyset$ be an arbitrary affine open subset of X. Then $V(f|_U)$ contains the dense open subset $W \cap U$, so $V(f|_U) = U$, and it follows that $f|_U = 0$ and hence $f = 0$.

Conversely, let us suppose that $\mathcal{O}_X(U)$ is an integral domain for every open subset U. Then X is reduced. It remains to show irreducibility. If X admits two distinct generic points ξ_1, ξ_2, there exist two affine open subsets U_1, U_2 with $\xi_i \in U_i$. As the U_i are integral and therefore irreducible, $U_1 \cap U_2 = \emptyset$ because otherwise ξ_1 and ξ_2 would both be generic points of $U_1 \cap U_2$. This leads to a contradiction because $\mathcal{O}_X(U_1 \cup U_2) = \mathcal{O}_X(U_1) \oplus \mathcal{O}_X(U_2)$ is not an integral domain. $\qquad\square$

Proposition 4.18. *Let X be an integral scheme, with generic point ξ.*

   (a) *Let V be an affine open subset of X; then $\mathcal{O}_X(V) \to \mathcal{O}_{X,\xi}$ induces an isomorphism $\operatorname{Frac}(\mathcal{O}_X(V)) \simeq \mathcal{O}_{X,\xi}$.*

   (b) *For any open subset U of X, and any point $x \in U$, the canonical homomorphisms $\mathcal{O}_X(U) \to \mathcal{O}_{X,x}$ and $\mathcal{O}_{X,x} \to \mathcal{O}_{X,\xi}$ (Remark 4.14) are injective.*

   (c) *By identifying $\mathcal{O}_X(U)$ and $\mathcal{O}_{X,x}$ to subrings of $\mathcal{O}_{X,\xi}$, we have $\mathcal{O}_X(U) = \bigcap_{x \in U} \mathcal{O}_{X,x}$.*

Proof (a) The point ξ is also the generic point of V and we have $\mathcal{O}_{X,\xi} = \mathcal{O}_{V,\xi}$, so $\operatorname{Frac}(\mathcal{O}_X(V)) \simeq \mathcal{O}_{X,\xi}$ by Lemma 3.3.

(b) Let $f \in \mathcal{O}_X(U)$ be such that $f_x = 0$. There exists an affine open set $W \ni x$ such that $f|_W = 0$. For any affine open subset V of U, we have $W \cap V \neq \emptyset$, so $f|_{W \cap V} = 0$ implies $f|_V = 0$ by (a), whence $f = 0$. Therefore $\mathcal{O}_X(U) \to \mathcal{O}_{X,x}$ is injective. The injectivity of $\mathcal{O}_{X,x} \to \mathcal{O}_{X,\xi}$ results from the fact that $\mathcal{O}_X(V) \to \mathcal{O}_{X,\xi}$ is injective for every open subset V containing x.

(c) By covering U with affine open subsets, we may assume that $U = \operatorname{Spec} A$ is affine. Let $f \in \operatorname{Frac}(A)$ be contained in all of the localizations $A_{\mathfrak{p}}$ for every $\mathfrak{p} \in \operatorname{Spec} A$. Let I be the ideal $\{g \in A \mid gf \in A\}$. Then I is not contained in any prime ideal, so $I = A$. It follows that $f \in A$. $\qquad\square$

Definition 4.19. Let X be an integral scheme, with generic point ξ. We denote the field $\mathcal{O}_{X,\xi}$ by $K(X)$. Sometimes, when X is an algebraic variety over a field k, one also denotes $K(X)$ by $k(X)$. An element of $K(X)$ is called a *rational function on X*. We call $K(X)$ the *field of rational functions* or *function field* of X.

We say that $f \in K(X)$ is *regular at $x \in X$* if $f \in \mathcal{O}_{X,x}$. Proposition 4.18(c) affirms that a rational function which is regular at every point of U is contained in $\mathcal{O}_X(U)$.

Exercises

4.1. Let k be a field and $P \in k[T_1, \ldots, T_n]$. Show that $\mathrm{Spec}(k[T_1, \ldots, T_n]/(P))$ is reduced (resp. irreducible; resp. integral) if and only if P has no square factor (resp. admits only one irreducible factor; resp. is irreducible).

4.2. Let X be a scheme and $x \in X$. Show that the image of the morphism $\mathrm{Spec}\,\mathcal{O}_{X,x} \to X$ is the set of points of X that specialize to x.

4.3. Let $\mathcal{O}_K$ be a discrete valuation ring with field of fractions K and *uniformizing parameter t* (i.e., a generator of the maximal ideal). Show that $\mathrm{Spec}\,K[T]$ can be identified with an open subscheme of $\mathrm{Spec}\,\mathcal{O}_K[T]$. Determine the set of closed points of $\mathrm{Spec}\,K[T]$ which specialize to the point corresponding to the maximal ideal (T, t) of $\mathcal{O}_K[T]$. See also Proposition 10.1.40(c).

4.4. We say that a scheme X is *connected* if the underlying topological space of X is connected. For example, an irreducible scheme is connected. Let X be a scheme having only a finite number of irreducible components $\{X_i\}_i$. Show that X is connected if and only if for any pair i, j, there exist indices $i_0 = i, i_1, \ldots, i_r = j$ such that $X_{i_l} \cap X_{i_{l+1}} \neq \emptyset$ for every $l < r$. Also show that X is integral if and only if X is connected, and integral at every point $x \in X$.

4.5. Let X be a scheme. Show that every irreducible component of X is contained in a *connected component* (in the topological sense). Show that if X is locally Noetherian, then the connected components are open. If X is Noetherian, then there are only finitely many connected components.

4.6. Let A be a ring. We say that $e \in A$ is *idempotent* if $e \neq 0$ and $e^2 = e$. We will say that e is *indecomposable* if it cannot be written as a sum of two idempotent elements. Let X be a scheme.

   (a) Show the equivalence of the following properties:

      (i) X is connected;

      (ii) $\mathcal{O}_X(X)$ has no other idempotent elements than 1;

      (iii) $\mathrm{Spec}\,\mathcal{O}_X(X)$ is connected.

   (b) Show that any *local scheme* (i.e., the spectrum of a local ring) is connected.

(c) Let us suppose that the connected components of X are open (e.g., X locally Noetherian). Let U be a connected component of X. Show that there exists a unique idempotent element e of $\mathcal{O}_X(X)$ such that $e|_U = 1$ and $e|_{X \setminus U} = 0$. Show that this establishes a canonical bijection from the set of connected components of X onto the set of indecomposable idempotent elements of $\mathcal{O}_X(X)$, the converse map being given by $e \mapsto V(1 - e)$.

4.7. We say that a topological space X verifies the separation axiom T_0 if for every pair of points $x \neq y$, there exists an open subset which contains one of the points and not the other one. Show that the underlying topological space of a scheme verifies T_0.

4.8. Let X be a quasi-compact scheme (Exercise 3.14). Show that X contains a closed point. See also Exercise 3.3.26 for a counterexample when X is not quasi-compact.

4.9. Let X be a Noetherian scheme. Show that the set of points $x \in X$ such that $\mathcal{O}_{X,x}$ is reduced (resp. is an integral domain) is open.

4.10. Let $f : X \to \operatorname{Spec} A$ be a quasi-compact morphism (Exercise 3.17). Let I be an ideal of A. Show that $f(X) \subseteq V(I)$ if and only if $f^{\#}(\operatorname{Spec} A)(I) \subseteq \mathcal{O}_X(X)$ is nilpotent.

4.11. Let $f : X \to Y$ be a morphism of irreducible schemes with respective generic points ξ_X, ξ_Y. We say that f is *dominant* if $f(X)$ is dense in Y. Let us suppose that X, Y are integral. Show that the following properties are equivalent:

(i) f is dominant;

(ii) $f^{\#} : \mathcal{O}_Y \to f_* \mathcal{O}_X$ is injective;

(iii) for every open subset V of Y and every open subset $U \subseteq f^{-1}(V)$, the map $\mathcal{O}_Y(V) \to \mathcal{O}_X(U)$ is injective;

(iv) $f(\xi_X) = \xi_Y$;

(v) $\xi_Y \in f(X)$.

4.12. Let B be a graded ring. Let Y be a reduced closed subscheme of $\operatorname{Proj} B$. Show that there exists a homogeneous ideal I of B such that $Y \simeq \operatorname{Proj} B/I$.

2.5 Dimension

In this section we study a numerical invariant attached to a scheme, the dimension. The fundamental result concerns the dimension of a hypersurface in a Noetherian scheme (Theorem 5.15, Corollary 5.26). As a consequence, we will see (Remark 5.27) that in the case of algebraic varieties, the abstract definition coincides with geometric intuition.

2.5.1   Dimension of schemes

Let X be a topological space. A *chain of irreducible closed subsets of X* is a strictly ascending sequence of irreducible closed subsets

$$Z_0 \subsetneq Z_1 \subsetneq \cdots \subsetneq Z_n \subseteq X.$$

The integer n is called the *length of the chain*.

Definition 5.1. Let X be a topological space. We define the (*Krull*) *dimension of X*, which we denote by $\dim X$, to be the supremum of the lengths of the chains of irreducible closed subsets of X. This number is not necessarily finite. By convention, the empty set is of dimension $-\infty$. We say that X is *pure of dimension n* if all irreducible components of X have the same dimension n.

Let X be a scheme. The *dimension of X* is the dimension of the underlying topological space of X.

Example 5.2. A discrete topological space is of dimension 0.

Definition 5.3. Let X be a topological space. Let $x \in X$. We set

$$\dim_x X = \inf\{\dim U \mid U \text{ open neighborhood of } x\}.$$

This is the *dimension of X at x*.

Example 5.4. Let $\mathcal{O}_K$ be a non-trivial discrete valuation ring. Let us keep the notation of Example 4.13. As $\mathcal{O}_K$ is an integral domain, $X := \operatorname{Spec} \mathcal{O}_K$ is irreducible. The chain of maximal length is $\{s\} \subset X$. Therefore

$$\dim X = \dim_s X = 1.$$

As $\{\eta\}$ is open, we have $\dim_\eta X = \dim_\eta \{\eta\} = 0$.

Proposition 5.5. *Let X be a topological space. The following properties are true.*

(a) *For any subset Y of X endowed with the induced topology, we have $\dim Y \leq \dim X$.*

(b) *Let us suppose that X is irreducible of finite dimension. Let Y be a closed subset of X. Then $Y = X$ if and only if $\dim Y = \dim X$.*

(c) *The dimension $\dim X$ is equal to the supremum of the dimensions of the irreducible components of X.*

(d) *We have $\dim X = \sup\{\dim_x X \mid x \in X\}$.*

Proof (a) Let $Y_1 \subsetneq Y_2$ be two irreducible closed subsets of Y, and let X_1, X_2 be their respective closures in X. Then the X_i are irreducible and $X_1 \subsetneq X_2$. It follows from this that $\dim Y \leq \dim X$.

(b) and (c) follow immediately.

(d) By (a), we have $\dim_x X \leq \dim X$. Let $X_0 \subsetneq X_1 \subsetneq \cdots \subsetneq X_n$ be a chain of irreducible closed subsets of X. Let $x \in X_0$. Then for any open neighborhood U of x, the $X_i \cap U$ form a chain of irreducible closed subsets of U; hence $\dim U \geq n$ and $\dim_x X \geq n$, which proves (d). $\square$

Remark 5.6. It follows from Proposition 5.5(d) that the notion of dimension is local: if $\{U_i\}$ is a covering of X by open subsets U_i, then $\dim X$ is the supremum of the $\dim U_i$.

Definition 5.7. Let Y be an irreducible closed subset of X, we define the *codimension of Y in X* to be the supremum of the lengths of the chains of irreducible closed subsets of X containing Y:

$$Y \subseteq Z_0 \subsetneq Z_1 \subsetneq \cdots \subsetneq Z_n.$$

Let Z be a closed subset of X; the *codimension of Z in X*, which we denote by $\mathrm{codim}(Z, X)$, is the infimum of the codimensions (in X) of its irreducible components.

It easily follows from the definition that $\mathrm{codim}(Z, X) + \dim Z \leq \dim X$. We will see further on a case of equality (Proposition 5.23), but in general, this is not true (see Exercise 5.3).

Let A be a ring and $\mathfrak{p}$ a prime ideal of A. Let us recall that $\mathrm{ht}(\mathfrak{p})$, the *height of $\mathfrak{p}$*, is the supremum of the lengths of the strictly ascending chains of prime ideals contained in $\mathfrak{p}$. For an arbitrary ideal I of A, $\mathrm{ht}(I)$ is the infimum of the heights of the prime ideals containing I.

The supremum of the $\mathrm{ht}(\mathfrak{p})$, when $\mathfrak{p}$ runs through $\mathrm{Spec}\, A$, is called the *Krull dimension of A*. We denote it $\dim A$. This is consequently the supremum of the lengths of the strictly ascending chains of prime ideals of A. By convention, $\dim\{0\} = -\infty$.

Proposition 5.8. *Let A be a ring. The following properties are true.*

(a) *Let N be the nilradical of A. We have $\dim \mathrm{Spec}\, A = \dim A = \dim A/N$.*

(b) *Let $\mathfrak{p}$ be a prime ideal of A; then $\dim A_{\mathfrak{p}} = \mathrm{ht}(\mathfrak{p}) = \mathrm{codim}(V(\mathfrak{p}), \mathrm{Spec}\, A)$.*

(c) *We have $\dim A = \sup\{\dim A_{\mathfrak{m}}\}$, the supremum being taken over the set of maximal ideals $\mathfrak{m}$ of A.*

Proof (a) By Proposition 4.7(a), we have a bijective correspondence between the irreducible closed subsets and the prime ideals: $V(I) \mapsto \sqrt{I}$, which implies that $\dim \mathrm{Spec}\, A = \dim A$. As $\mathrm{Spec}\, A/N$ is homeomorphic to $\mathrm{Spec}\, A$, we have the second equality.

(b) and (c) follow from the description given above of the irreducible closed subsets of $\mathrm{Spec}\, A$. □

Example 5.9. A field is of dimension 0. Any principal ideal domain which is not a field (e.g., $\mathbb{Z}$ or $\mathbb{Q}[T]$) is of dimension 1. Let K be a number field; then the ring of integers $\mathcal{O}_K$ of K is of dimension 1. Indeed, the localization of $\mathcal{O}_K$ at any maximal ideal is a principal ideal domain, so $\dim \mathcal{O}_K = 1$ by Proposition 5.8(c). We can also use the following proposition:

Proposition 5.10. *Let $\varphi : A \to B$ be an integral homomorphism. Let $\mathfrak{q} \in \mathrm{Spec}\, B$ and $\mathfrak{p} = \varphi^{-1}(\mathfrak{q}) \in \mathrm{Spec}\, A$.*

(a) *We have* $\operatorname{ht}(\mathfrak{q}) \leq \operatorname{ht}(\mathfrak{p})$. *In particular,* $\dim B \leq \dim A$.

(b) *Suppose, moreover, that* φ *is injective. Then* $\operatorname{Spec} B \to \operatorname{Spec} A$ *is surjective. Moreover, we have* $\dim V(\mathfrak{p}) = \dim V(\mathfrak{q})$ *and* $\dim A = \dim B$.

Proof (a) Let $\mathfrak{q}_0 \subsetneq \mathfrak{q}_1$ be two prime ideals of B. Let us show that $\varphi^{-1}(\mathfrak{q}_0) \subsetneq \varphi^{-1}(\mathfrak{q}_1)$. This will imply that $\operatorname{ht}(\mathfrak{q}) \leq \operatorname{ht}(\mathfrak{p})$. By replacing B by $B/\mathfrak{q}_0$ and A by $A/\varphi^{-1}(\mathfrak{q}_0)$, we may assume that A and B are integral domains, that $\mathfrak{q}_0 = 0$, and that φ is injective. We must then show that $\varphi^{-1}(\mathfrak{q}_1) \neq 0$. Let $b \in \mathfrak{q}_1$ be non-zero. Let $b^n + a_{n-1}b^{n-1} + \cdots + a_0 = 0$ be an integral equation for b over A, of minimal degree n. Then $a_0 \in \varphi^{-1}(\mathfrak{q}_1) \setminus \{0\}$.

(b) The surjectivity of $\operatorname{Spec} B \to \operatorname{Spec} A$ is shown in Exercise 1.8. Hence by (a), it is enough to show the first equality. Let $\mathfrak{p}_0 \subsetneq \mathfrak{p}_1$ be prime ideals of A. Let $\mathfrak{q}_0 \in \operatorname{Spec} B$ be such that $\varphi^{-1}(\mathfrak{q}_0) = \mathfrak{p}_0$. By considering the injective integral homomorphism $A/\mathfrak{p}_0 \hookrightarrow B/\mathfrak{q}_0$, we obtain a prime ideal $\mathfrak{q}_1$ of B such that $\varphi^{-1}(\mathfrak{q}_1) = \mathfrak{p}_1$ and $\mathfrak{q}_0 \subsetneq \mathfrak{q}_1$. By repeatedly applying this result to a chain of prime ideals of A containing $\mathfrak{p}$, we get $\dim V(\mathfrak{p}) \leq \dim V(\mathfrak{q})$, and hence the equality since $A/\mathfrak{p} \to B/\mathfrak{q}$ is integral. □

2.5.2   The case of Noetherian schemes

Let us start by characterizing the schemes of dimension 0. Let us recall that a ring A is called *Artinian* if every descending sequence of ideals of A is stationary.

Lemma 5.11. *Let* $(A, \mathfrak{m})$ *be a Noetherian local ring. The following conditions are equivalent:*

(i) $\dim A = 0$;

(ii) $\mathfrak{m} = \sqrt{0}$, *the nilradical of* A;

(iii) *there exists a* $q \geq 1$ *such that* $\mathfrak{m}^q = 0$;

(iv) A *is Artinian.*

Proof As A is local, $\dim A = 0$ if and only if $\mathfrak{m}$ is the unique prime ideal of A, which is equivalent to $\mathfrak{m} = \sqrt{0}$, whence (i) $\Longleftrightarrow$ (ii). Let us suppose that (ii) is true. Let $a_1, \ldots, a_s$ be a system of generators of $\mathfrak{m}$. There exists a $t \geq 1$ such that $a_i^t = 0$ for every i. We then have $\mathfrak{m}^{ts} = 0$, so (ii) implies (iii).

Let us show that (iii) implies (iv). Let $(I_n)_n$ be a descending sequence of ideals of A. For any $r \geq 0$, $\mathfrak{m}^r/\mathfrak{m}^{r+1}$ is, in a natural way, a vector space over the field $k = A/\mathfrak{m}$, of finite dimension because A is Noetherian. The $(I_n \cap \mathfrak{m}^r)/(I_n \cap \mathfrak{m}^{r+1})$ form a descending, and therefore stationary, sequence of subvector spaces of $\mathfrak{m}^r/\mathfrak{m}^{r+1}$. There then exists an n_0 such that for every $r \leq q$ and every $n \geq n_0$ we have $I_n \cap \mathfrak{m}^r \subseteq I_{n+1} + I_n \cap \mathfrak{m}^{r+1}$. This implies that for every $n \geq n_0$, we have $I_n \subseteq I_{n+1} + I_n \cap \mathfrak{m} \subseteq \cdots \subseteq I_{n+1} + I_n \cap \mathfrak{m}^q = I_{n+1}$. Consequently, $I_n = I_{n+1}$. Hence A is Artinian.

Finally, if A is Artinian, then the descending sequence of ideals $(\mathfrak{m}^n)_n$ is stationary. Hence there exists a q such that $\mathfrak{m}^q = \mathfrak{m}^{q+1}$. This implies that $\mathfrak{m}^q = 0$ by Nakayama's lemma. Hence (iv) implies (ii). □

Theorem 5.12 (Krull's principal ideal theorem). *Let A be a Noetherian ring, and $f \in A$ a non-invertible element. Then for any prime ideal $\mathfrak{p}$ that is minimal among those containing f, we have $\operatorname{ht}(\mathfrak{p}) \le 1$.*

Proof Localizing in $\mathfrak{p}$, if necessary, we may assume that A is local with maximal ideal $\mathfrak{p}$. Moreover, if necessary, replacing A by its quotient by a minimal prime ideal, we may assume that A is an integral domain. Let $\mathfrak{q} \subsetneq \mathfrak{p}$ be a prime ideal. We must show that $\mathfrak{q} = 0$.

By assumption, $\mathfrak{p}/fA$ is the unique prime ideal of A/fA; hence $\dim A/fA = 0$. By the lemma above, A/fA is Artinian. Let us consider the sequence of ideals $\mathfrak{q}_n := \mathfrak{q}^n A_f \cap A$. The image of this sequence in A/fA is stationary. Therefore there exists an n_0 such that for every $n \ge n_0$, we have $\mathfrak{q}_n \subseteq \mathfrak{q}_{n+1} + fA$.

Let $n \ge n_0$. Let $x \in \mathfrak{q}_n$. There exists a $y \in A$ such that $x - fy \in \mathfrak{q}_{n+1}$. It follows that $fy \in \mathfrak{q}_n$ and therefore $y \in \mathfrak{q}_n$. In other words, we have $\mathfrak{q}_n \subseteq \mathfrak{q}_{n+1} + f\mathfrak{q}_n \subseteq \mathfrak{q}_{n+1} + \mathfrak{p}\mathfrak{q}_n$. It follows from Nakayama's lemma that $\mathfrak{q}_n = \mathfrak{q}_{n+1}$. As $\mathfrak{q}_n A_f = \mathfrak{q}^n A_f$, we have

$$\mathfrak{q}^{n_0} A_f = \bigcap_{n \ge n_0} \mathfrak{q}^n A_f = 0,$$

by virtue of Krull's theorem (Corollary 1.3.13). It follows that $\mathfrak{q} = 0$. □

Lemma 5.13. *Let A be a Noetherian ring, $f \in A$. Then for any chain of prime ideals $\mathfrak{p}_0 \subsetneq \cdots \subsetneq \mathfrak{p}_n$ of A such that $f \in \mathfrak{p}_n$ and $n \ge 1$, there exists a chain of prime ideals $\mathfrak{q}_1 \subsetneq \cdots \subsetneq \mathfrak{q}_n$ with $\mathfrak{q}_n = \mathfrak{p}_n$ and $f \in \mathfrak{q}_1$.*

Proof Let us show this assertion by induction on n. It is trivial for $n = 1$. Let us suppose that $n \ge 2$. We may assume that $f \notin \mathfrak{p}_{n-1}$ (otherwise we apply the induction hypothesis to the sequence $\mathfrak{p}_0, \ldots, \mathfrak{p}_{n-1}$). Let $\mathfrak{q}_{n-1}$ be a minimal element among the prime ideals containing $\mathfrak{p}_{n-2} + fA$ and contained in $\mathfrak{p}_n$. Then $\mathfrak{p}_{n-2} \subsetneq \mathfrak{q}_{n-1}$. By applying the induction hypothesis to the sequence $\mathfrak{p}_0, \ldots, \mathfrak{p}_{n-2}, \mathfrak{q}_{n-1}$, we obtain a chain of ideals $\mathfrak{q}_1 \subsetneq \cdots \subsetneq \mathfrak{q}_{n-1}$ with $f \in \mathfrak{q}_1$. On the other hand, by virtue of Theorem 5.12, the image of $\mathfrak{q}_{n-1}$ in $A/\mathfrak{p}_{n-2}$ is of height 1, while that of $\mathfrak{p}_n$ is of height at least 2 by hypothesis. We therefore have $\mathfrak{q}_{n-1} \subsetneq \mathfrak{p}_n$, which shows the assertion at rank n. □

Corollary 5.14. *Let A be a Noetherian ring, and I an ideal of A generated by r elements.*

(a) *Let $\mathfrak{p}$ be a prime ideal of A, minimal among those containing I; then $\operatorname{ht}(\mathfrak{p}) \le r$. In particular $\operatorname{ht}(I) \le r$.*

(b) *If, moreover, A is local, with maximal ideal $\mathfrak{m}$, then $\dim A$ is finite and $\dim A \le \dim_{A/\mathfrak{m}} \mathfrak{m}/\mathfrak{m}^2$.*

Proof (a) We use induction on r. The case $r = 1$ is Theorem 5.12. Let us suppose that $r \ge 2$. Let $f_1, \ldots, f_r$ be generators of I. Then $I/f_r A$ is an ideal of $A/f_r A$ generated by $r - 1$ elements, and the image of $\mathfrak{p}$ in $A/f_r A$ is prime and minimal among those containing $I/f_r A$. Hence the induction hypothesis says that $\mathrm{ht}(\mathfrak{p}/f_r A) \leq r - 1$. Let $\mathfrak{p}_0 \subsetneq \cdots \subsetneq \mathfrak{p}_n$ be a chain of prime ideals of A such that $\mathfrak{p}_n = \mathfrak{p}$ and $n \geq 1$. By virtue of Lemma 5.13, there exists a chain of prime ideals $\mathfrak{q}_1 \subsetneq \cdots \subsetneq \mathfrak{q}_n$ with $\mathfrak{q}_n = \mathfrak{p}$ and $f_r \in \mathfrak{q}_1$. The image of this chain in $A/f_r A$ is a chain of prime ideals of $A/f_r A$. Hence we have $n - 1 \leq \mathrm{ht}(\mathfrak{p}/f_r A) \leq r - 1$, whence $\mathrm{ht}(\mathfrak{p}) \leq r$.

(b) Let $e = \dim_{A/\mathfrak{m}} \mathfrak{m}/\mathfrak{m}^2$. By Nakayama's lemma, $\mathfrak{m}$ is generated by e elements. It follows that $\dim A = \mathrm{ht}(\mathfrak{m}) \leq e$. □

Theorem 5.15. *Let $(A, \mathfrak{m})$ be a Noetherian local ring, $f \in \mathfrak{m}$. Then we have $\dim(A/fA) \geq \dim A - 1$. Moreover, equality holds if f is not contained in any minimal prime ideal of A.*

Proof For any chain of prime ideals $\mathfrak{p}_0 \subsetneq \cdots \subsetneq \mathfrak{p}_n$ of A with $\mathfrak{p}_n = \mathfrak{m}$, we have $f \in \mathfrak{p}_n$, and therefore there exists a chain of prime ideals $\mathfrak{q}_1 \subsetneq \cdots \subsetneq \mathfrak{q}_n$ with $\mathfrak{q}_n = \mathfrak{p}_n$ and $f \in \mathfrak{q}_1$ (Lemma 5.13). The images of the $\mathfrak{q}_i$ in A/fA form a chain of prime ideals of length $n - 1$, and it follows that $\dim A/fA \geq n - 1$. This shows that $\dim(A/fA) \geq \dim A - 1$. Let us moreover suppose that f is not contained in any minimal prime ideal of A. For any prime ideal $\mathfrak{p}$, minimal among those containing f, we have $\mathrm{ht}(\mathfrak{p}) = 1$ by Theorem 5.12 ($\mathrm{ht}(\mathfrak{p}) \neq 0$ since otherwise $\mathfrak{p}$ would be a minimal prime ideal). It follows that $\dim(A/fA) \leq \dim A - 1$, whence the equality. □

Lemma 5.16. *Let A be a Noetherian ring and let $\mathfrak{m}$ (resp. $\mathfrak{n}$) be a maximal ideal of A (resp. of $A[T_1, ..., T_n]$) such that $\mathfrak{n} \cap A = \mathfrak{m}$. Then $\mathrm{ht}(\mathfrak{n}) = \mathrm{ht}(\mathfrak{m}) + n$.*

Proof By Corollary 1.12, $A[T_1, ..., T_n]/\mathfrak{n}$ is a finite extension of $A/\mathfrak{m}$. If $\mathfrak{n}_1 = \mathfrak{n} \cap A[T_1, ..., T_{n-1}]$, then $A[T_1, ..., T_{n-1}]/\mathfrak{n}_1$ is a sub-$A/\mathfrak{m}$-algebra of $A[T_1, ..., T_n]/\mathfrak{n}$ and is therefore a field. Thus $\mathfrak{n}_1$ is maximal. We then see that it is enough to prove the lemma in the case when $n = 1$.

Let $\mathfrak{p}_0 \subsetneq \cdots \subsetneq \mathfrak{p}_r \subseteq \mathfrak{m}$ be a chain of prime ideals of A. Then the $\mathfrak{p}_i A[T]$ form a chain of prime ideals of $A[T]$. Moreover, $\mathfrak{p}_r A[T] \subsetneq \mathfrak{n}$ because $A[T]/\mathfrak{p}_r A[T]$ is not a field. Therefore $\mathrm{ht}(\mathfrak{n}) \geq \mathrm{ht}(\mathfrak{m}) + 1$.

We will show the inequality in the other direction by induction on $\mathrm{ht}(\mathfrak{m})$. If $\mathrm{ht}(\mathfrak{m}) = 0$, then $\mathfrak{m}$ is a minimal prime ideal of A, and any prime ideal of $A[T]$ contained in $\mathfrak{n}$ must contain $\mathfrak{m}$. Therefore $\mathrm{ht}(\mathfrak{n}) = \mathrm{ht}(\bar{\mathfrak{n}}) = 1$ where $\bar{\mathfrak{n}}$ is the image of $\mathfrak{n}$ in $A/\mathfrak{m}[T]$. Let us suppose that $\mathrm{ht}(\mathfrak{m}) \geq 1$. As A has only a finite number of minimal prime ideals (Proposition 4.9) and $\mathfrak{m}$ is not contained in any of them, there exists an $f \in \mathfrak{m}$ which does not belong to any minimal prime ideal (Exercise 5.4). Let $B = A/fA$, and let $\mathfrak{n}'$ be the image of $\mathfrak{n}$ in $B[T]$. We have

$$\mathrm{ht}(\mathfrak{n}) = \dim A[T]_{\mathfrak{n}} \leq \dim A[T]_{\mathfrak{n}}/(f) + 1 = \dim B[T]_{\mathfrak{n}'} + 1 = \mathrm{ht}(\mathfrak{n}') + 1.$$

Let $\mathfrak{m}'$ be the image of $\mathfrak{m}$ in B. Then $\mathrm{ht}(\mathfrak{m}') = \dim B_{\mathfrak{m}'} = \dim A_{\mathfrak{m}}/(f) = \mathrm{ht}(\mathfrak{m}) - 1$. The induction hypothesis implies that $\mathrm{ht}(\mathfrak{n}') \leq \mathrm{ht}(\mathfrak{m}') + 1$. It follows that $\mathrm{ht}(\mathfrak{n}) \leq \mathrm{ht}(\mathfrak{m}) + 1$, which completes the proof. □

Corollary 5.17. *Let A be a Noetherian ring. Then we have*

$$\dim A[T_1, \ldots, T_n] = \dim A + n.$$

Remark 5.18. If A is not necessarily Noetherian, the proof above shows that $\dim A[T] \geq \dim A + 1$. We can also show that $\dim A[T] \leq 1 + 2\dim A$, see [17], VIII, §2, n° 2, Corollaire 2.

2.5.3 Dimension of algebraic varieties

Let k be a field and K an extension of k. If there exists an integer d such that K is algebraic over a subextension isomorphic to $k(T_1, \ldots, T_d)$, we say that K is of finite transcendence degree over k, and d is called the *transcendence degree of K over k*. We know that d is then unique ([55], VIII, Section 1). We denote this degree by $\operatorname{trdeg}_k K$. The field of fractions of a finitely generated integral domain over k is clearly of finite transcendence degree over k (use Proposition 1.9).

Proposition 5.19. *Let X be an integral algebraic variety over a field k. Then for any non-empty open subvariety U of X, we have $\dim U = \dim X = \operatorname{trdeg}_k K(X)$.*

Proof Let ξ be the generic point of X. Then $K(U) = \mathcal{O}_{U,\xi} = \mathcal{O}_{X,\xi} = K(X)$, so it suffices to show the second equality. By Remark 5.6, we can reduce to $X = \operatorname{Spec} A$ affine.

By the Noether normalization lemma (Proposition 1.9), there exists a finite injective homomorphism $k[T_1, \ldots, T_d] \hookrightarrow A$. On the one hand we have $\dim A = d$ (Proposition 5.10 and Corollary 5.17), on the other hand $\operatorname{trdeg}_k A = d$ because $\operatorname{Frac}(A)$ is algebraic over $k(T_1, \ldots, T_d)$, whence the equality $\dim A = \operatorname{trdeg}_k \operatorname{Frac}(A)$. □

Example 5.20. Let $X = \operatorname{Proj} k[x, y, z, w]/(xz - y^2, yz - xw, z^2 - yw)$. Then $\dim X = 1$. Indeed, the function field $K(X)$ of X is $k(y/z)$.

Corollary 5.21. *Let B be a homogeneous algebra over a field k. Then*

$$\dim \operatorname{Spec} B = \dim \operatorname{Proj} B + 1.$$

Proof We may assume that B is an integral domain. Indeed, the minimal prime ideals $\mathfrak{p}_1, \ldots, \mathfrak{p}_n$ of B are homogeneous (Lemma 3.35(a)), and it suffices to compare the dimensions of $V(\mathfrak{p}_i)$ and $V_+(\mathfrak{p}_i)$. Let $f \in B_+$ be a homogeneous element of degree 1. Then the $B_{(f)}$-algebra homomorphism $B_{(f)}[T, 1/T] \to B_f$ which sends T to $1/f$ is an isomorphism. Hence

$$\dim \operatorname{Spec} B = \dim D(f) = \dim D_+(f) + 1 = \dim \operatorname{Proj} B + 1$$

(Corollary 5.17). □

Lemma 5.22. *Let A be a finitely generated integral domain over k, and let $\mathfrak{p}$ be a prime ideal of A of height 1. Then $\dim A/\mathfrak{p} = \dim A - 1$.*

Proof Let $f \in \mathfrak{p} \setminus \{0\}$. Then $\sqrt{fA_{\mathfrak{p}}} = \mathfrak{p}A_{\mathfrak{p}}$. As A is Noetherian, there exists an $h \in A \setminus \mathfrak{p}$ such that $\sqrt{fA_h} = \mathfrak{p}A_h$. As the dimension does not change when we restrict ourselves to a non-empty open subset (proposition above), we can replace A by A_h, and suppose that $\sqrt{fA} = \mathfrak{p}A$.

By Proposition 1.9, there exists a finite injective homomorphism $A_0 := k[T_1, \ldots, T_d] \to A$. Let $\mathfrak{q} = \mathfrak{p} \cap A_0$. By virtue of Proposition 5.10, we have $\dim A = d$ and $\dim A/\mathfrak{p} = \dim A_0/\mathfrak{q}$. As we have $\dim A/\mathfrak{p} \leq d-1$, it suffices to show that $\dim A_0/\mathfrak{q} \geq d - 1$. Let $F = \mathrm{Norm}_{\mathrm{Frac}(A)/\,\mathrm{Frac}(A_0)}(f) \in A_0$ (see Exercise 5.14). By the multiplicativity of the norm, we see that $fA \cap A_0 \subseteq \sqrt{FA_0}$. Let P be an irreducible factor of F. Then $\mathfrak{q} \subseteq PA_0$. Now in the proof of Proposition 1.9 we have seen that, if necessary applying an automorphism to A_0, we can make P monic in T_d with coefficients in $k[T_1, \ldots, T_{d-1}]$. This immediately implies that $A_0/(P)$ is finite over $k[T_1, \ldots, T_{d-1}]$, whence $\dim A_0/\mathfrak{q} \geq \dim A_0/(P) = d-1$.   □

Proposition 5.23. *Let A be a finitely generated integral domain over k. Let $\mathfrak{p}$ be a prime ideal of A.*

(a) *We have $\mathrm{ht}(\mathfrak{p}) + \dim A/\mathfrak{p} = \dim A$.*

(b) *If $\mathfrak{p}$ is maximal, then $\dim A_{\mathfrak{p}} = \dim A$.*

Proof (a) Let us use induction on $\mathrm{ht}(\mathfrak{p})$. The statement is trivial if $\mathrm{ht}(\mathfrak{p}) = 0$. Let us suppose that $\mathrm{ht}(\mathfrak{p}) \geq 1$. Let $0 = \mathfrak{p}_0 \subsetneq \mathfrak{p}_1 \subsetneq \cdots \subsetneq \mathfrak{p}_d = \mathfrak{p}$ be a chain of prime ideals of length $d = \mathrm{ht}(\mathfrak{p})$. Then $\mathfrak{p}/\mathfrak{p}_1$ is an ideal of $A/\mathfrak{p}_1$, of height $\mathrm{ht}(\mathfrak{p}) - 1$. Applying the induction hypothesis to $\mathfrak{p}/\mathfrak{p}_1$, we therefore obtain

$$(\mathrm{ht}(\mathfrak{p}) - 1) + \dim A/\mathfrak{p} = \dim A/\mathfrak{p}_1.$$

Using Lemma 5.22 with $\mathfrak{p}_1 \subset A$, we obtain the equality $\mathrm{ht}(\mathfrak{p}) + \dim A/\mathfrak{p} = \dim A$.

(b) Indeed, we then have $\dim A/\mathfrak{p} = 0$. It suffices to apply (a).   □

Corollary 5.24. *Let X be an irreducible algebraic variety over k, and let x be a closed point of X. Then $\dim \mathcal{O}_{X,x} = \dim X$.*

Proof We may assume that X is affine and integral. The statement then follows from Proposition 5.23(b).   □

Remark 5.25. This result is not true in general, even for Noetherian schemes. See Exercise 5.3(b).

Before stating the corollary that follows, we need some notation. Let X be a scheme, and let $f \in \mathcal{O}_X(X)$. We let $V(f) := \{x \in X \mid f_x \in \mathfrak{m}_x \mathcal{O}_{X,x}\}$. This is the complement of X_f (see Definition 3.11) in X. Such a set is called a *principal closed subset*. For any affine open subset U of X, $V(f) \cap U$ is the principal closed subset $V(f|_U)$ which we defined before (Section 2.1). The following corollary results from Lemma 5.22.

Corollary 5.26. *Let X be an irreducible algebraic variety, and $f \in \mathcal{O}_X(X)$ be non-nilpotent. Then every irreducible component of $V(f)$ is of dimension $\dim X - 1$.*

Remark 5.27. Let k be a field. Intuitively (see Corollary 1.15), the affine space $\mathbb{A}_k^d$ must be of dimension d, and an algebraic set of the form $f(x_1, \ldots, x_d) = 0$, where $f(x_1, \ldots, x_d)$ is a non-zero polynomial, of dimension $d - 1$ (a 'curve' if $d = 2$ and a 'surface' if $d = 3$). In fact, this indeed coincides with our abstract definition, by virtue of the corollary above and of Corollary 5.17.

Example 5.28. Let k be a field. Then $\dim \mathbb{P}^n_k = n$. Let $f \in k[T_0, \ldots, T_n]$ be a non-zero homogeneous polynomial. Then the irreducible components of $V_+(f)$ are of dimension $n - 1$. Thus the closed subscheme $x^p + y^p = z^p$ (where $p \in \mathbb{N}$ is fixed) of $\mathbb{P}^2_k$ is of dimension 1.

Definition 5.29. Let k be a field. An algebraic variety over k whose irreducible components are of dimension 1 (resp. dimension 2) is called an *algebraic curve* (resp. *algebraic surface*) over k.

Exercises

5.1. Let X be a topological space. Let $\{X_i\}_i$ be a covering of X by closed subsets X_i. We assume that it is a locally finite covering, that is to say that every point $x \in X$ admits an open neighborhood U which meets only a finite number of X_i. Show that $\dim X = \sup_i \dim X_i$.

5.2. Let X be a scheme and Z a closed subset of X. Show that for all $x \in X$, we have $\operatorname{codim}(\overline{\{x\}}, X) = \dim \mathcal{O}_{X,x}$ and $\operatorname{codim}(Z, X) = \min_{z \in Z} \dim \mathcal{O}_{X,z}$.

5.3. Show the following properties.

(a) Let Z be a closed subset of a topological space X. Show that we have $\operatorname{codim}(Z, X) = 0$ if and only if Z contains an irreducible component of X. Give an example (with X non-irreducible) where $\operatorname{codim}(Z, X) = 0$ and $\dim Z < \dim X$.

(b) Let $X = \operatorname{Spec} \mathcal{O}_K[T]$ where $\mathcal{O}_K$ is a discrete valuation ring, with uniformizing parameter $t \neq 0$. Let $f = tT - 1$. Show that the ideal generated by f is maximal. Let $x \in X$ be the corresponding point. Show that X is irreducible, $\dim \mathcal{O}_{X,x} = 1$, and that $\operatorname{codim}(\overline{\{x\}}, X) + \dim\overline{\{x\}} < \dim X$.

5.4. Let A be a ring and $\mathfrak{p}_1, \ldots, \mathfrak{p}_r$ be prime ideals of A. Let I be an ideal of A contained in none of the $\mathfrak{p}_i$. We want to show that I is not contained in $\mathfrak{p}_1 \cup \cdots \cup \mathfrak{p}_r$.

(a) Show that the property is true for $r \leq 2$.

(b) Assume that the property is true for $r-1$ and that $\mathfrak{p}_r$ does not contain any $\mathfrak{p}_i$, $i \leq r - 1$. Let $x \in I \setminus (\mathfrak{p}_1 \cup \cdots \cup \mathfrak{p}_{r-1})$. Show that there exists a $y \in (I\mathfrak{p}_1 \cdots \mathfrak{p}_{r-1}) \setminus \mathfrak{p}_r$.

(c) Show that either x or $x + y$ is not in $\mathfrak{p}_1 \cup \cdots \cup \mathfrak{p}_r$.

5.5. Let A be a graded ring, and let $\mathfrak{p}_1, \ldots, \mathfrak{p}_r$ be homogeneous prime ideals of A. Let I be a homogeneous ideal of A contained in none of the $\mathfrak{p}_i$. We want to show that there exists a homogeneous element of I not contained in $\cup_i \mathfrak{p}_i$. One can suppose that $\mathfrak{p}_r$ does not contain any $\mathfrak{p}_i$, $i < r$.

(a) Show that there exists a homogeneous element $a \in I\mathfrak{p}_1 \cdots \mathfrak{p}_{r-1} \setminus \mathfrak{p}_r$.

(b) Let $b \in I$ be a homogeneous element such that $b \notin \cup_{i \leq r-1}\mathfrak{p}_i$. Show that b or $a^{\deg b} + b^{\deg a}$ is not in $\mathfrak{p}_1 \cup \cdots \cup \mathfrak{p}_r$. Conclude.

5.6. Let A be a Noetherian ring, I an ideal of A.

    (a) Let $x \in I$. Show that $\operatorname{ht}(I/xA) + 1 \geq \operatorname{ht}(I)$, and that the equality holds if x does not belong to any minimal prime ideal of A. Show that any prime ideal minimal among those containing x has height equal to 1, if x is a regular element.

    (b) Show that $\operatorname{ht}(I) \geq 1$ if and only if I is not contained in the union of the minimal prime ideals of A (use (a) and Exercise 5.4).

    (c) Let $r = \operatorname{ht}(I)$. Show that I contains an ideal J generated by r elements such that $\operatorname{ht}(J) = r$ and that $\operatorname{ht}(I/J) = 0$.

5.7. Let $f : X \to Y$ be a morphism of algebraic varieties. We suppose that $f(X)$ is everywhere dense in Y. Show that $\dim X \geq \dim Y$ (use Proposition 5.19). Note a counterexample below for schemes in general.

5.8. Let $\mathcal{O}_K$ be as in Exercise 5.3, and let $K = \operatorname{Frac}(\mathcal{O}_K)$, $k = \mathcal{O}_K/t\mathcal{O}_K$ be the residue field of $\mathcal{O}_K$. Let us set $A = K \times k$ and let $\varphi : \mathcal{O}_K \to A$ be the homomorphism induced by $\mathcal{O}_K \to K$ and $\mathcal{O}_K \to k$. Show that $f_\varphi : \operatorname{Spec} A \to \operatorname{Spec} \mathcal{O}_K$ is surjective and that $\dim \mathcal{O}_K > \dim A$. Also show that A is a finitely generated $\mathcal{O}_K$-algebra (i.e., quotient of a polynomial ring over $\mathcal{O}_K$).

5.9. Let X be a scheme over a field k. Show that the points of $X(k)$ are closed in X. If X is an algebraic variety over k, then $x \in X$ is closed if and only if $k(x)$ is algebraic over k.

5.10. Let k be an non-countable infinite field. Let X be an algebraic variety over k with $\dim X \geq 1$, $(Y_n)_n$ a sequence of closed subsets of X with $\dim Y_n < \dim X$. We want to show that $\cup_n Y_n \neq X$.

    (a) Show the result for $X = \mathbb{A}^1_k$ and then for $\mathbb{A}^m_k$.

    (b) By using Noether's normalization lemma, show the result for an arbitrary affine variety. Deduce the general case from this.

5.11. (*Schemes of dimension* 0)

    (a) Let X be a scheme which is a (finite or not) disjoint union of open subschemes X_i. Show that $\mathcal{O}_X(X) \simeq \prod_i \mathcal{O}_X(X_i)$.

    (b) Show that any scheme of finite cardinal and dimension 0 is affine.

    (c) Let $X = \operatorname{Spec} A$ be a scheme of finite cardinal and dimension 0. Show that every point $x \in X$ is open. Deduce from this that $A \simeq \oplus_{\mathfrak{p} \in \operatorname{Spec} A} A\mathfrak{p}$.

    (d) Show that statement (c) is false if we do not suppose $\operatorname{Spec} A$ of dimension 0.

5.12. Let k be a field. We will determine the affine open subsets of $\mathbb{A}^n_k$ and of $\mathbb{P}^n_k$. See also Exercise 4.1.15.

    (a) Show that the principal open subsets of $\mathbb{A}^n_k$ and of $\mathbb{P}^n_k$ (not equal to $\mathbb{P}^n_k$) are affine.

(b) Let $X = \cup_i D(f_i)$ be a finite union of principal open subsets of $\mathbb{A}_k^n$. Show that $\mathcal{O}_{\mathbb{A}_k^n}(X) = k[T_1, \ldots, T_n]_f$, where $f = \gcd\{f_i\}_i$. Show that every affine open subset of $\mathbb{A}_k^n$ is principal.

(c) Show that every irreducible closed subset of $\mathbb{P}_k^n$ of dimension $n-1$ is principal.

(d) Let X be an affine open subset of $\mathbb{P}_k^n$. Show that the irreducible components of $\mathbb{P}_k^n \setminus X$ are of dimension $n-1$, and that X is a principal open subset of $\mathbb{P}_k^n$.

5.13. Let k be a field. A *function field in n variables* over k is a field K that is finitely generated over k (i.e., there exist $f_1, \ldots, f_r \in K$ such that $K = k(f_1, \ldots, f_r)$), of transcendence degree $\operatorname{trdeg}_k K = n$. Show that an extension K of k is a function field in n variables over k if and only if $K = K(X)$, where X is an integral algebraic variety over k, of dimension n. If this is the case, we can take X projective.

5.14. Let L/K be a finite field extension. Let $x \in L$. Then the multiplication by x is an endomorphism of L as K-vector space. We let $\operatorname{Norm}_{L/K}(x)$ denote the determinant of this endomorphism. We also call it the *norm* of x over K.

(a) Show that $\operatorname{Norm}_{L/K}$ is a multiplicative map from L to K.

(b) Let $A \subseteq B$ be rings such that $K = \operatorname{Frac}(A)$, $L = \operatorname{Frac}(B)$, and that B is integral over A. Show that for any $b \in B$, $\operatorname{Norm}_{L/K}(b)$ is integral over A.

(c) Let us moreover suppose that A is a polynomial ring over a field k. Show that $\operatorname{Norm}_{L/K}(B) \subseteq A$.

5.15. Let $X = \operatorname{Proj} B$ be a projective variety over a field k.

(a) Let $f \in B_+$ be a homogeneous element. Let $\mathfrak{p}$ be a prime ideal of B, minimal among those containing f. Show that $\mathfrak{p}$ is homogeneous, contained in B_+, and that $\operatorname{ht}(\mathfrak{p}) \leq 1$. Show that if $V_+(f) = \emptyset$, then $B_+ = \mathfrak{p}$ and $\dim X \leq 0$.

(b) Let Y be a closed subvariety of X, of dimension r. Show that for any sequence $f_1, \ldots, f_r$ of homogeneous elements of B_+, we have $V_+(f_1, \ldots, f_r) \cap Y \neq \emptyset$.

3

Morphisms and base change

In this chapter we study some properties of morphisms in relation to base change. The concept of base change is fundamental in the theory of schemes. It makes it possible to understand better the 'relative' (or functorial) nature of schemes, in the sense that it is sometimes more important to study the morphisms of schemes $X \to S$ than the schemes X themselves. Base change is also a very useful technique for reducing to a more suitable base scheme S. We will have many occasions to put it into practice.

In the first section, we introduce the notions of fibered product and base change. Next, we treat the particular case when the base scheme is the spectrum of a field. The chapter concludes with the study of a class of particularly important morphisms, that of the proper morphisms.

3.1 The technique of base change

3.1.1 Fibered product

Let X, Y, S be sets, $\varphi : X \to S$, $\psi : Y \to S$ maps. The fibered product $X \times_S Y$ of X and Y over S is the set of elements $(x, y) \in X \times Y$ such that $\varphi(x) = \psi(y)$. This set, defined very simply, can also be characterized by a universal property (Exercise 1.1). We use its analogue to define the fibered product of schemes.

Definition 1.1. Let S be a scheme, and let X, Y be two S-schemes. We define the *fibered product of X, Y over S* to be an S-scheme $X \times_S Y$, together with two morphisms of S-schemes $p : X \times_S Y \to X$, $q : X \times_S Y \to Y$ (called the *projections*), verifying the following universal property:

Let $f : Z \to X$, $g : Z \to Y$ be two morphisms of S-schemes. Then there exists a unique morphism of S-schemes $(f, g) : Z \to X \times_S Y$

making the following diagram commutative:

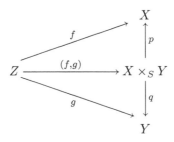

Proposition 1.2. *Let S be a scheme, and let X, Y be two S-schemes. Then the fibered product $(X \times_S Y, p, q)$ exists, and is unique up to isomorphism. If X, Y, and S are affine, then $X \times_S Y = \mathrm{Spec}(\mathcal{O}_X(X) \otimes_{\mathcal{O}_S(S)} \mathcal{O}_Y(Y))$, and the projection morphisms are induced by the canonical homomorphisms $\mathcal{O}_X(X), \mathcal{O}_Y(Y) \to \mathcal{O}_X(X) \otimes_{\mathcal{O}_S(S)} \mathcal{O}_Y(Y)$.*

Proof Let us first specify the meaning of an isomorphism of fibered products: if two triplets (Z, p, q) and (Z', p', q') satisfy the property of the fibered product, there exists a *unique* isomorphism $f : Z \to Z'$ such that $p = p' \circ f$ and $q = q' \circ f$. The uniqueness can easily be deduced from the universal property, in a similar manner as for Proposition 1.1.2. This uniqueness property will be used for the construction of the fibered product $X \times_S Y$.

The proof of the existence of the fibered product consists of constructing the fibered product in the case of affine schemes, and glueing affine pieces in the general case. Let us note that if $(X \times_S Y, p, q)$ exists, then for any open subscheme U of X, the fibered product of U and Y also exists. It suffices for this to take $U \times_S Y := p^{-1}(U)$, the projection morphisms being the restrictions of p and q to the open subset $p^{-1}(U)$ (use Exercise 2.2.12). Moreover, given the symmetry of the definition, if $(X \times_S Y, p, q)$ exists, then $(X \times_S Y, q, p)$ is the fibered product of Y and X.

Let us first consider the case when $S = \mathrm{Spec}\, A$, $X = \mathrm{Spec}\, B$, $Y = \mathrm{Spec}\, C$ are affine. Let us set $W = \mathrm{Spec}(B \otimes_A C)$. Let p (resp. q) be the morphism corresponding to the canonical homomorphism $B \to B \otimes_A C$ (resp. $C \to B \otimes_A C$). It follows from the universal property of $B \otimes_A C$ (Proposition 1.1.14) and from Proposition 2.3.25 that (W, p, q) is the fibered product of X and Y over S.

Let us now suppose S, Y affine, and X arbitrary. Let $\{X_i\}_i$ be a covering of X by affine open subsets. From the above, the fibered product $(X_i \times_S Y, p_i, q_i)$ exists for every i. For any pair i, j, $p_i^{-1}(X_i \cap X_j)$ and $p_j^{-1}(X_i \cap X_j)$ are canonically isomorphic to $(X_i \cap X_j) \times_S Y$, which gives an isomorphism of fibered products $f_{ij} : p_i^{-1}(X_i \cap X_j) \simeq p_j^{-1}(X_i \cap X_j)$. We then have $f_{ik} = f_{jk} \circ f_{ij}$ by the uniqueness of the isomorphism of fibered products $p_i^{-1}(X_i \cap X_j \cap X_k) \times_S Y \simeq p_k^{-1}(X_i \cap X_j \cap X_k) \times_S Y$. Consequently, we can glue the S-schemes $X_i \times_S Y$ to an S-scheme W (Lemma 2.3.33). As each $X_i \times_S Y$ can be considered as an X-scheme and a Y-scheme via the projection morphisms, and as the f_{ij} are compatible with the structures of X-schemes and of Y-schemes, we obtain projection morphisms

$p : W \to X$, $q : W \to Y$. The fact that (W, p, q) form the fibered product of X and Y can be verified immediately (use Exercise 2.2.11).

Let us now remove the assumption that Y is affine. We cover Y by affine open subsets Y_i. Then the fibered products of X and Y_i exist by what we have just seen and by the symmetry of the fibered product. By glueing the $X \times_S Y_i$ as above, we obtain the existence of the fibered product.

Let us conclude with the general case that S is not necessarily affine. Let $\{S_i\}_i$ be an affine open covering of S. Let us write $f : X \to S$ and $g : Y \to S$ for the structural morphisms, $X_i = f^{-1}(S_i)$, and $Y_i = g^{-1}(S_i)$. Any S_i-scheme is in a natural way an S-scheme. It immediately follows from the definition that the fibered product of X_i and Y_i over S_i (which exists by the preceding case) is also their fibered product over S. We leave it to the reader to verify that the schemes $X_i \times_S Y_i$ glue via the $(X_i \cap X_j) \times_S (Y_i \cap Y_j)$, and that the resulting scheme is the fibered product $X \times_S Y$.   □

In what follows, we will generally omit the projection morphisms p, q in the notation of the fibered product $(X \times_S Y, p, q)$. When $S = \operatorname{Spec} A$, we also write $X \times_A Y$ instead of $X \times_S Y$.

Example 1.3. Let k be a ring. By the construction above and Example 1.1.15, we have

$$\mathbb{A}_k^n \times_k \mathbb{A}_k^m = \mathbb{A}_k^{n+m}.$$

More generally, if X (resp. Y) is an affine closed subscheme of $\operatorname{Spec} k[T_1, \ldots, T_n]$ (resp. of $\operatorname{Spec} k[S_1, \ldots, S_m]$) defined by polynomials $P_1(T), \ldots, P_r(T)$ (resp. by polynomials $Q_1(S), \ldots, Q_l(S)$), then the fibered product $X \times_k Y$ is the closed subscheme of $\operatorname{Spec} k[T_i, S_j]_{i \le n, j \le m}$ defined by the polynomials $P_1(T), \ldots, P_r(T)$, $Q_1(S), \ldots, Q_l(S)$.

Let us note, on the other hand, that in general $\mathbb{P}_k^n \times_k \mathbb{P}_k^m \not\simeq \mathbb{P}_k^{n+m}$ (Exercise 3.21).

Proposition 1.4. *Let S be a scheme, and X, Y schemes over S. The following properties are true.*

  (a) *We have canonical isomorphisms of schemes: $X \times_S S \simeq X$, $X \times_S Y \simeq Y \times_S X$, $(X \times_S Y) \times_S Z \simeq X \times_S (Y \times_S Z)$.*

  (b) *Let Z be a Y-scheme, considered as an S-scheme via $Z \to Y \to S$. Then we have a canonical isomorphism of S-schemes $(X \times_S Y) \times_Y Z \simeq X \times_S Z$, where $X \times_S Y$ is endowed with the structure of a Y-scheme via the second projection.*

  (c) *Let $f : X \to X'$, $g : Y \to Y'$ be morphisms of S-schemes. There exists a unique morphism of S-schemes $f \times g : X \times_S Y \to X' \times_S Y'$ which makes* the following diagram commutative:

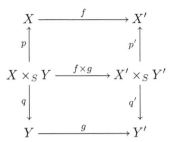

(d) Let $i : U \to X$, $j : V \to Y$ be open subschemes. Then the morphism $i \times j$ induces an isomorphism $U \times_S V \simeq p^{-1}(U) \cap q^{-1}(V) \subseteq X \times_S Y$.

Proof All these properties easily follow from the universal property of the fibered product. □

Remark 1.5. Let us denote the underlying topological space of X by $\operatorname{sp}(X)$. One must be careful when handling the space $\operatorname{sp}(X \times_S Y)$, which is not necessarily the fibered product of the topological spaces $\operatorname{sp}(X) \times_{\operatorname{sp}(S)} \operatorname{sp}(Y)$. See Exercise 1.10.

Remark 1.6. Keeping the notation of Proposition 1.4, let $\operatorname{Mor}_S(X, Y)$ denote the set of morphisms of S-schemes from X to Y. Let Z be an S-scheme. The projections p, q induce, by composition, maps

$$\operatorname{Mor}_S(Z, X \times_S Y) \to \operatorname{Mor}_S(Z, X), \quad \operatorname{Mor}_S(Z, X \times_S Y) \to \operatorname{Mor}_S(Z, Y).$$

This gives a map

$$\operatorname{Mor}_S(Z, X \times_S Y) \to \operatorname{Mor}_S(Z, X) \times \operatorname{Mor}_S(Z, Y), \quad (1.1)$$

which is bijective by the universal property. By taking $Z = S$, we obtain a canonical bijection of the sections

$$(X \times_S Y)(S) \simeq X(S) \times Y(S).$$

If we take $Y = Z$, we see that the bijection in (1.1) induces a bijection between the subsets

$$\operatorname{Mor}_Y(Y, X_Y) \simeq \operatorname{Mor}_S(Y, X), \quad (1.2)$$

where $X_Y = X \times_S Y$ is endowed with the structure of a Y-scheme via the second projection.

3.1.2 Base change

Definition 1.7. Let S be a scheme, and X an S-scheme. For any S-scheme S', the second projection $q : X \times_S S' \to S'$ endows $X \times_S S'$ with the structure of an S'-scheme. This process is called *base change by* $S' \to S$. We sometimes denote the S'-scheme $X \times_S S'$ by $X_{S'}$. If $f : X \to Y$ is a morphism of S-schemes, we let $f_{S'}$ denote the morphism $f \times \operatorname{Id}_{S'} : X \times_S S' \to Y \times_S S'$. If $S' = \operatorname{Spec} B$ is affine, we also denote $X_{S'}$ by X_B, and $f_{S'}$ by f_B.

Example 1.8. Let $X = \operatorname{Spec} A[T_1, \ldots, T_n]/(P_1, \ldots, P_m)$ be an affine scheme. Let B be an A-algebra. Then

$$X_B = \operatorname{Spec} B[T_1, \ldots, T_n]/(P_1, \ldots, P_m),$$

the P_i being considered as polynomials with coefficients in B.

Let us now consider base change for projective schemes. Let A be a ring. Let B be a graded A-algebra, C an A-algebra. Then we can canonically endow $E := B \otimes_A C$ with the structure of a graded C-algebra. Indeed, $B = \oplus_{d \geq 0} B_d$ implies that $E = \oplus_{d \geq 0} (B_d \otimes_A C)$. It then suffices to set $E_d = B_d \otimes_A C$. Let us note that E is still a graded A-algebra.

Proposition 1.9. *Let A be a ring. Let B be a graded A-algebra, and C an A-algebra. Let us endow $B \otimes_A C$ with the grading as above. Then we have a canonical isomorphism*

$$\operatorname{Proj}(B \otimes_A C) \simeq \operatorname{Proj} B \times_{\operatorname{Spec} A} \operatorname{Spec} C.$$

Proof Let $\varphi : B \to E = B \otimes_A C$ denote the canonical homomorphism ($\varphi(b) = b \otimes 1$). Then $\varphi(B_+)E = E_+$. We therefore obtain morphisms $g : \operatorname{Proj} E \to \operatorname{Proj} B$ (Lemma 2.3.40) and $\operatorname{Proj} E \to \operatorname{Spec} C$ (Proposition 2.3.38). These are morphisms of A-schemes. We therefore obtain a morphism of A-schemes

$$h : \operatorname{Proj} E \to \operatorname{Proj} B \times_{\operatorname{Spec} A} \operatorname{Spec} C.$$

For any $f \in B_+$, we have $h^{-1}(D_+(f) \times_{\operatorname{Spec} A} \operatorname{Spec} C) = g^{-1}(D_+(f)) = D_+(\varphi(f))$ (Lemma 2.3.40). It therefore suffices to show that

$$D_+(\varphi(f)) \to D_+(f) \times_{\operatorname{Spec} A} \operatorname{Spec} C$$

is an isomorphism, in other words that $\psi : B_{(f)} \otimes_A C \to E_{(\varphi(f))}$ is an isomorphism. It is clear that ψ, which is defined by $\psi((b/f^n) \otimes c) = (b \otimes c)/\varphi(f)^n$, is surjective. Moreover, as $B_{(f)}$ is a direct factor of

$$B_f = B_{(f)} \oplus (\oplus_{m \neq n \deg f} B_m/f^n),$$

the canonical homomorphism $B_{(f)} \otimes_A C \to B_f \otimes_A C \simeq E_{\varphi(f)}$ (Exercise 1.2.2) is injective. As this injection factors into ψ followed by the inclusion $E_{(\varphi(f))} \subseteq E_{\varphi(f)}$, we have ψ injective. Consequently, ψ is indeed an isomorphism. □

Example 1.10. Let $A \to C$ be a ring homomorphism. Then we have $(\mathbb{P}_A^n)_C = \mathbb{P}_C^n$. Let X be a closed subscheme of $\mathbb{P}_A^n$ defined by homogeneous polynomials $P_0(T), \ldots, P_m(T) \in A[T_0, \ldots, T_n]$. Then X_C is the closed subscheme of $\mathbb{P}_C^n$ defined by the same polynomials.

Notation. Let S be a scheme. It is in a unique way a $\mathbb{Z}$-scheme. We set

$$\mathbb{P}_S^n = \mathbb{P}_{\mathbb{Z}}^n \times_{\operatorname{Spec} \mathbb{Z}} S.$$

This is an S-scheme via the second projection $\mathbb{P}_S^n \to S$. If $S = \operatorname{Spec} A$ is affine, then $\mathbb{P}_S^n$ coincides with $\mathbb{P}_A^n$ (proposition above).

Definition 1.11. Let A be a ring. We say that a morphism $f : X \to \operatorname{Spec} A$ is *projective* if it factors into a closed immersion $X \to \mathbb{P}^n_A$ followed by the canonical morphism $\mathbb{P}^n_A \to \operatorname{Spec} A$. This coincides with Definition 2.3.42. We say that an A-scheme is *projective* if the structural morphism is projective.

In this book, we will essentially only use projective morphisms over affine schemes. The definition of a projective morphism over a general base scheme is more complicated (see [41], II.5.5.2). We adopt here the more restrictive definition of [43] (it coincides with that of [41] if the base scheme is affine, see also [41], II.5.5.4(ii)).

Definition 1.12. Let S be a scheme. We will say that a morphism $X \to S$ is *projective* if it factors into a closed immersion $X \to \mathbb{P}^n_S$ followed by the canonical morphism $\mathbb{P}^n_S \to S$.

Let us consider a particular case of base change, which is that of the morphism $\operatorname{Spec} k(y) \to Y$ (Example 2.3.18).

Definition 1.13. Let $f : X \to Y$ be a morphism of schemes. For any $y \in Y$, we set

$$X_y = X \times_Y \operatorname{Spec} k(y).$$

This is the *fiber of f over y*. The second projection $X_y \to \operatorname{Spec} k(y)$ makes X_y into a scheme over $k(y)$.

Example 1.14. Let $X = \mathbb{P}^n_Y$. Then for any $y \in Y$, we have $X_y = \mathbb{P}^n_{k(y)}$ by Proposition 1.4(b).

Definition 1.15. Let $f : X \to Y$ be a morphism with Y irreducible of generic point ξ; we call X_ξ the *generic fiber* of f.

Proposition 1.16. *Let $f : X \to Y$ be a morphism of schemes. Then for any $y \in Y$, the first projection $p : X_y = X \times_Y \operatorname{Spec} k(y) \to X$ induces a homeomorphism from X_y onto $f^{-1}(y)$.*

Proof We may assume that $Y = \operatorname{Spec} A$ is affine. Actually, if V is an affine neighborhood of y, then $X_y = (X \times_Y V) \times_V \operatorname{Spec} k(y) = f^{-1}(V)_y$. Moreover, for any open subset U of X, we have $p^{-1}(U) = U \times_Y \operatorname{Spec} k(y)$, and may therefore assume that $X = \operatorname{Spec} B$ is also affine. Let $\mathfrak{p}$ be the prime ideal of A corresponding to $y \in Y$. Let p_1 be the morphism associated to the canonical ring homomorphism $B \to B \otimes_A A_\mathfrak{p}$ (which is a localization), and let p_2 be the morphism associated to the canonical surjection $B \otimes_A A_\mathfrak{p} \to B \otimes_A k(\mathfrak{p})$. Then $p = p_2 \circ p_1$. Let us denote the homomorphism $f^\#(Y)$ by $\varphi : A \to B$. Lemma 2.1.7(b) and (c) imply that p induces a homeomorphism from X_y onto the set of prime ideals $\mathfrak{q}$ of B such that $\mathfrak{q} \supseteq \mathfrak{p}B$ and $\mathfrak{q} \cap \varphi(A \setminus \mathfrak{p}) = \emptyset$. Now this condition is equivalent to $\varphi^{-1}(\mathfrak{q}) = \mathfrak{p}$. This shows that the image of X_y in X is $f^{-1}(y)$. $\square$

Remark 1.17. Let $f : X \to Y$ be a morphism of schemes. Let $y \in Y$. The proposition above makes it possible to endow the set $f^{-1}(y)$ with the structure of a scheme over $k(y)$. Intuitively, we can consider X as a family of schemes $X_y \to \operatorname{Spec} k(y)$ parameterized by the points $y \in Y$.

Example 1.18. Let X be a scheme over a discrete valuation ring $\mathcal{O}_K$, and let s be the closed point of $\operatorname{Spec} \mathcal{O}_K$. We call X_s the *special fiber of X*. The topological space X is the disjoint union of the generic fiber X_η, which is a scheme over $\operatorname{Frac}(\mathcal{O}_K)$, and of the special fiber X_s, which is a scheme over the residue field $k(s)$ of $\mathcal{O}_K$. Moreover, X_η is open in X because $\{\eta\}$ is open in $\operatorname{Spec} \mathcal{O}_K$. The fiber X_s is closed in X because s is closed; hence it is also called the *closed fiber*.

Example 1.19. Let m be a non-zero integer. Let

$$f : X = \operatorname{Spec} \mathbb{Z}[T_1, T_2]/(T_1 T_2^2 - m) \to \operatorname{Spec} \mathbb{Z}$$

be the canonical morphism. For any prime number p, we let X_p denote the fiber of f over the point $p\mathbb{Z} \in \operatorname{Spec} \mathbb{Z}$. Then the generic fiber of X is

$$\operatorname{Spec} \mathbb{Q}[T_1, T_2]/(T_1 T_2^2 - m) = \operatorname{Spec} \mathbb{Q}[T_2, 1/T_2].$$

The fiber X_p is equal to $\operatorname{Spec} \mathbb{F}_p[T_1, T_2]/(T_1 T_2^2 - m)$. Thus X_p is integral (isomorphic to $\operatorname{Spec} \mathbb{F}_p[T, 1/T]$) if p does not divide m, while it has two irreducible components and is not reduced if p divides m. Let us note that X itself is integral. We have in a way 'cut' X 'up' into slices, most of these slices X_p (for $p \nmid m$) staying integral, but some become reducible. This is the phenomenon of 'degeneration'. We will return to it in later chapters.

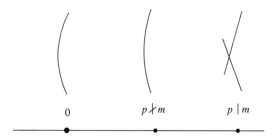

Figure 4. Family of varieties that degenerates.

Remark 1.20. Let $f : X \to Y$ be a morphism of S-schemes, and S' be an S-scheme. Then the morphism $f \times \operatorname{Id}_{S'} : X \times_S S' \to Y \times_S S'$ can be obtained from $f : X \to Y$ by the base change $Y \times_S S' \to Y$ because $X \times_S S' = X \times_Y (Y \times_S S')$.

Definition 1.21. A property $\mathcal{P}$ of morphisms of schemes $f : X \to Y$ is said to be *local on the base* Y if the following assertions are equivalent:

(i) f verifies $\mathcal{P}$;

(ii) for any $y \in Y$, there exists an affine neighborhood V of y such that $f|_{f^{-1}(V)}$ verifies $\mathcal{P}$.

For example, open and closed immersions are clearly properties which are local on the base.

Definition 1.22. We say that a property $\mathcal{P}$ of morphisms of schemes is *stable under base change* if for any morphism $X \to Y$ verifying $\mathcal{P}$, $X \times_Y Y' \to Y'$ also verifies $\mathcal{P}$ for every Y-scheme Y'.

Proposition 1.23. *Open immersions and closed immersions are stable under base change.*

Proof Let $X \to Y$ be an open immersion, and let $Y' \to Y$ be a morphism. By Proposition 1.4(d), $X \times_Y Y' \to Y' = Y \times_Y Y'$ induces an isomorphism from $X \times_Y Y'$ to $p^{-1}(X)$, where $p : Y \times_Y Y' \to Y$ is the first projection. Hence $X \times_Y Y' \to Y'$ is an open immersion.

As the property of closed immersion is local on Y, we may assume that $Y = \operatorname{Spec} A$, $Y' = \operatorname{Spec} B$. Then $X = \operatorname{Spec} A/I$, and the morphism $X \times_Y Y' \to Y'$ is induced by the surjective homomorphism $B \to B \otimes_A A/I = B/IB$; it is therefore a closed immersion. □

To conclude, let us study the behavior of the ring $\mathcal{O}_X(X)$ after base change. Let $A \to B$ be a ring homomorphism. Let X be a Noetherian A-scheme. The projection morphisms $X_B \to X$, $X_B \to \operatorname{Spec} B$ induce canonical homomorphisms $\mathcal{O}_X(X) \to \mathcal{O}_{X_B}(X_B)$, $B \to \mathcal{O}_{X_B}(X_B)$. By tensor product, this gives a canonical homomorphism $\mathcal{O}_X(X) \otimes_A B \to \mathcal{O}_{X_B}(X_B)$ (Proposition 1.1.14).

Proposition 1.24. *Let $A \to B$ be a flat ring homomorphism. Let X be a Noetherian A-scheme. Then the canonical homomorphism $\mathcal{O}_X(X) \otimes_A B \to \mathcal{O}_{X_B}(X_B)$ is an isomorphism.* (See also Lemma 5.2.26.)

Proof Let $\{X_i\}_i$ be a finite covering of X by affine open subsets. We have an exact sequence:

$$0 \to \mathcal{O}_X(X) \to \oplus_i \mathcal{O}_X(X_i) \to \oplus_{i,j} \mathcal{O}_X(X_i \cap X_j) \tag{1.3}$$

(see Lemma 2.2.7). As $A \to B$ is flat, this sequence tensored by B, which we denote by $(1.3) \otimes_A B$, stays exact. Moreover, $\{(X_i)_B\}_i$ is an affine open covering of X_B. We therefore have another exact sequence:

$$0 \to \mathcal{O}_{X_B}(X_B) \to \oplus_i \mathcal{O}_{X_B}((X_i)_B) \to \oplus_{i,j} \mathcal{O}_{X_B}((X_i)_B \cap (X_j)_B).$$

By comparing the latter with $(1.3) \otimes_A B$ as at the end of the proof of Proposition 2.3.12, we obtain the bijectivity of $\mathcal{O}_X(X) \otimes_A B \to \mathcal{O}_{X_B}(X_B)$. □

Exercises

1.1. Let S, X, Y be sets. We keep the notions of fibered product from the beginning of this section. Let $p : X \times_S Y \to X$, $q : X \times_S Y \to Y$ denote the restrictions of the projections $X \times Y \to X$ and $X \times Y \to Y$. Show that for any set Z and any pair of maps $f : Z \to X$, $g : Z \to Y$ such that $\varphi \circ f = \psi \circ g$, there exists a unique map $(f,g) : Z \to X \times_S Y$ such that $p \circ (f,g) = f$ and $q \circ (f,g) = g$.

1.2. Let $f : X \to Y$ be a morphism of schemes. For any scheme T, let $f(T) :$
$X(T) \to Y(T)$ denote the map defined by $f(T)(g) = f \circ g$. Show that
$f(T)$ is bijective for every T if and only if f is an isomorphism (use the
identity morphisms on X and Y).

1.3. Let $f : X \to Y$ be a dominant morphism of irreducible schemes. Show
that the generic fiber of f is dense in X.

1.4. Let $f : X \to Y$ be a morphism of affine schemes. Show that for any affine
open subscheme V of Y, $f^{-1}(V)$ is affine (see also Exercise 3.8).

1.5. Let X, Y be algebraic varieties over a field k. Show that

$$\dim(X \times_k Y) = \dim X + \dim Y$$

(use Proposition 2.1.9).

1.6. Let $\pi : T \to S$ be a morphism of schemes. Let us suppose that $\pi : T \to S$
is an open or closed immersion, or that $S = \operatorname{Spec} A$, and π is induced by
a localization $A \to F^{-1}A$ for a multiplicative set $F \subset A$.

   (a) Let $f, g : Z \to T$ be two morphisms of schemes such that $\pi \circ f = \pi \circ g$.
Show that $f = g$.

   (b) Let X, Y be T-schemes. Show that the canonical morphism $X \times_T Y \to$
$X \times_S Y$ is an isomorphism.

   (c) Show that (b) can be false if the hypothesis on $T \to S$ is not verified.

1.7. Let X, Y be S-schemes, p and q the projection morphisms from $X \times_S Y$
to X, Y. Let us fix $s \in S$. Show that for any $x \in X_s$ and any $y \in Y_s$,
there exists a natural homeomorphism

$$\operatorname{Spec}(k(x) \otimes_{k(s)} k(y)) \to \{z \in X \times_S Y \mid p(z) = x, q(z) = y\}.$$

1.8. Let $X \to S$ be a surjective morphism. Show that $X \times_S Y \to Y$ is surjective
for any S-scheme Y. In other words, surjective morphisms are stable under
base change. (Use Proposition 1.16.)

1.9. Let k be a field. We want to study $\operatorname{Spec} A$, where $A = k(u) \otimes_k k(v)$
is the tensor product of two purely transcendental extensions of k, of
transcendence degree 1.

   (a) We have $k(u) = T^{-1}k[u]$, where T is the multiplicative set made
up of the non-zero elements of $k[u]$. Deduce from this that A is the
localization of $k[u, v]$ with respect to the multiplicative set T' made
up of the non-zero elements of the form $P(u)Q(v) \in k[u, v]$.

   (b) Let $\mathfrak{m}$ be a maximal ideal of $k[u, v]$. Show that there exists a $P(u) \in$
$\mathfrak{m} \setminus \{0\}$. Deduce from this that $T' \cap \mathfrak{m} \neq \emptyset$.

   (c) Show that the maximal ideals of A are of the form gA, with $g \in$
$k[u, v] \setminus (k[u] \cup k[v])$ irreducible in $k[u, v]$.

   (d) Show that $\operatorname{Spec} A$ is an infinite set and that $\dim A = 1$.

1.10. Let S be a scheme, and let $\pi : X \to S$, $\rho : Y \to S$ be S-schemes. Let $\mathrm{sp}(X) \times_{\mathrm{sp}(S)} \mathrm{sp}(Y)$ be the fibered product of sets defined by π and ρ, endowed with the topology induced by the product topology on $\mathrm{sp}(X) \times \mathrm{sp}(Y)$. We are going to study some properties concerning the relation between $\mathrm{sp}(X \times_S Y)$ and $\mathrm{sp}(X) \times_{\mathrm{sp}(S)} \mathrm{sp}(Y)$.

(a) Show that we have a canonical continuous map $f : \mathrm{sp}(X \times_S Y) \to \mathrm{sp}(X) \times_{\mathrm{sp}(S)} \mathrm{sp}(Y)$.

(b) Show that f is surjective.

(c) Let us consider the example $X = Y = \mathrm{Spec}\,\mathbb{C}$ and $S = \mathrm{Spec}\,\mathbb{R}$. Show that $X \times_S Y \simeq \mathrm{Spec}(\mathbb{C} \oplus \mathbb{C})$ and that f is not injective.

(d) Show that in the case of Exercise 1.9, with $X = \mathrm{Spec}\,k(u)$, $Y = \mathrm{Spec}\,k(v)$, and $S = \mathrm{Spec}\,k$, the map f has infinite fibers.

(e) Let $S = \mathrm{Spec}\,k$ be the spectrum of an arbitrary field. By studying the example $X = Y = \mathbb{A}_k^1$, show that the image of an open subset under f is not necessarily an open subset.

1.11. Let k be a field and $z \in \mathbb{P}_k^n(k)$. We choose a homogeneous coordinate system such that $z = (1, 0, \ldots, 0)$.

(a) Show that there exists a morphism $p : \mathbb{P}_k^n \setminus \{z\} \to \mathbb{P}_k^{n-1}$ such that over $\overline{k}$, where $\overline{k}$ is the algebraic closure of k, we have

$$p_{\overline{k}}(a_0, a_1, \ldots, a_n) = (a_1, \ldots, a_n)$$

for every point of $\mathbb{P}_k^n(\overline{k})$. Such a morphism is called a *projection with center z*.

(b) Let X be a closed subset of $\mathbb{P}_k^n$ not containing z. Show that X cannot contain $p^{-1}(y)$ for any $y \in \mathbb{P}_k^{n-1}$.

3.2 Applications to algebraic varieties

3.2.1 Morphisms of finite type

Most interesting morphisms in algebraic geometry are of finite type and localizations of morphisms of finite type.

Definition 2.1. A morphism $f : X \to Y$ is said to be of *finite type* if f is quasi-compact (Exercise 2.3.17), and if for every affine open subset V of Y, and for every affine open subset U of $f^{-1}(V)$, the canonical homomorphism $\mathcal{O}_Y(V) \to \mathcal{O}_X(U)$ makes $\mathcal{O}_X(U)$ into a finitely generated $\mathcal{O}_Y(V)$-algebra. A Y-scheme is said to be of finite type if the structural morphism is of finite type.

Proposition 2.2. *Let $f : X \to Y$ be a morphism of schemes. Let us suppose that there exists a covering $\{V_i\}_i$ of Y by affine open subsets such that for every i, $f^{-1}(V_i)$ is a finite union of affine open subsets U_{ij}, and that $\mathcal{O}_X(U_{ij})$ is a finitely generated algebra over $\mathcal{O}_Y(V_i)$ for every j. Then f is of finite type.*

Proof Let V be an affine open subset of Y. For any i, $V \cap V_i$ is a union of open subsets V_{ik} which are principal both in V_i and in V. As V is quasi-compact, it is a finite union of open subsets of the form V_{ik}. We have $f^{-1}(V_{ik}) = \cup_j U_{ijk}$, where $U_{ijk} := U_{ij} \cap f^{-1}(V_{ik})$ is a principal open subset of U_{ij}. It follows that $f^{-1}(V)$ is a finite union of affine open subsets $\cup_\alpha U_\alpha$, where each U_α is of the form U_{ijk}. In particular, $f^{-1}(V)$ is quasi-compact.

Let us note, moreover, that $\mathcal{O}_X(U_\alpha) = \mathcal{O}_X(U_{ijk})$ is finitely generated over $\mathcal{O}_X(V_{ik})$, and hence finitely generated over $\mathcal{O}_Y(V)$, because $\mathcal{O}_X(V_{ik})$ is finitely generated over $\mathcal{O}_Y(V)$. Let U be an affine open subset of $f^{-1}(V)$. Then U is a finite union of open subsets $U_{\alpha\beta}$ which are principal both in U and in U_α. Each $\mathcal{O}_X(U_{\alpha\beta})$ is finitely generated over $\mathcal{O}_X(U_\alpha)$, and hence finitely generated over $\mathcal{O}_Y(V)$. Finally, we have reduced to the case when $V = \operatorname{Spec} A$, $U = \operatorname{Spec} B$, U is a finite union of principal open subsets $U_\gamma = D(b_\gamma)$ with $b_\gamma \in B$, and $\mathcal{O}_X(U_\gamma)$ finitely generated over A. It follows that there exists a finitely generated sub-A-algebra C of B, containing the b_γ and such that $C_{b_\gamma} = B_{b_\gamma}$ for every γ. We have an identity $1 = \sum_\gamma b_\gamma c_\gamma$ with $c_\gamma \in B$. It is easy to conclude that as an A-algebra, B is generated by C and the c_γ. Consequently, B is indeed finitely generated over A.   □

Example 2.3. Let k be a field. Then the schemes of finite type over $\operatorname{Spec} k$ are exactly the algebraic varieties over k (Definition 2.3.47).

Proposition 2.4. *We have the following properties:*

  (a) *Closed immersions are of finite type.*

  (b) *The composition of two morphisms $f : X \to Y$, $g : Y \to Z$ of finite type is of finite type.*

  (c) *Morphisms of finite type are stable under base change.*

  (d) *If $X \to Z$ and $Y \to Z$ are of finite type, then so is $X \times_Z Y \to Z$.*

  (e) *If the composition of two morphisms $f : X \to Y$ and $g : Y \to Z$ is of finite type, and if f is quasi-compact, then f is of finite type.*

Proof (a) results from Proposition 2.3.20; (b) let $h = g \circ f$. Let V be an affine open subset of Z. Then $g^{-1}(V)$ is a finite union of affine open subsets U_i of Y, and each $f^{-1}(U_i)$ is a finite union of affine open subsets W_{ij} of X. It is clear that the composition $W_{ij} \to U_i \to V$ is of finite type. Since $h^{-1}(V) = \cup_{i,j} W_{ij}$, it follows from Proposition 2.2 that h is of finite type. For (c) and (d), we use the construction of the fibered product of two affine schemes over an affine scheme (Proposition 1.2).

It remains to show (e). Using Proposition 2.2, we can reduce to affine Z. Let V be an affine open subset of Y. By hypothesis, $f^{-1}(V)$ is a finite union of affine open subsets U_i. As $\mathcal{O}_X(U_i)$ is finitely generated over $\mathcal{O}_Z(Z)$, it is also finitely generated over $\mathcal{O}_Y(V)$, which implies that f is of finite type.   □

Remark 2.5. Let $X \to Y$ be a morphism of finite type. Then for any $y \in Y$, X_y is an algebraic variety over $k(y)$, by Proposition 2.4(c). We can therefore consider X as a family of algebraic varieties (possibly over different fields) parameterized by the points of Y.

3.2.2 Algebraic varieties and extension of the base field

We fix a field k. In this subsection, we study some properties of algebraic varieties over k relative to the base change $\operatorname{Spec} K \to \operatorname{Spec} k$, for a field extension K/k. If A is a k-algebra, we denote by A_K the K-algebra $A \otimes_k K$. Let us note that as every k-module is free and therefore flat, any inclusions of k-modules $M \subseteq M'$, $N \subseteq N'$ induce an inclusion $M \otimes_k N \subseteq M' \otimes_k N'$.

Lemma 2.6. *Let X be an algebraic variety over k, and let K be an algebraic extension of k. Then for any reduced closed subvariety W of X_K, there exist a finite subextension K' of K, and a unique (for fixed K') reduced closed subvariety Z of $X_{K'}$ such that $W = Z_K$.*

Proof Let us first suppose that $X = \operatorname{Spec} A$. Then $W = V(I)$, where I is a radical ideal of A_K, generated by $f_1, \ldots, f_m$. There exists a finite subextension K' of K such that $f_i \in A_{K'}$ for every i. Let I' be the ideal of $A_{K'}$ generated by the f_i, and $Z = \operatorname{Spec} A_{K'}/I'$. Then $I' \otimes_{K'} K = I$, and hence $Z_K = W$. We have Z reduced, because $\mathcal{O}_Z(Z) \hookrightarrow \mathcal{O}_Z(Z) \otimes_{K'} K = \mathcal{O}_W(W)$. As the underlying space of Z is fixed (it is the image of W under the projection $X_K \to X_{K'}$, see Exercise 1.8), the uniqueness of the scheme structure on Z results from Proposition 2.4.2(e). Let us note that if (Z, K') fulfills the requirements, then so does (Z_L, L) for any finite subextension L/K' of K/K'.

In the general case, we cover X with a finite number of affine open subvarieties X_i. There then exist a finite subextension K'/k of K and reduced closed subvarieties Z_i of $(X_i)_{K'}$ such that $(Z_i)_K = W \cap (X_i)_K$. By the uniqueness in the affine case, the Z_i glue to a reduced closed subvariety Z of $X_{K'}$. We indeed have $Z_K = W$, and the uniqueness of Z comes from Proposition 2.4.2(e), as in the affine case. $\square$

Proposition 2.7. *Let X be an algebraic variety over k, and let K/k be an algebraic extension. The following properties are true.*

(a) *We have $\dim X_K = \dim X$.*

(b) *If X is reduced and K/k is separable, then X_K is reduced.*

(c) *If K/k is purely inseparable, then the projection $X_K \to X$ is a homeomorphism.*

Proof We may assume that $X = \operatorname{Spec} A$ is affine. As $A \to A_K$ is injective and integral, (a) follows from Proposition 2.5.10.

(b) Let $\mathfrak{p}_1, \ldots, \mathfrak{p}_n$ be the minimal prime ideals of A. Then A injects into $\oplus_i A/\mathfrak{p}_i$, and therefore A_K injects into $\oplus_i (A/\mathfrak{p}_i)_K$. We can therefore assume A to be integral. As A_K is a subring of $\operatorname{Frac}(A) \otimes_k K$, it suffices to show that $F \otimes_k K$ is reduced for any field F containing k. Every element of $F \otimes_k K$ is contained in $F \otimes_k K'$, with K' finite separable over k, and we can therefore assume K to be finite over k. It follows that $K \simeq k[T]/(P(T))$, where $P(T) \in k[T]$ is a separable polynomial. As $P(T)$ is still separable in $F[T]$, $F \otimes_k K \simeq F[T]/(P)$ is reduced.

(c) Let us denote the projection by $p : X_K \to X$. Let us first suppose that K is a simple extension of k (that is generated by a single element). We then have $K \simeq k[T]/(T^q - c)$, with $c \in k$ and q a power of $\operatorname{char}(k)$. Let $\mathfrak{p}$ be a prime ideal of A. Let $\mathfrak{q}$ be a prime ideal of A_K such that $\mathfrak{q} \cap A = \mathfrak{p}$. Then $\mathfrak{p} A_K \subseteq \mathfrak{q} A_K = \mathfrak{q}$. On the other hand, since $\alpha^q \in A \cap \mathfrak{q} = \mathfrak{p}$ for all $\alpha \in \mathfrak{q}$, we have $\mathfrak{q} \subseteq \sqrt{\mathfrak{p} A_K}$. Hence $\mathfrak{q} = \sqrt{\mathfrak{p} A_K}$ and p is injective. Let us now show that p is a closed map. Let I be an ideal of A_K. Let $J = I \cap A$. Then $I^q \subseteq J$. This implies that $p(V(I)) = V(J)$. As $X_K \to X$ is surjective (Proposition 2.5.10 or Exercise 1.8), p is a homeomorphism.

Let us now take K/k finite. The extension K/k is made up of successive simple extensions, and therefore $X_K \to X$ is a homeomorphism by the simple case. The general case immediately follows from the finite case and the lemma above. □

In the remainder of the book, we will sometimes deal with an algebraic closure of a given field k. Since all algebraic closures of k are isomorphic to each other, we will simply say 'the' algebraic closure. The same remark applies to the separable closures.

Definition 2.8. Let X be an algebraic variety over k. Let $\overline{k}$ be the algebraic closure of k. We say that X is *geometrically* (some authors prefer the adverb 'absolutely') *reduced* (resp. *geometrically integral*) if $X_{\overline{k}}$ is reduced (resp. integral). In a similar way, we define *geometrically connected* varieties and *geometrically irreducible* varieties.

Remark 2.9. If X is integral with function field $K(X)$, then for any open subset U of X, $\mathcal{O}(U_{\overline{k}})$ injects into $K(X) \otimes_k \overline{k}$. It follows that X is geometrically reduced (resp. geometrically integral) if $K(X) \otimes_k \overline{k}$ is reduced (resp. integral). The converse is also true because $K(X)$ is the direct limit of the $\mathcal{O}_X(U)$ over the non-empty open subsets U of X, and the tensor product by $\overline{k}$ commutes with the direct limit (Exercise 1.1.6).

Example 2.10. If k is perfect, then any reduced algebraic variety over k is geometrically reduced (Proposition 2.7(b)).

Remark 2.11. Let X be an algebraic variety over k. It is easy to see, using Lemma 2.6, that if $X_{\overline{k}}$ verifies one of the properties, reduced, connected, or irreducible, then X_K verifies the same property for any algebraic extension K/k.

Example 2.12. Let K be a non-trivial finite extension of k, and let $X = \operatorname{Spec} K$. If K/k is purely inseparable, then X is reduced but not geometrically reduced, because $X_K = \operatorname{Spec}(K \otimes_k K)$ is not reduced. If K/k is separable, then X is integral but not geometrically integral, nor even geometrically connected, because $X_{\overline{k}}$ is made up of $[K : k]$ isolated points.

Example 2.13. Let k be a field of characteristic different from 2, and let $a \in k$ which is not a square. Let us consider the projective variety

$$X = \operatorname{Proj} k[u, v, w]/(u^2 - av^2).$$

Let $\alpha \in \overline{k}$ be a square root of a and $K = k[\alpha]$. Then we easily verify that X is integral, while $X_K = \operatorname{Proj} K[u, v, w]/(u - \alpha v)(u + \alpha v)$ is not. See Figure 5 below.

X $\quad\quad\quad\quad\quad\quad\quad\quad\quad\quad\quad\quad$ X_K

Figure 5. A non-geometrically irreducible integral conic.

Corollary 2.14. *Let X be an integral algebraic variety over k, with function field $K(X)$.*

(a) *Let us suppose that char$(k) = p > 0$. Let $L := k^{p^{-\infty}}$ be the perfect closure of k. Then X is geometrically reduced if and only if X_L is reduced.*

(b) *Let k^s be the separable closure of k. Then X is geometrically connected (resp. irreducible) if and only if X_{k^s} is connected (resp. irreducible).*

(c) *The variety X is geometrically integral if and only if $K(X)$ and $\overline{k}$ are linearly disjoint over k. Moreover, in that case we have $K(X_{\overline{k}}) = K(X) \otimes_k \overline{k}$.*

(d) *X is geometrically irreducible if and only if $K(X) \cap k^s = k$.*

Proof (a) and (b) result from Proposition 2.7(b) (indeed, $\overline{k}$ is separable over $k^{p^{-\infty}}$) and (c).

(c) Let us suppose $K(X)$ and $\overline{k}$ are linearly disjoint over k. Then $K(X) \otimes_k \overline{k}$ is an integral domain. For any affine open subset U of X, we have $\mathcal{O}_{X_{\overline{k}}}(U_{\overline{k}}) = \mathcal{O}_X(U) \otimes_k \overline{k} \subseteq K(X) \otimes_k \overline{k}$. Therefore $U_{\overline{k}}$ is integral. This implies that $X_{\overline{k}}$ is integral. Conversely, let us suppose $X_{\overline{k}}$ is integral. Let $U = \operatorname{Spec} A$ be an affine open subset of X. Then $K(X) \otimes_k \overline{k} = \operatorname{Frac}(A) \otimes_k \overline{k}$ is a localization of $A \otimes_k \overline{k}$, and is therefore an integral domain. Moreover, it is integral over $K(X)$ because $\overline{k}$ is integral over k. It is therefore a field. In particular, $K(X)$ and $\overline{k}$ are linearly disjoint over k. Finally, $A \otimes_k \overline{k} \subseteq \operatorname{Frac}(A) \otimes_k \overline{k}$ implies that

$$K(X_{\overline{k}}) = \operatorname{Frac}(A \otimes_k \overline{k}) = \operatorname{Frac}(A) \otimes_k \overline{k} = K(X) \otimes_k \overline{k}.$$

(d) Let $K = K(X) \cap k^s$. Let us suppose X is geometrically irreducible. Then X_{k^s} is integral. The method above then shows that $K(X_{k^s}) = K(X) \otimes_k k^s$. Now $K(X) \otimes_k k^s$ contains $K \otimes_k K$, and hence $K = k$, because otherwise $K \otimes_k K$ is not an integral domain. Conversely, let us suppose that $K = k$. It suffices to show that $K(X) \otimes_k K'$ is an integral domain for any finite separable extension K'/k. We have $K' = k[T]/P(T)$. Let $Q(T) \in K(X)[T]$ be a monic polynomial dividing $P(T)$. Then $Q(T) \in k^s[T]$ because $P(T)$ splits over k^s. It follows that $Q(T) \in (K(X) \cap k^s)[T] = k[T]$, and hence $Q = 1$ or P. Consequently, $P(T)$ is irreducible in $K(X)[T]$, which proves that $K(X) \otimes_k K' = K(X)[T]/P(T)$ is an integral domain. $\square$

Proposition 2.15. *Let X be an integral algebraic variety over k. Then X is geometrically reduced if and only if $K(X)$ is a finite separable extension of a purely transcendental extension $k(T_1, \ldots, T_d)$.*

Proof If $K(X)$ is a finite separable extension of $L := k(T_1, \ldots, T_d)$, then $K(X) = L[S]/P(S)$ for an irreducible separable polynomial $P(S) \in L[S]$. Hence $K(X) \otimes_k \overline{k} = L'[S]/P(S)$, where $L' = L \otimes_k \overline{k} = \overline{k}(T_1, \ldots, T_n)$. As $P(S)$ stays separable in $L'[S]$, we have $K(X) \otimes_k \overline{k}$ reduced. It follows that X is geometrically reduced.

Conversely, let us suppose X is geometrically reduced. We have a presentation of $K(X)$ as a finite extension of $L := k(T_1, \ldots, T_d)$. If $K(X)/L$ is separable, there is nothing to prove. Let us suppose that $K(X)$ is not equal to the separable closure L_s of L in $K(X)$. Let $f \in K(X) \setminus L_s$ be such that $f^p \in L_s$. We will show that $L_s[f]$ is finite and separable over a purely transcendental extension. This will imply the proposition by decomposing $K(X)/L_s$ into a sequence of purely inseparable extensions of degree $p = \mathrm{char}(k)$.

Let $P(S) = S^r + Q_{r-1}S^{r-1} + \cdots + Q_0 \in L[S]$ be the minimal polynomial of f^p over L. Let us show that at least one $Q_i \notin k(T_1^p, \ldots, T_d^p)$. Indeed, in the opposite case, $P(S^p) = H(S)^p$ for an $H(S) \in \overline{k}[S]$. Therefore $L_s[f] \otimes_k \overline{k} = (L_s \otimes_k \overline{k})[S]/P(S^p)$ is not reduced. This is impossible because $K(X) \otimes_k \overline{k}$ is reduced (Remark 2.9) and contains $L_s[f] \otimes_k \overline{k}$ as a subalgebra. We can therefore assume, for example, that a power of T_1 prime to p appears in one of the Q_i. It follows that T_1 is algebraic and separable over $k(f, T_2, \ldots, T_d)$. As $L_s[f]$ is finite separable over $k(f, T_1, T_2, \ldots, T_d)$, we have $L_s[f]$ finite separable over $k(f, T_2, \ldots, T_d)$. Finally, this last extension is purely transcendental because its transcendence degree over k is equal to that of $L_s[f]$, which is d. □

Remark 2.16. Using differential forms (Remark 6.1.16), the construction of the subextension $k(T_1, \ldots, T_d)$ of $K(X)$ has a natural interpretation. Indeed, we can show that it suffices to take elements $T_1, \ldots, T_d \in K(X)$ such that the dT_i form a basis of $\Omega^1_{K(X)/k}$ as $K(X)$-vector space.

3.2.3 Points with values in an extension of the base field

Definition 2.17. Let X be an algebraic variety over a field k, and let K/k be a field extension. We let $X(K)$ denote the set of morphisms of k-schemes from $\mathrm{Spec}\, K$ to X. The elements of $X(K)$ are called the K-valued points of X.

In this subsection, we collect some elementary information on the set $X(K)$. It must be noted that in general, $X(K)$ is not the set of points $x \in X$ such that $k(x) \subseteq K$ (see (b) of the proposition below).

Proposition 2.18. Let X be an algebraic variety over k, and let K/k be a field extension. The following properties are true.

(a) We have a canonical bijection $X(K) \to X_K(K)$.

(b) Any element of $X(K)$ is uniquely determined by the data consisting of a point $x \in X$ and a homomorphism of k-algebras $k(x) \to K$.

(c) For any extension K' of K, we have a natural inclusion $X(K) \subseteq X(K')$.

(d) If X is a closed subvariety $V(I)$ of $\mathrm{Spec}\, k[T_1, \ldots, T_n]$, then we can identify $X(K)$ with $\{(t_1, \ldots, t_n) \in K^n \mid P(t_1, \ldots, t_n) = 0, \forall P \in I\}$.

(e) *If X is a closed subvariety $V_+(I)$ of $\operatorname{Proj} k[T_0, \ldots, T_n]$, then we can identify $X(K)$ with $\{(t_0, \ldots, t_n) \in \mathbb{P}(K^{n+1}) \mid P(t_0, \ldots, t_n) = 0, \forall P \in I\}$.*

Proof (a) This is just a translation of bijection (1.2) of Remark 1.6.

(b) Let $s \in X(K)$. Let $x \in X$ be the image of $s : \operatorname{Spec} K \to X$. Then $s_x^\# : \mathcal{O}_{X,x} \to K$ induces a homomorphism $k(x) \to K$. Conversely, taking $x \in X$ and a homomorphism $k(x) \to K$ makes it possible to define a morphism of k-schemes $\operatorname{Spec} K \to \operatorname{Spec} k(x)$. By composing with the canonical morphism $\operatorname{Spec} k(x) \to X$, we obtain an element of $X(K)$. One immediately sees that these two processes are mutual inverses.

(c) The composition with $\operatorname{Spec} K' \to \operatorname{Spec} K$ induces a map $X(K) \to X(K')$. The injectivity comes from (b). Finally, (d) and (e) result from (a), Example 2.3.32, and Corollary 2.3.44. □

Let K/k be a Galois extension with group G; then G naturally acts on $\operatorname{Spec} K$. Indeed, for any $\sigma \in G$, $\operatorname{Spec} \sigma : \operatorname{Spec} K \to \operatorname{Spec} K$ is an automorphism of k-schemes. Consequently, G acts (as a group of automorphisms of k-schemes) on $X_K = X \times_{\operatorname{Spec} k} \operatorname{Spec} K$, by the identity on the first component. Likewise, it naturally acts on $X(K)$. If X is a closed subvariety $V(I)$ of $\operatorname{Spec} k[T_1, \ldots, T_n]$, then for any $\sigma \in G$, and for any $x = (t_1, \ldots, t_n) \in X(K)$, we have $\sigma(x) = (\sigma(t_1), \ldots, \sigma(t_n))$. The action is similar for projective varieties.

Proposition 2.19. *Let X be an algebraic variety over k, and let K/k be a Galois extension with group G. Then the set of classes $G \backslash X(K)$ injects into X, and the set of fixed points $X(K)^G$ can be identified with $X(k)$.*

Proof We have a map $\rho : X(K) \to X$ deduced from Proposition 2.18(b). Let $x \in X$. The group G naturally acts on $\rho^{-1}(x)$ (if it is non-empty). Now $\rho^{-1}(x)$ can be identified with the set $\operatorname{Hom}_{k\text{-algebras}}(k(x), K)$, and the action of G on $\rho^{-1}(x)$ can be identified with its natural action on $\operatorname{Hom}_{k\text{-algebras}}(k(x), K)$. As this action is transitive, ρ induces an injective map from $G \backslash X(K)$ to X. Moreover, an element $\sigma : k(x) \to K$ is invariant under G if and only if $k(x) = k$. □

Proposition 2.20. *Let X be a geometrically reduced algebraic variety over a field k. Let k^s be the separable closure of k. Then $X(k^s) \neq \emptyset$.*

Proof By replacing X by X_{k^s} (which is geometrically reduced), we may assume that $k = k^s$. We must then show that $X(k) \neq \emptyset$. By replacing X by an irreducible affine open subset, we may assume X is affine and integral. By Proposition 2.15, $K(X)$ is finite separable over $k(T_1, \ldots, T_d)$. Hence $K(X) = k(T_1, \ldots, T_d)[f]$. Let us write $A = k[T_1, \ldots, T_d]$, $B = \mathcal{O}_X(X)$. Let $P(S) \in \operatorname{Frac}(A)[S]$ be the minimal polynomial of f. Then, localizing B, if necessary, we may assume that $A[f] \subseteq B$. As $\operatorname{Frac}(B) = \operatorname{Frac}(A)[f]$, there exists a $g \in A$ such that $B \subseteq A_g[f]$ and $P(S) \in A_g[S]$. It follows that $B_g = A_g[f] = A_g[S]/P(S)$. As $P(S)$ is a separable polynomial, its resultant $h := \operatorname{Res}(P(S), P'(S)) \in A_g$ is non-zero. As the field k is infinite since separably closed, there exists a $t \in k^d$ such that $g(t) \neq 0$ and $h(t) \neq 0$. Let $y \in \operatorname{Spec} A_g$ be the point corresponding to t. Then $k(y) = k$ and
$$B_g \otimes_{A_g} k(y) = k[S]/\tilde{P}(S),$$

where $\tilde{P}(S)$ is the image of $P(S)$ in $k(y)[S]$. The resultant of $\tilde{P}(S)$ is $h(t) \neq 0$. Therefore $\tilde{P}(S)$ is separable. It follows that $B_g \otimes_{A_g} k(y)$ is a direct sum of the ks. Therefore the points of $\operatorname{Spec} B_g$ over y (see Proposition 1.16) are rational over k. This shows that $X(k) \neq \emptyset$.   □

3.2.4 Frobenius

In this subsection we fix a prime number p. All considered schemes X will be schemes over $\mathbb{F}_p$. In other words, $\mathcal{O}_X$ is a sheaf of rings of characteristic p.

Definition 2.21. Let X be a scheme over $\mathbb{F}_p$. We call the morphism $F_X : X \to X$ induced by the ring homomorphism $\mathcal{O}_X \to \mathcal{O}_X : a \mapsto a^p$ the *absolute Frobenius of X*.

Lemma 2.22.

(a) *Let $f : X \to Y$ be a morphism of schemes over $\mathbb{F}_p$. Then $F_Y \circ f = f \circ F_X$.*

(b) *For any $x \in X$, we have $F_X(x) = x$.*

Proof (a) is immediate. Let $U = \operatorname{Spec} A$ be an affine open subset of X. Let $\rho : A \to A$ denote the homomorphism $\rho(a) = a^p$. For any $x \in U$ corresponding to a prime ideal $\mathfrak{p} \subset A$, we have $\rho^{-1}(\mathfrak{p}) = \mathfrak{p}$. Therefore $F_X(x) = F_U(x) = (\operatorname{Spec} \rho)(x) = x$.   □

Let S be a scheme over $\mathbb{F}_p$ and $\pi : X \to S$ an S-scheme. We let $X^{(p)}$ denote the fibered product $X \times_S S$, where the second factor S is endowed with the structure of an S-scheme via $F_S : S \to S$. We will endow $X^{(p)}$ with the structure of an S-scheme given by the second projection $q : X \times_S S \to S$. Let us denote the first projection by $\varphi : X^{(p)} \to X$. We have a commutative diagram

$$
\begin{array}{ccc}
X & \xrightarrow{\;F_X\;} & X \\
{\scriptstyle \pi}\downarrow & & \downarrow{\scriptstyle \pi} \\
S & \xrightarrow{\;F_S\;} & S
\end{array}
$$

There therefore exists a unique morphism of S-schemes $F_{X/S} : X \to X^{(p)}$ making the following diagram commutative:

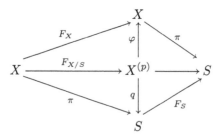

Definition 2.23. We call the morphism of S-schemes $F_{X/S} : X \to X^{(p)}$ as above the *relative Frobenius*.

Lemma 2.24. *The morphisms $F_{X/S}$ and φ are homeomorphisms.*

Proof Let us show that $f := F_{X/S} \circ \varphi$ is equal to $F_{X^{(p)}}$. As $F_{X^{(p)}}$ and $F_X = \varphi \circ F_{X/S}$ are both homeomorphisms, the lemma will be proved. With the notation of the previous commutative diagram, we have $q \circ f = (q \circ F_{X/S}) \circ \varphi = \pi \circ \varphi = F_S \circ q = q \circ F_{X^{(p)}}$ (the last equality comes from Lemma 2.22(a)), and $\varphi \circ f = (\varphi \circ F_{X/S}) \circ \varphi = F_X \circ \varphi = \varphi \circ F_{X^{(p)}}$. Therefore $F_{X^{(p)}} = F_X \times F_S = f$. $\qquad\square$

The construction of $F_{X/S}$ is functorial (Exercise 2.20(b)). If U is an open subset of X, then $U^{(p)}$ is an open subset of $X^{(p)}$ and $F_{U/S} = F_{X/S}|_U$. So most of local properties on $X^{(p)}$ can be reduced to the case when S and X are affine. Let $S = \operatorname{Spec} A$ and $X = \operatorname{Spec} B$. Then $X^{(p)} = \operatorname{Spec}(B \otimes_A A)$, where the second factor is endowed with the structure of an A-algebra via $a \mapsto a^p$, and $F_{X/S}$ corresponds to $B \otimes_A A \to B$, $b \otimes a \mapsto ab^p$. We then see easily the next lemma.

Lemma 2.25. *Let $S = \operatorname{Spec} A$ be an affine scheme over $\mathbb{F}_p$, and let $X = \operatorname{Spec} A[T_1, \ldots, T_n]/I$ be an affine S-scheme of finite type. Then $X^{(p)} = \operatorname{Spec} A[T_1, \ldots, T_n]/I^{(p)}$, where $I^{(p)}$ is the ideal generated by the elements of the form $\sum_{\nu \in \mathbb{N}^n} a_\nu^p T^\nu$, with $\sum_\nu a_\nu T^\nu \in I$. The relative Frobenius $F_{X/S} : X \to X^{(p)}$ is induced by the A-algebra homomorphism $T_i \mapsto T_i^p$.*

Corollary 2.26. *Let X be an algebraic variety over $k = \mathbb{F}_p$. Then the following properties are true.*

(a) *$X^{(p)} = X$, $F_X = F_{X/\operatorname{Spec} k}$.*

(b) *Let $\bar{k}$ be the algebraic closure of k. Let $\overline{F}_X : X(\bar{k}) \to X(\bar{k})$ be the map induced by the composition with F_X. If $X = V(I)$ is a closed subvariety of $\operatorname{Spec} k[T_1, \ldots, T_n]$, then we have the equality $\overline{F}_X(t_1, \ldots, t_n) = (t_1^p, \ldots, t_n^p)$ via the identification of Proposition 2.18(d).*

(c) *The set $X(k)$ can be identified with the fixed points of $\overline{F}_X$.*

Proof As $F_{\operatorname{Spec} k}$ is the identity morphism, we have (a). Property (b) follows from the preceding lemma. (c) is an immediate consequence of (b). $\qquad\square$

Corollary 2.27. *Let X be an integral geometrically reduced algebraic variety of dimension d over a field k of characteristic p. Then $X^{(p)}$ is integral, and $F_{X/\operatorname{Spec} k}$ induces a finite extension of function fields $K(X^{(p)}) \to K(X)$ of degree p^d and an isomorphism from $K(X^{(p)})$ onto the compositum $kK(X)^p$.*

Proof By definition, $X^{(p)} = X \times_{\operatorname{Spec} k} \operatorname{Spec} k'$ where $k \to k' = k$ is defined by $a \mapsto a^p$. So $X^{(p)}$ is reduced and irreducible (Lemma 2.24), hence integral. Moreover, as k'/k is an algebraic extension, we see immediately that $K(X^{(p)}) = K(X) \otimes_k k'$. Its image in $K(X)$ is $kK(X)^p$. The field $K(X)$ is finite of some degree m over $k(T_1, ..., T_d)$. Consider the following commutative diagram

$$
\begin{array}{ccc}
K(X) \otimes_k k' & \longrightarrow & K(X) \\
\uparrow & & \uparrow \\
k(T_1, ..., T_d) \otimes_k k' & \longrightarrow & k(T_1, ..., T_d)
\end{array}
$$

where the horizontal arrows are respectively induced by relative Frobenius on X and $\mathbb{A}_k^d$, and the vertical arrows are field extensions of degree m. The bottom arrow is a field extension of degree $[k(T_1, ..., T_d) : k(T_1^p, ..., T_d^p)] = p^d$. Therefore the top arrow have degree p^d. This achieves the proof.   □

Exercises

2.1. Let Y be a Noetherian scheme. Show that any Y-scheme X of finite type is Noetherian. Moreover, if Y is of finite dimension, then so is X.

2.2. Show that any open immersion into a locally Noetherian scheme is a morphism of finite type.

2.3. An *immersion* of schemes is a morphism which is an open immersion followed by a closed immersion.

  (a) Let $f : X \to Y$ be an immersion. Show that it can be decomposed into a closed immersion followed by an open immersion.

  (b) Show that the converse of (a) is true if f is moreover quasi-compact (e.g., if Y is locally Noetherian). Use the scheme-theoretic closure of f (Exercise 2.3.17).

  (c) Let $f : X \to Y$, $g : Y \to Z$ be two immersions with g quasi-compact. Show that $g \circ f$ is an immersion.

2.4. Let X, Y be schemes over a locally Noetherian scheme S, with Y of finite type over S. Let $x \in X$. Show that for any morphism of S-schemes $f_x :$ $\operatorname{Spec} \mathcal{O}_{X,x} \to Y$, there exist an open subset $U \ni x$ of X and a morphism of S-schemes $f : U \to Y$ such that $f_x = f \circ i_x$, where $i_x : \operatorname{Spec} \mathcal{O}_{X,x} \to U$ is the canonical morphism (in other words, the morphism f_x extends to an open neighborhood of x).

2.5. Let S be a locally Noetherian scheme. Let X, Y be S-schemes of finite type. Let us fix $s \in S$. Let $\varphi : X \times_S \operatorname{Spec} \mathcal{O}_{S,s} \to Y \times_S \operatorname{Spec} \mathcal{O}_{S,s}$ be a morphism of S-schemes. Show that there exist an open set $U \ni s$ and a morphism of S-schemes $f : X \times_S U \to Y \times_S U$ such that φ is obtained from f by the base change $\operatorname{Spec} \mathcal{O}_{S,s} \to U$. If φ is an isomorphism, show that there exists such an f which is moreover an isomorphism.

2.6. Let $f : X \to Y$ be a morphism of integral schemes. Let ξ be the generic point of X. We say that f is *birational* if $f_\xi^\# : K(Y) \to K(X)$ is an isomorphism. Let us suppose that X and Y are of finite type over a scheme S, and that f is a morphism of S-schemes. Show that f is birational if and only if there exist a non-empty open subset U of X and an open subset V of Y such that f induces an isomorphism from U onto V.

2.7. Let k be a field and $\overline{k}$ the algebraic closure of k. Let $\overline{X}, \overline{Y}$ be algebraic varieties over $\overline{k}$, $\overline{f}$ a morphism from $\overline{X}$ to $\overline{Y}$. Show that there exist a finite extension K of k, algebraic varieties X, Y over K, and a morphism $f : X \to Y$, such that $X_{\overline{k}} = \overline{X}$, $Y_{\overline{k}} = \overline{Y}$, and $\overline{f} = f \times_{\overline{k}}$.

2.8. Let $f : X \to Y$ be a morphism of schemes. We say that f has *finite fibers* if $f^{-1}(y)$ is a finite set for every $y \in Y$. We say that f is *quasi-finite* ([41], II.6.2.3 and Err_{III}, 20) if, moreover, $\mathcal{O}_{X_y, x}$ is finite over $k(y)$ for every $x \in X_y$. Show that a morphism of finite type with finite fibers is quasi-finite. Give an example of a morphism with finite fibers which is not quasi-finite.

2.9. Let X, Y be algebraic varieties over a field k, with X geometrically reduced. Let $f, g : X \to Y$ be two morphisms that canonically induce the same map $X(\overline{k}) \to Y(\overline{k})$. We want to show that $f = g$ as morphisms.

    (a) Show that $f(x) = g(x)$ for any closed point $x \in X$, and then show the same equality for any point x.

    (b) Reduce to the case X affine, $Y = \mathbb{A}_k^n$, and k algebraically closed.

    (c) Using Exercise 2.3.7 and Lemma 2.1.18, show that $f = g$.

2.10. Let k be a field and K/k a finite Galois extension with group G. Let X be an algebraic variety over k.

    (a) Let L be an extension of k. Show that G acts transitively on $\text{Spec}(L \otimes_k K)$.

    (b) Let us suppose X is irreducible. Show that G acts transitively on the irreducible components of X_K. Deduce from this that the irreducible components of $X_{\overline{k}}$ have the same dimension.

    (c) Let us suppose X is connected. Show that G acts transitively on the connected components of X_K.

2.11. Let X be an algebraic variety over a field k.

    (a) Show that if X is connected and if $X(k) \neq \emptyset$, then X is geometrically connected.

    (b) Show through an example that (a) is false if we replace 'connected' by 'irreducible'.

2.12. Let X be an irreducible algebraic variety over a field k. We call the irreducible components of $X_{\overline{k}}$ the *geometric irreducible components of X*. Let us assume X is integral. Show that the number of geometric irreducible components of X is equal to $[K(X) \cap k^s : k]$, where k^s is the separable closure of k.

2.13. Let X be an algebraic variety over a field k. Show that X is geometrically reduced (resp. irreducible; connected) if and only if $X_{k'}$ is reduced (resp. irreducible; connected) for every finite extension k' of k.

2.14. Let X be an algebraic variety over a field k. Let us suppose X is geometrically reduced (resp. integral; irreducible; connected). We want to show that for any field extension K/k, X_K is reduced (resp. integral; irreducible; connected).

(a) Show that we may assume that k is algebraically closed and K finitely generated over k.

(b) Let Y be an integral algebraic variety over k such that $K(Y) = K$. Applying Proposition 2.15 to Y, show that X_K is reduced.

(c) Let $P(S)$ be a monic polynomial in $k(T_1, \dots, T_d)[S]$. Let $P = Q_1 Q_2$ be a decomposition of P in $K(T_1, \dots, T_d)[S]$ in monic polynomials. Show that $Q_i \in k(T_1, \dots, T_d)[S]$ (consider the sub-k-algebra A of K generated by the elements of K which appear in the coefficients of the Q_i, and apply a homomorphism $A \to k$ to Q_i).

(d) Let us suppose X is integral. Applying Proposition 2.15 to X, show that $K(X) \otimes_k K$ is integral. Deduce from this that X_K is integral.

(e) Show assertion (d) for irreducible algebraic varieties and for connected algebraic varieties.

2.15. Let X be a geometrically integral algebraic variety over a field k. Show that for any integral algebraic variety Y over k, the fibered product $X \times_{\mathrm{Spec}\,k} Y$ is integral. Show the same assertion after subsequently replacing 'integral' everywhere in hypothesis for X and Y by 'reduced', 'irreducible', and 'connected'.

2.16. Let A be an integral domain with field of fractions K, and let B be a finitely generated A-algebra. We suppose that $B \otimes_A K$ is finite over K.

(a) For any $b \in B$, show that there exists a $g \in A \setminus \{0\}$ such that gb is integral over A.

(b) Show that there exists an $h \in A \setminus \{0\}$ such that B_h is finite over A_h.

2.17. Let Y be an irreducible scheme with generic point ξ, and let $f : X \to Y$ be a dominant morphism of finite type (Exercise 2.4.11).

(a) Let x be a closed point of X_ξ. Let Z be the reduced subscheme $\overline{\{x\}}$ of X. Show that there exists a dense open subset V of Y such that $f^{-1}(V) \cap Z \to V$ is surjective (use Exercises 2.16 and 2.1.8).

(b) Deduce from this that $f(X)$ contains a dense open subset of Y.

(c) Show that $f(X)$ is a *constructible subset* of Y, that is $f(X)$ is a finite disjoint union of sets Z_i, with each Z_i the intersection of an open subset and a closed subset (see [43], Exercise II.3.18).

2.18. Give an analogous version of Corollary 2.26(b) for projective varieties.

2.19. Show that assertions (b) and (c) of Corollary 2.26 remain true after replacing k by the field of $q = p^f$ elements and F_X by F_X^f.

2.20. Let S be a scheme over $\mathbb{F}_p$. Let X be an S-scheme.

(a) Show that the relative Frobenius $F_{X/S} : X \to X^{(p)}$ commutes with base change $T \to S$ (i.e., $F_{X/S} \times \mathrm{Id}_T = F_{X_T/T}$).

(b) Let $f : X \to Y$ be a morphism of S-schemes. Show that f canonically induces a morphism of S-schemes $f^{(p)} : X^{(p)} \to Y^{(p)}$ such that $f^{(p)} \circ F_{X/S} = F_{Y/S} \circ f$. Note that in the special case when $Y = S$, $f^{(p)}$ is just the structural morphism of $X^{(p)}$. Let us suppose that f satisfies a property $(\mathcal{P})$ which is stable by base change (e.g., closed immersion). Show that $f^{(p)}$ also satisfies $(\mathcal{P})$.

(c) Let us suppose that S is the spectrum of a field. Show that the morphism $\varphi : X^{(p)} \to X$ induces an injective homomorphism $\mathcal{O}_X \to \pi_* \mathcal{O}_{X^{(p)}}$. Deduce from this that if $X^{(p)}$ is integral, then so is X. Show that the converse is false in general.

2.21. Let B be a finitely generated algebra over a ring A. Show that there exist a subring A_0 of A that is finitely generated over $\mathbb{Z}$ and a finitely generated A_0-algebra B_0 such that $B = B_0 \otimes_{A_0} A$. If B is homogeneous, we can take B_0 as homogeneous.

Remark. We can show, more generally, that if X is a scheme of finite type over A with A Noetherian, then there exist A_0 as above and an A_0-scheme of finite type X_0 such that $X = X_0 \times_{\mathrm{Spec}\, A_0} \mathrm{Spec}\, A$. See [41], IV.8.8.2 (ii).

3.3 Some global properties of morphisms

Up to now, we have essentially been interested in local properties of schemes. In this section, we introduce some global notions of morphisms. Roughly speaking, the local study of schemes essentially depends on commutative algebra, while that of the global properties depends rather on purely geometric tools. The two aspects are of course inseparable. The core of this section is made up of the proper morphisms. This class of morphisms plays, in a way, the role of families of compact varieties.

3.3.1 Separated morphisms

Let X be a topological space. Let $\Delta : X \to X \times X$ denote the diagonal map: $x \mapsto (x, x)$. We endow $X \times X$ with the product topology.

Proposition 3.1. *Let X be a topological space. Then X is separated if and only if $\Delta(X)$ is closed.*

Proof Let us first suppose X is separated. If $(a, b) \in X \times X \setminus \Delta(X)$, then $a \neq b$. Let U, V be open subsets of X such that $a \in U, b \in V$, and $U \cap V = \emptyset$; then $U \times V$ is an open neighborhood of (a, b) which does not meet $\Delta(X)$, and hence $\Delta(X)$ is closed.

Conversely, let us suppose $\Delta(X)$ is closed. For any pair of points $a \neq b$, we have $(a, b) \in X \times X \setminus \Delta(X)$. Therefore there exists an open subset of the form

$U \times V$ with U, V open in X such that $(a, b) \in U \times V \subset X \times X \setminus \Delta(X)$. This implies that $a \in U, b \in V$, and that $U \cap V = \emptyset$. Hence X is separated.   □

As we have already seen, the underlying topological space of a scheme is almost never separated. Nevertheless, we will define the separatedness of schemes by drawing inspiration from the proposition above.

Definition 3.2. Let $f : X \rightarrow Y$ be a morphism of schemes. The morphism $(\mathrm{Id}_X, \mathrm{Id}_X) : X \rightarrow X \times_Y X$ is called the *diagonal morphism of f*; we denote it by $\Delta_{X/Y}$ or simply Δ if there is no confusion possible. We say that X is *separated over Y* if Δ is a closed immersion (of schemes). This is a local property on Y. We say that a scheme X is *separated* if it is separated over $\mathbb{Z}$.

Remark 3.3. Let p, q be the projections from $X \times_Y X$ onto X. If $s \in \Delta_{X/Y}(X)$, then $p(s) = q(s)$. But the converse is false. For example, let $Y = \mathrm{Spec}\,\mathbb{R}$, $X = \mathrm{Spec}\,\mathbb{C}$. Then $\Delta_{X/Y}(X)$ is reduced to a point, while for every $s \in X \times_Y X$ (which has two elements), we have $p(s) = q(s)$ since X is reduced to a point.

Proposition 3.4. *Any morphism of affine schemes $X \rightarrow Y$ is separated. In particular, any affine scheme is separated.*

Proof Let $X = \mathrm{Spec}\,B$ and $Y = \mathrm{Spec}\,A$. By construction of $X \times_Y X$, Δ is induced by the homomorphism $\rho : B \otimes_A B \rightarrow B$ defined by $\rho(b_1 \otimes b_2) = b_1 b_2$. It is clear that ρ is surjective, and therefore Δ is a closed immersion.   □

Corollary 3.5. *Let $f : X \rightarrow Y$ be a morphism of schemes such that $\Delta(X)$ is a closed subset of $X \times_Y X$. Then f is separated.*

Proof This is a consequence of the proposition above and Exercise 3.1.   □

Let X be a scheme. If U and V are open subsets of X, we have a canonical homomorphism $\mathcal{O}_X(U) \otimes_{\mathbb{Z}} \mathcal{O}_X(V) \rightarrow \mathcal{O}_X(U \cap V)$ defined by

$$f \otimes g \mapsto f|_{U \cap V} \cdot g|_{U \cap V}.$$

The following proposition gives a simple separatedness criterion using this type of homomorphism.

Proposition 3.6. *Let X be a scheme. Then the following properties are equivalent:*

(i) *X is separated;*

(ii) *for every pair of affine open subsets U, V of X, $U \cap V$ is affine and the canonical homomorphism $\mathcal{O}_X(U) \otimes_{\mathbb{Z}} \mathcal{O}_X(V) \rightarrow \mathcal{O}_X(U \cap V)$ is surjective;*

(iii) *there exists a covering of X by affine open subsets U_i such that for all i, j, $U_i \cap U_j$ is affine and $\mathcal{O}_X(U_i) \otimes_{\mathbb{Z}} \mathcal{O}_X(U_j) \rightarrow \mathcal{O}_X(U_i \cap U_j)$ is surjective.*

Proof Let Δ be the diagonal morphism $\Delta_{X/\mathbb{Z}}$. We have $\Delta^{-1}(U \times_{\mathbb{Z}} V) = U \cap V$. The homomorphism $\mathcal{O}_X(U) \otimes_{\mathbb{Z}} \mathcal{O}_X(V) \rightarrow \mathcal{O}_X(U \cap V)$ corresponds to the restriction $\Delta : U \cap V = \Delta^{-1}(U \times_{\mathbb{Z}} V) \rightarrow U \times_{\mathbb{Z}} V$.

If X is separated, Δ is a closed immersion. As $U \times_Z V$ is affine, this implies (ii). As (ii) implies (iii) is trivial, it remains to show that (iii) implies (i). Now (iii) implies that $\Delta : \Delta^{-1}(U_i \times_Z U_j) \to U_i \times_Z U_j$ is a closed immersion. As the open subsets $U_i \times_Z U_j$ cover $X \times_Z X$, Δ is a closed immersion. □

Example 3.7. The projective space $\mathbb{P}_{\mathbb{Z}}^n$ is separated. Indeed, we can cover $\mathbb{P}_{\mathbb{Z}}^n$ with the affine open subsets $D_+(T_i)$, $0 \le i \le n$ (Example 2.3.34). It is easy to verify that condition (iii) of Proposition 3.6 is satisfied. Therefore $\mathbb{P}_{\mathbb{Z}}^n$ is separated.

Example 3.8. Let p be a prime number. Let us set $X_1 = X_2 = \operatorname{Spec} \mathbb{Z}$, $X_{12} = D(p) \subset X_1$, and $X_{21} = D(p) \subset X_2$. Then the identity on $X_{12} = X_{21}$ makes it possible to glue the two schemes X_1 and X_2 (Lemma 2.3.33) to a scheme X. This scheme is not separated. Indeed, condition (ii) of Proposition 3.6 is not verified because the image of $\mathcal{O}_X(X_1) \otimes_{\mathbb{Z}} \mathcal{O}_X(X_2)$ in $\mathcal{O}_X(X_1 \cap X_2) = \mathcal{O}_X(X_{12}) = \mathbb{Z}[1/p]$ is $\mathbb{Z}$ (see also Exercise 3.18). In a way, we have enlarged X_{12} by adding the same point twice.

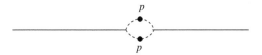

Figure 6. A line with a double point.

The following proposition gathers together some general properties of separated morphisms.

Proposition 3.9.

(a) *Open and closed immersions are separated morphisms.*

(b) *The composition of two separated morphisms is a separated morphism.*

(c) *Separated morphisms are stable under base change.*

(d) *If $X \to Z$ and $Y \to Z$ are separated, then so is $X \times_Z Y \to Z$.*

(e) *Let $f : X \to Y$, $g : Y \to Z$ be two morphisms such that $g \circ f$ is separated; then f is separated.*

(f) *Let Y be a separated Z-scheme. Then for any Y-schemes X_1, X_2, the canonical morphism $X_1 \times_Y X_2 \to X_1 \times_Z X_2$ is a closed immersion.*

Proof (a) Indeed, Δ is an isomorphism if $X \to Y$ is an open or closed immersion (Exercise 1.6(b)).

(b) The diagonal morphism $\Delta_{X/Z} : X \to X \times_Z X$ decomposes into $\Delta_{X/Y} : X \to X \times_Y X = X \times_Y Y \times_Y X$ and

$$\operatorname{Id}_X \times \Delta_{Y/Z} \times \operatorname{Id}_X : \ X \times_Y Y \times_Y X \to X \times_Y (Y \times_Z Y) \times_Y X = X \times_Z X.$$

Actually, the compositions of $(\operatorname{Id}_X \times \Delta_{Y/Z} \times \operatorname{Id}_X) \circ \Delta_{X/Y}$ with the projections $X \times_Z X \to X$ coincide with identity. Then (b) results from the fact that closed immersions are stable under base change (Proposition 1.23 and Remark 1.20) and by composition.

(c) Let $Y' \to Y$ be a morphism and $X' = X \times_Y Y'$. Then $\Delta_{X'/Y'}$ can be identified with a morphism $\Delta_{X/Y} \times \mathrm{Id}_{Y'} : X \times_Y Y' \to (X \times_Y X) \times_Y Y' = X' \times_{Y'} X'$. Hence $X' \to Y'$ is separated by the same reasoning as above.

(d) The morphism $X \times_Z Y \to Z$ decomposes into $X \times_Z Y \to Y$ and $Y \to Z$; it is therefore a separated morphism by (b) and (c).

(e) Let $h : X \times_Y X \to X \times_Z X$ be the canonical morphism. Then $\Delta_{X/Y}(X) \subseteq h^{-1}(\Delta_{X/Z}(X))$. Let us show the inclusion in the other direction (which will imply that $\Delta_{X/Y}(X)$ is closed, and therefore that f is separated). Let $s \in h^{-1}(\Delta_{X/Z}(X))$. Let $x \in X$ be such that $h(s) = \Delta_{X/Z}(x)$, and let $t = \Delta_{X/Y}(x)$. Let us take an affine open subset U (resp. V; resp. W) containing x (resp. $f(x)$; resp. $g(f(x))$), with $U \subseteq f^{-1}(V)$ and $V \subseteq g^{-1}(W)$. Then $s,t \in U \times_V U$ and $h|_{U \times_V U} : U \times_V U \to U \times_W U$ is a closed immersion because U, V, W are affine (using Proposition 1.2). Hence $s = t \in \Delta_{X/Y}(X)$ since $h(s) = h(t)$.

(f) We canonically have $X_1 \times_Y X_2 = X_1 \times_Y Y \times_Y X_2$ and $X_1 \times_Z X_2 = X_1 \times_Y (Y \times_Z Y) \times_Y X_2$. We then verify that the morphism $X_1 \times_Y X_2 \to X_1 \times_Z X_2$ is none other than

$$\mathrm{Id}_{X_1} \times \Delta_{Y/Z} \times \mathrm{Id}_{X_2} : X_1 \times_Y Y \times_Y X_2 \to X_1 \times_Y (Y \times_Z Y) \times_Y X_2.$$

It is therefore a closed immersion by Proposition 1.23 and Remark 1.20.   □

Corollary 3.10. *Any projective morphism is of finite type and separated.*

Proof Indeed, by construction, $\mathbb{P}^n_{\mathbb{Z}}$ is of finite type over $\mathbb{Z}$. The rest follows from Example 3.7 and Propositions 2.4 and 3.9.   □

We will conclude this subsection with a statement on the uniqueness of extensions of morphisms to separated schemes.

Proposition 3.11. *Let X be a reduced S-scheme, Y a separated S-scheme. Let us consider two morphisms of S-schemes f, g from X to Y. Let us suppose that $f|_U = g|_U$ for some everywhere dense open subset U of X; then $f = g$.*

Proof Let us write $\Delta = \Delta_{Y/S}$, $h = (f,g) : X \to Y \times_S Y$. We have $\Delta \circ f = (f,f)$ because the left-hand side verifies the universal property of the morphism (f,f) (Definition 1.1). It follows that on U, $\Delta \circ f$ coincides with h. Consequently, $U \subseteq h^{-1}(\Delta(Y))$.[1] By hypothesis, $\Delta(Y)$ is closed, and hence $X = h^{-1}(\Delta(Y))$. We therefore have $f(x) = g(x)$ for every $x \in X$.

It remains to show that $f = g$ as morphisms. We can assume that $X = \mathrm{Spec}\, A$ and $Y = \mathrm{Spec}\, B$. Let φ, ψ be the ring homomorphisms corresponding respectively to f and g (Lemma 2.3.23). Let $b \in B$. Let us write $a = \varphi(b) - \psi(b)$. Then $a|_U = 0$. It follows that $U \subseteq V(a)$, and hence $V(a) = \mathrm{Spec}\, A$ since U is dense. This implies that a is nilpotent, and as A is reduced (Proposition 4.2(a)), we have $a = 0$. Consequently, $\varphi = \psi$, and hence $f = g$.   □

---

[1]It is tempting here to reason in a set-theoretical manner. Let p and q be the projections from $Y \times_S Y$ onto its two factors. For any $x \in U$, we have $p(h(x)) = q(h(x))$. But we must be careful here because in general, this does not imply that $h(x) \in \Delta(Y)$. See Remark 3.3.

Remark 3.12. Proposition 3.11 is false if Y is not separated over S (or if X is not reduced). We can show a converse in the following form:

Let $Y \to S$ be a morphism of finite type to a Noetherian scheme S. Let us suppose that for any S-scheme $X = \operatorname{Spec} \mathcal{O}_K$, where $\mathcal{O}_K$ is a discrete valuation ring, two morphisms of S-schemes from X to Y which coincide on the open subset $\{\xi\}$ of X, where ξ is the generic point of X, are identical. Then $Y \to S$ is separated.

This is the valuative criterion of separatedness. See [41], II.7.2.3, or [43], Exercise II.4.11(b). This criterion will not be used in the rest of the book.

3.3.2 Proper morphisms

As for separated morphisms, the properness of a morphism is essentially a topological property.

Definition 3.13. We say that a morphism $f : X \to Y$ is *closed* if f maps any closed subset of X onto a closed subset of Y. We say that f is *universally closed* if for any base change $Y' \to Y$, $X \times_Y Y' \to Y'$ stays a closed morphism.

Let $f : X \to Y$ be a continuous map of topological spaces. We say that f is proper if the inverse image of a compact subset is compact. When X is separated and Y locally compact, we know that $f : X \to Y$ is proper if and only if for any topological space Z, the map $X \times Z \to Y \times Z$ defined by $(x, z) \mapsto (f(x), z)$ is closed (see [18], I, §10, n° 3, Proposition 7).

Definition 3.14. We say that a morphism of schemes $f : X \to Y$ is *proper* if it is of finite type, separated, and universally closed. We say that a Y-scheme is proper if the structural morphism is proper. This is clearly a local property on Y.

Closed immersions are proper morphisms (Proposition 3.16). But besides this case, there is no obvious example. We will see in the next subsection that every projective morphism is proper.

Lemma 3.15. *Let $\mathcal{P}$ be a property of morphisms of schemes. Let us suppose that:*

(1) *closed immersions verify $\mathcal{P}$;*
(2) *$\mathcal{P}$ is stable under composition;*
(3) *$\mathcal{P}$ is stable under base change.*

Then the following two properties are true.

(a) *$\mathcal{P}$ is stable under the fibered product in the sense of Proposition 2.4(d).*
(b) *Let $f : X \to Y$, $g : Y \to Z$ be morphisms such that g is separated and that $f \circ g$ verifies $\mathcal{P}$; then f verifies $\mathcal{P}$.*

Proof (a) Let $X \to Z$, $Y \to Z$ be two morphisms; then $X \times_Z Y \to Z$ can be decomposed into $X \times_Z Y \to Y$ (obtained from $X \to Z$ by the base change

$Y \to Z$), followed by $Y \to Z$. Therefore $X \times_Z Y \to Z$ verifies $\mathcal{P}$ by hypotheses (2) and (3).

(b) Let $q : X \times_Z Y \to Y$ be the second projection. Then q verifies $\mathcal{P}$ by hypothesis (3). On the other hand, $(\mathrm{Id}_X, f) : X \to X \times_Z Y$ is a closed immersion since $Y \to Z$ is separated (applying Proposition 3.9(f) with $X_1 = X$ and $X_2 = Y$). Therefore $f = q \circ (\mathrm{Id}_X, f)$ verifies $\mathcal{P}$ by hypotheses (1) and (2).   □

Proposition 3.16. *We have the following properties:*

(a) *Closed immersions are proper.*

(b) *The composition of two proper morphisms is a proper morphism.*

(c) *Proper morphisms are stable under base change.*

(d) *If $X \to Z$ and $Y \to Z$ are proper, then so is $X \times_Z Y \to Z$.*

(e) *If the composition of $X \to Y$ and $Y \to Z$ is proper and if $Y \to Z$ is separated, then $X \to Y$ is proper.*

(f) *Let $f : X \to Y$ be a surjective morphism of S-schemes. Let us suppose that Y is separated of finite type over S and that X is proper over S. Then Y is proper over S.*

Proof   (a), (b), and (c) immediately result from Propositions 1.23, 2.4, and 3.9. This implies (d) and (e) by virtue of Lemma 3.15. It remains to prove (f). Let $T \to S$ be an arbitrary morphism. Then $f_T : X_T \to Y_T$ is surjective (Exercise 1.8). We easily deduce from this that $Y_T \to T$ is closed. It follows that $Y \to S$ is proper.   □

One of the important consequences of properness is a finiteness property. We will give a weak version of it here (Proposition 3.18).

Lemma 3.17. *Let A be a ring. Let $Y = \mathrm{Spec}\, B$ be a proper affine scheme over A. Then B is finite over A.*

Proof   By hypothesis B is of finite type over A. It is enough to show that B is integral over A. Let us first suppose that B is generated by a single element over A, that is to say that we have a surjective homomorphism $\varphi : A[T] \to B$.

We consider $\mathrm{Spec}\, A[T]$ as an open subscheme of $\mathrm{Proj}\, A[T_1, T_2]$ by identifying it with the open subset $D_+(T_2)$ of $\mathrm{Proj}\, A[T_1, T_2]$ (we set $T = T_1/T_2$). Let $f : Y \to \mathrm{Proj}\, A[T_1, T_2]$ be the composition of the immersions $i : Y \to \mathrm{Spec}\, A[T]$ and $\mathrm{Spec}\, A[T] \to \mathrm{Proj}\, A[T_1, T_2]$. Then f is proper (Proposition 3.16(e) and Corollary 3.10); there therefore exists a homogeneous ideal $J \subset A[T_1, T_2]$ such that $f(Y) = V_+(J)$. It follows that $V_+(T_2) \cap V_+(J) = \emptyset$, which gives an inclusion of ideals $(T_1, T_2) \subseteq \sqrt{J + (T_2)}$. This implies the existence of a homogeneous polynomial $P(T_1, T_2) \in J$ of the form $T_1^n + T_2 Q(T_1, T_2)$ with $n \geq 1$. We have $i(Y) = V_+(J) \cap D_+(T_2) = V(J_{(T_2)})$, where $J_{(T_2)} = \{ H/R \in A[T_1, T_2]_{(T_2)} \mid H \in J \}$, and hence $\varphi(T_2^{-n} P)$ is nilpotent (Exercise 2.4.10). Now $\varphi(T_2^{-n} P) = \varphi(T)^n + Q(\varphi(T), 1)$ with $\deg Q(T, 1) \leq n - 1$; we therefore have $\varphi(T)$ integral over A. This shows that B is integral over A.

In the general case, we have $B = A[b_1, \ldots, b_m]$. Let $B_1 = A[b_1, \ldots, b_{m-1}] \subseteq B$. Then $\operatorname{Spec} B \to \operatorname{Spec} B_1$ is proper (Proposition 3.16(e)). It follows from the preceding case that B is integral over B_1. This implies, in particular, that $\operatorname{Spec} B \to \operatorname{Spec} B_1$ is surjective (Exercise 2.1.8). Hence $\operatorname{Spec} B_1 \to \operatorname{Spec} A$ is proper (Proposition 3.16(f)). Induction on m then completes the proof.   □

Proposition 3.18. *Let X be a proper scheme over a ring A. Then $\mathcal{O}_X(X)$ is integral over A.*

Proof We may assume X is reduced (Proposition 2.4.2(c)). Let $h \in \mathcal{O}_X(X)$. The homomorphism $\varphi : A[T] \to \mathcal{O}_X(X)$ defined by $\varphi(T) = h$ induces a morphism of A-schemes $f : X \to \operatorname{Spec} A[T]$ (Proposition 2.3.25). It follows that f is proper (Propositions 3.4 and 3.16(e)). Therefore there exists an ideal I of $A[T]$ such that $f(X) = V(I)$. By replacing I by its radical, the morphism f factors through a surjective morphism $X \to \operatorname{Spec} A[T]/I$. It follows from Proposition 3.16(f) that $\operatorname{Spec} A[T]/I \to \operatorname{Spec} A$ is proper. Therefore $A[T]/I$ is integral over A by virtue of the preceding lemma. Consequently, $h = \varphi(T)$ is integral over A because φ factors through $A[T]/I \to \mathcal{O}_X(X)$.   □

The following corollary in some sense illustrates the analogy between compact analytic varieties and proper schemes over a field.

Corollary 3.19. *Let X be a reduced algebraic variety, proper over a field k. Then $\mathcal{O}_X(X)$ is a k-vector space of finite dimension.*

Proof Let $X_1, \ldots, X_m$ be the irreducible components of X endowed with the reduced closed subscheme structure. Then the canonical homomorphism $\mathcal{O}_X(X) \to \oplus_i \mathcal{O}_{X_i}(X_i)$ is injective. This can be seen by restricting to an affine open covering of X. It then suffices to show that every $\mathcal{O}_{X_i}(X_i)$ is of finite dimension over k. We may therefore suppose X is integral. Now by Proposition 3.18, $\mathcal{O}_X(X)$ is an integral domain that is integral over k. It is therefore a field. Let $x \in X$ be an arbitrary closed point of X; then $\mathcal{O}_X(X) \to k(x)$ is injective since it is a field homomorphism. As $k(x)$ is finite over k, so is $\mathcal{O}_X(X)$.   □

Remark 3.20. More generally, if we replace k by an arbitrary Noetherian ring A, we can show that $\mathcal{O}_X(X)$ is finite over A, even without supposing X is reduced. This is a particular case of a finiteness theorem concerning coherent sheaves (Theorem 5.3.2 and Remark 5.3.3).

Corollary 3.21. *Let X be a reduced connected algebraic variety, proper over a field k. Then $\mathcal{O}_X(X)$ is a field, which is a finite extension of k. If X is geometrically connected (resp. geometrically connected and geometrically reduced), then $\mathcal{O}_X(X)$ is purely inseparable over k (resp. $\mathcal{O}_X(X) = k$).*

Proof Let us write $K = \mathcal{O}_X(X)$. By Corollary 3.19, K is a finite reduced k-algebra. It is therefore the direct sum (as k-algebra) of a finite number of fields. Now $\operatorname{Spec} K$ is connected (Exercise 2.4.6), it follows that K is a field which is finite over k.

Let $\bar{k}$ be the algebraic closure of k. Then $K \otimes_k \bar{k} = \mathcal{O}(X_{\bar{k}})$ by virtue of Proposition 2.24. Let us suppose X is geometrically connected. Let k^s be the separable closure of k. Then X_{k^s} is connected and reduced. Therefore $K \otimes_k k^s$ is a field and it contains $(K \cap k^s) \otimes_k (K \cap k^s)$. It follows that $K \cap k^s = k$. This completes the proof since K is purely inseparable over $K \cap k^s$. If, moreover, X is geometrically reduced, then $K \otimes_k \overline{k}$ is reduced, local (because $X_{\overline{k}}$ is connected), and finite over $\overline{k}$ and therefore equal to $\overline{k}$. Consequently, $K = k$.   □

In what follows, we will show another important property of proper schemes concerning the possibility of extending certain types of morphisms.

Definition 3.22. Let K be a field. We define a *valuation of K* to be a map ν from K^* to a totally ordered Abelian group Γ, verifying the following properties:

(a) $\nu(\alpha\beta) = \nu(\alpha) + \nu(\beta)$, (i.e., ν is a group homomorphism);

(b) $\nu(\alpha + \beta) \geq \min\{\nu(\alpha), \nu(\beta)\}$.

We will set, by convention, $\nu(0) = +\infty$. The set $\mathcal{O}_\nu = \{\alpha \in K \mid \nu(\alpha) \geq 0\}$ is a subring of K called the *valuation ring of ν* (or *of K* if there is no ambiguity). If there is no ambiguity over the valuation ν, we also denote the valuation ring by $\mathcal{O}_K$. A ring which is the valuation ring of a field is called a *valuation ring*. The set $\mathfrak{m}_\nu = \{\alpha \in K \mid \nu(\alpha) > 0\}$ is an ideal of $\mathcal{O}_\nu$, and it is the unique maximal ideal of $\mathcal{O}_\nu$. We call $\mathfrak{m}_\nu$ the *valuation ideal of ν*.

Example 3.23. Let $\mathcal{O}_K$ be a discrete valuation ring (Example 2.4.13), with field of fractions K and with maximal ideal $t\mathcal{O}_K$. We can define a valuation $\nu : K^* \to \mathbb{Z}$ in the following manner. For any non-zero $a \in \mathcal{O}_K$, we can write a in a unique way as $a = t^n u$, $u \in \mathcal{O}_K^*$. We then set $\nu(a) = n$. For any $\lambda \in K^*$, we write $\lambda = a/b$ with $a, b \in \mathcal{O}_K$. We then set $\nu(\lambda) = \nu(a) - \nu(b)$. We verify without difficulty that this integer is independent of the choice of a, b, and that ν is indeed a valuation, called the *normalized valuation of K*. The valuation ring of ν is $\mathcal{O}_K$, the valuation ideal is $t\mathcal{O}_K$. Such a valuation is called a *discrete valuation*. In this work, we will essentially use this type of valuation. Conversely, if a field K is endowed with a non-trivial valuation $K^* \to \mathbb{Z}$, it can immediately be seen that its valuation ring is a local principal ideal domain (hence a discrete valuation ring).

Let $A \subseteq B$ be local rings. We say that B *dominates* A if the inclusion $A \to B$ is a local ring homomorphism.

Lemma 3.24. *Let $\mathcal{O}_K$ be a valuation ring, $K = \mathrm{Frac}(\mathcal{O}_K)$, and A a local subring of K which dominates $\mathcal{O}_K$. Then $A = \mathcal{O}_K$.*

Proof Let us denote the valuation of K by ν and the maximal ideal of A by $\mathfrak{m}_A$. If there exists an $a \in A \setminus \mathcal{O}_K$, then $\nu(1/a) > 0$. Hence $1/a$ is contained in the maximal ideal of $\mathcal{O}_K$, and consequently $1/a \in \mathfrak{m}_A$. This implies that $1 = a \cdot (1/a) \in \mathfrak{m}_A$, which is impossible.   □

Theorem 3.25. *Let X be a proper scheme over a valuation ring $\mathcal{O}_K$, and let $K = \mathrm{Frac}(\mathcal{O}_K)$. Then the canonical map $X(\mathcal{O}_K) \to X_K(K)$ is bijective.*

Proof The injectivity of $X(\mathcal{O}_K) \to X_K(K)$ comes from the separatedness of $X \to \operatorname{Spec}\mathcal{O}_K$ because $\operatorname{Spec} K$ is dense in $\operatorname{Spec}\mathcal{O}_K$ (Proposition 3.11). Let us show the surjectivity.

Let $\pi : \operatorname{Spec} K \to X_K$ be a section of X_K. Let us denote the generic (resp. closed) point of $\operatorname{Spec}\mathcal{O}_K$ by η (resp. s), and let $x = \pi(\eta)$. Let $Z = \overline{\{x\}} \subseteq X$ be the closed subset endowed with the structure of a reduced (hence integral) subscheme. Then Z is proper over $\mathcal{O}_K$ (Proposition 3.16(a)–(b)). The point x is closed in X_K (Exercise 2.5.9), and dense in Z_K. Therefore $Z_K = \{x\}$. As the image of $Z \to \operatorname{Spec}\mathcal{O}_K$ is closed and contains η, it is equal to $\operatorname{Spec}\mathcal{O}_K$. Let $t \in Z_s$. Then $\mathcal{O}_{Z,t}$ is a local ring dominating $\mathcal{O}_K$, with field of fractions $\mathcal{O}_{Z,x} = K$. It follows from the preceding lemma that $\mathcal{O}_{Z,t} = \mathcal{O}_K$. We therefore have an extension of $\operatorname{Spec} K \to X$ to $\operatorname{Spec}\mathcal{O}_K = \operatorname{Spec}\mathcal{O}_{Z,t} \to Z \to X$. $\qquad\square$

Corollary 3.26. *Let $f : X \to Y$ be a proper morphism. Then for any Y-scheme $\operatorname{Spec}\mathcal{O}_K$, where $\mathcal{O}_K$ is a valuation ring with field of fractions K, the canonical map $X(\mathcal{O}_K) \to X(K)$ is bijective.*

Proof Let us set $Z = X \times_Y \operatorname{Spec}\mathcal{O}_K$. Then $X(\mathcal{O}_K) = Z(\mathcal{O}_K)$ and $X(K) = Z_K(K)$ (Remark 1.6, identity (1.2)). It now suffices to apply the theorem to $Z \to \operatorname{Spec}\mathcal{O}_K$. $\qquad\square$

Remark 3.27. The corollary above characterizes the proper morphisms among the morphisms of finite type (see [43], II.4.7). It is the *valuative criterion of properness*.

Remark 3.28. Let $f : X \to Y$ be a separated morphism of finite type. If f is proper, then for any $y \in Y$, the fiber $X_y \to \operatorname{Spec} k(y)$ is proper (Proposition 3.16(c)). We can ask ourselves whether the converse is true. It is trivially false if we do not take a minimum of precautions (we easily obtain a counter-example by taking Y Noetherian and $f : X \to Y$ an open immersion). But it is true under certain hypotheses. See [41], IV.15.7.10. Let us just state a particular case which we can easily draw from this reference:

Let $\mathcal{O}_K$ be a discrete valuation ring, X an irreducible scheme, $X \to \operatorname{Spec}\mathcal{O}_K$ a surjective morphism, separated and of finite type. If the fibers of $X \to \operatorname{Spec}\mathcal{O}_K$ are geometrically connected, and if the special fiber $X_s \to \operatorname{Spec} k(s)$ is proper, then $X \to \operatorname{Spec}\mathcal{O}_K$ is proper.

Remark 3.29. Let X be a separated scheme of finite type over a Noetherian scheme Y. A theorem of Nagata [73] says that there exists a proper scheme $\hat{X}$ over Y such that X is isomorphic to a dense open subset of $\hat{X}$. See [62] for a fairly recent proof. A proper algebraic variety over a field is also called a *complete variety*.

3.3.3 Projective morphisms

A fundamental example of proper morphisms is that of projective morphisms (Definition 1.12).

Theorem 3.30. *Let S be a scheme. Then any projective morphism to S is proper.*

Proof It suffices to see that for any n, $\mathbb{P}^n_{\mathbb{Z}}$ is proper over $\mathbb{Z}$ (Proposition 3.16(a)–(c)). We already know that it is separated and of finite type (Corollary 3.10). It remains to show that it is universally closed. Let Y be a scheme, and let $\pi : \mathbb{P}^n_Y \to Y$ be the canonical morphism. We must show that π is closed. As this property is local on Y, we can take $Y = \operatorname{Spec} A$ affine.

Let us write $B = A[T_0, \ldots, T_n]$. Let $V_+(I)$ be a closed subset of $\mathbb{P}^n_Y$. We want to show that $Y \setminus \pi(V_+(I))$ is open. Let $y \in Y$; we have $V_+(I) \cap \pi^{-1}(y) = V_+(I \otimes_A k(y))$ (Proposition 1.9). It follows that $y \in Y \setminus \pi(V_+(I))$ if and only if $B_+ \otimes_A k(y) \subseteq \sqrt{I \otimes_A k(y)}$ (Lemma 2.3.35). This inclusion is equivalent to $B_m \otimes_A k(y) \subseteq I \otimes_A k(y)$ for some m. It is also equivalent to $(B/I)_m \otimes_A k(y) = 0$.

Let us take $y \in Y \setminus \pi(V_+(I))$. Let $m \geq 1$ be such that $(B/I)_m \otimes_A k(y) = 0$. As $(B/I)_m$ is a finitely generated A-module, it follows from Nakayama's lemma that $(B/I)_m \otimes_A \mathcal{O}_{Y,y} = 0$. This implies that there exists an $f \in A$ such that $y \in D(f)$ and $f \cdot (B/I)_m = 0$, and hence $(B/I)_m \otimes_A A_f = 0$. Consequently, $y \in D(f) \subseteq Y \setminus \pi(V_+(I))$. This completes the proof. $\square$

Lemma 3.31. *Let S be a scheme. Then there exists a closed immersion*

$$\mathbb{P}^n_S \times_S \mathbb{P}^m_S \to \mathbb{P}^{nm+n+m}_S$$

(called the Segre embedding*).*

Proof It suffices to show the lemma for $S = \operatorname{Spec} A$, where $A = \mathbb{Z}$. Let us write $\mathbb{P}^n_S = \operatorname{Proj} A[T_i]_{0 \leq i \leq n}$, $\mathbb{P}^m_S = \operatorname{Proj} A[S_j]_{0 \leq j \leq m}$, and

$$\mathbb{P}^{nm+n+m}_S = \operatorname{Proj} A[U_{ij}]_{0 \leq i \leq n, 0 \leq j \leq m}.$$

For any pair (i, j), let

$$f_{ij} : D_+(T_i) \times_S D_+(S_j) \to D_+(U_{ij})$$

be the morphism induced by the homomorphism of A-algebras

$$U_{i'j'} U_{ij}^{-1} \mapsto (T_{i'} T_i^{-1}) \otimes (S_{j'} S_j^{-1}), \quad 0 \leq i' \leq n, 0 \leq j' \leq m.$$

Then it is clear that f_{ij} is a closed immersion. On the other hand, it is easy to verify that the morphisms f_{ij} glue to a morphism $f : \mathbb{P}^n_S \times_S \mathbb{P}^m_S \to \mathbb{P}^{nm+n+m}_S$, which is therefore a closed immersion. $\square$

Corollary 3.32. *The following properties are true.*

(a) *Closed immersions are projective morphisms.*

(b) *The composition of two projective morphisms is a projective morphism.*

(c) *Projective morphisms are stable under base change.*

(d) *Let $X \to S$, $Y \to S$ be projective morphisms; then $X \times_S Y \to S$ is projective.*

(e) *If the composition of $X \to Y$, $Y \to Z$ is projective, and if $Y \to Z$ is separated, then $X \to Y$ is projective.*

Proof (a) and (c) follow immediately; (b) and (d) result from the preceding lemma. Finally, (e) is a consequence of (a)–(c) and Lemma 3.15. $\square$

Remark 3.33. Let k be a field. Let X be a proper algebraic variety over k. We can show that: (1) if $\dim X \leq 1$, then X is projective over k (see Exercises 4.1.16 and 7.5.4); (2) if $\dim X = 2$ and X is smooth (Definition 4.3.28), then X is projective over k (Remark 9.3.5); (3) there exist normal proper varieties of dimension 2 and smooth proper algebraic varieties of dimension ≥ 3 which are not projective. See the bibliography in [43], II.4.10.2.

Remark 3.34. Proper morphisms are connected to projective morphisms by *Chow's lemma.* It can be stated as follows. Let Y be a Noetherian scheme. For any proper morphism $X \to Y$, there exists a commutative diagram

with f, g projective, and $f^{-1}(U) \to U$ is an isomorphism for some everywhere dense open subset U of X. This means that by modifying X slightly, we can make it projective over Y. This is a very useful method for extending certain results on projective morphisms to proper morphisms. This lemma will not be used directly in the rest of this work. One can consult [41], II.5.6.1 for the proof.

Definition 3.35. We say that a morphism $f : X \to Y$ is *quasi-projective* if it can be decomposed into an open immersion of finite type $X \to Z$ and a projective morphism $Z \to Y$. We also say that X is a *quasi-projective scheme over Y.*

Proposition 3.36. *Let A be a ring.*

(a) *Let F be a finite set of points of $\mathbb{P}_A^n$ ($n \geq 1$), and Z a closed subset of $\mathbb{P}_A^n$ which does not meet F. Then there exists a non-empty hypersurface $V_+(f) \subset \mathbb{P}_A^n$ such that $V_+(f) \cap F = \emptyset$ and $V_+(f) \supseteq Z$.*

(b) *Let X be a quasi-projective scheme over A, and F a finite set of points of X. Then F is contained in an affine open subset of X.*

Proof (a) We write $Z = V_+(I)$. The set F corresponds to a finite number of homogeneous prime ideals $\mathfrak{p}_1, \ldots, \mathfrak{p}_r$ of $A[T_0, \ldots, T_n]$, and I is contained in none of the $\mathfrak{p}_i$. Thus there exists a homogeneous element $f \in I \setminus \cup_{1 \leq i \leq r} \mathfrak{p}_i$ (Exercise 2.5.5). We then have $V_+(f) \cap F = \emptyset$ and $V_+(f) \supseteq Z$.

(b) By hypothesis, X is an open subscheme of some closed subscheme $\widehat{X}$ of $\mathbb{P}_A^n$. Let $Z = \widehat{X} \setminus X$. Let $V_+(f)$ be given by (a). Then $F \subseteq D_+(f) \cap \widehat{X}$, and the latter is an affine open subset of X. $\square$

Exercises

3.1. Let $f : X \to Y$ be a morphism of schemes. We suppose that there exist open subsets Y_i of Y such that $X = \cup_i f^{-1}(Y_i)$, and that the restrictions

$f : f^{-1}(Y_i) \to Y_i$ are closed immersions. Show that if $f(X)$ is closed in Y, then f is a closed immersion.

3.2. Let X be a scheme. Show that the following properties are equivalent:

 (i) X is separated;

 (ii) X is separated over an affine scheme;

 (iii) for any scheme Y, every morphism of schemes $X \to Y$ is separated.

3.3. Let $f : X \to Y$ be a surjective closed morphism between two Noetherian schemes (or, more generally, topological spaces). Show that we have the inequality $\dim X \ge \dim Y$ (reduce to the case when X is irreducible and use induction on $\dim X$). Compare with Exercises 2.5.8 and 4.3.3.

3.4. Let k be a field. Let $Z = \mathbb{A}^n_k$ or $\mathbb{P}^n_k$. Let us consider two closed subvarieties X, Y of Z, pure of respective dimensions q, r.

 (a) Show that the irreducible components of $X \times_k Y$ are of dimension $q + r$.

 (b) Show that $\dim(X \cap Y) \ge q + r - n$ if $X \cap Y \ne \emptyset$ (identify $X \cap Y$ with the subset $\Delta(Z) \cap (X \times_k Y)$ of $Z \times_k Z$ and use Corollary 2.5.26).

 (c) If $Z = \mathbb{P}^n_k$ and $q + r \ge n$, show that $X \cap Y \ne \emptyset$ (if $X = V_+(I)$, $Y = V_+(J)$, consider $V(I) \cap V(J)$ in $\mathbb{A}^{n+1}_k$ and use Corollary 2.5.21).

 (d) Show that every *projective plane curve* X (i.e., X is a closed subvariety of $\mathbb{P}^2_k$, pure of dimension 1) is connected.

3.5. Let $X \to S$ be a proper morphism. Let $f : X \to Y$ be an open immersion of S-schemes. Let us suppose that Y is separated over S and connected. Show that f is an isomorphism.

3.6. Let $f : X \to Y$ be a separated morphism. Show that any section of f is a closed immersion (apply Proposition 3.9(f) to $X \times_X Y \to X \times_Y Y$).

3.7. Let X be a Y-scheme. Show that X is separated over Y if and only if for every Y-scheme Y', the sections of $X \times_Y Y' \to Y'$ are closed immersions (*hint*: $\Delta_{X/Y}$ is a section of the second projection $X \times_Y X \to X$).

3.8. Let $f : X \to Y$ be a morphism with X affine and Y separated. Show that $f^{-1}(V)$ is affine for every affine open subset V of Y (use Proposition 3.9(f)).

3.9. Let S be a locally Noetherian scheme, $s \in S$. Let Y be a scheme of finite type over $\operatorname{Spec} \mathcal{O}_{S,s}$. Show that there exist an open neighborhood U of s and a scheme of finite type $X \to U$ such that $Y = X \times_U \operatorname{Spec} \mathcal{O}_{S,s}$. Moreover, if Y is separated, we can take $X \to U$ separated.

3.10. (*Graph of a morphism*) Let $f : X \to Y$ be a morphism of S-schemes. We define the *graph* Γ_f of f as being the image of $(\operatorname{Id}_X, f) : X \to X \times_S Y$.

(a) Show that (Id_X, f) is a section of the projection morphism $X \times_S Y \to X$. Deduce from this that Γ_f is a closed subset of $X \times_S Y$ if Y is separated over S.

(b) Let us suppose X is reduced and let us endow Γ_f with the structure of a reduced closed subscheme. Show that if Y is separated over S, then the projection $X \times_S Y \to X$ induces an isomorphism from Γ_f onto X.

3.11. Let X be a Y-scheme.

(a) Show that for any Y-scheme Z, $X_{\mathrm{red}} \times_Y Z \to X \times_Y Z$ is a closed immersion and a homeomorphism on the underlying topological spaces. Deduce from this that $X \to Y$ separated $\iff X_{\mathrm{red}} \to Y$ separated $\iff X_{\mathrm{red}} \to Y_{\mathrm{red}}$ separated.

(b) Let $\{F_i\}_i$ be a finite number of closed subsets of X such that $X = \cup_i F_i$. Let us endow the F_i with the structure of reduced closed subschemes. Show that $X \to Y$ is separated if and only if $F_i \to Y$ is separated for every i.

(c) Let us suppose X is of finite type over Y. Show that $X \to Y$ proper $\iff X_{\mathrm{red}} \to Y$ proper $\iff X_{\mathrm{red}} \to Y_{\mathrm{red}}$ proper. Also show that (b) is true for proper morphisms (by supposing $X \to Y$ of finite type).

3.12. Let $f : X \to Y$ be a surjective morphism. We suppose that Y is connected and that all of the fibers X_y are connected.

(a) Show that if X is proper over Y, then X is connected.

(b) Study the example $\operatorname{Spec} k[T_1, T_2]/T_1(T_1 T_2 - 1) \to \operatorname{Spec} k[T_1]$, where k is a field. Deduce from this that (a) is false if f is not proper.

3.13. (*Rational maps*) Let X, Y be schemes over S which are integral, with $Y \to S$ separated. A *rational map from X to Y*, denoted by $X \dashrightarrow Y$, is an equivalence class of morphisms of S-schemes from a non-empty open subscheme of X to Y. Two such morphisms $U \to Y$, $V \to Y$ are called equivalent if they coincide on $U \cap V$. Let us fix a rational map $X \dashrightarrow Y$.

(a) Show that in every equivalence class, there exists a unique element $f : U \to Y$ such that U is maximal for the inclusion relation, and that every element $g : V \to Y$ of the class verifies $g = f|_V$. We call U the *domain of definition of $X \dashrightarrow Y$*. We then denote the rational map associated to f by $f : X \dashrightarrow Y$.

(b) Let us suppose that f is dominant, that X, Y are of finite type over S, and that S is locally Noetherian. Let $x \in X$. Show that f is *defined* at x (i.e., $x \in U$) if and only if there exists a $y \in Y$ such that the image of $\mathcal{O}_{Y,y}$ under $K(Y) \to K(X)$ is a local ring dominated by $\mathcal{O}_{X,x}$.

(c) Let Γ_f be the closure in $X \times_S Y$ of the graph of $f : U \to Y$. We call it the *graph of the rational map f*. We endow Γ_f with the structure of a reduced closed subscheme. Show that Γ_f is integral and that the projection $p : X \times_S Y \to X$ induces a birational morphism from Γ_f to X.

(d) Show that there exist a birational morphism $g : Z \to X$ and a morphism $\tilde{f} : Z \to Y$ such that $\tilde{f} = f \circ g$ on $g^{-1}(U)$ (we say that we have eliminated the indetermination of $X \dashrightarrow Y$). Moreover, if Y is proper (resp. projective) over S, we can choose g proper (resp. projective).

3.14. Let X, Y be integral separated schemes over a Noetherian scheme S. Let $f : X \dashrightarrow Y$ be a *birational map* (i.e., f comes from an isomorphism from a non-empty open subset $U \subseteq X$ onto an open subset of Y). Show that f is defined everywhere if and only if $\Gamma_f \to X$ is an isomorphism.

3.15. We say that a morphism of schemes $f : X \to Y$ is *finite* (resp. *integral*) if for every affine open subset V of Y, $f^{-1}(V)$ is affine and $\mathcal{O}_X(f^{-1}(V))$ is finite (resp. integral) over $\mathcal{O}_Y(V)$. Let $f : X \to Y$ be a morphism such that there exists an affine open covering $Y = \cup_i Y_i$ with $f^{-1}(Y_i)$ affine and $\mathcal{O}_X(f^{-1}(Y_i))$ finite over $\mathcal{O}_Y(Y_i)$. We want to show that f is finite.

(a) Show that we can reduce to the case when Y is affine and that $Y_i = D(h_i)$ is a principal open subset.

(b) Show that X then verifies condition (3.2), Subsection 2.3.1. Deduce from this that the canonical morphism $X \to \operatorname{Spec} \mathcal{O}_X(X)$ is an isomorphism (hence X is affine).

(c) Show that $\mathcal{O}_X(X)$ is finite over $\mathcal{O}_Y(Y)$.

3.16. Present a statement analogous to that of Exercise 3.15 by replacing finite morphism with integral morphism.

3.17. Show that the following properties are true.

(a) Any finite morphism is of finite type and quasi-finite.

(b) The class of finite morphisms verifies hypotheses (1)–(3) of Lemma 3.15.

(c) Let $\rho : A \to B$ be a finite ring homomorphism. Then for any ideal I of B, we have $(\operatorname{Spec} \rho)(V(I)) = V(\rho^{-1}(I))$.

(d) Any finite morphism is proper (see also Exercise 3.22).

3.18. Let K be a number field. We know that the ring of integers $\mathcal{O}_K$ of K is finite over $\mathbb{Z}$. Show that if X is a separated connected scheme containing $\operatorname{Spec} \mathcal{O}_K$ as an open subscheme, then $X = \operatorname{Spec} \mathcal{O}_K$ (use Exercises 3.17(d) and 3.5).

3.19. Let X be a proper algebraic variety over a field k. Let $f : X \to Y$ be a morphism of k-schemes with Y affine. Show that $f(X)$ is a finite set of closed points (*hint: f factors into $X \to \operatorname{Spec} \mathcal{O}_X(X) \to Y$*).

3.20. (a) Show that Corollary 3.32 is true for quasi-projective morphisms (use Exercise 2.3(c)).

(b) Show that a morphism of finite type from an affine scheme to an affine scheme is quasi-projective.

3.21. Let k be a field. Show that two closed subsets of dimension 1 in $\mathbb{P}^2_k$ always have a non-empty intersection (use Exercise 3.4). Deduce from this that $\mathbb{P}^1_k \times_k \mathbb{P}^1_k \not\simeq \mathbb{P}^2_k$.

3.22. Let $f : \operatorname{Spec} B \to \operatorname{Spec} A$ be a finite morphism. We want to show that f is projective.

(a) Let us first suppose that $B = A[T]/(P(T))$, where $P(T) = \sum_{0 \le i \le n} a_i T^i$ with $a_n = 1$. Let $\tilde{P}(T, S) = \sum_{0 \le i \le n} a_i T^i S^{n-i}$. Show that the natural morphism $\operatorname{Spec} B \to \operatorname{Proj} A[T, S]/(\tilde{P}(T, S))$ is an isomorphism.

(b) Conclude in the general case.

3.23. Let X be a quasi-projective scheme over a scheme S, and let G be a finite group acting on the S-scheme X. Show that the quotient scheme X/G exists (use Exercise 2.3.21), and that G acts transitively on the fibers of $X \to X/G$. Moreover, the canonical morphism $X \to X/G$ is a finite morphism if S is locally Noetherian.

3.24. Let S be a locally Noetherian integral scheme with generic point ξ. Let X be a scheme of finite type over S. Show that if X_ξ is projective (resp. finite) over the function field $K(S) = k(\xi)$, then there exists a non-empty open subset U of S such that $X \times_S U \to U$ is projective (resp. finite). (Use Exercises 2.5 and 2.16.)

3.25. Let X, Y be schemes of finite type over a locally Noetherian scheme S. Let $f : X \to Y$ be a finite morphism of S-schemes. Let us fix a point $s \in S$ such that $X_s \neq \emptyset$.

(a) Let us suppose that $f_s : X_s \to Y_s$ is a closed immersion. Show that there exists an open neighborhood U of s such that $f_U : X_U \to Y_U$ is a closed immersion.

(b) Let us suppose that Y is reduced and that f is surjective. Show the analogue of assertion (a) for isomorphisms (compare to Exercise 2.5).

3.26. We will study the example (kindly communicated by Florian Pop) of a valuation ring having 'very many' prime ideals. Let k be a field. Let I be a totally ordered set. Let $K = k(T_i)_{i \in I}$ be the field of rational fractions whose variables are indexed by I.

(a) Let $\Gamma = \mathbb{Z}^{(I)}$ be the direct sum of copies of $\mathbb{Z}$ indexed by I. This is a group which is totally ordered by the lexicographical order. Let us denote the canonical basis of $\mathbb{Z}^{(I)}$ by $\{e_i\}_i$. Show that there exists a unique valuation (up to isomorphism) $\nu : K^* \to \Gamma$ such that $\nu(T_i) = e_i$ and $\nu(k^*) = 0$.

(b) Let $i \in I$. Show that the radical of the set $\mathfrak{p}(i) := \{a \in K \mid \nu(a) \ge e_i\}$ is a prime ideal of $\mathcal{O}_\nu$. Show that the map $I \to \operatorname{Spec} \mathcal{O}_\nu$ defined in this way is injective and increasing, if we endow $\operatorname{Spec} \mathcal{O}_\nu$ with the order relation determined by the inclusion.

(c) Let $\theta : \mathbb{Z}^{(I)} \setminus \{0\} \to I$ be the map defined by $\theta((n_i)_i) =$ the smallest i such that $n_i \neq 0$. Let $\mathfrak{p} \in \operatorname{Spec} \mathcal{O}_\nu$. Show that if $\theta(\nu(\mathfrak{p}))$ is bounded from above by an index i, then for every $j > i$, we have $\mathfrak{p} \subseteq \mathfrak{p}(j)$. Show that if $\theta(\nu(\mathfrak{p}))$ is not bounded, then $\mathfrak{p} = \mathfrak{m}_\nu$.

3.27. (*A scheme without closed point*) Let I be a totally ordered set without maximal element. Let $\mathcal{O}_\nu$ be the valuation ring defined in the preceding exercise, with maximal ideal $\mathfrak{m}_\nu$. Show that the scheme $\operatorname{Spec} \mathcal{O}_\nu \setminus \{\mathfrak{m}_\nu\}$ has no closed point.

4

Some local properties

In this chapter, we study some local aspects of schemes. First we define normal schemes and present an extension theorem for regular functions (Theorem 1.14). Next we study regular schemes. The important results are regularity criteria (Corollary 2.15 and the Jacobian criterion 2.19). After looking at flat morphisms, we arrive, in a natural way, at smooth morphisms, a relative version of (geometric) regularity. Section 4.4 is devoted to the proof of Zariski's 'Main Theorem', and to its application to proper birational morphisms.

4.1 Normal schemes

4.1.1 Normal schemes and extensions of regular functions

Let us recall that an integral domain A is called *normal* if it is integrally closed in $\mathrm{Frac}(A)$, that is, $\alpha \in \mathrm{Frac}(A)$ integral over A implies that $\alpha \in A$.

Definition 1.1. Let X be a scheme. We say that X is *normal at* $x \in X$ or that x is a *normal point* of X if $\mathcal{O}_{X,x}$ is normal. We say that X is *normal* if it is irreducible[2] and normal at all of its points.

Definition 1.2. We call a normal Noetherian integral domain of dimension 0 or 1 a *Dedekind domain*. We call a normal locally Noetherian scheme of dimension 0 or 1 a *Dedekind scheme*.

Remark 1.3. Usually, a Dedekind domain has dimension 1 by definition. Here we admit the dimension 0 because we want to make the class of Dedekind domains stable by localization. For Dedekind schemes, this makes an open sub-scheme of a Dedekind scheme into a Dedekind scheme.

Lemma 1.4.

    (a) *Let A be a unique factorization domain. Then A is normal.*

    (b) *Let A be a normal integral domain. Then $S^{-1}A$ is normal for any multiplicative subset S of A.*

---

[2] We require this additional condition of irreducibility to simplify the exposition.

Proof (a) Let $\alpha \in \mathrm{Frac}(A)$. We can write $\alpha = ab^{-1}$ with $a, b \in A$ without a common irreducible factor. Let us suppose that α verifies an integral equation

$$\alpha^n + c_{n-1}\alpha^{n-1} + \cdots + c_0 = 0, \quad c_i \in A.$$

Then

$$a^n + b(c_{n-1}a^{n-1} + \cdots + c_0 b^{n-1}) = 0.$$

Hence b divides a^n. It follows that b is invertible, and therefore that $\alpha \in A$.

(b) Immediate. $\qquad\square$

Proposition 1.5. *Let X be an irreducible scheme. The following properties are equivalent:*

(i) *The scheme X is normal.*

(ii) *For every open subset U of X, $\mathcal{O}_X(U)$ is a normal integral domain.*

If, moreover, X is quasi-compact, these properties are equivalent to:

(iii) *The scheme X is normal at its closed points.*

Proof Let us suppose X is normal. Let U be an open subset of X and $\alpha \in \mathrm{Frac}(\mathcal{O}_X(U))$ integral over $\mathcal{O}_X(U)$. Let V be an affine open subset of U. Then α is integral over $A := \mathcal{O}_X(V)$ and $\alpha \in \mathrm{Frac}(A)$. Let us set $I = \mathrm{Ann}((\alpha A + A)/A)$. If $I \neq A$, then $I \subseteq \mathfrak{p}$ for a prime ideal $\mathfrak{p}$ of A. As $A_\mathfrak{p}$ is normal, there exists an $s \in A \setminus \mathfrak{p}$ such that $s\alpha \in A$. It follows that $s \in I$, whence a contradiction. Hence $1 \in I$ and $\alpha \in \mathcal{O}_X(V)$. As this is true for every affine open subset V of U, we have $\alpha \in \mathcal{O}_X(U)$ (Proposition 2.4.18). This shows that (i) implies (ii).

Let us suppose (ii) is verified. Let $x \in X$, and let V be an affine open neighborhood of x. Then $\mathcal{O}_X(V)$ is normal by hypothesis. It follows by the lemma above that $\mathcal{O}_{X,x}$ is normal. Hence X is normal at x and we have (ii) $\Longrightarrow$ (i) by Proposition 2.4.17.

Let us suppose X is quasi-compact, irreducible, and normal at its closed points. Let $x \in X$ be an arbitrary point. There exists a closed point $y \in \overline{\{x\}}$ (Exercise 2.4.8). Let V be an affine open subset of X containing y. Then $x \in V$. It follows that $\mathcal{O}_{X,x}$ is a localization of $\mathcal{O}_{X,y}$, and therefore normal. Consequently, X is normal. $\qquad\square$

Example 1.6. Let k be a field. Then $\mathbb{A}^n_k$ and $\mathbb{P}^n_k$ are normal schemes because their local rings are factorial.

Example 1.7. Let X be a Noetherian integral scheme. Then X is a Dedekind scheme if and only if $\mathcal{O}_X(U)$ is a Dedekind domain for every open subset U of X. In particular, the spectrum of a Dedekind domain is a Dedekind scheme.

Example 1.8. Let $\mathcal{O}_K$ be a discrete valuation ring with uniformizing parameter t. Let $P(S) = S^n + a_{n-1}S^{n-1} + \cdots + a_0 \in \mathcal{O}_K[S]$ be an Eisenstein polynomial, that is, $a_i \in t\mathcal{O}_K$ and $a_0 \notin t^2\mathcal{O}_K$. Let us consider $\mathcal{O}_L := \mathcal{O}_K[S]/(P(S))$. Then $\mathcal{O}_L$ is normal. More precisely, $\mathcal{O}_L$ is a discrete valuation ring. Indeed, it is a classical result that $P(S)$ is an irreducible polynomial (if $P = P_1 P_2$, then

$P_1(0), P_2(0) \in t\mathcal{O}_K$; hence $P(0) \in t^2\mathcal{O}_K$, whence a contradiction). Let s be the image of S in $\mathcal{O}_L$. Then $L = \oplus_{0 \le i \le n-1} s^i K$. Let us denote the normalized valuation of $\mathcal{O}_K$ (Example 3.3.23) by ν_K. Let us set

$$\nu_L\left(\sum_{0 \le i \le n-1} a_i s^i\right) = \max_{0 \le i \le n-1}\{n\nu_K(a_i) + i\}, \quad a_i \in K.$$

It can then immediately be seen that $\nu_L : L^* \to \mathbb{Z}$ is a valuation of L, whose valuation ring is $\sum_i \mathcal{O}_K s^i = \mathcal{O}_L$.

Example 1.9. Let m be a square-free integer. Let us consider the scheme

$$X = \operatorname{Proj}\mathbb{Z}[S_0, S_1, S_2]/(S_2^2 - mS_1 S_0),$$

and let us show that X is normal. Let s_i denote the image of S_i in the quotient. We have $D_+(s_2) \subset D_+(s_0) \cup D_+(s_1)$, and the open subschemes $D_+(s_0)$ and $D_+(s_1)$ are isomorphic by symmetry. It therefore suffices to show that $D_+(s_0)$ is normal. Let $t_i = s_0^{-1} s_i \in A := \mathcal{O}_X(D_+(s_0))$. We have $A = \mathbb{Z}[t_1, t_2]$ with the relation $t_2^2 = mt_1$. This is clearly an integral domain, and the subring $\mathbb{Z}[t_1]$ of A is isomorphic to a polynomial ring. We moreover have $\operatorname{Frac}(A) = \mathbb{Q}(t_1)[t_2]$.

Let $f = g_0 + g_1 t_2 \in \operatorname{Frac}(A)$ with $g_i \in \mathbb{Q}(t_1)$ and let us suppose f is integral over A. Let us consider the automorphism (of $\mathbb{Z}$-algebras) $\sigma : A \to A$ defined by $\sigma(t_1) = t_1$, $\sigma(t_2) = -t_2$. Then $\sigma(f) = g_0 - g_1 t_2$ is also integral over A. This implies that $2g_0 = f + \sigma(f)$ is integral over A. Now, A is integral over $\mathbb{Z}[t_1]$. Since the latter is factorial ([55], V, Theorem 6.3), and hence normal, we have $2g_0 \in \mathbb{Z}[t_1]$. Consequently, $2g_1 t_2$ is integral over A. By considering its square, we deduce from this that $4g_1^2 mt_1 \in \mathbb{Z}[t_1]$. This immediately implies that $2g_1 \in \mathbb{Z}[t_1]$. Let us set $G_i = 2g_i \in \mathbb{Z}[t_1]$. As $g_0^2 + mt_1 g_1^2 = f\sigma(f)$ is integral over $\mathbb{Z}[t_1]$ and therefore belongs to $\mathbb{Z}[t_1]$, we have $G_0^2 + mt_1 G_1^2 = 4H$, with $H \in \mathbb{Z}[t_1]$. Finally, by considering this equality in $\mathbb{F}_2[t_1]$, we see that $G_0 \equiv 0 \mod 2$. This then implies that 4 divides $mt_1 G_1^2$. Hence 2 divides G_1. In other words, we have $f \in A$.

Remark 1.10. We see in the example above that testing the normality of a scheme is not always easy in practice, even if the scheme is given by a relatively simple equation. We will encounter other verification methods further on (Theorems 2.16(b) and 8.2.23).

Lemma 1.11. *Let A be a Noetherian integral domain, and I a non-zero ideal of A. Then $B := \{f \in \operatorname{Frac}(A) \mid If \subseteq I\}$ is a subring of $\operatorname{Frac}(A)$, finite over A.*

Proof It is clear that B is a ring containing A. We have an A-module homomorphism $\phi : B \to \operatorname{Hom}_A(I, I)$ defined by $\phi(f) : \alpha \mapsto \alpha f$. This is an injective homomorphism because A is an integral domain. Let $A^n \to I$ be a surjective A-module homomorphism. It induces an injective homomorphism $\operatorname{Hom}_A(I, I) \to \operatorname{Hom}_A(A^n, I) \simeq I^n$, which implies that $\operatorname{Hom}_A(I, I)$, and hence B, is finite over A (see also Exercise 1.1 for a more general statement). $\square$

Proposition 1.12. *Let X be a Dedekind scheme. Then for any $x \in X$, the ring $\mathcal{O}_{X,x}$ is a principal ideal domain.*

Proof We must show that any normal Noetherian local ring $(A, \mathfrak{m})$ of dimension 1 is a principal ideal domain (the dimension 0 case is trivial). Let $x \in \mathfrak{m} \setminus \mathfrak{m}^2$ (such an element exists because otherwise dim $A = 0$, see Lemma 2.5.11). Let us first show that $\mathfrak{m} = xA$. We have dim $A/xA = 0$ (Theorem 2.5.15), and hence $\mathfrak{m} = \sqrt{xA}$ (Lemma 2.5.11). Let $r \geq 2$ be an integer such that $\mathfrak{m}^r \subseteq xA$. Then for any $y \in \mathfrak{m}^{r-1}$, we have $(x^{-1}y)\mathfrak{m} \subseteq x^{-1}\mathfrak{m}^r \subseteq A$. Hence $(x^{-1}y)\mathfrak{m}$ is an ideal of A, contained in $\mathfrak{m}$ because otherwise $x \in \mathfrak{m}^r \subseteq \mathfrak{m}^2$. By Lemma 1.11, $x^{-1}y$ is integral over A, and hence $x^{-1}y \in A$, that is to say $y \in xA$. Consequently, $\mathfrak{m}^{r-1} \subseteq xA$. Repeating this, we see that $\mathfrak{m} = xA$.

Let I be a non-zero ideal of A. As $\cap_{n \geq 1}\mathfrak{m}^n = 0$ (using Lemma 1.11 and the fact that $x^{-1} \cdot (\cap_{n \geq 1}\mathfrak{m}^n) = \cap_{n \geq 1}\mathfrak{m}^n$; or Corollary 1.3.13), there exists an $n \geq 1$ such that $I \subseteq \mathfrak{m}^n = x^n A$ and $I \not\subseteq \mathfrak{m}^{n+1}$. It can then immediately be verified that $I = x^n A$. Consequently, A is indeed a principal ideal domain. □

Lemma 1.13. *Let A be a normal Noetherian ring of dimension ≥ 1. Then we have the equality*

$$A = \bigcap_{\mathfrak{p} \in \operatorname{Spec} A,\ \operatorname{ht}(\mathfrak{p})=1} A_{\mathfrak{p}}.$$

Proof Let A' denote the right-hand side in the equality we want to prove, and let us suppose that $A' \neq A$. For every $f \in A' \setminus A$, let I_f be the proper ideal $\{a \in A \mid af \in A\}$ of A. Since A is Noetherian, the set of I_f with $f \in A' \setminus A$ admits a maximal element $\mathfrak{q} := I_g$ for some $g \in A' \setminus A$. Let us first show that $\mathfrak{q}$ is a prime ideal. Let $a_1, a_2 \in A$ be such that $a_1 a_2 \in \mathfrak{q}$ and that $a_2 \notin \mathfrak{q}$. Then $a_2 g \in A' \setminus A$, $\mathfrak{q} \subseteq I_{a_2 g}$, and $a_1 \in I_{a_2 g}$. By the hypothesis on $\mathfrak{q}$, we have $a_1 \in I_{a_2 g} = \mathfrak{q}$, and hence $\mathfrak{q}$ is prime.

Let us now consider the ideal $g\mathfrak{q}A_{\mathfrak{q}}$ of $A_{\mathfrak{q}}$. If it is equal to $A_{\mathfrak{q}}$, then $\mathfrak{q}A_{\mathfrak{q}} = g^{-1}A_{\mathfrak{q}}$, and $\operatorname{ht}(\mathfrak{q}) = 1$ by Theorem 2.5.12. Therefore $g \in A_{\mathfrak{q}}$, and there exists an $s \in A \setminus \mathfrak{q}$ such that $sg \in A$. So $s \in I_g = \mathfrak{q}$, a contradiction. Hence $g\mathfrak{q}A_{\mathfrak{q}} \subseteq \mathfrak{q}A_{\mathfrak{q}}$. Applying Lemma 1.11 to the ideal $\mathfrak{q}A_{\mathfrak{q}}$ of $A_{\mathfrak{q}}$, we see that g is integral over, and therefore belongs to, $A_{\mathfrak{q}}$. This is impossible, as we saw above. Hence $A' = A$. □

Theorem 1.14. *Let X be a normal locally Noetherian scheme. Let F be a closed subset of X of codimension $\operatorname{codim}(F, X) \geq 2$. Then the restriction*

$$\mathcal{O}_X(X) \to \mathcal{O}_X(X \setminus F)$$

is an isomorphism. In other words, every regular function on $X \setminus F$ extends uniquely to a regular function on X.

Proof We may assume that $X = \operatorname{Spec} A$ is affine. For any prime ideal $\mathfrak{p} \subset A$ of height 1, we have $\mathfrak{p} \in X \setminus F$. The theorem immediately follows from Lemma 1.13.
□

Example 1.15. Let $X = \operatorname{Spec} k[x, y]$ be the affine plane over a field k. Let U be the complement of the origin in X. Then $\mathcal{O}_X(U) = \mathcal{O}_X(X)$. We can also verify this equality directly (Exercise 2.5.12(b)).

Proposition 1.16. *Let $Y \to S$ be a proper morphism over a locally Noetherian scheme. Let X be a normal S-scheme of finite type. Let us consider a morphism of S-schemes $f : U \to Y$ defined on a non-empty open subset U of X. Then f extends uniquely to a morphism $V \to Y$, where V is an open subset of X containing all points of codimension 1.*

Proof The uniqueness comes from the separatedness of $Y \to S$ and from the fact that X is reduced (Proposition 3.3.11). Let ξ be the generic point of X. Then $\xi \in U$ and f induces a morphism $f_\xi : \operatorname{Spec} K(X) \to Y$. Let $x \in X$ be a point of codimension 1. Then $\mathcal{O}_{X,x}$ is a discrete valuation ring with field of fractions $K(X)$. It follows from Corollary 3.3.26 that f_ξ extends to a morphism $f_x : \operatorname{Spec} \mathcal{O}_{X,x} \to Y$. As Y is of finite type over S, f_x extends to a morphism $g : U_x \to Y$ on an open neighborhood U_x of x (Exercise 3.2.4).

Let W be an affine open neighborhood of $g(x)$ and let us consider the restrictions of f and g to $U' := f^{-1}(W) \cap g^{-1}(W)$. The latter is a non-empty open subset of X because it contains ξ. The homomorphisms $\mathcal{O}_Y(W) \to \mathcal{O}_X(U')$ corresponding to $f|_{U'}$ and $g|_{U'}$ are identical because they coincide on $K(X) \supseteq \mathcal{O}_X(U')$. Consequently, $f|_{U'} = g|_{U'}$ (Proposition 2.3.25). By virtue of Proposition 3.3.11, f and g coincide on $U \cap U_x$. The same reasoning shows that if $x' \in X$ is another point of codimension 1, then the morphism $g' : U_{x'} \to Y$ coincides with f and g respectively on the intersections $U \cap U_{x'}$ and $U_x \cap U_{x'}$. Thus we see that f extends to a morphism on an open subset V containing all points of codimension 1. □

Corollary 1.17. *Let us keep the hypotheses of Proposition 1.16. Let us moreover suppose that X is of dimension 1. Then f extends uniquely to a morphism $X \to Y$.*

Lemma 1.18. *Let $\mathcal{O}_K$ be a discrete valuation ring, with field of fractions K and residue field k. Let X be an $\mathcal{O}_K$-scheme such that $\mathcal{O}_X(U)$ is flat over $\mathcal{O}_K$ for every affine open subset U of X. We suppose that X_K is normal and that X_k is reduced. Then X is normal.*

Proof We may suppose that $X = \operatorname{Spec} A$ is affine. The canonical homomorphism $A \to A \otimes_{\mathcal{O}_K} K$ is injective by the A flat over $\mathcal{O}_K$ hypothesis. Hence A is an integral domain. Let t be a uniformizing parameter for $\mathcal{O}_K$. Let $\alpha \in \operatorname{Frac}(A)$ be integral over A. Since $A \otimes_{\mathcal{O}_K} K$ is normal, there exist $a \in A$, $r \in \mathbb{Z}$ such that $\alpha = t^{-r}a$. We can suppose that $a \notin tA$. Let us show that $r \leq 0$. Let $\alpha^n + a_{n-1}\alpha^{n-1} + \cdots + a_0 = 0$ be an integral relation for α over A. If $r > 0$, by multiplying this relation by t^{rn}, we see that a is nilpotent in A/tA; hence, $a \in tA$, a contradiction. Thus $r \leq 0$ and $\alpha \in A$. (See also Exercise 8.2.11(b).) □

4.1.2 Normalization

Definition 1.19. Let X be an integral scheme. A morphism $\pi : X' \to X$ is called a *normalization morphism* if X' is normal, and if every dominant morphism

$f : Y \to X$ with Y normal factors uniquely through π:

   Note that if $\pi : X' \to X$ is a normalization of X, then for any open subscheme U of X, the restriction $\pi^{-1}(U) \to U$ is a normalization of U.

Definition 1.20. We say that a morphism $f : X' \to X$ is *integral* if for every affine open subset U of X, $f^{-1}(U)$ is affine and $\mathcal{O}_X(U) \to \mathcal{O}_{X'}(f^{-1}(U))$ is integral.

Lemma 1.21. *Let A be an integral domain. Let A' be the integral closure of A in $\mathrm{Frac}(A)$. Then the morphism $\mathrm{Spec}\, A' \to \mathrm{Spec}\, A$ induced by the canonical injection $A \to A'$ is a normalization morphism.*

Proof Let us recall that if A is a subring of a commutative ring C, the *integral closure of A in C* is the set of elements of C that are integral over A. This is a subring of C containing A (see [55], IX, Proposition 1.4).
   Let $\pi : Y \to \mathrm{Spec}\, A$ be a dominant morphism with Y normal. Then $A \to \mathcal{O}_Y(Y)$ is injective (Exercise 2.4.11). As $\mathcal{O}_Y(Y)$ is normal (Proposition 1.5), the homomorphism $A \to \mathcal{O}_Y(Y)$ factors through an injective homomorphism $A' \to \mathcal{O}_Y(Y)$. Hence π factors into $Y \to \mathrm{Spec}\, A'$ followed by $\mathrm{Spec}\, A' \to \mathrm{Spec}\, A$.   □

Proposition 1.22. *Let X be an integral scheme. Then there exists a normalization morphism $\pi : X' \to X$, and it is unique up to isomorphism (of X-schemes). Moreover, a morphism $f : Y \to X$ is the normalization morphism if and only if Y is normal, and f is birational and integral.*

Proof The uniqueness of the normalization is immediate. To show its existence, it suffices to cover X with affine open subsets U_i, to apply the lemma above to the U_i, and finally, to glue the normalizations $U_i' \to U_i$ owing to the uniqueness of the normalizations of the $U_i \cap U_j$. The rest of the properties also result directly from the lemma above.   □

Example 1.23. Let us consider Example 1.9 once more, but with $m = n^2 m'$, where m' is square-free. We have a homomorphism of graded algebras

$$\varphi : \mathbb{Z}[S_0, S_1, S_2]/(S_2^2 - m S_1 S_0) \to \mathbb{Z}[W_0, W_1, W_2]/(W_2^2 - m' W_1 W_0),$$

defined by $\varphi(S_2) = n W_2$, and $\varphi(S_i) = W_i$ for $i \leq 1$. The morphism of schemes f associated to φ is clearly birational and integral. It follows from Example 1.9 and from the proposition above that f is the normalization morphism.

   A more general notion than the normalization of a scheme is the following:

Definition 1.24. Let X be an integral scheme, and let L be an algebraic extension of the function field $K(X)$. We define the *normalization of X in L* to be an integral morphism $\pi : X' \to X$ with X' normal, $K(X') = L$, and such that π extends the canonical morphism $\mathrm{Spec}\, L \to X$.

In the same manner as in the proposition above, we easily see that the normalization $\pi : X' \to X$ of X in L exists and is unique. Moreover, for any affine open subset U of X, $\pi^{-1}(U)$ is affine and $\mathcal{O}_{X'}(\pi^{-1}(U))$ is the integral closure of $\mathcal{O}_X(U)$ in L. The normalization of X in $K(X)$ is none other than the normalization of X defined above.

Proposition 1.25. *Let X be a normal Noetherian scheme. Let L be a finite separable extension of $K(X)$. Then the normalization $X' \to X$ of X in L is a finite morphism.*

Proof It suffices to consider the affine case $X = \operatorname{Spec} A$. Let B be the integral closure of A in L. We want to show that B is finite over A as an A-module. As A is Noetherian, we can, if necessary, extend L in such a way that it is Galois over $K := K(X)$.

Let us consider the trace form $L \times L \to K$, $(x, y) \mapsto \operatorname{Tr}_{L/K}(xy)$. This is a nondegenerate bilinear form because L/K is separated (see [55], VI, Theorem 5.2). Let $\{e_1, \ldots, e_n\}$ be a basis of L/K made up of elements of B. There then exists a basis $\{e_1^*, \ldots, e_n^*\} \subset L$ dual to $\{e_1, \ldots, e_n\}$ (i.e., $\operatorname{Tr}_{L/K}(e_i e_j^*) = \delta_{ij}$). Let $b \in B$. We have $b = \sum_j \lambda_j e_j^*$ with $\lambda_j \in K$. It follows that $\lambda_j = \operatorname{Tr}_{L/K}(be_j) \in B \cap K = A$. Consequently, B is a sub-A-module of $\sum_j A e_j^*$, and is therefore finite over A. $\square$

Lemma 1.26. *Let k be a field, and let L be a finite extension of $k(T_1, \ldots, T_n)$. Then the integral closure of $k[T_1, \ldots, T_n]$ is finite over $k[T_1, \ldots, T_n]$.*

Proof Let $K = k(T_1, \ldots, T_n)$. There exists a finite extension of L (e.g., the normal closure of L/K) that is separable over a purely inseparable extension K' of K. By Proposition 1.25, it suffices to show that the integral closure A' of $k[T_1, \ldots, T_n]$ in K' is finite over $k[T_1, \ldots, T_n]$. We may assume that $\operatorname{char}(k) = p > 0$. Let $\{e_j\}_j$ be a finite system of generators of K' over K. There exists a $q = p^r$, $r \in \mathbb{N}$, such that $e_j^q \subseteq K$ for every j. Let us fix an algebraic closure $\overline{K}$ of K containing K'. Then $K' \subseteq k'[S_1, \ldots, S_n]$, where $S_i = T_i^{1/q} \in \overline{K}$, and where k' is the subextension of $\overline{K}$ over k generated by the qth roots of the coefficients of the e_j^q. Now $k'[S_1, \ldots, S_n]$ is normal and finite over $k[T_1, \ldots, T_n]$. It follows that $A' \subseteq k'[S_1, \ldots, S_n]$ is finite over $k[T_1, \ldots, T_n]$. $\square$

Proposition 1.27. *Let X be an integral algebraic variety over a field k. Let $L/K(X)$ be a finite extension. Then the normalization $X' \to X$ of X in L is a finite morphism. In particular, X' is an algebraic variety over k.*

Proof We may assume $X = \operatorname{Spec} A$ affine. Let $k[T_1, \ldots, T_n] \to A$ be a finite injective homomorphism (Noether's normalization lemma, Proposition 2.1.9). Then the integral closure of A in L is also the integral closure of $k[T_1, \ldots, T_n]$ in L. The proposition then follows from the lemma above. $\square$

Remark 1.28. See the end of Subsection 8.2.2 for more information about the finiteness of the normalization morphism.

Proposition 1.29. *Let X be an integral scheme such that normalization $X' \to X$ of X is a finite morphism. Then the set U of normal points of X is open in X.*

Proof To show that U is open, it suffices to show this in the affine case $X = \operatorname{Spec} A$. Let A' be the integral closure of A. Let $\mathfrak{p} \in \operatorname{Spec} A$. Then the integral closure of $A_{\mathfrak{p}}$ is $A' \otimes_A A_{\mathfrak{p}}$. Hence $\mathfrak{p} \in U$ if and only if $(A'/A) \otimes_A A_{\mathfrak{p}} = 0$. Let $I = \operatorname{Ann}(A'/A)$ be the annihilator of the A-module A'/A. Then we clearly have $\operatorname{Spec} A \setminus V(I) \subseteq U$. Conversely, let $\mathfrak{p} \in U$. Since A'/A is finitely generated over A, there exists an $a \in A \setminus \mathfrak{p}$ such that $a(A'/A) = 0$. Hence $\mathfrak{p} \in \operatorname{Spec} A \setminus V(I)$. This implies that $U = \operatorname{Spec} A \setminus V(I)$ is open.   □

Corollary 1.30. *Let X be an integral algebraic variety. Then the normalization $X' \to X$ of X is a finite morphism, and the set of normal points of X is open in X.*

Proof The finiteness of $X' \to X$ results from Proposition 1.27 by taking $L = K(X)$, and the openness of the set of normal points is a consequence of Proposition 1.29.   □

Proposition 1.31. *Let A be a Dedekind ring with field of fractions K, let L be a finite extension of K, and B the integral closure of A in L. Then B is a Dedekind ring and the canonical morphism $f : \operatorname{Spec} B \to \operatorname{Spec} A$ has finite fibers.*

Proof We can decompose the extension $K \to L$ into a separable extension followed by a purely inseparable extension. As the proposition is true for the separable extension (Propositions 2.5.10 and 1.25), we can suppose that L is purely inseparable over K. There therefore exists a power p^e of the characteristic $p = \operatorname{char}(K)$ such that $L^{p^e} \subseteq K$. Consequently, $B^{p^e} \subseteq A$. Let $\mathfrak{p}$ be a prime ideal of A. Then $\sqrt{\mathfrak{p}B}$ is the unique prime ideal of B lying above $\mathfrak{p}$. Hence f is bijective. Moreover, $\dim B = 1$ by Proposition 2.5.10.

Let $\mathfrak{q}$ be a maximal ideal of B. Then $\mathfrak{p} := \mathfrak{q} \cap A$ is a maximal ideal of A. Let $\nu : K \to \mathbb{Z}$ be a discrete valuation of K associated to $A_{\mathfrak{p}}$. Let us set $\nu_L(\beta) = \nu(\beta^{p^e})$. Then ν_L is a discrete valuation of L, and we immediately verify that $B_{\mathfrak{q}}$ is its valuation ring. Consequently, $B_{\mathfrak{q}}$ is a discrete valuation ring. It remains to show that B is Noetherian. Let I be a non-zero ideal of B. Let $b \in I$ be non-zero. Then B/bB is integral over $A/(bB \cap A)$. As $bB \cap A$ is non-zero because it contains b^{p^e}, we have $\dim B/bB = \dim A/(bB \cap A) = 0$. Moreover, as f is bijective, $V(b)$ is a finite set $\mathfrak{q}_1, \ldots, \mathfrak{q}_n$. Hence $B/bB \simeq \oplus_{1 \le i \le n} B_{\mathfrak{q}_i}/(b)$ (Exercise 2.5.11(c)) is Noetherian. It follows that I is finitely generated.   □

Exercises

1.1. Let A be a ring and B an A-algebra. We suppose that there exists a *faithful B-module M* (i.e., $bM = 0$ for $b \in B$ implies that $b = 0$) which is finitely generated over A. Using the Cayley–Hamilton identity on the matrices $\operatorname{Hom}_A(M, M)$ (i.e., any matrix cancels its characteristic polynomial), show that B is integral over A (this generalizes Lemma 1.11).

1.2. Let A be an integral domain, and let I be a finitely generated ideal of A. Show that for any $\varphi \in \operatorname{Hom}_A(I, I)$ and for any non-zero $a \in I$, $a^{-1}\varphi(a) \in \operatorname{Frac}(A)$ is integral over A (use Cayley–Hamilton identity).

1.3. Let X be a normal Noetherian local scheme of dimension 2, with closed point s. Show that $X \setminus \{s\}$ is a non-affine Dedekind scheme.

1.4. Let X be a connected algebraic variety. Show that X is reduced (resp. integral; resp. normal) if and only if this property is verified at the closed points of X.

1.5. Let A be a normal ring. We want to show that $A[T]$ is also normal. Let $f \in \mathrm{Frac}(A[T])$ be integral over A.

 (a) Show that $f \in \mathrm{Frac}(A)[T]$.

 (b) Show that there exists a subring A_0 of A such that A_0 is finitely generated over $\mathbb{Z}$, $f \in \mathrm{Frac}(A_0)[T]$, and f is integral over $A_0[T]$.

 (c) Show that there exists an $a \in A_0 \setminus \{0\}$ such that $af^n \in A_0[T]$ for every $n \geq 0$.

 (d) Let us write $f = \sum_{i \leq d} \alpha_i T^i$. Show that $a\alpha_0^n \in A_0$ for every $n \geq 0$, and that $a^{i+1}\alpha_i^n \in A_0$ for every $i \leq d$ and for every $n \geq 0$.

 (e) Show that $A_0[\alpha_i] \subseteq a^{-i-1}A_0$, and that $f \in A[T]$.

 (f) (A second method when A is Noetherian.) Show that $A[T]$ is normal if A is a discrete valuation ring; then prove the general case using Lemma 1.13.

1.6. Let $A \to B$ be a faithfully flat ring homomorphism, and let us suppose B is normal.

 (a) Let I, J be ideals of A such that $IB \subseteq JB$. Show that $I \subseteq J$.

 (b) Let $f = ab^{-1} \in \mathrm{Frac}(A)$ be integral over A. Show that $aB \subseteq bB$. Deduce from this that $f \in A$ and that A is normal.

1.7. Let k be a field. Determine the normalization of $\mathrm{Proj}\, k[x, y, z]/(x^3 - y^2 z)$.

1.8. Let $\mathcal{O}_K$ be a Dedekind domain, and let $a \in \mathcal{O}_K$ be non-zero. Show that the scheme $\mathrm{Proj}\, \mathcal{O}_K[x, y, z]/(xy - az^2)$ is normal.

1.9. Let $X = \mathrm{Spec}\, A$ be an affine Dedekind scheme. We are going to show that every open subscheme U of X is affine. We may assume that U is the complement of a closed point x_0.

 (a) Let $\mathfrak{m}_0$ be the maximal ideal of A corresponding to x_0. Let $t \in A$ be such that $t\mathcal{O}_{X,x_0} = \mathfrak{m}_0 \mathcal{O}_{X,x_0}$ and let $V(t) = \{x_0, x_1, \ldots, x_n\}$. Show that for every $1 \leq i \leq n$, there exists $t_i \in A$ such that $t_i \notin \mathfrak{m}_0$ and that t_i generates the maximal ideal of $\mathcal{O}_{X,x_i}$.

 (b) Show that there exists an $f \in \mathcal{O}_X(U) \setminus A$ (take $f = t^{-1} \prod_{1 \leq i \leq n} t_i^{d_i}$ with large d_i).

 (c) Let $Z = V(f) \subset U$ and $V = X \setminus Z$. Show that there exists a (unique) morphism $\varphi : X \to \mathbb{P}_X^1 = \mathrm{Proj}\, A[T_0, T_1]$ such that

$\varphi|_U$ is the morphism of A-schemes $U \rightarrow D_+(T_0)$ corresponding to $A[T_1/T_0] \rightarrow \mathcal{O}_X(U)$, $T_1/T_0 \mapsto f$; and $\varphi|_V$ is the morphism of A-schemes $V \rightarrow D_+(T_1)$ corresponding to $A[T_1/T_0] \rightarrow \mathcal{O}_X(V)$, $T_1/T_0 \mapsto f$.

(d) Show that φ is a section of the canonical morphism $\mathbb{P}^1_X \rightarrow X$, and that $U = \varphi^{-1}(D_+(T_0))$. Deduce from this that U is affine, and that $\mathcal{O}_X(U) = A[f]$ (use Exercise 3.3.6).

1.10. Let X be a quasi-projective scheme over a locally Noetherian scheme S. Let G be a finite group acting on X, so that the quotient scheme X/G exists (Exercise 3.3.23). Let us moreover suppose that X is normal.

(a) Show that $Y := X/G$ is normal.

(b) Show that the first projection $X \times_Y X \rightarrow X$ induces an isomorphism between any irreducible component Z of $X \times_Y X$, endowed with the reduced subscheme structure, and the scheme X. (Show that $Z \rightarrow X$ is finite and birational.)

1.11. Let X be an integral algebraic variety over a field k. Let $k' = K(X) \cap \bar{k}$.

(a) Let X' be the normalization of X. Show that X' is an algebraic variety over k'.

(b) Show that for any point $x \in X'$, k' is a subfield of $k(x)$. Deduce from this that k' is finite over k.

1.12. Let X be a normal and proper algebraic variety over a field k. Show that $\mathcal{O}_X(X) = K(X) \cap \bar{k}$.

1.13. Let X be a reduced algebraic variety. Show that X is normal at all of the generic points of X. Show that X contains an everywhere dense normal open subvariety.

1.14. Let X be an integral scheme of dimension 1, and let $f : X \rightarrow Y$ be a separated birational morphism of finite type from X to a Dedekind scheme. Show that f is an open immersion (this is a special case of a theorem of Zariski, see Corollary 4.6).

1.15. Let X be a locally Noetherian, separated, integral scheme, and U an affine open subset of X different from X. Let us suppose that the normalization X' of X is also locally Noetherian (e.g., if $X' \rightarrow X$ is finite). We want to show that the irreducible components of $X \setminus U$ are of codimension 1 in X. Let ξ be a generic point of $X \setminus U$.

(a) Show that there exists an affine open neighborhood V of ξ in X such that ξ is the unique generic point of $V \setminus (U \cap V)$.

(b) Show that $V \cap U$ is an affine open subset of V. Using Theorem 1.14, show that $\dim \mathcal{O}_{X,\xi} = 1$ if X is normal.

(c) Let $\pi : X' \to X$ be the normalization morphism. Show that any point of $\pi^{-1}(\xi)$ is a generic point of $X' \backslash U'$, where $U' = \pi^{-1}(U)$. Show that

$$\dim \mathcal{O}_{X,\xi} = \max_{\xi' \in \pi^{-1}(\xi)} \dim \mathcal{O}_{X',\xi'} = 1$$

(use Exercise 2.1.8 and Proposition 2.5.10).

(d) Show that the statement still holds if X is not necessarily integral, but is such that the normalization of its (reduced) irreducible components are locally Noetherian.

1.16. Let C be a normal proper curve over a field k. We are going to show that C is projective over k.

(a) Let $\{U_i\}_i$ be a finite covering of C by affine open subsets. Each U_i admits an open immersion into a projective variety Y_i over k (Exercise 2.3.18). Let Y be the fibered product of the Y_i over k. Show that the canonical morphism $\cap_i U_i \to Y$ extends uniquely to a proper morphism $h : C \to Y$.

(b) Let $Z = h(C)$ be the closed subvariety of Y endowed with the reduced structure. Show that g induces a surjective morphism $f : C \to Z$ and that the projection $Y \to Y_i$ induces a dominant morphism $Z \to Y_i$.

(c) Show that for any open subset U_i, the open immersion $U_i \to Y_i$ factors into $U_i \to C \to Z \to Y_i$. Deduce from this that $f_x^\# : \mathcal{O}_{Z,f(x)} \to \mathcal{O}_{C,x}$ is an isomorphism for every $x \in U_i$. Show that f is an isomorphism and that C is projective over k.

1.17. Let U be an integral affine algebraic curve over a field k.

(a) Show that there exists a proper curve $\hat{U}$ over k verifying the following properties:

   (1) there exists an open immersion $i : U \to \hat{U}$,

   (2) the points of $\hat{U} \setminus U$ are normal in $\hat{U}$.

(b) Show that $\hat{U}$ is the 'smallest proper completion' of U, in the sense that any morphism $U \to X$ into a proper variety over k factors uniquely through the open immersion $U \to \hat{U}$. Deduce from this that $\hat{U}$ is unique up to isomorphism.

(c) Show that there exists a finite morphism $U \to \mathbb{A}^1_k$, and that it extends to a finite morphism $f : \hat{U} \to \mathbb{P}^1_k$ such that $U = f^{-1}(\mathbb{A}^1_k)$.

1.18. Let k be a field, and let $P(x) \in k[x]$ be a monic polynomial.

(a) Show that $X = \text{Spec}\, k[x,y]/(y^2 - P(x))$ is integral if and only if $P(x)$ is not a square.

(b) Let us suppose $\text{char}(k) \neq 2$. Show that X is normal if and only if $P(x)$ is a polynomial without a square factor.

(c) Let us suppose $\text{char}(k) = 2$ and k is perfect. Then $P(x)$ can be written in a unique way as $P(x) = xQ_1(x)^2 + Q_2(x)^2$. Show that X is normal if and only if $P(x) \in k^*x + k[x^2]$. If k is not perfect, show that this condition is sufficient.

(d) Supposing X integral, and char$(k) \neq 2$ or k perfect, determine the normalization of X.

1.19. Let k be a field of characteristic $p > 0$. Let $f : X \to Y$ be a finite surjective morphism of normal algebraic varieties over k. We identify $K(Y)$ with a subfield of $K(X)$ via f. Use Lemma 3.2.25 to show the following properties:

(a) If $K(X)^p \subseteq K(Y)$, show that $F_{X/k}$ factors into $f : X \to Y$ followed by a morphism $Y \to X^{(p)}$.

(b) If $K(Y) \subseteq kK(X)^p$, show that f factors into $F_{X/k} : X \to X^{(p)}$ followed by a morphism $X^{(p)} \to Y$.

4.2 Regular schemes

In algebraic geometry, the 'simplest' schemes from the point of view of the local structure are the regular schemes X. They are (in some sense) close to affine spaces, either through the structure of the formal completions of the local rings $\mathcal{O}_{X,x}$ (Proposition 2.27), or through algebraic properties of the local rings $\mathcal{O}_{X,x}$ themselves (Theorem 2.16). We define the notion of tangent space in the first subsection. Then, the regularity of a scheme is defined, and the Jacobian criterion is proven (Theorem 2.19).

4.2.1 Tangent space to a scheme

Definition 2.1. Let X be a scheme and $x \in X$. Let $\mathfrak{m}_x$ be the maximal ideal of $\mathcal{O}_{X,x}$ and $k(x) = \mathcal{O}_{X,x}/\mathfrak{m}_x$ be the residue field. Then $\mathfrak{m}_x/\mathfrak{m}_x^2 = \mathfrak{m}_x \otimes_{\mathcal{O}_{X,x}} k(x)$ is in a natural way a $k(x)$-vector space. Its dual $(\mathfrak{m}_x/\mathfrak{m}_x^2)^{\vee}$ is called the *(Zariski) tangent space to X at x.* We denote it by $T_{X,x}$.

Let $f : X \to Y$ be a morphism of schemes, let $x \in X$, and $y = f(x)$. Then $f_x^{\#} : \mathcal{O}_{Y,y} \to \mathcal{O}_{X,x}$ canonically induces a $k(x)$-linear map $T_{X,x} \to T_{Y,y} \otimes_{k(y)} k(x)$ which we will denote by $T_{f,x}$, and call the *tangent map of f at x.*

Proposition 2.2. *Let X be a scheme. The following properties hold.*

(a) *If X is locally Noetherian, then for any $x \in X$, $\dim_{k(x)} T_{X,x} \geq \dim \mathcal{O}_{X,x}$.*

(b) *Let $g : Y \to Z$ be a morphism of schemes. Then $T_{g \circ f,x} = (T_{g,y} \otimes \mathrm{Id}_{k(x)}) \circ T_{f,x}$.*

Proof (a) is a direct translation of Corollary 2.5.14(b), and (b) results directly from the definition.   □

Lemma 2.3. *Let A be a ring, and $\mathfrak{m}$ a maximal ideal of A. Then the canonical homomorphism of A-modules $\mathfrak{m}/\mathfrak{m}^2 \to \mathfrak{m}A_{\mathfrak{m}}/\mathfrak{m}^2 A_{\mathfrak{m}}$ is an isomorphism.*

Proof It suffices to show that $\varphi : A/\mathfrak{m} \to A\mathfrak{m}/\mathfrak{m}A\mathfrak{m}$ is an isomorphism because we will then have the isomorphism of the lemma by taking the tensor product $\otimes_A \mathfrak{m}$. As $A/\mathfrak{m}$ is a field, it suffices to show that φ is surjective. Let $s \in A \setminus \mathfrak{m}$. There exist $a \in A$ and $m \in \mathfrak{m}$ such that $1 = as + m$. It follows that $s^{-1} = a$ modulo $\mathfrak{m}A\mathfrak{m}$. This immediately implies that φ is surjective. $\square$

In what follows, we are going to study an explicit case, where the notion of the Zariski tangent space takes the classical sense, namely, that of differential calculus (Proposition 2.5). Let $Y = \operatorname{Spec} k[T_1, \ldots, T_n]$ be the affine space over a field k, and let $y \in Y(k)$ be a rational point. Let E denote the vector space k^n. Let $D_y : k[T_1, \ldots, T_n] \to E^\vee$ (the dual of E) be the map defined by

$$D_y P : (t_1, \ldots, t_n) \mapsto \sum_{1 \le i \le n} \frac{\partial P}{\partial T_i}(y) t_i.$$

The linear form $D_y P$ is called the *differential of P in y*. It is clear that D_y is a k-linear map and that $D_y(PQ) = P(y)D_y Q + Q(y)D_y P$.

Lemma 2.4. *Let $\mathfrak{m}$ denote the maximal ideal of $k[T_1, \ldots, T_n]$ corresponding to y.*

(a) *The restriction of D_y to $\mathfrak{m}$ induces an isomorphism $\mathfrak{m}/\mathfrak{m}^2 \simeq E^\vee$.*

(b) *We have a canonical isomorphism $T_{Y,y} \simeq E$.*

Proof (a) We know that $\mathfrak{m}$ is of the form $(T_1 - \lambda_1, \ldots, T_n - \lambda_n)$, $\lambda_i \in k$ (Example 2.3.32). The assertion then follows immediately using the Taylor expansion of P in $(\lambda_1, \ldots, \lambda_n)$. Property (b) results from (a) and Lemma 2.3. $\square$

Let F be a subvector space of E. We denote the orthogonal of F in E by

$$F^\perp = \{\varphi \in E^\vee \mid \varphi(v) = 0, \forall v \in F\}.$$

We have a canonical exact sequence

$$0 \to F^\perp \to E^\vee \to F^\vee \to 0.$$

Proposition 2.5. *Let $X = V(I)$ be a closed subvariety of $Y = \mathbb{A}_k^n$, and let $x \in X(k)$ be a rational point. Let $f : X \to Y$ denote the closed immersion. Then, by identifying $T_{Y,x}$ with E as in Lemma 2.4, the linear map $T_{f,x} : T_{X,x} \to T_{Y,x}$ induces an isomorphism from $T_{X,x}$ onto $(D_x I)^\perp \subseteq E$. Consequently, $T_{X,x}$ can be identified with*

$$\left\{ (t_1, \ldots, t_n) \in E \ \middle| \ \sum_{1 \le i \le n} \frac{\partial P}{\partial T_i}(x) t_i = 0, \ \forall P \in I \right\}.$$

It is therefore the tangent space to X at x in the classical sense.

Proof We have an exact sequence of vector spaces over $k(x) = k$:

$$0 \to I/(I \cap \mathfrak{m}^2) \to \mathfrak{m}/\mathfrak{m}^2 \to \mathfrak{n}/\mathfrak{n}^2 \to 0, \tag{2.1}$$

where $\mathfrak{m}$ is the maximal ideal of $k[T_1, \ldots, T_n]$ corresponding to $x \in Y$, and $\mathfrak{n} = \mathfrak{m}/I$ is the maximal ideal of $\mathcal{O}_X(X)$ corresponding to $x \in X$. By the isomorphism $D_x : \mathfrak{m}/\mathfrak{m}^2 \to E^\vee$ of Lemma 2.4(a), we obtain an exact sequence

$$0 \to D_x I \to E^\vee \to \mathfrak{n}/\mathfrak{n}^2 \to 0.$$

By taking the dual, we obtain the isomorphism $(\mathfrak{n}/\mathfrak{n}^2)^\vee \simeq (D_x I)^\perp$. The proposition then results from Lemma 2.3. □

Example 2.6. Let $X = \operatorname{Spec} k[T, S]/(T^2 - S^3)$, and let $x_0 \in X \subset \mathbb{A}_k^2$ be the rational point corresponding to the maximal ideal generated by T and S. The differential form $D_x(T^2 - S^3)$ is zero if and only if $x = x_0$. Consequently, $\dim_k T_{X,x} = 1$ for any $x \in X(k) \setminus \{x_0\}$, and $\dim_k T_{X,x_0} = 2$.

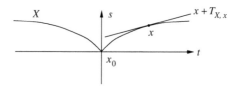

Figure 7. Tangent space to a curve X.

4.2.2   Regular schemes and the Jacobian criterion

Let $(A, \mathfrak{m})$ be a Noetherian local ring with residue field $k = A/\mathfrak{m}$. We have seen that $\dim_k \mathfrak{m}/\mathfrak{m}^2 \geq \dim A$ (Corollary 2.5.14(b)).

Definition 2.7. Let $(A, \mathfrak{m})$ be a Noetherian local ring. We say that A is *regular* if $\dim_k \mathfrak{m}/\mathfrak{m}^2 = \dim A$. By Nakayama's lemma, A is regular if and only if $\mathfrak{m}$ is generated by $\dim A$ elements.

Definition 2.8. Let X be a locally Noetherian scheme, and let $x \in X$ be a point. We say that X is *regular at* $x \in X$, or that x is a *regular point of X*, if $\mathcal{O}_{X,x}$ is regular, that is, $\dim \mathcal{O}_{X,x} = \dim_{k(x)} T_{X,x}$. We say that X is *regular* if it is regular at all of its points. A point $x \in X$ which is not regular is called a *singular point of X*. A scheme that is not regular is said to be *singular*.

Example 2.9. Any discrete valuation ring is regular. Consequently, any normal locally Noetherian scheme of dimension 1 is regular (Proposition 1.12). Conversely, let $(A, \mathfrak{m})$ be a Noetherian regular local ring of dimension 1; then by hypothesis $\mathfrak{m}$ is generated by one element, and A is a principal domain (see the second part of the proof of Proposition 1.12). Thus a Noetherian connected scheme of dimension 1 is a Dedekind scheme if and only if it is regular.

Example 2.10. In Example 2.6 above, the point x_0 is singular, while the other rational points of X are regular. We can see from Exercise 2.5 that x_0 is the unique singular point of X.

Proposition 2.11. Let $(A, \mathfrak{m})$ be a regular Noetherian local ring. Then A is an integral domain.

Proof We are going to use induction on $d := \dim A$. If $d = 0$, then $\mathfrak{m} = \mathfrak{m}^2$; hence $\mathfrak{m} = 0$. It follows that A is a field. Let $d \geq 1$. Let us suppose the proposition is true for $d - 1$. Let $k = A/\mathfrak{m}$ and let s be the closed point of $\operatorname{Spec} A$. Let us fix an $f \in \mathfrak{m} \setminus \mathfrak{m}^2$.

By Theorem 2.5.15 and Corollary 2.5.14, we have

$$d - 1 \leq \dim A/fA \leq \dim_k T_{V(f),s} = d - 1.$$

Therefore A/fA is regular of dimension $d - 1$. Let $\mathfrak{p}$ be a minimal prime ideal of A such that $\dim V(\mathfrak{p}) = d$. Let $B = A/\mathfrak{p}$, and let g be the image of f in B. Then

$$d - 1 \leq \dim B/gB \leq \dim_k T_{V(g),s} \leq \dim_k T_{V(f),s} = d - 1.$$

It follows that $A/(\mathfrak{p}+fA) = B/gB$ is regular of dimension $d-1$. By the induction hypothesis, A/fA and $A/(\mathfrak{p} + fA)$ are integral domains of the same dimension. Consequently, $\mathfrak{p} + fA = fA$, and hence $\mathfrak{p} \subseteq fA$. Therefore, for any $u \in \mathfrak{p}$, there exists a $v \in A$ such that $u = fv$. We have $f \notin \mathfrak{p}$ because $\dim V(f) < \dim V(\mathfrak{p})$, and hence $v \in \mathfrak{p}$. It follows that $\mathfrak{p} \subseteq f\mathfrak{p} \subseteq \mathfrak{m}\mathfrak{p}$. Therefore $\mathfrak{p} = 0$ and A is an integral domain. □

Corollary 2.12. Let $(A, \mathfrak{m})$ be a regular Noetherian local ring, and let $f \in \mathfrak{m} \setminus \{0\}$. Then A/fA is regular if and only if $f \notin \mathfrak{m}^2$.

Proof Indeed, since A is integral, A/fA is of dimension $\dim A-1$. Let $k = A/\mathfrak{m}$ and $\mathfrak{n} = \mathfrak{m}/fA$. Then A/fA is regular if and only if $\dim_k \mathfrak{n}/\mathfrak{n}^2 = \dim_k \mathfrak{m}/\mathfrak{m}^2 - 1$, which is equivalent to $f \notin \mathfrak{m}^2$. □

Example 2.13. Let $\mathfrak{m}$ be the maximal ideal $(p, T_1, \ldots, T_d)$ of $A := \mathbb{Z}[T_1, \ldots, T_d]$, where p is a prime number. Then $A_{\mathfrak{m}}$ is regular because it has dimension $d + 1$ (Lemma 2.5.16) and $\mathfrak{m}$ is generated by $d + 1$ elements. Let $F(T_1, \ldots, T_d) \in \mathfrak{m}$. Then $\operatorname{Spec}(A/FA)$ is regular at the point corresponding to $\mathfrak{m}/FA$ if and only if $F(T_1, \ldots, T_d) \notin (p, T_1, \ldots, T_d)^2$.

Definition 2.14. Let $(A, \mathfrak{m})$ be a regular Noetherian local ring of dimension d. Any system of generators of $\mathfrak{m}$ with d elements is called a *coordinate system* or *system of parameters* for A. If $d = 1$, a generator of $\mathfrak{m}$ is also called a *uniformizing parameter* for A.

Corollary 2.15. Let $(A, \mathfrak{m})$ be a regular Noetherian local ring. Let I be a proper ideal of A. Then A/I is regular if and only if I is generated by r elements of a coordinate system for A, with $r = \dim A - \dim A/I$.

Proof Let $\{f_1,\ldots,f_d\}$ be a coordinate system for A. Let $r \le d$. It can immediately be verified, by induction on r, that $A/(f_1,\ldots,f_r)$ is regular (hence an
integral domain) of dimension $d - r$.

Conversely, let A/I be a regular quotient of dimension $d - r$. By the exact
sequence (2.1) above, which holds in general, $\mathfrak{m}$ admits a coordinate system
$\{f_1,\ldots,f_d\}$ with $f_1,\ldots,f_r \subseteq I$. From the above, the two ideals I and $J :=
(f_1,\ldots,f_r)$ are prime, and $\dim A/J = \dim A/I = d - r$. Hence $I = J$ and I is
generated by a subset of a coordinate system. $\qquad\square$

In the following theorem, we gather together two results concerning regular
local rings. Their proof is difficult, and goes beyond the contents of this book.
Statement (a) is not really essential because we could restrict ourselves to regular
closed points, at least for algebraic varieties. Part (b), on the other hand, is
indispensable in several places in this book.

Theorem 2.16. *Let A be a regular Noetherian local ring. We have the following
properties:*

(a) *For any $\mathfrak{p} \in \operatorname{Spec} A$, $A_{\mathfrak{p}}$ is regular.*

(b) *The ring A is factorial (hence normal).*

Proof See [65], pages 139, 142. $\qquad\square$

Corollary 2.17. *Let X be a Noetherian scheme. Then X is regular if and only if
it is regular at its closed points. If this is the case, then any connected component
of X is normal.*

Proof By Theorem 2.16(a), the first assertion can be proven similarly to
Proposition 1.5. The second one comes from Theorem 2.16(b). $\qquad\square$

Example 2.18. Let $X = \mathbb{A}_k^n$ be the affine space of dimension n over a field k.
Then X is regular at its closed points. This results from Corollary 2.5.24 and
Exercise 2.1.5. Consequently, X is regular. It also follows from this that $\mathbb{P}_k^n$ is
regular, since it is a union of open subschemes isomorphic to $\mathbb{A}_k^n$.

Theorem 2.19 (Jacobian criterion). *Let k be a field. Let $X = V(I)$ be a closed
subvariety of $\mathbb{A}_k^n$, $F_1,\ldots,F_r$ a system of generators for I, and $x \in X(k)$ a rational
point. Let us consider the matrix*

$$J_x = \left(\frac{\partial F_i}{\partial T_j}(x)\right)_{1\le i\le r, 1\le j\le n}$$

in $M_{r\times n}(k)$. Then X is regular at x if and only if

$$\operatorname{rank} J_x = n - \dim \mathcal{O}_{X,x}.$$

Proof By Proposition 2.5, we have $\dim T_{X,x} = \dim(D_xI)^{\perp} = n - \dim D_xI$.
Now D_xI is generated by the line vectors of J_x, and hence $\dim D_xI = \operatorname{rank} J_x$,
whence the theorem. $\qquad\square$

Example 2.20. Let us consider the curve X of Example 2.6. Then for any $x \in X(k)$, we have $J_x = (2t, -3s^2)$, where $t = T(x), s = S(x) \in k$. Hence rank $J_x = 1 = 2 - \dim \mathcal{O}_{X,x}$ if and only if $x \neq x_0$. We therefore, once again, find the conclusions of Example 2.10.

Lemma 2.21. *Let X be a geometrically reduced algebraic variety over a field k. Then X contains a regular closed point.*

Proof We may assume $X = \operatorname{Spec} A$ is affine and integral. By Proposition 3.2.15, the field $K(X)$ is a finite separable extension of a purely transcendental extension $k(T_1, \ldots, T_d)$. Hence $K(X) = k(T_1, \ldots, T_d)[f]$. We may choose f such that $f \in A$ and the minimal polynomial $P(S)$ of f over $k(T_1, \ldots, T_d)$ has coefficients in $k[T_1, \ldots, T_d]$. Localizing A if necessary, we may assume that $k[T_1, \ldots, T_d][f] \subseteq A$. As A is finitely generated over k, there exists an $R(T) \in k[T_1, \ldots, T_d]$ such that $A \subseteq k[T_1, \ldots, T_d, R^{-1}][f]$. Hence, localizing A once again, if necessary, we may suppose that $A = k[T_1, \ldots, T_d, R^{-1}][f]$. Therefore $\operatorname{Spec} A$ is the principal open subset $D(R)$ of the variety $\operatorname{Spec} B$, where

$$B := k[T_1, \ldots, T_d, S]/(P(S)).$$

Let $\delta(T) \in k[T_1, \ldots, T_d]$ be the resultant of $P(S) \in k[T_1, \ldots, T_d][S]$. Then $\delta \neq 0$ because $K(X)$ is separable over $k[T_1, \ldots, T_d]$.

Let $\mathfrak{m}_x$ be a maximal ideal of $k[T_1, \ldots, T_d]$ not containing $R(T)\delta(T)$, and let $\mathfrak{m}_y$ be a maximal ideal of $k[T_1, \ldots, T_d, S]$ containing $P(S)$ and $\mathfrak{m}_x$. Then $\mathfrak{m}_y$ corresponds to a point y of $\operatorname{Spec} A \subset \operatorname{Spec} B = V(P) \subset \mathbb{A}_k^{d+1}$. Let us show that $\operatorname{Spec} B$ is regular at y. Let s be the image of S in $k(y)$. Then $k(y) = k(x)[s]$. There therefore exists a polynomial $Q(S) \in k[T_1, \ldots, T_d][S]$ such that $\mathfrak{m}_y = (\mathfrak{m}_x, Q(S))$. As $P'(s) \neq 0$, it is easy to verify that $P(S) \notin \mathfrak{m}_y^2 k[T_1, \ldots, T_d, S]_{\mathfrak{m}_y}$. It follows from Corollary 2.12 that $\operatorname{Spec} B$ (and therefore $\operatorname{Spec} A$) is regular at y. $\square$

Lemma 2.22. *Let $(A, \mathfrak{m})$ be a Noetherian local ring, and let $\mathfrak{p}$ be a prime ideal of A. Let us suppose that $A_{\mathfrak{p}}$ and $A/\mathfrak{p}$ are regular, and that $\mathfrak{p}$ is generated by e elements, where $e = \dim A_{\mathfrak{p}}$. Then A is regular.*

Proof Let us write $d = \dim A$. Then $\dim A/\mathfrak{p} \geq d - e$ (Theorem 2.5.15). As $\operatorname{ht}(\mathfrak{p}) + \dim A/\mathfrak{p} \leq \dim A$ and $\operatorname{ht}(\mathfrak{p}) = \dim A_{\mathfrak{p}} = e$, we have $\dim A/\mathfrak{p} = d - e$. Hence $\mathfrak{m}/\mathfrak{p}$ is generated by $d - e$ elements, and it follows that $\mathfrak{m}$ itself is generated by d elements. Consequently, A is regular. $\square$

Definition 2.23. Let X be a locally Noetherian scheme. We denote the set of regular points of X by $\operatorname{Reg}(X)$ and denote the set of singular points of a scheme X by $\operatorname{Sing}(X)$ or X_{sing}.

Proposition 2.24. *Let X be an algebraic variety over an algebraically closed field k. Then $\operatorname{Reg}(X)$ is an open subset of X. Moreover, if X is normal, then $\operatorname{codim}(X_{\operatorname{sing}}, X) \geq 2$.*

Proof As X is integral at the regular points, and the set of integral points is open in X (Exercise 2.4.9), we may reduce to X integral. As the question is local on X, we may suppose $X = \operatorname{Spec} A$ affine. Let $A = k[T_1, \ldots, T_n]/I$. Let $F_1, \ldots, F_r$ be a system of generators for I. Let us consider the matrix

$$M = \left(\frac{\partial F_i}{\partial T_j} \right)_{1 \le i \le r, 1 \le j \le n}$$

with coefficients in $k[T_1, \ldots, T_n]$. Let J be the ideal of A generated by the images of the minors of M of order $n - \dim X$. By Theorem 2.19, the singular closed points of X correspond to maximal ideals of A containing J. Let $\mathfrak{p} \in \operatorname{Spec} A$ correspond to a singular point of X. Then, by Theorem 2.16(a), any maximal ideal $\mathfrak{m}$ containing $\mathfrak{p}$ contains J. It follows that $J \subseteq \mathfrak{p}$. Consequently, $X_{\text{sing}} = X \setminus \operatorname{Reg}(X) \subseteq V(J)$. It remains to show that $\operatorname{Reg}(X) \subseteq X \setminus V(J)$.

Let $x \in X$ be a regular point. Let us consider the integral closed subvariety $Z = \overline{\{x\}}$ of X. Let $e = \dim \mathcal{O}_{X,x}$. Then $\mathfrak{m}_x \mathcal{O}_{X,x}$ is generated by e elements $f_1, \ldots, f_e$. We have $\mathfrak{m}_x \mathcal{O}_{X,y} = (f_1, \ldots, f_e) \mathcal{O}_{X,y}$ for every point y contained in a non-empty open subset of Z. Moreover, by Lemma 2.21, Z contains a regular closed point y contained in this open subset. It follows from the preceding lemma that X is regular at y. Now y is closed in X, and hence rational over k, whence $y \notin V(J)$ (Jacobian criterion). Consequently, $x \notin V(J)$.

Finally, let us suppose X is normal. For any point $\xi \in X$ of codimension 1 (i.e., $\operatorname{codim}(\overline{\{\xi\}}, X) = 1$), $\mathcal{O}_{X,\xi}$ is a normal local ring of dimension 1, and therefore a discrete valuation ring (Proposition 1.12). It follows that X is regular at every point of codimension 1. Therefore $\operatorname{codim}(X_{\text{sing}}, X) \ge 2$. □

Remark 2.25. One can show that Proposition 2.24 is true for an arbitrary field k. See Corollary 8.2.40(a).

We conclude with some information on the $\mathfrak{m}$-adic completion of regular Noetherian local rings.

Lemma 2.26. *Let $(A, \mathfrak{m})$ be a Noetherian local ring, and let $\hat{A}$ be the $\mathfrak{m}$-adic completion of A. Then $\dim \hat{A} = \dim A$. Moreover, A is regular if and only if $\hat{A}$ is regular.*

Proof Applying Theorem 3.12 to the flat morphism $\operatorname{Spec} \hat{A} \to \operatorname{Spec} A$ (Theorem 1.3.15) and to the closed fiber $\operatorname{Spec}(\hat{A} \otimes_A A/\mathfrak{m})$, which is of dimension 0 (Theorem 1.3.16), we see that $\dim \hat{A} = \dim A$. Since $\mathfrak{m}\hat{A}$ is the maximal ideal of $\hat{A}$ (Theorem 1.3.16) and $\mathfrak{m}/\mathfrak{m}^2 \simeq \mathfrak{m}\hat{A}/\mathfrak{m}^2\hat{A}$, we obtain the second assertion. □

Proposition 2.27. *Let X be an algebraic variety over k, and let $x \in X(k)$ be a regular point of X. Let $\mathfrak{m}_x$ denote the maximal ideal of $\mathcal{O}_{X,x}$, and $\hat{\mathcal{O}}_{X,x}$ the $\mathfrak{m}_x$-adic completion of $\mathcal{O}_{X,x}$. Then we have an isomorphism of k-algebras*

$$\hat{\mathcal{O}}_{X,x} \simeq k[[T_1, \ldots, T_d]],$$

with $d = \dim \mathcal{O}_{X,x}$.

Proof Let $A = \mathcal{O}_{X,x}$ and $\mathfrak{m} = \mathfrak{m}_x$. Let us fix a coordinate system $\{f_1, \ldots, f_d\}$ for A. Let $\varphi : k[T_1, \ldots, T_d] \to A$ be the k-algebra homomorphism defined by $\varphi(T_i) = f_i$. It is clear that for any $n \geq 1$, we have

$$A = \varphi(k[T_1, \ldots, T_d]) + \mathfrak{m}^n, \quad \mathfrak{m}^n = \varphi((T_1, \ldots, T_d)^n) + \mathfrak{m}^{n+1}.$$

Consequently, $k[T_1, \ldots, T_d]/(T_1, \ldots, T_d)^n \to A/\mathfrak{m}^n$ is surjective and the homomorphism

$$\hat{\varphi} : k[[T_1, \ldots, T_d]] = \varprojlim_n k[T_1, \ldots, T_d]/(T_1, \ldots, T_d)^n \to \hat{A}$$

induced by φ is surjective (Lemma 1.3.3). Now $k[[T_1, \ldots, T_d]]$ is a local, Noetherian (Proposition 1.3.7) integral domain of the same dimension as A (Lemma 2.26). It follows that $\hat{\varphi}$ is an isomorphism. $\qquad\square$

Remark 2.28. More generally, let $(A, \mathfrak{m})$ be a regular Noetherian local ring, containing a field k_0 (e.g., the local ring of an algebraic variety at a closed point), and let $\hat{A}$ be the $\mathfrak{m}$-adic completion of A. Then we have an isomorphism

$$\hat{A} \simeq k[[T_1, \ldots, T_d]],$$

where $k = A/\mathfrak{m}$ and $d = \dim A$ (Cohen structure theorem, see [65], page 206, Corollary 2). This means that, in a way, A is 'close' to the local ring $k[T_1, \ldots, T_d]_{(T_1, \ldots, T_d)}$ in the sense that they have the same formal completion. Note, however, that the above isomorphism is not an isomorphism of k_0-algebras in general. If A does not contain a field, the structure of $\hat{A}$ is more complex. See [17], VIII, Théorème 2.

Exercises

2.1. Let $f : X \to Y$ be a morphism of locally Noetherian schemes. Let $x \in X$, $y = f(x)$. Show that $T_{f,x}$ is injective if and only if the maximal ideal $\mathfrak{m}_x$ of $\mathcal{O}_{X,x}$ is generated by $f_x^\#(\mathfrak{m}_y)$. Show that if f is moreover of finite type, then x is discrete in X_y.

2.2. Let k be a field. Show that $\operatorname{Spec} k[x, y, z]/(x^2 - yz)$ is normal but singular.

2.3. Let $\mathcal{O}_K$ be a discrete valuation ring with uniformizing parameter t and field of fractions K. Let $n \geq 1$. We suppose that $\operatorname{char}(K) \neq 2, 3$.

(a) Show that $K[x, y]/(y^2 + x^3 + t^n)$ is regular.

(b) Let $A = \mathcal{O}_K[x, y]/(y^2 + x^3 + t^n)$. Show that the special fiber of $\operatorname{Spec} A \to \operatorname{Spec} \mathcal{O}_K$ is reduced. Deduce from this that A is normal (see Lemma 1.18).

(c) Let $\mathfrak{m}$ be the maximal ideal of A generated by x, y, and t. Show that $A_\mathfrak{m}$ is regular if and only if $n = 1$.

2.4. Let k be a field and let $\mathfrak{m}$ be a maximal ideal of $k[T_1,\ldots,T_n]$ corresponding to a closed point x of $\mathbb{A}_k^n$. Let us consider the $k(x)$-linear map $\varphi_x : \mathfrak{m}/\mathfrak{m}^2 \to E_x^\vee$ defined as before Lemma 2.4, with $E_x = k(x)^n$.

    (a) Let $P_1(T_1), P_2(T_1,T_2),\ldots, P_n(T_1,\ldots,T_n)$ be a system of generators of $\mathfrak{m}$ (see Exercise 2.1.5). Show that φ_x is an isomorphism if and only if $(\partial P_i/\partial T_i)(x) \neq 0$ for all $i \leq n$.

    (b) Deduce from (a) that φ_x is an isomorphism if and only if $k(x)$ is separable over k.

2.5. Show that the Jacobian criterion (with J_x as a matrix with coefficients in $k(x)$) remains valid at the closed points $x \in X$ for which $k(x)$ is separable over k. See also Proposition 3.30. Give an example with X regular at x and $\dim \mathcal{O}_{X,x} \neq n - \operatorname{rank} J_x$.

2.6. Let k be a field. Let $F_1,\ldots,F_n \in k[T_1,\ldots,T_n]$. Let us consider the homomorphism $\varphi : k[T_1,\ldots,T_n] \to k[T_1,\ldots,T_n]$ defined by $\varphi(T_i) = F_i(T)$ and let us suppose that the associated morphism $\mathbb{A}_k^n \to \mathbb{A}_k^n$ is an isomorphism. Show that

$$\det\left(\frac{\partial F_i}{\partial T_j}\right)_{1\leq i,j\leq n}$$

belongs to k^*. The converse assertion when $\operatorname{char}(k) = 0$ is called the Jacobian conjecture. It is an open problem.

2.7. Let X be a scheme over a field k, and let $x \in X(k)$. Show that there exists a canonical bijection between $T_{X,x}$ and the set of k-morphisms $\operatorname{Spec}(k[\epsilon]/\epsilon^2) \to X$ with image equal to x.

2.8. Let us once again consider the scheme $\operatorname{Proj} B$ of Exercise 2.3.13. Show that the fibers of $\operatorname{Proj} B \to \operatorname{Spec} A$ are all singular.

2.9. Let $X = V_+(I)$ be a closed subvariety of $\mathbb{P}_k^n$ over a field k. Let $x \in X(k)$, and let us consider the linear subvariety of $\mathbb{P}(k^{n+1})$

$$PT_{X,x} = \left\{(t_0,\ldots,t_n) \in \mathbb{P}(k^{n+1}) \;\middle|\; \sum_{0\leq i\leq n} \frac{\partial F}{\partial T_i}(x)t_i = 0,\ \forall F(T_0,\ldots,T_n) \in I\right\}.$$

We will abusively call this linear subvariety the *tangent space to X at x*. Show that $\dim_k T_{X,x} = \dim PT_{X,x}$.

2.10. Let X be a closed subvariety of $\mathbb{P}_k^n = \operatorname{Proj} k[T_0,\ldots,T_n]$ defined by some homogeneous polynomials $F_1,\ldots,F_r$. Let $x \in X(k)$. Let J_x denote the matrix

$$J_x = \left(\frac{\partial F_i}{\partial T_j}(x)\right)_{1\leq i\leq r,\ 0\leq j\leq n}.$$

Show that x_0 is regular in X if and only if $\operatorname{rank} J_x = n - \dim \mathcal{O}_{X,x}$. This is the Jacobian criterion for projective varieties.

2.11. Let k be an algebraically closed field, and let $n \geq 1$. Show that the curve $\text{Proj } k[x, y, z]/(x^n + y^n + z^n)$ is regular if and only if n is prime to $\text{char}(k)$.

2.12. Let X, Y be algebraic varieties over an algebraically closed field k. If X, Y are regular, show that $X \times_k Y$ is also regular. Show that this is false in general if k is not algebraically closed.

2.13. Let $(A, \mathfrak{m})$ be a Noetherian local ring with residue field k. We set

$$\text{gr}_{\mathfrak{m}}(A) = \oplus_{n \geq 0} \mathfrak{m}^n / \mathfrak{m}^{n+1}.$$

This is the *graded ring associated to the ideal* $\mathfrak{m}$. It naturally has the structure of a k-algebra. Let us suppose that A is regular of dimension d. We are going to show that $\text{gr}_{\mathfrak{m}}(A)$ is a polynomial ring in d variables over k. To simplify, we suppose that k is *infinite*.

(a) Let $f_1, \ldots, f_d$ be a coordinate system for $\mathfrak{m}$. Let $\phi : k[T_1, \ldots, T_d] \to \text{gr}_{\mathfrak{m}}(A)$ be the k-algebra homomorphism defined by $\phi(T_i) = $ the image of f_i in $\mathfrak{m}/\mathfrak{m}^2$. Show that ϕ is surjective, and that ϕ is an isomorphism if $d = 0$.

(b) Let $\tilde{\mathfrak{m}}$ be the maximal ideal of $\text{gr}_{\mathfrak{m}}(A)$ generated by the $\phi(T_i)$. Show that if $\dim \text{gr}_{\mathfrak{m}}(A)_{\tilde{\mathfrak{m}}} = 0$, then $\dim A = 0$.

(c) Let us suppose $d \geq 1$. Show that $\sum_i \phi(T_i) k$ is not contained in the union of the minimal prime ideals of $\text{gr}_{\mathfrak{m}}(A)$ (here we use the k infinite hypothesis).

(d) Let $f \in \mathfrak{m}$. Show that we have a canonical surjective homomorphism

$$\text{gr}_{\mathfrak{m}}(A) \to \text{gr}_{\mathfrak{m}/fA}(A/fA).$$

Show by induction on d that $\dim \text{gr}_{\mathfrak{m}}(A) = d$ and that ϕ is an isomorphism.

Remark. This result is true without hypothesis on k. See [65], 17.E, Theorem 35.

4.3 Flat morphisms and smooth morphisms

When we study families of varieties, the flatness is a basic property required in most of the important results. From a naive point of view, it guarantees some 'continuity' of objects attached to the fibers. This fact is illustrated for instance by Corollary 3.14. Flatness is treated in the first subsection. In the next subsection, we look at the notion of étale morphisms. This powerful tool plays, in some sense, the role of analytic neighborhood of a point. The last subsection deals with flat families of 'geometrically regular' varieties (smooth morphisms). We will return to étale morphisms and smooth morphisms in Section 6.2, equipped with the machinery of differential forms.

4.3.1   Flat morphisms

Definition 3.1. Let $f : X \to Y$ be a morphism of schemes. We say that f is *flat at the point* $x \in X$ if the homomorphism $f_x^\# : \mathcal{O}_{Y,f(x)} \to \mathcal{O}_{X,x}$ is flat. We say that f is *flat* if it is flat at every point of X. For such a morphism, the family of fibers X_y, $y \in Y$, is in some sense a 'continuous family'. We will consider examples which justify this assertion.

Remark 3.2. Let $y = f(x)$. Then saying that $X \to Y$ is flat at x comes down to saying that $g : X \times_Y \operatorname{Spec} \mathcal{O}_{Y,y} \to \operatorname{Spec} \mathcal{O}_{Y,y}$ is flat at x', where x' is the inverse image of x under the projection $X \times_Y \operatorname{Spec} \mathcal{O}_{Y,y} \to X$. Indeed, we have $f_x^\# = g_{x'}^\#$.

Proposition 3.3. *The following properties are true.*

    (a) *Open immersions are flat morphisms.*

    (b) *Flat morphisms are stable under base change.*

    (c) *The composition of flat morphisms is flat.*

    (d) *The fibered product of two flat morphisms (Proposition 3.1.4(c)) is flat.*

    (e) *Let $A \to B$ be a ring homomorphism. Then $\operatorname{Spec} B \to \operatorname{Spec} A$ is flat if and only if $A \to B$ is flat.*

Proof   This results from the general properties of flat homomorphisms (Proposition 1.2.2 and Corollary 1.2.15).   □

Example 3.4. The structural morphism $X \to \operatorname{Spec} k$ of an algebraic variety over k is flat. Indeed, any algebra over a field is flat over the field.

Example 3.5. Let X be an integral scheme, and let $\pi : X' \to X$ be the normalization morphism. Then π is flat if and only if it is an isomorphism (Exercise 1.2.10).

Example 3.6. Let X be a Noetherian scheme. Then a closed immersion into X is not flat, unless it is also open (Exercise 1.2.4).

Lemma 3.7. *Let $f : X \to Y$ be a flat morphism with Y irreducible. Then every non-empty open subset U of X dominates Y (i.e., $f(U)$ is dense in Y). If X has only a finite number of irreducible components, then every one of them dominates Y.*

Proof   We may suppose $Y = \operatorname{Spec} A$ and $U = \operatorname{Spec} B$ are affine. It follows from Proposition 3.3 that U is flat over Y. Let η be the generic point of Y, and N the nilradical of A. It follows from the flatness of B that

$$B/NB = B \otimes_A (A/N) \subseteq B \otimes_A \operatorname{Frac}(A/N) = B \otimes_A k(\eta) = \mathcal{O}(U_\eta).$$

If $U_\eta = \emptyset$, then $B = NB$ is nilpotent. Hence $U = \emptyset$, a contradiction. If X has only a finite number of irreducible components, then every one of them contains a non-empty open subset. They therefore dominate Y by what we have just proven.   □

Proposition 3.8. *Let Y be a scheme having only a finite number of irreducible components. Let $f : X \to Y$ be a flat morphism. Let us suppose that Y is reduced (resp. irreducible; resp. integral), and that the generic fibers of f are also reduced (resp. irreducible; resp. integral); then X is reduced (resp. irreducible; resp. integral).*

Proof Let us suppose Y is irreducible with generic point η, and X_η is irreducible. Then $Z := \overline{X_\eta}$ is a closed irreducible subset of X. Its complement $X \setminus Z$ is an open subset of X which does not dominate Y. It follows from Lemma 3.7 that $X \setminus Z = \emptyset$, and hence $X = Z$ is irreducible.

Let us suppose Y is reduced, and let us show that, under the hypotheses of the proposition, X is reduced. We may suppose $Y = \operatorname{Spec} A$, $X = \operatorname{Spec} B$ are affine. Let $\mathfrak{p}_1, \ldots, \mathfrak{p}_n$ be the minimal prime ideals of A. Then the canonical homomorphism $A \to \oplus_{1 \le i \le n} A/\mathfrak{p}_i$ is injective. Let η_i be the generic point of Y corresponding to $\mathfrak{p}_i$. By the flatness of B, there is an injection

$$B \to \oplus_i (B \otimes_A (A/\mathfrak{p}_i)) \subseteq \oplus_i (B \otimes_A k(\eta_i)) = \oplus_i \mathcal{O}(X_{\eta_i}).$$

Now, by hypothesis, X_{η_i} is reduced. It follows that B (and hence X) is reduced.

The integrality property can be shown in a similar way, or by using the fact that integral is equivalent to irreducible and reduced. $\qquad\square$

Proposition 3.9. *Let Y be a Dedekind scheme. Let $f : X \to Y$ be a morphism with X reduced. Then f is flat if and only if every irreducible component of X dominates Y.*

Proof Let us suppose that every irreducible component of X dominates Y. Let $x \in X$ and $y = f(x)$. If y is the generic point of Y, then $\mathcal{O}_{X,x}$ is flat over $\mathcal{O}_{Y,y} = K(Y)$, which is a field. Let us suppose y is closed, and π be a uniformizing parameter for $\mathcal{O}_{Y,y}$. Then π does not belong to any minimal prime ideal of $\mathcal{O}_{X,x}$. As X is reduced, this implies that π is not a zero divisor in $\mathcal{O}_{X,x}$ (Exercise 2.1.4), whence the flatness of $\mathcal{O}_{X,x}$ over $\mathcal{O}_{Y,y}$. The converse follows from Lemma 3.7. $\quad\square$

Corollary 3.10. *Let Y be a Dedekind scheme, and $f : X \to Y$ be a non-constant morphism (i.e., $f(X)$ is not reduced to a point) with X integral. Then f is flat.*

Proof Since f is not constant, $f(X)$ is dense in Y because the latter is irreducible and of dimension 1. Hence f is flat. $\qquad\square$

Remark 3.11. Let $f : X \to Y$ be a finite surjective morphism. Let us suppose X and Y are regular. Then f is flat (Example 6.3.18 and Exercise 6.3.5). See also [41], IV.15.4.2 for a more general statement.

Theorem 3.12. *Let $f : X \to Y$ be a morphism of locally Noetherian schemes. Let $x \in X$ and $y = f(x)$. Then*

$$\dim \mathcal{O}_{X_y,x} \ge \dim \mathcal{O}_{X,x} - \dim \mathcal{O}_{Y,y}.$$

If, moreover, f is flat, then we have equality.

Proof By the base change $\operatorname{Spec} \mathcal{O}_{Y,y} \to Y$, which preserves the local rings in the inequality, we may reduce to $Y = \operatorname{Spec} A$, where A is a Noetherian local ring, and y is the closed point of Y.

We are going to show the theorem by induction on $\dim Y$. If $\dim Y = 0$, then $X_{\mathrm{red}} = (X_y)_{\mathrm{red}}$, whence the equality. Let us suppose $\dim Y \geq 1$. We may replace Y by Y_{red}, and X by $X \times_Y Y_{\mathrm{red}}$. This does not change their dimensions in y and in x, respectively, nor that of X_y in x. Moreover, if f is flat, then so is $X \times_Y Y_{\mathrm{red}} \to Y_{\mathrm{red}}$. Let us therefore suppose Y is reduced. Let $t \in A$ which is neither a zero divisor nor invertible. Let us also denote the image of t in $B := \mathcal{O}_{X,x}$ by t. Then

$$\dim(A/tA) = \dim A - 1, \quad \dim(B/tB) \geq \dim B - 1$$

(Theorem 2.5.15). Moreover, if f is flat, then B is flat over A. It follows that t is not a zero divisor in B (consider the injective homomorphism $A \to A$ given by multiplication by t, and tensor by B), and hence that $\dim(B/tB) = \dim B - 1$.

Let us set $Y' = \operatorname{Spec} A/tA$ and $X' = X \times_Y Y'$. Then, by the induction hypothesis, we have

$$\dim \mathcal{O}_{X'_y,x} \geq \dim \mathcal{O}_{X',x} - \dim \mathcal{O}_{Y',y}$$

(with equality if f is flat, because $X' \to Y'$ is then flat). Now $X'_y = X_y$, whence

$$\dim \mathcal{O}_{X_y,x} \geq (\dim \mathcal{O}_{X,x} - 1) - (\dim \mathcal{O}_{Y,y} - 1) = \dim \mathcal{O}_{X,x} - \dim \mathcal{O}_{Y,y},$$

where the first inequality is an equality if f is flat. $\square$

Definition 3.13. We say that a scheme X is *equidimensional* or *pure* if all of its irreducible components have the same dimension.

Corollary 3.14. *Let $f : X \to Y$ be a flat surjective morphism of algebraic varieties. We suppose that Y is irreducible and that X is equidimensional. Then for every $y \in Y$, the fiber X_y is equidimensional, and we have*

$$\dim X_y = \dim X - \dim Y.$$

Proof Let x be a closed point of X_y. For every irreducible component X_i of X passing through x, we have

$$\dim \mathcal{O}_{X_i,x} = \dim X_i - \dim \overline{\{x\}} = \dim X - \dim \overline{\{x\}}$$

(Proposition 2.5.23(a)), which implies that

$$\dim \mathcal{O}_{X,x} = \dim X - \dim \overline{\{x\}}. \tag{3.2}$$

As x is the generic point of the algebraic variety $\overline{\{x\}}$ over k, we have $\dim \overline{\{x\}} = \operatorname{trdeg}_k k(x)$ (Proposition 2.5.19). In addition, since the point x is closed in X_y, $k(x)$ is algebraic over $k(y)$, and we therefore have

$$\dim \overline{\{x\}} = \operatorname{trdeg}_k k(x) = \operatorname{trdeg}_k k(y) = \dim \overline{\{y\}} = \dim Y - \dim \mathcal{O}_{Y,y}. \tag{3.3}$$

Combining equalities (3.2) and (3.3), we obtain

$$\dim \mathcal{O}_{X,x} - \dim \mathcal{O}_{Y,y} = \dim X - \dim Y.$$

The corollary then results from Theorem 3.12. $\square$

Remark 3.15. The corollary above says, in particular, that for a flat morphism of irreducible algebraic varieties, the dimension of the fibers is constant. See also Proposition 4.16, and Corollary 8.2.8. We will encounter other numerical invariants which are constant in the fibers (Proposition 5.3.28).

The following lemma is useful for constructing flat algebras.

Lemma 3.16. *Let $A \to B$ be a flat homomorphism of Noetherian local rings, and let $b \in B$ be such that its image in $B/\mathfrak{m}_A B$ is not a zero divisor. Then B/bB is flat over A.*

Proof This is a special case of [17], III, §5, n° 2, Théorème 1 (we apply the equivalence (i) $\Longleftrightarrow$ (iii) of this theorem with $M = B/bB$ and $I = \mathfrak{m}_A$). See also [66], Proposition 1.2.5, for a direct proof of the lemma. □

4.3.2 Étale morphisms

Definition 3.17. Let $f : X \to Y$ be a morphism of finite type of locally Noetherian schemes. Let $x \in X$ and $y = f(x)$. We say that f is *unramified* at x if the homomorphism $\mathcal{O}_{Y,y} \to \mathcal{O}_{X,x}$ verifies $\mathfrak{m}_y \mathcal{O}_{X,x} = \mathfrak{m}_x$ (in other words, $\mathcal{O}_{X,x}/\mathfrak{m}_y \mathcal{O}_{X,x} = k(x)$), and if the (finite) extension of residue fields $k(y) \to k(x)$ is separable. We say that f is *étale* at x if it is unramified and flat at x. We say that a homomorphism of Noetherian local rings $A \to B$ is *étale* if it is unramified, flat, and if B is a localization of a finitely generated A-algebra.

We say that f is *unramified* (resp. *étale*) if it is unramified (resp. étale) at every point of X.

Example 3.18. Let L/K be a finite field extension. Then $\operatorname{Spec} L \to \operatorname{Spec} K$ is unramified (and hence étale) if and only if the extension L/K is separable.

Example 3.19. Let L/K be an extension of number fields, and let $\mathcal{O}_L$, $\mathcal{O}_K$ be their respective rings of integers. Then $\operatorname{Spec} \mathcal{O}_L \to \operatorname{Spec} \mathcal{O}_K$ is flat (Proposition 3.9) and for any prime ideal $\mathfrak{q}$ of $\mathcal{O}_L$, setting $\mathfrak{p} = \mathfrak{q} \cap \mathcal{O}_K$, the extension $k(\mathfrak{q})$ of $k(\mathfrak{p})$ is separable. The morphism $\operatorname{Spec} \mathcal{O}_L \to \operatorname{Spec} \mathcal{O}_K$ is unramified (hence étale) at a prime ideal $\mathfrak{q}$ of $\mathcal{O}_L$ if and only if $\mathfrak{q}(\mathcal{O}_L)_\mathfrak{q}$ is generated by $\mathfrak{p} = \mathfrak{q} \cap \mathcal{O}_K$. It is therefore the usual definition of unramifiedness from number theory.

Lemma 3.20. *Let $f : X \to Y$ be a morphism of finite type between locally Noetherian schemes. Then f is unramified if and only if for every $y \in Y$, the fiber X_y is finite, reduced, and if $k(x)$ is separable over $k(y)$ for every $x \in X_y$.*

Proof Indeed, the quotient $\mathcal{O}_{X,x}/\mathfrak{m}_y \mathcal{O}_{X,x}$ remains unaltered when we pass from X to the fiber X_y. The condition is then clearly sufficient. Conversely, if f is unramified, X_y is of dimension 0 and reduced because $\mathfrak{m}_x \mathcal{O}_{X_y,x} = \mathfrak{m}_y \mathcal{O}_{X_y,x} = 0$ for every $x \in X_y$. The conclusion then follows immediately. □

Example 3.21. Let k be a field, $P(T) \in k[T]$ a monic polynomial, and $X = \operatorname{Spec} k[T]/(P)$. Then a point $x \in X$ corresponds to an irreducible factor $Q(T)$ of $P(T)$. The canonical morphism $X \to \operatorname{Spec} k$ is étale at x if and only if $Q(T)$ is a separable polynomial (i.e., without multiple root in the algebraic closure of k), and if $Q(T)$ is a simple factor of $P(T)$.

Proposition 3.22. *The following properties are true.*

  (a) *Any closed immersion of locally Noetherian schemes is an unramified morphism.*

  (b) *Any open immersion of locally Noetherian schemes is an étale morphism.*

  (c) *Unramified morphisms and étale morphisms are stable under base change, composition, and fibered products.*

Proof These properties are obvious for unramified morphisms. For étale morphisms, they are a consequence of Proposition 3.3. □

Let us recall that a morphism of finite type $X \to Y$ is called *quasi-finite* if X_y is a finite set for every $y \in Y$ (Exercise 3.2.8).

Proposition 3.23. *Let Y be a locally Noetherian scheme, and let $f : X \to Y$ be an étale morphism of finite type. Let $x \in X$ and $y = f(x)$. Then the following properties are true.*

  (a) *We have $\dim \mathcal{O}_{X,x} = \dim \mathcal{O}_{Y,y}$ and f is quasi-finite.*

  (b) *The tangent map $T_{X,x} \to T_{Y,y} \otimes_{k(y)} k(x)$ is an isomorphism.*

Proof Let $A = \mathcal{O}_{Y,y}$, $B = \mathcal{O}_{X,x}$, and let $\mathfrak{m}_y$, $\mathfrak{m}_x$ denote their respective maximal ideals. Then (a) results from Lemma 3.20, the flatness of $A \to B$, and Theorem 3.12. Let us show (b). We have

$$(\mathfrak{m}_y/\mathfrak{m}_y^2) \otimes_{k(y)} k(x) = (\mathfrak{m}_y \otimes_A k(y)) \otimes_{k(y)} k(x) = \mathfrak{m}_y \otimes_A (B/\mathfrak{m}_y B).$$

By flatness, we have $\mathfrak{m}_y \otimes_A B \simeq \mathfrak{m}_y B = \mathfrak{m}_x$. Therefore, the right-hand side is isomorphic to $(\mathfrak{m}_y \otimes_A B)/(\mathfrak{m}_y^2) = \mathfrak{m}_x/\mathfrak{m}_x^2$, which proves (b). □

Corollary 3.24. *Let Y be a locally Noetherian scheme, and let $f : X \to Y$ be a morphism of finite type, étale at $x \in X$. Then X is regular at x if and only if Y is regular at $f(x)$.*

Remark 3.25. The implication 'X regular at x' $\implies$ 'Y regular at y' is true assuming f is only flat at x instead of étale, see [65], 21.D, Theorem 51. A similar result is that if Y is normal and f is étale at x, then X is normal at x (Corollary 8.2.25).

Proposition 3.26. *Let Y be a locally Noetherian scheme, and let $f : X \to Y$ be a morphism of finite type. Let us a fix a point $x \in X$ such that $k(y) = k(x)$, where $y = f(x)$. Let $\widehat{\mathcal{O}}_{Y,y} \to \widehat{\mathcal{O}}_{X,x}$ be the morphism between the formal completions (for the topology defined by the respective maximal ideals) induced by $\mathcal{O}_{Y,y} \to \mathcal{O}_{X,x}$. Then f is étale at x if and only if $\widehat{\mathcal{O}}_{Y,y} \to \widehat{\mathcal{O}}_{X,x}$ is an isomorphism.*

Proof Let us keep the notation of the proof of the proposition above. Let us first suppose f is étale at x. As $A \to B$ is a flat homomorphism of local rings, it is faithfully flat. It follows that $A \to B$ is injective (Exercise 1.2.19(a)). We have $B = A + \mathfrak{m}_x$; hence $\mathfrak{m}_x = \mathfrak{m}_y B = \mathfrak{m}_y + \mathfrak{m}_x^2$. We deduce from this that $\mathfrak{m}_x = \mathfrak{m}_y + \mathfrak{m}_x^n$ for every $n \geq 1$. Hence $A/\mathfrak{m}_y^n \to B/\mathfrak{m}_x^n$ is surjective. Its kernel is $(\mathfrak{m}_x^n \cap A)/\mathfrak{m}_y^n = (\mathfrak{m}_y^n B \cap A)/\mathfrak{m}_y^n = 0$ (Exercise 1.2.6). Therefore $A/\mathfrak{m}_y^n \to B/\mathfrak{m}_x^n$ is an isomorphism for every $n \geq 1$, whence the proposition.

The converse results from Theorem 1.3.16 and Exercise 1.2.18.   □

Corollary 3.27. *Let us keep the hypotheses of Proposition 3.26. Let us suppose, moreover, that* $\dim \mathcal{O}_{X,x} = \dim \mathcal{O}_{Y,y}$, *and that* Y *is regular at* y. *If* $T_{X,x} \to T_{Y,y}$ *is injective, then* f *is étale at* x.

Proof The hypothesis on the tangent map implies that $\mathfrak{m}_y/\mathfrak{m}_y^2 \to \mathfrak{m}_x/\mathfrak{m}_x^2$ is surjective. Hence $\mathcal{O}_{X,x} = \mathcal{O}_{Y,y} + \mathfrak{m}_x^n$ and $\mathfrak{m}_x^n = \mathfrak{m}_y^n + \mathfrak{m}_x^{n+1}$ for every $n \geq 1$. It follows from Lemma 1.3.3 that $\widehat{\mathcal{O}}_{Y,y} \to \widehat{\mathcal{O}}_{X,x}$ is surjective. As these two rings are of the same dimension and $\widehat{\mathcal{O}}_{Y,y}$ is an integral domain (Lemma 2.26 and Proposition 2.11), $\widehat{\mathcal{O}}_{Y,y} \to \widehat{\mathcal{O}}_{X,x}$ is necessarily an isomorphism. Hence f is étale at x.   □

4.3.3   Smooth morphisms

Definition 3.28. Let X be an algebraic variety over a field k. Let $\overline{k}$ be the algebraic closure of k. We say that X is *smooth at* $x \in X$ if the points of $X_{\overline{k}}$ lying above x are regular points of $X_{\overline{k}}$. We say that X is *smooth over* k if it is smooth at all of its points (i.e., $X_{\overline{k}}$ is regular). For example, the varieties $\mathbb{A}_k^n$ and $\mathbb{P}_k^n$ are smooth over k.

Note that the smoothness of X can be verified by applying the Jacobian criterion (Theorem 2.19) and Corollary 2.17 to $X_{\overline{k}}$ (see Exercise 3.20).

Example 3.29. Let k be a field of characteristic char $k \neq 2$, and let $P(x) \in k[x]$ be a polynomial of degree $d \geq 1$. Let us consider $C = \operatorname{Spec} k[x,y]/(y^2 + P(x))$. Then C is smooth if and only if $C_{\overline{k}}$ is regular. As C is pure of dimension 1, the Jacobian criterion shows that this is equivalent to $P(x)$ being without multiple root, or to its discriminant $\operatorname{disc}(P(x))$ being non-zero.

Proposition 3.30. *Let* X *be an algebraic variety over* k, *and let* $x \in X$ *be a closed point. Let us suppose that* X *is smooth at* x; *then* X *is regular at* x. *Moreover, the converse is true if* $k(x)$ *is separable over* k.

Proof We may suppose X is affine. Let $x' \in X_{\overline{k}}$ be a point lying above x. The morphism $X_{\overline{k}} \to X$ has dimension 0 fibers. It results from Theorem 3.12 that $\dim \mathcal{O}_{X_{\overline{k}},x'} = \dim \mathcal{O}_{X,x}$. Let us write $X = V(I) \subseteq \mathbb{A}_k^n$, with $I = (F_1, \ldots, F_r)$, and let us consider the Jacobian J_x of X at x (see Theorem 2.19 and Exercise 2.5) and the Jacobian $J_{x'}$ of $X_{\overline{k}}$ at x'. We then clearly have $J_x = J_{x'}$ as matrices with coefficients in $\overline{k}$. On the other hand, using the notation of the proof of Proposition 2.5, we have a surjective map $I/(I \cap \mathfrak{m}^2) \to D_x I$, which implies that

$$\dim_{k(x)} T_{X,x} = n - \dim_{k(x)}(I/(I \cap \mathfrak{m}^2)) \leq n - \operatorname{rank} D_x I = n - \operatorname{rank} J_x. \quad (3.4)$$

Let us suppose X is smooth at x. Then we have

$$\dim_{k(x)} T_{X,x} \le n - \operatorname{rank} J_{x'} = \dim \mathcal{O}_{X_{\bar{k}},x'} = \dim \mathcal{O}_{X,x},$$

whence X is regular at x.

Let us now show the converse when $k(x)$ is separable over k. Let us first suppose x is rational over k. Inequality (3.4) is then an equality, it follows from the Jacobian criterion (applied to $X_{\bar{k}}$) that $X_{\bar{k}}$ is regular at x', and therefore that X is smooth at x. In the general case, let us set $k' = k(x)$. As $X_{k'} \to X$ is finite étale (because $\operatorname{Spec} k' \to \operatorname{Spec} k$ is finite étale), the smoothness of X at x results from Corollary 3.24, from the fact that the points of $X_{k'}$ lying above x are rational over k', and from the preceding case. $\qquad \square$

Remark 3.31. We have also just shown that if $X_{\bar{k}}$ is regular at *a point* x' lying above $x \in X$ (x closed), then X is smooth at x. Indeed, we have seen that $\operatorname{rank} J_{x'}$ and $\dim \mathcal{O}_{X_{\bar{k}},x'}$ are independent of the choice of x'. This is equally true for a point which is not necessarily closed, as we can see in the proof of the corollary below.

Corollary 3.32. *Let X be an algebraic variety over k, and let $x \in X$ be an arbitrary point. Let us suppose that X is smooth at x; then X is regular at x.*

Proof Let $x' \in X_{\bar{k}}$ be a point lying above x. By hypothesis, it is a regular point. We have seen in the proof of Proposition 2.24 that $\overline{\{x'\}}$ contains a closed point y' which is regular in $X_{\bar{k}}$. Let $y \in X$ be the image of y'. Then y is a regular point of X by the proposition above. Now y is a specialization of x, and hence x is a regular point (Theorem 2.16(a)). $\qquad \square$

Corollary 3.33. *Let X be an algebraic variety over a perfect field k. Then X is smooth over k if and only if it is regular.*

Remark 3.34. The converse of the corollary is not true if k is not perfect. For instance, for any finite inseparable extension k' of k, $X = \mathbb{P}^1_{k'}$ is a regular algebraic variety over k, but it is not smooth over k. See also Exercise 3.22 for an example with $H^0(X, \mathcal{O}_X) = k$.

Definition 3.35. Let Y be a locally Noetherian scheme, and let $f : X \to Y$ be a morphism of finite type. We say that f is *smooth* at a point $x \in X$ if it is flat at x, and if $X_y \to \operatorname{Spec} k(y)$ (with $y = f(x)$) is smooth at x. We say that f is *smooth* if it is smooth at every point $x \in X$. Note that it is enough to check the smoothness at closed points of fibers X_y over closed points $y \in Y$ (using Corollary 2.17 and Exercise 3.2). The set of points $x \in X$ such that f is smooth at x is called the *smooth locus* of f.

We say that f is *smooth of relative dimension n* if it is smooth and if all of its non-empty fibers are equidimensional of dimension n. For example, an étale morphism of finite type is smooth of relative dimension 0.

Theorem 3.36. *Let Y be a regular locally Noetherian scheme, and let $f : X \to Y$ be a smooth morphism. Then X is regular.*

Proof Let $x \in X$ and $y = f(x)$. Let $m = \dim \mathcal{O}_{X,x}$, $n = \dim \mathcal{O}_{Y,y}$. We have $\dim \mathcal{O}_{X_y,x} = m - n$ by Theorem 3.12. As X_y is regular at x (Corollary 3.32), the maximal ideal of $\mathcal{O}_{X_y,x}$ is generated by $m - n$ elements $b'_{n+1}, \ldots, b'_m$. Each b'_i is the image of an element b_i of the maximal ideal $\mathfrak{m}_x$ of $\mathcal{O}_{X,x}$ because $\mathcal{O}_{X_y,x} = \mathcal{O}_{X,x}/\mathfrak{m}_y \mathcal{O}_{X,x}$. Let $a_1, \ldots, a_n$ be generators of $\mathfrak{m}_y$; then $\mathfrak{m}_x$ is generated by the m elements $f_x^\#(a_i)$ and b_j. Consequently, X is regular at x. $\qquad\square$

Example 3.37. The schemes $\mathbb{A}^n_\mathbb{Z}$ (and therefore $\mathbb{P}^n_\mathbb{Z}$) are regular because they are smooth over $\mathrm{Spec}\,\mathbb{Z}$. This fact has partially been shown in Example 2.13. Using Exercise 2.11, we see that $\mathrm{Proj}\,\mathbb{Z}[1/n, x, y, z]/(x^n + y^n + z^n)$ is regular.

Proposition 3.38. *Smooth morphisms are stable under base change, composition, and fibered products.*

Proof The stability under base change follows immediately. Let $f : X \to Y$, $g : Y \to Z$ be two smooth morphisms. Then the composition $X \to Z$ is flat. Let $z \in Z$. Let $X_{\bar{z}} = X_z \times_{k(z)} \mathrm{Spec}\,\overline{k(z)}$. Then $X_{\bar{z}} \to Y_{\bar{z}}$ and $Y_{\bar{z}} \to \mathrm{Spec}\,\overline{k(z)}$ are smooth. It follows from Theorem 3.36 that $X_{\bar{z}}$ is regular; hence X_z is smooth over $\mathrm{Spec}\,k(z)$, which shows that $X \to Z$ is smooth. Finally, the stability under fibered products results from the first two properties. $\qquad\square$

Proposition 3.39. *Let $f : X \to Y$ be a morphism of finite type to a locally Noetherian scheme. Let us suppose f is smooth at $x \in X$. Let $y = f(x)$. Then we have an exact sequence of $k(x)$-vector spaces*

$$0 \to T_{X_y,x} \to T_{X,x} \to T_{Y,y} \otimes_{k(y)} k(x) \to 0.$$

Proof Let $\mathfrak{m}_x$, $\mathfrak{m}_y$ denote the respective maximal ideals of $\mathcal{O}_{X,x}$ and $\mathcal{O}_{Y,y}$. Then the maximal ideal $\mathfrak{n}_x$ of $\mathcal{O}_{X_y,x}$ is $\mathfrak{m}_x/\mathfrak{m}_y \mathcal{O}_{X,x}$. We therefore have an exact sequence of $k(x)$-vector spaces:

$$\mathfrak{m}_y/\mathfrak{m}_y^2 \otimes_{k(y)} k(x) \to \mathfrak{m}_x/\mathfrak{m}_x^2 \to \mathfrak{n}_x/\mathfrak{n}_x^2 \to 0.$$

It suffices to show that

$$\dim_{k(x)} T_{X,x} - \dim_{k(x)} T_{X_y,x} = \dim_{k(y)} T_{Y,y}.$$

For this, we will use induction on $d := \dim_{k(y)} T_{Y,y}$. Let us suppose that the canonical map $\mathfrak{m}_y/\mathfrak{m}_y^2 \to \mathfrak{m}_x/\mathfrak{m}_x^2$ is identically zero. Then $\dim_{k(x)} T_{X,x} = \dim_{k(x)} T_{X_y,x} \le \dim \mathcal{O}_{X,x}$, which implies that X is regular (and therefore integral) at x, and that $\dim \mathcal{O}_{X,x} = \dim \mathcal{O}_{X_y,x}$. Hence $\mathcal{O}_{Y,y}$ is of dimension 0 (Theorem 3.12) and contained in $\mathcal{O}_{X,x}$ (because f is flat at x). It follows that $\mathcal{O}_{Y,y}$ is a field, and $d = 0$. In this case, the equality which we want to show is trivial. Let us suppose that the map $\mathfrak{m}_y/\mathfrak{m}_y^2 \to \mathfrak{m}_x/\mathfrak{m}_x^2$ is not identically zero. Then there exists a $g \in \mathfrak{m}_y$ such that $g \notin \mathfrak{m}_x^2 \mathcal{O}_{X,x}$. When replacing Y, in a neighborhood of y, by the closed subscheme $Y' := V(g)$, and X by $X' := X \times_Y Y'$, the morphism $X' \to Y'$ remains smooth at x, and we do not change the fiber X_y. Moreover, $\dim_{k(x)} T_{X',x} = \dim_{k(x)} T_{X,x} - 1$ and $\dim_{k(y)} T_{Y',y} = \dim_{k(y)} T_{Y,y} - 1$. We can therefore apply the induction hypothesis and obtain the desired equality. $\qquad\square$

Remark 3.40. In Propositions 3.23 and 3.39, the results on tangent spaces are not sufficient to characterize étale morphisms and smooth morphisms. Indeed, they do not allow us to control the separability of the extension $k(x)/k(y)$. In Chapter 5, we will return to the study of smooth morphisms by using the sheaf of differential forms. In the next section, we give the local structure of étale morphisms (Proposition 4.11).

Exercises

3.1. Let X be a flat scheme of finite type over a Dedekind scheme S. Let $s \in S$ be a closed point and $x \in X_s$. Let t denote a generator of the maximal ideal of $\mathcal{O}_{S,s}$. We suppose that X is regular at x. Show that X_s is regular at x if and only if $t \in \mathfrak{m}_x \setminus \mathfrak{m}_x^2$, where $\mathfrak{m}_x$ is the maximal ideal of $\mathcal{O}_{X,x}$.

3.2. Let $f : X \to Y$ be a morphism of finite type. Show that f is flat if and only if for every closed point $y \in Y$ and for every closed point x of X_y, f is flat at x (use Proposition 1.2.13).

   Remark. Let $f : X \to Y$ be a morphism of finite type with Y locally Noetherian. We can show that the set of points $x \in X$ such that f is flat at x is open in X, see [41], IV.11.3.1.

3.3. Let $f : X \to Y$ be a morphism of locally Noetherian schemes. Show that $\dim X \le \dim Y$ if f is injective, and $\dim X \ge \dim Y$ if f is flat and surjective.

3.4. Let k be a field. Let A be the polynomial ring $k[x,y]$. Show, by computing the dimension of the fibers, that the canonical morphism $\operatorname{Proj} A[T_0, T_1]/(yT_0 - xT_1) \to \operatorname{Spec} A$ is not flat.

3.5. Let A be a reduced ring, and let $a \in A$ be a non-zero divisor. Let us consider a ring B containing $A[T]$ and an element $s \in B$ such that $s^n = aT$ for some $n \ge 1$. Show that the subalgebra $A[T][s]$ of B is free (and hence flat) over $A[T]$ (consider a linear relation between $1, \ldots, s^{n-1}$ and show that the coefficients are zero in $A_\mathfrak{p}[T]$ for every minimal prime ideal $\mathfrak{p}$ of A).

3.6. Let $f : X \to Y$ be a morphism of integral schemes. We suppose that the generic fiber of f is geometrically integral (the definition is the same as for algebraic varieties 3.2.8). Show that for any flat morphism $Z \to Y$ with Z integral, we have $X \times_Y Z$ integral. This partially generalizes Exercise 3.2.14.

3.7. Let $f : X \to Y$ be a morphism of schemes. Let X_1, X_2 be two closed subschemes of X, flat over Y. We suppose that there exists a dense open subscheme V of Y such that $X_1 \cap f^{-1}(V)$ and $X_2 \cap f^{-1}(V)$ are equal as closed subschemes of $f^{-1}(V)$. Let us suppose that Y is integral. We want to show that $X_1 = X_2$ as schemes.

(a) Show that one can suppose that $X = \operatorname{Spec} B$, $Y = \operatorname{Spec} A$, and V are affine.

(b) Let I, J be the ideals of B defining X_1 and X_2. Tensoring the exact sequence $0 \to A \to \mathcal{O}_Y(V)$ by B/I, show that $I = \operatorname{Ker}(B \to \mathcal{O}_X(f^{-1}(V))/(I))$. Deduce from this that $I = J$.

3.8. Let $f : X \to Y$ be a morphism of locally Noetherian schemes. Let Z be a closed subscheme of X. We suppose that there exists a point $y \in Y$ such that $Z_y = X_y$ as schemes.

(a) Show that if Z is flat over Y at $z \in X_y$, then Z equals to X in a neighborhood of z (use Exercise 1.2.14 and Nakayama's lemma).

(b) Show that if Z is flat over Y and if f is moreover of finite type (resp. proper) over Y, then there exists an open neighborhood $V \ni y$ such that $Z \cap f^{-1}(V) \to f^{-1}(V)$ is an open immersion (resp. an isomorphism) (use Exercise 3.2.5).

3.9. Let $f : X \to Y$ be a flat morphism of finite type with X, Y Noetherian. We will show that f is open.

(a) Use Exercise 3.2.17 to show that $f(X)$ contains a non-empty open subset V of Y.

(b) By considering the morphism $X \times_Y (Y \setminus V) \to Y \setminus V$, show that $f(X) \setminus V$ contains a non-empty open subset of $Y \setminus V$.

(c) Show that $Y \setminus f(X)$ is closed in Y by using the fact that the topological space Y is Noetherian; that is, every descending chain of closed subsets in Y is stationary.

3.10. Let k be a field of characteristic $\operatorname{char}(k) \neq 2$. Let $Y = \operatorname{Spec} k[u,v]/(v^2 - u^2(u+1))$, and let $f : X \to Y$ be the normalization morphism. Show that f is unramified, surjective, but not étale.

3.11. Let $f_1 : X_1 \to Y$, $f_2 : X_2 \to Y$ be morphisms of locally Noetherian schemes of finite type. Let us suppose that for every $y \in Y$, there exists $i = 1$ or 2 such that f_i is unramified (resp. étale; resp. smooth) at all points of $f_i^{-1}(y)$. Show that $X_1 \times_Y X_2 \to Y$ is unramified (resp. étale; resp. smooth).

3.12. Let $f : X \to Y$ be a flat morphism of algebraic variety. We suppose that X_y is irreducible for every closed point $y \in Y$ and that Y is irreducible. Show that X is irreducible.

3.13. Let X be a smooth morphism over a scheme S of characteristic $p > 0$.

(a) Show that $X^{(p)} \to S$ is smooth (use Proposition 3.38).

(b) Let us suppose that S is the spectrum of a field k. Let $F_{X/k} : X \to X^{(p)}$ be the relative Frobenius. Show that $F_{X/k}$ is flat at the rational points of X (use Proposition 2.27).

(c) Show that $F_{X/k}$ is *faithfully flat*, that is flat and surjective.

(d) Let us suppose $p = 2$. Let $Y = \operatorname{Spec} k[x, y]/(y^2 + x^3)$. Show that $F_{Y/k}$ is not flat.

3.14. Let us keep the notation and hypotheses of Lemma 3.16. Let $J = \operatorname{Ann}(b)$.

(a) Show that bB is flat over A.

(b) Using the exact sequence $0 \to J \to B \to B$, where the last map is multiplication by b, show that $J \otimes_A A/\mathfrak{m}_A = 0$. Deduce from this that $J = 0$ and that b does not divide zero in B.

3.15. Let X, Y be integral schemes, separated, and of finite type over a locally Noetherian scheme S. Let $f : X \dashrightarrow Y$ be a birational map. Let $T \to S$ be a faithfully flat morphism such that X_T and Y_T are integral. Suppose that $f_T : X_T \dashrightarrow Y_T$ is defined everywhere and that Y is affine. We want to show that f is also defined everywhere. See Exercise 5.2.14 for the Y arbitrary case.

(a) Show that we can suppose S, T are affine (and even local).

(b) Let us identify the function fields $K(X)$ and $K(Y)$. Show that f is defined everywhere if and only if $\mathcal{O}_Y(Y) \subseteq \mathcal{O}_X(X)$.

(c) Let $M = (\mathcal{O}_X(X) + \mathcal{O}_Y(Y))/\mathcal{O}_X(X)$. Show that $M \otimes_{\mathcal{O}_S(S)} \mathcal{O}_T(T) = 0$. Deduce from this that $M = 0$ and that f is defined everywhere.

3.16. Let B be a *semi-local* ring, that is, having only a finite number of maximal ideals $\mathfrak{m}_1, \ldots, \mathfrak{m}_n$. Let $\hat{B}$ be the formal completion of B for the topology defined by $\cap_i \mathfrak{m}_i$. Show that $\hat{B} \simeq \oplus_i \hat{B}_i$, where $\hat{B}_i$ is the formal completion of $B_{\mathfrak{m}_i}$ for the topology defined by its maximal ideal.

3.17. Let A be a Noetherian local ring, and let B be a finite A-algebra with maximal ideals $\mathfrak{m}_1, \ldots, \mathfrak{m}_n$. Show that $B \otimes_A \hat{A} \simeq \oplus_i \hat{B}_{\mathfrak{m}_i}$, where $\hat{\ }$ denotes the formal completion for the topology defined by the maximal ideal.

3.18. Let X be a quasi-projective scheme over a locally Noetherian scheme S. Let G be a finite group acting on X. Hence the quotient $Y := X/G$ exists (Exercise 3.3.23).

(a) Let G act on an A-algebra B, and C be a flat A-algebra. Then G acts on $B \otimes_A C$ by the identity on the second component. Let us denote A-linear map $g - \operatorname{Id}_B : B \to B$ by ϕ_g. Show that $\operatorname{Ker}(\phi_g \otimes \operatorname{Id}_C) = \operatorname{Ker}(\phi_g) \otimes_A C$ and that

$$\operatorname{Ker}(\phi_g \otimes \operatorname{Id}_C) \cap \operatorname{Ker}(\phi_{g'} \otimes \operatorname{Id}_C) = (\operatorname{Ker}(\phi_g) \cap \operatorname{Ker}(\phi_{g'})) \otimes_A C$$

in $B \otimes_A C$. Deduce from this that $(B \otimes_A C)^G = B^G \otimes_A C$. Show that the quotient morphism $f : X \to Y$ commutes with flat base change.

(b) Let us fix a point $x \in X$. Let us set $D := \{g \in G \mid gx = x\}$. This is the *decomposition group* of x. Show that D acts canonically on $\mathcal{O}_{X,x}$ as well as on the formal completion $\hat{\mathcal{O}}_{X,x}$. Let $y = f(x)$. Show that $(\hat{\mathcal{O}}_{X,x})^D = \hat{\mathcal{O}}_{Y,y}$ (use Exercise 3.17 and (a)). Show that $\mathcal{O}_{Y,y} = (\mathcal{O}_{X,x})^D$ if $G = D$.

(c) Show that the morphism $f : X \to Y$ factors into $X \to X/D$ and $X/D \to Y$. Let z be the image of x in X/D. Show that $X/D \to Y$ is étale at z.

3.19. We keep the notation and hypotheses of the preceding exercise. Let $x \in X$. Then the decomposition group D naturally acts on $k(x)$. This induces a group homomorphism $D \to \mathrm{Aut}(k(x)/k(y))$. The *inertia group* I of x is by definition the kernel of this homomorphism. We want to show that $f : X \to X/G$ is étale at x if and only if I acts trivially on $\mathcal{O}_{X,x}$.

(a) Show that $X \to X/G$ is étale at x if and only if $X \to X/D$ is étale at x. In what follows we will therefore reduce to the case that $G = D$, that is, $f^{-1}(y) = \{x\}$. Let us write $A = \mathcal{O}_{Y,y}$ and $B = \mathcal{O}_{X,x}$. Show that $B^D = A$ and that B is finite over A.

(b) Let k' be the separable closure of $k(y)$ in $k(x)$. For any element $b \in B$, let $\bar{b}$ denote its image in $k(x)$. Let $\theta \in B$ be such that $k' = k(y)[\bar{\theta}]$. Show that $P(T) := \prod_{g \in D}(T - g(\theta)) \in A[T]$ and that its image in $k(y)[T]$ vanishes in $\bar{\theta}$. Let $\tau \in \mathrm{Aut}(k'/k(y))$. Show that there exists a $g \in D$ such that $g(\bar{\theta}) = \tau(\bar{\theta})$. Deduce from this that $\tau = g$ and that $D \to \mathrm{Aut}(k(x)/k(y)) = \mathrm{Aut}(k'/k(y))$ is surjective. Show that $k'/k(y)$ is Galois, and that we have an exact sequence of groups

$$1 \to I \to D \to \mathrm{Gal}(k'/k(y)) \to 1.$$

(c) Let $C = A[T]/(P(T))$. This is a finite étale local A-algebra. Let D act on $C \otimes_A B$ as the identity on the first component. Show that $\mathrm{Spec}\, C = \mathrm{Spec}(C \otimes_A B)/D$, and that B is étale over A if and only if $C \otimes_A B$ is étale over C. Show that the decomposition group of every point of $\mathrm{Spec}(C \otimes_A B)_z$, where z is the closed point of $\mathrm{Spec}\, C$, is equal to I. Hence, by (c), we have reduced to the case when $D = I$, and $k(x)$ is purely inseparable over $k(y)$.

(d) Using Proposition 3.26, show that $X \to Y$ is étale at x if and only if I acts trivially on $\mathcal{O}_{X,x}$. Show that in general, $X/I \to Y$ is étale at the image of x in X/I.

3.20. Let $X = \mathrm{Spec}\, k[T_1, \ldots, T_n]/(F_1, \ldots, F_r)$ be an affine variety over a field k, and x a closed point of X. Let us consider $J_x = ((\partial F_i/\partial T_j)(x))_{i,j}$ as a matrix with coefficients in $k(x)$. Show that X is smooth at x if and only if $\mathrm{rank}\, J_x = n - \dim \mathcal{O}_{X,x}$. Hence the Jacobian criterion is a criterion for smoothness.

3.21. Let $F \in k[T_0, \ldots, T_n]$ be a homogeneous polynomial of degree d over a field k. We suppose that d is prime to the characteristic of k. Show that the closed subscheme $V_+(F)$ of $\mathbb{P}^n_k$ is smooth over k if and only if the $\frac{\partial F}{\partial T_i}$ are without common zero in $\bar{k}$.

3.22. Let C be a *conic* over a field k, that is $C = V_+(F) \subset \mathbb{P}^2_k$, where $F(x, y, z)$ is a homogeneous polynomial of degree 2.

(a) Show that C is reduced (resp. irreducible) if and only if F is not a square modulo k^* (resp. F is irreducible).

(b) Let us suppose that C contains a rational point P.

    (1) Show that if C is singular at P, then (up to isomorphism) C is of the form $V_+(G)$, with $G(x,y) \in k[x,y]$ homogeneous of degree 2.

    (2) Let us suppose C is regular at P. Let D be a line in $\mathbb{P}_k^2$ passing through P and different from $PT_{C,P}$ (see the notation in Exercise 2.9). Show that D meets C at another rational point Q. Show that $C \simeq V_+(ax^2 + yz)$ with $a \in k$ (let R be the intersection point of the lines $PT_{C,P}$ and $PT_{C,Q}$, and reduce to $P = (1,0,0)$, $Q = (0,1,0)$ and $R = (0,0,1)$). Deduce from this that either C is the union of two distinct lines, or $C \simeq \mathbb{P}_k^1$.

(c) Let us suppose char$(k) \neq 2$. Show that $C \simeq V_+(ax^2 + by^2 + cz^2)$, and that C is smooth if and only if $abc \neq 0$.

(d) Let us suppose char$(k) = 2$, $C(k) = \emptyset$, and that C is not smooth.

    (1) Show that there exist $a, b, c \in k$ such that $C \simeq V_+(ax^2 + by^2 + cz^2)$. Show that C does not contain any smooth point.

    (2) Show that $\mathcal{O}_C(D_+(z))$ is equal to the integral closure of $k[x/z]$ in $K(C)$. Deduce from this that C is normal, and therefore regular.

3.23. Let k be a field and let $F(x,y,z) \in k[x,y,z]$ be a homogeneous polynomial of degree 2 defining a smooth conic $V_+(F)$ over k.

(a) If char$(k) \neq 2$, show that after a change of variables, one can write $F = ax^2 + by^2 + cz^2$ for some $a,b,c \in k^*$. Let us suppose moreover that k is finite with q elements. Show that the sets $\{ax^2 \mid x \in k\}$ and $\{-c - by^2 \mid y \in k\}$ both have $(q+1)/2$ elements. Deduce from this that there exists a rational point in $V_+(F)$.

(b) Let us suppose that char$(k) = 2$ and that k is perfect. Show that after a suitable change of variables, one has $F = xy + z^2$. Deduce from this that $V_+(F)$ contains a rational point.

3.24. Let X be an algebraic variety over a field k. Show that the following properties are equivalent:

  (i) X is smooth;

  (ii) for every finite extension l of k, X_l is regular;

  (iii) for every purely inseparable extension l of k, X_l is regular;

  (iv) for every (not necessarily algebraic) extension l of k, X_l is regular.

3.25. Let $f : X \to Y$ be a morphism of schemes, let $x \in X$, and $y = f(x)$.

(a) Show that we have an exact sequence of $k(x)$-vector spaces:

$$0 \to T_{X_y,x} \to T_{X,x} \to T_{Y,y} \otimes_{k(y)} k(x).$$

(b) Let us suppose that f admits a section $g : Y \to X$, and that $x = g(y)$.

Show that $k(x) = k(y)$, and that we have a decomposition $T_{X,x} \simeq T_{X_y,x} \oplus T_{Y,y}$.

(c) Let us suppose that X, Y are locally Noetherian and regular, and that f is of finite type. Show that under the hypothesis of (b), X_y is smooth at x.

(d) Let us keep the hypothesis above. Let $d = \dim_x X_y$. Show that we have an isomorphism $\hat{\mathcal{O}}_{X,x} \simeq \hat{\mathcal{O}}_{Y,y}[[T_1, \ldots, T_d]]$, and that f is smooth at x.

3.26. Let X be a scheme of finite type over a locally Noetherian regular scheme S. Let $x \in X$.

(a) Show that in a neighborhood of x, there exists a closed immersion $i : X \to Z$ into a scheme Z that is of finite type over S and regular at x (locally embed X into an affine space over S and use Theorem 3.36).

(b) Using exact sequence (2.1) in the proof of Proposition 2.5, show that one can successively reduce the dimension of $T_{Z,i(x)}$ in such a way that the canonical map $T_{X,x} \to T_{Z,i(x)}$ is an isomorphism.

4.4 Zariski's 'Main Theorem' and applications

There are several variants of what we call Zariski's 'Main Theorem'. We are going to present a version concerning quasi-finite morphisms (Corollary 4.6).

Lemma 4.1. *Let Y be the spectrum of a Noetherian complete local ring, with closed point y. Let X be a Y-scheme of finite type such that X_y consists of a single point x. Then x admits an affine neighborhood that is finite over Y.*

Proof Restricting X if necessary, we may assume that it is affine, and that all of its irreducible components pass through x. We are going to show that $X \to Y$ is finite. As it is a morphism of finite type, it suffices to show that it is integral. We may therefore suppose X is reduced. Let $B = \mathcal{O}_X(X)$, $A = \mathcal{O}_Y(Y)$, and let $\mathfrak{m}$ be the maximal ideal of A. Then $B/\mathfrak{m}B = \mathcal{O}(X_y)$ is a finite $A/\mathfrak{m}$-algebra. There therefore exists a finite sub-A-algebra N of B such that $B = N + \mathfrak{m}B$. It follows that $B = N + \mathfrak{m}^n B$ for every $n \geq 1$. Let us consider the inverse systems of A-modules $N_n = N/\mathfrak{m}^n N$, $B_n = B/\mathfrak{m}^n B$, and $M_n = (\mathfrak{m}^n B) \cap N/\mathfrak{m}^n N$. We have an exact sequence of these systems:

$$0 \to M_n \to N_n \to B_n \to 0.$$

Moreover, $M_{n+1} \to M_n$ is surjective because

$$(\mathfrak{m}^n B) \cap N = (\mathfrak{m}^n N + \mathfrak{m}^{n+1} B) \cap N = \mathfrak{m}^n N + (\mathfrak{m}^{n+1} B \cap N),$$

whence an exact sequence of inverse limits (Lemma 1.3.1)

$$0 \to \varprojlim_n M_n \to \varprojlim_n N_n \to \varprojlim_n B_n \to 0.$$

As A is Noetherian and complete, and N finitely generated, the limit in the middle is equal to N (Corollary 1.3.14). Let $\mathfrak{n}$ denote the maximal ideal of B

corresponding to the point x. The hypothesis $X_y = \{x\}$ implies that $\sqrt{\mathfrak{m}B} = \mathfrak{n}$. There therefore exists an $r \geq 1$ such that $\mathfrak{n}^r \subseteq \mathfrak{m}B$. Consequently,

$$\mathrm{Ker}(B \to \varprojlim_n B_n) = \cap_{n \geq 1}\mathfrak{m}^n B = \cap_{m \geq 1}\mathfrak{n}^m.$$

Let $b \in \cap_{m \geq 1}\mathfrak{n}^m$. By Krull's theorem (Corollary 1.3.13), there exists an $\epsilon \in \mathfrak{n}$ such that $(1 + \epsilon)b = 0$. The hypothesis on the irreducible components of X translates into the fact that $\mathfrak{n}$ contains all minimal prime ideals of B. We deduce from this that b belongs to the intersection of the minimal prime ideals; hence $b = 0$ and $B \to \varprojlim B_n$ is injective. This implies that $B = N$ is finite over A. $\square$

Proposition 4.2. *Let Y be a locally Noetherian scheme, and let $f : X \to Y$ be a separated morphism of finite type. Let us suppose that the canonical homomorphism $\mathcal{O}_Y \to f_*\mathcal{O}_X$ is an isomorphism. Then there exists a (possibly empty) open subset V of Y such that $f^{-1}(V) \to V$ is an isomorphism, and X_y has no isolated points for every $y \in Y \setminus V$.*

Proof It suffices to show that for every $y \in Y$ such that X_y has an isolated point x, there exists an open subset $W \ni y$ such that $f^{-1}(W) \to W$ is an isomorphism. Let us fix such x and y.

Let us first suppose that Y is the spectrum of a complete local ring A, with closed point y. It follows from the lemma above that there exists an affine (and connected) open neighborhood U of x, finite over Y. As U is proper over Y (Exercise 3.3.17), $U \to X$ is proper (Proposition 3.3.16(e)), and hence closed. This implies that U is a connected component of X. Now $\mathcal{O}_X(X) = A$, and hence X is connected. It follows that $X = U$ is affine. Consequently, $X \to Y$ is an isomorphism.

In the general case, the property which we want to show being local on Y, we may suppose Y is affine, and even local with closed point y (Exercise 3.2.5). Let us therefore write $Y = \mathrm{Spec}\,A$, with A local with maximal ideal $\mathfrak{m}$. Let B be the $\mathfrak{m}$-adic completion of A. Let us also denote the closed point of $\hat{Y} := \mathrm{Spec}\,B$ by y. Then the morphism $f_{\hat{Y}} : X_{\hat{Y}} \to \hat{Y}$ obtained by the base change $\hat{Y} \to Y$ is separated of finite type, and has the same closed fiber as $X \to Y$. Moreover, as B is flat over A (Theorem 1.3.16), we have $B \simeq \mathcal{O}_X(X) \otimes_A B = \mathcal{O}(X_{\hat{Y}})$ (Proposition 3.1.24). Hence $f_{\hat{Y}} : X_{\hat{Y}} \to \hat{Y}$ is an isomorphism. Let U be an affine open subset of X which meets X_y. Then $f_{\hat{Y}}(U \times_Y \hat{Y})$ is an open subset of $\hat{Y}$ containing the closed point; it is therefore equal to $\hat{Y}$. Consequently, $U \times_Y \hat{Y} = X_{\hat{Y}}$. Now $\hat{Y} \to Y$ is surjective since $A \to B$ is faithfully flat, and hence the projection $X_{\hat{Y}} \to X$ is surjective. We deduce from this that $X = U$ is affine, and that f is an isomorphism. $\square$

Corollary 4.3. *Let Y be a normal locally Noetherian scheme, let X be an integral scheme, and $f : X \to Y$ a proper birational morphism. Then the following properties are true.*

(a) *The canonical homomorphism $\mathcal{O}_Y \to f_*\mathcal{O}_X$ is an isomorphism.*

(b) *There exists an open subset V of Y such that $f^{-1}(V) \to V$ is an isomorphism, and X_y has no isolated points if $y \notin V$. Moreover, the complement of V has codimension ≥ 2. (See also Lemma 9.2.10.)*

Proof (a) We identify the function fields $K(Y)$ and $K(X)$ via $f^\#$. For every affine open subset W of Y, $f^{-1}(W) \to W$ is proper, and hence $f_*\mathcal{O}_X(W) = \mathcal{O}_X(f^{-1}(W))$ is integral over $\mathcal{O}_Y(W)$ (Proposition 3.3.18) and contained in $K(Y)$. It follows that $f_*\mathcal{O}_X(W) = \mathcal{O}_Y(W)$, whence $f_*\mathcal{O}_X = \mathcal{O}_Y$.

(b) results from (a) and Proposition 4.2. We just have to show the assertion concerning the codimension of $Y \setminus V$. Let $y \in Y$ be a point of codimension 1. Then $\mathcal{O}_{Y,y}$ is a discrete valuation ring (Proposition 1.12). Let $x \in X_y$; then $\mathcal{O}_{X,x} = \mathcal{O}_{Y,y}$ (Lemma 3.3.24). It follows from Proposition 3.3.11 that $X_y = \{x\}$. Consequently, $y \in V$. Hence codim$(Y \setminus V, Y) \geq 2$. $\square$

Theorem 4.4. *Let $Y = \operatorname{Spec} A$ be an Noetherian affine scheme, and let $f : X \to Y$ be a separated morphism of finite type. We suppose that $\mathcal{O}_X(X)$ is a finitely generated A-module.[3] Let X_f be the set of points $x \in X$ that are isolated in $X_{f(x)}$. Then*

(a) *X_f is open in X,*

(b) *the canonical morphism $g : X \to \operatorname{Spec} \mathcal{O}_X(X)$ induces an open immersion from X_f to $\operatorname{Spec} \mathcal{O}_X(X)$.*

Proof Let us set $Z = \operatorname{Spec} \mathcal{O}_X(X)$. Let us consider the decomposition of f into $g : X \to Z$ and $h : Z \to Y$. The latter is finite by assumption. By virtue of Proposition 4.2, there exists an open subset V of Z such that $g^{-1}(V) \to V$ is an isomorphism and $g^{-1}(y)$ has no isolated points for every $y \in Y \setminus V$. It now suffices to show that $X_f = g^{-1}(V)$. For any $y \in Y$, X_y is the finite union of the fibers X_z, $z \in Z_y$. Each fiber X_z is closed in X_y because z is closed in Z_y. It follows that $X_f = X_g$. Finally, it is clear that $X_g = g^{-1}(V)$. $\square$

Lemma 4.5. *Let X, Y be irreducible schemes and $f : X \to Y$ a separated morphism which is locally an open immersion. Then f is an open immersion.*

Proof Let us first show that $X \times_Y X$ is irreducible. Let $\{U_i\}_i$ be a covering of X by open subsets U_i such that $f|_{U_i}$ is an open immersion for all i. Then $X \times_Y X$ is covered by the $U_i \times_Y U_j \simeq f(U_i) \times_Y f(U_j) = f(U_i) \cap f(U_j)$. Hence $U_i \times_Y U_j$ is irreducible. Since these open subsets have non-empty intersection with each other, $X \times_Y X$ is irreducible.

The diagonal morphism $\Delta_{X/Y} : X \to X \times_Y X$ is a closed immersion because f is separated, and it is locally an open immersion. Since $X \times_Y X$ is irreducible, $\Delta_{X/Y}$ is surjective. Let $x_1, x_2 \in X$ be such that $f(x_1) = f(x_2)$. Then there exists a point $z \in X \times_Y X$ whose projections on X are x_1 and x_2 (Exercise 3.1.7). Since $z \in \Delta_{X/Y}(X)$, we have $x_1 = x_2$. Hence f is injective, and therefore an open immersion. $\square$

---

[3]We will see in the next chapter that this condition is verified as soon as f is projective.

Corollary 4.6 (Zariski's 'Main Theorem'). *Let Y be a normal, locally Noetherian, integral scheme. Let $f : X \to Y$ be a separated birational morphism of finite type. We suppose that f is quasi-finite; then f is an open immersion.*

Proof We may suppose Y is affine. Let us show that f is locally an open immersion. Let $x \in X$, and U be an affine open neighborhood of x. There exists an integral projective scheme $Z \to Y$ such that U is isomorphic to an open subscheme of Z. This morphism is birational. We have $\mathcal{O}_Z(Z)$ integral over $\mathcal{O}_Y(Y)$ (Proposition 3.3.18) and contained in $K(Y)$; hence $\mathcal{O}_Z(Z) = \mathcal{O}_Y(Y)$ since Y is normal. Let $y = f(x)$. By hypothesis, $\{x\}$ is open in U_y (hence open in Z_y). It follows that x is an isolated point of Z_y. By Proposition 4.2, $Z \to Y$ is an isomorphism over an open neighborhood of y, and hence f is an open immersion in an open neighborhood of x. By Lemma 4.5, f is an open immersion.   □

Let $f : X \to Y$ be a projective morphism. Anticipating the next chapter (by admitting that $f_* \mathcal{O}_X(W)$ is finite over $\mathcal{O}_Y(W)$ for any affine open subset W of Y, Theorem 5.3.2(a)), the following corollary is an immediate consequence of Theorem 4.4 (see also an elementary proof in Exercise 4.2).

Corollary 4.7. *Let Y be a locally Noetherian scheme, and let $f : X \to Y$ be a quasi-finite projective morphism. Then f is a finite morphism.*

Still anticipating Theorem 5.3.2, we have another variant of Zariski's 'Main Theorem':

Corollary 4.8. *Let Y be a locally Noetherian scheme, and let $f : X \to Y$ be a quasi-projective morphism (e.g., a morphism of finite type with X, Y affine). Let X_y be a finite fiber. Then there exists an open neighborhood U of X_y such that $f|_U : U \to Y$ factors into an open immersion $U \to Z$ followed by a finite morphism $Z \to Y$.*

Proof We embed X as an open subscheme of a projective scheme $\overline{f} : \overline{X} \to Y$. Since X_y is open in $\overline{X}_y$, every point of X_y is isolated in $\overline{X}_y$. Hence $X_y \subseteq \overline{X}_{\overline{f}}$. By Theorem 4.4, $\overline{X}_{\overline{f}} \to Y$ is finite. Now it suffices to take $Z = \overline{X}_{\overline{f}}$ and $U = Z \cap X$.   □

Remark 4.9. One can find a proof using methods from commutative algebra in [77]. The schemes are not necessarily locally Noetherian, and we must replace finite morphism by integral morphism. See also [69], III.9, for a discussion on the different forms of the 'Main Theorem'.

As an application of Zariski's Main Theorem, we are going to describe the local structure of étale morphisms.

Example 4.10. Let A be a Noetherian ring, let $P(T) \in A[T]$ be a monic polynomial, and $P'(T)$ its derivative. Let $B = A[T, P'(T)^{-1}]/(P(T))$. Then B is flat over A because it is the localization of the flat A-algebra $A[T]/(P(T))$. Let $k(s)$ denote the residue field of a point $s \in \operatorname{Spec} A$; then

$$B \otimes_A k(s) = k(s)[T, \tilde{P}'(T)^{-1}]/(\tilde{P}(T)),$$

where $\tilde{P}$ is the image of P in $k(s)[T]$. Thus it is the direct sum of the decomposition fields of the simple roots of $\tilde{P}(T)$. Consequently, either $\operatorname{Spec} B$ is empty, or Spec $B \to$ Spec A is étale. In the second case, the morphism Spec $B \to$ Spec A is called a *standard étale morphism*.

Proposition 4.11. *Let Y be a locally Noetherian scheme. Let $f : X \to Y$ be a morphism of finite type, unramified at a point x. Let $y = f(x)$. Then, if necessary replacing X (resp. Y) by an open neighborhood of x (resp. of y), there exists a standard étale morphism $h : Z \to Y$ such that we have the following commutative diagram*

where g is an immersion. If f is étale, then g is an open immersion.

Proof As the property that we want to show is local, we may suppose X and Y are affine. Moreover, as X is of finite type over Y, we may suppose Y is local with closed point y (Exercise 3.2.5). Let us set $A = \mathcal{O}_Y(Y)$ and $B = \mathcal{O}_X(X)$. By Corollary 4.8, we may even suppose B is finite over A. The fiber X_y is the disjoint union of Spec $k(x)$ and of an open subset U. As $k(x)$ is separable over $k(y)$, there exists a $\tilde{b} \in \mathcal{O}(X_y)$ such that $k(x) = k(y)[\tilde{b}]$, $\tilde{b} \neq 0$, and $\tilde{b}|_U = 0$. We can lift $\tilde{b}$ to an element $b \in B$. Let us consider the subalgebra $C := A[b]$ of B. Let $\mathfrak{m}$ be the prime ideal of B corresponding to the point x, and $\mathfrak{q} = \mathfrak{m} \cap C$. Let us show that $C_\mathfrak{q} \to B \otimes_C C_\mathfrak{q}$ is an isomorphism. First, $\mathfrak{m}$ is the unique prime ideal of B lying above $\mathfrak{q}$ because $\mathfrak{m}$ does not contain b while the other prime ideals of B above $\mathfrak{m}_y$ all contain b. It follows that $B \otimes_C C_\mathfrak{q}$ is a local ring. As $C \to C_\mathfrak{q}$ is flat, the morphism $C_\mathfrak{q} \to B \otimes_C C_\mathfrak{q}$ is finite and injective. Finally, $(B \otimes_C C_\mathfrak{q}) \otimes_{C_\mathfrak{q}} k(\mathfrak{q}) = k(\mathfrak{q})$. It follows from Nakayama's lemma that $C_\mathfrak{q} = B \otimes_C C_\mathfrak{q}$.

As B and C are finitely generated over A, this isomorphism extends to a neighborhood of x. We can therefore replace B by C, and suppose that B is generated, as an A-algebra, by a single element b. Let $n = \dim_{k(y)} B \otimes_A k(y)$. Then $\{1, \tilde{b}, \ldots, \tilde{b}^{n-1}\}$ is a basis of $B \otimes_A k(y)$ over $k(y)$. Once more by Nakayama, $\{1, b, \ldots, b^{n-1}\}$ is a system of generators for the A-module B. There therefore exists a monic polynomial $P(T) \in A[T]$ of degree n which vanishes in b, and we have a surjective homomorphism of A-algebras $A[T]/(P(T)) \to B$ which sends T to b. The image of $b' := P'(b)$ in $k(x)$ is non-zero because $k(x)$ is separable over $k(y)$, and hence $x \in D(b')$. Replacing X by the open subset $D(b')$ if necessary, we have a surjective homomorphism

$$D := A[T, P'(T)^{-1}]/(P(T)) \to B.$$

This shows the proposition for unramified morphisms. Let us now suppose that f is étale at x. Let $\mathfrak{n}$ be the inverse image of $\mathfrak{m}$ in D. Let us show that $\psi : D_\mathfrak{n} \to B_\mathfrak{m}$ is an isomorphism. We know that $\psi \otimes k(y)$ is an isomorphism. Let $I = \operatorname{Ker} \psi$. As $B_\mathfrak{p}$ is flat over A, and ψ is surjective, we have $I \otimes_A k(\mathfrak{p}) = 0$

(Proposition 1.2.6); hence $I = 0$ by Nakayama. As D is Noetherian, there exists an open neighborhood U of x such that the closed immersion $\operatorname{Spec} B \to \operatorname{Spec} D$ is an isomorphism over U. □

Corollary 4.12. *Let Y be a locally Noetherian scheme. Let $f : X \to Y$ be a morphism of finite type. Then the set of points $x \in X$ such that f is étale at x is open in X.*

Proof If f is étale at x, there exists an open neighborhood U of x such that U is isomorphic to an open subscheme of an étale scheme W over Y. Hence $U \to Y$ is étale. □

Let us conclude with one more application of Zariski's theorem to the study of the dimensions of the fibers of a morphism. The finiteness Theorem 5.3.2 is necessary because we use Corollary 4.8.

Lemma 4.13. *Let Y be a locally Noetherian irreducible scheme with generic point η, and let $f : X \to Y$ be a morphism of finite type with X irreducible. If there exists a $y \in Y$ such that X_y is non-empty and finite, then the generic fiber X_η contains at most one point. Moreover, there exists a non-empty open subset V of Y such that $f^{-1}(V) \to V$ is quasi-finite.*

Proof We may suppose f is dominant and X, Y are affine. Then f is quasi-projective. By Corollary 4.8, there exists an open subset $U \supseteq X_y$ of X which is isomorphic to an open subscheme of a finite Y-scheme Z. Since X is irreducible, $U \cap X_\eta \neq \emptyset$. Therefore $U \cap X_\eta$ is a finite set. As X_η is irreducible, it is therefore reduced to a point. Finally, the existence of V comes from Exercise 3.2.16. □

Remark 4.14. More generally, if X_y is not necessarily finite, we still have the inequality $\dim X_y \geq \dim X_\eta$ (Exercise 4.6). Moreover, equality holds for the points of a non-empty open subset of Y (Chevalley's theorem, see [41], IV.13.1.3).

Lemma 4.15. *Let A be a discrete valuation ring with uniformizing element t. Let B be a flat A-algebra such that B/tB is finitely generated over A/tA and of dimension d. Then there exists an injective homomorphism of A-algebras $\varphi : A[T_1, \ldots, T_d] \hookrightarrow B$ such that the homomorphism $A/tA[T_1, \ldots, T_d] \to B \otimes_A A/tA$ is finite injective.*

Proof By the Noether normalization lemma (2.1.9), there exists an injective finite homomorphism

$$\overline{\varphi} : A/\mathfrak{m}[T_1, \ldots, T_d] \hookrightarrow B/\mathfrak{m}B = B \otimes_A A/\mathfrak{m}.$$

Let $b_i \in B$ be with image $\overline{\varphi}(T_i)$ in $B/\mathfrak{m}B$. Then we can define a homomorphism of A-algebras $\varphi : A[T_1, \ldots, T_d] \to B$ by sending T_i to b_i. We then immediately verify that φ is injective. □

Proposition 4.16. *Let S be a Dedekind scheme with generic point η, and let $X \to S$ be a dominant morphism of finite type with X irreducible. Then for any $s \in S$ such that $X_s \neq \emptyset$, we have X_s equidimensional of dimension $\dim X_\eta$. (See also Corollary 8.2.8.)*

Proof We may suppose that S is local and X is integral, and therefore flat over S (Proposition 3.9). Let F be an irreducible component of X_s. We want to show that $\dim F = \dim X_\eta$. As X is irreducible, we can replace X by an affine open subscheme which meets F but not the other irreducible components of X_s. We then have X affine and $\dim X_s = \dim F$. Let $f : X \to \mathbb{A}_S^d$ be the morphism induced by a homomorphism $\mathcal{O}_S(S)[T_1, \ldots, T_d] \to \mathcal{O}_X(X)$ defined as in the lemma above, with $d = \dim X_s$. Then, by construction, $f_s : X_s \to \mathbb{A}_{k(s)}^d$ is a finite morphism. In particular, for any $y \in \mathbb{A}_S^d$ lying above s, the fiber X_y is finite. It follows that the function field $K(X_\eta) = K(X)$ is finite over $K(\mathbb{A}_S^d) = K(S)(T_1, \ldots, T_d)$ (Lemma 4.13), and hence X_η is of dimension d (Proposition 2.5.19). □

Exercises

4.1. Let $f : X \to Y$ be a projective morphism of algebraic varieties over an algebraically closed field. Let us suppose that f is injective and $T_{f,x}$ is injective for every closed point $x \in X$. Show that f is a finite morphism, and, using Nakayama's lemma, show that f is a closed immersion.

4.2. Let $f : X \to Y$ be a projective morphism. Let $y \in Y$ be such that X_y is finite. We want to show that there exists a neighborhood V of y such that $f^{-1}(V) \to V$ is finite, without using Zariski's 'Main Theorem'. We may suppose Y is affine.

    (a) Let us fix an embedding $X \to \mathbb{P}_Y^n$. Show that there exists a non-empty principal closed subset $V_+(P)$ of $\mathbb{P}_Y^n$ such that $V_+(P) \cap X_y = \emptyset$ (use Proposition 3.3.36).

    (b) Show that there exists an affine neighborhood V of y such that $X \cap V_+(P) \cap \mathbb{P}_V^n = \emptyset$.

    (c) Show that $f^{-1}(V) = X \cap D_+(P) \cap \mathbb{P}_V^n$ and that $f^{-1}(V)$ is affine. Conclude using Proposition 3.3.18.

4.3. Let Y be a locally Noetherian scheme, and let $f : X \to Y$ be a quasi-finite and quasi-projective morphism. Show that f factors into an open immersion $X \to Z$ followed by a finite morphism $Z \to Y$.

4.4. Let A be a discrete valuation ring, and let B be a finitely generated flat A-algebra. Show that there exists a subalgebra C of B with a finite injective homomorphism $A[T_1, \ldots, T_d] \to C$, and such that the morphism $\operatorname{Spec} B \to \operatorname{Spec} C$ is an open immersion. This is a sort of generalization of Noether's normalization lemma. Show that in general we cannot have $C = B$.

4.5. Let Y be a locally Noetherian integral scheme, and let $f : X \to Y$ be a projective morphism with geometrically connected fibers. Show that for any affine open subset V of Y, there exists an $n \geq 1$ such that $(f_*\mathcal{O}_X(V))^{p^n}$ is contained in the image of $\mathcal{O}_Y(V) \to f_*\mathcal{O}_X(V)$, where

p is the characteristic of $K(Y)$ (by convention, we set $p^n = 1$ if $p = 0$). In particular, if char $K(Y) = 0$, then $\mathcal{O}_Y \to f_*\mathcal{O}_X$ is an isomorphism.

4.6. Let Y be a locally Noetherian irreducible scheme, and let $f : X \to Y$ be a dominant morphism of finite type with X irreducible. We are going to generalize Lemma 4.13 and show that $\dim X_\eta \le \dim X_y$ if $X_y \ne \emptyset$. By this lemma, we may suppose that $\dim X_y \ge 1$. We will use induction on $\dim X_y + \dim \mathcal{O}_{Y,y}$.

   (a) Let $x \in X_y$ be such that $\dim \mathcal{O}_{X_y,x} = \dim X_y$. Show that there exists a $b \in \mathfrak{m}_x \mathcal{O}_{X,x}$ which belongs to none of the minimal prime ideals containing $\mathfrak{m}_y \mathcal{O}_{X,x}$. Restricting X if necessary, we may suppose X is affine and $b \in \mathcal{O}_X(X)$.

   (b) Let b be as above. Show that there exists an irreducible component Z of $V(b)$ passing through x and not contained in X_y. Let z' be the generic point of Z, and $y' = f(z')$ the generic point of $\overline{f(Z)}$ (Exercise 2.4.11). Show that

$$\dim \mathcal{O}_{Y,y'} < \dim \mathcal{O}_{Y,y}, \quad \dim X_{y'} - \dim X_y \le \dim Z_{y'} - \dim Z_y.$$

   By applying the induction hypothesis to $Z \to \overline{f(Z)}$ and $X \times_Y \operatorname{Spec} \mathcal{O}_{Y,y'} \to \operatorname{Spec} \mathcal{O}_{Y,y'}$, show that $\dim X_\eta \le \dim X_y$.

5

Coherent sheaves and Čech cohomology

The notion of quasi-coherent sheaves in algebraic geometry was introduced by J.-P. Serre ([85]). These are sheaves which are in some sense 'locally constant' with respect to the structural sheaf. They are the analogues of modules over a ring. The theory of quasi-coherent sheaves and their cohomology groups is a powerful tool for studying schemes. Among the first applications of the theory, we can cite the determination of the closed subschemes of a projective scheme (Proposition 1.30) and Serre's affineness criterion (Theorem 2.23). The first fundamental theorem on projective morphisms is the direct image theorem (Theorem 3.2). We will apply this theorem to show Zariski's connectedness principle. In the next chapter, we will apply the theory to sheaves of differentials on a scheme.

5.1 Coherent sheaves on a scheme

After a brief introduction to the notion of sheaves of modules on a ringed topological space, we will present some general properties of quasi-coherent sheaves. In particular, we will determine them explicitly on affine schemes (Theorem 1.7). In the last subsection, we will look at coherent sheaves on a projective scheme. The principal results are Theorem 1.27, which is sort of equivalent to Theorem 1.7, and Theorem 1.34, which characterizes ample sheaves.

5.1.1 Sheaves of modules

Definition 1.1. Let $(X, \mathcal{O}_X)$ be a ringed topological space. A *sheaf of $\mathcal{O}_X$-modules* or an *$\mathcal{O}_X$-module* is a sheaf $\mathcal{F}$ on X such that $\mathcal{F}(U)$ is an $\mathcal{O}_X(U)$-module for every open subset U, and that if $V \subseteq U$, then $(af)|_V = a|_V f|_V$ for every $a \in \mathcal{O}_X(U)$ and every $f \in \mathcal{F}(U)$. We define the notion of homomorphisms of $\mathcal{O}_X$-modules in an obvious way.

Starting with two $\mathcal{O}_X$-modules $\mathcal{F}$, $\mathcal{G}$, the usual operations on modules over a ring make it possible to construct other $\mathcal{O}_X$-modules. We define the *tensor product* $\mathcal{F} \otimes_{\mathcal{O}_X} \mathcal{G}$ to be the sheaf associated to the presheaf

$$U \mapsto \mathcal{F}(U) \otimes_{\mathcal{O}_X(U)} \mathcal{G}(U).$$

It is an $\mathcal{O}_X$-module. Let us note that for any $x \in X$, we have

$$(\mathcal{F} \otimes_{\mathcal{O}_X} \mathcal{G})_x = \mathcal{F}_x \otimes_{\mathcal{O}_{X,x}} \mathcal{G}_x$$

([16], II, Section 6, Proposition 7, or Proposition 1.5(d) for quasi-coherent sheaves). The direct sum of a family of $\mathcal{O}_X$-modules is an $\mathcal{O}_X$-module. The kernel and cokernel of a homomorphism of $\mathcal{O}_X$-modules are $\mathcal{O}_X$-modules.

Definition 1.2. Let $(X, \mathcal{O}_X)$ be a ringed topological space. We say that an $\mathcal{O}_X$-module $\mathcal{F}$ is *generated by its global sections at* $x \in X$ if the canonical homomorphism $\mathcal{F}(X) \otimes_{\mathcal{O}_X(X)} \mathcal{O}_{X,x} \to \mathcal{F}_x$ is surjective. We say that $\mathcal{F}$ is *generated by its global sections* if this is true at every point of X. Let S be a subset of $\mathcal{F}(X)$. We say that $\mathcal{F}$ is generated by S if $\{s_x\}_{s \in S}$ generates $\mathcal{F}_x$ for every $x \in X$.

Lemma 1.3. *Let X be a ringed topological space. Then an $\mathcal{O}_X$-module $\mathcal{F}$ is generated by its global sections if and only if there exist a set I and a surjective homomorphism of $\mathcal{O}_X$-modules $\mathcal{O}_X^{(I)} \to \mathcal{F}$, where $\mathcal{O}_X^{(I)}$ is the direct sum of those $\mathcal{O}_X$ indexed by I. Moreover, if $\mathcal{F}$ is generated by a set S of global sections, then we can take $I = S$.*

Proof If there exists a surjective homomorphism $\mathcal{O}_X^{(I)} \to \mathcal{F}$, as $\mathcal{O}_X^{(I)}$ is generated by its global sections, we easily see that the same is true for $\mathcal{F}$. Let us suppose $\mathcal{F}$ is generated by its global sections. Let S be a subset of $\mathcal{F}(X)$ which generates $\mathcal{F}$ (e.g., $S = \mathcal{F}(X)$). Let $\{\varepsilon_s\}_{s \in S}$ be the canonical basis of $\mathcal{O}_X(X)^{(S)}$. For any open subset U of X, the $\varepsilon_s|_U$ form a basis of $\mathcal{O}_X(U)^{(S)}$ over $\mathcal{O}_X(U)$. Let $\psi : \mathcal{O}_X^{(S)} \to \mathcal{F}$ be the $\mathcal{O}_X$-module homomorphism defined by

$$\psi(U) : \sum_{s \in S} f_s \cdot \varepsilon_s|_U \mapsto \sum_{s \in S} f_s \cdot s|_U$$

for any $f_s \in \mathcal{O}_X(U)$. Then ψ_x is surjective for every $x \in X$ and hence ψ is surjective.   □

Definition 1.4. Let $(X, \mathcal{O}_X)$ be a ringed topological space. Let $\mathcal{F}$ be an $\mathcal{O}_X$-module. We say that $\mathcal{F}$ is *quasi-coherent* if for every $x \in X$, there exist an open neighborhood U of x and an exact sequence of $\mathcal{O}_X$-modules

$$\mathcal{O}_X^{(J)}|_U \to \mathcal{O}_X^{(I)}|_U \to \mathcal{F}|_U \to 0.$$

The structural sheaf $\mathcal{O}_X$, for example, is quasi-coherent.

5.1.2 Quasi-coherent sheaves on an affine scheme

We are going to classify quasi-coherent sheaves on an affine scheme $X = \operatorname{Spec} A$. Let M be an A-module. We define an $\mathcal{O}_X$-module $\widetilde{M}$ as follows. For any principal open subset $D(f)$ of X, we set $\widetilde{M}(D(f)) = M_f$. We verify, as for Proposition 2.3.1, that this indeed defines a sheaf on X, that $\widetilde{M}(X) = M$, and that $\widetilde{M}_{\mathfrak{p}} = M_{\mathfrak{p}}$ for every $\mathfrak{p} \in \operatorname{Spec} A$. In addition to which it is clearly an $\mathcal{O}_X$-module.

Proposition 1.5. *Let $X = \operatorname{Spec} A$ be an affine scheme. The following properties are true.*

(a) *Let $\{M_i\}_i$ be a family of A-modules. Then $(\oplus_i M_i)^\sim \simeq \oplus_i \widetilde{M_i}$.*

(b) *A sequence of A-modules $L \to M \to N$ is exact if and only if the associated sequence of $\mathcal{O}_X$-modules $\widetilde{L} \to \widetilde{M} \to \widetilde{N}$ is exact.*

(c) *For any A-module M, the sheaf $\widetilde{M}$ is quasi-coherent.*

(d) *Let M, N be two A-modules. Then we have a canonical isomorphism*

$$(M \otimes_A N)^\sim \simeq \widetilde{M} \otimes_{\mathcal{O}_X} \widetilde{N}.$$

Proof (a) results from the definition. (b) Let us suppose $L \to M \to N$ is exact. Then for every $\mathfrak{p} \in \operatorname{Spec} A$, the sequence $L_{\mathfrak{p}} \to M_{\mathfrak{p}} \to N_{\mathfrak{p}}$ is exact because $A_{\mathfrak{p}}$ is flat over A. It follows that $\widetilde{L} \to \widetilde{M} \to \widetilde{N}$ is exact (Proposition 2.2.18). Conversely, let us suppose that we have an exact sequence

$$\widetilde{L} \xrightarrow{\;\alpha\;} \widetilde{M} \xrightarrow{\;\beta\;} \widetilde{N}. \tag{1.1}$$

For any prime ideal $\mathfrak{p} \in \operatorname{Spec} A$, we have commutative diagram

$$
\begin{array}{ccccc}
L & \xrightarrow{\;\alpha(X)\;} & M & \xrightarrow{\;\beta(X)\;} & N \\
\downarrow & & \downarrow & & \downarrow \\
L_{\mathfrak{p}} & \longrightarrow & M_{\mathfrak{p}} & \longrightarrow & N_{\mathfrak{p}}
\end{array}
$$

where the vertical arrows are localizations, and where the second horizontal sequence is exact (it is the fiber at $\mathfrak{p} \in \operatorname{Spec} A$ of exact sequence (1.1)). We deduce from this that $(\operatorname{Ker} \beta(X)/\operatorname{Im} \alpha(X))_{\mathfrak{p}} = 0$. It follows that $\operatorname{Ker} \beta(X) = \operatorname{Im} \alpha(X)$ (Lemma 1.2.12), and hence $L \to M \to N$ is exact. Finally, (c) results from (a) and (b) because we have an exact sequence of A-modules

$$K \to L \to M \to 0$$

with K, L free, and therefore of the form $A^{(J)}$ and $A^{(I)}$.

(d) Let $L = M \otimes_A N$. For any principal open subset $D(f)$ of X, we have a canonical isomorphism of $\mathcal{O}_X(D(f))$-modules

$$
\begin{aligned}
\widetilde{L}(D(f)) &= (M \otimes_A N) \otimes_A A_f \simeq (M \otimes_A A_f) \otimes_{A_f} (N \otimes_A A_f) \\
&= \widetilde{M}(D(f)) \otimes_{\mathcal{O}_X(D(f))} \widetilde{N}(D(f)),
\end{aligned}
$$

compatible with the restriction homomorphisms. This therefore induces an isomorphism of $\mathcal{O}_X$-modules $\widetilde{L} \simeq \widetilde{M} \otimes_{\mathcal{O}_X} \widetilde{N}$ because the principal open subsets form a base for the topology of X (Exercise 2.2.7). $\qquad\square$

Proposition 1.6. Let $\mathcal{F}$ be a quasi-coherent sheaf on a scheme X. Let us suppose X is Noetherian or separated and quasi-compact. Then for any $f \in \mathcal{O}_X(X)$, the canonical homomorphism

$$\mathcal{F}(X)_f = \mathcal{F}(X) \otimes_{\mathcal{O}_X(X)} \mathcal{O}_X(X)_f \to \mathcal{F}(X_f),$$

where $X_f := \{x \in X \mid f_x \in \mathcal{O}_{X,x}^*\}$, is an isomorphism.

Proof Let us first show that every point $x \in X$ has an affine open neighborhood U such that the canonical homomorphism $\mathcal{F}(U)^{\sim} \to \mathcal{F}|_U$ is an isomorphism. By hypothesis, there exist an affine open neighborhood U of x and an exact sequence of $\mathcal{O}_X$-modules

$$\mathcal{O}_X^{(J)}|_U \to \mathcal{O}_X^{(I)}|_U \xrightarrow{\alpha} \mathcal{F}|_U \to 0,$$

Let $M = \mathrm{Im}(\alpha(U))$. By Proposition 1.5(a) and (b), we have an exact sequence

$$\mathcal{O}_X^{(J)}|_U \to \mathcal{O}_X^{(I)}|_U \to \widetilde{M} \to 0,$$

which implies that $\mathcal{F}|_U \simeq \widetilde{M}$. We have $M = \widetilde{M}(U) \simeq \mathcal{F}(U)$. As X is quasi-compact, we can cover X with a finite number of affine open subsets U_i such that $\mathcal{F}|_{U_i} \simeq \mathcal{F}(U_i)^{\sim}$. Let $V_i = U_i \cap X_f = D(f|_{U_i})$. By an argument similar to that of Proposition 2.3.12, we have a commutative diagram

$$\begin{array}{ccccccc}
0 & \longrightarrow & \mathcal{F}(X)_f & \longrightarrow & \oplus_i \mathcal{F}(U_i)_f & \longrightarrow & \oplus_{i,j} \mathcal{F}(U_i \cap U_j)_f \\
& & \downarrow & & \downarrow{\gamma} & & \downarrow \\
0 & \longrightarrow & \mathcal{F}(X_f) & \longrightarrow & \oplus_i \mathcal{F}(V_i) & \longrightarrow & \oplus_{i,j} \mathcal{F}(V_i \cap V_j)
\end{array}$$

where the horizontal sequences are exact. The homomorphism γ is an isomorphism because $\mathcal{F}|_{U_i} \simeq \mathcal{F}(U_i)^{\sim}$. Still by the same arguments of Proposition 2.3.12, we obtain that $\mathcal{F}(X)_f \to \mathcal{F}(X_f)$ is an isomorphism (let us note that our hypotheses on X assure that condition (3.2), Subsection 2.3.1, is satisfied). □

Theorem 1.7. Let X be a scheme, and $\mathcal{F}$ an $\mathcal{O}_X$-module. Then $\mathcal{F}$ is quasi-coherent if and only if for every affine open subset U of X, we have $\mathcal{F}(U)^{\sim} \simeq \mathcal{F}|_U$.

Proof Let us suppose $\mathcal{F}$ is quasi-coherent. Let U be an affine open subset of X. For any $f \in \mathcal{O}_X(U)$, we have $\mathcal{F}(U)_f \simeq \mathcal{F}(D(f))$ by virtue of Proposition 1.6. Hence $\mathcal{F}(U)^{\sim} \simeq \mathcal{F}|_U$. The converse is just Proposition 1.5(c). □

Proposition 1.8. Let X be an affine scheme. Let $0 \to \mathcal{F} \to \mathcal{G} \to \mathcal{H} \to 0$ be an exact sequence of $\mathcal{O}_X$-modules with $\mathcal{F}$ quasi-coherent. Then the sequence

$$0 \to \mathcal{F}(X) \to \mathcal{G}(X) \to \mathcal{H}(X) \to 0$$

is exact.

Proof It suffices to show that $\mathcal{G}(X) \to \mathcal{H}(X)$ is surjective, the remainder of the sequence always being exact. We identify $\mathcal{F}$ with a sub-$\mathcal{O}_X$-module of $\mathcal{G}$. Let $s \in \mathcal{H}(X)$. There exists a covering of X by a finite number of principal open subsets U_i such that $s|_{U_i}$ lifts to a section $t_i \in \mathcal{G}(U_i)$. For any pair i, j, the image of $t_i|_{U_{ij}} - t_j|_{U_{ij}}$ in $\mathcal{H}(U_{ij})$, where $U_{ij} = U_i \cap U_j$, is zero. It follows that

$$f_{ij} := t_i|_{U_{ij}} - t_j|_{U_{ij}} \in \mathcal{F}(U_{ij}).$$

For any triplet (i, j, k), we have

$$f_{ij}|_{U_{ijk}} - f_{ik}|_{U_{ijk}} + f_{jk}|_{U_{ijk}} = [(t_i - t_j) - (t_i - t_k) + (t_j - t_k)]|_{U_{ijk}} = 0,$$

where $U_{ijk} = U_i \cap U_j \cap U_k$. It results from Lemma 2.17 (the case $p = 1$; one can verify that its proof does not use any results of this subsection other than Theorem 1.7) that there exist $f_i \in \mathcal{F}(U_i)$ such that $f_i|_{U_{ij}} - f_j|_{U_{ij}} = f_{ij}$. The $t_i - f_i$ therefore glue to a section $t \in \mathcal{G}(X)$, whose image in $\mathcal{H}(X)$ is equal to s. □

Remark 1.9. This proposition characterizes affine schemes among Noetherian or separated and quasi-compact schemes. See the proof of Theorem 2.23.

5.1.3 Coherent sheaves

Definition 1.10. Let $(X, \mathcal{O}_X)$ be a ringed topological space, and let $\mathcal{F}$ be an $\mathcal{O}_X$-module. We say that $\mathcal{F}$ is *finitely generated* if for every $x \in X$, there exist an open neighborhood U of x, an integer $n \geq 1$, and a surjective homomorphism $\mathcal{O}_X^n|_U \to \mathcal{F}|_U$. We say that $\mathcal{F}$ is *coherent* if it is finitely generated, and if for every open subset U of X, and for every homomorphism $\alpha : \mathcal{O}_X^n|_U \to \mathcal{F}|_U$, the kernel $\mathrm{Ker}\,\alpha$ is finitely generated. The notion of finitely generated (resp. coherent) sheaf is of local nature on X.

To simplify, we will essentially consider coherent sheaves in the case of locally Noetherian schemes.

Proposition 1.11. *Let X be a scheme. Let $\mathcal{F}$ be a quasi-coherent $\mathcal{O}_X$-module. Let us consider the following properties:*

(i) *$\mathcal{F}$ is coherent;*

(ii) *$\mathcal{F}$ is finitely generated;*

(iii) *for every affine open subset U of X, $\mathcal{F}(U)$ is finitely generated over $\mathcal{O}_X(U)$.*

Then (i) $\Longrightarrow$ (ii) $\Longrightarrow$ (iii). Moreover, if X is locally Noetherian, then these properties are equivalent.

Proof By definition, (i) implies (ii). Let us suppose $\mathcal{F}$ is finitely generated. Let U be an affine open subset of X. Then U can be covered with a finite number of principal open subsets U_i such that there exists an exact sequence $\mathcal{O}_X^{n_i}|_{U_i} \to \mathcal{F}|_{U_i} \to 0$. It follows that $\mathcal{O}_X(U_i)^{n_i} \to \mathcal{F}(U_i) \to 0$ is exact (Proposition 1.5(b)). In particular, $\mathcal{F}(U_i)$ is finitely generated over $\mathcal{O}_X(U_i)$. As we have $\mathcal{F}(U_i) =$

$\mathcal{F}(U) \otimes_{\mathcal{O}_X(U)} \mathcal{O}_X(U_i)$, there exists a finitely generated sub-$\mathcal{O}_X(U)$-module M of $\mathcal{F}(U)$ such that $M \otimes_{\mathcal{O}_X(U)} \mathcal{O}_X(U_i) = \mathcal{F}(U_i)$. Enlarging M if necessary, we may suppose this equality is true for every i. The sequence $\widetilde{M} \to \mathcal{F}|_U \to 0$ is then exact because it is exact on every U_i. Consequently, $M \to \mathcal{F}(U)$ is surjective, and (ii) implies (iii).

Let us now suppose (iii) is true and X locally Noetherian. Let us show that $\mathcal{F}$ is coherent. Let V be an open subset of X and $\alpha : \mathcal{O}_X^n|_V \to \mathcal{F}|_V$ a homomorphism. We must show that $\mathrm{Ker}(\alpha)$ is finitely generated. As this is a local property, we may suppose that V is affine, and hence that $\mathcal{F}|_V = \widetilde{N}$. Then $\mathrm{Ker}\,\alpha = (\mathrm{Ker}\,\alpha(V))^{\sim}$ (Proposition 1.5(b)). Now $\mathrm{Ker}\,\alpha(V)$ is finitely generated because $\mathcal{O}_X(V)$ is Noetherian; hence $\mathrm{Ker}\,\alpha$ is finitely generated.   □

Proposition 1.12. *Let X be a scheme. We have the following properties for $\mathcal{O}_X$-modules.*

(a) *A direct sum of quasi-coherent sheaves is quasi-coherent; a finite direct sum of finitely generated quasi-coherent sheaves is finitely generated.*

(b) *If $\mathcal{F}$, $\mathcal{G}$ are quasi-coherent (resp. finitely generated quasi-coherent) sheaves, then so is $\mathcal{F} \otimes_{\mathcal{O}_X} \mathcal{G}$. Moreover, for any affine open subset U of X, we have $(\mathcal{F} \otimes_{\mathcal{O}_X} \mathcal{G})(U) = \mathcal{F}(U) \otimes_{\mathcal{O}_X(U)} \mathcal{G}(U)$.*

(c) *Let $u : \mathcal{F} \to \mathcal{G}$ be a morphism of quasi-coherent sheaves; then $\mathrm{Ker}\,u$, $\mathrm{Im}\,u$, and $\mathrm{Coker}\,u$ are quasi-coherent.*

(d) *Let $0 \to \mathcal{F} \to \mathcal{G} \to \mathcal{H} \to 0$ be an exact sequence of $\mathcal{O}_X$-modules. If two of them are quasi-coherent, then so is the third.*

(e) *If X is locally Noetherian, then properties (c) and (d) are true for coherent sheaves.*

Proof Properties (a)–(c) easily follow from all the above, and (e) is trivially true. The least obvious part of (d) is that where $\mathcal{F}$ and $\mathcal{H}$ are quasi-coherent. We may suppose X is affine. By Proposition 1.8, the sequence $0 \to \mathcal{F}(X) \to \mathcal{G}(X) \to \mathcal{H}(X) \to 0$ is exact. We then have a commutative diagram of exact sequences

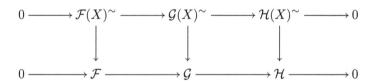

where the first and last vertical arrows are isomorphisms. It is then easy to see that $\mathcal{G}(X)^{\sim} \to \mathcal{G}$ is an isomorphism.   □

Definition 1.13. Let $f : X \to Y$ be a morphism of schemes. The homomorphism $f^{\#} : \mathcal{O}_Y \to f_*\mathcal{O}_X$ induces a homomorphism of sheaves of rings $f^{-1}\mathcal{O}_Y \to \mathcal{O}_X$ (Exercise 2.2.13(a)). Let $\mathcal{G}$ be an $\mathcal{O}_Y$-module. On the ringed topological space $(X, f^{-1}\mathcal{O}_Y)$, we can consider the tensor product of $f^{-1}\mathcal{O}_Y$-modules

$$f^*\mathcal{G} := f^{-1}\mathcal{G} \otimes_{f^{-1}\mathcal{O}_Y} \mathcal{O}_X.$$

It is an $\mathcal{O}_X$-module by the multiplication on the right. We call it the *inverse image* (or *pull-back*) of $\mathcal{G}$ by f. For example, $f^*\mathcal{O}_Y = \mathcal{O}_X$.

Proposition 1.14. *Let $f : X \to Y$ be a morphism of schemes.*

(a) *Let $\mathcal{G}$ be an $\mathcal{O}_Y$-module. Then for any $x \in X$, we have a canonical isomorphism*

$$(f^*\mathcal{G})_x \simeq \mathcal{G}_{f(x)} \otimes_{\mathcal{O}_{Y,f(x)}} \mathcal{O}_{X,x}.$$

(b) *Let us suppose $\mathcal{G}$ is quasi-coherent. Let U be an affine open subset of X such that $f(U)$ is contained in an affine open subset V of Y. Then*

$$f^*\mathcal{G}|_U \simeq (\mathcal{G}(V) \otimes_{\mathcal{O}_Y(V)} \mathcal{O}_X(U))^{\sim}.$$

In particular, $f^*\mathcal{G}$ is quasi-coherent. It is finitely generated if $\mathcal{G}$ is finitely generated.

(c) *Let $\mathcal{F}$ be a quasi-coherent sheaf on X. If X is Noetherian, or if f is separated and quasi-compact, then $f_*\mathcal{F}$ is quasi-coherent on Y.*

(d) *If f is finite and $\mathcal{F}$ quasi-coherent and finitely generated, then $f_*\mathcal{F}$ is quasi-coherent and finitely generated on Y.*

Proof (a) results from the fact that $(f^{-1}\mathcal{G})_x = \mathcal{G}_{f(x)}$ (see Subsection 2.2.1, page 37).

(b) Let $g : U \to V$ be the restriction of f to U. Then $(f^*\mathcal{G})|_U = g^*(\mathcal{G}|_V)$. We may therefore suppose that $X = U = \operatorname{Spec} B$ and $Y = V = \operatorname{Spec} A$. The property is trivially true if $\mathcal{G} = \mathcal{O}_Y^{(I)}$ because $f^*\mathcal{O}_Y = \mathcal{O}_X$ and f^* commutes with taking direct sums. In the general case, let us write an exact sequence

$$K \xrightarrow{\alpha} L \to \mathcal{G}(Y) \to 0$$

with K, L free over A. We therefore have exact sequences

$$\tilde{K} \to \tilde{L} \to \mathcal{G} \to 0 \quad \text{and} \quad f^*\tilde{K} \xrightarrow{\beta} f^*\tilde{L} \to f^*\mathcal{G} \to 0$$

(Proposition 1.1.12). As β is associated to $\alpha_B : K \otimes_A B \to L \otimes_A B$, we have

$$f^*\mathcal{G} = \operatorname{Coker} \beta = (\operatorname{Coker} \alpha_B)^{\sim} = ((\operatorname{Coker} \alpha) \otimes_A B)^{\sim} = (\mathcal{G}(Y) \otimes_A B)^{\sim},$$

which proves (b).

(c) As the property is of local nature on Y, we may suppose Y is affine. Then X is Noetherian or separated and quasi-compact. For any $g \in \mathcal{O}_Y(Y)$, let g' denote its image in $\mathcal{O}_X(X)$. We have

$$f_*\mathcal{F}(D(g)) = \mathcal{F}(f^{-1}(D(g))) = \mathcal{F}(D(g')) \simeq \mathcal{F}(X)_{g'} = \mathcal{F}(X) \otimes_{\mathcal{O}_Y(Y)} \mathcal{O}_Y(D(g))$$

(the isomorphism in the middle comes from Proposition 1.6). This shows that $f_*\mathcal{F} \simeq \mathcal{F}(X)^{\sim}$ on Y.

(d) We may suppose Y is affine. As f is affine, hence separated and quasi-compact, we have just seen that $f_*\mathcal{F} = \mathcal{F}(X)^\sim$. As X is affine and finite over Y, $\mathcal{F}(X)$ is finitely generated over $\mathcal{O}_X(X)$ and consequently finitely generated over $\mathcal{O}_Y(Y)$. □

Let us now consider coherent sheaves of ideals on a scheme X. They allow us to classify the closed subschemes of X.

Proposition 1.15. *Let X be a scheme, and let Z be a closed subscheme of X. Let $i : Z \to X$ denote the canonical closed immersion. Then $\operatorname{Ker} i^\#$ is a quasi-coherent sheaf of ideals on X. The correspondence $Z \mapsto \operatorname{Ker} i^\#$ establishes a bijection between the closed subschemes of X and the quasi-coherent sheaves of ideals on X.*

Proof Using Proposition 2.2.24, we see that the correspondence $\rho : Z \mapsto \operatorname{Ker} i^\#$ establishes a bijection between the closed ringed topological subspaces of X and the sheaves of ideals of $\mathcal{O}_X$. If Z is a closed subscheme, then $\operatorname{Ker} i^\#$ is quasi-coherent by Proposition 2.3.20 and Lemma 2.3.17. Conversely, let us suppose $\mathcal{J} := \operatorname{Ker} i^\#$ is quasi-coherent. Let j be the closed topological immersion $V(\mathcal{J}) \to X$. Then $\mathcal{J}$ corresponds to $(V(\mathcal{J}), j^{-1}(\mathcal{O}_X/\mathcal{J}))$ via ρ. We must therefore show that $(V(\mathcal{J}), j^{-1}(\mathcal{O}_X/\mathcal{J}))$ is a scheme. We may suppose that $X = \operatorname{Spec} A$ is affine, and therefore that $\mathcal{J} = \tilde{J}$ for some ideal J of A. Now by Lemma 2.3.17, the closed subscheme $\operatorname{Spec} A/J$ corresponds, via ρ^{-1}, to the sheaf of ideals $\tilde{J}$. Hence $V(\mathcal{J}) \simeq \operatorname{Spec} A/J$ is a scheme, which completes the proof. □

Remark 1.16. Let X be a scheme. For any quasi-coherent sheaf of ideals $\mathcal{J}$, the closed subset $V(\mathcal{J})$ will implicitly be endowed with the closed subscheme structure as above.

5.1.4   Quasi-coherent sheaves on a projective scheme

In this subsection, we are interested in quasi-coherent sheaves on a projective scheme. On such a scheme X, the sheaves of the form $\mathcal{O}_X(n)$ (see Definition 1.23) play an essential role. An illustration of this is given by Theorem 1.27. Next, we show that closed subschemes of a projective scheme are also projectives schemes (Proposition 1.30). Finally, we study ample sheaves, which are a sort of generalization of the sheaves $\mathcal{O}_X(n)$ (Theorem 1.34).

Let $B = \oplus_{n\geq0}B_n$ be a graded ring, and let $M = \oplus_{n\in\mathbb{Z}}M_n$ be a graded B-module (i.e., $B_nM_m \subseteq M_{n+m}$ for every $n \geq 0$ and every $m \in \mathbb{Z}$). We are going to construct a quasi-coherent sheaf on $X = \operatorname{Proj} B$ in the following way. For any non-nilpotent homogeneous $f \in B_+$, we let

$$M_{(f)} = \{mf^{-d} \in M_f \mid m \in M_{d\deg f}\}.$$

These are the elements of degree 0 in M_f. This is, in a natural way, a $B_{(f)}$-module.

Proposition 1.17. With the notation above, there exists a unique quasi-coherent $\mathcal{O}_X$-module $\widetilde{M}$ (do not confuse this with the affine case) such that:

(i) for any non-nilpotent homogeneous $f \in B_+$, $\widetilde{M}|_{D_+(f)}$ is the quasi-coherent sheaf $(M_{(f)})^{\sim}$ on $D_+(f) = \operatorname{Spec} B_{(f)}$;

(ii) for any $\mathfrak{p} \in \operatorname{Proj} B$, $\widetilde{M}_{\mathfrak{p}}$ is isomorphic to $M_{(\mathfrak{p})}$, the set of elements of degree 0 in $M_{\mathfrak{p}}$.

Proof We define $\widetilde{M}$ on the principal open sets $D_+(f)$ by property (i). We must verify, which is routine, that the glueing conditions (Exercise 2.2.8) are satisfied. The second property (ii) can be verified just as easily.   □

Another way to construct $\widetilde{M}$ is by considering the canonical injection $\rho :$ $\operatorname{Proj} B \to \operatorname{Spec} B$ (which is a continuous map), and taking the inverse image $\rho^{-1}\mathcal{F}$, where $\mathcal{F}$ is the quasi-coherent sheaf on $\operatorname{Spec} B$ associated to M. For any homogeneous prime ideal $\mathfrak{p} \in \operatorname{Proj} B$, we have $(\rho^{-1}\mathcal{F})_{\mathfrak{p}} = M_{\mathfrak{p}}$. Then $\widetilde{M}$ is the subsheaf of $\rho^{-1}\mathcal{F}$ of homogeneous elements of degree 0.

Remark 1.18. Let $M = \oplus_{n \geq 0} M_n$ be a graded B-module. Let $N = \oplus_{n \geq n_0} M_n$ for some $n_0 \geq 0$. Then $\widetilde{M} = \widetilde{N}$ because $M_{(f)} = N_{(f)}$ for every homogeneous element $f \in B$. This implies, in particular, that $\widetilde{M}$ does not determine M, contrary to the affine case.

Example 1.19. Let A be a ring, and B be a graded A-algebra. For any $n \in \mathbb{Z}$, we let $B(n)$ denote the graded B-module defined by $B(n)_d = B_{n+d}$. This is a *'twist'* of B. Let $X = \operatorname{Proj} B$, and let $\mathcal{O}_X(n)$ denote the $\mathcal{O}_X$-module $B(n)^{\sim}$. These sheaves play an essential role in the theory of projective schemes.

For any homogeneous $f \in B$ of degree 1, we have $B(n)_{(f)} = f^n B_{(f)}$. Thus on the affine open subset $D_+(f)$, we have $\mathcal{O}_X(n)|_{D_+(f)} = f^n \mathcal{O}_X|_{D_+(f)}$. We easily verify that $\mathcal{O}_X(n) \otimes_{\mathcal{O}_X} \mathcal{O}_X(m) = \mathcal{O}_X(n+m)$ (Proposition 1.12(b)).

Definition 1.20. Let S be a scheme. Let $f : \mathbb{P}_S^d \to \mathbb{P}_{\mathbb{Z}}^d$ be the canonical morphism. For any $n \in \mathbb{Z}$, we let

$$\mathcal{O}_{\mathbb{P}_S^d}(n) = f^*\mathcal{O}_{\mathbb{P}_{\mathbb{Z}}^d}(n).$$

If S is affine, we verify without difficulty that this coincides with the notation of the example above (see also Exercise 1.20).

Definition 1.21. Let X be a scheme. We say that an $\mathcal{O}_X$-module $\mathcal{L}$ is *invertible* if for every point $x \in X$, there exist an open neighborhood U of x and an isomorphism of $\mathcal{O}_U$-modules $\mathcal{O}_X|_U \simeq \mathcal{L}|_U$. If X is locally Noetherian, this comes down to saying that $\mathcal{L}$ is coherent and that $\mathcal{L}_x$ is free of rank 1 over $\mathcal{O}_{X,x}$ for every $x \in X$ (see also Exercise 1.12).

See Exercise 1.12(d) for a justification of the word 'invertible'. We easily see that the tensor product of two invertible sheaves is invertible, and that the inverse image of an invertible sheaf by a morphism is an invertible sheaf. For example, the sheaves $\mathcal{O}_X(n)$ on a projective space $X = \mathbb{P}_S^d$ are invertible.

Lemma 1.22. *Let* $B = A[T_0, \ldots, T_n]$. *With the notation of Example 1.19, we have* $\mathcal{O}_X(n)(X) = B_n$ *if* $n \geq 0$ *and* $\mathcal{O}_X(n) = 0$ *if* $n < 0$. *In other words,* $\oplus_{n \in \mathbb{Z}} \mathcal{O}_X(n)(X) = B$.

Proof We may suppose $d \geq 1$. The data of a global section $f \in \mathcal{O}_X(n)(X)$ consists of the data of local sections of $\mathcal{O}_X(n)(D_+(T_i))$ which coincide on the intersections. The A-modules $\mathcal{O}_X(n)(D_+(T_i))$ and $\mathcal{O}_X(n)(D_+(T_i) \cap D_+(T_j))$ are canonically sub-A-modules of $A[T_0, \ldots, T_d, T_0^{-1}, \ldots, T_d^{-1}]$. We may therefore consider f as an element of the latter. Let $1 \leq i \leq d$. The fact that $f \in T_i^n \mathcal{O}_X(D_+(T_i))$ implies that f does not contain T_0 in its denominator. As $f \in T_0^n \mathcal{O}_X(D_+(T_0))$, this implies that f is homogeneous of degree n if $n \geq 0$, and that $f = 0$ if $n < 0$. Conversely, it can immediately be verified that $B_n \subseteq \mathcal{O}_X(n)(X)$.   □

Definition 1.23. Let $X = \mathbb{P}_A^d$. For any integer $n \in \mathbb{Z}$ and any quasi-coherent sheaf $\mathcal{F}$ on X, we denote the tensor product $\mathcal{F} \otimes_{\mathcal{O}_X} \mathcal{O}_X(n)$ by $\mathcal{F}(n)$. The $\mathcal{F}(n)$ are called the *twists* of $\mathcal{F}$. For any affine open subset U of X, we have $\mathcal{F}(n)(U) = \mathcal{F}(U) \otimes_{\mathcal{O}_X(U)} \mathcal{O}_X(n)(U)$ (Proposition 1.12(b)).

Definition 1.24. Let X be a scheme. We say that an $\mathcal{O}_X$-module $\mathcal{F}$ is *generated* by $s_0, \ldots, s_d \in \mathcal{F}(X)$ if $\mathcal{F}_x = \sum_i (s_i)_x \mathcal{O}_{X,x}$ for every $x \in X$ (see Definition 1.2). Let $\mathcal{L}$ be an invertible sheaf. Let $s \in \mathcal{L}(X)$. Let us set

$$X_s := \{ x \in X \mid \mathcal{L}_x = s_x \mathcal{O}_{X,x} \}.$$

This is an open subset of X. When $\mathcal{L} = \mathcal{O}_X$, X_s coincides with Definition 2.3.11.

The following lemma generalizes Proposition 1.6 by replacing $\mathcal{O}_X$ by an invertible sheaf $\mathcal{L}$. For simplicity, we denote the tensor power $\mathcal{L}^{\otimes n}$ by $\mathcal{L}^n$ and the tensor power $s^{\otimes n}$, $s \in \mathcal{L}(X)$, by s^n.

Lemma 1.25. *Let* X *be a Noetherian or separated and quasi-compact scheme, let* $\mathcal{F}$ *be a quasi-coherent sheaf on* X, *and* $\mathcal{L}$ *an invertible sheaf on* X. *Let us fix sections* $f \in \mathcal{F}(X)$ *and* $s \in \mathcal{L}(X)$.

(a) *If* $f|_{X_s} = 0$, *then there exists an* $n \geq 1$ *such that* $f \otimes s^n = 0$ *in* $(\mathcal{F} \otimes \mathcal{L}^n)(X)$.

(b) *Let* $g \in \mathcal{F}(X_s)$. *Then there exists an* $n_0 \geq 1$ *such that* $g \otimes (s^n|_{X_s})$ *lifts to a section of* $(\mathcal{F} \otimes \mathcal{L}^n)(X)$ *for all* $n \geq n_0$.

Proof (See also Exercise 1.18.) We cover X by a finite number of affine open subsets $X_1, \ldots, X_r$ such that $\mathcal{L}|_{X_i}$ is free for every $i \leq r$. Let e_i be a generator of $\mathcal{L}|_{X_i}$, and let $h_i \in \mathcal{O}_X(X_i)$ be such that $s|_{X_i} = e_i h_i$. Then $X_s \cap X_i$ is the principal open subset $D(h_i)$ of X_i.

(a) The $f|_{X_s} = 0$ hypothesis implies that $f|_{X_s \cap X_i} = 0$. Hence there exists an integer $n \geq 1$ such that $f|_{X_i} h_i^n = 0$. We can take n sufficiently large so that it is independent of i. On X_i, the isomorphism $\mathcal{O}_{X_i} \to \mathcal{L}^n|_{X_i}$ defined as the multiplication by e_i^n induces an isomorphism $\varphi_{i,n} : \mathcal{F}|_{X_i} \to (\mathcal{F} \otimes \mathcal{L}^n)|_{X_i}$. We have

$$0 = \varphi_{i,n}(f|_{X_i} h_i^n) = f|_{X_i} h_i^n \otimes e_i^n = (f \otimes s^n)|_{X_i}.$$

Hence $(f \otimes s^n)|_{X_i} = 0$ for every i. Consequently, $f \otimes s^n = 0$.

(b) There exists an $m \geq 1$ such that for every $i \leq r$, there exists an $f_i \in \mathcal{F}(X_i)$ with

$$g|_{X_s \cap X_i} h_i^m|_{X_s \cap X_i} = f_i|_{X_s \cap X_i}.$$

Let us set $t_i = \varphi_{i,m}(f_i) \in (\mathcal{F} \otimes \mathcal{L}^{\otimes m})(X_i)$. Then $t_i|_{X_s \cap X_i} = g|_{X_s \cap X_i} \otimes s^m|_{X_s \cap X_i}$. For any pair i, j, we then have

$$t_i|_{X_s \cap (X_i \cap X_j)} = t_j|_{X_s \cap (X_i \cap X_j)}.$$

As $X_i \cap X_j$ is Noetherian or separated and quasi-compact, by (a) there exists an integer $q \geq 1$ such that

$$(t_i|_{X_i \cap X_j} - t_j|_{X_i \cap X_j}) \otimes s^q|_{X_i \cap X_j} = 0 \in ((\mathcal{F} \otimes \mathcal{L}^m) \otimes \mathcal{L}^q)(X_i \cap X_j).$$

Let $n_0 = m + q$. Then for all $n \geq n_0$, the sections $t_i \otimes s^{q+n-n_0}|_{X_i} \in (\mathcal{F} \otimes \mathcal{L}^n)(X_i)$ coincide on the intersections $X_i \cap X_j$, and therefore glue to a section $t \in (\mathcal{F} \otimes \mathcal{L}^n)(X)$ whose restriction to X_s is equal to $g \otimes (s^n|_{X_s})$. $\square$

Definition 1.26. Let $f : X \to \operatorname{Spec} A$ be a scheme over a ring A. Let $i : X \to \mathbb{P}_A^d$ be an immersion (Exercise 3.2.3). We let $\mathcal{O}_X(n) = i^* \mathcal{O}_{\mathbb{P}_A^d}(n)$. This is an invertible sheaf on X which depends on i. The sheaf $\mathcal{O}_X(1)$ is called a *very ample sheaf* (relative to f). Let $\mathcal{F}$ be a quasi-coherent $\mathcal{O}_X$-module. We denote the tensor product $\mathcal{F} \otimes_{\mathcal{O}_X} \mathcal{O}_X(n)$ by $\mathcal{F}(n)$.

Theorem 1.27. *Let X be a projective scheme over a ring A. Then for any finitely generated quasi-coherent sheaf $\mathcal{F}$ on X, there exists an $n_0 \geq 0$ such that $\mathcal{F}(n)$ is generated by its global sections for every $n \geq n_0$.*

Proof Let $f : X \to \mathbb{P}_A^d$ be a closed immersion. Then $f_* \mathcal{F}$ is a finitely generated quasi-coherent sheaf on $\mathbb{P}_A^d$ (Proposition 1.14(d)), and we have $f_*(\mathcal{F}(n)) = f_* \mathcal{F}(n)$ (Exercise 1.16(c)). As $f_* \mathcal{F}(n)(\mathbb{P}_A^d) = \mathcal{F}(n)(X)$ by definition of f_*, it suffices to show the theorem with $X = \mathbb{P}_A^d$.

Let $U_i = D_+(T_i)$, $0 \leq i \leq d$. Then $\mathcal{F}(U_i)$ is generated by a finite number of elements s_{ij}, $j \leq m$. It follows from Lemma 1.25 that there exists an $n_0 \geq 0$ such that for every $i \leq d$, $j \leq m$, and $n \geq n_0$, $T_i^n \otimes s_{ij}$ is the restriction to U_i of a global section of $\mathcal{F}(n)$. Since $\mathcal{F}(n)(U_i) = T_i^n \otimes \mathcal{F}(U_i)$, this immediately implies that $\mathcal{F}(n)$ is generated by its global sections. $\square$

Corollary 1.28. *With the hypotheses of the theorem, there exist integers $m \in \mathbb{Z}$, $r \geq 1$ such that $\mathcal{F}$ is a quotient sheaf of $\mathcal{O}_X(m)^r$.*

Proof Let $n \in \mathbb{Z}$ be such that $\mathcal{G} := \mathcal{F}(n)$ is generated by its global sections. We cover X with a finite number of affine open subsets U_i, $1 \leq i \leq d$. On each U_i, $\mathcal{G}(U_i)$ is generated by a finite number of global sections. Therefore there exist a finite number of global sections $s_1, \ldots, s_r \in \mathcal{G}(X)$ which generate $\mathcal{G}$ on every open subset U_i. We immediately deduce from this a surjective homomorphism $\mathcal{O}_X^r \to \mathcal{G}$ (Lemma 1.3). By tensoring with $\mathcal{O}_X(-n)$, we obtain a surjective homomorphism $\mathcal{O}_X(-n)^r \to \mathcal{F}$. This proves the corollary. $\square$

On an affine scheme, a quasi-coherent sheaf $\mathcal{F}$ is determined by its global sections $\mathcal{F}(X)$. The following lemma is an analogue of this result for projective schemes, but by taking the global sections of all twists of $\mathcal{F}$.

Lemma 1.29. *Let A be a ring, $B = A[T_0, \ldots, T_d]$, and $X = \operatorname{Proj} B$. Then for any quasi-coherent sheaf $\mathcal{F}$ on X, the direct sum $\oplus_{n \geq 0} \mathcal{F}(n)(X)$ is naturally a graded B-module, and we have*

$$(\oplus_{n \geq 0} \mathcal{F}(n)(X))^\sim \simeq \mathcal{F}.$$

Proof Let $M = \oplus_{n \geq 0} \mathcal{F}(n)(X)$. For any $b \in B_m = \mathcal{O}_X(m)(X)$ and for any $t \in \mathcal{F}(n)(X)$, the tensor product $b \otimes t \in \mathcal{O}_X(m)(X) \otimes \mathcal{F}(n)(X)$ is naturally sent to an element of $(\mathcal{O}_X(m) \otimes \mathcal{F}(n))(X) = \mathcal{F}(n+m)(X)$. This defines the structure of a graded B-module on M.

Let $f \in B$ be one of the elements T_i. Let $U = D_+(f)$. Let us show that we have a canonical isomorphism $\varphi : M_{(f)} \simeq \mathcal{F}(U)$. Let $f^{-n}t \in M_{(f)}$ with $t \in \mathcal{F}(n)(X)$. Then $t|_U \in \mathcal{F}(n)(U) = f^n \otimes \mathcal{F}(U)$. There exists a unique $s \in \mathcal{F}(U)$ such that $t|_U = f^n \otimes s$, because f^n is a basis of $\mathcal{O}_X(n)(U)$ over $\mathcal{O}_X(U)$. Let us set $\varphi(f^{-n}t) = s$. It follows from Lemma 1.25(b) that φ is surjective. If $s = 0$, then $t|_U = 0$. Hence there exists an $m \geq 1$ such that $f^m \otimes t = 0 \in \mathcal{O}_X(m)(X) \otimes \mathcal{F}(n)(X)$ (Lemma 1.25(a)). It follows that $f^m t = 0$ in M. Hence $f^{-n}t = 0$ and φ is bijective. When we let f vary, this defines an isomorphism $M^\sim \simeq \mathcal{F}$. $\square$

Proposition 1.30. *Let A be a ring, $B = A[T_0, \ldots, T_d]$, and $X = \operatorname{Proj} B$. Then any closed subscheme Z of X is of the form $\operatorname{Proj} B/I$ for a homogeneous ideal I of B. In particular, any projective scheme over A is isomorphic to $\operatorname{Proj} C$, where C is a homogeneous A-algebra.*

Proof Let $\mathcal{I}$ be the sheaf of ideals of $\mathcal{O}_X$ which defines the closed subscheme Z. Then $\mathcal{I}$ is quasi-coherent by Proposition 1.15. Let us set $I = \oplus_{n \geq 0} \mathcal{I}(n)(X)$. The canonical homomorphism $\mathcal{I}(n) \to \mathcal{O}_X(n)$ is injective because it is injective at every point $x \in X$ owing to the fact that $\mathcal{O}_X(n)_x \simeq \mathcal{O}_{X,x}$ is flat over $\mathcal{O}_{X,x}$. Therefore I is an ideal of $\oplus_{n \geq 0} B_n = B$ (Lemma 1.22). By the lemma above, we have the equality of sheaves of ideals $\tilde{I} = \mathcal{I}$. It is now easy to see that $V(\tilde{I}) = \operatorname{Proj} B/I$ as schemes. This proves the proposition. $\square$

To conclude, we are going to consider the relation between morphisms to a projective space and invertible sheaves. Let $\mathcal{L}$ be an invertible sheaf on X. Let $s \in \mathcal{L}(X)$. Then the multiplication by s induces an isomorphism

$$\mathcal{O}_X|_{X_s} \xrightarrow{\cdot s} s\mathcal{O}_X|_{X_s} = \mathcal{L}|_{X_s}$$

because s_x is a basis of $\mathcal{L}_x$ over $\mathcal{O}_{X,x}$ for every $x \in X_s$. In particular, for every $t \in \mathcal{L}(X)$, we can, without ambiguity, write t/s as an element of $\mathcal{L}(X_s)$.

Proposition 1.31. *Let* $Y = \operatorname{Proj} A[T_0, \ldots, T_d]$ *be a projective space over a ring* A, *and let* X *be a scheme over* A.

(a) *Let* $f : X \to Y$ *be a morphism of* A-*schemes. Then* $f^*\mathcal{O}_Y(1)$ *is an invertible sheaf on* X, *generated by* $d+1$ *of its global sections.*

(b) *Conversely, for any invertible sheaf* $\mathcal{L}$ *on* X *generated by* $d+1$ *global sections* $s_0, \ldots, s_d$, *there exists a morphism* $f : X \to Y$ *such that* $\mathcal{L} \simeq f^*\mathcal{O}_Y(1)$ *and* $f^*T_i = s_i$ *via this isomorphism.*

Proof (a) By the computations of Example 1.19, $\mathcal{O}_Y(1)$ is generated by $d+1$ global sections $T_0, \ldots, T_d$. These sections canonically induce global sections $s_0, \ldots, s_d$ of $f^*\mathcal{O}_Y(1)$. Let $x \in X$, let $y = f(x)$. Then we have

$$(f^*\mathcal{O}_Y(1))_x = \mathcal{O}_Y(1)_y \otimes_{\mathcal{O}_{Y,y}} \mathcal{O}_{X,x} = \sum_i (T_i)_y \mathcal{O}_{Y,y} \otimes_{\mathcal{O}_{Y,y}} \mathcal{O}_{X,x} = \sum_i (s_i)_x \mathcal{O}_{X,x}.$$

Hence $f^*\mathcal{O}_Y(1)$ is generated by the global sections $s_0, \ldots, s_d$.

(b) The open subsets X_{s_i} cover X. For every $i \le d$, let us consider the morphism $f_i : X_{s_i} \to D_+(T_i)$ corresponding to the ring homomorphism

$$\mathcal{O}_Y(D_+(T_i)) \to \mathcal{O}_X(X_{s_i}), \quad T_j/T_i \mapsto s_j/s_i \in \mathcal{O}_X(X_{s_i}).$$

It is clear that the morphisms f_i glue to a morphism $f : X \to Y$, and that $\mathcal{L} \simeq f^*\mathcal{O}_Y(1)$ by Proposition 1.14(b) and the description in Example 1.19. □

Remark 1.32. Let us fix a set of sections $\{s_0, \ldots, s_d\}$ of $\mathcal{L}(X)$ which generates $\mathcal{L}$. Then the morphism f as in (b) is unique. Moreover, let us consider an isomorphism $\varphi : \mathcal{L} \to \mathcal{L}'$ of invertible sheaves on X, and let t_i be the image of s_i under $\varphi(X)$. By examining the proof above, we easily see that the morphism $X \to Y$ associated to $\mathcal{L}'$ and to the sections $t_0, \ldots, t_d$ is identical to f.

A more flexible notion than that of a very ample sheaf is the following.

Definition 1.33. Let X be a quasi-compact scheme. Let $\mathcal{L}$ be an invertible sheaf on X. We say that $\mathcal{L}$ is *ample* if for any finitely generated quasi-coherent sheaf $\mathcal{F}$ on X, there exists an integer $n_0 \ge 1$ such that for every $n \ge n_0$, $\mathcal{F} \otimes \mathcal{L}^{\otimes n}$ is generated by its global sections. Theorem 1.27 says that a very ample sheaf (on a projective scheme over an affine scheme) is ample. Note that the notion of ample sheaf is an absolute notion (independent of a base scheme).

Theorem 1.34. *Let* $f : X \to \operatorname{Spec} A$ *be a morphism of finite type to an affine scheme* $\operatorname{Spec} A$. *Let us suppose that* X *is Noetherian or that* f *is separated. Let* $\mathcal{L}$ *be an ample sheaf on* X. *Then there exists an* $m \ge 1$ *such that* $\mathcal{L}^{\otimes m}$ *is very ample for* f.

Proof Let us denote $\mathcal{L}^{\otimes n}$ by $\mathcal{L}^n$ if there is no confusion possible. Let x be a point of X. Let us show that there exist an $n = n(x)$ and a section $s \in \mathcal{L}^n(X)$ such that

X_s is an affine neighborhood of x. Let U be an affine open neighborhood of x such that $\mathcal{L}|_U$ is free, and let $\mathcal{J}$ be a sheaf of ideals of $\mathcal{O}_X$ such that $V(\mathcal{J}) = X \setminus U$. For any $n \geq 1$, $\mathcal{J} \otimes \mathcal{L}^n$ can be identified with $\mathcal{J}\mathcal{L}^n \subseteq \mathcal{L}^n$ because $\mathcal{L}_x^n$ is flat over $\mathcal{O}_{X,x}$. By hypothesis, there exists an $n \geq 1$ such that $\mathcal{J}\mathcal{L}^n$ is generated by its global sections. There therefore exists a section $s \in (\mathcal{J}\mathcal{L}^n)(X) \subseteq \mathcal{L}^n(X)$ such that s_x is a basis of $\mathcal{L}_x^n = \mathcal{J}_x\mathcal{L}_x^n$. This means that $x \in X_s \subseteq U$. Let us write $\mathcal{L}|_U = e\mathcal{O}_U$ and $s|_U = eh$ with $h \in \mathcal{O}_X(U)$. Then $X_s = D_U(h)$. This implies that X_s is affine. We will note that the rest of the proof no longer uses the $\mathcal{L}$ ample hypothesis.

As X is quasi-compact, we can cover X with a finite number of open subsets X_{s_i} as above. Moreover, taking positive powers of the s_i if necessary, we may assume that $s_i \in \mathcal{L}^n(X)$ for some n that does not depend on i. We have $\mathcal{O}_X(X_{s_i}) = A[f_{ij}]_j$ with a finite number of f_{ij} (Definition 3.2.1). By Lemma 1.25, there exists an $r \geq 1$ such that $s_i^r \otimes f_{ij}$ is the restriction to X_{s_i} of a section $s_{ij} \in \mathcal{L}^{nr}(X)$. We can choose the same r for all of the f_{ij}.

The sections $\{s_i^r, s_{ij}\}_{i,j}$ of $\mathcal{L}^{nr}$ generate it, because the s_i^r generate $\mathcal{L}^{nr}$ owing to the fact that $X = \cup_i X_{s_i}$. Let $\pi : X \to \operatorname{Proj} A[S_i, S_{ij}]_{i,j}$ be the morphism from X to a projective space associated to the sections s_i, s_{ij} (Proposition 1.31). Let $U_i = D_+(S_i)$. Then $X_{s_i} = \pi^{-1}(U_i)$ and $\mathcal{O}(U_i) \to \mathcal{O}_X(X_{s_i})$ is surjective because it sends S_{ij}/S_i to f_{ij}. It follows that π induces a closed immersion from X into $U := \cup_i U_i$. Hence π is an immersion (Exercise 3.2.3). This proves that $\mathcal{L}^{nr}$ is very ample for f. □

Lemma 1.35. *Let X be a Noetherian or separated and quasi-compact scheme and $\mathcal{L}$ an invertible sheaf on X.*

(a) *We suppose that there exist $s_1, \ldots, s_r \in \mathcal{L}(X)$ such that X_{s_i} is affine for every i and that $X = \cup_{1 \leq j \leq r} X_{s_j}$. Then $\mathcal{L}$ is ample.*

(b) *Let U be an open quasi-compact subscheme of X. If $\mathcal{L}$ is ample, then $\mathcal{L}|_U$ is ample.*

Proof (a) Let $\mathcal{F}$ be a finitely generated quasi-coherent sheaf on X. Let $f_1, \ldots, f_q$ be generators of $\mathcal{F}(X_{s_1})$. By Lemma 1.25, there exists an $n \geq 1$ such that $f_i \otimes s_1^n|_{X_{s_1}}$ lifts to a section in $(\mathcal{F} \otimes \mathcal{L}^n)(X)$ for every $i \leq q$. As s_1^n is a basis of $\mathcal{L}^n|_{X_{s_1}}$, $(\mathcal{F} \otimes \mathcal{L}^n)(X_{s_1})$ is generated by the $f_i \otimes s_1^n|_{X_{s_1}}$. Hence, at the points of X_{s_1}, $\mathcal{F} \otimes \mathcal{L}^n$ is generated by its global sections. And this remains true when we replace n by a larger integer. Therefore, by applying the result to the other s_j, we see that $\mathcal{F} \otimes \mathcal{L}^n$ is generated by its global sections for every sufficiently large n. Hence $\mathcal{L}$ is ample.

(b) Let us suppose $\mathcal{L}$ is ample. It suffices to show that a positive tensor power of $\mathcal{L}|_U$ is ample. The beginning of the proof of Theorem 1.34 (that part does not use the hypothesis that X is of finite type over an affine scheme) shows that, replacing $\mathcal{L}$ by a positive tensor power if necessary, there exist sections $s_1, \ldots, s_r \in \mathcal{L}(X)$ satisfying the hypotheses of (a). Any open subset $U \cap X_{s_j}$ of X_{s_j} is a finite union of principal open subsets $D(h_{1j}), \ldots, D(h_{mj})$, $h_{ij} \in \mathcal{O}_X(X_{s_j})$. By Lemma 1.25, if necessary replacing $\mathcal{L}$ by a positive tensor power and the s_j by the same power, the $s_j h_{ij}$ lift to sections $t_{ij} \in \mathcal{L}(X)$. Then the sections $s_{ij} := t_{ij}|_U \in \mathcal{L}(U)$, $1 \le i \le m, 1 \le j \le r$, verify the hypotheses of (a) because $U_{s_{ij}} = D(h_{ij})$. Consequently, $\mathcal{L}|_U$ is ample. $\square$

Corollary 1.36. *Let $X \to \operatorname{Spec} A$ be a morphism as in Theorem 1.34. Then $X \to \operatorname{Spec} A$ is quasi-projective if and only if there exists an ample sheaf on X.*

Proof Let us suppose $X \to \operatorname{Spec} A$ is quasi-projective. Then X is an open subscheme of a projective scheme Y over A. There exists an ample sheaf on Y by Theorem 1.27. The restriction of this ample sheaf to X gives an ample sheaf on X by Lemma 1.35(b). The converse is an immediate consequence of Theorem 1.34. $\square$

Proposition 1.37. *Let $f : X \to Y$ be a proper morphism of locally Noetherian schemes. Let $\mathcal{L}$ be an invertible sheaf on X. Let us fix a point $y \in Y$ and let $\varphi : X \times_Y \operatorname{Spec} \mathcal{O}_{Y,y} \to X$ be the canonical morphism. Then the following properties are true.*

(a) *If $\varphi^*\mathcal{L}$ is generated by its global sections, then there exists an open neighborhood V of y such that $\mathcal{L}|_{f^{-1}(V)}$ is generated by its global sections.*

(b) *If $\varphi^*\mathcal{L}$ is ample, then there exists an open neighborhood V of y such that $\mathcal{L}|_{f^{-1}(V)}$ is ample.*

Proof We can suppose $Y = \operatorname{Spec} A$ is affine.

(a) Let $Z = X \times_Y \operatorname{Spec} \mathcal{O}_{Y,y}$. In a way similar to Proposition 3.1.24, we show that $\varphi^*\mathcal{L}(Z) = \mathcal{L}(X) \otimes_A \mathcal{O}_{Y,y}$. We easily deduce from this that for any $x \in f^{-1}(y)$, we have

$$\mathcal{L}(X) \otimes \mathcal{O}_{X,x} = (\varphi^*\mathcal{L})(Z) \otimes \mathcal{O}_{Z,x}, \quad (\varphi^*\mathcal{L})_{\varphi^{-1}(x)} = \mathcal{L}_x.$$

We therefore see that $\mathcal{L}$ is generated by its global sections at the points $x \in f^{-1}(y)$. The set of points where $\mathcal{L}$ is not generated by its global sections is a closed subset F that does not meet $f^{-1}(y)$. Let $V = Y \setminus f(F)$. Then $\mathcal{L}|_{f^{-1}(V)}$ is generated by its global sections.

(b) If necessary replacing $\mathcal{L}$ by a tensor power, we can suppose $\varphi^*\mathcal{L}$ is very ample and generated by its global sections. By (a), we can suppose that $\mathcal{L}$ is generated by its global sections. As X is quasi-compact, $\mathcal{L}$ is generated by a finite number of sections $s_0, \ldots, s_n \in \mathcal{L}(X)$. Let $g : X \to \mathbb{P}^n_Y$ be the morphism associated to these sections. Let Z be the scheme-theoretic closure of $g(X)$ in $\mathbb{P}^n_Y$. Let us also denote by g the morphism $X \to Z$ induced by g. By hypothesis, g is an isomorphism above $\operatorname{Spec} \mathcal{O}_{Y,y}$. Hence g is an isomorphism above an open subscheme $V \ni y$ (Exercise 3.2.5). This shows that $\mathcal{L}|_{f^{-1}(V)}$ is very ample. $\square$

Exercises

1.1. Let $f : X \to Y$ be a continuous map of ringed topological spaces. Let $\mathcal{F}$ be an $\mathcal{O}_X$-module and $\mathcal{G}$ an $\mathcal{O}_Y$-module.

(a) Show that there exists a canonical homomorphism $f^*f_*\mathcal{F} \to \mathcal{F}$ (resp. $\mathcal{G} \to f_*f^*\mathcal{G}$) of $\mathcal{O}_X$-modules (resp. of $\mathcal{O}_Y$-modules) (use

Exercise 2.2.13(a)). Show that the first homomorphism is an isomorphism if f is an immersion.

(b) Show that there exists a canonical isomorphism

$$\mathrm{Hom}_{\mathcal{O}_X}(f^*\mathcal{G}, \mathcal{F}) \simeq \mathrm{Hom}_{\mathcal{O}_Y}(\mathcal{G}, f_*\mathcal{F}).$$

(c) Show that there exists a canonical homomorphism $\mathcal{G} \otimes_{\mathcal{O}_Y} f_*\mathcal{O}_X \to f_*f^*\mathcal{G}$ and that it is an isomorphism if f is an affine morphism of schemes (see Exercise 2.2) and if $\mathcal{G}$ is quasi-coherent (note that $f_*\mathcal{O}_X$ is quasi-coherent).

1.2. Let A be an integral domain with field of fractions K, and let $x_0 \in X = \mathrm{Spec}\, A$. We construct a sheaf $\mathcal{F}$ on X by setting $\mathcal{F}(U) = K$ if $x_0 \in U$ and $\mathcal{F}(U) = 0$ otherwise. The restriction maps are the obvious ones. Show that $\mathcal{F}$ is an $\mathcal{O}_X$-module. Under what condition is $\mathcal{F}$ quasi-coherent?

1.3. Let X be a scheme and $f \in \mathcal{O}_X(X)$. Show that the sheaf $f\mathcal{O}_X$ (Exercise 2.3.4) is finitely generated quasi-coherent. Show that in general it is not an invertible sheaf.

1.4. Let $X = \mathrm{Spec}\, A$ be an affine scheme. Let $\mathcal{F}$ be a quasi-coherent $\mathcal{O}_X$-module. Show that for any affine open subset U of X, we have a canonical isomorphism $\mathcal{F}(X) \otimes_A \mathcal{O}_X(U) \simeq \mathcal{F}(U)$. Show that this is false if U is not affine (consider Example 2.3.6 with $M = \mathbb{Z}[T]/(T, p)$).

1.5. Let X be a scheme, and let $\mathcal{F}, \mathcal{G}$ be $\mathcal{O}_X$-modules.

(a) Show that $U \mapsto \mathrm{Hom}_{\mathcal{O}_X|_U}(\mathcal{F}|_U, \mathcal{G}|_U)$ defines an $\mathcal{O}_X$-module. This sheaf will be denoted $\mathcal{H}om_{\mathcal{O}_X}(\mathcal{F}, \mathcal{G})$.

(b) Let us suppose $X = \mathrm{Spec}\, A$. Show that the canonical map

$$\mathrm{Hom}_{\mathcal{O}_X}(\mathcal{F}, \mathcal{G}) \to \mathrm{Hom}_A(\mathcal{F}(X), \mathcal{G}(X))$$

is bijective if $\mathcal{F}$ is quasi-coherent.

1.6. Let X be a locally Noetherian scheme, and let $\mathcal{F}, \mathcal{G}$ be $\mathcal{O}_X$-modules.

(a) Show that if $\mathcal{F}$ is coherent and $\mathcal{G}$ quasi-coherent, then $\mathcal{H}om_{\mathcal{O}_X}(\mathcal{F}, \mathcal{G})$ is quasi-coherent (use Exercise 1.2.8).

(b) Show that if $\mathcal{F}$ and $\mathcal{G}$ are coherent, then so is $\mathcal{H}om_{\mathcal{O}_X}(\mathcal{F}, \mathcal{G})$.

1.7. Let X be a scheme. Show that any finitely generated sheaf of ideals $\mathcal{J}$ is quasi-coherent. If X is locally Noetherian, show that any quasi-coherent sheaf of ideals is coherent.

1.8. Let X be a scheme. Let $\mathcal{F}, \mathcal{G}$ be two quasi-coherent subsheaves of an $\mathcal{O}_X$-module $\mathcal{H}$. Show that $\mathcal{F} \cap \mathcal{G}$ and $\mathcal{F} + \mathcal{G}$ are quasi-coherent, where $\mathcal{F}+\mathcal{G}$ is the sheaf associated to the presheaf $U \mapsto \mathcal{F}(U)+\mathcal{G}(U)$. Show that these properties are true for coherent sheaves if X is locally Noetherian.

1.9. Let $\mathcal{F}$ be a coherent sheaf on a locally Noetherian scheme X.

(a) Show that the support $\operatorname{Supp} \mathcal{F}$ of $\mathcal{F}$ (Exercise 2.2.5) is closed.

(b) Show that the multiplication by scalars defines a canonical homomorphism of $\mathcal{O}_X$-modules $\alpha : \mathcal{O}_X \to \mathcal{H}om_{\mathcal{O}_X}(\mathcal{F}, \mathcal{F})$. Show that $\operatorname{Ker}\alpha$ is quasi-coherent. We denote it by $\operatorname{Ann}\mathcal{F}$. This sheaf is called the *annihilator of $\mathcal{F}$*.

(c) Show that for any affine open subset U of X, we have

$$(\operatorname{Ann}\mathcal{F})(U) = \{a \in \mathcal{O}_X(U) \mid a\mathcal{F}(U) = 0\}.$$

(d) Show that $\operatorname{Supp}\mathcal{F}$ coincides with the closed subset $V(\operatorname{Ann}\mathcal{F})$. Show that $\mathcal{F}$ is the direct image of a coherent sheaf on the closed subscheme $V(\operatorname{Ann}\mathcal{F})$ under the closed immersion $V(\operatorname{Ann}\mathcal{F}) \to X$.

1.10. Let X be a scheme.

(a) Let $\mathcal{N}$ be the sheaf of nilpotent elements of $\mathcal{O}_X$ (defined in the proof of Proposition 2.4.2). Show that it is quasi-coherent.

(b) Let us suppose X locally Noetherian. Show that the set of $x \in X$ such that $\mathcal{O}_{X,x}$ is reduced is open in X.

(c) Let $\mathcal{I}$ be a quasi-coherent sheaf of ideals of $\mathcal{O}_X$. Show that $U \mapsto \sqrt{\mathcal{I}(U)}$, for every affine open U, defines a quasi-coherent sheaf of ideals of $\mathcal{O}_X$.

1.11. Let X be a locally Noetherian scheme, $\phi : \mathcal{F} \to \mathcal{G}$ a homomorphism of coherent sheaves. Show that $\{x \in X \mid \phi_x$ is an isomorphism$\}$ is an open subset of X.

1.12. Let X be a scheme. A sheaf $\mathcal{F}$ on X is called *free of rank n* if it is isomorphic to $\mathcal{O}_X^n$. We say that $\mathcal{F}$ is *locally free of rank n* if there exists a covering $\{X_i\}_i$ of X such that $\mathcal{F}|_{X_i}$ is free of rank n for every i. We say that it is *locally free of finite rank* if there exist a covering $X = \{X_i\}_i$ and integers n_i such that $\mathcal{F}|_{X_i}$ is locally free of rank n_i. We will suppose in what follows that X is locally Noetherian.

(a) Let $\mathcal{F}$ be a coherent sheaf on X. Show that if $\mathcal{F}_x$ is free of rank n on $\mathcal{O}_{X,x}$, then there exists an open neighborhood U of x such that $\mathcal{F}|_U$ is free of rank n. Deduce from this that $\mathcal{F}$ is locally free of rank n if and only if $\mathcal{F}_x$ is free of rank n on $\mathcal{O}_{X,x}$ for every $x \in X$. We suppose in the remainder of the exercise that $\mathcal{F}$ satisfies this condition.

(b) Show that $\mathcal{F}^\vee := \mathcal{H}om_{\mathcal{O}_X}(\mathcal{F}, \mathcal{O}_X)$ is locally free of rank n.

(c) Show that the canonical homomorphism $\mathcal{F} \otimes_{\mathcal{O}_X} \mathcal{F}^\vee \to \mathcal{O}_X$ is an isomorphism if $n = 1$.

(d) Let $\operatorname{Pic}(X)$ denote the set of isomorphism classes of invertible sheaves on X. Show that the tensor product makes $\operatorname{Pic}(X)$ into a commutative group, whose unit element is the class of $\mathcal{O}_X$. Show that the inverse of the class of $\mathcal{L}$ is the class of $\mathcal{L}^\vee$. We will return to the group $\operatorname{Pic}(X)$ (called the *Picard group of X*) in Chapter 7.

1.13. Let X be a locally Noetherian scheme, and let $\varphi : \mathcal{F} \to \mathcal{G}$ be a surjective morphism between two locally free sheaves of equal (finite) rank. Show that φ is an isomorphism (use Proposition 1.2.6).

1.14. Let X be an integral scheme, $\mathcal{F}$ a quasi-coherent sheaf on X.

(a) Let $\mathcal{F}(U)_{\mathrm{tors}}$ be the set of $x \in \mathcal{F}(U)$ such that there exists a non-zero $a \in \mathcal{O}_X(U)$ with $ax = 0$. Show that this is a submodule of $\mathcal{F}(U)$.

(b) Show that there exists a unique subsheaf $\mathcal{F}_{\mathrm{tors}}$ of $\mathcal{F}$ such that $\mathcal{F}_{\mathrm{tors}}(U) = \mathcal{F}(U)_{\mathrm{tors}}$ for every affine open subset U of X. Show that $\mathcal{F}_{\mathrm{tors}}$ is quasi-coherent.

(c) We say that $\mathcal{F}$ is *torsion-free* if $\mathcal{F}_{\mathrm{tors}} = 0$. Show that $\mathcal{F}/\mathcal{F}_{\mathrm{tors}}$ is always torsion-free.

(d) Let us suppose, moreover, that X is locally Noetherian and that $\mathcal{F}$ is coherent and non-zero. Show that $\mathrm{Supp}\,\mathcal{F} = X$ if and only if $\mathcal{F} \neq \mathcal{F}_{\mathrm{tors}}$. Show that there exist an $n \geq 0$ and a non-empty open subset U of X such that $\mathcal{O}_X^n|_U \simeq \mathcal{F}|_U$ (by convention, $\mathcal{O}_X^n = \mathcal{O}_X$ if $n = 0$).

1.15. Let $\mathcal{F}$ be a coherent sheaf on a locally Noetherian scheme X. For any $x \in X$, we set
$$\phi(x) = \dim_{k(x)}(\mathcal{F}_x \otimes_{\mathcal{O}_{X,x}} k(x)).$$

(a) Let $n \in \mathbb{N}$. Using Nakayama's lemma, show that the set $\{x \in X \mid \phi(x) \leq n\}$ is open in X. In particular, if X is irreducible with generic point ξ, then $\phi(x) \geq \phi(\xi)$ for all $x \in X$.

(b) Let A be a ring, and let $\alpha : A^n \to M$ be a homomorphism of A-modules. We suppose that for every $\mathfrak{p} \in \mathrm{Spec}\,A$, the homomorphism
$$\alpha \otimes \mathrm{Id}_{k(\mathfrak{p})} : k(\mathfrak{p})^n \to M_{\mathfrak{p}} \otimes_{A_{\mathfrak{p}}} k(\mathfrak{p})$$
is injective. Show that $\mathrm{Ker}\,\alpha \subseteq N^n$, where N is the nilradical of A.

(c) Let us suppose that ϕ is constant of value n on X, and that X is reduced. Show that $\mathcal{F}$ is locally free of rank n.

(d) Show through an example that (c) is false if X is not reduced.

1.16. Let $f : X \to S$ be a morphism of schemes. Let us do a base change

$$
\begin{array}{ccc}
X_T & \xrightarrow{\;f_T\;} & T \\[2pt]
{\scriptstyle p}\big\downarrow & & \big\downarrow{\scriptstyle \pi} \\[2pt]
X & \xrightarrow{\;f\;} & S
\end{array}
$$

(a) Let $\mathcal{F}$ be an arbitrary $\mathcal{O}_X$-module. Show that we have a canonical homomorphism $\pi^* f_* \mathcal{F} \to (f_T)_* p^* \mathcal{F}$ (use Exercise 1.1).

(b) Let us suppose that $T \to S$ is flat and that $\mathcal{F}$ is quasi-coherent. Show that $\pi^* f_* \mathcal{F} \to (f_T)_* p^* \mathcal{F}$ is an isomorphism if X is Noetherian or if f is separated and quasi-compact (see also Proposition 3.1.24).

(c) Let X and f be as in (b). Let $\mathcal{G}$ be a locally free sheaf on S. Show that we have a canonical isomorphism

$$f_*\mathcal{F} \otimes_{\mathcal{O}_S} \mathcal{G} \simeq f_*(\mathcal{F} \otimes_{\mathcal{O}_X} f^*\mathcal{G})$$

(see also Proposition 2.32).

1.17. Let X be a scheme. Let $\mathcal{A}$ be a sheaf of (commutative) $\mathcal{O}_X$-algebras. We say that $\mathcal{A}$ is *quasi-coherent* if it is quasi-coherent as an $\mathcal{O}_X$-module.

(a) Let us suppose $\mathcal{A}$ is quasi-coherent. Show that there exists a unique X-scheme $\pi : Z \to X$ with π affine and $\pi_*\mathcal{O}_Z = \mathcal{A}$. We denote Z by Spec $\mathcal{A}$.

(b) Show that Spec $\mathcal{A} \to X$ is finite if and only if $\mathcal{A}$ is quasi-coherent and finitely generated as an $\mathcal{O}_X$-module.

(c) Let $f : W \to X$ be an affine morphism. Show that $f_*\mathcal{O}_W$ is a quasi-coherent $\mathcal{O}_X$-algebra and that $W \simeq$ Spec $f_*\mathcal{O}_W$ as X-schemes.

1.18. Let X be a Noetherian or separated and quasi-compact scheme. Let $\mathcal{L}$ be an invertible sheaf on X. Let us fix a section $s \in \mathcal{L}(X)$.

(a) For any $n \geq 1$, we denote the tensor power $\mathcal{L}^{\otimes n}$ by $\mathcal{L}^n$, and let $\mathcal{L}^{\otimes 0} = \mathcal{O}_X$, by convention. Let us set $\mathcal{A} = \oplus_{n \geq 0}\mathcal{L}^{\otimes n}$. Show that $\mathcal{A}$ has a natural structure of an $\mathcal{O}_X$-algebra. Let $Z = $ Spec $\mathcal{A}$. Let $\pi : Z \to X$ denote the canonical morphism. Show that for any $x \in X$, there exists an open neighborhood $U \ni x$ such that $\pi^{-1}(U) \simeq \mathbb{A}^1_U$. Deduce from this that Z is Noetherian or separated and quasi-compact.

(b) Let $\mathcal{F}$ be a quasi-coherent $\mathcal{O}_X$-module. Let $\mathcal{G} = \pi^*\mathcal{F}$. Show that

$$\mathcal{G}(\pi^{-1}(X_s)) = \oplus_{n \geq 0}(\mathcal{L}^n \otimes_{\mathcal{O}_X} \mathcal{F})(X_s)$$

(use Exercise 1.1(c)).

(c) Let us consider s as an element of $\mathcal{O}_Z(Z)$. Show that $\mathcal{G}(Z_s) = \mathcal{G}(Z)_s$ and that $Z_s \subset \pi^{-1}(X_s)$.

(d) Give a new proof of Lemma 1.25.

1.19. By considering (if necessary infinite) direct sums of sheaves of the form $\mathcal{O}_X(-m)$, show that there does not exist an integer n_0 as in Theorem 1.27 which suits all coherent sheaves on X, and that this theorem is false for quasi-coherent sheaves on X.

1.20. Let B be a graded algebra over a ring A, C an A-algebra. Let $f :$ Proj$(B \otimes_A C) \to$ Proj B be the projection morphism (Proposition 3.1.9). For any graded B-module M, we endow $M \otimes_A C$ with the grading $(M \otimes_A C)_n = M_n \otimes C$. Show that it is a graded $B \otimes_A C$-module and that we have $f^*(\widetilde{M}) = (M \otimes_A C)^{\sim}$.

1.21. Let B be a graded Noetherian ring, and let M be a finitely generated graded B-module. Show that $\widetilde{M}$ is coherent on Proj B.

1.22. Let B be a homogeneous algebra over a ring A, quotient of some $A[T_0, T_1, \ldots, T_m]$. Let M be a graded B-module and let us fix $n \in \mathbb{Z}$. Let $\mathcal{O}_X(1)$ denote the sheaf on $X = \operatorname{Proj} B$ given by the closed immersion $X \to \mathbb{P}_A^m$. Let us consider the graded B-module $M(n)$ given by $M(n)_d = M_{d+n}$ for every $d \in \mathbb{Z}$. Show that

$$M(n)^\sim = \widetilde{M}(n) := M^\sim \otimes \mathcal{O}_X(n).$$

1.23. (*Automorphisms of $\mathbb{P}_k^n$*) Let k be a field, $B = k[T_0, \ldots, T_n]$ with $n \geq 1$ and $X = \operatorname{Proj} B$. We are going to show that the automorphisms of X are linear; that is, that they come from k-linear transformations of B. We admit that the group $\operatorname{Pic}(X)$ (see Exercise 1.12) is generated by the class of $\mathcal{O}_X(1)$ (Proposition 7.2.9). Let σ be an automorphism of X.

(a) Show that we have an isomorphism $\lambda : \sigma^* \mathcal{O}_X(1) \simeq \mathcal{O}_X(1)$.

(b) Let $\tau : \mathcal{O}_X(1)(X) \to \sigma^* \mathcal{O}_X(1)(X) \to \mathcal{O}_X(1)(X)$ be the composition of the canonical homomorphism followed by $\lambda(X)$. Show that τ is an automorphism of the k-vector space $B_1 = \mathcal{O}_X(1)(X)$.

(c) Show that τ canonically induces an automorphism $\tilde\tau$ of the graded k-algebra B and that σ is the automorphism of X induced by $\tilde\tau$.

1.24. Let X be an algebraic variety over a field k. Let $f : X \to Y := \mathbb{P}_k^n$ be a morphism of algebraic k-varieties. Let $\mathcal{L} = f^* \mathcal{O}_Y(1)$. Let σ be an automorphism of X such that $\sigma^* \mathcal{L} \simeq \mathcal{L}$. Show that there exists an automorphism σ' of Y such that $\sigma' \circ f = f \circ \sigma$.

1.25. Let $f : X \to Y$ be a finite dominant morphism of integral locally Noetherian schemes. The integer $n = [K(X) : K(Y)]$ is called the *degree of f*.

(a) Show that for any $y \in Y$, $\mathcal{O}(X_y)$ is a $k(y)$-vector space of finite dimension and that $\dim_{k(y)} \mathcal{O}(X_y) \geq n$. Show that f is flat if and only if there is equality for every $y \in f(X)$ (use Exercise 1.15).

(b) Let us suppose f is flat above y and that X_y is made up of exactly n points. Show that f is étale at every point of $f^{-1}(y)$.

1.26. Let $\mathcal{L}$ be an invertible sheaf on a scheme X. We call a point $x \in X$ such that $H^0(X, \mathcal{L}) \otimes \mathcal{O}_{X,x} \subseteq \mathfrak{m}_x \mathcal{L}_x$ a *base point* of $\mathcal{L}$. Let $x_1, \ldots, x_n \in X$ be points that are not base points. Let us suppose that X is an algebraic variety over a field k with k infinite or $\operatorname{Card} k \geq n$. Show that there exists a section $s \in H^0(X, \mathcal{L})$ such that $\mathcal{L}_{x_i} = s_{x_i} \mathcal{O}_{X,x_i}$ for every $i \leq n$.

1.27. (*d-uple embedding*) Let A be a ring, and $X = \operatorname{Proj} A[T_0, \ldots, T_n]$ with $n \geq 1$. Let us fix an integer $d \geq 1$.

(a) Let $\{M_\alpha\}_\alpha$ denote the set of monomials of degree d in $A[T_0, \ldots, T_n]$. This set is of cardinal $N := \binom{n+d}{n}$. Let us consider the homomorphism $\varphi : A[S_\alpha]_\alpha \to A[T_0, \ldots, T_n]$ which sends S_α to $M_\alpha(T_0, \ldots, T_n)$. Show that φ induces a closed immersion $f : X \to \mathbb{P}_A^{N-1} = \operatorname{Proj} A[S_\alpha]_\alpha$ (Lemma 2.3.40).

(b) Show that $\{M_\alpha\}_\alpha$ generates $H^0(X, \mathcal{O}_X(d))$, that $\mathcal{O}_X(d) \simeq f^*\mathcal{O}_{\mathbb{P}^{N-1}_A}(1)$, and that f is the morphism associated to $\mathcal{O}_X(d)$ and the global sections $\{M_\alpha\}_\alpha$ as in Proposition 1.31(b).

(c) Show that $\mathcal{O}_X(d)$ is very ample if and only if $d \geq 1$.

1.28. Let X' be a projective scheme over a ring A, and let $\mathcal{L}, \mathcal{M}$ be two invertible sheaves generated by their global sections. Let $f : X \to \mathbb{P}^n_A$, $g : X \to \mathbb{P}^m_A$ be morphisms such that $\mathcal{L} \simeq f^*\mathcal{O}_{\mathbb{P}^n_A}(1)$ and $\mathcal{M} \simeq g^*\mathcal{O}_{\mathbb{P}^m_A}(1)$. Let $Y = \mathbb{P}^n_A \times_A \mathbb{P}^m_A$, and let p, q denote the respective projections of Y onto $\mathbb{P}^n_A$ and $\mathbb{P}^m_A$.

(a) Show that $\mathcal{L} \otimes \mathcal{M} \simeq h^*\mathcal{N}$, where $h = (f, g)$ and $\mathcal{N} := p^*\mathcal{O}_{\mathbb{P}^n_A}(1) \otimes q^*\mathcal{O}_{\mathbb{P}^m_A}(1)$.

(b) Let $i : Y \to \mathbb{P}^{nm+n+m}_A$ be the Segre embedding (Lemma 3.3.31). Show that $\mathcal{N} \simeq i^*\mathcal{O}_{\mathbb{P}^{nm+n+m}_A}(1)$.

(c) Show that if f is a closed immersion then so is h (use Lemma 3.3.15(b)). Deduce from this that $\mathcal{L} \otimes \mathcal{M}$ is very ample if $\mathcal{L}$ is very ample.

1.29. Let $\pi : X' \to X$ be a faithfully flat morphism of schemes.

(a) Let $\mathcal{F}$ be a quasi-coherent sheaf on X. Show that there exists a homomorphism $\phi : \mathcal{O}_X^{(I)} \to \mathcal{F}$ such that $\phi(X)$ is surjective. Show that $\mathcal{F}$ is generated by its global sections if and only if ϕ is surjective.

(b) Show that $\mathcal{F}$ is generated by its global sections if and only if $\pi^*\mathcal{F}$ is generated by its global sections.

(c) Let us suppose X, X' are quasi-compact. Let $\mathcal{L}$ be an invertible sheaf on X. Show that if $\pi^*\mathcal{L}$ is ample, then $\mathcal{L}$ is ample.

1.30. Let X be a proper scheme over a Noetherian ring A, and let $\mathcal{L}$ be an invertible sheaf on X. Let us suppose that there exists a faithfully flat base change $\operatorname{Spec} B \to \operatorname{Spec} A$ such that B is Noetherian and that $\pi^*\mathcal{L}$ is very ample relative to B, where $\pi : X_B \to X$ is the canonical projection. We are going to show that $\mathcal{L}$ is very ample relative to A.

(a) Show that $\mathcal{L}$ is ample and generated by some global sections $s_0, \ldots, s_n$. Let $f : X \to Y := \mathbb{P}^n_A$ be the morphism associated to $\mathcal{L}$ and to the s_i. Show that $f_B : X_B \to Y_B$ is a closed immersion.

(b) Let $y \in Y$ and let $y' \in Y_B$ be a point lying above y. Show that the fiber of X_B over y' is isomorphic to $X_y \times_{\operatorname{Spec} k(y)} \operatorname{Spec} k(y')$. Deduce from this that $f^{-1}(y)$ consists of at most one point, and that f is a topological closed immersion.

(c) Show that $\mathcal{O}_Y \to f_*\mathcal{O}_X$ is surjective and that f is a closed immersion.

1.31. Let $\mathcal{L}$ be an ample sheaf and $\mathcal{M}$ an invertible sheaf on a Noetherian scheme. Show that $\mathcal{L} \otimes \mathcal{M}$ is ample if $\mathcal{M}$ is ample or if $\mathcal{M}$ is generated by its global sections.

1.32. Let $\mathcal{L}$ be an invertible sheaf on a quasi-compact scheme. Show that the following properties are equivalent:

(i) $\mathcal{L}$ is ample;

(ii) $\mathcal{L}^n$ is ample for every $n \geq 1$;

(iii) there exists an $n \geq 1$ such that $\mathcal{L}^n$ is ample.

1.33. Let X be a Noetherian or separated and quasi-compact scheme, and let $f : X \to Y = \operatorname{Spec} A$ be a morphism.

(a) Let $\mathcal{F}$ be a finitely generated quasi-coherent sheaf on X. Let M be a sub-A-module of $\mathcal{F}(X)$, and let $x \in X$ be such that $M \otimes \mathcal{O}_{X,x} \to \mathcal{F}_x$ is surjective. Show that this is then true in an open neighborhood of x.

(b) Let $\mathcal{F}$ be as in (a). Show that $\mathcal{F}$ is generated by its global sections if and only if for every $y \in Y$, the sheaf $\rho^* \mathcal{F}$, where ρ is the projection $X \times_Y \operatorname{Spec} \mathcal{O}_{Y,y} \to \operatorname{Spec} \mathcal{O}_{Y,y}$, is generated by its global sections.

(c) Let $\mathcal{L}$ be an invertible sheaf on X. Let us suppose that $\rho^* \mathcal{L}$ is ample for every $y \in Y$. Show that $\mathcal{L}$ is ample.

5.2   Čech cohomology

In this section we consider the theory of cohomology in algebraic geometry. It is an extremely rich and varied theory. In this book we are interested in one of the most elementary cohomology theories, the Čech cohomology of quasi-coherent sheaves. We will show Serre's criterion for affine schemes (Theorem 2.23) and we will conclude with properties of flat base change (Corollary 2.27 and Proposition 2.32). In passing, we present several results on the vanishing of cohomology (Theorem 2.18, Propositions 2.24 and 2.34).

5.2.1   Differential modules and cohomology with values in a sheaf

Let us begin with some general formalism. Let A be a ring. We define a *differential A-module* to be an A-module M endowed with a homomorphism $d : M \to M$ such that $d^2 = 0$. We call d the *differential* of M. If there is no ambiguity possible, we omit d in the notation of a differential module. We let

$$H(M) = \operatorname{Ker} d / \operatorname{Im} d.$$

If $M = \oplus_{n \in \mathbb{Z}} M_n$ is a graded A-module, then we will say that d is of degree $r \in \mathbb{Z}$ if $dM_n \subseteq M_{n+r}$ for all $n \geq 0$. For example, let

$$\cdots \to M_{-1} \xrightarrow{d_{-1}} M_0 \xrightarrow{d_0} M_1 \xrightarrow{d_1} \cdots \to M_n \xrightarrow{d_n} \cdots \qquad (2.2)$$

be a complex of A-modules. Then $M := \oplus_{n \in \mathbb{Z}} M_n$ endowed with $d := \oplus_n d_n$ is a graded differential A-module, with d of degree 1.

A *homomorphism of differential A-modules* $(M, d) \to (M', d')$ is the data of a homomorphism of A-modules $\varphi : M \to M'$ such that $d' \circ \varphi = \varphi \circ d$. The kernel $\operatorname{Ker} \varphi$, as well as $\operatorname{Im} \varphi$, are clearly endowed with the structure of a differential module. To a homomorphism of differential A-modules $\varphi : M \to M'$, we associate, in a natural way, a homomorphism of A-modules $H(\varphi) : H(M) \to H(M')$. If M is graded and d is of degree 1, then $H(M)$ is endowed with a grading $\oplus_{n \in \mathbb{Z}} H^n(M)$ with

$$H^n(M) := \operatorname{Ker}(d : M_n \to M_{n+1})/\operatorname{Im}(d : M_{n-1} \to M_n).$$

In the case of the differential module coming from the complex (2.2), we have

$$H^n(M) = \operatorname{Ker} d_n / \operatorname{Im} d_{n-1}.$$

Lemma 2.1. *Let A be a ring, and let*

$$0 \to M'' \xrightarrow{\psi} M \xrightarrow{\varphi} M' \to 0$$

be an exact sequence of differential A-modules. Then there exists a homomorphism $\partial : H(M') \to H(M'')$ such that

$$H(M) \xrightarrow{H(\varphi)} H(M') \xrightarrow{\partial} H(M'') \xrightarrow{H(\psi)} H(M) \xrightarrow{H(\varphi)} H(M')$$

is exact. Moreover, if these modules are graded with differentials of degree 1, then $H(\varphi)$, $H(\psi)$ are of degree 0 and ∂ is of degree 1.

Proof For any differential module (M, d), we let $Z(M) = \operatorname{Ker} d$. We define ∂ in the following way: let $x' \in H(M')$ be the image of an element $y' \in Z(M')$. There exists a $y \in M$ such that $\varphi(y) = y'$. As $\varphi(dy) = d'y' = 0$, there exists a $y'' \in M''$ such that $\psi(y'') = dy$. We have $\psi(d''y'') = d^2 y = 0$. Hence $y'' \in Z(M'')$. We let $\partial(x')$ be the image of y'' in $H(M'')$. It is now easy to verify the stated properties. □

Remark 2.2. In the case of graded differential modules, we then obtain an exact sequence

$$\cdots \to H^{p-1}(M') \xrightarrow{\partial} H^p(M'') \to H^p(M) \to H^p(M') \xrightarrow{\partial} H^{p+1}(M'') \to \cdots.$$

From here to the end of the subsection, we fix a topological space X, a commutative ring A, and a sheaf of A-modules $\mathcal{F}$ on X. Let us note that a sheaf of Abelian groups is a sheaf of $\mathbb{Z}$-modules. We are going to define the cohomology groups of X with values in $\mathcal{F}$. We essentially follow the article of Serre [85]. Moreover, we refer to this article for certain proofs. Let $\mathcal{U} = \{U_i\}_{i \in I}$

be an (open) covering of X. For any integer $p \geq 0$ and for any sequence of indices $i_0, \ldots, i_p \in I$, we let

$$U_{i_0 \ldots i_p} = U_{i_0} \cap \cdots \cap U_{i_p}.$$

We set

$$C^p(\mathcal{U}, \mathcal{F}) = \prod_{(i_0, \ldots, i_p) \in I^{p+1}} \mathcal{F}(U_{i_0 \ldots i_p}).$$

An element of $C^p(\mathcal{U}, \mathcal{F})$ is called a *p-cochain* (of $\mathcal{U}$ in $\mathcal{F}$). We say that a *p*-cochain $f \in C^p(\mathcal{U}, \mathcal{F})$ is *alternating* if $f_{i_0 \ldots i_p} = 0$ as soon as two indices are equal and if for every permutation σ of the indices, we have $f_{\sigma(i_0) \ldots \sigma(i_p)} = \varepsilon(\sigma) f_{i_0 \ldots i_p}$, where $\varepsilon(\sigma)$ is the signature of σ. The set of alternating elements clearly forms a sub-A-module of $C^p(\mathcal{U}, \mathcal{F})$. We denote this submodule by $C'^p(\mathcal{U}, \mathcal{F})$. Let us set

$$C(\mathcal{U}, \mathcal{F}) = \oplus_{p \geq 0} C^p(\mathcal{U}, \mathcal{F}), \quad C'(\mathcal{U}, \mathcal{F}) = \oplus_{p \geq 0} C'^p(\mathcal{U}, \mathcal{F}).$$

These are graded A-modules. Let us define a differential d on $C(\mathcal{U}, \mathcal{F})$. Let $f \in C^p(\mathcal{U}, \mathcal{F})$. Then $df \in C^{p+1}(\mathcal{U}, \mathcal{F})$ will be the element given by

$$(df)_{i_0 \ldots i_{p+1}} = \sum_{k=0}^{p+1} (-1)^k f_{i_0 \ldots \hat{i}_k \ldots i_{p+1}} \big|_{U_{i_0 \ldots i_{p+1}}}$$

where '$\hat{i}_k$' means that we remove the index i_k. For example, if $p = 1$, then

$$(df)_{i_0 i_1 i_2} = f_{i_1 i_2} \big|_{U_{i_0 i_1 i_2}} - f_{i_0 i_2} \big|_{U_{i_0 i_1 i_2}} + f_{i_0 i_1} \big|_{U_{i_0 i_1 i_2}}.$$

A direct computation shows that $d^2 = 0$. If f is alternating, we verify that df is equally alternating. Hence $C(\mathcal{U}, \mathcal{F})$ and $C'(\mathcal{U}, \mathcal{F})$ are differential A-modules with a differential d of degree 1. The following proposition says that $H(C(\mathcal{U}, \mathcal{F}))$ can be computed with alternating cochains.

Proposition 2.3. *The canonical injection* $i : C'(\mathcal{U}, \mathcal{F}) \to C(\mathcal{U}, \mathcal{F})$ *induces an isomorphism* $H(i) : H(C'(\mathcal{U}, \mathcal{F})) \simeq H(C(\mathcal{U}, \mathcal{F}))$ *of graded* A-*modules.*

Proof   See [85], n° 20, Proposition 2.   □

Let us endow the set of indices I with a total ordering. Then we have a direct summand

$$C''^p(\mathcal{U}, \mathcal{F}) := \prod_{i_0 < \cdots < i_p} \mathcal{F}(U_{i_0 \ldots i_p})$$

of $C^p(\mathcal{U}, \mathcal{F})$. It can immediately be verified that the differential d of $C(\mathcal{U}, \mathcal{F})$ induces a differential on $C''(\mathcal{U}, \mathcal{F})$, by restriction.

Corollary 2.4. *Given a total ordering on* I, *there exists a canonical isomorphism* $H(C''(\mathcal{U}, \mathcal{F})) \simeq H(C(\mathcal{U}, \mathcal{F}))$.

Proof Let $C''^p(\mathcal{U}, \mathcal{F}) \to C'''^p(\mathcal{U}, \mathcal{F})$ be the projection onto the coordinates $(i_0, \ldots, i_p)$ such that $i_0 < i_1 < \cdots < i_p$. This homomorphism is compatible with the differentials d and is an isomorphism, as we can easily verify with the definition of alternating cochains. It therefore induces an isomorphism of A-modules $H(C'(\mathcal{U}, \mathcal{F})) \simeq H(C''(\mathcal{U}, \mathcal{F}))$. The corollary then follows from Proposition 2.3. □

By construction, $C(\mathcal{U}, \mathcal{F})$ is graded and the differential d is of degree 1. For any $p \geq 0$, we set

$$H^p(\mathcal{U}, \mathcal{F}) := H^p(C(\mathcal{U}, \mathcal{F}))$$
$$= \operatorname{Ker}(C^p(\mathcal{U}, \mathcal{F}) \to C^{p+1}(\mathcal{U}, \mathcal{F})) / \operatorname{Im}(C^{p-1}(\mathcal{U}, \mathcal{F}) \to C^p(\mathcal{U}, \mathcal{F})).$$

By convention, $C^{-1}(\mathcal{U}, \mathcal{F}) = 0$, and hence $H^{-1}(\mathcal{U}, \mathcal{F}) = 0$.

Corollary 2.5. *If $\mathcal{U}$ is made up of n open subsets, then $H^p(\mathcal{U}, \mathcal{F}) = 0$ for every $p \geq n$.*

Proof Indeed, if $p \geq n$, there does not exist any strictly increasing $(p+1)$-uple of indices $i_0, \ldots, i_p$. Hence $C'''^p(\mathcal{U}, \mathcal{F}) = 0$, whence $H^p(\mathcal{U}, \mathcal{F}) = 0$. □

Proposition 2.6. *We have $H^0(\mathcal{U}, \mathcal{F}) = \mathcal{F}(X)$.*

Proof By definition, $H^0(\mathcal{U}, \mathcal{F})$ is the kernel of $d : C^0(\mathcal{U}, \mathcal{F}) \to C^1(\mathcal{U}, \mathcal{F})$. This homomorphism is

$$\prod_{i \in I} \mathcal{F}(U_i) \to \prod_{i,j \in I} \mathcal{F}(U_i \cap U_j)$$
$$(f_i)_{i \in I} \mapsto (f_i|_{U_i \cap U_j} - f_j|_{U_j \cap U_i})_{i,j}.$$

The proposition then follows from the definition of sheaves (see Lemma 2.2.7). □

Example 2.7. Let A be a ring. Let $X = \operatorname{Proj} A[T_0, T_1]$. Let us consider the covering $\mathcal{U} = \{U_0, U_1\}$ of X with $U_i = D_+(T_i)$. Then $H^p(\mathcal{U}, \mathcal{O}_X) = 0$ for every $p \geq 1$. Indeed, by Corollary 2.5, it suffices to verify this for $p = 1$. We have to show that the complex

$$\mathcal{O}_X(U_0) \oplus \mathcal{O}_X(U_1) \xrightarrow{d} \mathcal{O}_X(U_{01}) \to 0$$

corresponding to $C''(\mathcal{U}, \mathcal{O}_X)$ is exact. Let us set $t = T_1/T_0$. Then this complex is

$$A[t] \oplus A[1/t] \xrightarrow{d} A[t, 1/t] \to 0$$

with $d(f, g) = f - g$. It is clear that d is surjective. This shows that $H^1(\mathcal{U}, \mathcal{O}_X) = 0$.

Passage to a refinement

We say that a covering $\mathcal{V} = \{V_j\}_{j \in J}$ of X is a *refinement* of another covering $\mathcal{U} = \{U_i\}_{i \in I}$ if there exists a map $\sigma : J \to I$ such that $V_j \subseteq U_{\sigma(j)}$ for every $j \in J$. We then have a homomorphism, which we also denote by σ:

$$\sigma : C^p(\mathcal{U}, \mathcal{F}) \to C^p(\mathcal{V}, \mathcal{F})$$

defined by

$$\sigma(f)_{j_0 \dots j_p} = f_{\sigma(j_0) \dots \sigma(j_p)}|_{V_{j_0 \dots j_p}}.$$

This homomorphism commutes with the differentials and therefore induces a homomorphism

$$\sigma^* : H^p(\mathcal{U}, \mathcal{F}) \to H^p(\mathcal{V}, \mathcal{F}).$$

Lemma 2.8. *The homomorphism σ^* does not depend on the choice of σ.*

Proof   See [85], n° 21, Proposition 3.   □

Corollary 2.9. *If $\mathcal{V}$ is a refinement of $\mathcal{U}$ and if $\mathcal{U}$ is a refinement of $\mathcal{V}$, then $H^p(\mathcal{U}, \mathcal{F}) \to H^p(\mathcal{V}, \mathcal{F})$ is an isomorphism.*

Proof   Let us keep the notation above. If $\mathcal{U}$ is a refinement of a covering $\mathcal{W}$ with a map τ such that $U_i \subseteq W_{\tau(i)}$, then $\mathcal{V}$ is a refinement of $\mathcal{W}$ because $V_j \subseteq W_{\tau \circ \sigma(j)}$. Moreover, $(\tau \circ \sigma)^* = \tau^* \circ \sigma^*$ in an obvious way. Let us now take $\mathcal{W} = \mathcal{V}$. Then $\tau^* \circ \sigma^*$ coincides with Id^* by the lemma above. Hence τ^* is injective and σ^* is surjective. By symmetry, τ^* is also surjective, and hence bijective.   □

The 'fineness' property defines a partial ordering on the family of coverings of X. This family is filtering: two coverings $\{U_i\}_i$, $\{V_j\}_j$ always admit a common refinement; it suffices to take their intersection $\{U_i \cap V_j\}_{i,j}$. Two coverings are called *equivalent* if each one is a refinement of the other one. The corollary above says that $H^p(\mathcal{U}, \mathcal{F})$ depends only on the equivalence class of $\mathcal{U}$ for this equivalence relation.

Definition 2.10. Let X be a topological space, and let $\mathcal{F}$ be a sheaf on X. We set

$$H^p(X, \mathcal{F}) = \varinjlim_{\mathcal{U}} H^p(\mathcal{U}, \mathcal{F}),$$

where the $\mathcal{U}$ run through the classes of open coverings of X. The group $H^p(X, \mathcal{F})$ is called the *pth (Čech) cohomology group of $\mathcal{F}$*. For example, by Proposition 2.6, we have $H^0(X, \mathcal{F}) = \mathcal{F}(X)$. If $\mathcal{F}$ is a sheaf of modules over a ring A, then $H^p(X, \mathcal{F})$ is canonically an A-module.

Remark 2.11. Even though we cannot talk about the set of open coverings of X, the direct limit above does indeed have a meaning because we can restrict ourselves to coverings which are subsets of the set of open subsets of X, and these coverings do form a set. Indeed, if $\mathcal{U} = \{U_i\}_i$ is an arbitrary covering of X, let $\mathcal{U}'$ be the covering defined by taking the set of the U_i (i.e., we omit possible repetitions in the family of the U_i). Then $\mathcal{U}$ and $\mathcal{U}'$ are equivalent.

In addition to this, we can restrict ourselves in this direct limit to a cofinal subset of coverings $\mathcal{U}$. If X is quasi-compact, the finite coverings form a cofinal system (i.e., any covering admits a finite refinement). If X is a scheme, the affine coverings form a cofinal system.

In the following theorem, we are going to compare the cohomology of a covering $\mathcal{U} = \{U_i\}_{i \in I}$ with that of X. Let $\mathcal{V} = \{V_j\}_{j \in J}$ be another covering of X. For any $\alpha = (i_0, \ldots, i_n) \in I^{n+1}$, we let $U_\alpha := U_{i_0 \ldots i_n}$ and let $\mathcal{V}_\alpha$ denote the covering $\{U_\alpha \cap V_j\}_{j \in J}$ of U_α.

Theorem 2.12 (Leray acyclicity theorem). *Let $\mathcal{F}$ be a sheaf on a topological space X. Let $\mathcal{U} = \{U_i\}_{i \in I}$ be a covering of X. Let us suppose that there exists a family of coverings $(\mathcal{V}^\lambda)_{\lambda \in \Lambda}$, cofinal in the family of coverings of X, such that*

$$H^q(\mathcal{V}^\lambda_\alpha, \mathcal{F}|_{U_\alpha}) = 0$$

for every $\lambda \in \Lambda$, $\alpha \in I^{n+1}$ ($n \geq 0$), and for every $q \geq 1$. Then the canonical homomorphism

$$H^p(\mathcal{U}, \mathcal{F}) \to H^p(X, \mathcal{F})$$

is an isomorphism for every $p \geq 0$.

Proof See [85], n° 29, Théorème 1.   □

Remark 2.13. The construction of $H^p(X, \mathcal{F})$ is functorial in $\mathcal{F}$: if $\mathcal{F} \to \mathcal{G}$ is a homomorphism of sheaves of A-modules on X, then we have a natural homomorphism of cohomology groups $H^p(X, \mathcal{F}) \to H^p(X, \mathcal{G})$. Moreover, if

$$\mathcal{F}'' \to \mathcal{F} \to \mathcal{F}'$$

is a complex, then so is

$$H^p(X, \mathcal{F}'') \to H^p(X, \mathcal{F}) \to H^p(X, \mathcal{F}').$$

But in general, this does not preserve the exactness, see the remark below.

Remark 2.14. Let $0 \to \mathcal{F}'' \to \mathcal{F} \to \mathcal{F}' \to 0$ be an exact sequence of sheaves of A-modules on X. We would like to compare the cohomology groups of these sheaves. Let $\mathcal{U}$ be a covering of X. We first canonically have a complex of graded differential modules

$$0 \to C(\mathcal{U}, \mathcal{F}'') \to C(\mathcal{U}, \mathcal{F}) \to C(\mathcal{U}, \mathcal{F}') \to 0. \qquad (2.3)$$

If this complex is exact, then we will have a long exact sequence (Remark 2.2)

$$\cdots \to H^{p-1}(\mathcal{U}, \mathcal{F}') \xrightarrow{\partial} H^p(\mathcal{U}, \mathcal{F}'') \to H^p(\mathcal{U}, \mathcal{F})$$
$$\to H^p(\mathcal{U}, \mathcal{F}') \xrightarrow{\partial} H^{p+1}(\mathcal{U}, \mathcal{F}'') \to \cdots.$$

Moreover, if (2.3) is exact for a cofinal system of coverings $\mathcal{U}$, then the exact sequence above transforms into an exact sequence of cohomology on X:

$$\cdots \to H^{p-1}(X, \mathcal{F}') \xrightarrow{\partial} H^p(X, \mathcal{F}'') \to H^p(X, \mathcal{F})$$
$$\to H^p(X, \mathcal{F}') \xrightarrow{\partial} H^{p+1}(X, \mathcal{F}'') \to \cdots .$$

(2.4)

The default of Čech cohomology is that in general, the complex (2.3) may not be exact; hence we cannot say whether the complex of cohomology groups above is exact. However, see Corollary 2.22 for a positive result.

Proposition 2.15. *Let X be a topological space. Let*

$$0 \to \mathcal{F}'' \xrightarrow{\alpha} \mathcal{F} \xrightarrow{\beta} \mathcal{F}' \to 0$$

be an exact sequence of sheaves on X. Then there exists a canonical homomorphism $\partial : \mathcal{F}'(X) \to H^1(X, \mathcal{F}'')$ such that the sequence

$$0 \to \mathcal{F}''(X) \to \mathcal{F}(X) \to \mathcal{F}'(X) \to H^1(X, \mathcal{F}'') \to H^1(X, \mathcal{F}) \to H^1(X, \mathcal{F}')$$

is exact.

Proof Let $s \in \mathcal{F}'(X)$. Then there exist a covering $\{U_i\}_i$ of X and sections $t_i \in \mathcal{F}(U_i)$ such that $s|_{U_i} = \beta(U_i)(t_i)$. Hence there exists a unique section $z_{ij} \in \mathcal{F}''(U_{ij})$ such that $\alpha(U_{ij})(z_{ij}) = t_i|_{U_{ij}} - t_j|_{U_{ij}}$. Let $z_s = (z_{ij})_{ij} \in C^1(\mathcal{U}, \mathcal{F}'')$. It is then clear that $z_s \in \mathrm{Ker}(C^1(\mathcal{U}, \mathcal{F}'') \to C^2(\mathcal{U}, \mathcal{F}''))$. Let $\bar{z}_s$ denote the image of z_s in $H^1(\mathcal{U}, \mathcal{F}'')$ and $\tilde{z}_s$ the image of $\bar{z}_s$ in $H^1(X, \mathcal{F}'')$. We easily verify that $\bar{z}_s$ is independent of the choice of the t_i. Let $\mathcal{V} = \{V_j\}_j$ be a covering of X, and finer than $\mathcal{U}$. Then for any j, $s|_{V_j}$ lifts to a section of $\mathcal{F}(V_j)$. This allows us to define, as above, an element $\bar{w}_s \in H^1(\mathcal{V}, \mathcal{F}'')$. It is just as easy to verify that $\bar{w}_s$ is equal to the image of $\bar{z}_s$ in $H^1(\mathcal{V}, \mathcal{F}'')$. Thus we see that $\tilde{z}_s \in H^1(X, \mathcal{F}'')$ is independent of the choice of $\mathcal{U}$.

Let $\partial(s) = \tilde{z}_s \in H^1(X, \mathcal{F}'')$. We can note that $\bar{z}_s = 0$ if and only if there exist $z_i \in \mathcal{F}''(U_i)$ such that $z_{ij} = z_i|_{U_{ij}} - z_j|_{U_{ij}}$ for every i, j. Therefore if $\bar{z}_i = 0$, the sections $t_i - \alpha(U_i)(z_i) \in \mathcal{F}(U_i)$ glue to a section $t \in \mathcal{F}(X)$ whose image in $\mathcal{F}'(X)$ is equal to s. Conversely, if s is the image of $t \in \mathcal{F}(X)$, we can take $t_i = t|_{U_i}$, which implies that $\bar{z}_s = 0$. Thus we have just shown that

$$0 \to \mathcal{F}''(X) \to \mathcal{F}(X) \to \mathcal{F}'(X) \to H^1(X, \mathcal{F}'')$$

is exact (see Exercise 2.2.4).

Let $\mathcal{U}$ be a covering of X, and let $z = (z_{ij})_{ij} \in \mathrm{Ker}(C^1(\mathcal{U}, \mathcal{F}'') \to C^2(\mathcal{U}, \mathcal{F}''))$ be such that its image in $H^1(\mathcal{U}, \mathcal{F})$ is zero. Then there exist $t_i \in \mathcal{F}(U_i)$ such that $\alpha(U_{ij})(z_{ij}) = t_i|_{U_{ij}} - t_j|_{U_{ij}}$. It follows that the sections $s_i := \beta(U_i)(t_i)$ glue to a section $s \in \mathcal{F}'(X)$. With the notation above, we immediately have $z_s = z$. This implies that

$$\mathcal{F}'(X) \xrightarrow{\partial} H^1(X, \mathcal{F}'') \to H^1(X, \mathcal{F})$$

is exact. Finally, we leave it to the reader to show the exactness of

$$H^1(X, \mathcal{F}'') \to H^1(X, \mathcal{F}) \to H^1(X, \mathcal{F}')$$

using a similar method. $\square$

Remark 2.16. The most correct way to define the cohomology group $H^p(X, \mathcal{F})$ is to use the notion of derived functors (see a summary in [43], III.1). With this construction, sequence (2.4) is always exact, without assuming that (2.3) also is. In the context of this book, we will usually restrict ourselves to quasi-coherent sheaves on a separated scheme. The two notions of cohomology are then equivalent (Theorem 2.18 and [40], Théorème II.5.9.2). With the help of Proposition 2.15, we can show that the equivalence stays valid in all of its generality if we restrict ourselves to the first groups $H^p(X, \mathcal{F})$, $p = 0, 1$. One of the advantages of Čech cohomology is that it is convenient for direct computations (see Lemma 3.1).

5.2.2 Čech cohomology on a separated scheme

Lemma 2.17. *Let $X = \operatorname{Spec} A$ be an affine scheme, $\mathcal{F}$ a quasi-coherent $\mathcal{O}_X$-module. Then for any finite covering $\mathcal{U} = \{U_i\}_{i \in I}$ of X by principal open subsets $U_i = D(g_i)$, we have $H^p(\mathcal{U}, \mathcal{F}) = 0$ for every $p \geq 1$.*

Proof We have $\mathcal{F} = \mathcal{F}(X)^{\sim}$ by Theorem 1.7. Let $f \in C^p(\mathcal{U}, \mathcal{F}) \cap \operatorname{Ker} d$. We must construct an $f' \in C^{p-1}(\mathcal{U}, \mathcal{F})$ such that $df' = f$. The construction will be similar to the proof of Proposition 2.3.1. As I is finite, there exists an integer $r \geq 1$ such that

$$f_{i_0 \ldots i_p} = \frac{x_{i_0 \ldots i_p}}{(g_{i_0} \cdots g_{i_p})^r}, \quad x_{i_0 \ldots i_p} \in \mathcal{F}(X). \tag{2.5}$$

Let $i \in I$; then $(df)_{ii_0 \ldots i_p} = 0$ by hypothesis. We can write this relation as

$$\left(\frac{x_{i_0 \ldots i_p}}{(g_{i_0} \cdots g_{i_p})^r} + \sum_{k=0}^{p} (-1)^{k+1} \frac{g_{i_k}^r x_{ii_0 \ldots \hat{i}_k \ldots i_p}}{g_i^r (g_{i_0} \cdots g_{i_p})^r} \right) \Bigg|_{U_{ii_0 \ldots i_p}} = 0.$$

There therefore exists an $l \geq 1$ such that for every $i \in I$ and every $(i_0, \ldots, i_p) \in I^{p+1}$, we have the following identity in $\mathcal{F}(U_{i_0 \ldots i_p})$:

$$\frac{g_i^{r+l} x_{i_0 \ldots i_p}}{(g_{i_0} \cdots g_{i_p})^r} = \sum_{k=0}^{p} (-1)^k \frac{g_i^l g_{i_k}^r x_{ii_0 \ldots \hat{i}_k \ldots i_p}}{(g_{i_0} \cdots g_{i_p})^r}. \tag{2.6}$$

As the open subsets $D(g_i^{r+l})$ cover X, we have an identity $1 = \sum_{i \in I} h_i g_i^{r+l}$ with $h_i \in A$. Let

$$f'_{i_0 \ldots i_{p-1}} = \sum_{i \in I} h_i g_i^l \frac{x_{ii_0 \ldots i_{p-1}}}{(g_{i_0} \cdots g_{i_{p-1}})^r} \in \mathcal{F}(U_{i_0 \ldots i_{p-1}}).$$

This defines an element $f' \in C^{p-1}(\mathcal{U}, \mathcal{F})$. For any $0 \leq k \leq p$, we have

$$f'_{i_0 \ldots \hat{i}_k \ldots i_p} \big|_{U_{i_0 \ldots i_p}} = \sum_{i \in I} h_i \frac{g_i^l g_{i_k}^r x_{ii_0 \ldots \hat{i}_k \ldots i_p}}{(g_{i_0} \cdots g_{i_p})^r}.$$

Keeping relations (2.5)–(2.6) in mind, we obtain

$$(df')_{i_0 \ldots i_p} = \sum_{k=0}^{p} (-1)^k f'_{i_0 \ldots \hat{i}_k \ldots i_p} \big|_{U_{i_0 \ldots i_p}} = \sum_{i \in I} h_i g_i^{r+l} f_{i_0 \ldots i_p} = f_{i_0 \ldots i_p}.$$

Therefore $df' = f$. $\qquad \square$

Theorem 2.18. *Let X be an affine scheme. Then for any quasi-coherent sheaf $\mathcal{F}$ on X and for any integer $p \geq 1$, we have $H^p(X, \mathcal{F}) = 0$.*

Proof Indeed, the family of coverings by principal open subsets is cofinal in the family of open coverings of X. The theorem then results from Lemma 2.17 and Remark 2.11. ☐

Theorem 2.19. *Let X be a separated scheme (Definition 3.3.2), let $\mathcal{F}$ be a quasi-coherent sheaf on X, and $\mathcal{U}$ an affine covering of X. Then the canonical homomorphism*

$$H^p(\mathcal{U}, \mathcal{F}) \to H^p(X, \mathcal{F})$$

is an isomorphism for every $p \geq 0$.

Proof We are going to show that this is a consequence of Theorem 2.12, whose notation we also use here. Let us first suppose X is affine. Then the theorem results from Lemma 2.17 and Theorem 2.12 by taking for $(\mathcal{V}^\lambda)_{\lambda \in \Lambda}$ the family of coverings by principal open subsets of X. In the general case, we take for $(\mathcal{V}^\lambda)_{\lambda \in \Lambda}$ the family of coverings of X by affine open subsets. As $\mathcal{V}^\lambda_\alpha$ is an affine covering of U_α (Proposition 3.3.6), we have $H^q(\mathcal{V}^\lambda_\alpha, \mathcal{F}|_{U_\alpha}) = H^q(U_\alpha, \mathcal{F}|_{U_\alpha}) = 0$ by the affine case that we have just shown and Theorem 2.18. We therefore have the desired result, by once more applying Theorem 2.12. ☐

Example 2.20. Let A be a ring. Let $X = \mathbb{P}^1_A$. Then $H^p(X, \mathcal{O}_X) = 0$ for every $p \geq 1$ (Example 2.7). See also Lemma 3.1.

Remark 2.21. Let X be a separated scheme which has a covering by m affine open subsets. Then for any quasi-coherent sheaf $\mathcal{F}$ on X, we have $H^p(X, \mathcal{F}) = 0$ for every $p \geq m$. This results from the theorem above and Corollary 2.5.

Corollary 2.22. *Let X be a separated scheme, and let*

$$0 \to \mathcal{F}'' \to \mathcal{F} \to \mathcal{F}' \to 0$$

be an exact sequence of sheaves on X with $\mathcal{F}''$ quasi-coherent. Then we have a long exact sequence

$$0 \to H^0(X, \mathcal{F}'') \to H^0(X, \mathcal{F}) \to H^0(X, \mathcal{F}')$$

$$\xrightarrow{\partial} H^1(X, \mathcal{F}'') \to H^1(X, \mathcal{F}) \to H^1(X, \mathcal{F}') \xrightarrow{\partial} H^2(X, \mathcal{F}'') \to \cdots .$$

Proof Let $\mathcal{U}$ be an affine covering of X. Then the complex

$$0 \to C(\mathcal{U}, \mathcal{F}'') \to C(\mathcal{U}, \mathcal{F}) \to C(\mathcal{U}, \mathcal{F}'') \to 0$$

is exact by Proposition 1.8. The proposition then results from Remark 2.14 and Theorem 2.19. ☐

Theorem 2.23 (Serre's criterion). *Let X be a scheme that is either Noetherian or separated and quasi-compact. Then the following conditions are equivalent:*

(i) *X is affine.*

(ii) *For every quasi-coherent sheaf $\mathcal{F}$ on X, we have $H^p(X, \mathcal{F}) = 0$ for every $p \geq 1$.*

(iii) *For every quasi-coherent (resp. coherent if X is Noetherian) sheaf $\mathcal{F}$ on X, we have $H^1(X, \mathcal{F}) = 0$.*

Proof (i) implies (ii): this is Theorem 2.18. (ii) trivially implies (iii). It remains to show that (iii) implies (i).

Let $A = \mathcal{O}_X(X)$. We must show that the canonical morphism $\varphi : X \to \operatorname{Spec} A$ is an isomorphism. Let $f \in A$. Then $\varphi^{-1}(D(f)) = X_f$ (see Subsection 2.3.1 for the definition of X_f). Under the hypotheses of the theorem, condition (3.2) of Subsection 2.3.1, is satisfied. Consequently, $\mathcal{O}_X(X_f) = A_f$ (Proposition 2.3.12). In particular, the restriction of φ to X_f is an isomorphism if X_f is affine. It therefore suffices to show that we can cover X with affine open subsets of the form $X_{f_1}, \ldots, X_{f_n}$ with $\operatorname{Spec} A = \cup_i D(f_i)$. The proof will be similar to that of Theorem 1.34.

Let x be a point of X. We want to show that it is contained in an affine open subset of the form X_f. The closure $\overline{\{x\}}$ of x in X is a closed subset of X, and hence quasi-compact. Hence $\overline{\{x\}}$ contains a closed point of X (Exercise 2.4.8). It therefore suffices to consider the case when x itself is closed. Let $\mathcal{M}$ be the sheaf of ideals of $\mathcal{O}_X$ which defines the reduced closed subscheme of X whose underlying space is $\{x\}$. Let us consider an affine open neighborhood $U \ni x$ and a sheaf of ideals $\mathcal{J}$ such that $V(\mathcal{J}) = X \setminus U$. We have an exact sequence of quasi-coherent sheaves

$$0 \to \mathcal{M}\mathcal{J} \to \mathcal{J} \to \mathcal{J}/\mathcal{M}\mathcal{J} \to 0.$$

We have $(\mathcal{J}/\mathcal{M}\mathcal{J})_y = 0$ if $y \neq x$; hence $\mathcal{J}/\mathcal{M}\mathcal{J}$ is a skyscraper sheaf, whose fiber at x is $\mathcal{O}_{X,x}/\mathfrak{m}_x = k(x)$. By hypothesis, $H^1(X, \mathcal{M}\mathcal{J}) = 0$. It follows from Proposition 2.15 that $\mathcal{J}(X) \to (\mathcal{J}/\mathcal{M}\mathcal{J})(X) = k(x)$ is surjective. There therefore exists an $f \in \mathcal{J}(X)$ such that $f_x \notin \mathfrak{m}_x \mathcal{O}_{X,x}$. Consequently $x \in X_f \subseteq U$. As $X_f = D(f|_U)$, it is therefore affine.

As X is quasi-compact, we can cover it with a finite number of affine open subsets of the form $X_{f_1}, \ldots, X_{f_n}$. It remains to show that $\operatorname{Spec} A = \cup_i D(f_i)$, in other words, that $\sum_i f_i A = A$. Let us consider the homomorphism $\psi : \mathcal{O}_X^n \to \mathcal{O}_X$ defined by $(a_1, \ldots, a_n) \mapsto \sum_i f_i a_i$. It is surjective because it is surjective in all of the fibers. Applying Proposition 2.15 to the exact sequence

$$0 \to \operatorname{Ker} \psi \to \mathcal{O}_X^n \to \mathcal{O}_X \to 0$$

(the kernel $\operatorname{Ker} \psi$ is quasi-coherent by Proposition 1.12(c)), we obtain the surjectivity of $\psi(X) : A^n \to A$ owing to the $H^1(X, \operatorname{Ker} \psi) = 0$ hypothesis, and therefore $A = \sum_i f_i A$. $\square$

Proposition 2.24. *Let X be a quasi-projective scheme of dimension d over a Noetherian ring A. Then X admits a covering by $d+1$ affine open subsets. In particular, $H^p(X, \mathcal{F}) = 0$ for every $p > d$.*

Proof By definition, X is an open subscheme of a closed subscheme of $\mathbb{P}^n_A$. Let $\overline{X}$ be the closure of X in $\mathbb{P}^n_A$, and $Z = \overline{X} \setminus X$. By Proposition 3.3.36(a), there exists a principal closed subset $V_+(P_1)$ of $\mathbb{P}^n_A$ such that $V_+(P_1) \supseteq Z$ and that $V_+(P_1)$ does not contain any generic point of X (as X is Noetherian, it only has a finite number of generic points). Thus we construct, by induction, a sequence of principal closed subsets $V_+(P_i)$ containing Z and such that $V_+(P_{i+1})$ does not contain any generic point of $X \cap (\cap_{j \leq i} V_+(P_j))$. As the dimension of this intersection decreases strictly every time, we have $X \cap (\cap_{j \leq d+1} V_+(P_j)) = \emptyset$. This is equivalent to

$$X = \cup_{i \leq d+1} (D_+(P_i) \cap \overline{X}).$$

As $D_+(P_i) \cap \overline{X}$ is an affine open subset of $\overline{X}$, and hence of X, we see that X is covered by $d+1$ affine open subsets. This proves the proposition (Remark 2.21). $\square$

Remark 2.25. For any Noetherian separated scheme X of dimension d, we have $H^p(X, \mathcal{F}) = 0$ for any quasi-coherent sheaf $\mathcal{F}$ and any $p > d$. This is a particular case of a vanishing theorem of Grothendieck. See [43], Theorem III.2.7, or [40], Théorème II.4.15.2.

5.2.3   Higher direct image and flat base change

Let $f : X \to Y$ be a separated morphism of schemes. Let $\mathcal{F}$ be a quasi-coherent sheaf on X. We want to have a notion of 'relative cohomology' of $\mathcal{F}$. Let us recall that a morphism f is called quasi-compact if the inverse image of any affine open subset is quasi-compact. For example, if X is Noetherian or if f is of finite type, then f is quasi-compact.

Lemma 2.26. *Let $f : X \to \operatorname{Spec} A$ be a separated quasi-compact morphism. Let $\mathcal{F}$ be a quasi-coherent sheaf on X and M an A-module. Let $\mathcal{F} \otimes_A M$ denote the quasi-coherent sheaf $U \mapsto \mathcal{F}(U) \otimes_A M$ for every affine open set U. Then for any $p \geq 0$, we have a canonical homomorphism*

$$H^p(X, \mathcal{F}) \otimes_A M \to H^p(X, \mathcal{F} \otimes_A M).$$

Moreover, this homomorphism is an isomorphism if M is flat over A.

Proof Let us first note that $\mathcal{F} \otimes_A M$ is indeed a quasi-coherent sheaf because $(\mathcal{F} \otimes_A M)|_U = (\mathcal{F}(U) \otimes_A M)^\sim$ for every affine open subset U of X (Exercise 1.4(a)). For any complex of A-modules $N'' \xrightarrow{\alpha} N \xrightarrow{\beta} N'$, we have a canonical homomorphism

$$(\operatorname{Ker} \beta / \operatorname{Im} \alpha) \otimes_A M \to \operatorname{Ker}(\beta \otimes \operatorname{Id}_M) / \operatorname{Im}(\alpha \otimes \operatorname{Id}_M) \tag{2.7}$$

which is an isomorphism if M is flat. Let us take a finite affine covering $\mathcal{U} = \{U_i\}_i$ of X. Then $(\mathcal{F} \otimes_A M)(U_{i_0\ldots i_p}) = \mathcal{F}(U_{i_0\ldots i_p}) \otimes_A M$ because, since X is separated, $U_{i_0\ldots i_p}$ is affine, whence the identity of complexes

$$C(\mathcal{U}, \mathcal{F}) \otimes_A M = C(\mathcal{U}, \mathcal{F} \otimes_A M)$$

(note that the covering $\mathcal{U}$ is finite, and hence $C(\mathcal{U}, \mathcal{F})$ is a direct sum). We easily verify that the differential on $C(\mathcal{U}, \mathcal{F} \otimes_A M)$ is equal to $d \otimes \mathrm{Id}_M$. We therefore obtain a canonical homomorphism

$$H(C(\mathcal{U}, \mathcal{F})) \otimes_A M \to H(C(\mathcal{U}, \mathcal{F} \otimes_A M))$$

in the manner of (2.7) above. Moreover, it is an isomorphism if M is flat. We conclude by Theorem 2.19 because X is a separated scheme (Propositions 3.3.4 and 3.3.9(b)). $\square$

Corollary 2.27. *Let $f : X \to \operatorname{Spec} A$ be a separated and quasi-compact morphism. Let B be a flat A-algebra. Let $X_B = X \times_{\operatorname{Spec} A} \operatorname{Spec} B$ and let $\rho : X_B \to X$ be the projection morphism. Then for any quasi-coherent sheaf $\mathcal{F}$ on X and for any $p \geq 0$, we have a canonical isomorphism*

$$H^p(X, \mathcal{F}) \otimes_A B \simeq H^p(X_B, \rho^* \mathcal{F}).$$

Proof Let $\mathcal{U} = \{U_i\}_i$ be a finite affine covering of X. Then $\mathcal{U}_B := \{(U_i)_B\}_i$ is an affine covering of X_B and $C(\mathcal{U}_B, \rho^* \mathcal{F}) = C(\mathcal{U}, \mathcal{F} \otimes_A B)$, with the notation of Lemma 2.26. The corollary then results from Lemma 2.26 and Theorem 2.19 applied to X_B, which is separated (Proposition 3.3.9(d)). $\square$

Proposition 2.28. *Let $f : X \to Y$ be a separated and quasi-compact morphism of schemes. Let $\mathcal{F}$ be a quasi-coherent sheaf on X. Then for every $p \geq 0$, there exists a unique quasi-coherent sheaf $R^p f_* \mathcal{F}$ on Y such that for every affine open subset V of Y, we have*

$$R^p f_* \mathcal{F}(V) = H^p(f^{-1}(V), \mathcal{F}|_{f^{-1}(V)}).$$

Proof Let $W \subseteq V$ be two affine open subsets of Y. As the restriction homomorphism $\mathcal{O}_Y(V) \to \mathcal{O}_Y(W)$ is flat, it follows from Corollary 2.27 (applied to $f^{-1}(V) \to V$ and $\operatorname{Spec} B = W$) that

$$H^p(f^{-1}(W), \mathcal{F}|_{f^{-1}(W)}) \simeq H^p(f^{-1}(V), \mathcal{F}|_{f^{-1}(V)}) \otimes_{\mathcal{O}_Y(V)} \mathcal{O}_Y(W).$$

This immediately implies the proposition. $\square$

Remark 2.29. We have $R^0 f_* \mathcal{F} = f_* \mathcal{F}$. The sheaves $R^p f_* \mathcal{F}$, for $p \geq 1$, are called *higher direct images* of $\mathcal{F}$.

Definition 2.30. Let X be a scheme, and let $\mathcal{F}$ be an $\mathcal{O}_X$-module. We say that $\mathcal{F}$ is *flat* at a point $x \in X$ if $\mathcal{F}_x$ is a flat $\mathcal{O}_{X,x}$-module. We say that $\mathcal{F}$ is flat if it is flat at every point $x \in X$. Let $f : X \to Y$ be a morphism of schemes. Then $\mathcal{F}_x$ is endowed with the structure of an $\mathcal{O}_{Y,f(x)}$-module via the canonical homomorphism $\mathcal{O}_{Y,f(x)} \to \mathcal{O}_{X,x}$. We say that $\mathcal{F}$ is *flat over Y* at x if $\mathcal{F}_x$ is flat over $\mathcal{O}_{Y,f(x)}$. We say that $\mathcal{F}$ is flat over Y if $\mathcal{F}$ is flat over Y at every point $x \in X$.

Lemma 2.31. *Let $\mathcal{F}$ be a quasi-coherent sheaf on a scheme X. Then the following properties are true.*

(a) *$\mathcal{F}$ is flat over X if and only if for every affine open subset U of X, $\mathcal{F}(U)$ is flat over $\mathcal{O}_X(U)$.*

(b) *Let us suppose X is locally Noetherian and $\mathcal{F}$ coherent. Then $\mathcal{F}$ is flat over X if and only if it is locally free of finite rank (Exercise 1.12).*

Proof (a) follows from Proposition 1.2.13 and (b) is a consequence of Theorem 1.2.16. □

Proposition 2.32. *Let $f : X \to Y$ be a separated and quasi-compact morphism of schemes. Let $\mathcal{F}$ be a quasi-coherent sheaf on X. Let $\mathcal{G}$ be a quasi-coherent sheaf on Y. Then for any $p \geq 0$, we have a canonical homomorphism*

$$(R^p f_* \mathcal{F}) \otimes_{\mathcal{O}_Y} \mathcal{G} \to R^p f_*(\mathcal{F} \otimes_{\mathcal{O}_X} f^* \mathcal{G}) \tag{2.8}$$

that is an isomorphism if $\mathcal{G}$ is flat over Y. In the latter case, the isomorphism is called the projection formula.

Proof For any affine open subset V of Y, the left-hand side of (2.8) is equal to

$$H^p\left(f^{-1}(V), \mathcal{F}|_{f^{-1}(V)}\right) \otimes_{\mathcal{O}_Y(V)} \mathcal{G}(V)$$

while the right-hand side is

$$H^p\left(f^{-1}(V), \mathcal{F}|_{f^{-1}(V)} \otimes_{\mathcal{O}_Y(V)} \mathcal{G}(V)\right).$$

The proposition is therefore a consequence of Lemma 2.26. □

Remark 2.33. In general, the hypothesis that $\mathcal{G}$ is flat over Y cannot be removed. See Example 3.23.

Proposition 2.34. *Let $f : X \to Y$ be a quasi-projective morphism to a locally Noetherian scheme. Let r be the supremum of the dimensions $\dim X_y$, $y \in Y$. Then $R^p f_* \mathcal{F} = 0$ for every $p > r$ and for every quasi-coherent sheaf $\mathcal{F}$ on X.*

Proof This is an immediate consequence of Proposition 2.24. □

Exercises

2.1. (*Cohomology of flasque sheaves*) Let X be a topological space. We say that a sheaf $\mathcal{F}$ on X is *flasque* if for every pair of open subsets $V \subseteq U$, the restriction $\mathcal{F}(U) \to \mathcal{F}(V)$ is surjective. Let $\mathcal{U} = \{U_i\}_{i \in I}$ be a covering of X. We are going to show that $H^{p+1}(\mathcal{U}, \mathcal{F}) = 0$ for every $p \geq 0$. Let us endow I with a total ordering with minimal element i_0. Let us fix an $f \in C^{p+1}(\mathcal{U}, \mathcal{F})$ such that $df = 0$.

(a) For any $i \in I$, let $\mathcal{U}_i$ denote the covering $U_i \cap \mathcal{U}$ of U_i. Note that $C^p(\mathcal{U}, \mathcal{F}) = C^{p-1}(\mathcal{U}_i, \mathcal{F}) \times \prod_{i' \neq i} C^{p-1}(\mathcal{U}_{i'}, \mathcal{F})$. Let J be a subset of

I containing i_0. Let us consider the following hypothesis (H_J): there exist

$$g_i = (g_{i\alpha})_{\alpha \in I^p} \in C^{p-1}(\mathcal{U}_i, \mathcal{F}|_{U_i}), \quad i \in J$$

such that for every $i, j \in J$ with $i < j$, and for every $\alpha \in I^p$, we have

$$f_{ij\alpha} = g_{j\alpha}|_{U_{ij\alpha}} - (dg_i)_{j\alpha}.$$

Show that (H_J) is satisfied if $J = \{i_0\}$.

(b) Let us suppose H_J is true. Let $k \in I$ be such that $k > i$ for every $i \in J$. Show that for any $i, j \in J$, $i \neq j$, we have

$$f_{ik\alpha} + (dg_i)_{k\alpha} = f_{jk\alpha} + (dg_j)_{k\alpha}$$

on $U_{ijk\alpha}$ (develop $(df)_{ijk\alpha}$). Show that there exist $h_{k\alpha} \in \mathcal{F}(\cup_{i \in J} U_{ik\alpha})$ such that $h_{k\alpha}|_{U_{ik\alpha}} = f_{ik\alpha} + (dg_i)_{k\alpha}$ for every $i \in J$. Deduce from this that $(H_{J \cup \{k\}})$ is true (lift $h_{k\alpha}$ to $g_{k\alpha} \in \mathcal{F}(U_{k\alpha})$).

(c) Show that (H_I) is true (consider the set of couples $(J, (g_i)_{i \in J})$ verifying (H_J), ordered in a natural manner, and apply Zorn's lemma). Deduce from this that $H^{p+1}(\mathcal{U}, \mathcal{F}) = 0$ (use Corollary 2.4).

2.2. Let $f : X \to Y$ be an *affine morphism* (i.e., for every affine open subset V of Y, $f^{-1}(V)$ is affine). Show that for any quasi-coherent sheaf $\mathcal{F}$ on X, and for any $p \geq 1$, we have $R^p f_* \mathcal{F} = 0$.

2.3. Let $f : X \to Y$ be a morphism of schemes. Let $\mathcal{F}$ be a sheaf on X.

(a) Show that we have a canonical homomorphism $H^p(Y, f_* \mathcal{F}) \to H^p(X, \mathcal{F})$ for every $p \geq 0$.

(b) Let us suppose $\mathcal{F}$ is quasi-coherent. Show that the homomorphism in (a) is an isomorphism in each of the following cases: (1) X is separated and f is affine (use Theorem 2.19); (2) f is a closed immersion.

2.4. Let X be an integral scheme. Let $\mathcal{K}_X$ denote the constant sheaf $K(X)$ on X.

(a) Show that $\mathcal{K}_X$ is a quasi-coherent sheaf.

(b) For any point $x \in X$, let Q_x denote the skyscraper sheaf with support $\{x\}$ and fiber $K(X)/\mathcal{O}_{X,x}$ over x. Let us suppose X is of dimension 1 and quasi-compact. Show that we have an exact sequence of quasi-coherent sheaves on X:

$$0 \to \mathcal{O}_X \to \mathcal{K}_X \to \oplus_{x \in X} Q_x \to 0$$

(show that every section s of $\mathcal{K}_X/\mathcal{O}_X$ over an open subset U verifies $s_x = 0$ for all but finitely many points x).

(c) Let $X = \mathbb{P}_k^1$ be the projective line over a field k. Let $x_1, \ldots, x_n$ be points of X that are pairwise distinct, and let $f_1, \ldots, f_n \in K(X)$.

Show that there exists an $f \in K(X)$ such that $f - f_i \in \mathcal{O}_{X,x_i}$ for every i (apply Proposition 2.15 to the exact sequence in (b)).

2.5. Let X be an open subset of the affine plane $\mathbb{A}_k^2$ over a field k.

(a) Let F be a finite set of closed points $p_1, \dots, p_n$ of $\mathbb{A}_k^2$. Show that p_i corresponds to a maximal ideal of $k[T, S]$ generated by two irreducible polynomials $f_i \in k[T], g_i \in k[T, S]$. Changing the numbering if necessary, we may assume that $f_1, \dots, f_m$ are pairwise prime to each other, and that f_i is equal to one of the $f_1, \dots, f_m$ if $i \geq m + 1$. Let us set

$$f = \prod_{1 \leq i \leq m} f_i, \quad g = \sum_{1 \leq i \leq m} \left(\left(\prod_{1 \leq j \leq m, j \neq i} f_j \right) \prod_{f_l = f_i} g_l \right).$$

Show that $V(f, g) = \{p_1, \dots, p_n\}$.

(b) Show that X is the union of two affine open sets (more generally, one can show that any non-empty open subset of $\mathbb{A}_k^d$ is the union of d affine open subsets. See [53], Corollary V.1.5). Deduce from this that for any quasi-coherent sheaf $\mathcal{F}$ and for any $p \geq 2$, we have $H^p(X, \mathcal{F}) = 0$.

2.6. Let X be a Noetherian topological space. Let $(\mathcal{F}_\lambda)_{\lambda \in \Lambda}$ be a direct system of sheaves on X.

(a) Show that the correspondence $U \mapsto \varinjlim_\lambda \mathcal{F}_\lambda(U)$, for every open set U, defines a sheaf $\mathcal{F}$ on X.

(b) Let us fix $p \geq 0$. Let $\mathcal{U}$ be an open covering of X. Show that the cohomology groups $H^p(C(\mathcal{U}, \mathcal{F}_\lambda))$, $\lambda \in \Lambda$, naturally form a direct system and that we have a canonical isomorphism

$$\varinjlim_\lambda H^p(C(\mathcal{U}, \mathcal{F}_\lambda)) \simeq H^p(C(\mathcal{U}, \mathcal{F})).$$

(c) Show property (b) when replacing $H^p(C(\mathcal{U}, *))$ by $H^p(X, *)$.

2.7. Let X be a scheme. Let $\mathcal{O}_X^*$ denote the sheaf of Abelian groups $U \mapsto \mathcal{O}_X(U)^*$.

(a) Let $\mathcal{L}$ be an invertible sheaf on X. Let $\mathcal{U} = \{U_i\}_{i \in I}$ be a covering of X such that for every $i \in I$, $\mathcal{L}|_{U_i}$ is free, generated by a section $e_i \in \mathcal{L}(U_i)$. Show that there exist $f_{ij} \in \mathcal{O}_X^*(U_{ij})$ such that $e_i|_{U_{ij}} = e_j|_{U_{ij}} f_{ij}$ for every $i, j \in I$. Let $f := (f_{ij})_{i,j \in I} \in C'^1(\mathcal{U}, \mathcal{O}_X^*)$. Show that the image of f in $H^1(\mathcal{U}, \mathcal{O}_X^*)$ is independent of the choice of the e_i. Show that the image $\phi(\mathcal{L})$ of f in $H^1(X, \mathcal{O}_X^*)$ is independent of the choice of $\mathcal{U}$.

(b) Show that $\phi(\mathcal{L}) = 1$ if and only if $\mathcal{L}$ is free.

(c) Show that the correspondence $\mathcal{L} \mapsto \phi(\mathcal{L})$ induces a canonical group isomorphism $\mathrm{Pic}(X) \simeq H^1(X, \mathcal{O}_X^*)$ (to prove the surjectivity, use

Exercise 2.2.8 by taking $\mathcal{F}_i = \mathcal{O}_{U_i}$, and φ_{ij} to be the isomorphism defined by multiplication by f_{ij}).

(d) Let $f : X \to Y$ be a morphism of schemes. Then $\mathcal{L} \mapsto f^*\mathcal{L}$ induces a group homomorphism $\mathrm{Pic}(Y) \to \mathrm{Pic}(X)$. In addition, the homomorphism $\mathcal{O}_Y^* \to f_*\mathcal{O}_X^*$ canonically induces a homomorphism $H^1(Y, \mathcal{O}_Y^*) \to H^1(X, \mathcal{O}_X^*)$. Show that these two homomorphisms are identical via the isomorphisms of (c).

2.8. Let X be a quasi-compact open subscheme of an affine scheme $\mathrm{Spec}\, A$. We suppose that $H^p(X, \mathcal{O}_X) = 0$ for every $p \geq 1$.

(a) Show that there exist $f_1, \ldots, f_n \in A$ such that $U_i := D(f_i) \subseteq X$ and that $X = \cup_{1 \leq i \leq n} U_i$.

(b) Let us set $B = \mathcal{O}_X(X)$ and $\mathcal{U} = \{U_i\}_{1 \leq i \leq n}$. Show that

$$0 \to B \to C^0(\mathcal{U}, \mathcal{O}_X) \to C^1(\mathcal{U}, \mathcal{O}_X) \to \ldots \qquad (2.9)$$

is an exact sequence of flat B-modules (use Proposition 2.3.12), zero from a finite rank on. Show that for any B-module M, $(2.9) \otimes_B M$ stays exact (use Proposition 1.2.6).

(c) Let $\mathcal{F}$ be a quasi-coherent sheaf on X. Show that $H^p(X, \mathcal{F}) = 0$ for every $p \geq 1$ (use Proposition 1.6 and (b)). Deduce from this that X is affine.

2.9. Show that in Serre's criterion (Theorem 2.23), we can replace condition (iii) by $H^1(X, \mathcal{J}) = 0$ for every quasi-coherent sheaf of ideals $\mathcal{J}$.

2.10. Let us keep the notation and hypotheses of Exercise 1.16. Show that we have a canonical isomorphism $\pi^*(R^n f_*\mathcal{F}) \to R^n f_{T*}(p^*\mathcal{F})$ for every $n \geq 0$.

2.11. Let $f : X \to Y$ be a morphism. Let $\mathcal{F}$ be an $\mathcal{O}_X$-module, flat over Y.

(a) Show that for any affine open subset V of Y and any affine open subset U of $f^{-1}(V)$, $\mathcal{F}(U)$ is a flat $\mathcal{O}_Y(V)$-module (draw inspiration from the proof of Corollary 1.2.15).

(b) Let us suppose that f is separated and quasi-compact, and that $R^p f_*\mathcal{F} = 0$ for every $p \geq 1$. Show that $f_*\mathcal{F}$ is a flat $\mathcal{O}_Y$-module (use Exercise 1.2.15).

2.12. Let S be a scheme and $f : X \to S$ a morphism with X Noetherian or f separated and quasi-compact. We suppose that there exists a faithfully flat S-scheme T such that $X_T \to T$ is affine. We want to show that f is an affine morphism.

(a) Let us suppose S is affine. Let us fix a quasi-coherent sheaf $\mathcal{F}$ on X. Show that for any $s \in S$, we have $H^1(X, \mathcal{F}) \otimes_{\mathcal{O}_S(S)} \mathcal{O}_{S,s} = 0$ (use Corollary 2.27).

(b) Deduce from this that X is affine if S is affine (use Lemma 1.2.12). Conclude.

2.13. Let S be a scheme and $g : X \to Y$ a morphism of S-schemes such that X is Noetherian or g is separated and quasi-compact. We suppose that there exists a faithfully flat S-scheme T such that g_T is an isomorphism. We want to show that g is an isomorphism.

(a) Show that we can reduce to S and Y affine.

(b) Show that under the hypotheses of (a), we have X affine and f is an isomorphism (use Lemma 1.2.12).

2.14. Let $f : X \dashrightarrow Y$ be a birational map as in Exercise 4.3.15 but without the Y affine hypothesis. Using Exercises 2.13 and 3.3.14, show that f is defined everywhere.

2.15. Let S be a locally Noetherian scheme, $f : X \to S$ a proper morphism.

(a) Let us suppose that $S = \operatorname{Spec} A$, with A a complete local ring. Let s be the closed point of S. Show that if X_s is finite, then X is finite over S (use Lemma 4.4.1).

(b) Show that (a) is true without the A complete hypothesis (use Exercise 2.13 and Lemma 3.3.17).

(c) Let us suppose that X_s is finite for every $s \in S$. Show that f is a finite morphism (compare with Exercise 4.4.2).

2.16. Let $f : X \to Y$ be an affine morphism.

(a) Let $\mathcal{F}$ be a quasi-coherent sheaf on X. Show that the canonical homomorphism $f^* f_* \mathcal{F} \to \mathcal{F}$ (Exercise 1.1) is surjective.

(b) Let us suppose Y is quasi-compact and f finite. Let $\mathcal{L}$ be an ample sheaf on Y. Show that $f^* \mathcal{L}$ is ample on X (use Propositions 1.14(d) and 2.32).

2.17. Let $(X, \mathcal{O}_X)$ be a ringed topological space, and let $\mathcal{F}$, $\mathcal{G}$ be two $\mathcal{O}_X$-modules. We are going to define an operation called the *cup product*

$$H^p(X, \mathcal{F}) \times H^q(X, \mathcal{G}) \to H^{p+q}(X, \mathcal{F} \otimes_{\mathcal{O}_X} \mathcal{G})$$

for any $p, q \geq 0$. Let $\mathcal{U} = \{U_i\}_i$, $\mathcal{V} = \{V_j\}_j$ be two coverings of X. Let $\mathcal{W} = \{U_i \cap V_j\}_{(i,j)}$ be the covering $\mathcal{U} \cap \mathcal{V}$ of X. We will denote the index (i, j) by α.

(a) Let $f \in C^p(\mathcal{U}, \mathcal{F})$, $g \in C^q(\mathcal{V}, \mathcal{G})$, and set $f \smile g \in C^{p+q}(\mathcal{W}, \mathcal{F} \otimes_{\mathcal{O}_X} \mathcal{G})$ equal to the cochain defined by

$$(f \smile g)_{\alpha_0 \dots \alpha_{p+q}} = f_{i_0 \dots i_p}|_{W_\alpha} \otimes g_{j_p \dots j_{p+q}}|_{W_\alpha}, \quad W_\alpha = U_i \cap V_j.$$

Show that $d(f \smile g) = df \smile g + (-1)^p (f \smile dg)$.

(b) Show that the operation $\smile$ in (a) induces a homomorphism

$$H^p(\mathcal{U}, \mathcal{F}) \times H^q(\mathcal{V}, \mathcal{G}) \to H^{p+q}(\mathcal{W}, \mathcal{F} \otimes_{\mathcal{O}_X} \mathcal{G}).$$

By taking the direct limit, we thus obtain a homomorphism

$$\smile : H^p(X, \mathcal{F}) \times H^q(X, \mathcal{G}) \to H^{p+q}(X, \mathcal{F} \otimes_{\mathcal{O}_X} \mathcal{G})$$

called the cup product.

(c) Show that the cup product is associative in an evident sense. We can show that it is anti-commutative, that is to say that

$$\xi \smile \eta = (-1)^{pq} \eta \smile \xi$$

if $\xi \in H^p(X, \mathcal{F})$, $\eta \in H^q(X, \mathcal{G})$, and if we identify $\mathcal{F} \otimes_{\mathcal{O}_X} \mathcal{G}$ with $\mathcal{G} \otimes_{\mathcal{O}_X} \mathcal{F}$ (see [40], Section II.6.6).

5.3 Cohomology of projective schemes

In this section we are going to study more closely the cohomology groups of a projective scheme. The first fundamental theorem is the finiteness of these groups (Theorem 3.2). A first application of this theorem is the connectedness principle of Zariski (Theorem 3.15).

5.3.1 Direct image theorem

Let us begin by giving the cohomology groups of certain sheaves on a projective space.

Lemma 3.1. *Let A be a ring, $B = A[T_0, \ldots, T_d]$, and $X = \operatorname{Proj} B$. Then for any $n \in \mathbb{Z}$, we have:*

(a) $H^0(X, \mathcal{O}_X(n)) = B_n$ *(if $n < 0$, we set $B_n = 0$ by convention);*

(b) $H^p(X, \mathcal{O}_X(n)) = 0$ *if $p \neq 0, d$;*

(c) $H^d(X, \mathcal{O}_X(n)) \simeq H^0(X, \mathcal{O}_X(-n - d - 1))^{\vee}$ *(where $^{\vee}$ means dual as an A-module). In particular, $H^d(X, \mathcal{O}_X(n)) = 0$ if $n \geq -d$.*

Proof (a) This is Lemma 1.22. (c) will be done in Exercise 3.2. The proof of part (b) also uses explicit computations, but is much more complex. See [41], III.2.1.12 or [43], III.5.1. □

The following theorem is fundamental for the study of projective schemes.

Theorem 3.2 (Serre). *Let A be a Noetherian ring, and X be a projective scheme over A. Let $\mathcal{F}$ be a coherent sheaf on X. Then we have the following properties.*

(a) *For any integer $p \geq 0$, the A-module $H^p(X, \mathcal{F})$ is finitely generated.*

(b) *There exists an integer n_0 (which depends on $\mathcal{F}$) such that for every $n \geq n_0$ and for every $p \geq 1$, we have $H^p(X, \mathcal{F}(n)) = 0$.*

Proof Let $f : X \to \mathbb{P}^d_A$ be a closed immersion. Then we have

$$H^p(X, \mathcal{F}(n)) = H^p(\mathbb{P}^d_A, f_* \mathcal{F}(n))$$

(Exercise 2.3). As $f_*(\mathcal{F}(n))$ is coherent (Propositions 1.11 and 1.14(d)) and isomorphic to $(f_* \mathcal{F})(n)$ (Exercise 1.16(c)), it suffices to restrict ourselves to the case when $X = \mathbb{P}^d_A$. As X is covered by $d + 1$ affine open subsets (Example 2.3.34), we have $H^p(X, \mathcal{F}) = 0$ for every $p \geq d+1$ (Remark 2.21). The theorem is therefore true for every $p \geq d+1$. Let us show the general case by descending induction on p. Let us suppose that we have proved the result at rank $p+1$. By Corollary 1.28, we have an exact sequence of coherent sheaves

$$0 \to \mathcal{G} \to \mathcal{O}_X(m)^r \to \mathcal{F} \to 0$$

for some $m, r \in \mathbb{Z}$. As $\mathcal{O}_X(n)$ is invertible and therefore flat over $\mathcal{O}_X$, the sequence

$$0 \to \mathcal{G}(n) \to \mathcal{O}_X(m+n)^r \to \mathcal{F}(n) \to 0$$

is exact for every $n \in \mathbb{Z}$. By taking the exact cohomology sequence (Corollary 2.22), we obtain an exact sequence

$$H^p(X, \mathcal{O}_X(m+n)^r) \to H^p(X, \mathcal{F}(n)) \to H^{p+1}(X, \mathcal{G}(n)).$$

It is now clear that the theorem is true at rank p by the induction hypothesis and the lemma above.   □

Remark 3.3. Theorem 3.2(a) is true for every proper scheme over A (see [41], Théorème III.3.2.1). The proof consists of first reducing to the case when X is integral and showing the theorem for a single coherent sheaf with support X. Next we use the result in the projective case with the help of Chow's lemma (Remark 3.3.34) as well as that of the theory of spectral sequences (or, as pointed out by C. Ivorra, of a flasque resolution of the sheaf).

Remark 3.4. Theorem 3.2(a) generalizes Corollary 3.3.19 to the case of projective varieties (proper ones if we take into account the remark above).

Corollary 3.5. *Let* $f : X \to Y$ *be a projective morphism with* Y *locally Noetherian, and let* $\mathcal{F}$ *be a coherent sheaf on* X. *Then for any* $p \geq 0$, $R^p f_* \mathcal{F}$ *(Proposition 2.28) is a coherent sheaf on* Y. *In particular,* $f_* \mathcal{F}$ *is coherent.*

Proposition 3.6. *Let* $f : X \to \operatorname{Spec} A$ *be a proper morphism over a Noetherian affine scheme. Let* $\mathcal{L}$ *be an invertible sheaf on* X. *Then the following properties are equivalent:*

  (i) *The sheaf* $\mathcal{L}$ *is ample.*

 (ii) *For any coherent sheaf* $\mathcal{F}$ *on* X, *there exists an integer* n_0 *such that for every* $n \geq n_0$ *and for every* $p \geq 1$, *we have* $H^p(X, \mathcal{F} \otimes \mathcal{L}^n) = 0$.

(iii) *For any coherent sheaf of ideals* $\mathcal{J}$ *of* $\mathcal{O}_X$, *there exists an integer* n_0 *such that for every* $n \geq n_0$, *we have* $H^1(X, \mathcal{J} \otimes \mathcal{L}^n) = 0$.

If one of these conditions is satisfied, then f *is projective and* $\mathcal{L}^n$ *is very ample for* f *for a sufficiently large* n.

Proof (i) $\implies$ (ii). Some tensor power $\mathcal{L}^n$ of $\mathcal{L}$ is very ample for f (Theorem 1.34); therefore there exists an immersion $i : X \to \mathbb{P}^N_A$ such that $\mathcal{L}^n \simeq i^* \mathcal{O}_{\mathbb{P}^N_A}(1)$. As f is proper, i is a closed immersion (Proposition 3.3.16(e)), and hence f is projective. Consequently (ii) is true by virtue of Theorem 3.2(b).

(ii) $\implies$ (iii) being trivial, it remains to show that (iii) implies (i). Let x be a point of X. We first want to show that it is contained in an affine open set of the form X_s, where s is a section of $\mathcal{L}^n(X)$ for some n (see Definition 1.24). As in the proof of Theorem 2.23, it suffices to consider the case that x itself is closed. Let U be an affine open neighborhood of x such that $\mathcal{L}|_U$ is free, and let

$$0 \to \mathcal{MJ} \to \mathcal{J} \to \mathcal{J}/\mathcal{MJ} \to 0$$

be the exact sequence of quasi-coherent sheaves as in the proof of Theorem 2.23, where $\mathcal{J}/\mathcal{MJ}$ is a skyscraper sheaf with support $\{x\}$ whose fiber at x is equal to $\mathcal{O}_{X,x}/\mathfrak{m}_x = k(x)$. As $\mathcal{L}^n$ is flat over X for every n, the sequence

$$0 \to \mathcal{MJL}^n \to \mathcal{JL}^n \to (\mathcal{J}/\mathcal{MJ}) \otimes \mathcal{L}^n \to 0$$

remains exact. By hypothesis, there exists an n such that $H^1(X, \mathcal{MJL}^n) = 0$. Hence

$$H^0(X, \mathcal{JL}^n) \to H^0(X, (\mathcal{J}/\mathcal{MJ}) \otimes \mathcal{L}^n) = k(x) \otimes_{\mathcal{O}_{X,x}} \mathcal{L}^n_x \to 0$$

is exact (Proposition 2.15). By Nakayama, this implies that

$$H^0(X, \mathcal{JL}^n) \otimes_A \mathcal{O}_{X,x} \to \mathcal{L}^n_x$$

is surjective. Let $s \in H^0(X, \mathcal{JL}^n)$ be such that s_x is a basis of $\mathcal{L}^n_x$. Then $x \in X_s \subseteq U$. The rest of the proof is exactly as for Theorem 1.34, which shows that $\mathcal{L}^m$ is very ample (hence ample by Theorem 3.27) for some positive power $\mathcal{L}^m$. Hence $\mathcal{L}$ is ample (Exercise 1.32). $\qquad\square$

Lemma 3.7. *Let X be a proper scheme over a Noetherian ring A. Let $\mathcal{L}$ be an invertible sheaf on X. We suppose that there exist two coherent sheaves of ideals $\mathcal{I}, \mathcal{J}$ such that $\mathcal{IJ} = 0$, and that $\mathcal{L}|_Y$, $\mathcal{L}|_Z$ are ample respectively on $Y := V(\mathcal{I})$ and on $Z := V(\mathcal{J})$. Then $\mathcal{L}$ is ample.*

Proof Let $\mathcal{F}$ be a coherent sheaf on X. We have an exact sequence

$$0 \to (\mathcal{JF}) \otimes \mathcal{L}^{\otimes n} \to \mathcal{F} \otimes \mathcal{L}^{\otimes n} \to (\mathcal{F} \otimes \mathcal{L}^{\otimes n}) \otimes \mathcal{O}_Z \to 0,$$

and hence an exact sequence

$$H^1(X, (\mathcal{JF}) \otimes \mathcal{L}^{\otimes n}) \to H^1(X, \mathcal{F} \otimes \mathcal{L}^{\otimes n}) \to H^1(X, (\mathcal{F} \otimes \mathcal{L}^{\otimes n}) \otimes \mathcal{O}_Z).$$

As $\mathcal{JF}$ is killed by $\mathcal{I}$, it is a coherent sheaf on Y and we have the equality $H^1(X, (\mathcal{JF}) \otimes \mathcal{L}^{\otimes n}) = H^1(Y, (\mathcal{JF}) \otimes_{\mathcal{O}_Y} (\mathcal{L}|_Y)^{\otimes n})$. Likewise

$$H^1(X, (\mathcal{F} \otimes \mathcal{L}^{\otimes n}) \otimes \mathcal{O}_Z) = H^1(Z, \mathcal{F}|_Z \otimes_{\mathcal{O}_Z} (\mathcal{L}|_Z)^{\otimes n}).$$

It now suffices to apply Proposition 3.6. $\qquad\square$

Corollary 3.8. *Let X be a proper scheme over a Noetherian ring A, with (reduced) irreducible components $X_1, \ldots, X_n$. Let $\mathcal{L}$ be an invertible sheaf on X. Then $\mathcal{L}$ is ample if and only if $\mathcal{L}|_{X_i}$ is ample on X_i for every i.*

Proof As $X_i \to X$ is a closed immersion, we immediately verify that $\mathcal{L}|_{X_i}$ is ample if $\mathcal{L}$ is. Let us now suppose $\mathcal{L}|_{X_i}$ is ample for every $i \leq n$. By induction on n and using Lemma 3.7, we see that $\mathcal{L}|_{X'}$ is ample on $X' := X_{\mathrm{red}}$. Let $\mathcal{N}$ be the coherent sheaf of nilpotent elements of $\mathcal{O}_X$. There exists an $r \geq 1$ such that $\mathcal{N}^r = 0$. We can once again reason by induction on r by applying Lemma 3.7 with $\mathcal{I} = \mathcal{N}^{r-1}$ and $\mathcal{J} = \mathcal{N}$.   □

Remark 3.9. Let X be as in the corollary above, let $f : X' \to X$ be a finite surjective morphism, and $\mathcal{L}$ an invertible sheaf on X. We can show that $\mathcal{L}$ is ample if and only if $f^*\mathcal{L}$ is ample. The 'only if' part is in Exercise 2.16. The converse is more difficult, see [43], Exercise III.5.7, or [41], III.2.6.2. See also Exercise 7.5.3 for a particular case.

5.3.2   Connectedness principle

As an application of the finiteness theorem (Theorem 3.2(a)), we are going to show Zariski's connectedness principle (Theorem 3.15). For this purpose, we first show a special case of the theorem on formal functions (Corollary 3.11). Let M be a module over a ring A. For any $\alpha \in A$, we let

$$M[\alpha] := \{x \in M \mid \alpha x = 0\}$$

denote the submodule of α-torsion elements. It is the kernel of the multiplication by α. Let $\mathcal{F}$ be a quasi-coherent sheaf on an A-scheme X; then $\mathcal{F}[\alpha] := \mathrm{Ker}(\mathcal{F} \xrightarrow{\cdot \alpha} \mathcal{F})$ is also quasi-coherent.

Lemma 3.10. *Let $X \to \mathrm{Spec}\, A$ be a projective scheme over a Noetherian affine scheme. Let $\alpha \in A$, $\mathcal{F}$ a coherent sheaf on X, and $p \geq 0$. For any $n \geq 0$, let us consider the exact sequence*

$$H^p(X, \alpha^n \mathcal{F}) \xrightarrow{\lambda_n} H^p(X, \mathcal{F}) \xrightarrow{\rho_n} H^p(X, \mathcal{F}/\alpha^n \mathcal{F}) \xrightarrow{\partial_n} H^{p+1}(X, \alpha^n \mathcal{F})$$

induced by the exact sequence $0 \to \alpha^n \mathcal{F} \to \mathcal{F} \to \mathcal{F}/\alpha^n \mathcal{F} \to 0$.

  (a) *There exists an $n_0 \geq 0$ such that for every $n \geq n_0$, we have*

$$\alpha^n H^p(X, \mathcal{F}) \subseteq \mathrm{Ker}\, \rho_n \subseteq \alpha^{n-n_0} H^p(X, \mathcal{F}).$$

  (b) *Let $m \geq n_0$, and $n \geq m$. Let us consider the homomorphism*

$$\pi_{n,m} : H^p(X, \mathcal{F}/\alpha^n \mathcal{F}) \to H^p(X, \mathcal{F}/\alpha^m \mathcal{F})$$

  induced by the canonical homomorphism $\mathcal{F}/\alpha^n \mathcal{F} \to \mathcal{F}/\alpha^m \mathcal{F}$. Then if n is sufficiently large with respect to m, we have $\mathrm{Im}(\pi_{n,m}) \subseteq \mathrm{Im}\, \rho_m$.

Proof We have an ascending sequence of sub-$\mathcal{O}_X$-modules $(\mathcal{F}[\alpha^n])_n$ of $\mathcal{F}$. As $\mathcal{F}$ is coherent, this sequence is stationary, starting let us say at rank n_0. If $n \geq m \geq n_0$, we have a commutative diagram

$$
\begin{array}{ccc}
\alpha^m \mathcal{F} & \longrightarrow & \mathcal{F} \\
{\scriptstyle \cdot \alpha^{n-m}} \big\downarrow & & \big\downarrow {\scriptstyle \cdot \alpha^{n-m}} \\
\alpha^n \mathcal{F} & \longrightarrow & \mathcal{F}
\end{array}
\tag{3.10}
$$

where the horizontal arrows are canonical inclusions and the vertical arrows are the multiplication by α^{n-m}. As $\mathcal{F}[\alpha^n] = \mathcal{F}[\alpha^m]$, the first vertical arrow is an isomorphism. Taking $H^p(X, .)$ of this diagram, we immediately deduce from it that $\operatorname{Im} \lambda_n = \alpha^{n-m} \operatorname{Im} \lambda_m$, whence the second inclusion of (a) by taking $m = n_0$. The first inclusion is always true because α^n kills $\mathcal{F}/\alpha^n \mathcal{F}$.

Let us fix $m \geq n_0$. Let $M = H^{p+1}(X, \alpha^m \mathcal{F})$ and $N = H^{p+1}(X, \alpha^m \mathcal{F})[\alpha^m]$. Let $\delta_{n,m}$ be the canonical homomorphism $H^{p+1}(X, \alpha^n \mathcal{F}) \to H^{p+1}(X, \alpha^m \mathcal{F})$ deduced from the inclusion $\alpha^n \mathcal{F} \to \alpha^m \mathcal{F}$. Diagram (3.10) stays valid when replacing $\mathcal{F}$ by $\alpha^m \mathcal{F}$ in the second column. We deduce from this that $\operatorname{Im} \delta_{n,m} = \alpha^{n-m} M$. As $\mathcal{F}/\alpha^m \mathcal{F}$, and therefore $H^p(X, \mathcal{F}/\alpha^m \mathcal{F})$, are killed by α^m, we have $\operatorname{Im} \partial_m \subseteq N$. Hence

$$
\operatorname{Im}(\partial_m \circ \pi_{n,m}) = \operatorname{Im}(\delta_{n,m} \circ \partial_n) \subseteq \operatorname{Im} \delta_{n,m} \cap \operatorname{Im} \partial_m \subseteq (\alpha^{n-m} M) \cap N.
$$

As M is finitely generated over A (Theorem 3.2), we can apply the Artin–Rees lemma (Corollary 1.3.12). There exists an $n_m \geq m$ such that for every $n \geq n_m$, we have $(\alpha^{n-m} M) \cap N = \alpha((\alpha^{n-m-1} M) \cap N)$. For any $n \geq n_m + m$, we then have $(\alpha^{n-m} M) \cap N \subseteq \alpha^m N = 0$, which shows that $\operatorname{Im} \pi_{n,m} \subseteq \operatorname{Ker} \partial_m = \operatorname{Im} \rho_m$. $\square$

Let us keep the notation of the lemma. Let I be an ideal of A. Let us fix $p \geq 0$. We have two natural inverse systems

$$
H^p(X, \mathcal{F}) \otimes_A (A/I^n A), \quad H^p(X, \mathcal{F}/I^n \mathcal{F}).
$$

The canonical surjection $\mathcal{F} \to \mathcal{F}/I^n \mathcal{F}$ then induces a homomorphism of the inverse limits

$$
\varphi_p : \varprojlim_n (H^p(X, \mathcal{F}) \otimes_A A/I^n A) \to \varprojlim_n H^p(X, \mathcal{F}/I^n \mathcal{F}).
\tag{3.11}
$$

The term on the left is isomorphic to $H^p(X, \mathcal{F}) \otimes_A \hat{A}$, where $\hat{A}$ is the (separated) completion of A for the I-adic topology (Subsection 1.3.1) because $H^p(X, \mathcal{F})$ is finitely generated over A (Theorem 3.2, Corollary 1.3.14).

Corollary 3.11. *Homomorphism* (3.11) *is an isomorphism if* $I = \alpha A$.

Proof Let us keep the notation of Lemma 3.10. Part (a) of the lemma implies that

$$
\varprojlim_n (H^p(X, \mathcal{F}) \otimes_A A/\alpha^n A) \simeq \varprojlim_n \operatorname{Im}(\rho_n).
$$

Let $M_n = H^p(X, \mathcal{F}/\alpha^n \mathcal{F})$. Part (b) of the lemma says that for any $m \geq n_0$, there exists an $n \geq m$ such that $\pi_{n,m}(M_n) \subseteq \operatorname{Im} \rho_m \subseteq M_m$, which implies that $\varprojlim_n (\operatorname{Im} \rho_n) = \varprojlim_n M_n$, whence the corollary. $\square$

Remark 3.12 (*Theorem on formal functions*). We can show that for any ideal I of A, homomorphism (3.11) is an isomorphism. This is the theorem on formal functions. See [43], III.11.1 (and [41], III.4.1.7, for the more general case of proper morphisms). This theorem will play an important role in the proof of Castelnuovo's theorem 9.3.8 and of Theorem 9.4.15.

Definition 3.13. We say that a morphism $f : X \to Y$ is *purely inseparable* if it is injective and if for every $x \in X$, the extension of residue fields $k(f(x)) \to k(x)$ is a purely inseparable extension. This is equivalent to $X_{\mathrm{red}} \to Y_{\mathrm{red}}$ purely inseparable (see also Exercise 3.8). For example, an immersion is a purely inseparable morphism. We will say that a ring homomorphism is purely inseparable if the corresponding morphism of schemes is purely inseparable. Finally, we will say that $\mathcal{O}_Y \to f_*\mathcal{O}_X$ is purely inseparable if the morphism $\operatorname{Spec} f_*\mathcal{O}_X \to Y$ (Exercise 1.17) is purely inseparable.

Lemma 3.14. *Let A be a Noetherian ring, and let X be a projective scheme over A. Let us suppose that $A \to H^0(X, \mathcal{O}_X)$ is purely inseparable. Then for any ideal I of A, there exists an ideal J such that $J \subseteq I \subseteq \sqrt{J}$ and*

$$A/J \to H^0(X, \mathcal{O}_X/J\mathcal{O}_X)$$

is purely inseparable.

Proof Let $\alpha \in I$. With the notation of Lemma 3.10 (with $p = 0$ and $\mathcal{F} = \mathcal{O}_X$), there exist integers $n > m > 0$ such that $\operatorname{Im} \pi_{n,m} \subseteq \operatorname{Im} \rho_m$. Since $\rho_m = \pi_{n,m} \circ \rho_n$, we have

$$H^0(X, \mathcal{O}_X/\alpha^n\mathcal{O}_X) \subseteq \operatorname{Im} \rho_n + \operatorname{Ker} \pi_{n,m}.$$

Now $\operatorname{Ker} \pi_{n,m} = H^0(X, \alpha^m\mathcal{O}_X/\alpha^n\mathcal{O}_X)$, which is a nilpotent ideal. We immediately deduce from this that $\rho_n : H^0(X, \mathcal{O}_X) \to H^0(X, \mathcal{O}_X/\alpha^n\mathcal{O}_X)$ is purely inseparable. By composition, $A \to H^0(X, \mathcal{O}_X/\alpha^n\mathcal{O}_X)$ is also purely inseparable. Let us consider the projective scheme $X' = X \times_{\operatorname{Spec} A} \operatorname{Spec}(A/\alpha^n A)$ on $A' := A/\alpha^n A$. Then

$$A' \to H^0(X', \mathcal{O}_{X'})$$

is purely inseparable because the right-hand side is equal to $H^0(X, \mathcal{O}_X/\alpha^n\mathcal{O}_X)$. We can then repeat the argument with X'. Applying this result successively to a system of generators of I, we obtain an ideal $J \subseteq I$ generated by the powers of a system of generators of I (hence $I \subseteq \sqrt{J}$), such that $A/J \to H^0(X, \mathcal{O}_X/J\mathcal{O}_X)$ is purely inseparable. □

Theorem 3.15 (Zariski's connectedness principle). *Let $f : X \to Y$ be a projective scheme over a locally Noetherian scheme. Let us suppose that $\mathcal{O}_Y \to f_*\mathcal{O}_X$ is an isomorphism. Then X_y is connected for every $y \in Y$.*

Proof We may suppose Y is affine, and even local with closed point y because the construction of f_* commutes with localization (Proposition 3.1.24). Let us apply the lemma above to $A = \mathcal{O}_{Y,y}$, $I = \mathfrak{m}_y\mathcal{O}_{Y,y}$. Let $Y' = \operatorname{Spec} A/J$, and $X' = X \times_Y Y'$. Then $\operatorname{Spec} H^0(X', \mathcal{O}_{X'})$ is reduced to a point because $Y' = \{y\}$. Consequently, X' is connected (Exercise 2.4.6). Now X' has X_y as the underlying topological space, and hence the latter is also connected. □

Corollary 3.16. *Let $f : X \to Y$ be a projective birational morphism with Y normal, locally Noetherian, and integral. Then all of the fibers X_y are connected.*

Proof Indeed, $\mathcal{O}_Y \to f_*\mathcal{O}_X$ is an isomorphism (Corollary 4.4.3(a)). The corollary then follows from Theorem 3.15. □

Corollary 3.17. *Under the hypotheses of Theorem 3.15, the fibers X_y are geometrically connected.*

Proof We may suppose $Y = \operatorname{Spec} A$ is affine, local, with closed point y. Let $E = k(y)[T]/(\tilde{P}(T))$ be a simple extension of $k(y)$, where $\tilde{P}(T)$ is a monic polynomial. Let us lift $\tilde{P}(T)$ to a monic polynomial $P(T) \in A[T]$ (i.e., P is monic and its image in $k(y)[T]$ is equal to $\tilde{P}$) and let us consider $B = A[T]/(P(T))$. This is a local A-algebra, flat and finitely generated. Let $q : X_B = X \times_Y \operatorname{Spec} B \to \operatorname{Spec} B$ be the projection morphism. Then q is projective and we have $\mathcal{O}_{\operatorname{Spec} B} \simeq q_*\mathcal{O}_{X_B}$ (Proposition 3.1.24). By Theorem 3.15, the fiber of q over a closed point of $\operatorname{Spec} B$ is connected. Now this fiber is precisely $X_y \times_{\operatorname{Spec} k(y)} \operatorname{Spec} E$. It is therefore connected. As every finite extension of $k(y)$ is composed of successive simple extensions, this indeed shows that X_y is geometrically connected. □

5.3.3 Cohomology of the fibers

Let $f : X \to S$ be a projective morphism to a locally Noetherian scheme S, let $\mathcal{F}$ be a coherent sheaf on X. For any $s \in S$, we have a coherent sheaf and $\mathcal{F}_s := i^*\mathcal{F}$, where $i : X_s \to X$ is the canonical morphism. We are going to study the cohomology groups $H^p(X_s, \mathcal{F}_s)$ with respect to $R^p f_*\mathcal{F}$. We are going to treat the particular case when S is regular of dimension 1. Since the problem is of local nature, and as the construction of $R^p f_*$ commutes with flat base change over S (Corollary 2.27), we can reduce to the case when S is local.

Lemma 3.18. *Let $\mathcal{O}_K$ be a discrete valuation ring, with field of fractions K, uniformizing parameter t, and residue field $k := \mathcal{O}_K/t\mathcal{O}_K$. Let M be a finitely generated $\mathcal{O}_K$-module. Then the dimensions of the k-vector spaces $M[t]$ and $M \otimes_{\mathcal{O}_K} k$ verify the identity*

$$\dim_k(M \otimes_{\mathcal{O}_K} k) - \dim_k M[t] = \dim_K(M \otimes_{\mathcal{O}_K} K).$$

Proof As M is finitely generated, we have an exact sequence

$$0 \to M' \to L \xrightarrow{\phi} M \to 0,$$

where L is free of finite rank. The submodule M' of L is torsion-free and therefore free over $\mathcal{O}_K$. By tensoring by k, we obtain an exact sequence

$$0 \to (tL \cap M')/tM' \to M'/tM' \to L/tL \to M/tM \to 0.$$

The homomorphism $tL \to M$ defined by $tx \mapsto \phi(x)$ induces an isomorphism

$$(tL \cap M')/tM' \simeq M[t].$$

It follows that

$$\dim_k(M/tM) - \dim_k M[t] = \dim_k(L/tL) - \dim_k(M'/tM')$$
$$= \dim_K(L \otimes K) - \dim_K(M' \otimes K)$$
$$= \dim_K(M \otimes K),$$

which proves the lemma.   □

Lemma 3.19. *Let $\mathcal{O}_K$ be a discrete valuation ring. Let X be a scheme over $\mathcal{O}_K$ and $\mathcal{F}$ a quasi-coherent sheaf on X, flat over $\mathcal{O}_K$. Then for any $p \geq 0$ and any $\alpha \in \mathcal{O}_K \setminus \{0\}$, we have an exact sequence*

$$0 \to H^p(X, \mathcal{F})/\alpha H^p(X, \mathcal{F}) \to H^p(X, \mathcal{F}/\alpha \mathcal{F}) \to H^{p+1}(X, \mathcal{F})[\alpha] \to 0. \quad (3.12)$$

Proof As $\mathcal{F}[\alpha] = 0$ (Corollary 1.2.5), the commutative diagram (3.10) is valid with $m = 0$ and $n = 1$. It suffices to apply $H^p(X, .)$ and $H^{p+1}(X, .)$ to this diagram.   □

Theorem 3.20. *Let $S = \operatorname{Spec} \mathcal{O}_K$ be the spectrum of a discrete valuation ring $\mathcal{O}_K$, with generic point η and closed point s. Let $f : X \to S$ be a projective morphism and $\mathcal{F}$ a coherent sheaf on X that is flat over S. Let us fix $p \geq 0$.*

(a) *(Upper semi-continuity) We have the inequality*

$$\dim_{k(s)} H^p(X_s, \mathcal{F}_s) \geq \dim_{k(\eta)} H^p(X_\eta, \mathcal{F}_\eta). \quad (3.13)$$

(b) *Inequality (3.13) is an equality if and only if $H^p(X, \mathcal{F})$ is free over $\mathcal{O}_K$ and the canonical homomorphism*

$$H^p(X, \mathcal{F}) \otimes_{\mathcal{O}_K} k(s) \to H^p(X_s, \mathcal{F}_s) \quad (3.14)$$

is an isomorphism.

(c) *Let us suppose that (3.14) is surjective. Then the canonical homomorphism*

$$\rho_M : H^p(X, \mathcal{F}) \otimes_{\mathcal{O}_K} M \to H^p(X, \mathcal{F} \otimes_{\mathcal{O}_K} M)$$

(Lemma 2.26) is an isomorphism for every $\mathcal{O}_K$-module M.

(d) *The homomorphism (3.14) is an isomorphism if and only if $H^{p+1}(X, \mathcal{F})$ is free over $\mathcal{O}_K$.*

Proof We will omit $\mathcal{O}_K$ in the index of the tensor product. Let $k = k(s)$ and $K = k(\eta)$. Let t be a uniformizing parameter for $\mathcal{O}_K$. Taking $\alpha = t$ in exact sequence (3.12), we get

$$\dim_k H^p(X_s, \mathcal{F}_s) \geq \dim_k(H^p(X, \mathcal{F}) \otimes k)$$
$$\geq \dim_K H^p(X, \mathcal{F}) \otimes K = \dim_K H^p(X_\eta, \mathcal{F}_\eta)$$

(the second inequality comes from Lemma 3.18 and the last equality from the fact that $\operatorname{Spec} K \to S$ is flat), whence (a). Properties (b) and (d) also follow from Lemma 3.18 and exact sequence (3.12). It remains to show (c).

As M is a direct limit of finitely generated $\mathcal{O}_K$-modules, and $\otimes$ and H^p commute with direct limits (Exercises 1.1.6 and 2.6), we may suppose that M is finitely generated. Let us first show that ρ_M is surjective. The sequence of submodules $(M[t^n])_{n\geq 0}$ of M ($M[t^0] = 0$ by convention) is ascending and therefore stationary. Let m be the smallest integer ≥ 0 such that $M[t^m] = M[t^r]$ for every $r \geq m$. Let us set $N = M/M[t]$. As $\mathcal{F}$ is flat over $\mathcal{O}_K$, we have an exact sequence

$$0 \to \mathcal{F} \otimes M[t] \to \mathcal{F} \otimes M \to \mathcal{F} \otimes N \to 0.$$

Taking the cohomology, we obtain a commutative diagram

$$
\begin{array}{ccccccc}
H^p(X,\mathcal{F}) \otimes M[t] & \longrightarrow & H^p(X,\mathcal{F}) \otimes M & \longrightarrow & H^p(X,\mathcal{F}) \otimes N & \longrightarrow & 0 \\
\downarrow{\scriptstyle \rho_{M[t]}} & & \downarrow{\scriptstyle \rho_M} & & \downarrow{\scriptstyle \rho_N} & & \\
H^p(X,\mathcal{F} \otimes M[t]) & \longrightarrow & H^p(X,\mathcal{F} \otimes M) & \longrightarrow & H^p(X,\mathcal{F} \otimes N) & &
\end{array}
$$

where the lines are exact. As $M[t]$ is a k-vector space, $\rho_{M[t]}$ decomposes into

$$H^p(X,\mathcal{F}) \otimes M[t] \to H^p(X,\mathcal{F} \otimes k) \otimes_k M[t] \to H^p(X,(\mathcal{F} \otimes k) \otimes_k M[t]).$$

The first part is surjective by hypothesis ($H^p(X,\mathcal{F} \otimes k) = H^p(X_s,\mathcal{F}_s)$ by Exercise 2.3(b)) and the second is an isomorphism (Proposition 2.32). Hence $\rho_{M[t]}$ is surjective. Consequently, if ρ_N is surjective, then so is ρ_M. Now $(N[t^n])_{n\geq 0}$ is stationary starting at rank $m-1$ if $m \geq 1$. Hence by induction on m, we reduce to $m = 0$, that is the case when M is torsion-free and hence flat over $\mathcal{O}_K$. But in that case ρ_M is an isomorphism (Proposition 2.32), whence the surjectivity of ρ_M in the general case.

To conclude, let us show that ρ_M is injective. Let us write an exact sequence

$$0 \to M' \to L \to M \to 0$$

as in the proof of Lemma 3.18. We have a commutative diagram of exact sequences

$$
\begin{array}{ccccccc}
H^p(X,\mathcal{F}) \otimes M' & \longrightarrow & H^p(X,\mathcal{F}) \otimes L & \longrightarrow & H^p(X,\mathcal{F}) \otimes M & \longrightarrow & 0 \\
\downarrow{\scriptstyle \rho_{M'}} & & \downarrow{\scriptstyle \rho_L} & & \downarrow{\scriptstyle \rho_M} & & \\
H^p(X,\mathcal{F} \otimes M') & \longrightarrow & H^p(X,\mathcal{F} \otimes L) & \longrightarrow & H^p(X,\mathcal{F} \otimes M) & &
\end{array}
$$

The first two vertical arrows are isomorphisms because M' and L are flat over $\mathcal{O}_K$ (Proposition 2.32), and hence we indeed have ρ_M injective by some diagram chasing. $\qquad\square$

Remark 3.21. If we replace S by a locally Noetherian base scheme Y, then the theorem reads as follows. Let $f : X \to Y$ be a projective morphism. Let $\mathcal{F}$ be a coherent sheaf on X, flat over Y. Let us fix a $p \geq 0$.

(a) The function $h^p : y \mapsto \dim_{k(y)} H^p(X_y, \mathcal{F}_y)$ is upper semi-continuous, that is for any n, the set $\{y \in Y \mid h^p(y) \geq n\}$ is closed in Y.

(b) Let us suppose Y is integral. Then h^p is constant on Y if and only if $R^p f_* \mathcal{F}$ is locally free on Y and the canonical homomorphism

$$R^p f_* \mathcal{F} \otimes_{\mathcal{O}_Y} k(y) \to H^p(X_y, \mathcal{F}_y) \tag{3.15}$$

is an isomorphism for every $y \in Y$.

(c) Let us suppose that (3.15) is surjective at a point y. It is then bijective. If this is true at every point $y \in Y$, then for any quasi-coherent sheaf $\mathcal{G}$ on Y, the canonical homomorphism

$$(R^p f_* \mathcal{F}) \otimes_{\mathcal{O}_Y} \mathcal{G} \to R^p f_*(\mathcal{F} \otimes_{\mathcal{O}_X} f^* \mathcal{G})$$

is an isomorphism.

See [43], III, 12.5–12.11, [70], Section 5, pp. 50–51.

Corollary 3.22. *Let us keep the hypotheses of* Theorem 3.20. *Then the following properties are equivalent:*

(i) For any base change $\pi : T \to S$, let $\rho : X_T \to X$ denote the first projection. Then the canonical homomorphism

$$\pi^*(R^p f_* \mathcal{F}) \to R^p f_{T*}(\rho^* \mathcal{F})$$

(in the manner of Corollary 2.27) is an isomorphism, that is the construction of $R^p f_* \mathcal{F}$ commutes with base change.

(ii) The canonical homomorphism $H^p(X, \mathcal{F}) \otimes_{\mathcal{O}_K} k(s) \to H^p(X_s, \mathcal{F}_s)$ is bijective.

(iii) The canonical homomorphism $H^p(X, \mathcal{F}) \otimes_{\mathcal{O}_K} k(s) \to H^p(X_s, \mathcal{F}_s)$ is surjective.

(iv) The $\mathcal{O}_K$-module $H^{p+1}(X, \mathcal{F})$ is torsion-free.

Proof We have (i) implies (ii) by taking $T = \operatorname{Spec} k(s)$. (ii) trivially implies (iii). The implication (iii) $\Longrightarrow$ (i) results from Theorem 3.20(c). Finally, (iv) is equivalent to (ii) by exact sequence (3.12). □

Example 3.23. Let $\mathcal{O}_K$ be a discrete valuation ring, $X = \mathbb{P}^1_{\mathcal{O}_K}$. We are going to give an example of a sheaf such that homomorphism (3.14) of Theorem 3.20 is not an isomorphism. Let $Z = V(\mathcal{I})$ be the reduced closed subscheme of X whose underlying space consists of two closed points $x_1, x_2 \in X_s$. From the exact sequence $0 \to \mathcal{I} \to \mathcal{O}_X \to \mathcal{O}_Z \to 0$ we deduce the exact sequence

$$\mathcal{O}_K = H^0(X, \mathcal{O}_X) \xrightarrow{\phi} H^0(Z, \mathcal{O}_Z) \to H^1(X, \mathcal{I}) \to H^1(X, \mathcal{O}_X) = 0.$$

As $H^0(Z, \mathcal{O}_Z) = k(x_1) \oplus k(x_2)$, this implies that ϕ is not surjective; hence $H^1(X, \mathcal{I})$ is non-zero and torsion. The sheaf $\mathcal{I}$ is torsion-free because it is contained in $\mathcal{O}_X$ and therefore flat over $\mathcal{O}_K$. By the corollary above,

$$H^0(X, \mathcal{I}) \otimes k \to H^0(X_s, \mathcal{I}_s)$$

is not bijective. We can also give such an example with $\mathcal{F} = \mathcal{O}_X$ (Exercise 3.15).

Corollary 3.24. *Let S be an affine Dedekind scheme and $f : X \to S$ a projective morphism. Let $\mathcal{L}$ be an invertible sheaf on X such that $\mathcal{L}_s$ is ample for every closed point $s \in S$. Then $\mathcal{L}$ is ample.*

Proof Let us first suppose that S is the spectrum of a discrete valuation ring $\mathcal{O}_K$ with uniformizing parameter t. Let $\mathcal{I}$ be a coherent sheaf of ideals of $\mathcal{O}_X$ (hence flat over S). We have $(\mathcal{J} \otimes \mathcal{L}^n)_s = \mathcal{J}_s \otimes \mathcal{L}_s^n$. There therefore exists an $n_0 \in \mathbb{Z}$ such that $H^1(X_s, (\mathcal{J} \otimes \mathcal{L}^n)_s) = 0$ for every $n \geq n_0$. It follows from Theorem 3.20(a)–(b) that $H^1(X, \mathcal{J} \otimes \mathcal{L}^n) = 0$ for every $n \geq n_0$. Hence $\mathcal{L}$ is ample by Proposition 3.6.

   Let us show the general case. By the above and Proposition 1.37(b), there exists an affine open covering $\{V_i\}_{1 \leq i \leq m}$ of S such that $\mathcal{L}|_{X_{V_i}}$ is ample for every $i \leq m$. Let $\mathcal{F}$ be a coherent sheaf on X. There exists an n_0 such that for every $n \geq n_0$, and for every $i \leq m$, $H^1(X, (\mathcal{F} \otimes \mathcal{L}^n)|_{X_{V_i}}) = 0$. Now $H^1(X, (\mathcal{F} \otimes \mathcal{L}^n)|_{X_{V_i}}) = H^1(X, \mathcal{F} \otimes \mathcal{L}^n) \otimes_{\mathcal{O}_S(S)} \mathcal{O}_S(V_i)$. We deduce from this that $H^1(X, \mathcal{F} \otimes \mathcal{L}^n) = 0$, whence $\mathcal{L}$ is ample. $\square$

Remark 3.25. We can show that the corollary above remains true if S is only a Noetherian affine scheme. Indeed, [41], III.4.7.1, says that $\mathcal{L}|_{X_{V_i}}$ is ample on X_{V_i} for an affine open covering $\{V_i\}_i$ of S. The method above then implies that $\mathcal{L}$ is ample on X.

Definition 3.26. Let X be a projective variety over a field k. Let $\mathcal{F}$ be a coherent sheaf on X. We call the alternating sum

$$\chi_k(\mathcal{F}) := \sum_{i \geq 0} (-1)^i \dim_k H^i(X, \mathcal{F})$$

the *Euler–Poincaré characteristic of $\mathcal{F}$.* It is a finite integer by Theorem 3.2 and because $H^i(X, \mathcal{F}) = 0$ if $i > \dim X$ (Proposition 2.24). If no confusion is possible, we will simply write $\chi(\mathcal{F})$.

Example 3.27. Let X be the projective space $\mathbb{P}_k^d$ over a field k. By the computations of Lemma 3.1, we have $\chi(\mathcal{O}_X(n)) = \binom{n+d}{d}$ if $n \geq 0$. In particular, $\chi(\mathcal{O}_X) = 1$.

Proposition 3.28. *Let us keep the notation of Theorem 3.20. Then we have the equality*

$$\chi_{k(s)}(\mathcal{F}_s) = \chi_{k(\eta)}(\mathcal{F}_\eta)$$

(Invariance of the Euler–Poincaré characteristic).

Proof Exact sequence (3.12) implies that

$$\chi_{k(s)}(\mathcal{F}_s) = \sum_{p \geq 0} (-1)^p \left(\dim_{k(s)} H^p(X, \mathcal{F}) \otimes k(s) + \dim_{k(s)} H^{p+1}(X, \mathcal{F})[t] \right)$$

$$= \sum_{p \geq 0} (-1)^p \left(\dim_{k(s)} H^p(X, \mathcal{F}) \otimes k(s) - \dim_{k(s)} H^p(X, \mathcal{F})[t] \right)$$

(note that $H^0(X, \mathcal{F})[t] = 0$). The proposition then follows from Lemma 3.18. $\square$

Remark 3.29. Proposition 3.28 is true over a locally Noetherian and connected base scheme S. See [70], Section 5.

Remark 3.30. Let us note that in the last two subsections, even though the considered morphisms are always supposed projective, we have only used the finiteness theorem (Theorem 3.2(a)). In particular, all of the results are true if we assume the finiteness theorem for proper morphisms (Remark 3.3).

Exercises

3.1. Let k be a field, and let X be the open subvariety $D(x) \cup D(y)$ of $\operatorname{Spec} k[x, y]$. We want to show that $H^1(X, \mathcal{O}_X)$ is of infinite dimension over k.

   (a) Show that the $x^i y^j$, $i, j \in \mathbb{Z}$, form a basis of $k[x, 1/x, y, 1/y]$ as a k-vector space.

   (b) Show that $k[x, 1/x, y] + k[x, y, 1/y]$ is generated by the vectors $x^i y^j$, with $\max\{i, j\} \geq 0$, and that

$$H^1(X, \mathcal{O}_X) \simeq k[x, 1/x, y, 1/y]/(k[x, 1/x, y] + k[x, y, 1/y])$$

   has infinite dimension over k.

3.2. We are going to show Lemma 3.1(c). Let A be a ring, and let $X = \operatorname{Proj} A[T_0, \ldots, T_d]$.

   (a) Let $\mathcal{U} = \{U_i\}_{0 \leq i \leq d}$ denote the covering of X with $U_i = D_+(T_i)$. Let $q \in \mathbb{Z}$. Show that $\mathcal{O}_X(q)(U_{01\ldots d})$ admits as a basis over A the elements of the form D/N, where D, N are monomials without common variable and such that $\deg D = \deg N + q$. Show that if a T_i does not divide N, then D/N belongs to $\mathcal{O}_X(q)(U_{01\ldots \hat{i}\ldots d})|_{U_{01\ldots d}}$.

   (b) Show that the images under $\mathcal{O}_X(q)(U_{01\ldots d}) \to H^d(X, \mathcal{O}_X(q))$ of the elements of the form $T_0^{r_0} \cdots T_d^{r_d}$ with $r_i \leq -1$ and $\sum_i r_i = q$ form a basis of A. In particular, the module $H^d(X, \mathcal{O}_X(-d-1))$ is free of rank 1 over A, with $e := [(T_0 \cdots T_d)^{-1}]$ as a basis.

   (c) Let $n \geq 0$. Show that the cup product (Exercise 2.17)

$$\smile : H^0(X, \mathcal{O}_X(n)) \times H^d(X, \mathcal{O}_X(-d-1-n)) \to H^d(X, \mathcal{O}_X(-d-1))$$

   induces a non-degenerate bilinear map (let M, N be monomials of degree n, show that $[M] \smile [(T_0 \cdots T_d N)^{-1}] = e$ if $M = N$ and 0 otherwise).

3.3. Let $P = \mathbb{P}_k^d$ be a projective space over a field k. Let X be a closed subvariety of P of dimension r. We say that X is a *complete intersection* if it is defined (as a variety) by $d - r$ homogeneous polynomials $F_1, \ldots, F_{d-r}$. This notion depends on the embedding $X \to P$.

(a) Let $Z = V_+(F_1, \ldots, F_{d-r-1})$ ($Z = P$ if $d - r = 1$). Show that Z is a complete intersection. Let $\mathcal{J}$ be the sheaf of ideals of $\mathcal{O}_Z$ that defines the subscheme X of Z. Show that $\mathcal{J} \simeq \mathcal{O}_Z(-\deg F_{d-r})$.

(b) Let $q \in \mathbb{Z}$. Show the following properties by induction on $d - r$:

(1) $H^p(X, \mathcal{O}_X(q)) = 0$ if $1 \le p \le r - 1$.

(2) $H^0(P, \mathcal{O}_P(q)) \to H^0(X, \mathcal{O}_X(q))$ is surjective if $r \ge 1$.

(3) $\dim_k H^p(X, \mathcal{O}_X(q))$ is an integer that only depends on p, q, d, and the $\deg F_i$.

(c) Show that X is geometrically connected if $r \ge 1$.

(d) Let $d = 2$, and let $X = V_+(F)$ with $\deg F = n$. Show that

$$\dim_k H^1(X, \mathcal{O}_X) = (n - 1)(n - 2)/2.$$

3.4. Let Y be a Noetherian local scheme with closed point y_0, and $X = V_+(F_1, \ldots, F_{d-r})$ be a closed subscheme of $\mathbb{P}^d_Y$. Let us suppose that X_{y_0} is a complete intersection in $\mathbb{P}^d_{k(y_0)}$, of dimension r. Show that for any $y \in Y$, X_y is a complete intersection in $\mathbb{P}^d_{k(y)}$, of dimension r (use induction on $d - r$). Likewise, show that X is flat over Y (use Lemma 4.3.16).

3.5. Let Y be a locally Noetherian scheme. Let Z be a closed subscheme of $\mathbb{P}^d_Y$, flat over Y. Let $y_0 \in Y$ be a point such that Z_{y_0} is a complete intersection in $\mathbb{P}^d_{k(y_0)}$. Let us set $r = \dim Z_{y_0}$.

(a) Let us suppose Y is affine. Let us write $Z = V_+(I)$. Show that there exist $F_1, \ldots, F_{d-r} \in I$ such that if we set $X = V_+(F_1, \ldots, F_{d-r})$, then $X_{y_0} = Z_{y_0}$ as a subscheme of $\mathbb{P}^d_{k(y_0)}$.

(b) Show that there exists an open neighborhood V of y_0 such that $Z \times_Y V = X \times_Y V$ (Exercise 4.3.8). Show that for V small enough, Z_y is a complete intersection in $\mathbb{P}^d_{k(y)}$ of dimension r for every $y \in V$.

(c) Let us suppose that Y is connected and that Z_y is a complete intersection in $\mathbb{P}^d_{k(y)}$ for all $y \in Y$. Let $p \ge 0, q \in \mathbb{Z}$. Show that the function $y \mapsto \dim_{k(y)} H^p(Z_y, \mathcal{O}_{Z_y}(q))$ is constant on Y.

3.6. Let S be a scheme over a finite field $\mathbb{F}_p$, and let $X \to S$ be a scheme of finite type. Show that the relative Frobenius $X^{(p)} \to X$ (Definition 3.2.23) is a purely inseparable morphism.

3.7. Let k be an algebraically closed field. Let $f : X \to Y$ be a finite surjective morphism of integral algebraic varieties over k.

(a) Let us suppose that $K(X)/K(Y)$ is separable of degree n. Show that there exists a non-empty open subset V of Y such that $f^{-1}(V) \to V$ decomposes into a finite birational morphism $f^{-1}(V) \to W$ followed by a finite étale morphism $W \to V$. Show that for any $y \in V(k)$, $f^{-1}(y)$ consists of at least n points.

(b) Use (a) to show that if f induces an injective map $X(k) \to Y(k)$, then it is purely inseparable.

3.8. Let $f : X \to Y$ be a morphism of schemes. Show that the following properties are equivalent:

(i) f is purely inseparable.

(ii) For every Y-scheme Z, the projection morphism $X \times_Y Z \to Z$ is injective (in other words, f is *universally injective*).

(iii) For every field K, the canonical map $X(K) \to Y(K)$ is injective.

3.9. Let Y be a normal, locally Noetherian, integral scheme, and let $f : X \to Y$ be a projective morphism with X integral.

(a) Let us suppose f is finite. Show that f is purely inseparable if and only if the extension $K(Y) \to K(X)$ is purely inseparable. Moreover, if f is purely inseparable, it will be a homeomorphism.

(b) Let us suppose that the algebraic closure of $K(Y)$ in $K(X)$ is purely inseparable over $K(Y)$. Show that the fibers of $X \to Y$ are geometrically connected.

3.10. Let Y be a locally Noetherian scheme, and $f : X \to Y$ be a projective morphism such that $\mathcal{O}_Y \to f_* \mathcal{O}_X$ is purely inseparable. Show that for any morphism $Z \to Y$ with Z connected, the fibered product $X \times_Y Z$ is connected. Also show that $f^{-1}(F)$ is connected for any connected subset F of Y.

3.11. (*Stein factorization*). Let $f : X \to Y$ be a projective morphism to a locally Noetherian scheme Y.

(a) Show that f factors into $g : X \to Z$ and $h : Z \to Y$ such that g has only geometrically connected fibers, and h is finite.

(b) Show that we can factor f as in (a) with moreover $\mathcal{O}_Z \simeq g_* \mathcal{O}_X$. Show that this condition implies that Z is unique up to isomorphism (Exercise 1.17).

3.12. Let $f : X \to Y$ be a projective morphism. Let us suppose that for every $y \in Y$, X_y is geometrically connected (resp. geometrically integral). Show that $\mathcal{O}_Y \to f_* \mathcal{O}_X$ is purely inseparable (resp. an isomorphism).

3.13. Let $X \to S$ be a projective morphism to a locally Noetherian scheme. Let us suppose that the fibers of $X \to S$ are of dimension $\leq d$.

(a) Show that if $\mathcal{F} \to \mathcal{G}$ is a surjective homomorphism of quasi-coherent sheaves on X, then $H^d(X, \mathcal{F}) \to H^d(X, \mathcal{G})$ is surjective (use Proposition 2.34). Show that the canonical homomorphism

$$R^d f_* \mathcal{F} \otimes_{\mathcal{O}_S} k(s) \to H^d(X_s, \mathcal{F}_s) \qquad (3.16)$$

is surjective for all $s \in S$.

(b) Let us suppose that S is a Dedekind scheme, and that $\mathcal{F}$ is coherent and flat over S. Show that (3.16) is bijective.

3.14. Let Y be a locally Noetherian scheme, and let $f : X \to Y$ be a flat projective morphism. We say that f is *cohomologically flat* (in dimension 0) if the construction of $f_*\mathcal{O}_X$ commutes with base change (Corollary 3.22(i)).

(a) Show that f is cohomologically flat if and only if for every $y \in Y$, the morphism $X \times_Y \operatorname{Spec} \mathcal{O}_{Y,y} \to \operatorname{Spec} \mathcal{O}_{Y,y}$ is cohomologically flat.

(b) Let $Z \to Y$ be a faithfully flat morphism. Show that f is cohomologically flat if and only if $X \times_Y Z \to Z$ is cohomologically flat.

(c) Let us suppose that Y is local with closed point y, and that $H^0(X_y, \mathcal{O}_{X_y})$ is a direct sum of fields isomorphic to $k(y)$. Show that

$$H^0(X, \mathcal{O}_X) \otimes_{\mathcal{O}_{Y,y}} k(y) \to H^0(X_y, \mathcal{O}_{X_y}) \qquad (3.17)$$

is surjective (use Lemma 3.14 with $A = H^0(X, \mathcal{O}_X)$).

(d) Let Y be local with closed point y. Let R be a finite étale algebra over $k(y)$. Show that there exists a finite free (hence faithfully flat) $\mathcal{O}_{Y,y}$-algebra A such that $A \otimes_{\mathcal{O}_{Y,y}} k(y) = R$. Deduce from this that if X_y is geometrically reduced, then (3.17) is surjective (consider $R = H^0(X_y, \mathcal{O}_{X_y})$ and $X_A \to \operatorname{Spec} A$).

(e) Let us suppose Y is regular of dimension 1 and that either X_y is geometrically reduced for every closed point $y \in Y$, or X is a complete intersection in some $\mathbb{P}_Y^d$. Show that f is cohomologically flat (note that this is true without restriction on Y if we assume Remark 3.21(c)).

3.15. (*A non-cohomologically flat scheme*). Let $S = \operatorname{Spec} A$ be a Noetherian integral affine scheme, and let $t \in A$ be a non-zero non-invertible element. Let us consider the S-scheme

$$X = \operatorname{Proj} A[w, x, y, z]/(x^2 - t^2 wy, xz - ty^2, xy - twz, z^2 w - y^3),$$

and let $f : X \to S$ denote the structural morphism.

(a) Show that over the open subset $T := D(t)$ of S, there exists a morphism $\mathbb{P}_T^1 \to \mathbb{P}_T^3$ that induces an isomorphism $\mathbb{P}_T^1 \simeq X \times_S T$ (consider the 3-uple embedding $\mathbb{P}_T^1 \to \mathbb{P}_T^3$ (Exercise 1.27) followed by the automorphism $(w, x, y, z) \to (w, tx, y, z)$).

(b) Show that $D_+(z) \cap X \simeq \mathbb{A}_S^1$ and that $\mathcal{O}_X(D_+(w) \cap X)$ is free over A as an A-module. Conclude that X is flat over S.

(c) Let $s \in V(t)$. Show that X_s is an irreducible curve, reduced except at the point $(1, 0, 0, 0)$. Show that we have $\dim_{k(s)} H^0(X_s, \mathcal{O}_{X_s}) = 2$ and that $A \to H^0(X, \mathcal{O}_X)$ is an isomorphism. Deduce from this that $X \to S$ is not cohomologically flat.

(d) Show that X is not normal, and that its normalization is isomorphic to $\mathbb{P}_S^1$ if S is normal (Exercise 4.1.5).

See [81], Section 9.4, for examples of non-cohomologically flat schemes $X \to S$ with S regular of dimension 1 and X regular of dimension 2.

6

Sheaves of differentials

Differential calculus is a very useful tool in algebraic geometry. We approached it in Section 4.2, where we studied the tangent space $T_{X,x}$ of a scheme X at a point x. The sheaf of Kähler differentials on X is in some sense the union, indexed by $x \in X$, of the dual vector spaces $T_{X,x}^\vee$. In the first section, we introduce basic notions. We then apply them to the study of smooth morphisms in the second section. When we are dealing with not necessarily smooth morphisms, Kähler differentials become less easy to handle. For local complete intersections (Definition 3.17), another sheaf, the canonical sheaf, plays a very interesting role and advantageously replaces Kähler differentials in certain situations (for Grothendieck duality, for example). The last section is therefore devoted to local complete intersections and to the associated canonical sheaves.

6.1 Kähler differentials

6.1.1 Modules of relative differential forms

Definition 1.1. Let us fix a ring A. Let B be an A-algebra and M a B-module. An *A-derivation of B into M* is an A-linear map $d : B \to M$ such that the Leibniz rule

$$d(b_1 b_2) = b_1 db_2 + b_2 db_1, \quad b_i \in B$$

is verified, and that $da = 0$ for every $a \in A$ (i.e., the elements of A are 'constant'). Let us note that the last property is in fact an immediate consequence of the Leibniz rule. We denote the set of these derivations by $\mathrm{Der}_A(B, M)$.

Definition 1.2. Let B be an A-algebra. The *module of relative differential forms of B over A* is a B-module $\Omega^1_{B/A}$ endowed with an A-derivation $d : B \to \Omega^1_{B/A}$ having the following universal property:

> For any B-module M and for any A-derivation $d' : B \to M$, there exists a unique homomorphism of B-modules $\phi : \Omega^1_{B/A} \to M$ such that $d' = \phi \circ d$.

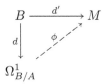

Proposition 1.3. *The module of relative differential forms* $(\Omega^1_{B/A}, d)$ *exists and is unique up to unique isomorphism.*

Proof The uniqueness follows from the definition, as for any solution to a universal problem (see, for example, the proof of Proposition 1.1.2). Let us therefore show existence. Let F be the free B-module generated by the symbols db, $b \in B$. Let E be the submodule of F generated by the elements of the form da, $a \in A$; $d(b_1 + b_2) - db_1 - db_2$, and $d(b_1 b_2) - b_1 db_2 - b_2 db_1$ with $b_i \in B$. Let us set $\Omega^1_{B/A} = F/E$ and $d : B \to \Omega^1_{B/A}$ which sends b to the image of db in $\Omega^1_{B/A}$. It is clear that $(\Omega^1_{B/A}, d)$ has the required properties. □

Remark 1.4. In what follows, if there is no confusion possible, we will simply denote the module of differential forms by $\Omega^1_{B/A}$, the derivation d being implicitly contained in the definition. For any $b \in B$, we will denote its image in $\Omega^1_{B/A}$ by db. By construction, $\Omega^1_{B/A}$ is generated as a B-module by dB.

Remark 1.5. For any B-module M, we have a map

$$\mathrm{Hom}_B(\Omega^1_{B/A}, M) \to \mathrm{Der}_A(B, M)$$

defined by $\phi \mapsto \phi \circ d$. The universal property of $\Omega^1_{B/A}$ comes down to saying that this map is an isomorphism of A-modules.

Example 1.6. Let us consider the following example which is at the basis of every computation of modules of differentials. Let A be a ring, and let B be the polynomial ring $A[T_1, \ldots, T_n]$. Let us show that $\Omega^1_{B/A}$ is the free B-module generated by the dT_i.

Let $F \in B$, and let $d' : B \to M$ be an A-derivation into a B-module M. Using the definition of a derivation, we immediately obtain that $d'F = \sum_i (\partial F / \partial T_i) d' T_i$, where $(\partial F / \partial T_i)$ is the partial derivative in the usual sense. Therefore d' is entirely determined by the images of the T_i. Let Ω be the free B-module generated by the symbols dT_i, $1 \le i \le n$. Let $d : B \to \Omega$ be the map defined by $d(F) = \sum_i \frac{\partial F}{\partial T_i} dT_i$. It can then immediately be verified that (Ω, d) fulfills the conditions of the universal property of the module $\Omega^1_{B/A}$. Hence $\Omega^1_{B/A} \simeq \Omega$.

Example 1.7. Let B be a localization or a quotient of A; then $\Omega^1_{B/A} = 0$. Indeed, if $A \to B$ is surjective, $d(b) = ad(1) = 0$ for $a \in A$ an inverse image of b. Let us suppose that $B = T^{-1}A$ is a localization of A. For any $b \in B$, there exists a $t \in T$ such that $tb \in A$, and hence $tdb = d(tb) = 0$, whence $db = 0$ since t is invertible in B.

Let $\rho : B \to C$ be a homomorphism of A-algebras. Then it follows from the universal property that there exist canonical homomorphisms of C-modules

$$\alpha : \Omega^1_{B/A} \otimes_B C \to \Omega^1_{C/A}, \quad \beta : \Omega^1_{C/A} \to \Omega^1_{C/B}.$$

By definition $\alpha(db \otimes c) = cd\rho(b)$. In the following proposition, we enumerate some properties of modules of differentials.

Proposition 1.8. *Let B be an algebra over a ring A.*

(a) *(Base change) For any A-algebra A', let us set $B' = B \otimes_A A'$. There exists a canonical isomorphism of B'-modules $\Omega^1_{B'/A'} \simeq \Omega^1_{B/A} \otimes_B B'$.*

(b) *Let $B \to C$ be a homomorphism of A-algebras. Let α, β be as above. Then*

$$\Omega^1_{B/A} \otimes_B C \xrightarrow{\alpha} \Omega^1_{C/A} \xrightarrow{\beta} \Omega^1_{C/B} \to 0$$

is an exact sequence.

(c) *Let S be a multiplicative subset of B; then $S^{-1}\Omega^1_{B/A} \simeq \Omega^1_{S^{-1}B/A}$.*

(d) *If $C = B/I$, we have an exact sequence*

$$I/I^2 \xrightarrow{\delta} \Omega^1_{B/A} \otimes_B C \xrightarrow{\alpha} \Omega^1_{C/A} \to 0$$

where for any $b \in I$, if we let $\bar{b}$ denote the image of b in I/I^2, then $\delta(\bar{b}) := db \otimes 1$.

Proof (a) The canonical derivation $d : B \to \Omega^1_{B/A}$ induces an A'-derivation $d' = d \otimes \mathrm{Id}_{A'} : B' \to \Omega^1_{B/A} \otimes_A A' = \Omega^1_{B/A} \otimes_B B'$. We easily see that $(\Omega^1_{B/A} \otimes_B B', d')$ verifies the universal property of $\Omega^1_{B'/A'}$.

(b) It suffices to show that for any C-module N, the dual sequence

$$0 \to \mathrm{Hom}_C(\Omega^1_{C/B}, N) \to \mathrm{Hom}_C(\Omega^1_{C/A}, N) \to \mathrm{Hom}_C(\Omega^1_{B/A} \otimes_B C, N)$$

is exact. We have $\mathrm{Hom}_C(\Omega^1_{B/A} \otimes_B C, N) = \mathrm{Hom}_B(\Omega^1_{B/A}, N)$ (Exercise 1.1.5). By Remark 1.5, this sequence is canonically isomorphic to the sequence

$$0 \to \mathrm{Der}_B(C, N) \to \mathrm{Der}_A(C, N) \to \mathrm{Der}_A(B, N),$$

the last homomorphism being the composition with $B \to C$. It follows from the definition of a derivation that this sequence is exact, which proves (b).

(c) results from (b) by taking $C = S^{-1}B$ and using Example 1.7.

(d) We have $I/I^2 = I \otimes_B C$. As in (b), it suffices to show that the sequence

$$0 \to \mathrm{Der}_A(C, N) \to \mathrm{Der}_A(B, N) \to \mathrm{Hom}_C(I/I^2, N) = \mathrm{Hom}_B(I, N) \quad (1.1)$$

is exact for any C-module N. The last homomorphism associates to any A-derivation $d' : B \to N$ its restriction to I. The latter is indeed a homomorphism of B-modules because $IN = 0$. It can immediately be seen that sequence (1.1) is exact. $\qquad \square$

Corollary 1.9. *Let A be a ring, and let B be a finitely generated A-algebra or a localization of such an algebra. Then $\Omega^1_{B/A}$ is finitely generated over B.*

Proof This follows from Proposition 1.8(d) and (c), and Examples 1.6, 1.7. □

Example 1.10. Let $B = A[T_1, \ldots, T_n]$, let $F \in B$, and let $C = B/FB$. Then by Example 1.6 and Proposition 1.8(d), we have

$$\Omega^1_{C/A} = (\oplus_{1 \leq i \leq n} C dT_i)/C dF,$$

with $dF = \sum_i (\partial F/\partial T_i) dT_i$. This is the quotient of a free C-module of rank n by a *simple* submodule (i.e., generated by one element).

Next, we study the kernel of the homomorphism α of Proposition 1.8(b) when C is a finitely generated B-algebra.

Lemma 1.11. *Let A be a ring.*

(a) *Let B_1, B_2 be two A-algebras and $R = B_1 \otimes_A B_2$. Then we have a canonical isomorphism*

$$\varphi : (\Omega^1_{B_1/A} \otimes_{B_1} R) \oplus (\Omega^1_{B_2/A} \otimes_{B_2} R) \simeq \Omega^1_{R/A}$$

defined by $(db_1 \otimes r_1) + (db_2 \otimes r_2) \mapsto r_1 d(b_1 \otimes 1) + r_2 d(1 \otimes b_2)$.

(b) *Let B be an A-algebra and $C = B[T_1, \ldots, T_n]/I$. Let*

$$\alpha : \Omega^1_{B/A} \otimes_B C \to \Omega^1_{C/A}, \quad \delta : I/I^2 \to \Omega^1_{B[T_1,\ldots,T_n]/B} \otimes_{B[T_1,\ldots,T_n]} C$$

be the canonical homomorphisms as defined in Proposition 1.8(b) and (d). Then there exists a surjective homomorphism $\operatorname{Ker} \delta \to \operatorname{Ker} \alpha$ of B-modules.

Proof (a) This results from the canonical exact sequences

$$\Omega^1_{B_1/A} \otimes_{B_1} R \xrightarrow{\varphi_1} \Omega^1_{R/A} \xrightarrow{\psi_2} \Omega^1_{R/B_1} = \Omega^1_{B_2/A} \otimes_{B_2} R \to 0$$

and

$$\Omega^1_{B_2/A} \otimes_{B_2} R \xrightarrow{\varphi_2} \Omega^1_{R/A} \xrightarrow{\psi_1} \Omega^1_{R/B_2} = \Omega^1_{B_1/A} \otimes_{B_1} R \to 0$$

(we verify that φ_i is a section of ψ_i). The isomorphism φ is simply $\varphi_1 \oplus \varphi_2$.

(b) Let us abbreviate $B[T_1, \ldots, T_n]$ to $B[T]$. Since $B[T] = A[T] \otimes_A B$, by (a) we have a canonical isomorphism

$$\Omega^1_{B[T]/A} \simeq (\Omega^1_{A[T]/A} \otimes_{A[T]} B[T]) \oplus (\Omega^1_{B/A} \otimes_B B[T]) = \Omega^1_{B[T]/B} \oplus (\Omega^1_{B/A} \otimes_B B[T]),$$

and hence an isomorphism

$$\rho : \Omega^1_{B[T]/A} \otimes_{B[T]} C \simeq (\Omega^1_{B[T]/B} \otimes_{B[T]} C) \oplus (\Omega^1_{B/A} \otimes_B C). \qquad (1.2)$$

We have a commutative diagram

$$
\begin{array}{ccc}
I/I^2 & \xrightarrow{\ \ \delta\ \ } & \Omega^1_{B[T]/B} \otimes_{B[T]} C \\[2mm]
\delta_1 \downarrow & & \uparrow p \\[2mm]
\Omega^1_{B[T]/A} \otimes_{B[T]} C & \xrightarrow{\ \ \rho\ \ } & (\Omega^1_{B[T]/B} \otimes_{B[T]} C) \oplus (\Omega^1_{B/A} \otimes_B C) \\[2mm]
\alpha_1 \downarrow & & \uparrow i \\[2mm]
\Omega^1_{C/A} & \xleftarrow{\ \ \alpha\ \ } & \Omega^1_{B/A} \otimes_B C
\end{array}
$$

where δ_1 and α_1 are canonical homomorphisms as defined in Proposition 1.8(d) (with B replaced by $B[T]$), i is the canonical injection, and p is the projection. Now it is easy to check that $i(\operatorname{Ker}\alpha) = \rho \circ \delta_1(\operatorname{Ker}\delta)$, whence (b). $\qquad\square$

Example 1.12. Let us suppose $n = 1$ and I is generated by a monic polynomial $P(T)$. We have an exact sequence

$$
0 \to \operatorname{Ker}\alpha \to \Omega^1_{B/A} \otimes_B C \xrightarrow{\ \alpha\ } \Omega^1_{C/A} \to \Omega^1_{C/B} = C/(P'(t))dT \to 0,
$$

where t is the image of T in C, and where $P'(T)$ is the derivative of $P(T)$. The kernel of $\delta : I/I^2 \to \Omega^1_{B[T]/B} \otimes C = C\,dT$ is made up of the images in I/I^2 of the polynomials $Q(T)P(T)$ such that $Q(t)P'(t) = 0$. Let $C' = C_{P'(t)}$. Then $\operatorname{Ker}\delta \otimes_C C' = 0$ and $\Omega^1_{C/B} \otimes_C C' = 0$. It follows from Lemma 1.11(b) that

$$
\Omega^1_{B/A} \otimes_B C' \to \Omega^1_{C'/A}
$$

is an isomorphism.

To conclude, let us study differentials on function fields.

Lemma 1.13. Let k be a field. Let E be an extension of k and $K = E[T]/(P(T))$ a simple algebraic extension of E.

(a) If K/E is separable, then $\Omega^1_{K/E} = 0$ and $\Omega^1_{E/k} \otimes_E K \to \Omega^1_{K/k}$ is an isomorphism.

(b) If K/E is inseparable, then $\Omega^1_{K/E} \simeq K$ and

$$
\dim_E \Omega^1_{E/k} \le \dim_K \Omega^1_{K/k} \le \dim_E \Omega^1_{E/k} + 1.
$$

(c) Suppose that K is finite over k. Then K/k is separable if and only if $\Omega^1_{K/k} = 0$.

Proof Let $P'(T)$ denote the derivative of the polynomial $P(T)$ and t the image of T in K. We saw in Example 1.10 that

$$
\Omega^1_{K/E} = K\,dT/(P'(t))dT \simeq K/(P'(t)).
$$

In case (a), we have $P'(t) \in K^*$, and hence $\Omega^1_{K/E} = 0$. And $\Omega^1_{E/k} \otimes_E K \simeq \Omega^1_{K/k}$ by Example 1.12.

(b) If K/E is inseparable, then $P'(t) = 0$ and $\Omega^1_{K/E}$ is isomorphic to K. With the notation of the proof of Lemma 1.11, we have

$$\Omega^1_{K/k} \simeq (\Omega^1_{E[T]/k} \otimes_{E[T]} K)/(\delta_1(P)),$$

and using isomorphism (1.2),

$$\dim_K(\Omega^1_{E[T]/k} \otimes_{E[T]} K) = \dim_K(KdT) + \dim_K(\Omega^1_{E/k} \otimes_E K) = 1 + \dim_E \Omega^1_{E/k}.$$

This implies the desired inequalities.

(c) If K/k is separable, then it is simple and $\Omega^1_{K/k} = 0$ by (a). Suppose that K/k is inseparable. Then K is a simple inseparable extension of some subfield E. By Proposition 1.8(b), $\Omega^1_{K/k}$ maps onto $\Omega^1_{K/E} \neq 0$, whence $\Omega^1_{K/k} \neq 0$.   □

Definition 1.14. Let K be a function field over a field k. We say that the extension K/k is *separable* if K is finite separable over a purely transcendental extension of k. If K is the function field of an integral algebraic variety X, then this is equivalent to saying that X is geometrically reduced (Proposition 3.2.15).

Proposition 1.15. *Let K be a function field over a field k, of transcendence degree $n = \mathrm{trdeg}_k K$. Then $\Omega^1_{K/k}$ is a K-vector space of finite dimension and $\dim_K \Omega^1_{K/k} \geq n$. Moreover, the equality holds if and only if K/k is separable.*

Proof As K is a finite extension of a purely transcendental extension $F := k(T_1, \ldots, T_n)$, Example 1.6 and Proposition 1.8(c) show that $\dim_F \Omega^1_{F/k} = n$. By repeating Lemma 1.13(a)–(b) a finite number of times, we obtain $n \leq \dim_K \Omega^1_{K/k} < +\infty$.

Let us suppose that K/k is separable. Then K is a simple separable extension of F. It results from Lemma 1.13(a) that $\dim_K \Omega^1_{K/k} = \dim_F \Omega^1_{F/k} = n$. Let us now suppose that $\dim_K \Omega^1_{K/k} = n$. Let $f_1, \ldots, f_n \in K$ be such that the df_i form a basis of $\Omega^1_{K/k}$ over K. Let $E = k(f_1, \ldots, f_n)$ be the subextension of K generated by the f_i. The exact sequence

$$\Omega^1_{E/k} \otimes_E K \to \Omega^1_{K/k} \to \Omega^1_{K/E} \to 0$$

implies that $\Omega^1_{K/E} = 0$. Hence K/E is finite. It follows from Lemma 1.13(c) that K/E is separable, and hence K is separable over k.   □

Remark 1.16. In the proof above, we have seen that if K/k is separable, then K is finite separable over $k(f_1, \ldots, f_n)$, provided that the df_i form a basis of $\Omega^1_{K/k}$ over K. This explains the construction of the f_i in Proposition 3.2.15.

6.1.2   Sheaves of relative differentials (of degree 1)

On a scheme, we can glue the differential forms on affine open subsets and thus define the sheaf of relative differentials.

Proposition 1.17. *Let* $f : X \to Y$ *be a morphism of schemes. Then there exists a unique quasi-coherent sheaf* $\Omega^1_{X/Y}$ *on* X *such that for any affine open subset* V *of* Y, *any affine open subset* U *of* $f^{-1}(V)$, *and any* $x \in U$, *we have*

$$\Omega^1_{X/Y}|_U \simeq (\Omega^1_{\mathcal{O}_X(U)/\mathcal{O}_Y(V)})^{\sim}, \quad (\Omega^1_{X/Y})_x \simeq \Omega^1_{\mathcal{O}_{X,x}/\mathcal{O}_{Y,f(x)}}. \qquad (1.3)$$

Proof Let $\rho : A \to B$ be a ring homomorphism. Let $\mathfrak{q} \in \operatorname{Spec} B$ and $\mathfrak{p} = \rho^{-1}(\mathfrak{q}) \in \operatorname{Spec} A$. Then we have canonical isomorphisms

$$\Omega^1_{B/A} \otimes_B B_{\mathfrak{q}} \simeq \Omega^1_{B_{\mathfrak{q}}/A} \simeq \Omega^1_{B_{\mathfrak{q}}/A_{\mathfrak{p}}}$$

by Proposition 1.8(c) and (b). To simplify, we let $\Omega_x = \Omega^1_{\mathcal{O}_{X,x}/\mathcal{O}_{Y,f(x)}}$ for any $x \in X$. If V is an affine open subset of Y, U an affine open subset of $f^{-1}(V)$, $\omega \in \Omega^1_{\mathcal{O}_X(U)/\mathcal{O}_Y(V)}$, and $x \in U$, we let ω_x denote the canonical image of ω in $\Omega^1_{\mathcal{O}_X(U)/\mathcal{O}_Y(V)} \otimes \mathcal{O}_{X,x} \simeq \Omega_x$.

For any open subset U of X, let us set $\Omega^1_{X/Y}(U)$ equal to the set of maps $s : U \to \coprod_{x \in U} \Omega_x$ (disjoint union) such that for any $x \in U$, there exist an affine open neighborhood V_y of $y := f(x)$, an affine open neighborhood $U_x \subseteq f^{-1}(V_y)$ of x, and $\omega \in \Omega^1_{\mathcal{O}_X(U_x)/\mathcal{O}_Y(V_y)}$ with $\omega_{x'} = s(x')$ for every $x' \in U_x$. It is easy to verify that $\Omega^1_{X/Y}$ has a natural structure of sheaf of $\mathcal{O}_X$-modules, the restriction homomorphisms being restrictions of maps. Moreover, for any $x \in X$, the stalk $(\Omega^1_{X/Y})_x$ is canonically isomorphic to Ω_x. Let $U \subseteq f^{-1}(V)$ be affine open subsets as above. By construction, we have a natural homomorphism of modules $\Omega^1_{\mathcal{O}_X(U)/\mathcal{O}_Y(V)} \to \Omega^1_{X/Y}(U)$, whence a homomorphism of $\mathcal{O}_U$-modules $(\Omega^1_{\mathcal{O}_X(U)/\mathcal{O}_Y(V)})^{\sim} \to \Omega^1_{X/Y}|_U$. As this homomorphism is an isomorphism on the stalks, it is an isomorphism of $\mathcal{O}_U$-modules. This proves the isomorphisms (1.3), and in particular that $\Omega^1_{X/Y}$ is a quasi-coherent sheaf on X. $\qquad \square$

Remark 1.18. The diagonal morphism $\Delta : X \to X \times_Y X$ is locally a closed immersion, and hence $\Delta(X)$ is closed in an open subset U of $X \times_Y X$. Let $\mathcal{I} = \operatorname{Ker} \Delta^{\#}$ be the sheaf of ideals defining the closed subset $\Delta(X)$ in U. Then we can show that $\Omega^1_{X/Y} \simeq \Delta^*(\mathcal{I}/\mathcal{I}^2)$ (Exercise 1.1). We could have taken this as a definition of $\Omega^1_{X/Y}$.

Definition 1.19. Let $f : X \to Y$ be a morphism of schemes. The quasi-coherent sheaf $\Omega^1_{X/Y}$ is called the *sheaf of relative differentials (or differential forms) of degree* 1 *of* X *over* Y. If $Y = \operatorname{Spec} A$, we also denote this sheaf by $\Omega^1_{X/A}$. If there is no ambiguity, we also denote $\Omega^1_{X/Y}$ by Ω^1_X.

Proposition 1.20. *Let* $f : X \to Y$ *be a morphism of finite type with* Y *Noetherian. Then* $\Omega^1_{X/Y}$ *is a coherent sheaf on* X.

Proof This results from Corollary 1.9. $\qquad \square$

Example 1.21. If $X = \mathbb{A}^n_Y$, then $\Omega^1_{X/Y} \simeq \mathcal{O}^n_X$. This comes from the computations of Example 1.6.

Example 1.22. Let A be a ring and $X = \mathbb{P}^1_A$; then we have $\Omega^1_{X/A} \simeq \mathcal{O}_X(-2)$. Indeed, on $D_+(T_0) \cap D_+(T_1)$, we have $T_0^2 d(T_1/T_0) = -T_1^2 d(T_0/T_1)$. This induces a global section of $\mathcal{O}_X(2) \otimes \Omega^1_{X/A}$. It can then immediately be seen that this global section induces an isomorphism $\mathcal{O}_X \simeq \mathcal{O}_X(2) \otimes \Omega^1_{X/A}$, whence $\Omega^1_{X/A} \simeq \mathcal{O}_X(-2)$ (Exercise 5.1.12(d)).

Example 1.23. Let $X = \operatorname{Spec} k[T, S]/(F(T, S))$ be a smooth affine curve over a field k. Let t, s denote the respective images of T, S in $k[T, S]/(F)$, and $\partial_t F$ the image in $k[t, s]$ of the partial derivative $\partial F(T, S)/\partial T$. Then $\Omega^1_{X/k}$ is free of rank 1 over $\mathcal{O}_X$, and $ds/(\partial_t F)$ and $dt/(\partial_s F)$ are both bases of $\Omega^1_{X/k}$ over $\mathcal{O}_X$. Indeed, we have the relation $\partial_t F dt = -\partial_s F ds$. On the principal open subset $D(\partial_s F)$ of X, the sheaf $\Omega^1_{D(\partial_s F)/k}$ is free with basis $dt/(\partial_s F)$, by Example 1.10, and $\Omega^1_{D(\partial_t F)/k}$ is free with basis $ds/(\partial_t F)$. By the Jacobian criterion, X is the union of the open subsets $D(\partial_s F)$ and $D(\partial_t F)$, which implies the result since $dt/(\partial_s F) = -ds/(\partial_t F)$.

In the setting of morphisms of schemes, Proposition 1.8 translates into the following proposition:

Proposition 1.24. *Let $f : X \to Y$ be a morphism of schemes.*

(a) *(Base change) Let Y' be a Y-scheme. Let $X' = X \times_Y Y'$ and let $p : X' \to X$ be the first projection. Then $\Omega^1_{X'/Y'} \simeq p^* \Omega^1_{X/Y}$.*

(b) *Let $Y \to Z$ be a morphism of schemes. We have an exact sequence*

$$f^* \Omega^1_{Y/Z} \to \Omega^1_{X/Z} \to \Omega^1_{X/Y} \to 0.$$

(c) *Let U be an open subset of X; then $\Omega^1_{X/Y}|_U \simeq \Omega^1_{U/Y}$. For any $x \in X$, we have $(\Omega^1_{X/Y})_x \simeq \Omega^1_{\mathcal{O}_{X,x}/\mathcal{O}_{Y,f(x)}}$.*

(d) *If Z is a closed subscheme of X defined by a quasi-coherent sheaf of ideals $\mathcal{I}$, we have a canonical exact sequence*

$$\mathcal{I}/\mathcal{I}^2 \xrightarrow{\delta} \Omega^1_{X/Y} \otimes_{\mathcal{O}_X} \mathcal{O}_Z \to \Omega^1_{Z/Y} \to 0.$$

Definition 1.25. Let k be a field. We define an *elliptic curve* over k to be a smooth projective curve E over k, isomorphic to a closed subvariety of $\mathbb{P}^2_k$ defined by a homogeneous polynomial $F(u, v, w)$ of the form

$$F(u, v, w) = v^2 w + (a_1 u + a_3 w) vw - (u^3 + a_2 u^2 w + a_4 uw^2 + a_6 w^3) \quad (1.4)$$

with the privileged rational point $o = (0, 1, 0)$.

The fact that E is smooth expresses itself in the non-vanishing of the discriminant of equation (1.4), see the formula for the discriminant in [92], Section III.1. An elliptic curve is connected, like any projective plane curve (Exercise 5.3.3(c)), and hence integral since it is regular (Corollary 4.3.32). As an example of a computation, we are going to determine $\Omega^1_{E/k}$ in the following proposition.

Proposition 1.26. *Let E be an elliptic curve over a field k, given by equation (1.4) above. Let $x = u/w, y = v/w \in K(E)$, and*

$$\omega := \frac{dx}{(2y + a_1 x + a_3)} \in \Omega^1_{K(E)/k}.$$

Then $\omega \in H^0(E, \Omega^1_{E/k})$ and $\Omega^1_{E/k} = \omega \mathcal{O}_E$.

Proof (See also Proposition 7.4.26.) The curve E is the union of the two principal open subsets $U = D_+(w) \cap E$ and $V = D_+(v) \cap E$. We have

$$\mathcal{O}_E(U) = k[x, y]/(y^2 + (a_1 x + a_3)y - (x^3 + a_2 x^2 + a_4 x + a_6))$$

and

$$\mathcal{O}_E(V) = k[t, z]/(z + a_1 tz + a_3 z^2 - (t^3 + a_2 t^2 z + a_4 tz^2 + a_6 z^3))$$

with $t := u/v = x/y$ and $z := w/v = 1/y$. By Example 1.23, $\Omega^1_{U/k}$ is free, with basis ω; and $\Omega^1_{V/k}$ is free, with basis

$$\omega' := \frac{dz}{a_1 z - (3t^2 + 2a_2 tz + a_4 z^2)}.$$

Now in $\Omega^1_{K(E)/k}$, we have $dz = -dy/y^2$. A direct computation shows that

$$\omega' = \frac{-dy}{a_1 y - (3x^2 + 2a_2 x + a_4)} = \omega,$$

whence the proposition.   □

Exercises

1.1. There is a more concrete way to construct $\Omega^1_{B/A}$ using the diagonal homomorphism. Let B be an A-module. Let $\rho : B \otimes_A B \to B$ be the 'diagonal' homomorphism: $\rho(b_1 \otimes b_2) = b_1 b_2$. Let $I = \operatorname{Ker} \rho$. We can consider $B \otimes_A B$ as a B-module via the multiplication on the right $(b \cdot (b_1 \otimes b_2) = b_1 \otimes bb_2)$; I/I^2 is thus a B-module. Let $d : B \to I/I^2$ be defined by $db = b \otimes 1 - 1 \otimes b$.

   (a) Show that d is an A-derivation.

   (b) Show that the B-module I/I^2 endowed with the A-derivation d verifies the universal property of the module of differentials of B over A.

   (c) Let $f : X \to Y$ be a morphism of schemes. Let $\Delta : X \to X \times_Y X$ be the diagonal morphism. Let $\mathcal{I}$ denote the kernel of $\Delta^\# : \mathcal{O}_{X \times_Y X} \to \Delta_* \mathcal{O}_X$. Show that $\Omega^1_{X/Y} \simeq \Delta^*(\mathcal{I}/\mathcal{I}^2)$.

1.2. Let A be a ring. Show that in general $\Omega^1_{A[[T]]/A}$ is not finitely generated over $A[[T]]$. Show that we have an exact sequence

$$0 \to \cap_{n \geq 1} T^n \Omega^1_{A[[T]]/A} \to \Omega^1_{A[[T]]/A} \xrightarrow{\varphi} A[[T]]dT,$$

with φ defined by $\varphi(dF(T)) = F'(T)dT$.

1.3. Let A be a ring and B a A-algebra. Let I be an ideal of B and $\widehat{B}$ the I-adic completion of B. Let $\rho : \Omega^1_{B/A} \to \Omega^1_{\widehat{B}/A}$ denote the canonical homomorphism.

(a) Show that we have $\Omega^1_{\widehat{B}/A} = \rho(\Omega^1_{B/A}) + I\Omega^1_{\widehat{B}/A}$. Deduce from this that for any $n \geq 1$, ρ induces an exact sequence

$$0 \to K_n \to \Omega^1_{B/A}/I^n\Omega^1_{B/A} \to \Omega^1_{\widehat{B}/A}/I^n\Omega^1_{\widehat{B}/A} \to 0$$

for some B-module K_n. Moreover, $K_{n+1} \to K_n$ is surjective.

(b) Let us suppose A is Noetherian, and that B is a localization of a finitely generated A-algebra. Show that

$$\varprojlim_n \Omega^1_{\widehat{B}/A}/I^n\Omega^1_{\widehat{B}/A} \simeq \Omega^1_{B/A} \otimes_B \widehat{B}.$$

1.4. Let $X \to Y$ be a morphism, and let $Y \to Z$ be an open or closed immersion. Show that we have a canonical isomorphism $\Omega^1_{X/Z} \to \Omega^1_{X/Y}$.

1.5. Let X, Y be S-schemes. Let p, q be the projections from $X \times_S Y$ onto X and Y, respectively. Show that $\Omega^1_{X \times_S Y} \simeq p^*\Omega^1_{X/S} \oplus q^*\Omega^1_{Y/S}$.

1.6. Let $K \subseteq L$ be two function fields that are separable over a field k. Show that the canonical map $\Omega^1_{K/k} \otimes_K L \to \Omega^1_{L/k}$ is injective if and only if L is separable over K.

1.7. Let K be a perfect field of characteristic $p > 0$. Let L be an extension of K. Show that $K \subseteq L^p$ and that the canonical homomorphism $\Omega^1_{L/K} \to \Omega^1_{L/L^p}$ is an isomorphism.

1.8. Let X be a smooth algebraic variety of dimension d over a field K.

(a) Let L be a function field over K. Let $r = \dim_L \Omega^1_{L/K}$. Show that L can be generated, as a function field over K, by $r + 1$ elements.

(b) Let $x \in X$. Show that the residue field $k(x)$ can be generated by $d + 1$ elements over K.

1.9. Let p be a prime number, let $S = \operatorname{Spec} \mathbb{Z}_p$, and $X = \operatorname{Spec} \mathbb{Z}_p[T_1, T_2]/(T_1T_2^2 + p)$. Show that X is regular, integral, and that $(\Omega^1_{X/S})_{\mathrm{tors}}$ (see Exercise 5.1.14) is a non-zero coherent sheaf, with support $V(T_2) \cap X$.

6.2   Differential study of smooth morphisms

In this section we are going to use the sheaf of differentials to extend the study
of smooth morphisms (Definition 4.3.35). All considered schemes will be locally
Noetherian. For a morphism $X \to S$, we denote $\Omega^1_{X/S}$ by Ω^1_X if there is no
ambiguity concerning the base scheme.

6.2.1   Smoothness criteria

Lemma 2.1. *Let A be a finitely generated algebra over a field k. Let $x \in \operatorname{Spec} A$
be a rational point corresponding to a maximal ideal $\mathfrak{m}$ of A. Then the canonical
homomorphism*

$$\delta : \mathfrak{m}/\mathfrak{m}^2 \to \Omega^1_{A/k} \otimes_A k(x)$$

deduced from Proposition 1.8(d) is an isomorphism.

Proof   The cokernel of δ is $\Omega^1_{k(x)/k} = 0$. It suffices to show that δ is injective.
Let us write $A = B/I$, where $B = k[T_1, \ldots, T_n]$. Let $\mathfrak{n}$ be the inverse image of $\mathfrak{m}$
in B. We have a commutative diagram of exact sequences

$$
\begin{array}{ccccccccc}
I & \longrightarrow & \mathfrak{n}/\mathfrak{n}^2 & \longrightarrow & \mathfrak{m}/\mathfrak{m}^2 & \longrightarrow & 0 \\
\| & & \downarrow{\scriptstyle\delta'} & & \downarrow{\scriptstyle\delta} & & \\
I & \xrightarrow{\;\gamma\;} & \Omega^1_{B/k} \otimes k(x) & \longrightarrow & \Omega^1_{A/k} \otimes k(x) & \longrightarrow & 0
\end{array}
$$

where γ is given by the composition of $I \to I/I^2 \to \Omega^1_{B/k} \otimes_B A$ tensored by
$k(x)$, and where δ' is given by Proposition 1.8(d). It therefore suffices to show
that δ' is injective. But this results from Example 1.6 and by using the Taylor
expansion in x. See also Exercise 2.5.   □

Proposition 2.2. *Let X be an algebraic variety over a field k, and let $x \in X$.
Then the following properties are equivalent:*

  (i) *X is smooth in a neighborhood of x;*

  (ii) *$\Omega^1_{X,x}$ is free of rank $\dim_x X$;*

  (iii) *X is smooth at x.*

Proof   (ii) $\Longrightarrow$ (i). There exists an open neighborhood U of x such that $\Omega^1_{U/k} =
\Omega^1_{X/k}|_U$ is free of rank $n := \dim_x X$ (Exercise 5.1.12). By taking U sufficiently
small, we may suppose U is connected and $\dim U = n$. Let $\bar{k}$ be an algebraic
closure of k. Let $\overline{U} = U_{\bar{k}}$. Then $\overline{U}$ is of dimension n (Proposition 3.2.7(a)). By
Proposition 1.24(a), $\Omega^1_{\overline{U}/\bar{k}}$ is also locally free of rank n. For any closed point
$y \in \overline{U}$, we have, by virtue of Lemma 2.1,

$$\dim_{k(y)} T_{\overline{U},y} = \dim_{k(y)} \Omega^1_{\overline{U},y} \otimes k(y) = n.$$

Consequently, $\overline{U}$ is regular at the closed points y such that $\dim_y \overline{U} = n$. Let us
show that U is irreducible. Otherwise, there exist two irreducible components of

U that meet at a closed point x_0 with $\dim_{x_0} U = n$. But we have just shown that such a point is smooth, and hence integral (Proposition 4.2.11). Consequently, U is irreducible, and hence $\overline{U}$ is of dimension n at all of the closed points (Exercise 3.2.10(b)). It follows that $\overline{U}$ is regular (Corollary 4.2.17) and therefore that U is smooth.

(iii) $\Longrightarrow$ (ii). Let $x' \in X_{\bar{k}}$ be a point lying over x. Then x' has an irreducible regular open neighborhood W in $X_{\bar{k}}$ (Proposition 4.2.24), and W is defined over a finite extension k' of k: $W'_{\bar{k}} = W$ for some open subset W' of $X_{k'}$ (apply Lemma 3.2.6 to $X_{\bar{k}} \setminus W$). Let $p : X_{k'} \to X$ be the first projection. Then p is open (Exercise 4.3.9), and $U := p(W')$ is contained in the smooth locus of X (Remark 4.3.31). Thus U is irreducible of some dimension n. Let $z \in U$ be a closed point, and let z' be a point of $\overline{U}$ lying above z. Then

$$\dim_{k(z)} \Omega^1_{U,z} \otimes k(z) = \dim_{k(z')} \Omega^1_{U,z} \otimes k(z) \otimes_{k(z)} k(z') = \dim_{k(z')} \Omega^1_{\overline{U},z'} \otimes k(z') = n.$$

As U is regular (Corollary 4.3.32), hence reduced (Proposition 4.2.11), it follows from Exercise 5.1.15(a) and (c) that $\Omega^1_{U/k}$ is locally free of rank n. $\qquad\square$

Corollary 2.3. *Let $f : X \to S$ be a morphism of finite type of locally Noetherian schemes. Then f is unramified at a point $x \in X$ if and only if $\Omega^1_{X/S,x} = 0$.*

Proof Let us denote $\Omega^1_{X/S}$ by Ω^1_X. Let $s = f(x)$. Then $\Omega^1_{X,x} \otimes k(s) = \Omega^1_{X_s,x}$ (Proposition 1.24(a)). The morphism f is unramified at x if and only if X_s is smooth at x and x is an isolated point in X_s (Lemma 4.3.20). Let us suppose f is unramified at x; and then $\Omega^1_{X,x} \otimes_{\mathcal{O}_{S,s}} k(s) = \Omega^1_{X_s,x} = 0$. So $\Omega^1_{X,x} \subseteq \mathfrak{m}_s \Omega^1_{X,x} \subseteq \mathfrak{m}_x \Omega^1_{X,x}$. As Ω^1_X is coherent, $\Omega^1_{X,x} = 0$ by Nakayama.

Let us now suppose that $\Omega^1_{X,x} = 0$; and thus $\Omega^1_{X_s,x} = 0$. As $\Omega^1_{X_s/k(s)}$ is a coherent sheaf, there exists an open neighborhood U of x in X_s such that $\Omega^1_{X_s,y} = 0$ for every $y \in U$. Consequently, $\Omega^1_{U/k(s)} = 0$. Let ξ be a generic point of U. Then $\Omega^1_{k(\xi)/k(s)}$ is a localization of $\Omega^1_{U,\xi} = 0$, and hence $k(\xi)$ is algebraic over $k(s)$. It follows that U is of dimension 0 (Proposition 1.15). By Proposition 2.2, U is smooth over $k(s)$. Hence f is unramified at x. $\qquad\square$

Lemma 2.4. *Let $X \to S$ be a morphism of finite type over a locally Noetherian scheme S. Fix $s \in S$, $x \in X_s$, and let*

$$d = \dim_{k(x)} \Omega^1_{X_s/k(s),x} \otimes_{\mathcal{O}_{X_s,x}} k(x).$$

Then in a neighborhood of x, there exists a closed immersion $X \to Z$ into a scheme Z that is smooth over S at x, and such that $\dim_x Z_s = d$ and that $\Omega^1_{Z/S,x}$ is free of rank d over $\mathcal{O}_{Z,x}$.

Proof We may assume that X and S are affine. We will denote $\Omega^1_{Y/T}$ by Ω^1_Y for any morphism $Y \to T$. The scheme X is a closed subscheme of an affine S-scheme Y, smooth at x, and such that $\Omega^1_{Y,x}$ is free (e.g., $Y = \mathbb{A}^r_S$ for a large enough r). Let $\mathcal{I} \subseteq \mathcal{O}_Y$ be the sheaf of ideals defining X. We have an exact sequence

$$\mathcal{I}_x/\mathcal{I}_x^2 \to \Omega^1_{Y,x} \otimes_{\mathcal{O}_{Y,x}} \mathcal{O}_{X,x} \to \Omega^1_{X,x} \to 0$$

(Proposition 1.8(d)). Tensoring by $\otimes_{\mathcal{O}_{S,s}} k(s)$, and then by $\otimes_{\mathcal{O}_{X_s,x}} k(x)$, we get an exact sequence

$$\mathcal{I}_x/\mathcal{I}_x^2 \to \Omega^1_{Y_s,x} \otimes_{\mathcal{O}_{Y_s,x}} k(x) \to \Omega^1_{X_s,x} \otimes_{\mathcal{O}_{X_s,x}} k(x) \to 0$$

(Proposition 1.8(a)). If $\dim_x Y_s = d$ then we are done by taking $Z = Y$. Suppose that $m := \dim_x Y_s > d$.

By Proposition 2.2, the dimension of $\Omega^1_{Y_s,x} \otimes_{\mathcal{O}_{Y_s,x}} k(x)$ over $k(x)$ is equal to $\dim_x Y_s$. Hence there exists an $f \in \mathcal{I}_x$ such that its image $d\bar{f}$ in $\Omega^1_{Y_s,x} \otimes \mathcal{O}_{X_s,x}$ can be completed to a basis $\{d\bar{f}_1 = d\bar{f}, d\bar{f}_2, \ldots, d\bar{f}_m\}$ of $\Omega^1_{Y_s,x} \otimes \mathcal{O}_{X_s,x}$. Restricting Y, if necessary, we can suppose $f \in \mathcal{O}_Y(Y)$. Let us consider the closed subscheme $Z := V(f)$ of Y. Then X is a closed subscheme of Z. Let us show that Z is smooth over S at x and that $\Omega^1_{Z,x}$ is free. By Nakayama, $\mathcal{E} := \{df_1, \ldots, df_m\}$ is a system of generators for $\Omega^1_{Y,x}$. As $\Omega^1_{Y,x}$ is free of rank m, $\mathcal{E}$ is a basis of $\Omega^1_{Y,x}$. Hence

$$\Omega^1_{Z,x} = (\Omega^1_{Y,x}/(df)) \otimes_{\mathcal{O}_{Y,x}} \mathcal{O}_{Z,x}$$

is free of rank $m-1$. In particular, $\Omega^1_{Z_s,x}$ is free of rank $m-1$. Now the irreducible components of Z_s passing through x are of dimension $m-1$ because $\mathcal{O}_{Y_s,x}$ is an integral domain and the image of f in $\mathcal{O}_{Y_s,x}$ is non-zero. Hence Z_s is smooth at x by Proposition 2.2. Applying Lemma 4.3.16 to the homomorphism $\mathcal{O}_{S,s} \to \mathcal{O}_{Y,x}$, we deduce from this that $\mathcal{O}_{Z,x} = \mathcal{O}_{Y,x}/(f)$ is flat over $\mathcal{O}_{S,s}$. Hence Z is smooth at x. We can now conclude the proof by descending induction on $\dim_x Z_s$.   □

Proposition 2.5. *Let S be a locally Noetherian scheme, and let $X \to S$ be a morphism of finite type, smooth at a point x. Then $\Omega^1_{X/S}$ is free of rank $\dim_x X_s$ in a neighborhood of x, where $s \in S$ is the image of x.*

Proof Let d and $X \to Z$ be as in Lemma 2.4. Then $\dim_x X_s = d = \dim_x Z_s$ by Proposition 2.2. Thus $X_s = Z_s$ in a neighborhood of x. By Exercise 4.3.8(a), $X = Z$ in a neighborhood of x. Hence $\Omega^1_{X/S,x} = \Omega^1_{Z/S,x}$ is free of rank d. As $\Omega^1_{X/S}$ is coherent, it is free of rank d in a neighborhood of x (Exercise 5.1.12(a)).   □

Corollary 2.6. *Let S be a locally Noetherian scheme, and let $X \to S$ be a flat morphism of finite type with equidimensional fibers of dimension n. Then $X \to S$ is smooth if and only if $\Omega^1_{X/S}$ is locally free of rank n.*

Proof This results from Propositions 2.2, 2.5, and the base change property for sheaves of differentials (Proposition 1.24).   □

The condition that $\Omega^1_{X/S}$ is locally free of rank n does not suffice to characterize smoothness. Indeed, let $X = S_{\mathrm{red}}$. Then $\Omega^1_{X/S} = 0$. But $X \to S$ is not smooth if S is not reduced. Let us, however, show a partial converse of Proposition 2.5. It also generalizes Proposition 2.2.

Theorem 2.7. *Let S be a regular locally Noetherian connected scheme, and let $f : X \to S$ be a morphism of finite type with X irreducible. Let $x \in X$ be a point such that*

$$\dim \mathcal{O}_{X,x} = \dim \mathcal{O}_{X_s,x} + \dim \mathcal{O}_{S,s}, \quad \text{where } s = f(x) \qquad (2.5)$$

Then f is smooth at x if and only if $\Omega^1_{X/S,x}$ is free of rank $\dim_x X_s$.

Proof We already know that f smooth at x implies that $\Omega^1_{X,x}$ is free of rank $\dim_x X_s$ (Proposition 2.5). Conversely, let us suppose that $\Omega^1_{X/S}$ is locally free of rank n. Proceeding in a manner similar to the proof of Proposition 2.5, we can find a neighborhood U of x in X and a closed immersion $U \to Z$ into a scheme $Z \to S$, the latter being smooth at x and such that $\Omega^1_{Z,x}$ is free of rank n. In particular, $\dim_x Z_s = n$. The dimension formula (Theorem 4.3.12) implies that $\dim \mathcal{O}_{Z,x} = \dim \mathcal{O}_{X,x}$. Now $\mathcal{O}_{Z,x}$ is regular (Theorem 4.3.36), and hence integral (Proposition 4.2.11); it follows that $\mathcal{O}_{Z,x} \to \mathcal{O}_{X,x}$ is an isomorphism. Consequently, f is smooth at x.   □

Remark 2.8. In this theorem, the $\Omega^1_{X/S}$ locally free condition essentially guarantees that X_s is smooth at x. The S regular hypothesis serves uniquely to show that $\mathcal{O}_{Z,x}$ is an integral domain. In fact, this condition is verified if S is normal, or more generally if S is reduced and geometrically unibranch. See [41], IV.17.5.6.

Remark 2.9. Equality (2.5) of the theorem is not very restrictive. Indeed, it is verified if the following conditions are fulfilled: (a) X, S are locally Noetherian, irreducible, and f is dominant of finite type; (b) S is universally catenary (Definition 8.2.1, e.g., if S is an algebraic variety, or, more generally, a scheme of finite type over a regular scheme, see Corollary 8.2.16); (c) X_s is equidimensional of dimension $\dim X_\eta$, where η is the generic point of S. See [41], IV.5.6.5.

Proposition 2.10. *Let X, Y be schemes of finite type over a locally Noetherian scheme S. Let $f : X \to Y$ be a morphism of S-schemes. Then the canonical homomorphism*

$$\varphi : (f^*\Omega^1_{Y/S})_x \to (\Omega^1_{X/S})_x$$

is an isomorphism if f is étale at x. The converse is true if X (resp. Y) is smooth over S at x (resp. y).

Proof We have $(f^*\Omega^1_{Y/S})_x = (\Omega^1_{Y/S})_y \otimes_{\mathcal{O}_{Y,y}} \mathcal{O}_{X,x}$. Let us first suppose that f is étale. As the proposition is of local nature on X, we can suppose that f is a standard étale morphism (Proposition 4.4.11). The isomorphism then results from Example 1.12.

    Conversely, let us suppose that φ is an isomorphism. Then $(\Omega^1_{X/Y})_x = 0$. Hence f is unramified at x. By Proposition 4.4.11, in a neighborhood of x, we can decompose $f : X \to Y$ into a closed immersion $g : X \to Z$ followed by an étale morphism $h : Z \to Y$. From the above, $h^*\Omega^1_{Z/S} \to \Omega^1_{Y/S}$ is an isomorphism. We can therefore replace Y by Z and suppose that f is a closed immersion. By Corollary 2.6, X and Y are smooth over S of the same relative dimension in a neighborhood of x; thus f is an open immersion, and hence étale in a neighborhood of x.   □

6.2.2   Local structure and lifting of sections

The preceding results let us refine the local structure of smooth schemes over S. Such a scheme is 'locally, for the étale topology' an open subset of $\mathbb{A}^n_S$. More precisely we have:

Corollary 2.11. *Let S be a locally Noetherian scheme, and let $f : X \to S$ be a morphism, smooth at a point x. Then there exists an open neighborhood U of x such that $f|_U : U \to S$ decomposes into*

with g étale at x, and where the vertical arrow is the structural morphism.

Proof We may suppose $S = \operatorname{Spec} A$ is affine. Let $\{df_1, \ldots, df_n\}$ be a basis of $\Omega^1_{X,x}$ over $\mathcal{O}_{X,x}$. Restricting X if necessary, we can suppose that the functions f_i are regular on X. They define a homomorphism $A[T_1, \ldots, T_n] \to \mathcal{O}_X(X)$ which sends T_i to f_i. Let $Y = \mathbb{A}^n_S$ and let $g : X \to Y$ denote the corresponding morphism. Let us show that g is étale at x. We have $g^*(dT_i) = df_i$. It follows that

$$(g^*\Omega^1_{Y/S})_x \to (\Omega^1_{X/S})_x$$

is surjective and therefore bijective because the two factors are free of the same rank. It follows from Proposition 2.10 that g is étale at x.   □

Corollary 2.12. *Let S be a locally Noetherian scheme, and let $f : X \to S$ be a morphism of finite type. Then the set of points $x \in X$ such that f is smooth at x is open in X.*

Proof By the preceding corollary, it suffices to show the assertion for étale morphisms. But this has already been done (Corollary 4.4.12).   □

Corollary 2.13. *Let $S = \operatorname{Spec} A$ be the spectrum of a complete Noetherian local ring, with closed point s. Let $X \to S$ be a smooth morphism. Then the canonical map*

$$X(S) \to X_s(k(s))$$

is surjective. In other words, any rational point of X_s lifts to a section of $X \to S$.

Proof Let $x \in X_s$ be a rational point. Let $g : U \to \mathbb{A}^n_S$ be as in Corollary 2.11. It is clear that $g(x)$ lifts to a section $S \to \mathbb{A}^n_S$. The latter induces, by base change, an étale morphism $Z := U \times_{\mathbb{A}^n_S} S \to S$. Let $p : Z \to U$ be the first projection. Then $x = p(z)$, where $z = (x, f(x)) \in Z_s(k(s))$. It is enough to show that z lifts to a section of $Z \to S$. By Proposition 4.3.26, the homomorphism $A \to \mathcal{O}_{Z,z}$ induces an isomorphism $A = \widehat{A} \simeq \widehat{\mathcal{O}}_{Z,z}$, and hence a homomorphism of A-algebras $\mathcal{O}_{Z,z} \to A$ which induces a section $S \to \operatorname{Spec} \mathcal{O}_{Z,z} \to Z$.   □

Remark 2.14. The proof above shows that the A complete hypothesis is not necessary for the surjectivity. It suffices that A be a Henselian local ring. See [15], Proposition 2.3.5.

Let us conclude with a property of infinitesimal lifting of smooth or unramified morphisms.

Proposition 2.15. *Let S be a locally Noetherian scheme and $f : X \to S$ a smooth (resp. étale; resp. unramified) morphism. Let $Y = \operatorname{Spec} A$ be a Noetherian local affine scheme. Then for any closed subscheme Y_0 of Y defined by a nilpotent ideal I, the canonical morphism*

$$\theta : \operatorname{Mor}_S(Y, X) \to \operatorname{Mor}_S(Y_0, X)$$

induced by the closed immersion $Y_0 \to Y$ is surjective (resp. bijective; resp. injective).

Proof Let us start by reducing the general case. First of all, by replacing $X \to S$ by $X \times_S Y \to Y$, we can suppose $Y = S$. Next we may suppose $I^2 = 0$. Indeed, we have $I^r = 0$ for some $r \geq 2$. Let $S_i = V(I^{i+1})$, $0 \leq i \leq r-1$. Then S_i is a closed subscheme of S_{i+1} defined by an ideal of square zero. As $\operatorname{Mor}_S(S, X) \to \operatorname{Mor}_S(S_0, X)$ is the composition of the maps $\operatorname{Mor}_S(S_{i+1}, X) \to \operatorname{Mor}_S(S_i, X)$, and $\operatorname{Mor}_S(S_i, X) = \operatorname{Mor}_{S_i}(S_i, X \times_S S_i)$ (this is bijection (1.2) of Remark 3.1.6), we see that it is sufficient to consider the case $I^2 = 0$.

Let $\sigma : S_0 \to X$. Let s be the closed point of S, and $x = \sigma(s)$. For any open subscheme U of X containing x, we have $\sigma \in \operatorname{Mor}_S(S_0, U)$ and $\theta^{-1}(\operatorname{Mor}_S(S_0, U)) = \operatorname{Mor}_S(S, U)$. Hence the problem of lifting σ is of local nature in X. Let us suppose f is smooth. After, if necessary, shrinking X, we can suppose that f decomposes into an étale morphism $g : X \to \mathbb{A}_S^n$ followed by the projection $\mathbb{A}_S^n \to S$ (Corollary 2.11). Moreover, we may suppose that X is an open subscheme of a scheme that is standard étale over an open subset of $\mathbb{A}_S^n$ (Proposition 4.4.11). Hence $\mathcal{O}_{X,x}$ is the localization of

$$B := A[T_0, T_1, \ldots, T_n,, P'(T)^{-1}]/(P(T))$$

at a prime ideal $\mathfrak{p}$, where $P(T)$ is a monic polynomial in T_0 and $P'(T)$ denotes the derivative with respect to T_0. The morphism σ corresponds to a homomorphism of A-algebras $\sigma^{\#} : \mathcal{O}_{X,x} \to A/I$. Let t_i denote the image of T_i in $B_{\mathfrak{p}}$. For any $a \in A$, let $\tilde{a}$ denote its image in A/I. We have $\sigma^{\#}(t_i) = \tilde{a}_i$. Let $\alpha = (a_0, \ldots, a_n)$. Then $P(\alpha) \in I$ and $P'(\alpha) \in A^*$. Let $\varepsilon = -P(\alpha)/P'(\alpha) \in I$. The Taylor expansion of P (as a polynomial in T_0) gives

$$P(a_0 + \varepsilon, a_1, \ldots, a_n) = P(\alpha) + P'(\alpha)\varepsilon = 0$$

(note that $I^2 = 0$). If necessary replacing α by $(a_0 + \varepsilon, a_1, \ldots, a_n)$ (which does not modify $\sigma^{\#}$), we can suppose that $P(\alpha) = 0$. It is now clear that $\sigma^{\#} : B_{\mathfrak{p}} \to A/I$ lifts to a homomorphism of A-algebras $B_{\mathfrak{p}} \to A$ which sends t_i to a_i. The morphism $S \to \operatorname{Spec} B_{\mathfrak{p}} \to X$ induced by this homomorphism is a lift of $\sigma : S_0 \to X$.

If $\beta = (b_0, \ldots, b_n) \in A^{n+1}$ is another possible lift, then $P(\beta) = 0$, and $b_i - a_i \in I$. Therefore

$$P(\beta) = P(\alpha) + D_{\alpha}P \cdot (\beta - \alpha),$$

which implies that $\beta - \alpha = 0$ because $D_{\alpha}P$ is an invertible linear map. In other words, if f is étale (hence $n = 0$), then $\operatorname{Mor}_S(S, X) \to \operatorname{Mor}_S(S_0, X)$ is bijective. The same method shows that if f is unramified, then this map is injective. $\quad\square$

Remark 2.16. We can show that the proposition above is true if we take an arbitrary affine scheme for Y, and that in that case, the converse is also true. See [15], Proposition 2.2.6, or [41], IV.17.4.1, IV.17.5.1, and IV.17.6.1.

Exercises

2.1. Let $X \to S$ be a smooth morphism of locally Noetherian schemes. Let $s \in S$ and $x \in X_s$ be a $k(s)$-rational point. Show that there exists an isomorphism of $\widehat{\mathcal{O}}_{S,s}$-algebras

$$\widehat{\mathcal{O}}_{X,x} \simeq \widehat{\mathcal{O}}_{S,s}[[T_1, \ldots, T_n]],$$

where $n = \dim \mathcal{O}_{X_s,x}$.

2.2. Let X be an algebraic variety over a field k. Show that X is smooth if and only if $\Omega^1_{X/k}$ is locally free and for any generic point ξ of X, $k(\xi)$ is a separable extension of k (Definition 1.14).

2.3. Let S be a locally Noetherian scheme. Show that if $X \to S$ is smooth, then $s \mapsto \dim X_s$ is locally constant.

2.4. Let X be a smooth connected algebraic variety over a field k of characteristic $p > 0$. Let $F : X \to X^{(p)}$ be the Frobenius. We want to compute $\Omega^1_{X/X^{(p)}}$.

    (a) Let us suppose $X = \operatorname{Spec} \mathbb{A}^n_k$. Show that $\Omega^1_{X/X^{(p)}}$ is free of rank n on X.

    (b) Show that in the general case, $\Omega^1_{X/X^{(p)}}$ is locally free of rank $\dim X$.

2.5. Let $X \to S$ be a morphism of finite type, and let $x \in X_s$ for some $s \in S$.

    (a) Using Lemma 2.4, show that $\dim_{k(x)} T_{X_s,x} \le \dim_{k(x)} \Omega^1_{X_s,x} \otimes k(x)$.

    (b) Let $\mathfrak{m}_x$ be the maximal ideal of $\mathcal{O}_{X,x}$. Show that we have a canonical exact sequence

$$\mathfrak{m}_x/\mathfrak{m}_x^2 \to \Omega^1_{X_s,x} \otimes k(x) \to \Omega^1_{k(x)/k(s)} \to 0.$$

    (c) Let us suppose that x is closed in X_s and that $k(x)$ is separable over $k(s)$. Show that $(\Omega^1_{X/S,x} \otimes k(x))^\vee \simeq T_{X_s,x}$.

2.6. Let $X \to Y$ be a morphism of finite type of locally Noetherian regular schemes. Let $y \in Y$, and let $x \in X_y$ be a closed point such that $k(x)$ is separable over $k(y)$.

    (a) Show that in a neighborhood of x, there exists a closed immersion $i : X \to Z$ into a scheme Z that is smooth of relative dimension $\dim \mathcal{O}_{X,x}$ over Y.

    (b) Show that locally in x, X is a closed subscheme of codimension $\dim \mathcal{O}_{Y,y}$ of a smooth scheme over Y.

2.7. Let S be a locally Noetherian scheme, and let $X \to S$ be a morphism of finite type, smooth at a closed point $x \in X_s$.

    (a) Let us suppose $\dim X_s > 0$. Show that there exists a closed subscheme Z of X containing x, smooth over S at x, with $\dim Z_s < \dim X_s$.

    (b) Using (a) or Corollary 2.11, show that there exists an étale morphism $T \to S$ such that $X \times_S T \to T$ admits a section, whose image $T \to X \times_S T \to X$ contains x. Such a section is called an *étale quasi-section* of $X \to S$.

2.8. Let S be a Noetherian local scheme, with closed point s. Let $f : X \to S$ be a proper flat morphism. We suppose that X_s is smooth over $k(s)$. Show that X is smooth over S.

2.9. Let X, Y be locally Noetherian integral schemes. Let $f : X \to Y$ be a dominant morphism of finite type. We say that f is *generically separable* (resp. *generically étale*) if $k(Y)/k(X)$ is a separable (resp. finite separable) extension.

    (a) Show that if f is generically separable (resp. generically étale), then there exists a non-empty open subset U of X such that $f|_U : U \to Y$ is smooth (resp. étale).

    (b) Let us suppose that f is proper and that the generic fiber of f is smooth. Show that under the hypothesis of (a), there exists a non-empty open subset V of Y such that $f^{-1}(V) \to V$ is smooth (resp. étale).

2.10. Let S be a locally Noetherian scheme, and let $f : X \to Y$ be a morphism of smooth S-schemes.

    (a) Show that if f is smooth, then the complex

$$0 \to f^* \Omega^1_{Y/S} \to \Omega^1_{X/S} \to \Omega^1_{X/Y} \to 0$$

    is exact.

    (b) Let us suppose X, Y are integral. Show that the complex above stays exact if $X \to Y$ is generically separable. See also Exercise 3.8.

2.11. Let $A \to B$ be a finite homomorphism of Noetherian rings. We suppose that A is a complete local ring, with residue field k. Let k' be the largest sub-k-algebra of $B \otimes_A k$ that is separable over k. Show that $A \to B$ decomposes into a finite étale homomorphism $A \to C$ followed by a finite homomorphism $C \to B$, with C simple over A (i.e., $C = A[c]$ for some $c \in C$) and $C \otimes_A k = k'$.

6.3   Local complete intersection

In this section, we study a more general class of morphisms than that of the smooth morphisms, namely that of the local complete intersections

(Definition 3.17). It turns out that any morphism $X \to Y$ of finite type between regular Noetherian schemes is a local complete intersection (Example 3.18). To stay at an elementary level, we have tried to avoid homological algebra, which would have rendered the local study of these morphisms more elegant, and which, above all, is more powerful. We therefore encourage the curious reader to pursue this matter by consulting, for example, [33], Chapter III.

Unless explicitly mentioned otherwise, we will suppose all schemes in this section to be *locally Noetherian*. This condition is not useful in all of the statements, but we prefer thus to keep a uniform hypothesis.

6.3.1   Regular immersions

Definition 3.1. Let A be a ring, and let $a_1, \ldots, a_n$ be a sequence of elements of A. We say that it is a *regular sequence* if a_1 is not a zero divisor and if for any $i \geq 2$, a_i is not a zero divisor in $A/(a_1, \ldots, a_{i-1})$.

Example 3.2. Let A be a regular Noetherian local ring; then we easily verify, using Corollary 4.2.15 and Proposition 4.2.11, that any coordinate system forms a regular sequence.

Remark 3.3. Let $a_1, \ldots, a_n$ be a sequence of elements in a ring A. The regularity property depends, in general, on the order of the a_i. However, we can show that if A is a Noetherian local ring and the a_i belong to the maximal ideal, then the regularity of the sequence is independent of the order. See Exercise 3.1.

Definition 3.4. Let Y be a locally Noetherian scheme, and let $f : X \to Y$ be an immersion (Exercise 3.2.3; we can restrict ourselves to closed immersions if we want to, which will not modify the results). We say that f is a *regular immersion at $x \in X$* (resp. *regular immersion of codimension n at x*) if the ideal $\mathrm{Ker}(\mathcal{O}_{Y,f(x)} \to \mathcal{O}_{X,x})$ is generated by a regular sequence (resp. regular sequence of n elements) in $\mathcal{O}_{Y,f(x)}$. We say that f is a regular immersion (resp. regular immersion of codimension n) if the property holds at every point of X.

Example 3.5. Let $Y = \mathbb{P}_k^d$ be a projective space over a field k. Let X be a closed subvariety of Y. If X is a complete intersection of dimension r (Exercise 5.3.3), then the canonical injection $X \to Y$ is a regular immersion of codimension $d - r$.

Lemma 3.6. *Let A be a ring. Let I be an ideal generated by a regular sequence $a_1, \ldots, a_n$. Then the images of the a_i in I/I^2 form a basis of I/I^2 over A/I. In particular, I/I^2 is a free A/I-module of rank n.*

Proof Let us first show the following property by induction on n:

$$\sum_{1 \leq i \leq n} a_i x_i = 0, x_i \in A \Longrightarrow x_i \in I \text{ for every } i.$$

This is obvious if $n = 1$. Let us suppose this is true for any regular sequence of $n - 1$ elements. As a_n does not divide zero in $A/(a_1, \ldots, a_{n-1})$, we have $x_n \in (a_1, \ldots, a_{n-1})$. Let us write $x_n = \sum_{1 \leq i \leq n-1} a_i y_i$ with $y_i \in A$. It follows that

$\sum_{1 \leq i \leq n-1} a_i(x_i + a_n y_i) = 0$. By the induction hypothesis, we have $x_i + a_n y_i \in (a_1, \ldots, a_{n-1})$, hence $x_i \in I$ for $i \leq n - 1$, and we have already shown that $x_n \in I$.

Now let $x_1, \ldots, x_n \in A$ be such that

$$\sum_{1 \leq i \leq n} a_i x_i \in I^2 = \sum_{1 \leq i \leq n} a_i I.$$

We then have $\sum_{1 \leq i \leq n} a_i(x_i - z_i) = 0$ for some $z_i \in I$. From the above, we have $x_i - z_i \in I$, whence $x_i \in I$. This shows that the images of the a_i in I/I^2 form a free family over A/I. It is therefore a basis since the a_i generate I. $\square$

Definition 3.7. Let $f : X \to Y$ be an immersion into a scheme. Let V be an open subscheme of Y such that f factors through a closed immersion $i : X \to V$, and let $\mathcal{J}$ be the sheaf of ideals defining i. We call the sheaf $i^*(\mathcal{J}/\mathcal{J}^2)$ on X the *conormal sheaf of X in Y*, and we denote it by $\mathcal{C}_{X/Y}$ (we easily verify that $\mathcal{C}_{X/Y}$ does not depend on the choice of V). As $\mathcal{J}/\mathcal{J}^2$ is killed by $\mathcal{J}$, we have $i_* \mathcal{C}_{X/Y} = \mathcal{J}/\mathcal{J}^2$. The dual of $\mathcal{C}_{X/Y}$ is called the *normal sheaf of X in Y*, and we denote it by $\mathcal{N}_{X/Y}$.

Corollary 3.8. *Let $f : X \to Y$ be a regular immersion. Then $\mathcal{C}_{X/Y}$ is a locally free sheaf on X, of rank n if f is a regular immersion of codimension n.*

Proof For any $x \in X$, we have $f^*(\mathcal{J}/\mathcal{J}^2)_x = \mathcal{J}_y/\mathcal{J}_y^2$, where $y = f(x)$, and the kernel of the surjective homomorphism $\mathcal{O}_{Y,y} \to \mathcal{O}_{X,x}$ is $\mathcal{J}_y$. The lemma above implies that $f^*(\mathcal{J}/\mathcal{J}^2)_x$ is free (of rank n if f is of codimension n) over $\mathcal{O}_{X,x}$. $\square$

Remark 3.9. Let $f : X \to Y$ be a regular immersion. If X is connected, then $f^*(\mathcal{J}/\mathcal{J}^2)$ is locally free of some rank n because $x \mapsto \mathrm{rank}(\mathcal{J}/\mathcal{J}^2)_x$ is a locally constant map on X (Exercise 5.1.12(a)). Consequently, f is a regular immersion of codimension n.

Lemma 3.10. *Let $\phi : A \to B$ be a flat homomorphism of rings, let $a_1, \ldots, a_n$ be a regular sequence in A. Then $\phi(a_1), \ldots, \phi(a_n)$ is a regular sequence in B.*

Proof By induction on n, and because $A/(a_1, \ldots, a_i) \to B/(\phi(a_1), \ldots, \phi(a_i))$ is flat for every $i \leq n$, we can reduce to the case when $n = 1$. Now a_1 not dividing zero is equivalent to saying that the map $A \to A$ given by multiplication by a_1 is an injective linear homomorphism. By tensoring with B, the map becomes the multiplication by $\phi(a_1)$ on B. It is injective by the flatness hypothesis. Therefore $\phi(a_1)$ does not divide zero. $\square$

Proposition 3.11. *Let X, Y, Y', Z be locally Noetherian schemes. The following properties are true.*

(a) *Let $f : X \to Y$, $g : Y \to Z$ be regular immersions (resp. of respective codimensions n and m). Then $g \circ f$ is a regular immersion (resp. of codimension $n + m$). Moreover, we have a canonical exact sequence*

$$0 \to f^* \mathcal{C}_{Y/Z} \to \mathcal{C}_{X/Z} \to \mathcal{C}_{X/Y} \to 0. \tag{3.6}$$

(b) *Let X be a closed subscheme of Y. We suppose that the canonical injection $X \to Y$ is a regular immersion of codimension n. Then for any irreducible component Y' of Y, we have $\operatorname{codim}(X \cap Y', Y') = n$ if $X \cap Y' \neq \emptyset$. Moreover, $\dim \mathcal{O}_{X,x} = \dim \mathcal{O}_{Y,x} - n$ for all $x \in X$.*

(c) *Let $f : X \to Y$ be a regular immersion. Let $Y' \to Y$ be a morphism, and let us consider the base change $X' := X \times_Y Y' \to Y'$. Then we have a canonical surjective homomorphism $p^* \mathcal{C}_{X/Y} \to \mathcal{C}_{X'/Y'}$, where $p : X' \to X$ is the projection.*

(d) *Let us keep the hypotheses of (c) and let us suppose that $Y' \to Y$ is flat. Then $X' \to Y'$ is a regular immersion, of codimension n if f is of codimension n. Moreover, $p^* \mathcal{C}_{X/Y} \to \mathcal{C}_{X'/Y'}$ is an isomorphism.*

Proof   (a) Let us show exact sequence (3.6), the rest following immediately from the definition. By construction of the conormal sheaves, we immediately have a canonical exact sequence

$$f^* \mathcal{C}_{Y/Z} \to \mathcal{C}_{X/Z} \xrightarrow{\alpha} \mathcal{C}_{X/Y} \to 0.$$

The sheaves in the sequence are locally free and coherent. It follows that $\operatorname{Ker} \alpha$ is flat (Exercise 1.2.15) and coherent, and hence locally free (Theorem 1.2.16). The homomorphism $f^* \mathcal{C}_{Y/Z} \to \operatorname{Ker} \alpha$ is surjective, and the two sheaves have stalks of the same rank; it is therefore an isomorphism (Exercise 5.1.13).

(b) As the property is local (Exercise 2.5.2), we can reduce to the case when Y is the spectrum of a Noetherian local ring A, with closed point y. Then $\mathcal{J}_y$ is generated by a regular sequence $a_1, \ldots, a_n$. By induction on n, it suffices to show the property when $n = 1$ (hence $X = V(a_1)$). Let $\mathfrak{p}$ be the minimal prime ideal of A corresponding to Y'. Let b be the image of a_1 in $A/\mathfrak{p}$. Then $b \neq 0$ (Exercise 2.1.4(a)). It follows that $\operatorname{codim}(X \cap Y', Y') = 1$ (Theorem 2.5.12) and $\dim X \cap Y' = \dim Y' - 1$ (Theorem 2.5.15). This last equality implies that $\dim X = \dim Y - 1$, by varying Y'.

As the proof of (c) poses no difficulty, it remains to show assertion (d). The fact that $X' \to Y'$ is a regular immersion results from Lemma 3.10. Finally, $p^* \mathcal{C}_{X/Y} \to \mathcal{C}_{X'/Y'}$ is an isomorphism because it is surjective and the two sheaves are, locally on X, free of the same rank. $\square$

Remark 3.12. Let $X \to Y$ be an immersion. Let $Y' \to Y$ be a faithfully flat morphism with Y' locally Noetherian. If $X \times_Y Y' \to Y'$ is a regular immersion, then so is $X \to Y$ (Exercise 3.2(c)). This is a partial converse of Proposition 3.11(d).

Proposition 3.13. *Let S be a locally Noetherian scheme. Let X, Y be smooth schemes over S. Then any immersion $f : X \to Y$ of S-schemes is a regular immersion, and we have a canonical exact sequence on X:*

$$0 \to \mathcal{C}_{X/Y} \to f^* \Omega^1_{Y/S} \to \Omega^1_{X/S} \to 0. \tag{3.7}$$

Proof   Let $x \in X$, $y = f(x)$, and let s be the image of x in S. We are going to show by induction on $e := \dim_y Y_s - \dim_x X_s$ that f is a regular immersion of codimension e at x. We have seen in the proof of Proposition 2.5 that if $e = 0$, then f is an isomorphism (hence a regular immersion) at x, and that if $e \geq 1$, then there exists an $f_1 \in \mathcal{J}_y \setminus \{0\}$ such that $Z := V(f_1)$, which is defined in a neighborhood of y, is smooth over S at x. The canonical injection $Z \to Y$ is clearly a regular immersion at x, and we have $\dim_y Z_s - \dim_x X_s = e - 1$. By applying the induction hypothesis and Proposition 3.11(a), we conclude that f is a regular immersion.

Let us now show exact sequence (3.7). Replacing Y if necessary, by an open subset, we can suppose that f is a closed immersion. The complex (3.7) is given in Proposition 1.24(d), and we know that it is exact except perhaps on the left. The sheaves of the complex are locally free, and for any $x \in X$, we have

$$\mathrm{rank}(f^*\Omega^1_{Y/S})_x - \mathrm{rank}(\Omega^1_{X/S})_x = \dim_y Y_s - \dim_x X_s = e = \mathrm{rank}(\mathcal{C}_{X/Y})_x$$

(Corollary 3.8). By an argument similar to that of Proposition 3.11(a), we conclude that (3.7) is exact on the left.   □

Corollary 3.14. *Let Y be a locally Noetherian scheme, and let $f : X \to Y$ be a separated smooth morphism. Then any section $\pi : Y \to X$ of f is a regular closed immersion. Moreover, we have a canonical isomorphism $\mathcal{C}_{Y/X} \simeq \pi^*\Omega^1_{X/Y}$.*

Proof Indeed, π is a closed immersion (Exercise 3.3.6) of smooth schemes over Y. Applying exact sequence (3.7) to the morphism of Y-schemes $\pi : Y \to X$, we obtain the desired isomorphism.   □

Lemma 3.15. *Let S be a locally Noetherian scheme, and let $i : X \to Y$ be an immersion of locally Noetherian S-schemes. Let us fix $s \in S$ and $x \in X_s$ and let us suppose that $i_s : X_s \to Y_s$ is a regular immersion at x and that X and Y are flat over S at x. Then i is a regular immersion at x.*

Proof We may suppose that i is a closed immersion. Let $\mathcal{J} \subseteq \mathcal{O}_Y$ be the sheaf of ideals defining i. From the exact sequence

$$0 \to \mathcal{J}_x \to \mathcal{O}_{Y,x} \to \mathcal{O}_{X,x} \to 0$$

we deduce the exact sequence

$$0 \to \mathcal{J}_x \otimes_{\mathcal{O}_{S,s}} k(s) = \mathcal{J}_x \mathcal{O}_{Y_s,x} \to \mathcal{O}_{Y_s,x} \to \mathcal{O}_{X_s,x} \to 0$$

(Proposition 1.2.6). By hypothesis, there exist $a_1, \dots, a_n \in \mathcal{J}_x$ whose images $a'_1, \dots, a'_n$ in $\mathcal{J}_x \mathcal{O}_{Y_s,x}$ form a regular sequence and generate $\mathcal{J}_x \mathcal{O}_{Y_s,x}$. We have

$$(\mathcal{J}_x/(a_1, \dots, a_n)) \otimes_{\mathcal{O}_{S,s}} k(s) = 0.$$

By Nakayama's lemma on $\mathcal{O}_{Y,x}$, this implies that $\mathcal{J}_x = (a_1, \dots, a_n)$. As a'_1 does not divide zero, $\mathcal{O}_{Y,x}/(a_1)$ is flat over $\mathcal{O}_{S,s}$ (Lemma 4.3.16). Hence a_1 does not divide zero (Exercise 4.3.14). The scheme $V(a_1)$, which is defined in a neighborhood of x, is flat over S at x. Moreover, $V(a_1) \to Y$ is a regular immersion. We can therefore replace Y by $V(a_1)$. Thus we see, little by little, that $X \to Y$ is a regular immersion at x.   □

Remark 3.16. The converse of the lemma is true. It follows from Corollary 3.24 and Lemma 3.21.

6.3.2   Local complete intersections

Definition 3.17. Let Y be a locally Noetherian scheme. Let $f : X \to Y$ be a morphism of finite type. We say that f is a *local complete intersection at x* if there exist a neighborhood U of x and a commutative diagram

where i is a regular immersion and g is a smooth morphism. We say that f is a *local complete intersection* (l.c.i., to abbreviate) if it is a local complete intersection at all of its points.

Example 3.18. Let $f : X \to Y$ be a morphism of finite type of regular locally Noetherian schemes. Then f is an l.c.i. Indeed, as the property is local, we may suppose that X is a closed subscheme of some $\mathbb{A}^n_Y$. Now since $\mathbb{A}^n_Y$ is regular, it follows from Corollary 4.2.15 that $X \to \mathbb{A}^n_Y$ is a regular immersion. As $\mathbb{A}^n_Y \to Y$ is smooth, f is indeed an l.c.i.

Remark 3.19. Let $f : X \to Y$ be a morphism of locally Noetherian schemes that is smooth at a point x. Then locally in x, we can always decompose f into a regular immersion $X \to \mathbb{A}^{n+1}_Y$ followed by the projection $\mathbb{A}^{n+1}_Y \to Y$. This immediately follows from Proposition 3.13. We can therefore replace the smooth morphism $Z \to Y$ in Definition 3.17 by the projection $\mathbb{A}^m_Y \to Y$ for some $m \geq 1$.

Proposition 3.20. *The following properties are true.*

(a) *Regular immersions and smooth morphisms are l.c.i.*

(b) *Local complete intersections are stable under composition.*

(c) *Let $f : X \to Y$ be an l.c.i., and let $Y' \to Y$ be a flat morphism. Then $f_{Y'} : X \times_Y Y' \to Y'$ is an l.c.i.*

Proof (a) immediately follows from the definition. (b) Let $f : X \to Y$ and $g : Y \to Z$ be two l.c.i.s. We want to show that $g \circ f$ is an l.c.i. As the composition of regular immersions (resp. smooth morphisms) is still a regular immersion (resp. a smooth morphism), it suffices to consider the case when f is smooth and g is a regular immersion. As the property is local, we may suppose that f factors into a closed immersion $i : X \to \mathbb{A}^n_Y$ followed by the projection onto Y. It follows from Proposition 3.13 that i is a regular immersion. Moreover, g induces a regular immersion $\mathbb{A}^n_Y \to \mathbb{A}^n_Z$ (Proposition 3.11(d)), and hence $g \circ f$ factors into a regular immersion $X \to \mathbb{A}^n_Z$ followed by the projection $\mathbb{A}^n_Z \to Z$ which is a smooth morphism. This proves that $g \circ f$ is an l.c.i. Finally, (c) results from Propositions 3.11(d) and 4.3.38. $\square$

Lemma 3.21. *Let $f : X \to Y$ be an immersion into a locally Noetherian scheme. Let us suppose that f is an l.c.i. Then f is a regular immersion.*

Proof We decompose f locally into a regular immersion $i : X \to \mathbb{A}_Y^m$ followed by the projection $\mathbb{A}_Y^m \to Y$ (see Remark 3.19). The lemma then immediately results from Exercise 3.2(c). □

Corollary 3.22. *Let $f : X \to Y$ be an l.c.i. Let us suppose that f decomposes into an immersion $i : X \to Z$ followed by a smooth morphism $g : Z \to Y$. Then i is a regular immersion. Moreover, if f is a regular immersion, then we have a canonical exact sequence*

$$0 \to \mathcal{C}_{X/Y} \to \mathcal{C}_{X/Z} \to i^*\Omega^1_{Z/Y} \to 0.$$

Proof Replacing Z by an open subscheme if necessary, we may suppose that i is a closed immersion. Let us consider $W := X \times_Y Z$ and the commutative fibered product diagram

Then the projection $q : W \to Z$ is an l.c.i. by Proposition 3.20(c). The immersion $i : X \to Z$ induces a section $\pi : X \to W$ of the projection $p : W \to X$. As the latter is smooth, π is a regular immersion (Corollary 3.14); hence $i = q \circ \pi$ is an l.c.i., and a regular immersion (Lemma 3.21). Let us suppose that f is a regular immersion. Then so is q, and we have the exact sequence

$$0 \to \pi^*\mathcal{C}_{W/Z} \to \mathcal{C}_{X/Z} \to \mathcal{C}_{X/W} \to 0$$

(Proposition 3.11(a)). Moreover,

$$\pi^*\mathcal{C}_{W/Z} = \pi^*p^*\mathcal{C}_{X/Y} = \mathcal{C}_{X/Y}, \quad \mathcal{C}_{X/W} = \pi^*\Omega^1_{W/X} = \pi^*q^*\Omega^1_{Z/Y} = i^*\Omega^1_{Z/Y}$$

(Corollary 3.14). This completes the proof of the corollary. □

Remark 3.23. The corollary above and Remark 3.12 imply the following partial converse of Proposition 3.20(c): if $f : X \to Y$ is of finite type and if $X \times_Y Y' \to Y'$ is an l.c.i. for a locally Noetherian scheme Y' that is faithfully flat over Y, then f is an l.c.i.

Corollary 3.24. *Let Y be a locally Noetherian scheme, and let $f : X \to Y$ be a flat morphism of finite type. Then f is an l.c.i. if and only if for every $y \in Y$, $f_y : X_y \to \operatorname{Spec} k(y)$ is an l.c.i.*

Proof As the property is local, we may suppose that f decomposes into an immersion $i : X \to Z = \mathbb{A}_Y^n$ followed by the projection $Z \to Y$. Let us fix $y \in Y$ and $x \in X_y$. Let us suppose that f_y is an l.c.i. By Corollary 3.22, $X_y \to Z_y$ is a regular immersion, and hence $X \to Z$ is a regular immersion (Lemma 3.15).

This proves that f is an l.c.i. Conversely, let us suppose that f is an l.c.i., hence i is a regular immersion, defined by a regular sequence $b_1, \ldots, b_m \in \mathcal{O}_{Z,x}$. In particular, we have $\dim \mathcal{O}_{X,x} = \dim \mathcal{O}_{Z,x} - m$ (Exercise 3.4(a)). In addition,

$$\dim \mathcal{O}_{X_y,x} = \dim \mathcal{O}_{X,x} - \dim \mathcal{O}_{Y,y}, \quad \dim \mathcal{O}_{Z_y,x} = \dim \mathcal{O}_{Z,x} - \dim \mathcal{O}_{Y,y}$$

(Theorem 4.3.12). Hence $\dim \mathcal{O}_{X_y,x} = \dim \mathcal{O}_{Z_y,x} - m$. It follows from Exercise 3.4(c) that the images of the b_i in $\mathcal{O}_{Z_y,x}$ form a regular sequence, and hence $X_y \to Z_y$ is a regular immersion.   □

Exercises

3.1. Let $(A, \mathfrak{m})$ be a Noetherian local ring, and let $(a_1, a_2, \ldots, a_n)$ be a regular sequence of elements of $\mathfrak{m}$. We want to show that $(a_{\sigma(1)}, \ldots, a_{\sigma(n)})$ is a regular sequence for any permutation σ.

   (a) Show that it suffices to consider the case $n = 2$ (decompose σ as a product of transpositions $(i, i+1)$ and consider the ring $A/(a_1, \ldots, a_{i-1})$).

   (b) Let $J = \text{Ann}(a_2)$ be the annihilator of a_2. Show that $J = a_1 J$. Deduce from this that $J = 0$ (Theorem 1.3.13). Here we need the A Noetherian and $a_1 \in \mathfrak{m}$ hypothesis. Show that (a_2, a_1) is a regular sequence.

3.2. Let us keep the notation and hypotheses of Exercise 3.1. Let $(b_1, \ldots, b_m)$ be a minimal system of generators of $I = (a_1, \ldots, a_n)$.

   (a) Show that, permuting the a_i if necessary (which, by Exercise 3.1, does not change the fact that $a_1, \ldots, a_n$ is a regular sequence), there exists an $a \in A^*$ such that $b_m \in aa_n + (a_1, \ldots, a_{n-1})$. Show that $(a_1, \ldots, a_{n-1}, b_m)$ is a regular sequence.

   (b) Show that $m = n$ and that $(b_1, \ldots, b_m)$ is a regular sequence.

   (c) Let T be a variable, and B the localization of $A[T]$ at the maximal ideal $(\mathfrak{m}, T)$. Show that if an ideal J of A is such that $(J, T) \subseteq B$ is generated by a regular sequence, then J is generated by a regular sequence in A. Show the same result with more than one variable.

   (d) Let C be a faithfully flat Noetherian local A-algebra. Let J be an ideal of A such that JC is generated by a regular sequence. Show that J is generated by a regular sequence in A.

3.3. Let $f : X \to Y$, $g : Y \to Z$ be immersions of locally Noetherian schemes. Show that if f and $g \circ f$ are regular immersions, then so is g. Give an example where $g \circ f$ and g are regular immersions without f being one.

3.4. Let $(A, \mathfrak{m})$ be a Noetherian local ring that is *equidimensional* (i.e., the irreducible components of $\text{Spec } A$ are of the same dimension).

   (a) Let $a_1, \ldots, a_r$ be a regular sequence in $\mathfrak{m}$. Show that $A/(a_1, \ldots, a_r)$ is equidimensional of dimension $\dim A - r$ (use Theorem 2.5.15).

(b) Let $a_{r+1}, \ldots, a_m \in \mathfrak{m}$ be such that a_{r+1} is a zero divisor in $A/(a_1, \ldots, a_r)$. Show that $\dim A/(a_1, \ldots, a_m) \geq \dim A - m + 1$.

(c) Let $b_1, \ldots, b_m \in \mathfrak{m}$ be such that $A/(b_1, \ldots, b_m)$ is of dimension $\dim A - m$. Show that $b_1, \ldots, b_m$ is a regular sequence.

3.5. Let Y be a locally Noetherian scheme, and let $f : X \to Y$ be a finite surjective l.c.i.

(a) Let $x \in X$ and $y = f(x)$. Show that $\dim \mathcal{O}_{X,x} = \dim \mathcal{O}_{Y,y}$.

(b) Show that there exists an affine open neighborhood U of x, isomorphic to a closed subscheme $V(I)$ of an affine space $\mathbb{A}_V^n$ over an affine open neighborhood V of y, with I generated by n elements $F_1, \ldots, F_n \in \mathcal{O}_Y(V)[T_1, ..., T_n]$ (use Exercise 3.4(a)).

(c) Let $\overline{F}_1, \ldots, \overline{F}_n$ be the images of the F_i in $k(y)[T_1, ..., T_n]$. Show that the $\overline{F}_i$ form a regular sequence (use Exercise 3.4(b)). Deduce from this that f is flat (use Lemma 4.3.16).

3.6. Let Y be a locally Noetherian scheme, and let $f : X \to Y$ be a closed immersion defined by a sheaf of ideals $\mathcal{J} \subseteq \mathcal{O}_Y$. Let $x \in X$. Show that if $\mathcal{J}_{f(x)}$ is generated by a regular sequence, then there exists an affine open neighborhood V of $f(x)$ such that $\mathcal{J}(V)$ is generated by a regular sequence.

3.7. (*Lifting of the formal structure*). Let $X \to Y$ be a flat morphism of finite type between locally Noetherian schemes. Let $y \in Y$ and let $x \in X_y$ be such that $k(x) = k(y)$. We suppose that $X_y \to \operatorname{Spec} k(y)$ is an l.c.i. at x.

(a) Show that there exists an exact sequence

$$0 \to \overline{I} \to k(y)[[T_1, \ldots, T_n]] \to \widehat{\mathcal{O}}_{X_y,x} \to 0$$

where $\overline{I}$ is an ideal generated by a regular sequence $f_1, \ldots, f_r$, and where $\widehat{\mathcal{O}}_{X_y,x}$ is the completion for the $\mathfrak{m}_x$-adic topology. Show that we can choose $n = \dim_{k(x)} \Omega^1_{X_y/k(y),x} \otimes k(x) + 1$.

(b) Show that for any exact sequence as in (a), there exists an exact sequence

$$0 \to I \to \widehat{\mathcal{O}}_{Y,y}[[T_1, \ldots, T_n]] \to \widehat{\mathcal{O}}_{X,x} \to 0$$

where I is an ideal generated by a regular sequence $F_1, \ldots, F_r$, and where the canonical image of F_i in $k(y)[[T_1, \ldots, T_n]]$ is equal to f_i.

3.8. Let X, Y be Noetherian integral schemes over a scheme S. Let $f : X \to Y$ be a dominant l.c.i., and suppose that $K(X)/K(Y)$ is a (not necessarily algebraic) separable extension.

(a) Let us suppose that there exists a regular immersion $i : X \to Z := \mathbb{A}_Y^n$. Show that the canonical homomorphism $\delta : \mathcal{C}_{X/Z} \to \Omega^1_{Z/Y}|_X$ is injective (show that the support of $\operatorname{Ker} \delta$ is different from X).

(b) Show that the complex

$$0 \to f^*\Omega^1_{Y/S} \to \Omega^1_{X/S} \to \Omega^1_{X/Y} \to 0$$

is exact (use Lemma 1.11(b)).

6.4   Duality theory

An important tool for studying the cohomology of coherent sheaves on a projective scheme is Serre duality (generalized by Grothendieck), see Definition 4.18. To stay within the limits of this book, we will not prove the duality theorem. This section is rather devoted to the canonical sheaf $\omega_{X/Y}$ for a quasi-projective l.c.i. $f : X \to Y$. This sheaf coincides with the determinant of $\Omega^1_{X/Y}$ when $X \to Y$ is smooth. It turns out that this sheaf is also the dualizing sheaf (Theorem 4.32) if f is moreover flat.

6.4.1   Determinant

Let us first of all recall some notions from tensor algebra. Let A be a ring and M an A-module. We define the tensor algebra $T(M)$ of M as the direct sum

$$T(M) = \bigoplus_{n \geq 0} M^{\otimes n}.$$

By convention, $M^{\otimes 0} = A$ even if $M = 0$. We have a natural structure of an (in general non-commutative) A-algebra on $T(M)$. Let

$$\wedge(M) = \bigoplus_{n \geq 0} \wedge^n M$$

be the quotient of $T(M)$ by the two-sided ideal generated by the elements of the form $x \otimes x$, $x \in M$. We call $\wedge(M)$ the exterior algebra of M. We let $x_1 \wedge \cdots \wedge x_n$ denote the image of $x_1 \otimes \cdots \otimes x_n$ in $\wedge^n M$. We have $x_1 \wedge \cdots \wedge x_n = 0$ if there exist two distinct indices $i \neq j$ such that $x_i = x_j$. Moreover, we have $x \wedge y = -y \wedge x$ (expand $(x + y) \wedge (x + y)$). We deduce from this that $x_{\sigma(1)} \wedge \cdots \wedge x_{\sigma(n)} = \varepsilon(\sigma)x_1 \wedge \cdots \wedge x_n$ for any permutation σ (where ε is the signature of σ).

   Let M be a free module of finite rank r over A; then $\det M := \wedge^r M$ is free of rank 1 over A. It is called the *determinant of M*. If $M = 0$, we have $T(M) = \wedge(M) = \det M = A$. Let $\{e_1, \dots, e_r\}$ be a basis of M; then $e_1 \wedge \cdots \wedge e_r$ is a basis of $\det M$. Let $e'_1, \dots, e'_r$ be elements of M; we have $(e'_1, \dots, e'_r) = (e_1, \dots, e_r) \cdot P$ for some matrix $P \in M_r(A)$. A direct computation shows that

$$e'_1 \wedge \cdots \wedge e'_r = (\det P) \cdot (e_1 \wedge \cdots \wedge e_r). \tag{4.8}$$

In particular, we deduce from this that $\{e'_1, \dots, e'_r\}$ is a basis of M if and only if $e'_1 \wedge \cdots \wedge e'_r$ is a basis of $\det M$. If $\phi : M \to N$ is a homomorphism of A-modules, then we canonically have a homomorphism $\wedge^n M \to \wedge^n N$ for every $n \geq 0$. If M and N are free of the same finite rank, then ϕ canonically induces a homomorphism $\det \phi : \det M \to \det N$.

Lemma 4.1. *Let M be a free A-module of finite rank.*

   (a) *Let B be an A-algebra. Then we have a canonical isomorphism $\det(M \otimes_A B) \simeq (\det M) \otimes_A B$, the left-hand side being the determinant as a B-module.*

(b) Let $0 \to L \xrightarrow{\alpha} M \xrightarrow{\beta} N \to 0$ be an exact sequence of free A-modules of finite rank. Then we have a canonical isomorphism:

$$(\det L) \otimes_A (\det N) \simeq \det M.$$

(c) We have a canonical isomorphism $\det(M^\vee) \simeq (\det M)^\vee$ (the sign $^\vee$ denotes the dual).

(d) The isomorphisms of (b) and (c) commute with extension of scalars.

Proof We may suppose that the considered modules are non-zero. (a) Let $\{e_1, \ldots, e_r\}$ be a basis of M over A. Let $e_i' = e_i \otimes 1 \in M \otimes_A B$. Then the homomorphism $(\det M) \otimes_A B \to \det(M \otimes_A B)$ defined by $(e_1 \wedge \cdots \wedge e_r) \otimes b \mapsto (e_1' \wedge \cdots \wedge e_r')b$ is clearly an isomorphism independent of the choice of the basis.

(b) Let $\{e_1, \ldots, e_r\}$, $\{f_1, \ldots, f_q\}$ be respective bases of L and N. Let $g_j \in \beta^{-1}(f_j)$, $j \le q$. Then the set $\{\alpha(e_i), g_j\}_{i \le r, j \le q}$ is a basis of M. Let us set

$$\psi((e_1 \wedge \cdots \wedge e_r) \otimes (f_1 \wedge \cdots \wedge f_q)) = \alpha(e_1) \wedge \cdots \wedge \alpha(e_r) \wedge g_1 \wedge \cdots \wedge g_q.$$

Then ψ defines an isomorphism $\det L \otimes \det N \simeq \det M$. It is easy to verify first that ψ does not depend on the choice of the $g_j \in \beta^{-1}(f_j)$, and next that it does not depend on the choice of the bases of L and N either.

(c) Let $e_1^*, \ldots, e_n^*$ be the dual basis of $M^\vee$. We define a bilinear map $\psi :$ $(\det M) \times \det(M^\vee) \to A$ by $\psi(e_1 \wedge \cdots \wedge e_n, e_1^* \wedge \cdots \wedge e_n^*) = 1$. Then ψ is clearly non-degenerate and we verify without trouble that its definition does not depend on the choice of a basis. Finally, (d) comes from the construction of the isomorphisms. $\qquad\square$

The notions that we have just looked at carry over to quasi-coherent sheaves $\mathcal{F}$ on a scheme X. Let U be an affine open subset of X, and let $r \ge 0$. Then $(\wedge^r \mathcal{F})|_U := (\wedge^r \mathcal{F}(U))^\sim$. In particular, if $\mathcal{F}$ is locally free of finite rank, we obtain an invertible sheaf $\det \mathcal{F}$ on X (on each connected component X_i of X, $\mathcal{F}$ is of constant rank r_i, and we set $(\det \mathcal{F})|_{X_i} = \wedge^{r_i}(\mathcal{F}|_{X_i})$).

Corollary 4.2. Let X be a scheme. Let $\mathcal{F}$ be a locally free sheaf on X.

(a) For any morphism of schemes $p : Y \to X$, we have a canonical isomorphism $\det(p^*\mathcal{F}) \simeq p^*(\det \mathcal{F})$.

(b) Let $0 \to \mathcal{E} \to \mathcal{F} \to \mathcal{G} \to 0$ be an exact sequence of locally free sheaves of finite rank on X. Then we have a canonical isomorphism

$$\det \mathcal{E} \otimes_{\mathcal{O}_X} \det \mathcal{G} \simeq \det \mathcal{F}.$$

(c) We have a canonical isomorphism $\det(\mathcal{F}^\vee) \simeq (\det \mathcal{F})^\vee$.

(d) The isomorphisms in (b) and (c) commute with base change.

Proof Let us show (b), the rest can be deduced in the same manner. We cover X with open subsets $\{X_i\}_{i\in I}$ such that the sheaves in question are all free on each X_i. The lemma above gives an isomorphism

$$\psi_i : \det(\mathcal{E}|_{X_i}) \otimes \det(\mathcal{G}|_{X_i}) \to \det(\mathcal{F}|_{X_i})$$

in terms of a suitable choice of bases of the sheaves. On the intersections $X_i \cap X_j$, we have $\psi_i|_{X_i\cap X_j} = \psi_j|_{X_i\cap X_j}$ by the independence of ψ with respect to the choice of the bases. We therefore obtain an isomorphism $\det \mathcal{E} \otimes \det \mathcal{G} \simeq \det \mathcal{F}$. □

6.4.2   Canonical sheaf

Definition 4.3. Let $f : X \to Y$ be a morphism of schemes. For any $r \geq 1$, we call the quasi-coherent sheaf $\Omega^r_{X/Y} := \wedge^r \Omega^1_{X/Y}$ (do not confuse this with the direct product of r copies of $\Omega^1_{X/Y}$) the *sheaf of differentials of order r*. If Y is locally Noetherian and f is smooth, then $\Omega^1_{X/Y}$ is locally free on X (Proposition 2.5). We then have the notion of the determinant $\det \Omega^1_{X/Y}$.

Example 4.4. Let $X = \operatorname{Spec} A[T_1, \dots, T_n]$. Then $\det \Omega^1_{X/A} = \Omega^n_{X/A}$ is free over $\mathcal{O}_X$, generated by $dT_1 \wedge \cdots \wedge dT_n$.

Lemma 4.5. *Let $f : X \to Y$ be a morphism of finite type of locally Noetherian schemes. Let us suppose that we have a commutative diagram*

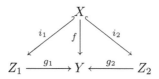

where g_1, g_2 are smooth, and i_1, i_2 are regular immersions. Then we have a canonical isomorphism of sheaves on X:

$$\phi_{12} : \det(\mathcal{C}_{X/Z_1})^\vee \otimes i_1^*(\det \Omega^1_{Z_1/Y}) \simeq \det(\mathcal{C}_{X/Z_2})^\vee \otimes i_2^*(\det \Omega^1_{Z_2/Y}).$$

Proof Let $W = Z_1 \times_Y Z_2$, and $h = (i_1, i_2) : X \to W$. By Corollary 3.22, we have the following canonical exact sequences:

$$0 \to \mathcal{C}_{X/Z_1} \to \mathcal{C}_{X/W} \to h^*\Omega^1_{W/Z_1} \to 0,$$

$$0 \to \mathcal{C}_{X/Z_2} \to \mathcal{C}_{X/W} \to h^*\Omega^1_{W/Z_2} \to 0.$$

Now $\Omega^1_{W/Z_1} = p_2^*\Omega^1_{Z_2/Y}$, where $p_2 : W \to Z_2$ is the second projection. Hence $h^*\Omega^1_{W/Z_1} = i_2^*\Omega^1_{Z_2/Y}$. In the same manner, $h^*\Omega^1_{W/Z_2} = i_1^*\Omega^1_{Z_1/Y}$. We now only have to apply Corollary 4.2 to the two exact sequences above. □

Remark 4.6. Let $X \to Z_3 \to Y$ be a third decomposition of f into a regular immersion followed by a smooth morphism. Then we have two other isomorphisms ϕ_{13} and ϕ_{23}. We can show that we have $\phi_{13} = \phi_{23} \circ \phi_{12}$ (Exercise 4.5).

Definition 4.7. Let Y be a locally Noetherian scheme, and let $f : X \to Y$ be a quasi-projective l.c.i. Let $i : X \to Z$ be an immersion into a scheme Z that is smooth over Y (e.g., $Z = \mathbb{P}^n_Y$ for a suitable n). We define the *canonical sheaf of* $X \to Y$ to be the invertible sheaf

$$\omega_{X/Y} := \det(\mathcal{C}_{X/Z})^\vee \otimes_{\mathcal{O}_X} i^*(\det \Omega^1_{Z/Y}).$$

By Lemma 4.5, this sheaf is independent of the choice of the decomposition $X \to Z \to Y$, up to isomorphisms. See also Exercises 4.5 and 4.6. It is clear from the definition that for any open subset U of X, we have $\omega_{X/Y}|_U = \omega_{U/Y}$. If $f : X \to Y$ is a regular immersion, then $\omega_{X/Y} = \det(\mathcal{C}_{X/Y})^\vee$. Let us also note that $\omega_{X/X} = \mathcal{O}_X$, while $\Omega^1_{X/X} = 0$.

Example 4.8. Let A be a Noetherian ring, $B = A[T_1, \dots, T_n]/I$, where I is an ideal generated by a regular sequence $F_1, \dots, F_r$. Then $\omega_{\operatorname{Spec} B / \operatorname{Spec} A}$ is generated by

$$(\overline{F}_1 \wedge \cdots \wedge \overline{F}_r)^\vee \otimes ((dT_1 \wedge \cdots \wedge dT_n) \otimes 1_B),$$

where $\overline{F}_i$ denotes the image of F_i in I/I^2. Actually, the $\overline{F}_i$ form a basis of I/I^2 over B (Lemma 3.6).

Theorem 4.9. *Let* $f : X \to Y$, $g : Y \to Z$ *be quasi-projective l.c.i.s.*

(a) (Adjunction formula) *We have a canonical isomorphism*

$$\omega_{X/Z} \simeq \omega_{X/Y} \otimes_{\mathcal{O}_X} f^* \omega_{Y/Z}.$$

(b) (Base change) *Let* $Y' \to Y$ *be a morphism. Let* $X' := X \times_Y Y'$ *and let* $p : X' \to X$ *be the first projection. If either* $Y' \to Y$ *or* $X \to Y$ *is flat, then* $X' \to Y'$ *is an l.c.i., and we have a canonical isomorphism*

$$\omega_{X'/Y'} \simeq p^* \omega_{X/Y}.$$

Proof (a) Let us choose two immersions $i : X \to U = \mathbb{P}^n_Y$ and $j : Y \to W = \mathbb{P}^n_Z$. They are both regular immersions by Corollary 3.22. Let $V = \mathbb{P}^n_W$. Then $U = Y \times_Z W$ and we have a commutative diagram

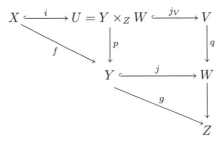

where the vertical arrows are the canonical projections. Using Proposition 3.11 (a) and (d) and Corollary 4.2(b), we have

$$\omega_{X/V} = \omega_{X/U} \otimes i^* \omega_{U/V} = \omega_{X/U} \otimes i^* p^* \omega_{Y/W} = \omega_{X/U} \otimes f^* \omega_{Y/W}. \qquad (4.9)$$

In addition, Exercise 2.10(a) shows that $\omega_{V/Z} = \omega_{V/W} \otimes q^* \omega_{W/Z}$. Now we have $j_V^* \Omega^1_{V/W} = \Omega^1_{U/Y}$ by the base change $j : Y \to W$, and hence $j_V^* \omega_{V/W} = \omega_{U/Y}$, which implies that

$$(j_V \circ i)^* (\omega_{V/Z}) = i^* \omega_{U/Y} \otimes (j_V \circ i)^* q^* \omega_{W/Z} = i^* \omega_{U/Y} \otimes f^* j^* \omega_{W/Z}. \quad (4.10)$$

Taking the tensor product of identities (4.9) and (4.10), we obtain (a).

(b) The $Y' \to Y$ flat case results from Propositions 3.11(d) and 1.24(a). Let us suppose $X \to Y$ is flat. We decompose $X \to Y$ locally into a closed immersion $X \to W := \mathbb{A}_Y^n$ followed by the projection onto Y. Let $x \in X$ and $y = f(x)$, and let m be the codimension of X in W at x. In the proof of Corollary 3.24, we have seen that $X_y \to W_y$ is also a regular immersion of codimension m at x. It easily follows that at every point $x' \in X'$ above x, $X' \to W' := W \times_Y Y'$ is a regular immersion of codimension m. The surjective homomorphism $p^* \mathcal{C}_{X/W} \to \mathcal{C}_{X'/W'}$ of Proposition 3.11(c) is bijective because the two terms are locally free of the same rank. As $\omega_{W/Y}$ obviously commutes with any base change, we have shown the isomorphism of (b).   □

By definition, $\omega_{X/Y} = \wedge^d \Omega^1_{X/Y}$ for a smooth morphism of relative dimension d (see the definition below). It is natural to compare the two sheaves in a more general situation.

Definition 4.10. Let $f : X \to Y$ be a morphism of finite type. Let $x \in X$. We call the integer $\dim_x X_{f(x)}$ the *relative dimension* of f at x. Corollary 4.3.14 computes the relative dimension in a particular case. For a smooth morphism, the relative dimension is locally constant because it coincides with the rank of $\Omega^1_{X/Y}$ at x.

Definition 4.11. Let $f : X \to Y$ be an l.c.i. with Y locally Noetherian. Let $x \in X$. Let us consider a decomposition of f in a neighborhood $U \ni x$ into a regular immersion $i : U \to Z$ of codimension e, followed by a smooth morphism $Z \to Y$ of relative dimension d at $i(x)$. We call the integer $d - e$ the *virtual relative dimension* of f at x. The exact sequences in the proof of Lemma 4.5 show that this integer is independent of the choice of the decomposition into $U \to Z \to Y$. The relative virtual dimension is smaller than or equal to the relative dimension (Theorem 2.5.15). Let us suppose f is flat. Let $y = f(x)$. Then $X_y \to \operatorname{Spec} k(y)$ is an l.c.i. (Corollary 3.24), and the relative dimension at x coincides with the virtual relative dimension. This results from Theorem 2.5.15 and Corollary 2.5.17.

Lemma 4.12. *Let A be a Noetherian ring, $B = A[T_1, \ldots, T_n]/I$, where I is an ideal generated by a regular sequence $F_1, \ldots, F_r$. Let t_i denote the image of T_i in B. Let us suppose $n \geq r$. Then there exists a canonical homomorphism*

$$c : \Omega^{n-r}_{B/A} \to \omega_{B/A} := H^0(\operatorname{Spec} B, \omega_{\operatorname{Spec} B / \operatorname{Spec} A})$$

such that for any (ordered) subset $S = \{j_{r+1}, \ldots, j_n\}$ of $\{1, \ldots, n\}$, we have

$$c\left(dt_{j_{r+1}} \wedge \cdots \wedge dt_{j_n}\right) = \Delta_S \cdot (\overline{F}_1 \wedge \cdots \wedge \overline{F}_r)^\vee \otimes ((dT_1 \wedge \cdots \wedge dT_n) \otimes 1_B),$$

where Δ_S is the determinant of the Jacobian matrix $(\partial F_i / \partial T_j)_{i,j}$ where we have removed the columns $j_{r+1}, \ldots, j_n$, and where $\overline{F}_i$ is the image of F_i in I/I^2. Moreover, if $\operatorname{Spec} B \to \operatorname{Spec} A$ is smooth, then c is an isomorphism.

Proof Let $C = A[T_1, \ldots, T_n]$, $M = \Omega^1_{C/A}$, and let N be the submodule of M generated by the dF_i. Then $M/N \simeq \Omega^1_{B/A}$.

We have a commutative diagram

$$\begin{array}{ccccc} \wedge^r N \otimes \wedge^{n-r} M & \xrightarrow{\phi} & \wedge^r M \otimes \wedge^{n-r} M & \xrightarrow{\rho} & \wedge^n M \\ \downarrow & & & \nearrow \psi & \\ \wedge^r N \otimes \wedge^{n-r}(M/N) & & & & \end{array}$$

where ρ is defined by $\rho((x_1 \wedge \cdots \wedge x_r) \otimes (x_{r+1} \wedge \cdots \wedge x_n)) = x_1 \wedge \cdots \wedge x_n$, ψ is defined in a similar way, and the other homomorphisms are canonical. The fact that $\rho \circ \phi$ factors through ψ can easily be verified owing to the fact that $\wedge^{r+1} N = 0$. By tensoring ψ by B and by composing with $\delta : I/I^2 \to N \otimes_C B$, we obtain a canonical homomorphism

$$\wedge^r(I/I^2) \otimes_B \wedge^{n-r}(\Omega^1_{B/A}) \to \wedge^n \Omega^1_{C/A} \otimes_C B.$$

By tensoring with the dual of $\wedge^r(I/I^2)$, we obtain a homomorphism

$$c \;:\; \wedge^{n-r}(\Omega^1_{B/A}) \to \omega_{B/A}.$$

By definition of ρ, we immediately have

$$\rho\left((dF_1 \wedge \cdots \wedge dF_r) \otimes (dT_{j_{r+1}} \wedge \cdots \wedge dT_{j_n})\right) =$$
$$dF_1 \wedge \cdots \wedge dF_r \wedge dT_{j_{r+1}} \wedge \cdots \wedge dT_{j_n} = \Delta_S \cdot (dT_1 \wedge \cdots \wedge dT_n).$$

This shows the equality concerning $c(dt_{j_{r+1}} \wedge \cdots \wedge dt_{j_n})$. Finally, if $\operatorname{Spec} B \to \operatorname{Spec} A$ is smooth, then the Jacobian criterion (Theorem 4.2.19) immediately implies that $\{\Delta_S\}_S$ generates the unit ideal in B. Hence c is surjective by Example 4.8. As it is a homomorphism of locally free modules of same rank 1, it is therefore an isomorphism. Note that the construction of c is independent of the choice of the F_i and of the T_j. $\qquad\square$

Corollary 4.13. *Let Y be a locally Noetherian scheme, and let $f : X \to Y$ be a quasi-projective l.c.i. of constant virtual relative dimension equal to $d \geq 0$. We then have a canonical homomorphism $c_{X/Y} : \Omega^d_{X/Y} \to \omega_{X/Y}$. Moreover, $c_{X/Y}$ coincides with the identity on the (possibly empty) smooth locus of f.*

Lemma 4.12 makes it possible to have an explicit expression for the sheaf $\omega_{X/Y}$ in certain situations. Let $f : X \to Y$ be a quasi-projective l.c.i. of locally Noetherian integral schemes. We suppose that f is dominant and that the function field extension $K(X)/K(Y)$ is separable (Definition 1.14), in order that f be smooth at the generic point ξ of X (Exercise 2.9). As $\omega_{X/Y}$ is invertible and X is integral, $\omega_{X/Y}$ is canonically a subsheaf of the constant sheaf $\omega_{X/Y,\xi} = \det \Omega^1_{K(X)/K(Y)}$.

Corollary 4.14. *Let* $Y = \operatorname{Spec} A$ *be a Noetherian integral scheme, and let* X *be an integral closed subscheme of* $Z = \operatorname{Spec} A[T_1, \ldots, T_n]$ *defined by an ideal generated by a regular sequence* $F_1, \ldots, F_r$ *with* $r \leq n$. *Let us suppose that*

$$\Delta := \det \left(\frac{\partial F_i}{\partial T_j} \right)_{1 \leq i,j \leq r}$$

is non-zero in $K(X)$. *Let* ξ *be the generic point of* X.

(a) *Let* t_i *be the image of* T_i *in* $\mathcal{O}_X(X)$. *Then*

$$\omega_{X/Y,\xi} = (dt_{r+1} \wedge \cdots \wedge dt_n)\mathcal{O}_{X,\xi}.$$

(b) *As a subsheaf of* $\omega_{X/Y,\xi}$, *we have*

$$\omega_{X/Y} = \Delta^{-1} \cdot (dt_{r+1} \wedge \cdots \wedge dt_n)\mathcal{O}_X.$$

Example 4.15. *Let* A *be a Noetherian integral domain with field of fractions* K, *and let*

$$X = \operatorname{Spec} A[x, y]/(y^2 + Q(x)y - P(x)), \quad P, Q \in A[x].$$

Let us suppose that $Q(x) \neq 0$ *if* $\operatorname{char}(K) = 2$. *Then* $K(X)$ *is separable over* K, *and* $\omega_{X/\operatorname{Spec} A}$ *is generated by* $dx/(2y + Q(x))$.

Let us conclude with the description of the canonical sheaf on the projective space.

Lemma 4.16. *Let* $Y = \operatorname{Spec} \mathbb{Z}$, *and let* $X = \mathbb{P}_Y^n$. *Then we have*

$$\omega_{X/Y} \simeq \mathcal{O}_X(-n-1).$$

Proof Let us write $X = \operatorname{Proj} \mathbb{Z}[T_0, \ldots, T_n]$. Let $t_i = T_i/T_0 \in K(X)$ for each $0 \leq i \leq n$. Then $K(X) = \mathbb{Q}(t_1, \ldots, t_n)$. Hence $\Omega^1_{X/Y,\xi}$, where ξ is the generic point of X, is free over $K(X)$ with basis $dt_1, \ldots, dt_n$. Consequently, $\omega_0 := dt_1 \wedge \cdots \wedge dt_n$ is a basis of $\omega_{X/Y,\xi}$ over $K(X)$. Let $U_i = D_+(T_i)$. For any $j \neq i$, we have

$$d(T_j/T_i) = d(t_j/t_i) = -t_j t_i^{-2} dt_i + t_i^{-1} dt_j.$$

Taking the exterior product of the $d(T_j/T_i)$ for $j \neq i$, we deduce from this that $\omega_{U_i/Y}$ has $\omega_i := t_i^{-n-1}\omega_0$ as a basis over $\mathcal{O}_X|_{U_i}$. This shows (see Example 5.1.19) that

$$\omega_{X/Y} = \omega_0 T_0^{n+1} \mathcal{O}_X(-n-1) \simeq \mathcal{O}_X(-n-1),$$

whence the lemma. $\square$

Corollary 4.17. *Let* Y *be a locally Noetherian scheme; then for any* $n \geq 1$, *we have* $\omega_{\mathbb{P}_Y^n/Y} \simeq \mathcal{O}_{\mathbb{P}_Y^n}(-n-1)$.

Proof This results from the lemma above, Proposition 1.24(a), and Corollary 4.2(a). □

6.4.3 Grothendieck duality

Let $f : X \to Y$ be a proper morphism to a locally Noetherian scheme Y, with fibers of dimension $\leq r$. Let $\mathcal{F}, \mathcal{G}$ be quasi-coherent sheaves on X. For any affine open subset V of Y, each homomorphism $\phi : \mathcal{F}|_{f^{-1}(V)} \to \mathcal{G}|_{f^{-1}(V)}$ induces a homomorphism

$$H^r(f^{-1}(V), \mathcal{F}|_{f^{-1}(V)}) \xrightarrow{H^r(\phi)} H^r(f^{-1}(V), \mathcal{G}|_{f^{-1}(V)}).$$

This defines a canonical bilinear map

$$f_* \mathcal{H}om_{\mathcal{O}_X}(\mathcal{F}, \mathcal{G}) \times R^r f_* \mathcal{F} \to R^r f_* \mathcal{G}. \tag{4.11}$$

Definition 4.18. We define the *dualizing sheaf* (or *r-dualizing sheaf*) for f to be a quasi-coherent sheaf ω_f on X, endowed with a homomorphism of $\mathcal{O}_Y$-modules (the *trace*)

$$\mathrm{tr}_f : R^r f_* \omega_f \to \mathcal{O}_Y$$

such that for any quasi-coherent sheaf $\mathcal{F}$ on X, the natural bilinear map

$$f_* \mathcal{H}om_{\mathcal{O}_X}(\mathcal{F}, \omega_f) \times R^r f_* \mathcal{F} \to R^r f_* \omega_f \xrightarrow{\mathrm{tr}_f} \mathcal{O}_Y$$

induces an isomorphism

$$f_* \mathcal{H}om_{\mathcal{O}_X}(\mathcal{F}, \omega_f) \simeq \mathcal{H}om_{\mathcal{O}_Y}(R^r f_* \mathcal{F}, \mathcal{O}_Y). \tag{4.12}$$

Lemma 4.19. *Let $f : X \to Y$ be as above. Let us suppose that a dualizing sheaf $(\omega_f, \mathrm{tr}_f)$ exists. Then it is unique in the following sense: if $(\omega'_f, \mathrm{tr}'_f)$ is another dualizing sheaf for f, then there exists a unique isomorphism $\rho : \omega'_f \to \omega_f$ such that $\mathrm{tr}'_f = \mathrm{tr}_f \circ \rho_r$, where ρ_r denotes the canonical homomorphism $R^r f_* \omega'_f \to R^r f_* \omega_f$ induced by ρ.*

Proof Indeed, by taking $\mathcal{F} = \omega'_f$ in (4.12), we obtain an isomorphism

$$\mathrm{Hom}_{\mathcal{O}_X}(\omega'_f, \omega_f) \simeq \mathrm{Hom}_{\mathcal{O}_Y}(R^r f_* \omega'_f, \mathcal{O}_Y).$$

Then ρ is necessarily the inverse image of tr'_f by this isomorphism (hence ρ is unique). By symmetry we have a homomorphism $\rho' : \omega_f \to \omega'_f$ such that $\mathrm{tr}'_f = \mathrm{tr}_f \circ \rho'_r$. Then $\rho \circ \rho' : \omega_f \to \omega_f$ verifies $\mathrm{tr}_f = \mathrm{tr}_f \circ (\rho \circ \rho')_r$. By the uniqueness, this implies that $\rho \circ \rho'$ is the identity, and hence ρ is an isomorphism. □

Remark 4.20. Taking once more the definition of $R^r f_*$, on a scheme with affine base $Y = \mathrm{Spec}\, A$, isomorphism (4.12) translates into an isomorphism

$$\mathrm{Hom}_{\mathcal{O}_X}(\mathcal{F}, \omega_f) \simeq H^r(X, \mathcal{F})^{\vee} \tag{4.13}$$

(the right-hand side is the dual seen as an A-module).

Remark 4.21. Let $\mathcal{F}, \mathcal{G}$ be quasi-coherent sheaves on a scheme X. Then we have a canonical homomorphism

$$\mathcal{F}^{\vee} \otimes_{\mathcal{O}_X} \mathcal{G} \to \mathcal{H}om_{\mathcal{O}_X}(\mathcal{F}, \mathcal{G}) \tag{4.14}$$

which on any affine open subset U sends $\phi \otimes y \in \mathcal{F}^{\vee}(U) \otimes \mathcal{G}(U)$ to the homomorphism $x \mapsto \phi(x)y$. It is easy to verify that it is an isomorphism if $\mathcal{F}$ or $\mathcal{G}$ is locally free. In particular, we will then have a canonical isomorphism

$$f_*(\mathcal{F}^{\vee} \otimes \mathcal{G}) \simeq f_* \mathcal{H}om_{\mathcal{O}_X}(\mathcal{F}, \mathcal{G}). \tag{4.15}$$

When Y is affine, then isomorphism (4.13) becomes

$$H^0(X, \mathcal{F}^{\vee} \otimes \omega_f) \simeq H^r(X, \mathcal{F})^{\vee}. \tag{4.16}$$

Proposition 4.22. *Let* Y *be a locally Noetherian scheme, let* $X = \mathbb{P}_Y^d$, *and let* $f : X \to Y$ *be the canonical morphism. Then the* (*d*-)*dualizing sheaf* ω_f *exists and is isomorphic to* $\omega_{X/Y}$.

Proof By Corollary 4.17, $\omega_{X/Y} \simeq \mathcal{O}_X(-d-1)$. We know that there exists a canonical isomorphism

$$\mathrm{tr}_f : R^d f_* \mathcal{O}_X(-d-1) \simeq f_* \mathcal{O}_X = \mathcal{O}_Y$$

(see Lemma 5.3.1, where this result is shown above any affine open subset V of Y; these local isomorphisms coincide on the intersections because they all come from the isomorphism over $\mathrm{Spec}\,\mathbb{Z}$, whence a global isomorphism). Let us therefore take $\omega_f = \mathcal{O}_X(-d-1)$ and let us show isomorphism (4.12).

We may assume that $Y = \mathrm{Spec}\,A$ is affine. We leave it to the reader to verify the commutativity of the following diagram:

$$
\begin{array}{ccc}
H^0(X, \mathcal{F}^{\vee} \otimes \omega_f) \times H^d(X, \mathcal{F}) & \xrightarrow{c} & H^d(X, \omega_f) \\
{\scriptstyle \rho \times \mathrm{Id}} \downarrow & & \| \\
\mathrm{Hom}_{\mathcal{O}_X}(\mathcal{F}, \omega_f) \times H^d(X, \mathcal{F}) & \longrightarrow & H^d(X, \omega_f)
\end{array}
$$

where the first horizontal arrow c is the cup product (Exercise 5.2.17), the second horizontal arrow comes from the bilinear map (4.11), and where the vertical arrow ρ comes from (4.14). Now we have seen in Exercise 5.3.2(c) that if $\mathcal{F}$ is of the form $\mathcal{O}_X(m)$, $m \in \mathbb{Z}$, then c is a non-degenerate bilinear form. This shows isomorphism (4.12) in that case. The case of an arbitrary quasi-coherent sheaf results from Lemma 4.23 below. $\square$

Lemma 4.23. *Let* $f : X \to \mathrm{Spec}\,A$ *be a projective scheme over an affine Noetherian scheme, with fibers of dimension* $\leq r$. *Let us fix a very ample sheaf* $\mathcal{O}_X(1)$. *Let* ω_f *be a quasi-coherent sheaf on* X. *Let us suppose that we have a trace homomorphism* $\mathrm{tr}_f : H^r(X, \omega_f) \to A$ *and that the homomorphism*

$$\psi_{\mathcal{F}} : \mathrm{Hom}_{\mathcal{O}_X}(\mathcal{F}, \omega_f) \to H^r(X, \mathcal{F})^{\vee}$$

induced by tr_f *is an isomorphism for every sheaf* $\mathcal{F}$ *of the form* $\mathcal{O}_X(m)$, $m \in \mathbb{Z}$. *Then* $\psi_{\mathcal{F}}$ *is an isomorphism for every quasi-coherent sheaf* $\mathcal{F}$.

Proof Let $\mathcal{F}$ be a quasi-coherent sheaf on X. Then we have an exact sequence

$$0 \to \mathcal{G} \to \mathcal{L} \to \mathcal{F} \to 0,$$

where $\mathcal{L}$ is a direct sum of sheaves of the form $\mathcal{O}_X(m)$ (Corollary 5.1.28). In particular, $\psi_{\mathcal{L}}$ is an isomorphism. We then have a canonical commutative diagram

$$
\begin{array}{ccccccc}
0 & \longrightarrow & \mathrm{Hom}_{\mathcal{O}_X}(\mathcal{F}, \omega_f) & \longrightarrow & \mathrm{Hom}_{\mathcal{O}_X}(\mathcal{L}, \omega_f) & \longrightarrow & \mathrm{Hom}_{\mathcal{O}_X}(\mathcal{G}, \omega_f) \\
& & \downarrow{\psi_{\mathcal{F}}} & & \downarrow{\psi_{\mathcal{L}}} & & \downarrow{\psi_{\mathcal{G}}} \\
0 & \longrightarrow & H^r(X, \mathcal{F})^{\vee} & \longrightarrow & H^r(X, \mathcal{L})^{\vee} & \longrightarrow & H^r(X, \mathcal{G})^{\vee}
\end{array}
$$

because $H^{r+1}(X, \mathcal{F}) = 0$ (Proposition 5.2.34). This implies that $\psi_{\mathcal{F}}$ is injective for every quasi-coherent sheaf. Applying this result to $\mathcal{G}$, we have the injectivity of $\psi_{\mathcal{G}}$. The commutative diagram then shows that $\psi_{\mathcal{F}}$ is also surjective. $\square$

Proposition 4.22 establishes the existence of the dualizing sheaf for projective spaces. Let us now consider the case of finite morphisms.

Lemma 4.24. *Let $\rho : A \to B$ be a ring homomorphism. Let M be an A-module, and N a B-module.*

(a) *The A-module $\rho^! M := \mathrm{Hom}_A(B, M)$ also admits the structure of a B-module defined by $(b \cdot \phi)(b') = \phi(bb')$ for every $\phi \in \rho^! M$ and $b, b' \in B$.*

(b) *Let $\mathrm{tr}_M : \rho^! M \to M$ denote the A-linear homomorphism 'evaluation in 1': $\theta \mapsto \theta(1)$. Then tr_M induces a canonical isomorphism*

$$\lambda : \mathrm{Hom}_B(N, \rho^! M) \to \mathrm{Hom}_A(N, M).$$

Proof (a) follows immediately. (b) By construction, we have $\lambda(\psi)(bx) = \psi(x)(b)$ for every $\psi \in \mathrm{Hom}_B(N, \rho^! M)$, $x \in N$, and $b \in B$. We immediately deduce from this that λ is an isomorphism. $\square$

Proposition 4.25. *Let $f : X \to Y$ be a finite morphism of locally Noetherian schemes. Let $\mathcal{F}$ (resp. $\mathcal{G}$) be a quasi-coherent sheaf on X (resp. on Y). Let us set $f^! \mathcal{G} = \mathcal{H}om_{\mathcal{O}_Y}(f_* \mathcal{O}_X, \mathcal{G})$.*

(a) *The sheaf $f^! \mathcal{G}$ is canonically endowed with the structure of a quasi-coherent $\mathcal{O}_X$-module and with a homomorphism $\mathrm{tr}_{\mathcal{G}} : f_*(f^! \mathcal{G}) \to \mathcal{G}$.*

(b) *The canonical homomorphism $f_* \mathcal{H}om_{\mathcal{O}_X}(\mathcal{F}, f^! \mathcal{G}) \to \mathcal{H}om_{\mathcal{O}_Y}(f_* \mathcal{F}, \mathcal{G})$ induced by $\mathrm{tr}_{\mathcal{G}}$ is an isomorphism.*

(c) *Let $\omega_f = f^! \mathcal{O}_Y$ and $\mathrm{tr}_f = \mathrm{tr}_{\mathcal{O}_Y}$. Then $(\omega_f, \mathrm{tr}_f)$ is the (0-)dualizing sheaf for f.*

Proof By Proposition 5.1.14(c), $f_* \mathcal{O}_X$ is a coherent $\mathcal{O}_Y$-module. Hence $\mathcal{H}om_{\mathcal{O}_Y}(f_* \mathcal{O}_X, \mathcal{G})$ is a quasi-coherent $\mathcal{O}_Y$-module (Exercise 5.1.6). Then (a) and (b) follow from the lemma above, and (c) is a consequence of (a) and (b). $\square$

Lemma 4.26. *Let Y be a locally Noetherian scheme. Let $f : X \to Y$ be a projective morphism with fibers of dimension $\leq r$. Let us suppose that f decomposes into a finite morphism $\pi : X \to Z$ followed by a projective morphism $g : Z \to Y$ with fibers of dimension $\leq r$ and that g admits a r-dualizing sheaf $(\omega_g, \mathrm{tr}_g)$. Then the following properties are true.*

(a) *There exists an r-dualizing sheaf $(\omega_f, \mathrm{tr}_f)$ for f with $\omega_f = \pi^! \omega_g$.*

(b) *If, moreover, π is flat or if ω_g is locally free, then $\omega_f = \omega_\pi \otimes_{\mathcal{O}_X} \pi^* \omega_g$.*

Proof We have a trace homomorphism $\mathrm{tr}_{\omega_g} : \pi_* \pi^! \omega_g \to \omega_g$. Thus by taking $R^r g_*$ and by composing with tr_g, a homomorphism $\mathrm{tr}_f : R^r f_*(\pi^! \omega_g) \to \mathcal{O}_Y$. Let $\mathcal{F}$ be a quasi-coherent sheaf on X. Then $R^r f_* \mathcal{F} = R^r g_*(\pi_* \mathcal{F})$ (use Exercise 5.2.3). Now

$$f_* \mathcal{H}om_{\mathcal{O}_X}(\mathcal{F}, \pi^! \omega_g) \simeq g_* \mathcal{H}om_{\mathcal{O}_Z}(\pi_* \mathcal{F}, \omega_g) \simeq \mathcal{H}om_{\mathcal{O}_Y}(R^r g_*(\pi_* \mathcal{F}), \mathcal{O}_Y).$$

We verify that the composition of the two isomorphisms is the homomorphism induced by tr_f. This proves (a). Finally, (b) results from the definition of $\pi^! \omega_g$ and the isomorphism (4.15) because either $\pi_* \mathcal{O}_X$ or ω_g is locally free on Z. □

Lemma 4.27. *Let k be an infinite field. Let X be a projective variety of dimension r over k. Then there exists a finite morphism $X \to \mathbb{P}^r_k$.*

Proof We embed X in a projective space $X \to \mathbb{P}^n_k$. Let us suppose that $r < n$ (otherwise there is nothing to show). Then there exists a point $z \in \mathbb{P}^n_k(k) \setminus X$. Let $p : \mathbb{P}^n_k \setminus \{z\} \to \mathbb{P}^{n-1}_k$ be a projection with center z (Exercise 3.1.11). Then $p|_X : X \to \mathbb{P}^{n-1}_k$ is a quasi-finite morphism (indeed, every fiber $p^{-1}(y)$ is of dimension 1 and not contained in X), and hence finite (Corollary 4.4.7). If $r = n-1$, we are done. Otherwise, as $p(X)$ is of dimension r, we can continue the reasoning with $p(X)$ (the subvariety structure does not intervene) and we obtain a finite morphism $X \to \mathbb{P}^r_k$ after a finite number of steps. □

Remark 4.28. Let $Y = \mathrm{Spec}\, A$ be an affine scheme. For any section $\sigma : Y \to \mathbb{P}^n_Y$, we can define a projection $\mathbb{P}^n_Y \setminus \sigma(Y) \to \mathbb{P}^{n-1}_Y$ centered in $\sigma(Y)$ in a way similar to that of Exercise 3.1.11. The lemma above then remains valid when replacing k by a local ring A with an infinite residue field.

Corollary 4.29. *Let $f : X \to \mathrm{Spec}\, k$ be a projective morphism over an infinite field k, of dimension $\leq r$. Then there exists a coherent r-dualizing sheaf for f.*

Proof This is an immediate consequence of Lemmas 4.26 and 4.27. □

Remark 4.30. We have in fact a much more general statement:

Let $f : X \to Y$ be a projective morphism to a locally Noetherian scheme Y, with fibers of dimension $\leq r$. Then the r-dualizing sheaf ω_f exists.

We will not show this theorem. See [43], Proposition III.7.5, for the case of projective varieties, and [51], Theorem 4, for the general case (we take $\omega_f := f^! \mathcal{O}_Y$). Let us note that in [51], Definition 6, the notion of dualizing sheaf imposes stronger conditions. The proof of [43] uses the Ext functors, while [51] uses the theory of adjoint functors. See also [44] and [24].

However, if Y has an infinite residue field (e.g., if Y is an algebraic variety over an infinite field), then we can show the result in an elementary way with the help of Lemma 4.26 and Remark 4.28.

Remark 4.31. The theory of duality which we have presented here is the duality of order 0. The higher order duality concerns, more generally, the sheaves $R^{r-m}f_*\mathcal{F}$, $0 \le m \le r$. See [43], III.7, or [51]. The higher order duality is needed to show the following result, which will be used to prove the Riemann–Roch theorem in the following chapter:

Theorem 4.32. *Let Y be a locally Noetherian scheme, and let $f : X \to Y$ be a flat projective l.c.i. of relative dimension r. Then the r-dualizing sheaf ω_f is isomorphic to $\omega_{X/Y}$. In particular, if f is smooth, then $\omega_f \simeq \Omega^r_{X/Y}$.*

Proof   We apply [51], Corollary 19, and Proposition 4.22 above. See also Exercise 4.11 for the case of global complete intersections.   □

Exercises

4.1. Let Y be a locally Noetherian scheme. Let $f : X \to Y$ be a morphism of finite type. We suppose that there exists an integer $r \ge 1$ such that $\Omega^r_{X/Y} = 0$.

   (a) Let M be a finitely generated module over a local ring $(A, \mathfrak{m})$. Let us suppose that $\wedge^r M = 0$. Show that $\wedge^r(M/\mathfrak{m}M) = 0$ (exterior product as an $A/\mathfrak{m}$-module). Deduce from this that $M/\mathfrak{m}M$, and therefore M, are generated by $r - 1$ elements.

   (b) Let B be a finitely generated algebra over a Noetherian ring A. Let $f_1, \ldots, f_n \in B$ be such that $df_1, \ldots, df_n$ generate $\Omega^1_{B/A}$ as a B-module. Let $\varphi : A[T_1, \ldots, T_n] \to B$ be the homomorphism of A-algebras defined by $\varphi(T_i) = f_i$. Show that $\Omega^1_{B/A[T_1,\ldots,T_n]} = 0$.

   (c) Show that locally, X admits an immersion into $\mathbb{A}^r_Y$. (Use Corollary 2.3 and Proposition 4.4.11.)

4.2. Let k be a field, and let $F = x^n + y^n + z^n \in k[x, y, z]$, where $n \ge 1$ is prime to $\mathrm{char}(k)$. Determine $H^0(X, \Omega^1_{X/k})$ for $X = V_+(F) \subset \mathbb{P}^2_k$.

4.3. Let $\mathcal{F}, \mathcal{G}$ be locally free sheaves of respective ranks m and n on a scheme X. Show that

$$\det(\mathcal{F} \otimes \mathcal{G}) \simeq (\det \mathcal{F})^{\otimes n} \otimes (\det \mathcal{G})^{\otimes m}.$$

4.4. Let $f : X \to \operatorname{Spec} k$ be a quasi-projective algebraic variety and an l.c.i. over a perfect field k. Let us moreover suppose that X is normal.

   (a) Let U be an affine open subset of X, $A = \mathcal{O}_X(U)$ and I an ideal of A. Show that any linear form $\phi : I \to A$ is a homothety $\alpha \mapsto \theta\alpha$ for some

$\theta \in \mathrm{Frac}(A)$ (if $I \neq 0$, take $\theta = \phi(\alpha)/\alpha$ for some arbitrary $\alpha \in I \backslash \{0\}$). Show that $\theta \in A_{\mathfrak{p}}$ for every prime ideal $\mathfrak{p}$ not containing I. Deduce from this that if I is of height ≥ 2, then $\theta \in A$ (Lemma 4.1.13) and that the inclusion $I \rightarrow A$ induces an isomorphism $A^{\vee} \rightarrow I^{\vee}$.

(b) Show that the homomorphism $c_{X/k} : \Omega^d_{X/k} \rightarrow \omega_{X/k}$ of Corollary 4.13 induces an isomorphism of the biduals $(\Omega^d_{X/k})^{\vee\vee} \simeq (\omega_{X/k})^{\vee\vee} = \omega_{X/k}$.

4.5. Let Y be a locally Noetherian scheme, let $f : X \rightarrow Y$ be an l.c.i. admitting a decomposition into a regular immersion $i : X \rightarrow Z$ followed by a smooth morphism $g : Z \rightarrow Y$ (e.g., if f is a quasi-projective l.c.i.). We will then temporarily write $\omega_{i,g} := \det(\mathcal{C}_{X/Z})^{\vee} \otimes i^* \det \Omega^1_{Z/Y}$. Let

$$X \xrightarrow{i_k} Z_k \xrightarrow{g_k} Y, \quad k = 1, 2, 3$$

be decompositions of f with i_k regular immersions, g_k smooth. With the notation of Lemma 4.5, we will show that $\phi_{13} = \phi_{23} \circ \phi_{12}$ and give a more precise definition of $\omega_{X/Y}$.

(a) Show, using the fibered products of the Z_k, that we can reduce to the case when for any pair $3 \geq l > k \geq 1$, there exists a smooth morphism $h_{lk} : Z_l \rightarrow Z_k$ such that $g_l = g_k \circ h_{lk}$ and $i_k = h_{lk} \circ i_l$. We will denote this relation by $(i_k, g_k) \geq (i_l, g_l)$. Let us make this hypothesis in (b)–(d).

(b) Let $x \in X$. We will still denote its images in the Z_k by x. Let $\alpha_1, \ldots, \alpha_q \in \mathcal{O}_{Z_1,x}$ be such that the $d\alpha_i \in \Omega^1_{Z_1/Y,x}$ form a basis of the latter over $\mathcal{O}_{Z_1,x}$. Show that there exist $\beta_1, \ldots, \beta_r \in \mathrm{Ker}\, i^{\#}_{2,x}$ such that $\{d\alpha_1, \ldots, d\alpha_q, d\beta_1, \ldots, d\beta_r\}$ is a basis of $\Omega^1_{Z_2/Y,x}$.

(c) Let $a_1, \ldots, a_e \in \mathrm{Ker}\, i^{\#}_{1,x}$ be such that their images in $\mathcal{C}_{X/Z_1,x}$ form a basis over $\mathcal{O}_{X,x}$. Show that the images of the f_i, β_j in $\mathcal{C}_{X/Z_2,x}$ form a basis and that the isomorphism $\phi_{12} : \omega_{i_1,g_1} \rightarrow \omega_{i_2,g_2}$ is given by

$$(\phi_{12})_x : a^* \otimes i_1^* d\alpha \mapsto (a^* \wedge \beta^*) \otimes i_2^*(d\alpha \wedge d\beta),$$

where $a^* = a_1^* \wedge \cdots \wedge a_e^*$ is the exterior product of the basis dual to the a_i; $d\alpha = d\alpha_1 \wedge \cdots \wedge d\alpha_q$; and β^* and $d\beta$ are defined in a similar fashion.

(d) Let $\gamma_1, \ldots, \gamma_s \in \mathrm{Ker}\, i^{\#}_{3,x}$ be such that the $d\gamma_i$ form a basis of $\Omega^1_{Z_3/Z_2,x}$. Show that

$$(\phi_{13})_x : a^* \otimes i_1^* d\alpha \mapsto (a^* \wedge \beta^* \wedge \gamma^*) \otimes i_3^*(d\alpha \wedge d\beta \wedge d\gamma),$$

$$(\phi_{23})_x : (a^* \wedge \beta^*) \otimes i_2^*(d\alpha \wedge d\beta) \mapsto ((a^* \wedge \beta^*) \wedge \gamma^*) \otimes i_3^*((d\alpha \wedge d\beta) \wedge d\gamma).$$

Deduce from this that $\phi_{13} = \phi_{23} \circ \phi_{12}$.

(e) Let $f : X \rightarrow Y$ be a quasi-projective l.c.i. Let $\mathcal{D}$ denote the set of pairs $d = (i, g)$ where $i : X \rightarrow Z$ is a regular immersion into a projective space $Z = \mathbb{P}^n_Y$ over Y and where $g : Z \to Y$ is the projection. Note that $\mathcal{D}$ is indeed a set. If $d_1 \geq d_2$, we consider the isomorphism $\phi_{12} : \omega_{d_1} \to \omega_{d_2}$. Show that $(\omega_d, \phi)_{d \in \mathcal{D}}$ forms a direct system. Forgetting Definition 4.7, we will set

$$\omega_{X/Y} := \varinjlim_{(i,g) \in \mathcal{D}} \omega_{i,g}.$$

Show that for any open subscheme U of X, $U \to Y$ is a quasi-projective l.c.i. and that $\omega_{X/Y}|_U = \omega_{U/Y}$.

4.6. Let Y be a locally Noetherian scheme, and let $f : X \to Y$ be a local complete intersection. We keep the notation of Exercise 4.5.

(a) Show that there exists a unique invertible sheaf $\omega_{X/Y}$ on X such that for any open subscheme U of X that is quasi-projective over Y (e.g., U affine with image in an affine open subscheme of Y), we have $\omega_{X/Y}|_U = \omega_{U/Y}$.

(b) Show that for any open subscheme U of X, we have $\omega_{X/Y}|_U = \omega_{U/Y}$.

(c) Let us suppose that f decomposes into a regular immersion $i : X \to Z$ followed by a smooth morphism $g : Z \to Y$. Show that

$$\omega_{X/Y} \simeq \det(\mathcal{C}_{X/Z})^\vee \otimes i^* \det \Omega^1_{Z/Y}.$$

(d) Show that Theorem 4.9 is true without the f quasi-projective hypothesis.

4.7. Let $f : X \to Y$ be a finite flat projective morphism of locally Noetherian schemes. Hence $f_* \mathcal{O}_X$ is a locally free sheaf of finite rank on Y.

(a) Let V be an affine open subset of Y, and $A = \mathcal{O}_Y(V)$, $B = f_* \mathcal{O}_X(V)$. Let $\rho(V) : B \to (f_* \omega_f)(V) = \mathrm{Hom}_A(B, A)$ be the homomorphism which sends $b \in B$ to $b' \mapsto \mathrm{Tr}_{B/A}(bb')$. Show that the $\rho(V)$ canonically induce a homomorphism of $\mathcal{O}_Y$-modules $\rho : f_* \mathcal{O}_X \to f_* \omega_f$, and a homomorphism of $\mathcal{O}_X$-modules $\lambda : \mathcal{O}_X \to \omega_f$.

(b) Let k be a field, E a finite algebra over k. Show that E is étale over $k \iff$ there exists an $e \in E$ such that $\mathrm{Tr}_{E/k}(e) = 1 \iff$ the map $e \mapsto \mathrm{Tr}_{E/k}(e\cdot)$ from E to $E^\vee$ is an isomorphism (see [16], V, Section 8, Proposition 1).

(c) Let us suppose that f is an étale morphism. Show that for any $y \in Y$, the canonical homomorphism $(f_* \mathcal{O}_X) \otimes_{\mathcal{O}_Y} k(y) \to (f_* \omega_f) \otimes_{\mathcal{O}_Y} k(y)$ induced by ρ is an isomorphism; that ρ is surjective (use Nakayama's lemma); and that it is also injective (Exercise 5.1.13). Deduce from this that $\lambda : \mathcal{O}_X \to \omega_f$ is an isomorphism. This shows Theorem 4.32 for finite, projective, étale morphisms.

(d) Show the converse of (c): if λ is an isomorphism, then f is étale.

(e) Let us suppose X, Y are integral and f generically separable. Show that $\lambda : \mathcal{O}_X \to \omega_f$ is injective and that $\mathrm{Coker}\,\lambda$ has support in a proper closed subset of X which is empty or pure of codimension 1.

4.8. Let $A \to B$ be a finite injective homomorphism of Noetherian integral domains. We want to identify the B-module $\mathrm{Hom}_A(B, A)$ with a submodule of $L := \mathrm{Frac}(B)$. We suppose that L is separable over $K := \mathrm{Frac}(A)$. Let us set

$$W_{B/A} = \{\beta \in L \mid \mathrm{Tr}_{L/K}(\beta B) \subseteq A\}.$$

Show that the homomorphism $W_{B/A} \to \mathrm{Hom}_A(B, A)$ which sends β to the map $b \mapsto \mathrm{Tr}_{L/K}(\beta b)$ is an isomorphism. When A and B are Dedekind domains, the fractional ideal $W_{B/A}$ is called the *codifferent* of the extension $A \to B$. So we have $\omega_{\mathrm{Spec}\, B/\,\mathrm{Spec}\, A} \simeq (W_{B/A})^\sim$.

4.9. (*Dualizing sheaf of a finite birational morphism.*)

(a) Let A be a ring and B a subring of the total ring of fractions $\mathrm{Frac}(A)$ (Definition 7.1.11.) Show that the homomorphism $\mathrm{Hom}_A(B, A) \to B$ defined by $\phi \mapsto \phi(1)$ is an isomorphism from $\mathrm{Hom}_A(B, A)$ to $\{a \in A \mid aB \subseteq A\}$. It is an ideal of A and of B. We call it the *conductor of B in A*.

(b) Let $f : X \to Y$ be a finite, projective, birational morphism of locally Noetherian integral schemes. Show that we have a canonical injective homomorphism $\omega_f \to \mathcal{O}_X$.

4.10. Let $f : X \to Y$ be a projective morphism with r-dualizing sheaf ω_f. Show that for any flat morphism $Y' \to Y$, the dualizing sheaf of $X \times_Y Y' \to Y'$ is isomorphic to $p^*\omega_f$, where p is the projection $X \times_Y Y' \to X$.

4.11. (*Duality for a global complete intersection*) Let $Y = \mathrm{Spec}\, A$ be an affine Noetherian scheme, and X be a closed subscheme $V_+(F_1, \ldots, F_{d-r})$ of $P := \mathbb{P}_Y^d$, where $0 \le r \le d$. We suppose that X_y is pure of dimension r for every $y \in Y$. We are going to show, without using Theorem 4.32, that $\omega_{X/Y}$ verifies the properties of the r-dualizing sheaf for $f : X \to Y$. We may assume $r < d$.

(a) Show that X is flat over Y (proceed as in Exercise 3.5) and that X_y is a complete intersection in P_y (Exercise 5.3.3). We will say that X is a (global) complete intersection in P.

(b) Let $Z = V_+(F_2, \ldots, F_{d-r})$. Show that Z is a complete intersection in P. Let $\mathcal{J} \subseteq \mathcal{O}_P$ denote the sheaf of ideals corresponding to $V_+(F_1)$, and $d_1 = \deg F_1$. Show that $\mathcal{J} \simeq \mathcal{O}_P(-d_1)$ and that

$$\omega_{X/Y} = \omega_{Z/Y}|_X \otimes (\mathcal{J}/\mathcal{J}^2)^\vee|_X \simeq \omega_{Z/Y}|_X \otimes \mathcal{O}_X(d_1).$$

Deduce from this that $\omega_{X/Y} \simeq \mathcal{O}_X(s - n - 1)$, where $s = \sum_i \deg F_i$. Show that for any $m \in \mathbb{Z}$, we have the exact sequences

$$0 \to \mathcal{O}_Z(m - d_1) \to \mathcal{O}_Z(m) \to \mathcal{O}_X(m) \to 0, \qquad (4.17)$$

$$0 \to \omega_{Z/Y}(-m) \to \omega_{Z/Y}(-m + d_1) \to \omega_{X/Y}(-m) \to 0. \qquad (4.18)$$

(c) Show the following properties by descending induction on r:

(1) $R^p f_* \mathcal{O}_X(m)$ is locally free on Y if $p = 0$ or r;

(2) $R^p f_* \mathcal{O}_X(m) = 0$ if $0 < p < r$;

(3) $H^0(X, \mathcal{O}_X) = A$ if $r \geq 1$.

(d) Let us henceforth suppose that $\omega_{Z/Y}$ is isomorphic to the $(r+1)$-dualizing sheaf ω_g, where $g : Z \to Y$ is the canonical morphism. Taking $m = 0$ in exact sequence (4.18), show that we have a canonical homomorphism $R^r f_* \omega_{X/Y} \to R^{r+1} g_* \omega_{Z/Y}$. Let tr_f be the composition of this homomorphism with tr_g. For simplicity, we write ω_X and ω_Z, omitting the reference to Y. Show that the exact sequences (4.17) and (4.18) induce a commutative diagram

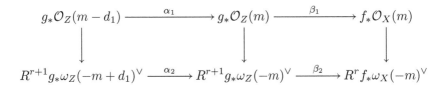

where the vertical arrows are induced by tr_f and tr_g, and show that α_1, α_2 are injective. Moreover, the β_i are surjective if $r \geq 1$. Deduce from this that $(\omega_X, \mathrm{tr}_f)$ is the dualizing sheaf if $r \geq 1$. If $r = 0$, show the same result using this commutative diagram as well as its dual (changing m).

7

Divisors and applications to curves

One way to study a scheme X is to study all closed subschemes of dimension strictly smaller than that of X. The linear combinations of these closed subsets with coefficients in $\mathbb{Z}$ are called *cycles* (Definition 2.1). To begin, we are going to examine the simplest among these cycles, the cycles of codimension 1. On a regular Noetherian scheme X, these cycles are locally defined by a rational function (Proposition 2.16). If X is not necessarily regular, it is more convenient to use Cartier divisors (Definition 1.17) which are, to some extent, locally principal cycles of codimension 1. On general Noetherian schemes, it turns out that Cartier divisors are often equivalent to invertible sheaves (Corollary 1.19 and Proposition 1.32). This gives an additional tool for studying the Picard group.

The second part of this chapter is devoted to the study of algebraic curves over a field, using mainly the notion of divisors, but also the cohomology of coherent sheaves. A fundamental result in the study of algebraic curves is the Riemann–Roch theorem (Theorem 3.26) when the curve is regular, or, more generally, a local complete intersection (l.c.i.). As singular curves can naturally occur as fibers of arithmetic surfaces, we will consider them in the last section.

7.1 Cartier divisors

We introduce the notion of Cartier divisors after constructing the sheaf of meromorphic functions in the first subsection. Then, we study the possibility of constructing the inverse image of a Cartier divisor by a morphism of schemes.

7.1.1 Meromorphic functions

We are going to generalize the notion of rational functions to schemes that are not necessarily reduced. This essentially consists of replacing the field of fractions of an integral domain A by the total ring of fractions of A in the general case. Let A be a ring. A non-zero divisor $f \in A$ is called a *regular element*. This is equivalent to saying that the annihilator $\mathrm{Ann}(f)$ of f is zero. In Exercise 2.1.4, we have seen that the elements of the minimal prime ideals of A are zero divisors, and that the converse is true if A is reduced. We will first rapidly review the notion of associated prime ideals, which makes it possible, among other things, to characterize the set of zero divisors in a Noetherian ring (Corollary 1.3).

Definition 1.1. Let A be a Noetherian ring, M an A-module. For any $x \in M$, we let $\mathrm{Ann}(x) := \{a \in A \mid ax = 0\}$ denote the annihilator of x. A prime ideal $\mathfrak{p}$ of A of the form $\mathrm{Ann}(x)$ with $x \in M \setminus \{0\}$ is called an *associated prime ideal of M*. The set of these ideals is denoted by $\mathrm{Ass}_A(M)$, or $\mathrm{Ass}(M)$ if there is no ambiguity concerning A.

Lemma 1.2. *Let M be a module over a Noetherian ring A.*

(a) *If $\mathrm{Ass}(M) = \emptyset$, then $M = 0$.*

(b) *Let S be a multiplicative subset of A. Then the elements of $\mathrm{Ass}_{S^{-1}A}(S^{-1}M)$ are exactly the primes of the form $S^{-1}\mathfrak{p}$ with $\mathfrak{p} \in \mathrm{Ass}_A(M)$ and $\mathfrak{p} \cap S = \emptyset$.*

Proof (a) Let us suppose $M \neq 0$. Let us consider the set of ideals $\{\mathrm{Ann}(x)\}$, where x runs through the non-zero elements of M. As A is Noetherian, this set contains a maximal element (for the inclusion) $I := \mathrm{Ann}(x_0)$. Let $a, b \in A$ be such that $ab \in I$ and $b \notin I$. As $\mathrm{Ann}(x_0) \subseteq \mathrm{Ann}(bx_0)$ and $bx_0 \neq 0$, since I is maximal we have $a \in \mathrm{Ann}(bx_0) = \mathrm{Ann}(x_0)$. This proves that I is prime, and hence $I \in \mathrm{Ass}(M)$.

(b) If $\mathfrak{p} = \mathrm{Ann}(x) \in \mathrm{Ass}_A(M)$, then $\mathrm{Ann}_{S^{-1}A}(x) = S^{-1}\mathfrak{p}$, because $A \to S^{-1}A$ is flat. Hence $S^{-1}\mathfrak{p} \in \mathrm{Ass}_{S^{-1}A}(S^{-1}M)$. Conversely, every prime ideal $\mathrm{Ann}(y)$ of $\mathrm{Ass}_{S^{-1}A}(S^{-1}M)$ is of the form $S^{-1}\mathfrak{p}$ with $\mathfrak{p} \in \mathrm{Spec}\, A$. Multiplying y by an invertible element of $S^{-1}A$ if necessary, we may suppose $y \in M$. This implies that $\mathfrak{p} = \mathrm{Ann}(y)$, and hence $\mathfrak{p} \in \mathrm{Ass}(M)$. □

Corollary 1.3. *Let A be a Noetherian ring. Then the following properties are true.*

(a) *The set of zero divisors of A is equal to the union of the associated prime ideals of A.*

(b) *The minimal prime ideals of A belong to $\mathrm{Ass}(A)$.*

Proof (a) Let $a \in A$ be a zero divisor. We have $ab = 0$ for some $b \neq 0$. Hence there exists a prime ideal $\mathfrak{p} = \mathrm{Ann}(bc) \in \mathrm{Ass}(bA)$. Now $abc = 0$, and hence $a \in \mathrm{Ann}(bc) = \mathfrak{p}$. By definition, we have $\mathfrak{p} \in \mathrm{Ass}(A)$. (b) Let $\mathfrak{p}$ be a minimal prime ideal of A. Localizing at $\mathfrak{p}$ if necessary, we may assume that it is the unique prime ideal of A. Now $\mathrm{Ass}(A) \neq \emptyset$, so $\mathfrak{p} \in \mathrm{Ass}(A)$. □

Lemma 1.4. *Let M be a finitely generated module over a Noetherian ring A. Then there exists a chain of submodules*

$$0 = M_0 \subset M_1 \subset \cdots \subset M_n = M$$

of M such that each successive quotient M_{i+1}/M_i is isomorphic to $A/\mathfrak{p}_i$, where $\mathfrak{p}_i$ is a prime ideal of A.

Proof If $M = 0$, there is nothing to show. Otherwise, there exists an associated prime ideal $\mathfrak{p}_1 = \text{Ann}(x_1)$ (Lemma 1.2). Let us set $M_1 = x_1 A \simeq A/\mathfrak{p}_1$. If $M = M_1$, we are done. Otherwise, we consider $\text{Ass}(M/M_1)$ and thus construct a strictly ascending chain $(M_i)_i$ of submodules of M whose successive quotients are of the form $A/\mathfrak{p}$ with $\mathfrak{p}$ prime. This chain is finite because M is Noetherian. $\square$

Corollary 1.5. *Let A be a Noetherian ring, and let M be a finitely generated A-module. Then $\text{Ass}(M)$ is a finite set.*

Proof Let N be a submodule of M. Let us show that

$$\text{Ass}(M) \subseteq \text{Ass}(N) \cup \text{Ass}(M/N).$$

The corollary will then be an immediate consequence of Lemma 1.4. Let $\mathfrak{p} = \text{Ann}(x) \in \text{Ass}(M)$. If $xA \cap N = 0$, we immediately verify that $\mathfrak{p} = \text{Ann}(\overline{x}) \in \text{Ass}(M/N)$, where $\overline{x} \neq 0$ is the image of x in M/N. Let us suppose that $xA \cap N \neq 0$. Let $y \in xA \cap N$ be non-zero. Then $\text{Ass}(yA) \subseteq \text{Ass}(xA) = \{\mathfrak{p}\}$. Hence $\text{Ass}(yA) = \{\mathfrak{p}\}$. As $y \in N$, we have $\mathfrak{p} \in \text{Ass}(N)$. $\square$

Definition 1.6. Let X be a locally Noetherian scheme, and let

$$\text{Ass}(\mathcal{O}_X) := \{x \in X \mid \mathfrak{m}_x \in \text{Ass}_{\mathcal{O}_{X,x}}(\mathcal{O}_{X,x})\}.$$

The points of $\text{Ass}(\mathcal{O}_X)$ are called the *associated points* of X. For any open subset U of X, we have $\text{Ass}(\mathcal{O}_X) \cap U = \text{Ass}(\mathcal{O}_U)$. If X is affine, then $\text{Ass}(\mathcal{O}_X) = \text{Ass}(\mathcal{O}_X(X))$. The generic points of X are associated points of X. An associated point that is not a generic point of X is called an *embedded point* of X. By Corollary 1.5, $\text{Ass}(\mathcal{O}_X)$ is a locally finite set.

Example 1.7. If X is locally Noetherian and reduced, then X has no embedded points (Exercise 2.1.4). Meanwhile $\text{Spec}\, k[u, v]/(u^2, uv)$, where k is a field, admits an embedded point corresponding to the maximal ideal (u, v).

Remark 1.8. An l.c.i. scheme (see Definition 6.3.17) over a regular locally Noetherian scheme has no embedded points. See Proposition 8.2.15(b).

Lemma 1.9. *Let X be a locally Noetherian scheme, U an open subset of X, and $i : U \to X$ the canonical inclusion. Then the canonical homomorphism $\mathcal{O}_X \to i_*\mathcal{O}_U$ is injective if and only if $\text{Ass}(\mathcal{O}_X) \subseteq U$.*

Proof As the property is local on X, we may suppose $X = \text{Spec}\, A$ and restrict ourselves to showing that $A \to \mathcal{O}_X(U)$ is injective if and only if $\text{Ass}(A) \subseteq U$. Let us first suppose that $\text{Ass}(A) \subseteq U$. Let $a \in A$ be such that $a|_U = 0$. If $a \neq 0$, then there exists a $\mathfrak{p} = \text{Ann}(ab) \in \text{Ass}(aA)$ (Lemma 1.2(a)). As $a = 0$ in $A_\mathfrak{p}$, there exists an $s \in A \setminus \mathfrak{p}$ such that $sa = 0$. Hence $s \in \text{Ann}(ab) = \mathfrak{p}$, which is absurd. Consequently, $a = 0$ in A.

Let us now suppose that there exists a $\mathfrak{p} = \text{Ann}(a) \in \text{Ass}(A)$ with $\mathfrak{p} \notin U$. Then for any point $x \in U$, we have $\text{Ann}(a)\mathcal{O}_{X,x} = \mathfrak{p}\mathcal{O}_{X,x} = \mathcal{O}_{X,x}$; hence $a_x = 0$. Consequently, $a|_U = 0$. $\square$

Remark 1.10. If X has no embedded points, then $\mathrm{Ass}(\mathcal{O}_X) \subseteq U$ if and only if U is everywhere dense in X.

Definition 1.11. Let A be a ring. We let $\mathrm{Frac}(A)$ denote the *total ring of fractions of A*, the localization of A with respect to the multiplicative subset of regular elements of A. It is a ring containing A as a subring.

For any commutative ring A, let us denote by $R(A)$ the (multiplicative) group of the regular elements of A. Then $\mathrm{Frac}(A)$ is the localization $R(A)^{-1}A$.

Lemma 1.12. *Let X be a scheme. For any open subset U of X, let $\mathcal{R}_X(U) := \{a \in \mathcal{O}_X(U) \mid a_x \in R(\mathcal{O}_{X,x}), \forall x \in U\}$. Then $\mathcal{R}_X$ is a sheaf on X, and $\mathcal{R}_X(U) = R(\mathcal{O}_X(U))$ if U is affine. Moreover, there exists a unique presheaf of algebras $\mathcal{K}'_X$ on X containing $\mathcal{O}_X$, verifying the following properties:*

(a) *For any open subset U of X, we have $\mathcal{K}'_X(U) = \mathcal{R}_X(U)^{-1}\mathcal{O}_X(U)$ (localization). In particular, $\mathcal{K}'_X(U) = \mathrm{Frac}(\mathcal{O}_X(U))$ if U is affine.*

(b) *For any open subset U of X, the canonical homomorphism $\mathcal{K}'_X(U) \to \prod_{x \in U} \mathcal{K}'_{X,x}$ is injective.*

(c) *If X is locally Noetherian, then for any $x \in X$, $\mathcal{K}'_{X,x} \simeq \mathrm{Frac}(\mathcal{O}_{X,x})$.*

Proof Obviously, $\mathcal{R}_X$ is a sheaf and we have $\mathcal{R}_X(U) \subseteq R(\mathcal{O}_X(U))$. If U is affine, then for any regular element $a \in R(\mathcal{O}_X(U))$, the restriction $a|_V$ to any affine open subset V of U is still a regular element by virtue of Lemma 6.3.10, because $\mathcal{O}_X(U) \to \mathcal{O}_X(V)$ is flat. This implies that $a \in \mathcal{R}_X(U)$. The existence and the uniqueness of $\mathcal{K}'_X$ are given by (a) and the fact that it contains $\mathcal{O}_X$ as a subsheaf of algebras (which determines the restriction maps).

Let us prove the property (b). Let $s = a/b \in \mathcal{K}'_X(U)$, with $a \in \mathcal{O}_X(U)$ and $b \in \mathcal{R}_X(U)$. Let us suppose that $s_x = 0$ for every $x \in U$. Then $a_x = 0$ for every $x \in U$. It follows that $a = 0$, and hence $s = 0$, whence (b).

(c) We have $\mathcal{K}'_{X,x} = \mathcal{R}^{-1}_{X,x}\mathcal{O}_{X,x}$ and $\mathcal{R}_{X,x} \subseteq R(\mathcal{O}_{X,x})$. We have to show that the latter inclusion is an equality when X is locally Noetherian. Let $b_x \in R(\mathcal{O}_{X,x})$. The stalk b_x comes from a section $b \in \mathcal{O}_X(W)$ with W affine. Let $I := \mathrm{Ann}(b)$, the annihilator of b. Then $I\mathcal{O}_{X,x} = 0$. As X is locally Noetherian, $\mathcal{O}_X(W)$ is Noetherian, and hence I is finitely generated. We at once deduce from this the existence of an affine open neighborhood $V \subseteq W$ of x such that $I\mathcal{O}_X(V) = 0$. Hence $b|_V \in R(\mathcal{O}_X(V)) = \mathcal{R}_X(V)$, and $b_x \in \mathcal{R}_{X,x}$. This completes the proof. $\square$

Definition 1.13. Let X be a scheme. We denote the sheaf of algebras associated to the presheaf $\mathcal{K}'_X$ by $\mathcal{K}_X$, and we call it the *sheaf of stalks of meromorphic functions* on X. By Lemma 1.12(b), $\mathcal{K}'_X$ is a sub-presheaf of $\mathcal{K}_X$, and therefore $\mathcal{O}_X$ is a subsheaf of $\mathcal{K}_X$. If X is locally Noetherian, then $\mathcal{K}_{X,x} = \mathcal{K}'_{X,x} = \mathrm{Frac}(\mathcal{O}_{X,x})$. An element of $\mathcal{K}_X(X)$ is called a *meromorphic function* on X. Let us note that if X is integral, then $\mathcal{K}_X$ is the constant sheaf $K(X)$. We denote the subsheaf of invertible elements of $\mathcal{K}_X$ by $\mathcal{K}^*_X$.

Remark 1.14. If X is locally Noetherian, or if it is reduced with only a finite number of irreducible components, then for any affine open subset U of X, we have $\mathcal{K}_X(U) = \mathcal{K}'_X(U) = \mathrm{Frac}(\mathcal{O}_X(U))$. See the end of the next proof.

Proposition 1.15. *Let X be a locally Noetherian scheme, and let U be an open subset of X containing $\mathrm{Ass}(\mathcal{O}_X)$. Let $i : U \to X$ denote the canonical injection. Then the canonical homomorphism $\mathcal{K}_X \to i_*\mathcal{K}_U$ is an isomorphism.*

Proof As $U \supseteq \mathrm{Ass}(\mathcal{O}_X)$, the canonical homomorphism $\mathcal{O}_X \to i_*\mathcal{O}_U$ is injective (Lemma 1.9); hence $\mathcal{K}'_X \to i_*\mathcal{K}'_U$ is also injective. Thus $\mathcal{K}_X \to i_*\mathcal{K}_U$ is injective, because $\mathcal{K}_{X,x} = \mathcal{K}'_{X,x}$ for all $x \in X$ and $i_*\mathcal{K}'_U$ is a sub-presheaf of $i_*\mathcal{K}_U$.

To prove the surjectivity of $\mathcal{K}_X \to i_*\mathcal{K}_U$, we can suppose that $X = \mathrm{Spec}\, A$ is affine. Let us first show that $\mathcal{K}'_X(X) \to \mathcal{K}'_X(U)$ is surjective. Write $X \setminus U = V(I)$. Then $V(I) \cap \mathrm{Ass}(A) = \emptyset$. Consequently, I is not contained in any prime ideal in $\mathrm{Ass}(A)$. As $\mathrm{Ass}(A)$ is finite (Corollary 1.5), I is not contained in $\cup_{\mathfrak{p} \in \mathrm{Ass}(A)}\mathfrak{p}$ (Exercise 2.5.4). Let $a \in I \setminus \cup_{\mathfrak{p} \in \mathrm{Ass}(A)}\mathfrak{p}$. Then a is regular, $D(a) \subseteq U$, and $\mathrm{Ass}(A) \subseteq D(a)$. As the composition of the injective canonical homomorphisms $\mathcal{K}'_X(X) \hookrightarrow \mathcal{K}'_X(U) \hookrightarrow \mathcal{K}'_X(D(a))$ is clearly an isomorphism, $\mathcal{K}'_X(X) \to \mathcal{K}'_X(U)$ is an isomorphism.

Let $s \in i_*\mathcal{K}_U(X) = \mathcal{K}_X(U)$. There exists an affine covering $\{U_i\}_i$ of U such that $s|_{U_i} \in \mathcal{K}'_X(U_i)$. Write $s|_{U_i} = a_i/b_i$ with $b_i \in R(\mathcal{O}_X(U_i))$. Let $V_i = D(b_i) \subseteq U_i$ and let $V = \cup_i V_i$. Then $s|_{V_i} \in \mathcal{O}_X(V_i)$ and $s|_V \in \mathcal{O}_X(V)$. As $\mathrm{Ass}(\mathcal{O}_X(U_i)) \subseteq V_i$ by Corollary 1.3, we have $\mathrm{Ass}(A) = \mathrm{Ass}(\mathcal{O}_U) \subseteq V$. From the above, $s|_V = t|_V$ for some $t \in \mathcal{K}'_X(X)$. Since $\mathcal{K}_X(U) \to \mathcal{K}_X(V)$ is injective, $s = t|_U$. So $\mathcal{K}'_X(X) \to \mathcal{K}_X(U)$ is surjective. Note that taking $U = X$, we obtain $\mathcal{K}'_X(X) = \mathcal{K}_X(X)$. □

Remark 1.16. For any reduced scheme X with only a finite number of irreducible components, the set of generic points of X plays the same role as $\mathrm{Ass}(\mathcal{O}_X)$ (Exercise 2.1.4(b)). By a similar method, we can show that the proposition stays valid for X by taking for U any everywhere dense open subset.

7.1.2   Cartier divisors

Definition 1.17. Let X be a scheme. We denote the group $H^0(X, \mathcal{K}^*_X/\mathcal{O}^*_X)$ by $\mathrm{Div}(X)$. The elements of $\mathrm{Div}(X)$ are called *Cartier divisors* on X. Let $f \in H^0(X, \mathcal{K}^*_X)$; its image in $\mathrm{Div}(X)$ is called a *principal Cartier divisor* and denoted by $\mathrm{div}(f)$. The group law on $\mathrm{Div}(X)$ is noted additively. We say that two Cartier divisors D_1 and D_2 are *linearly equivalent* if $D_1 - D_2$ is principal. We then write $D_1 \sim D_2$. A Cartier divisor D is called *effective* if it is in the image of the canonical map $H^0(X, \mathcal{O}_X \cap \mathcal{K}^*_X) \to H^0(X, \mathcal{K}^*_X/\mathcal{O}^*_X)$. We then write $D \geq 0$. The set of effective Cartier divisors is denoted by $\mathrm{Div}_+(X)$. Let U be an open subset of X, and let $D|_U$ denote the restriction of D (as a section of the sheaf $\mathcal{K}^*_X/\mathcal{O}^*_X$) to U.

By definition, we can represent a Cartier divisor D by a system $\{(U_i, f_i)_i\}$, where the U_i are open subsets of X forming a covering of X, f_i is the quotient of two regular elements of $\mathcal{O}_X(U_i)$, and $f_i|_{U_i \cap U_j} \in f_j|_{U_i \cap U_j}\mathcal{O}_X(U_i \cap U_j)^*$ for every i, j. Two systems $\{(U_i, f_i)_i\}$ and $\{(V_j, g_j)_j\}$ represent the same Cartier divisor if on $U_i \cap V_j$, f_i and g_j differ by a multiplicative factor in $\mathcal{O}_X(U_i \cap V_j)^*$. Let D_1, D_2 be two Cartier divisors, represented by $\{(U_i, f_i)_i\}$ and $\{(V_j, g_j)_j\}$, respectively. Then $D_1 + D_2$ is represented by $\{(U_i \cap V_j, f_ig_j)_{i,j}\}$. We have $D \geq 0$ if and only if it can be represented by $\{(U_i, f_i)_i\}$ with $f_i \in \mathcal{O}_X(U_i)$. It is principal if it can be represented by a system $\{(X, f)\}$.

We denote by $\mathrm{CaCl}(X)$ the group of isomorphism classes of Cartier divisors modulo the linear equivalence relation. We are going to compare this group to the Picard group $\mathrm{Pic}(X)$ (Exercise 5.1.12). To a Cartier divisor D represented by $\{(U_i, f_i)_i\}$, we can associate a subsheaf $\mathcal{O}_X(D) \subset \mathcal{K}_X$ defined by $\mathcal{O}_X(D)|_{U_i} = f_i^{-1}\mathcal{O}_X|_{U_i}$. It is clearly independent of the choice of a representing system $\{(U_i, f_i)_i\}$, and we see that it is an invertible sheaf on X. By construction, $D \geq 0$ if and only if $\mathcal{O}_X(-D) \subseteq \mathcal{O}_X$. If U is an open subset of X, then $\mathcal{O}_X(D)|_U = \mathcal{O}_U(D|_U)$.

Proposition 1.18. *Let X be a scheme. The following properties are true.*

(a) *The map $\rho : D \mapsto \mathcal{O}_X(D)$ is additive, that is*

$$\rho(D_1 + D_2) = \mathcal{O}_X(D_1)\mathcal{O}_X(D_2) \simeq \mathcal{O}_X(D_1) \otimes_{\mathcal{O}_X} \mathcal{O}_X(D_2).$$

(b) *The map ρ induces an injective homomorphism*

$$\mathrm{CaCl}(X) \to \mathrm{Pic}(X).$$

(c) *The image of ρ corresponds to the invertible sheaves contained in $\mathcal{K}_X$.*

Proof (a) follows immediately from the construction of ρ. (b) It is clear that ρ sends a principal Cartier divisor to a free sheaf of rank 1. Hence ρ induces a group homomorphism $\mathrm{CaCl}(X) \to \mathrm{Pic}(X)$. Let $D \in \mathrm{Ker}\,\rho$. Then there exists an $f \in H^0(X, \mathcal{O}_X(D)) \subseteq H^0(X, \mathcal{K}_X)$ such that $\mathcal{O}_X(D) = f\mathcal{O}_X$. The fact that $\mathcal{O}_X(D)$ is locally generated by elements of $\mathcal{K}_X^*$ implies that $f \in H^0(X, \mathcal{K}_X^*)$. Finally, it can immediately be verified that the equality $\mathcal{O}_X(D) = f\mathcal{O}_X$ implies that $D = \mathrm{div}(f)$. This shows the injectivity of $\mathrm{CaCl}(X) \to \mathrm{Pic}(X)$. (c) By construction, $\rho(D) = \mathcal{O}_X(D) \subseteq \mathcal{K}_X$. Let $\mathcal{L} \subseteq \mathcal{K}_X$ be an invertible subsheaf. Let $\{U_i\}_i$ be an open covering of X such that $\mathcal{L}|_{U_i}$ is free and generated by an $f_i \in \mathcal{K}_X'(U_i)$ for each i. Then $f_i \in \mathcal{K}_X'(U_i)^* \subseteq \mathcal{K}_X^*(U_i)$. And the Cartier divisor D associated to $\{(U_i, f_i)\}_i$ verifies $\rho(D) = \mathcal{L}$. $\square$

Corollary 1.19. *Let X be a Noetherian scheme without embedded point (e.g., reduced); then the canonical homomorphism $\mathrm{CaCl}(X) \to \mathrm{Pic}(X)$ is an isomorphism.*

Proof It is a matter of showing that every invertible sheaf $\mathcal{L}$ on X is isomorphic to a sub-$\mathcal{O}_X$-module of $\mathcal{K}_X$. Let $\xi_1, \ldots, \xi_n$ be the generic points of X (which are also the associated points of X, by hypothesis). Each ξ_j has an open neighborhood U_j contained in $\overline{\{\xi_j\}} \setminus \cup_{k \neq j} \overline{\{\xi_k\}}$. Moreover, by taking U_j sufficiently small, we may suppose $\mathcal{L}|_{U_j}$ is free. Let us set $U = \cup_{1 \leq j \leq n} U_j$. Let $i : U \to X$ be the canonical injection. Then $\mathcal{L} \to i_*(\mathcal{L}|_U)$ is injective by Lemma 1.9, because this is a property of local nature. As $\mathcal{L}|_U$ is free, we have

$$\mathcal{L} \hookrightarrow i_*(\mathcal{L}|_U) \simeq i_*\mathcal{O}_U \hookrightarrow i_*\mathcal{K}_U \simeq \mathcal{K}_X,$$

by Proposition 1.15, which proves the corollary. $\square$

Remark 1.20. More generally, $\mathrm{CaCl}(X) \to \mathrm{Pic}(X)$ is an isomorphism if X is locally Noetherian such that $\mathrm{Ass}(\mathcal{O}_X)$ is contained in an affine open subset (e.g., if X is quasi-projective over an affine Noetherian scheme) or if X is reduced with only a finite number of irreducible components. See [41], IV.21.3.5 and Proposition 1.32.

Length of modules

On a Noetherian scheme, Cartier divisors can be seen as cycles of codimension 1 (see the next section). To do this, we are going to recall the notion of length of a module. It will also serve to define intersection numbers on a regular surface (Section 9.1). Let A be a ring and M an A-module. We say that M is *simple* if $M \neq 0$, and if the only sub-A-modules of M are 0 and M. We say that M is *of finite length* if there exists a chain

$$0 = M_0 \subset \cdots \subset M_n = M \tag{1.1}$$

of sub-A-modules of M such that M_{i+1}/M_i is simple for every $i \leq n-1$ (in particular, $M_{i+1} \neq M_i$). The chain (1.1) is called a *composition series of M*. We can show that n is independent of the choice of such a composition series (Jordan–Hölder theorem, see for instance [33], Theorem 2.13(a)). We then call n the *length of M*, and we denote it by $\mathrm{length}_A(M)$, or if there is no confusion possible, $\mathrm{length}(M)$. By convention, $\mathrm{length}(0) = 0$.

Example 1.21. A vector space V is simple if and only if it is of dimension 1. We immediately deduce from this that V is of finite length if it is of finite dimension, in which case the length coincides with the dimension.

Example 1.22. Let $\mathcal{O}_K$ be a discrete valuation ring with uniformizing parameter t. Let $M = \mathcal{O}_K/t^n\mathcal{O}_K$ with $n \geq 1$. Then $\mathrm{length}_{\mathcal{O}_K} M = n$. Indeed,

$$0 \subset t^{n-1}\mathcal{O}_K/t^n\mathcal{O}_K \subset \cdots \subset t\mathcal{O}_K/t^n\mathcal{O}_K \subset \mathcal{O}_K/t^n\mathcal{O}_K$$

is a composition series of M, because the successive quotients are isomorphic to $\mathcal{O}_K/t\mathcal{O}_K$, which is simple.

Lemma 1.23. *Let A be a ring, M an A-module, and N a submodule of M. Then M is of finite length if and only if N and M/N are. Moreover, we will then have the equality*

$$\mathrm{length}\, M = \mathrm{length}\, N + \mathrm{length}(M/N).$$

Proof Let us suppose N and M/N of finite length. Let

$$N_0 \subset \cdots \subset N_q, \quad L_0 \subset \cdots \subset L_r$$

be composition series of N and M/N, respectively. For each $i \leq r$, L_i is the quotient of a unique sub-A-module M_{i+q} of M by N. Let us set $M_j = N_j$ if $j \leq q-1$ (note that $M_q = N = N_q$). It can immediately be verified that the M_i,

$0 \leq i \leq q + r$, form a composition series of M. This proves that M is of finite length, equal to $q + r$.

Conversely, let us suppose that M admits a composition series $(M_i)_{0 \leq i \leq n}$. Then the $N \cap M_i$ form an ascending sequence of submodules of N with $N \cap M_0 = 0$ and $N \cap M_n = N$. Moreover, $(N \cap M_{i+1})/(N \cap M_i)$ is isomorphic to a submodule of M_{i+1}/M_i, and is therefore zero or isomorphic to this module. Therefore, by leaving out the repetitions in the sequence, we extract a composition series of N from $(N \cap M_i)_i$. Similarly, by considering the images of the M_i in M/N, we show that M/N is of finite length. $\qquad\square$

Definition 1.24. We say that a module M is *Artinian* if every descending sequence of submodules of M is stationary. Any submodule and any quotient module of an Artinian module is Artinian.

Proposition 1.25. *Let M be a module over a ring A. Then M is of finite length if and only if it is Artinian and Noetherian.*

Proof Let us first suppose M is of finite length. Let $(M_n)_n$ be a monotonic sequence of submodules of M. By virtue of Lemma 1.23, $(\text{length } M_n)_n$ is a monotonic sequence of positive integers, bounded from above by length M, and therefore stationary. Therefore the chain $(M_n)_n$ is stationary, again by the preceding lemma, which shows that M is Artinian and Noetherian.

For the converse, let us first show that $M = 0$ or contains a simple submodule. Let us suppose the contrary. Then M contains a proper non-zero submodule L_1. Similarly, L_1 contains a proper non-zero submodule L_2. We thus construct a strictly descending chain $(L_n)_{n \geq 1}$ of submodules of M, which contradicts the hypothesis that M is Artinian. Let us suppose $M \neq 0$. Let M_1 be a simple submodule of M. Either $M/M_1 = 0$, and M is of length 1, or, as M/M_1 is Artinian, there exists a simple submodule M_2/M_1 of M/M_1. We thus construct a chain of modules $M_1 \subset M_2 \subset \cdots$ whose successive quotients are simple modules. As M is Noetherian, the chain is finite, and hence M is of finite length. $\qquad\square$

Lemma 1.26. *Let A be a Noetherian local ring of dimension 1, and let $f \in A$ be a regular element. Then A/fA is of finite length. Let $g \in A$ be regular; then we have*

$$\text{length}(A/fgA) = \text{length}(A/fA) + \text{length}(A/gA).$$

Proof No minimal prime ideal $\mathfrak{p}$ of A contains f (Corollary 1.3). It follows from Theorem 2.5.15 that A/fA is of dimension 0, and therefore Noetherian and Artinian (Lemma 2.5.11). By Proposition 1.25, A/fA is of finite length. We have an exact sequence of A-modules

$$0 \to gA/fgA \to A/fgA \to A/gA \to 0.$$

As $gA/fgA \simeq A/fA$, the equality of the lengths results from Lemma 1.23. $\quad\square$

Definition 1.27. Let A be a Noetherian local ring of dimension 1. For any regular element $f \in A$, $\text{length}_A(A/fA)$ is a finite integer. The lemma above shows that the map $f \mapsto \text{length}_A(A/fA)$ extends to a group homomorphism

$\mathrm{Frac}(A)^* \to \mathbb{Z}$. Moreover, its kernel contains the invertible elements of A. We thus obtain a group homomorphism $\mathrm{mult}_A : \mathrm{Frac}(A)^*/A^* \to \mathbb{Z}$.

Let X be a locally Noetherian scheme. Let $D \in \mathrm{Div}(X)$ be a Cartier divisor. For any point $x \in X$ of codimension 1, the stalk of D at x belongs to $(\mathcal{K}_X^*/\mathcal{O}_X^*)_x = \mathrm{Frac}(\mathcal{O}_{X,x})^*/\mathcal{O}_{X,x}^*$. We can therefore define

$$\mathrm{mult}_x(D) := \mathrm{mult}_{\mathcal{O}_{X,x}}(D_x) \qquad (1.2)$$

as above. Let U be an open and everywhere dense subset of X such that $D|_U = 0$. Then any $x \in X$ of codimension 1 such that $\mathrm{mult}_x(D) \neq 0$ is a generic point of $X \setminus U$. This implies that in any affine open subset of X, there are only a finite number of points x of codimension 1 such that $\mathrm{mult}_x(D) \neq 0$.

7.1.3 Inverse image of Cartier divisors

Let $f : X \to Y$ be a morphism of schemes. It is natural to want to transport Cartier divisors on X to Y, and conversely. In this subsection, we are going to construct the inverse images of Cartier divisors for morphisms or divisors fulfilling certain conditions. If for invertible sheaves $\mathcal{L}$ on Y, the inverse image $f^*\mathcal{L}$ is always defined and is an invertible sheaf, this is not always the case for Cartier divisors.

Definition 1.28. Let X be a scheme, and let $D \in \mathrm{Div}(X) = H^0(X, \mathcal{K}_X^*/\mathcal{O}_X^*)$. The *support of* D, which we denote by $\mathrm{Supp}\, D$, is defined to be the set of points $x \in X$ such that $D_x \neq 1$. This comes down to saying that $\mathcal{O}_X(D)_x \neq \mathcal{O}_{X,x}$. The set $\mathrm{Supp}\, D$ is a closed subset of X.

Lemma 1.29. *Let X be a closed subscheme of a locally Noetherian scheme Y. Let $i : X \to Y$ be the canonical injection.*

(a) *The set $G_{X/Y}$ of Cartier divisors E on Y such that*

$$(\mathrm{Supp}\, E) \cap \mathrm{Ass}(\mathcal{O}_X) = \emptyset$$

is a subgroup of $\mathrm{Div}(Y)$.

(b) *There exists a natural homomorphism $G_{X/Y} \to \mathrm{Div}(X)$, denoted by $E \mapsto E|_X$, compatible with the homomorphism $\mathcal{O}_Y \to i_*\mathcal{O}_X$. Moreover we have a canonical isomorphism $\mathcal{O}_Y(E)|_X \simeq \mathcal{O}_X(E|_X)$, and*

$$\mathrm{Supp}(E|_X) = (\mathrm{Supp}\, E) \cap X,$$

If $E > 0$, then $E|_X \geq 0$. The image of a principal divisor is a principal divisor.

Proof (a) Let $E \in \mathrm{Div}(Y) = H^0(Y, \mathcal{K}_Y^*/\mathcal{O}_Y^*)$. Then $E \in G_{X/Y}$ if and only if $E_x = 1$ for every $x \in \mathrm{Ass}(\mathcal{O}_X)$. This shows that $G_{X/Y}$ is a subgroup.

(b) Let $E \in G_{X/Y}$ be represented by a system $\{(U_i, f_i)\}_{i \in I}$. There exists an open subset V of Y such that $\mathrm{Ass}(\mathcal{O}_X) \subseteq V$ and $f_i|_{U_i \cap V} \in \mathcal{O}_Y(U_i \cap V)^*$ for every i. Let $\overline{U}_i = X \cap U_i$, and $\overline{f}_i$ be the image of $f_i|_{U_i \cap V}$ in $\mathcal{O}_X^*(\overline{U}_i \cap V) \subseteq \mathcal{K}_X^*(\overline{U}_i \cap V)$. Now

$$\overline{U}_i \cap V \supseteq \operatorname{Ass}(\mathcal{O}_X) \cap \overline{U}_i = \operatorname{Ass}(\mathcal{O}_{\overline{U}_i}),$$

and hence $\mathcal{K}_X^*(\overline{U}_i \cap V) = \mathcal{K}_X^*(\overline{U}_i)$, by Proposition 1.15. Let $E|_X \in \operatorname{Div}(X)$ be the Cartier divisor represented by $\{(\overline{U}_i, \overline{f}_i)\}_{i \in I}$. We verify without difficulty that $E|_X$ does not depend on the choice of representatives of E. The properties of $E|_X$ can immediately be deduced from the construction. □

Remark 1.30. If X is reduced, then the $(\operatorname{Supp} E) \cap \operatorname{Ass}(\mathcal{O}_X) = \emptyset$ condition of the lemma above is equivalent to saying that $\operatorname{Supp} E$ does not contain any irreducible component of X.

Lemma 1.31. *Let X be a quasi-projective scheme over an affine Noetherian scheme. Let $\mathcal{L}$ be an invertible sheaf on X. Then there exist very ample sheaves $\mathcal{L}_1$ and $\mathcal{L}_2$ such that $\mathcal{L} \simeq \mathcal{L}_1 \otimes \mathcal{L}_2^\vee$.*

Proof Let $\mathcal{M}$ be a very ample sheaf on X (Corollary 5.1.36). There exists an $n \geq 1$ such that $\mathcal{L} \otimes \mathcal{M}^{\otimes n}$ is generated by its global sections (Theorem 5.1.27). Hence $\mathcal{L} \otimes \mathcal{M}^{\otimes(n+1)}$ and $\mathcal{M}^{\otimes(n+1)}$ are very ample (Exercise 5.1.28), whence the lemma by taking $\mathcal{L}_1 = \mathcal{L} \otimes \mathcal{M}^{\otimes(n+1)}$ and $\mathcal{L}_2 = \mathcal{M}^{\otimes(n+1)}$. □

Proposition 1.32. *Let X be quasi-projective over a Noetherian affine scheme. Then the canonical homomorphism $\operatorname{CaCl}(X) \to \operatorname{Pic}(X)$ is an isomorphism.*

Proof We have to find an injective homomorphism $\mathcal{L} \hookrightarrow \mathcal{K}_X$ for any invertible sheaf $\mathcal{L}$ on X. Let U be an affine open subset of X containing $\operatorname{Ass}(\mathcal{O}_X)$ (Proposition 3.3.36(b)). Let R be the set of regular elements of $\mathcal{O}_X(U)$, $B = R^{-1}\mathcal{O}_X(U)$, and $N = R^{-1}\mathcal{L}(U)$. Then $\widetilde{N}$ is an invertible sheaf on $\operatorname{Spec} B$. We first show that $N \simeq B$. There are only a finite number of maximal ideals $\mathfrak{m}_1, \ldots, \mathfrak{m}_n$ in B because they all correspond to associated prime ideals of $\mathcal{O}_X(U)$. Let $I = \cap_i \mathfrak{m}_i$. Then $V(I)$ is a reduced closed subscheme of $\operatorname{Spec} B$ of dimension 0. Therefore $B/I = \oplus_i B/\mathfrak{m}_i$ (apply Exercise 2.5.11(c) to $\operatorname{Spec} B/I$). Tensoring by N, we obtain $N/IN \simeq \oplus_i N/\mathfrak{m}_i N \simeq \oplus_i B/\mathfrak{m}_i$. Therefore there exists $v \in N$ such that $N = Bv + IN$. By Corollary 1.2.9, we have $N = Bv$. Since $\widetilde{N}$ is invertible on $\operatorname{Spec} B$, we have $N \simeq B$.

We then see immediately that $\mathcal{L}|_V \simeq \mathcal{O}_V$ for some $V := D(t) \subseteq U$ with $t \in R$. So $\operatorname{Ass}(\mathcal{O}_X) = \operatorname{Ass}(\mathcal{O}_U) \subseteq V$. Let $i : V \to X$ be the canonical injection. Then $\mathcal{L} \hookrightarrow i_*(\mathcal{L}|_V) \simeq i_*\mathcal{O}_V \subseteq i_*\mathcal{K}_V \simeq \mathcal{K}_X$ (Lemma 1.9, Proposition 1.15). □

If we want to define the inverse image by $f : X \to Y$ of any Cartier divisor on Y, it suffices that we have a natural homomorphism $\mathcal{K}_Y^* \to f_*\mathcal{K}_X^*$.

Lemma 1.33. *Let $f : X \to Y$ be a morphism. We suppose that one of the following hypotheses is verified:*

(1) *f is flat;*

(2) *X is reduced, having only a finite number of irreducible components, and each of these dominates an irreducible component of Y (e.g., if X and Y are integral and f is dominant).*

Then the canonical homomorphism $\mathcal{O}_Y \to f_*\mathcal{O}_X$ extends to a homomorphism $\mathcal{K}_Y \to f_*\mathcal{K}_X$.

Proof It suffices to show that $\mathcal{O}_Y \to f_*\mathcal{O}_X$ extends to a homomorphism of presheaves $\mathcal{K}'_Y \to f_*\mathcal{K}'_X$ (see the notation of Lemma 1.12). Or to show that if $U \subseteq X$ and $V \subseteq Y$ are affine open subsets such that $f(U) \subseteq V$, then $\mathcal{O}_Y(V) \to \mathcal{O}_X(U)$ sends a regular element to a regular element. In case (1), this results from the fact that $\mathcal{O}_Y(V) \to \mathcal{O}_X(U)$ is flat (see the proof of Lemma 1.12). Let us now assume to be in case (2). Then $\mathcal{O}_X(U)$ is reduced and every irreducible component of U dominates an irreducible component of V (use Proposition 2.4.5(b)). Let us suppose that $\phi := f^\#(U) : \mathcal{O}_Y(V) \to \mathcal{O}_X(U)$ sends an element $b \in \mathcal{O}_Y(V)$ to a zero divisor. Then $\phi(b)$ belongs to a minimal prime ideal $\mathfrak{q}$ of $\mathcal{O}_Y(V)$ (Corollary 1.3). By hypothesis, $\phi^{-1}(\mathfrak{q})$ is a minimal prime ideal of $\mathcal{O}_X(U)$. Hence $b \in \phi^{-1}(\mathfrak{q})$ divides zero. This proves the lemma. $\square$

Definition 1.34. Let $f : X \to Y$ be as in Lemma 1.33. As we always have $f^\#(\mathcal{O}_Y^*) \subseteq f_*\mathcal{O}_X^*$, we have a homomorphism $\mathcal{K}_Y^*/\mathcal{O}_Y^* \to f_*(\mathcal{K}_X^*/\mathcal{O}_X^*)$. For any Cartier divisor $D \in \mathrm{Div}(Y) = H^0(Y, \mathcal{K}_Y^*/\mathcal{O}_Y^*)$ on Y, we will denote its image in $\mathrm{Div}(X)$ by f^*D. By construction, f^*D is principal if D is principal. Hence f^* induces a canonical homomorphism $\mathrm{CaCl}(Y) \to \mathrm{CaCl}(X)$.

Remark 1.35. Let $f : X \to Y$ be a morphism verifying one of the conditions of Lemma 1.33. Let $D \in \mathrm{Div}_+(Y)$ be an effective divisor. Then, by construction, it is easy to see that $f^*D \in \mathrm{Div}_+(X)$ and that the sheaf of ideals $\mathcal{O}_X(-f^*D)$ on X is the image of $f^*(\mathcal{O}_Y(-D))$ in $\mathcal{O}_X$.

Lemma 1.36. Let $(A, \mathfrak{m})$ be a local ring and B an A-algebra.

(a) Suppose B is local with maximal ideal $\mathfrak{n}$ and $[B/\mathfrak{n} : A/\mathfrak{m}] < +\infty$. Let M be a B-module of finite length. Then M is of finite length over A, and we have

$$\mathrm{length}_A(M) = [B/\mathfrak{n} : A/\mathfrak{m}]\,\mathrm{length}_B(M).$$

(b) Let us suppose B is finite over A, with maximal ideals $\mathfrak{n}_1, \dots, \mathfrak{n}_q$, and that A is Noetherian of dimension 1. Let $b \in B$ be a regular element. Let $B_i := B_{\mathfrak{n}_i}$. Then we have

$$\mathrm{length}_A(B/bB) = \sum_{1 \le i \le q} [B_i/\mathfrak{n}_i B_i : A/\mathfrak{m}]\,\mathrm{length}_{B_i}(B_i/bB_i).$$

(c) Let us suppose B is free of finite rank n over A, and let us keep the notation of (b). Then we have

$$n = \sum_{1 \le i \le q} [B_i/\mathfrak{n}_i B_i : A/\mathfrak{m}]\,\mathrm{length}_{B_i}(B_i/\mathfrak{m}B_i).$$

Proof (a) Let $(M_j)_{0 \le j \le d}$ be a composition series of M. Then M_{j+1}/M_j is a simple B-module, and hence isomorphic to $B/\mathfrak{n}$ (Exercise 1.6(a)). Using

Lemma 1.23, we can reduce to showing (a) when $M \simeq B/\mathfrak{n}$. The A-module structure on $B/\mathfrak{n}$ is induced by its structure of vector space over $A/\mathfrak{m}$; hence $\mathrm{length}_A(B/\mathfrak{n}B) = \mathrm{length}_{A/\mathfrak{m}}(B/\mathfrak{n}) = [B/\mathfrak{n} : A/\mathfrak{m}]$.

(b) As B is Noetherian of dimension ≤ 1 (Proposition 2.5.10), B/bB is of dimension 0. We deduce from this that $B/bB = \oplus_{1 \leq i \leq q}(B_i/bB_i)$ (Exercise 2.5.11). It follows that $\mathrm{length}_A(B/bB) = \sum_i \mathrm{length}_A(B_i/bB_i)$. We then obtain (b) by applying (a). Finally, the proof of (c) is similar to that of (b). □

Remark 1.37. Lemma 1.36(c) generalizes the well-known formula $n = \sum_i e_i f_i$ for finite extensions of discrete valuation rings.

Proposition 1.38. Let $f : X \to Y$ be a finite dominant morphism of Noetherian integral schemes. Let $y \in Y$ be a point of codimension 1. Let $D \in \mathrm{Div}(Y)$. Then we have

$$[K(X) : K(Y)]\,\mathrm{mult}_y\,D = \sum_{x \in f^{-1}(y)} [k(x) : k(y)]\,\mathrm{mult}_x(f^*D).$$

Proof Let us first note that every $x \in f^{-1}(y)$ has codimension at most 1 by Proposition 2.5.10(a). Hence x is of codimension 1 (it is not of codimension 0 because f is dominant), and $\mathrm{mult}_x(f^*D)$ indeed has a meaning. As the property to be shown is clearly of local nature, we can suppose that $Y = \mathrm{Spec}\,A$ is affine and local, and that y is the closed point of Y. Let $B = \mathcal{O}_X(X)$. We can suppose that D is defined by a regular element $a \in A$.

Let $n = [K(X) : K(Y)]$. Taking into account Lemma 1.36(b), the proposition reduces to showing that $\mathrm{length}_A(B/aB) = n\,\mathrm{length}_A(A/aA)$. Let $b_1, \dots, b_n$ be elements of B that form a basis of $K(X)$ over $K(Y)$. Let us consider the sub-A-module $M := \sum_{1 \leq i \leq n} Ab_i$ of B. As B/M is a torsion A-module, it is of finite length over A. We moreover have two exact sequences of A-modules

$$0 \to aB/aM \to B/aM \to B/aB \to 0,$$

$$0 \to M/aM \to B/aM \to B/M \to 0,$$

whose terms are of finite length over A. The multiplication by a induces an isomorphism $B/M \simeq aB/aM$; hence Lemma 1.23 implies that $\mathrm{length}_A(B/aB) = \mathrm{length}_A(M/aM)$. Now M is free of rank n over A, so we have $\mathrm{length}_A(M/aM) = n\,\mathrm{length}_A(A/aA)$, which completes the proof. □

Exercises

1.1. Let X be a locally Noetherian scheme without embedded point, or reduced with only a finite number of irreducible components. We want to show that $\mathcal{K}_X$ is a quasi-coherent sheaf on X. Let U be an affine open subset of X.

(a) Let $A = \mathcal{O}_X(U)$, and let $f \in A$ be non-nilpotent. Let us consider an element $a \in A$ that is a zero divisor and regular in A_f.

(1) Show that a is not nilpotent. Let $\mathfrak{p}_1,\dots,\mathfrak{p}_r$ be the minimal prime ideals of A containing a, and $\mathfrak{p}_{r+1},\dots,\mathfrak{p}_n$ those which do not contain a. Show that $n > r \geq 1$.

(2) Let $b \in (\cap_{r+1\leq i\leq n}\mathfrak{p}_i)\setminus\cup_{1\leq j\leq r}\mathfrak{p}_j$. Show that there exists an $m \geq 1$ such that $b^m a^m = 0$ and $a^m + b^m$ is a regular element.

(3) Show that the image of $a^{m-1}/(a^m + b^m) \in \mathrm{Frac}(A)$ in $\mathrm{Frac}(A_f)$ is equal to $1/(a|_{D(f)})$. Deduce from this that the canonical homomorphism

$$\mathrm{Frac}(A) \otimes_A A_f \to \mathrm{Frac}(A_f)$$

is an isomorphism.

(b) Show that $\mathcal{K}'_X$ is a quasi-coherent sheaf on X. Deduce from this that $\mathcal{K}_X$ is quasi-coherent, and that $\mathcal{K}_X(U) = \mathrm{Frac}(\mathcal{O}_X(U))$.

Remark. There exists a Noetherian scheme X such that $\mathcal{K}_X$ is not quasi-coherent ([41], IV.20.2.13.(iii)).

1.2. Let X be a locally Noetherian scheme without embedded point. Show that X is reduced if and only if it is reduced at the generic points (use Lemma 1.9).

1.3. Let X be a locally Noetherian scheme. We suppose that there exists a unique point $x \in X$ such that $\mathcal{O}_{X,x}$ is not reduced. Show that $x \in \mathrm{Ass}(\mathcal{O}_X)$.

1.4. Let X be a Noetherian scheme without embedded point.

(a) Let $\xi_1,\dots,\xi_n$ be its generic points. Show that for any open subset V of X, we canonically have

$$\mathcal{K}_X(V) = \oplus_{\xi_j\in V}\mathcal{K}_{X,\xi_j}.$$

(b) Deduce from this that $\mathcal{K}_X$ and $\mathcal{K}_X^*$ are flasque sheaves (Exercise 5.2.1), and that $H^1(X,\mathcal{K}_X) = 0$, $H^1(X,\mathcal{K}_X^*) = 1$.

(c) ([5], VII, 3.8) Using Exercise 5.2.7, Proposition 5.2.15, and the exact sequence $1 \to \mathcal{O}_X^* \to \mathcal{K}_X^* \to \mathcal{K}_X^*/\mathcal{O}_X^* \to 1$, show that $\mathrm{CaCl}(X) \to \mathrm{Pic}(X)$ is an isomorphism. This gives another proof of Corollary 1.19.

1.5. Let $A \to B$ be a surjective ring homomorphism. Show that for any B-module M, we have $\mathrm{length}_B(M) = \mathrm{length}_A(M)$. Show through an example that this is false if $A \to B$ is not surjective.

1.6. Let $(A, \mathfrak{m})$ be a Noetherian local ring. Let M be a finitely generated A-module.

(a) Show that M is simple if and only if $M \simeq A/\mathfrak{m}$.

(b) Show that M is of finite length if and only if there exists an $r \geq 1$ such that $\mathfrak{m}^r M = 0$.

(c) Let us suppose M is of finite length. Show that we have

$$\text{length}_A(M) = \sum_{i \geq 0} \dim_{A/\mathfrak{m}}(\mathfrak{m}^i M/\mathfrak{m}^{i+1}M).$$

(d) Let us suppose that A is an algebra over a field k. Show that

$$\text{length}_A(M) \dim_k A/\mathfrak{m} = \dim_k M.$$

1.7. Let A be a ring and M a finitely generated A-module. Let $a \in A$. When M/aM and $M[a]$ are of finite length over A, we set

$$e_A(a, M) := \text{length}_A(M/aM) - \text{length}_A(M[a]).$$

We have already come across this number in Lemma 5.3.18.

(a) Show that if M is of finite length, then $e_A(a, M)$ is well defined and is zero.

(b) If M is killed by an ideal I of A, show that $e_A(a, M) = e_{A/I}(\bar{a}, M)$, where $\bar{a}$ is the image of a in A/I.

(c) Let $0 \to M' \to M \to M'' \to 0$ be an exact sequence of A-modules. Show that we have the following exact sequences:

$$0 \to N \to M'/aM' \to M/aM \to M''/aM'' \to 0,$$

$$0 \to M'[a] \to M[a] \to M''[a] \to N \to 0,$$

where $N = (aM \cap M')/aM'$. Show that if two of the integers $e_A(a, M')$, $e_A(a, M)$, $e_A(a, M'')$ are well defined, then so is the third, and that in that case we have

$$e_A(a, M) = e_A(a, M') + e_A(a, M'').$$

1.8. Let $A \to B$ be a homomorphism of local rings. We call the (not necessarily finite) integer $e_{B/A} := \text{length}_B(B/\mathfrak{m}_A B)$ the *ramification index* of B over A. We have $e_{B/A} < +\infty$ if, for example, $\text{Spec } B \to \text{Spec } A$ is quasi-finite (see Exercise 3.2.8).

(a) Let $\mathcal{O}_K \subseteq \mathcal{O}_L$ be an extension of discrete valuation rings. Show that we have $e_{\mathcal{O}_L/\mathcal{O}_K} = \nu_L(t_K)$, where t_K is a uniformizing parameter for K, and ν_L is the normalized valuation (Example 3.3.23) of L. Thus, in this case, the definition of the ramification index coincides with the classical definition.

(b) Let M be an A-module of finite length. Let us suppose $e_{B/A}$ is finite. Show that

$$\text{length}_B(M \otimes_A B) \leq e_{B/A} \text{length}_A(M),$$

and that the equality holds if B is flat over A.

1.9. Let X be a quasi-projective scheme over a Noetherian ring A. Let $\mathcal{L}$ be an ample sheaf on X. Show that there exists an $m \geq 1$ such that $\mathcal{L}^{\otimes m} \simeq \mathcal{O}_X(D)$ for some effective Cartier divisor D.

1.10. Let X be an integral projective variety over a field k, of dimension ≥ 2. A Cartier divisor D on X is called an *ample divisor* if $\mathcal{O}_X(D)$ is an ample sheaf. Let D be an ample effective Cartier divisor on X.

(a) Suppose that X is normal. Show that for a sufficiently large n, the canonical homomorphism of k-algebras

$$H^0(X, \mathcal{O}_X) \to H^0(nD, \mathcal{O}_{nD}),$$

where we identify nD with the closed subvariety $V(\mathcal{O}_X(-nD))$, is surjective (use [Har], Corollary III.7.8), hence bijective. Deduce from this that the support of D is connected.

(b) Let $\pi : X' \to X$ be the normalization morphism. Show that $\pi^* D$ is ample and that X' is projective (use Exercise 5.2.16(b)). Show that $\operatorname{Supp} D$ is connected.

(c) Show that (b) is false without the assumption X irreducible.

1.11. Let X be a locally Noetherian scheme. Show that we have canonical isomorphisms

$$\varinjlim_{U \supseteq \operatorname{Ass}(\mathcal{O}_X)} \mathcal{O}_X(U) \simeq H^0(X, \mathcal{K}_X), \qquad \varinjlim_{U \supseteq \operatorname{Ass}(\mathcal{O}_X)} \mathcal{O}_X^*(U) \simeq H^0(X, \mathcal{K}_X^*),$$

of rings and groups, respectively.

1.12. Let X be a locally Noetherian scheme, and $\mathcal{F}$ be a coherent sheaf on X. We set

$$\operatorname{Ass}(\mathcal{F}) = \{x \in X \mid \mathfrak{m}_x \in \operatorname{Ass}_{\mathcal{O}_{X,x}}(\mathcal{F}_x)\}.$$

Let U be an open subset of X containing $\operatorname{Ass}(\mathcal{F})$.

(a) Let $i : U \to X$ be the canonical injection. Show that the canonical homomorphism $\mathcal{F} \to i_*(\mathcal{F}|_U)$ is injective.

(b) Let $\phi : \mathcal{F} \to \mathcal{G}$ be a homomorphism to a quasi-coherent sheaf. Show that if $\phi|_U$ is injective, then ϕ is injective.

1.13. Let $\mathcal{L}$ be an invertible sheaf on an integral scheme X. Let $s \in H^0(X, \mathcal{L} \otimes_{\mathcal{O}_X} \mathcal{K}_X)$ be a non-zero *rational section* of $\mathcal{L}$.

(a) Let $\{U_i\}_i$ be a covering of X such that $\mathcal{L}|_{U_i}$ is free and generated by an element e_i. Show that there exist $f_i \in K(X)^*$ such that $s|_{U_i} = e_i f_i$. Show that $\{(U_i, f_i)\}_i$ defines a Cartier divisor on X. We denote it by $\operatorname{div}(s)$. Show that $\mathcal{O}_X(\operatorname{div}(s)) = \mathcal{L}$.

(b) If $\mathcal{L} = \mathcal{O}_X$, show that $\operatorname{div}(s)$ is the principal Cartier divisor associated to s.

(c) Show that $\operatorname{div}(s) \geq 0$ if and only if $s \in H^0(X, \mathcal{L})$.

(d) Let $D \in \operatorname{Div}(X)$. For any open subset U of X, show that

$$\mathcal{O}_X(D)(U) = \{f \in \mathcal{K}_X^*(U) \mid \operatorname{div}(f) + D|_U \geq 0\} \cup \{0\}.$$

7.2 Weil divisors

This section is devoted to the study of Weil divisors (cycles of codimension 1 on a Noetherian integral scheme). These are geometrically more intuitive than Cartier divisors. However, in favorable situations, the two types of divisors are equivalent (Proposition 2.16). At the end of the section, we apply the technique of divisors to the study of birational morphisms (van der Waerden, Theorem 2.22).

All considered schemes are Noetherian.

7.2.1 Cycles of codimension 1

Definition 2.1. Let X be a Noetherian scheme. A *prime cycle* on X is an irreducible closed subset of X. A *cycle* on X is an element of the direct sum $\mathbb{Z}^{(X)}$. Thus, any cycle Z can be written in a unique way as a finite sum

$$Z = \sum_{x \in X} n_x[x].$$

By definition, the sum of two cycles is done component-wise, and we have $Z = 0$ if and only if $n_x = 0$ for every $x \in X$.

As we have a canonical bijection between X and the set of its irreducible closed subsets via the map $x \mapsto \overline{\{x\}}$, we rather write Z as a finite sum

$$Z = \sum_{x \in X} n_x[\overline{\{x\}}].$$

By this identification, a prime cycle is a cycle. The coefficient n_x is called the *multiplicity* of Z at x, and is denoted by $\mathrm{mult}_x(Z)$. We say that a cycle Z is *positive* if $\mathrm{mult}_x(Z) \geq 0$ for every $x \in X$. By gathering the (strictly) positive coefficients on the one hand, and the (strictly) negative coefficients on the other hand, we can always write Z as the difference of two positive cycles

$$Z = Z_0 - Z_\infty. \tag{2.3}$$

The (finite) union of the $\overline{\{x\}}$ such that $n_x \neq 0$ is called the *support* of Z, and is denoted by $\mathrm{Supp}\, Z$. It is a closed subset of X. By convention, the support of 0 is the empty set. We say that a cycle Z is (pure) of *codimension* 1 if the irreducible components of $\mathrm{Supp}\, Z$ are of codimension 1 (Definition 2.5.7) in X. Let us note that $\overline{\{x\}}$ is of codimension 1 if and only if $\dim \mathcal{O}_{X,x} = 1$ (Exercise 2.5.2). We will then say that x is a *point of codimension* 1. The cycles of codimension 1 form a subgroup $Z^1(X)$ of the group of cycles on X.

Example 2.2. Let X be a curve over a field k (Definition 2.5.29). Then a cycle of codimension 1 on X is simply a finite sum $\sum_i n_i[x_i]$ with $n_i \in \mathbb{Z}$, and where the x_i are closed points of X.

Example 2.3. Let $X = \mathbb{P}^n_k$ be a projective space over a field. Then any prime cycle of codimension 1 is of the form $V_+(P)$, with P homogeneous and irreducible (Exercise 2.5.12(c)). Hence every element of $Z^1(X)$ is a finite sum $Z = \sum_i n_i[V_+(P_i)]$.

Definition 2.4. Let X be a Noetherian integral scheme. A cycle of codimension 1 on X is called a *Weil divisor* on X.

Lemma 2.5. *Let X be a Noetherian integral scheme, and let $f \in K(X)$ be non-zero. Then for every point $x \in X$ of codimension 1 except possibly a finite number, we have $f \in \mathcal{O}_{X,x}^*$.*

Proof Let U be a dense affine open subset of X. Then $f = a/b$ with $a, b \in \mathcal{O}_X(U)$. For any point $x \in X$ of codimension 1 such that

$$x \notin Y := V(a) \cup V(b) \cup (X \setminus U),$$

we have $f \in \mathcal{O}_{X,x}^*$. As at the end of Definition 1.27, there are only a finite number of such x in Y.   □

Example 2.6. Let X be a normal Noetherian scheme. Let $f \in K(X)$ be a non-zero rational function. Let $x \in X$ be a point of codimension 1. Then $\mathcal{O}_{X,x}$ is local of dimension 1 and normal; it is therefore a discrete valuation ring (Proposition 4.1.12). Let $\mathrm{mult}_x : K(X) \to \mathbb{Z} \cup \{\infty\}$ be the normalized valuation of $K(X)$ associated to $\mathcal{O}_{X,x}$ (Example 3.3.23). Let us set

$$(f) := \sum_{x \in X, \ \dim \mathcal{O}_{X,x}=1} \mathrm{mult}_x(f)[\overline{\{x\}}].$$

This is a Weil divisor (we have seen in Lemma 2.5 that this sum is finite). Such a divisor is called a *principal divisor* (we will give later a more general construction based on Cartier divisors, see Proposition 2.14). As mult_x is additive, we have $(fg) = (f) + (g)$. Therefore the set of principal divisors is a subgroup of $Z^1(X)$.

Definition 2.7. Let X be a normal Noetherian scheme. We denote the quotient of $Z^1(X)$ by the subgroup of principal divisors by $\mathrm{Cl}(X)$. We say that two Weil divisors Z_1, Z_2 are *linearly equivalent*, and write $Z_1 \sim Z_2$, if they have the same class in $Z^1(X)$.

Example 2.8. Let K be a number field, $\mathcal{O}_K$ its ring of integers. Let $X = \mathrm{Spec}\,\mathcal{O}_K$. To every Weil divisor $D = \sum_i n_i[x_i]$ on X, we can associate the fractional ideal $\prod_i \mathfrak{p}_i^{n_i}$, where $\mathfrak{p}_i$ is the maximal ideal of $\mathcal{O}_K$ corresponding to the point x_i (x_i is closed because it is of codimension 1 in a scheme of dimension 1). This correspondence establishes, in an evident manner, an isomorphism between $\mathrm{Cl}(X)$ and the class group of K.

Proposition 2.9. *Let $X = \mathbb{P}_k^n$ be a projective space over a field k, with $n \geq 1$. Then there exists a group homomorphism $\delta : Z^1(X) \to \mathbb{Z}$ which induces an isomorphism $\mathrm{Cl}(X) \to \mathbb{Z}$. Moreover, any hyperplane in X is a generator of $\mathrm{Cl}(X)$.*

Proof Let us consider the homomorphism $\delta : Z^1(X) \to \mathbb{Z}$ defined by

$$\sum_i n_i[V_+(P_i)] \mapsto \sum_i n_i \deg P_i$$

(see Example 2.3). It is clearly surjective, because $\delta([H]) = 1$ if H is a hyperplane. Let $f \in K(X)$ be a non-zero rational function. We can write f as a finite product

$f = \prod_i P_i^{n_i}$, where the P_i are irreducible homogeneous polynomials that are pairwise prime to each other, and $n_i \in \mathbb{Z}$. Let us first show that

$$(f) = \sum_i n_i [V_+(P_i)]. \qquad (2.4)$$

Let $x \in X$ be a point of codimension 1. Then $\overline{\{x\}} = V_+(P)$ for an irreducible homogeneous polynomial P of some degree d. It is easy to see that for any homogeneous polynomial Q of degree d, prime to P, the function P/Q is a generator of $\mathfrak{m}_x \mathcal{O}_{X,x}$. We can write $f = (P/Q)^r g$, with $r = n_i$ if P_i divides P ($r = 0$ if no P_i divides P), and $g \in K(X)$ containing P neither in its numerator nor in its denominator. It follows that $\mathrm{mult}_x(f) = r$. This immediately implies (2.4). Let us note that $\sum_i n_i \deg P_i = 0$, because f is a rational function on X. Thus $\delta((f)) = 0$. Hence δ induces a surjective homomorphism $\tilde{\delta} : \mathrm{Cl}(X) \to \mathbb{Z}$.

Let $Z = \sum n_i [V_+(F_i)] \in \mathrm{Ker}\, \delta$. Then $\sum_i n_i \deg F_i = 0$. It follows that $h := \prod_i F_i^{n_i} \in K(X)$. By formula (2.4), we have $Z = (h)$. This shows that $\tilde{\delta}$ is an isomorphism. Let H be a hyperplane in X. Then $\delta(H) = 1$. This shows that the class of H in $\mathrm{Cl}(X)$ is a basis. $\qquad \square$

Definition 2.10. Let f be a non-zero rational function on a normal Noetherian scheme. The divisor $(f)_0$ (see formula (2.3)) is called the *divisor of zeros of f*, while $(f)_\infty$ is called the *divisor of poles of f*. Let us take $X = \mathbb{P}_k^n$. Let $f = P/Q \in K(X)$, where P and Q are two homogeneous polynomials of the same degree that are prime to each other. Then $\mathrm{Supp}(f)_0 = V_+(P)$, and $\mathrm{Supp}(f)_\infty = V_+(Q)$, by formula (2.4).

Proposition 2.11. *Let X be a normal Noetherian scheme. Let $f \in K(X)^*$. Let U be an open subset of X. Then the following properties are true.*

(a) $f \in \mathcal{O}_X(U)$ *if and only if* $U \cap \mathrm{Supp}(f)_\infty = \emptyset$.

(b) *Let $x \in X$. Then* $f \in \mathfrak{m}_x \mathcal{O}_{X,x}$ *if and only if* $x \in \mathrm{Supp}(f)_0 \setminus \mathrm{Supp}(f)_\infty$.

Proof (a) Let us suppose $f \in \mathcal{O}_X(U)$. For any $x \in \mathrm{Supp}(f)_\infty$ of codimension 1 in X, we have $x \in X \setminus U$, because otherwise $\mathrm{mult}_x(f) \geq 0$. Hence the irreducible components of $\mathrm{Supp}(f)_\infty$ are contained in $X \setminus U$, which shows that $U \cap \mathrm{Supp}(f)_\infty = \emptyset$. Conversely, let us suppose this condition is verified. For any point $x \in U$ of codimension 1, x is of codimension 1 in X, because $\mathcal{O}_{X,x} = \mathcal{O}_{U,x}$. Hence $\mathrm{mult}_x(f) \geq 0$. In other words, $f \in \mathcal{O}_{X,x}$. It follows from Theorem 4.1.14 that $f \in \mathcal{O}_X(U)$.

(b) Let us suppose $f \in \mathfrak{m}_x \mathcal{O}_{X,x}$. Then $x \notin \mathrm{Supp}(f)_\infty$ by (a). We have $x \in \mathrm{Supp}(f)_0$, because otherwise $f \in \mathcal{O}_{X,x}^*$. Conversely, let us suppose $x \in \mathrm{Supp}(f)_0 \setminus \mathrm{Supp}(f)_\infty$; then $f \in \mathcal{O}_{X,x}$. We have $f \in \mathfrak{m}_x \mathcal{O}_{X,x}$, because otherwise f would be invertible in a neighborhood of x and $\mathrm{Supp}(f)_0$ would not pass through x. $\qquad \square$

Definition 2.12. Let X be a Noetherian scheme and U an open subset of X. Let $Z \in Z^1(X)$. Then we can take the restriction of Z to U:

$$Z|_U = \sum_{x \in U, \dim \mathcal{O}_{X,x} = 1} \mathrm{mult}_x(Z) \overline{[\{x\}]}.$$

We have $\mathrm{Supp}(Z|_U) = (\mathrm{Supp}\,Z) \cap U$. If X is normal and $f \in K(X)^*$, then $(f)|_U = (f|_U)$.

For any $D \in \mathrm{Div}(X)$, let us set

$$[D] = \sum_{x \in X, \dim \mathcal{O}_{X,x}=1} \mathrm{mult}_x(D)[\overline{\{x\}}] \in Z^1(X).$$

Thus $[D]$ is a cycle of codimension 1 such that $\mathrm{mult}_x([D]) = \mathrm{mult}_x(D)$ at every point x of codimension 1. For any open subset U of X, we have $[D|_U] = [D]|_U$ (see Definition 2.12).

Example 2.13. Let B be an integral homogeneous algebra over a ring A, generated by elements $t_0, \ldots, t_n$ of degree 1. Let $b \in B$ be a homogeneous element of degree $d > 0$. Then we have an effective Cartier divisor D on $\mathrm{Proj}\,B$ given by the system $\{(D_+(t_i), b/t_i^d)\}_{1 \le i \le n}$. It can immediately be verified that $\mathrm{Supp}[D] = V_+(b)$ as subsets of $\mathrm{Proj}\,B$.

Proposition 2.14. *Let X be a Noetherian scheme. Then the following properties are true:*

  (a) *The correspondence $D \mapsto [D]$ establishes a group homomorphism*

$$\mathrm{Div}(X) \to Z^1(X)$$

   which sends effective divisors to positive cycles.

  (b) *Let us suppose X is normal. Let $f \in H^0(X, \mathcal{K}_X^*) = K(X)^*$. Then the Weil divisor $(f) \in Z^1(X)$ coincides with the image of the principal Cartier divisor $\mathrm{div}(f)$.*

  (c) *Let us suppose X is normal. Then the homomorphism in (a) is injective and induces an injective homomorphism $\mathrm{CaCl}(X) \to \mathrm{Cl}(X)$. Moreover, we have $D \ge 0$ if and only if $[D] \ge 0$.*

Proof   (a) follows from the definition. Let us suppose X is normal. Then (b) results from Example 1.22. Let $D \in \mathrm{Div}(X)$ be a Cartier divisor represented by $\{(U_i, f_i)\}_i$. If $[D] \ge 0$, then the Weil divisor $(f_i) = [D|_{U_i}]$ on U_i is positive. It follows from Proposition 2.11 that $f_i \in \mathcal{O}_X(U_i)$; hence $D \ge 0$. If $[D] = 0$, then $f_i, 1/f_i \in \mathcal{O}_X(U_i)$; hence $f_i \in \mathcal{O}_X(U_i)^*$. In other words, $D = 0$, whence the injectivity of $\mathrm{Div}(X) \to Z^1(X)$. This implies the injectivity of $\mathrm{CaCl}(X) \to \mathrm{Cl}(X)$, by virtue of (b).   □

Example 2.15. The normality hypothesis cannot be omitted in Proposition 2.14. Let us consider the integral curve $X = \mathrm{Spec}\,k[s, t]$ over a field k, with the relation $s^2 - t^3 = 0$. Let $p \in X$ be the point $s = t = 0$. A simple computation shows that

$$\mathrm{length}_{\mathcal{O}_{X,p}}(\mathcal{O}_{X,p}/(t)) = \mathrm{length}_{\mathcal{O}_{X,p}}(\mathcal{O}_{X,p}/(t-s)) = 2.$$

Let D be the principal Cartier divisor associated to $f := (t - s)/t \in K(X)^*$. For any $x \ne p$, we have $f \in \mathcal{O}_{X,x}^*$. We therefore have $\mathrm{mult}_x(D) = 0$ for every closed point $x \in X$. In other words, $[D] = 0$. However, $D \ne 0$, because $f_p \notin \mathcal{O}_{X,p}$.

Proposition 2.16. *Let X be a Noetherian regular integral (hence normal, by Theorem 4.2.16(b)) scheme. Then the homomorphisms*

$$\mathrm{Div}(X) \to Z^1(X), \quad \mathrm{CaCl}(X) \to \mathrm{Cl}(X)$$

are isomorphisms.

Proof Owing to Proposition 2.14, it suffices to show that $\mathrm{Div}(X) \to Z^1(X)$ is surjective. Let E be an irreducible closed subset of X of codimension 1. Let $\mathcal{I}$ be the sheaf of ideals of $\mathcal{O}_X$ defining E (endowed with the reduced subscheme structure). Let $x \in X$. If $x \in E$, $\mathcal{I}_x$ is a prime ideal of height 1, and hence principal, because $\mathcal{O}_{X,x}$ is factorial (Theorem 4.2.16(b)). If $x \notin E$, then $\mathcal{I}_x = \mathcal{O}_{X,x}$. Therefore there always exist an open neighborhood U_x of x and $f_x \in \mathcal{O}_X(U_x)$, such that $\mathcal{I}|_{U_x} = f_x\mathcal{O}_X|_{U_x}$. Let $y \in X$. Then on $U_x \cap U_y$, the elements f_x and f_y differ by an invertible multiplicative factor, because they generate the same sheaf of ideals. Let D be the Cartier divisor represented by $\{(U_x, f_x)_{x\in X}\}$ on X. Then $\mathrm{mult}_\xi(D) = 1$ if ξ is the generic point of E, and $\mathrm{mult}_x(D) = 0$ if x is a point of codimension 1 distinct from ξ, because $x \notin E$, and therefore f_x is invertible. Consequently, $[D]$ is equal to the prime cycle E. As the prime cycles of codimension 1 generate $Z^1(X)$, this shows that $\mathrm{Div}(X) \to Z^1(X)$ is surjective. □

Definition 2.17. Let $f : X \to Y$ be a finite dominant morphism of Noetherian integral schemes. Let Z be a cycle on X. We define the *direct image of Z by f* to be the cycle f_*Z given by

$$\mathrm{mult}_y(f_*Z) = \sum_{x \in f^{-1}(y)} [k(x) : k(y)]\,\mathrm{mult}_x(Z)$$

for every $y \in Y$. In particular, if Z is a prime cycle, we have

$$f_*Z = [K(Z) : K(W)]W, \quad \text{where } W = f(Z).$$

It can immediately be verified that $Z \mapsto f_*Z$ is a group homomorphism, and that $g_*(f_*Z) = (g \circ f)_*Z$ if $g : Y \to S$ is a finite dominant morphism to a Noetherian integral scheme S. Note that if Z is a cycle of codimension 1, then f_*Z is not necessarily a cycle of codimension 1. See, however, Corollary 8.2.6.

Theorem 2.18. *Let $f : X \to Y$ be a finite dominant morphism of Noetherian integral schemes. Let $D \in \mathrm{Div}(Y)$. Then we have the following equality:*

$$f_*[f^*D] = n[D],$$

where $n = [K(X) : K(Y)]$. (See also Proposition 9.2.11.)

Proof Since the equality is local on Y and linear on D, we can assume that D is a principal and effective divisor. It is then clear that the support of $f_*[f^*D]$ is equal to that of $[D]$. It now suffices to apply Proposition 1.38. □

Remark 2.19. Let $b \in K(X)^*$ and let $N(b) := \mathrm{Norm}_{K(X)/K(Y)}(b)$ be the norm of b in $K(Y)$. Then one can show that

$$f_*[\mathrm{div}(b)] = [\mathrm{div}(N(b))]$$

(see Exercise 2.6 when Y is normal, and [37], Proposition 1.4.(b), for the general case). This is a generalization of Theorem 2.18.

7.2.2　Van der Waerden's purity theorem

We are going to use the notion of divisors to show a theorem of van der Waerden concerning the exceptional locus of a birational morphism.

Lemma 2.20. *Let X, Y be Noetherian integral schemes. Let $f : X \to Y$ be a separated birational morphism of finite type.*

    (a) *There exists a non-empty open subset V of Y such that $f^{-1}(V) \to V$ is an isomorphism.*

    (b) *Let W be the union of the open sets V as in (a). Then $x \in f^{-1}(W)$ if and only if $\mathcal{O}_{Y,f(x)} \to \mathcal{O}_{X,x}$ is an isomorphism.*

Proof　There exist non-empty open subsets U, V of X and Y, respectively, such that $f|_U : U \to V$ is an isomorphism (Exercise 3.2.6). We have $U \subseteq f^{-1}(V)$. The morphism $f^{-1}(V) \to V$ is separated and admits a section $V \simeq U \subseteq f^{-1}(V)$. Hence U is open and closed in $f^{-1}(V)$ (Exercise 3.3.6). Now $f^{-1}(V)$ is irreducible, and hence $f^{-1}(V) = U$ is isomorphic to V. This shows (a).

    (b) Let $x \in X$ be such that $\mathcal{O}_{Y,y} \to \mathcal{O}_{X,x}$, where $y = f(x)$, is an isomorphism. Then an argument similar to that for (a) shows that $X \times_Y \mathrm{Spec}\,\mathcal{O}_{Y,y} \to \mathrm{Spec}\,\mathcal{O}_{Y,y}$ is an isomorphism. This isomorphism extends to an open neighborhood of y (Exercise 3.2.5). Hence $y \in W$.　　　　　　　　　　　　　　　　　□

Definition 2.21. Let us keep the hypotheses and notation of the lemma above. We call the closed subset $E := X \setminus f^{-1}(W)$ the *exceptional locus* of f. The definition of E is local on X in the sense that if U is an open subset of X, then the exceptional locus of $f|_U : U \to Y$ is $U \cap E$. This can be seen by using Lemma 2.20(b).

Theorem 2.22 (van der Waerden's purity theorem). *Let X, Y be Noetherian integral schemes, and let $f : X \to Y$ be a separated birational morphism of finite type. Let us suppose that Y is regular. Then the exceptional locus E of f is empty or pure of codimension 1 in X.*

Proof　Let us suppose $E \neq \emptyset$. As the theorem is of local nature, we may suppose X, Y are affine. Hence $X \to Y$ is quasi-projective. Let $X \to \mathbb{P}^n_Y$ be an immersion. Let X' be the closure of X in $\mathbb{P}^n_Y$, endowed with the reduced closed subscheme structure. Then X' is integral and f extends to a projective birational morphism $X' \to Y$ (which is the composition of the closed immersion into $\mathbb{P}^n_Y$ followed by the projection onto Y). It then suffices to show that the exceptional locus of $X' \to Y$ is pure of codimension 1. We may therefore assume that X is itself projective over Y.

Let $H = V_+(P)$ be a hypersurface in $\mathbb{P}_Y^n$ containing no generic point of E (Proposition 3.3.36). Let us write $X = \operatorname{Proj} A[T_0, \ldots, T_n]/\mathfrak{p}$. As H does not contain X, the image $\overline{P}$ of P in $A[T_0, \ldots, T_n]/\mathfrak{p}$ is non-zero. Let $D \in \operatorname{Div}_+(X)$ be the effective Cartier divisor associated to $\overline{P}$ (see Example 2.13). Let us write $D = \sum_i n_i \Gamma_i$, where the Γ_i are prime cycles of codimension 1. Then the generic point of Γ_i does not belong to E, because $\operatorname{Supp}[D] = H \cap X$. It follows that $f(\Gamma_i)$ is a cycle of codimension 1 on Y. As Y is regular, there exists a $\Delta_i \in \operatorname{Div}_+(Y)$ such that $[\Delta_i] = f(\Gamma_i)$ (Propositions 2.16 and 2.14(c)). Let us set $\Delta = \sum_i n_i \Delta_i$ and $Z = [f^*\Delta] - [D]$. Let us show that $E = \operatorname{Supp} Z$, which will prove the theorem.

First, let $W = X \setminus E$. As $f|_W$ is an isomorphism from W to $f(W)$, we have $Z|_W = [f^*\Delta]|_W - [D]|_W = [f^*\Delta|_W] - [D|_W] = 0$, which proves that $\operatorname{Supp} Z \subseteq E$. Conversely, for any $x \in E$, let us set $y = f(x)$. As Y is normal, we have $f_*\mathcal{O}_X = \mathcal{O}_Y$ (Corollary 4.4.3). It follows that X_y has no isolated points (Theorem 4.4.4), and hence $\dim X_y \geq 1$. Consequently, $H \cap X_y \neq \emptyset$ (Exercise 3.3.4(c)). It follows that $y \in f(H \cap X)$. Therefore

$$E \subseteq f^{-1}(f(H \cap X)) = \operatorname{Supp}[f^*\Delta] \subseteq (H \cap X) \cup \operatorname{Supp} Z.$$

As $H \cap X$ does not contain any generic point of E, this implies that $E \subseteq \operatorname{Supp} Z$.

$\square$

Remark 2.23. See [69], III.9, Example 0, for a counterexample when Y is not regular.

Remark 2.24. If we only suppose that Y is normal, then the end of the proof shows that E does not have any isolated point.

Exercises

2.1. Let A be a normal Noetherian ring, let $X = \operatorname{Spec} A$. Let us suppose that $\operatorname{Cl}(X) = 0$.

(a) Let $\mathfrak{p}$ be a prime ideal of height 1 in A. Let $f \in \operatorname{Frac}(A)$ be such that $V(\mathfrak{p}) = (f)$ as Weil divisors on X. Show that $\mathfrak{p} = fA$ (use Proposition 2.11 and Theorem 4.1.14).

(b) Let $f \in A$ be an irreducible element (i.e., if $f = gh$ with $g, h \in A$, then g or h is invertible). Let $\mathfrak{p}$ be a prime ideal, minimal among those containing fA. Show that $fA = \mathfrak{p}$.

(c) Show that A is a unique factorization domain.

(d) Show that if B is a unique factorization domain, then $\operatorname{Cl}(Y) = \operatorname{Pic}(Y) = \operatorname{CaCl}(Y) = 0$, where $Y = \operatorname{Spec} B$.

2.2. Let L/K be an extension of number fields, let $\mathcal{O}_L, \mathcal{O}_K$ be their respective rings of integers, and $f : \operatorname{Spec} \mathcal{O}_L \to \operatorname{Spec} \mathcal{O}_K$ the canonical morphism.

Let $Z \in Z^1(\operatorname{Spec}\mathcal{O}_L)$ be a cycle corresponding to a fractional ideal $\mathfrak{p}$. Show that f_*Z corresponds to the norm of $\mathfrak{p}$ on K. Give an interpretation of Theorem 2.18 in this concrete case.

2.3. Let $f : X \to Y$ be a morphism of Noetherian schemes. We suppose that either f is flat, or X, Y are integral and f is finite surjective.

(a) Let $x \in X$ be a point of codimension 1, and $y = f(x)$. Show that $\dim \mathcal{O}_{Y,y} = 1$ if f is finite surjective, and that $\dim \mathcal{O}_{Y,y} \le 1$ if f is flat. Show that when $\dim \mathcal{O}_{Y,y} = 1$, $e_{x/y} := e_{\mathcal{O}_{X,x}/\mathcal{O}_{Y,y}}$ is finite.

(b) Let $D \in \operatorname{Div}(Y)$. Show that $\operatorname{mult}_x(f^*D) = e_{x/y}\operatorname{mult}_y(D)$ or 0, according to whether y is of codimension 1 or 0. Deduce from this the following equality in $Z^1(X)$:

$$[f^*D] = \sum_x e_{x/y}\operatorname{mult}_y(D)[\overline{\{x\}}],$$

where the sum is taken over the set of points $x \in X$ of codimension 1 such that $y := f(x)$ is of codimension 1.

2.4. Let X be a scheme. Let $D \in \operatorname{Div}(X)$ be a Cartier divisor.

(a) If D is effective, show that $\operatorname{Supp} D = V(\mathcal{O}_X(-D))$.

(b) Let us suppose X is Noetherian. Show that $\operatorname{Supp}[D] \subseteq \operatorname{Supp} D$, and that we have equality if X is regular or D is effective.

(c) Compare $\operatorname{Supp} D$ and $\operatorname{Supp}[D]$ in Example 2.15.

(d) Let $f : X \to Y$ be a morphism verifying one of the conditions of Lemma 1.33. Let $E \in \operatorname{Div}_+(Y)$. Show that $\operatorname{Supp}[f^*E] = f^{-1}(\operatorname{Supp}[E])$.

2.5. Let X be a Noetherian integral scheme. Let us suppose that the normalization morphism $\pi : X' \to X$ is finite (e.g., if X is an algebraic variety, Proposition 4.1.27). Let $D \in \operatorname{Div}(X)$.

(a) Show that $[D] \ge 0 \iff [\pi^*D] \ge 0 \iff \pi^*\mathcal{O}_X(-D) \subseteq \mathcal{O}_{X'}$.

(b) Show that $[D] \ge 0$ if and only if $\mathcal{O}_X(-D) \subseteq \pi_*\mathcal{O}_{X'}$, that is to say that D is locally defined by rational functions that are integral over $\mathcal{O}_X$.

2.6. Let $f : X \to Y$ be a finite dominant morphism of integral Noetherian schemes. Let us suppose that Y is normal. Let $b \in K(X)^*$ and $a = \operatorname{Norm}_{K(X)/K(Y)}(b)$ be the norm of b in $K(Y)$.

(a) Let ξ be a point of codimension 1 of Y, $A = \mathcal{O}_{Y,\xi}$, and $B = (f_*\mathcal{O}_X)_\xi$. Show that B is finite and free over A.

(b) Let $N \subseteq M$ be free A-modules of the same finite rank. Show that there exist a basis $e_1, \ldots, e_n$ of M over A, and $a_1, \ldots, a_n \in A$ such that $a_1e_1, \ldots, a_ne_n$ form a basis of N over A. Taking $M = B$ and $N = bB$, show that

$$a = \operatorname{Norm}_{B/A}(b) = a_1 \cdots a_n, \quad \operatorname{length}_A(A/aA) = \operatorname{length}_A(B/bB).$$

(c) Using Lemma 1.36(b), show that $f_*[\operatorname{div}(b)] = [\operatorname{div}(a)]$.

7.3 Riemann–Roch theorem

Let k be a field. Let us recall that a *curve* over k is an algebraic variety over k whose irreducible components are of dimension 1 (Definition 2.5.29). We will be interested not only in smooth or normal curves, but also, when this is possible, in singular, possibly non-reduced curves. This type of curve can appear when we study the fibers of a regular fibered surface over a Dedekind scheme.

7.3.1 Degree of a divisor

In this subsection, we fix a field k and a curve X over k.

Definition 3.1. Let D be a Cartier divisor on X. The *degree of D*, which we denote by $\deg_k D$, is defined to be the integer

$$\deg_k D := \sum_x \mathrm{mult}_x(D)[k(x) : k],$$

where x runs through the closed points of X (see Definition 1.27). This integer depends on the base field k, but if there is no ambiguity, we will omit the mention of k. It is clear that $D \to \deg_k D$ establishes a group homomorphism.

Remark 3.2. A *0-cycle* on a Noetherian scheme Y is an element of the free Abelian group generated by the closed points of Y. On X, the 0-cycles coincide with the cycles of codimension 1. For any 0-cycle $Z = \sum_x n_x[x]$ on X; we define its degree by $\deg_k Z := \sum_x n_x[k(x) : k]$. For any Cartier divisor D on X, we then have $\deg_k D = \deg_k[D]$.

Example 3.3. Let $X = \mathbb{P}^1_k$ and $D \in \mathrm{Div}(X)$. Then $\deg_k D = \deg_k[D] = \delta([D])$, where δ is defined in Proposition 2.9.

Definition 3.4. Let E be an effective Cartier divisor on a scheme Y. We let $(E, \mathcal{O}_E)$ denote the closed subscheme of Y associated to the invertible sheaf of ideals $\mathcal{O}_Y(-E) \subseteq \mathcal{O}_Y$.

Lemma 3.5. *Let X be a curve over a field k, and let D be a non-zero effective Cartier divisor on X. Then $\deg_k D = \dim_k H^0(D, \mathcal{O}_D)$.*

Proof The scheme $(D, \mathcal{O}_D)$ is finite, and hence affine. For any $x \in D$, $\{x\}$ is a connected component of D. Let $A = H^0(D, \mathcal{O}_D)$. We have $A = \oplus_{x \in D} \mathcal{O}_D(\{x\}) = \oplus_{x \in D} \mathcal{O}_{D,x}$. Hence $\dim_k A = \sum_{x \in D} \dim_k \mathcal{O}_{D,x}$. For any $x \in D$, we have

$$\mathrm{mult}_x(D) := \mathrm{length}_{\mathcal{O}_{X,x}}(\mathcal{O}_{D,x}) = [k(x) : k]^{-1} \dim_k \mathcal{O}_{D,x}$$

(Exercise 1.6(d)). Taking the sum over the points $x \in D$ (note that $\mathrm{mult}_x(D) = 0$ if $x \notin D$), we obtain the lemma.   □

Lemma 3.6. *Let X be a Noetherian scheme of dimension 1. Let D be a Cartier divisor on X. Then there exist two non-zero effective Cartier divisors E and F on X such that $D = E - F$.*

Proof We represent D by a system $\{(U_i, f_i)_i\}$, with $f_i = a_i/b_i$. Then $V(b_i) \subset U_i$ is a finite set. As b_i is a regular element, it does not contain generic points of U_i. Hence $V(b_i)$ is closed in X. Let D_i be the Cartier divisor on X defined by the system $(X \setminus V(b_i), 1)$, (U_i, b_i). Then $F := \sum_i D_i$ is effective, as is $E := D + F$. Adding a non-zero effective divisor to F and E if necessary, we have E, F are non-zero.   □

Proposition 3.7. *Let X be a curve over k, and let K be a field extension of k. Let $p : X_K \to X$ denote the projection morphism.*

(a) *Let $D_K = p^*D$ for every $D \in \mathrm{Div}(X)$ (Definition 1.34). Then $\deg_K D_K = \deg_k D$.*

(b) *Let k' be a subfield of k of finite index. Consider X as a curve over k'. Then $\deg_{k'} D = [k : k'] \deg_k D$.*

Proof (a) It suffices to show the equality for effective D (Lemma 3.6). Let $(D, \mathcal{O}_D)$ be the subvariety of X associated to the divisor D as in Lemma 3.5. Then $(D_K, \mathcal{O}_{D_K})$ is none other than $D \times_k \mathrm{Spec}\, K$. As K is flat over k, we have $H^0(D_K, \mathcal{O}_{D_K}) = H^0(D, \mathcal{O}_D) \otimes_k K$ (Proposition 3.1.24). This proves the proposition, by virtue of Lemma 3.5.

(b) For all $x \in X$, $\mathrm{mult}_x(D)$ does not depend on the base field. So the equality comes from the fact that $[k(x) : k'] = [k : k'][k(x) : k]$.   □

Proposition 3.8. *Let $\pi : X \to Y$ be a finite morphism of integral curves over k. Then for any $D \in \mathrm{Div}(Y)$, we have*

$$\deg \pi^*D = [K(X) : K(Y)] \deg D.$$

In particular, we have $\deg D = \deg \pi^*D$ if π is a normalization morphism (Definition 4.1.19).

Proof Note that condition (2) of Lemma 1.33 is verified; hence π^*D is well defined. Let Z be a 0-cycle on X. Then it immediately follows from the definition that $\deg \pi_* Z = \deg Z$ (Remark 3.2). It now suffices to apply Theorem 2.18.   □

Corollary 3.9. *Let $\pi : X \to Y$ be a finite morphism of normal curves with $Y = \mathbb{P}^1_k$. Let us consider a generator t of $K(Y)$ over k, and let f be the image of t in $K(X)$. Then we have*

$$[K(X) : K(Y)] = \deg(f)_0 = \deg(f)_\infty.$$

Proof As X, Y are regular, we will identify Weil divisors with Cartier divisors (Proposition 2.16). By the choice of t, we have $(t) = [P] - [Q]$, where P, Q are two distinct rational points. It follows that

$$(f) = \pi^*(t) = \pi^*[P] - \pi^*[Q].$$

As $\pi^*[P]$ and $\pi^*[Q]$ are non-zero effective divisors with disjoint support, we have $\pi^*[P] = (f)_0$ and $\pi^*[Q] = (f)_\infty$. As $[P]$ is of degree 1, the corollary results from Proposition 3.8.   □

Morphisms to $\mathbb{P}^1_k$

Let X be a normal curve over a field k, and let $f \in K(X)$ be a rational function. Then f induces a morphism $\pi : X \to \mathbb{P}^1_k = \operatorname{Proj} k[T_0, T_1]$ in the following way. Let $U = X \setminus \operatorname{Supp}(f)_\infty$, $V = X \setminus \operatorname{Supp}(f)_0$ be such that $f \in \mathcal{O}_X(U)$ and $1/f \in \mathcal{O}_X(V)$. Let us consider the morphism $\pi_0 : U \to D_+(T_0)$ corresponding to the homomorphism $k[T_1/T_0] \to \mathcal{O}_X(U)$ which sends T_1/T_0 to f, and $\pi_1 : V \to D_+(T_1)$ corresponding to $T_0/T_1 \mapsto 1/f$. Then we immediately verify that π_0, π_1 coincide on $U \cap V$ and therefore define a morphism $\pi : X \to \mathbb{P}^1_k$. Clearly, π is dominant if and only if f is transcendent over k. If this is the case, π is then quasi-finite.

Lemma 3.10. *Let X, Y be projective curves over k, and $\pi : X \to Y$ be a morphism. Then the following conditions are equivalent:*

(i) *π is finite.*

(ii) *For any irreducible component X_i of X, $\pi(X_i)$ is not reduced to a point.*

(iii) *The image of any generic point of X is a generic point of Y.*

Proof We clearly have (i) implies (ii). (ii) implies (iii) because $\pi(X_i)$ is an irreducible closed subset of dimension 1. If (iii) is verified, as X is of dimension 1, π is quasi-finite. Now π is projective (Corollary 3.3.32(e)), and hence finite (Corollary 4.4.7). □

Remark 3.11. Let X be a normal curve over k, let $f \in K(X)$, and $D \in \operatorname{Div}_+(X)$ be the Cartier divisor such that $[D] = (f)_\infty$, and let $\mathcal{L} = \mathcal{O}_X(D)$. Then it is easy to see that $\{1, f\}$ are sections of $H^0(X, \mathcal{L})$ which generate $\mathcal{L}$ (see the definition before Proposition 5.1.31). The morphism $\pi : X \to \mathbb{P}^1_k$ is none other than the morphism associated to $\mathcal{L}$ and to the sections $\{1, f\}$ (see the construction of Proposition 5.1.31). Moreover, the morphism π is of degree (Exercise 5.1.25) $\deg D$ by Corollary 3.9.

Corollary 3.12. *Let X be a normal projective curve over k. We suppose that there exists an $f \in K(X)$ such that $(f) = [x_0] - [x_1]$, where x_0, x_1 are two distinct rational points. Then the morphism $\pi : X \to \mathbb{P}^1_k$ induced by f is an isomorphism.*

Proof Indeed, π is finite because it is non-constant, and of degree 1 by the above, and hence birational. As $\mathbb{P}^1_k$ is normal, π is an isomorphism. □

Let us conclude with the following proposition which characterizes normal projective curves by their function fields. Let us recall that a function field in d variables over k is a finitely generated extension of k, of transcendence degree d over k.

Proposition 3.13. *Let k be a field.*

(a) *For any function field in one variable K/k, there exists, up to isomorphism, a unique normal projective curve X such that $K(X) = K$.*

(b) *For any normal projective curves X, Y over k, there exists a canonical injection*

$$\rho : \mathrm{Hom}_{k-\mathrm{alg}}(K(Y), K(X)) \hookrightarrow \mathrm{Mor}_k(X, Y)$$

whose image is the set of dominant morphisms, and which is compatible with the composition of morphisms. In particular, isomorphisms correspond to isomorphisms.

(c) *The image by ρ of a homomorphism $\varphi : K(Y) \to K(X)$ is the normalization morphism of Y in $K(X)$.*

Proof (a) The existence of X is true in every dimension. Indeed, K is the field of fractions of a finitely generated algebra A. The integral affine variety $\mathrm{Spec}\, A$ is a dense open subscheme of an integral projective variety Y. Let X be the normalization of Y (Definition 4.1.19). Then X is a normal variety, finite over Y (Proposition 4.1.27), of dimension $\mathrm{trdeg}_k K$. Finally, X is projective over k, by Exercise 5.2.16. The uniqueness of X (in dimension 1) results from part (b), whose proof is independent of (a).

(b) Let $\varphi : K(Y) \to K(X)$ be a homomorphism of k-algebras. Let η, ξ be the generic points of X and Y, respectively. Then $K(X) = \mathcal{O}_{X,\eta}$, $K(Y) = \mathcal{O}_{Y,\xi}$, and φ induces a morphism $\mathrm{Spec}\, \mathcal{O}_{X,\eta} \to \mathrm{Spec}\, \mathcal{O}_{Y,\xi}$. This morphism extends to a morphism $U \to Y$ for some non-empty open subscheme U (Exercise 3.2.4), and therefore to a morphism $f : X \to Y$ (Corollary 4.1.17). We will set $\rho(\varphi) = f$. This is a dominant morphism, by construction. Conversely, let $f : X \to Y$ be a dominant morphism. Then f induces a homomorphism $\varphi : \mathcal{O}_{Y,\xi} \to \mathcal{O}_{X,\eta}$. It is then easy to verify that $\rho(\varphi) = f$, as well as the other stated properties.

(c) Let $Y' \to Y$ be the normalization morphism of Y in $K(X)$. Then φ decomposes into $K(Y) \to K(Y') = K(X)$. It follows from (b) that $\rho(\varphi)$ is the morphism $Y' \to Y$.   □

Remark 3.14. This proposition can be stated as follows: there exists an equivalence between the category of normal projective curves over k with finite morphisms and the category of function fields in one variable over k with homomorphisms of k-algebras.

Example 3.15. There exist normal projective curves that are geometrically integral and not smooth over k. Indeed, let $p > 2$ be a prime number, and let k be a non-perfect field of characteristic p. Let us consider the extension $K = k(t, y)$ of $k(t)$ defined by $y^2 = t^p - \alpha$, where $\alpha \in k \setminus k^p$. Then the normal projective curve X associated to K is an example. So Proposition 3.13(a) is not true if we replace normal by smooth.

7.3.2 Riemann–Roch for projective curves

Riemann's theorem expresses the Euler–Poincaré characteristic $\chi(\mathcal{L})$ (Definition 5.3.26) of an invertible sheaf $\mathcal{L}$ associated to a Cartier divisor D as a function of the degree of D. We fix a field k. The algebraic varieties considered in this subsection are projective varieties over k.

Lemma 3.16. *Let X be a projective variety over a field k. Let*
$$0 \to \mathcal{F} \to \mathcal{G} \to \mathcal{H} \to 0 \tag{3.5}$$
be an exact sequence of coherent sheaves on X. Then we have
$$\chi(\mathcal{G}) = \chi(\mathcal{F}) + \chi(\mathcal{H}).$$

Proof If we have an exact sequence of vector spaces over k
$$0 \to E_0 \to E_1 \to \cdots \to E_n \to 0,$$
then $\sum_{0 \le i \le n} (-1)^i \dim_k E_i = 0$. It suffices to apply this results to the long exact cohomology sequence deduced from exact sequence (3.5). $\qquad\square$

Theorem 3.17 (Riemann). *Let X be a projective curve over a field k. Let D be a Cartier divisor on X. Then we have*
$$\chi(\mathcal{O}_X(D)) = \deg D + \chi(\mathcal{O}_X).$$

Proof We omit k in the indices. By Lemma 3.6, we can write $D = E - F$ with E and F non-zero effective. We have an exact sequence of sheaves
$$0 \to \mathcal{O}_X(-F) \to \mathcal{O}_X \to \mathcal{O}_F \to 0.$$
It stays exact when tensoring over $\mathcal{O}_X$ with $\mathcal{O}_X(E)$, because the latter is flat over $\mathcal{O}_X$:
$$0 \to \mathcal{O}_X(D) \to \mathcal{O}_X(E) \to \mathcal{O}_X(E)|_F \to 0.$$
As F is a finite scheme, $\mathcal{O}_X(E)|_F \simeq \mathcal{O}_F$; moreover, $H^q(F, \mathcal{O}_F) = 0$ for every $q \ge 1$. It follows from Lemmas 3.5, 3.16 that
$$\chi(\mathcal{O}_X(D)) = \chi(\mathcal{O}_X(E)) - \deg F.$$
Applying this equality to $D = 0$, we obtain $\chi(\mathcal{O}_X(E)) = \chi(\mathcal{O}_X) + \deg E$, whence $\chi(\mathcal{O}_X(D)) = \deg E + \chi(\mathcal{O}_X) - \deg F = \deg D + \chi(\mathcal{O}_X)$. $\qquad\square$

Corollary 3.18. *Let X be a projective curve over a field. Let $\mathrm{div}(f)$ be a principal Cartier divisor on X. Then $\deg(\mathrm{div}(f)) = 0$.*

Proof The invertible sheaf $f\mathcal{O}_X$ associated to $\mathrm{div}(f)$ is isomorphic to $\mathcal{O}_X$. It follows that $\chi(f\mathcal{O}_X) = \chi(\mathcal{O}_X)$, whence $\deg(\mathrm{div}(f)) = 0$ by Theorem 3.17. If X is normal, this also follows from Corollary 3.9. $\qquad\square$

Definition 3.19. Let X be a projective curve over a field k. The *arithmetic genus of X* is defined to be the integer
$$p_a(X) := 1 - \chi_k(\mathcal{O}_X).$$
It depends on k. For any extension k'/k, we have $p_a(X_{k'}) = p_a(X)$ (Corollary 5.2.27) where $p_a(X_{k'})$ is computed over k'. If X is geometrically connected and geometrically reduced, so that we have $H^0(X, \mathcal{O}_X) = k$ (Corollary 3.3.21), then $p_a(X) = \dim_k H^1(X, \mathcal{O}_X)$. For a smooth projective variety Y over k, we define the *geometric genus of Y* to be
$$p_g(Y) := \dim_k H^0(Y, \omega_{Y/k}).$$
If Y is a curve, we usually denote the geometric genus by $g(Y)$. See Remark 3.28. For any Cartier divisor $D \in \mathrm{Div}(X)$, we let
$$L(D) := H^0(X, \mathcal{O}_X(D)), \quad l(D) := \dim_k L(D).$$

Remark 3.20. Let X be a normal projective curve over k. Let $Z = \sum_x n_x[x]$ be a 0-cycle on X, and $D \in \mathrm{Div}(X)$ be such that $[D] = Z$. Then

$$L(D) = \{f \in K(X)^* \mid \mathrm{mult}_x(f) + n_x \geq 0, \ \forall x\} \cup \{0\},$$

where x runs through the closed points of X (Exercise 1.13 and Example 2.6). It is therefore the set of rational functions in which we have a certain control over the zeros and the poles. This is what motivated the study of these vector spaces in the first place. If X is only integral, we have (Exercise 1.13)

$$L(D) = \{f \in K(X)^* \mid \mathrm{div}(f) + D \geq 0\} \cup \{0\}.$$

Example 3.21. The projective line $\mathbb{P}^1_k$ is of genus $p_a = p_g = 0$. This follows from Lemma 5.3.1 and Corollary 6.4.17.

Example 3.22. Let X be a projective plane curve over k (Exercise 5.3.3(c)), defined by a homogeneous polynomial F of degree $n \geq 1$. Then $H^0(X, \mathcal{O}_X) = k$ and $p_a(X) = (n-1)(n-2)/2$ (Exercise 5.3.3(b.2) and (d)). In particular, an elliptic curve (6.1.25) is of arithmetic genus 1.

Corollary 3.23. *Let X be a projective curve over k, and let $D \in \mathrm{Div}(X)$ be a Cartier divisor. Then we have*

$$l(D) \geq \deg D + 1 - p_a(X).$$

Proof This follows from Theorem 3.17.   □

Corollary 3.24. *Let X be a normal projective curve over k. Then $X \simeq \mathbb{P}^1_k$ if and only if there exists a Cartier divisor D such that $\deg D = 1$ and $l(D) \geq 2$. Moreover, for such a divisor D the sheaf $\mathcal{O}_X(D)$ is very ample.*

Proof Let us suppose that there exists such a divisor D. Let $g \in L(D)$. Then $D \sim \mathrm{div}(g) + D \geq 0$. Hence we may suppose $D \geq 0$. It follows that D is the divisor associated to a rational point $x_1 \in X(k)$. In particular, $H^0(X, \mathcal{O}_X) = k$ (the canonical homomorphism $H^0(X, \mathcal{O}_X) \to k(x_1) = k$ is a field homomorphism, and hence injective). Let $f \in L(D) \setminus k$; then $\mathrm{div}(f) + D$ is an effective Cartier divisor, distinct from D. It follows that $(f) = [x_0] - [x_1]$ with $x_0 \in X(k) \setminus \{x_1\}$. It follows from Corollary 3.12 and the remark preceding it that $\mathcal{O}_X(D)$ induces an isomorphism $X \simeq \mathbb{P}^1_k$. See Example 3.34 for the remaining part of the corollary.   □

Proposition 3.25. *Let X be a projective curve over a field k.*

(a) *Let $D' \leq D$ be Cartier divisors on a projective curve X. Then we have*

$$l(D') \leq l(D) \leq l(D') + \deg(D - D').$$

In particular, if $D \geq 0$, then

$$l(D) \leq \deg D + \dim_k H^0(X, \mathcal{O}_X).$$

(b) *Let us suppose X is integral. If $\deg D = 0$, then $l(D) \neq 0$ if and only if $D \sim 0$. If $\deg D < 0$, then $l(D) = 0$.*

(c) Let X be as in (b). Let $\mathcal{L}$ be an invertible sheaf on X. Then $\mathcal{L} \simeq \mathcal{O}_X$ if and only if $\deg \mathcal{L} = 0$ and $H^0(X, \mathcal{L}) \neq 0$.

Proof (a) As $\mathcal{O}_X(D')$ is a subsheaf of $\mathcal{O}_X(D)$, we have $l(D') \leq l(D)$. Suppose $D' < D$. Let us write $D = D' + E$ with E non-zero effective. Then by tensoring the exact sequence

$$0 \to \mathcal{O}_X(-E) \to \mathcal{O}_X \to \mathcal{O}_E \to 0$$

(see Definition 3.4 for $\mathcal{O}_E$) with the invertible (hence flat) sheaf $\mathcal{O}_X(D)$, we obtain an exact sequence

$$0 \to \mathcal{O}_X(D') \to \mathcal{O}_X(D) \to \mathcal{O}_X(D)|_E \to 0.$$

As E is a finite scheme, we have $\mathcal{O}_X(D)|_E \simeq \mathcal{O}_E$, whence an exact sequence

$$0 \to L(D') \to L(D) \to H^0(E, \mathcal{O}_E).$$

As the term on the right is of dimension $\deg E$ over k, we obtain the desired inequality.

(b) Let us suppose $l(D) \neq 0$. Let $f \in L(D) \setminus \{0\}$. Then $D \sim \mathrm{div}(f) + D \geq 0$. This implies first that $\deg D \geq 0$, and then that if $\deg D = 0$, $\mathrm{div}(f) + D = 0$, and hence $D \sim 0$.

(c) Let us suppose that $\deg \mathcal{L} = 0$ and that there exists a non-zero global section $s \in H^0(X, \mathcal{L})$. Then $\mathcal{L}' := \mathcal{L} \otimes (s\mathcal{O}_X)^\vee$ is an invertible sheaf containing $\mathcal{O}_X$ and contained in the constant sheaf $K_{X/k}$ (here we use the X integral hypothesis). It is therefore associated to a unique effective Cartier divisor D. As $\mathcal{L}' \simeq \mathcal{L}$, we have $\deg D = \deg \mathcal{L} = 0$. This implies that $D = 0$; hence $\mathcal{L} \simeq \mathcal{O}_X$. The converse is trivial. □

To have more precise information on $L(D)$, we need to determine the term $H^1(X, \mathcal{O}_X(D))$. Let $f : X \to \mathrm{Spec}\, k$ be the structural morphism. Let ω_f be the 1-dualizing sheaf for f (Definition 6.4.18, Corollary 6.4.29, and Remark 6.4.30).

Theorem 3.26 (Riemann–Roch). *Let $f : X \to \mathrm{Spec}\, k$ be a projective curve over a field k. Then for any Cartier divisor $D \in \mathrm{Div}(X)$, we have*

$$\dim_k H^0(X, \mathcal{O}_X(D)) - \dim_k H^0(X, \omega_f \otimes \mathcal{O}_X(-D)) = \deg D + 1 - p_a.$$

Proof Let $\mathcal{L}$ be an invertible sheaf on X. We have

$$H^0(X, \mathcal{L}^\vee \otimes \omega_f) \simeq \mathrm{Hom}_{\mathcal{O}_X}(\mathcal{L}, \omega_f) \simeq H^1(X, \mathcal{L})^\vee$$

(Remarks 6.4.20–21). It suffices to apply this result to $\mathcal{L} = \mathcal{O}_X(D)$, and the theorem then follows from Theorem 3.17. □

Remark 3.27. If $X \to \mathrm{Spec}\, k$ is an l.c.i. (6.3.17), for example if f is regular, we know that ω_f coincides with the canonical sheaf $\omega_{X/k}$ $(= \Omega^1_{X/k}$ if X is smooth). This result can be shown in a relatively elementary, but not immediate, manner, when X is a smooth projective curve over k, either by using the method of residues ([86], II, n° 4–9), or by Noether's method, using the fact that the curve is birational to a projective plane curve having only ordinary double points as singular points, see [38], Section 8.6.

Remark 3.28. Let $X \to \operatorname{Spec} k$ be as in the remark above. Then

$$H^0(X, \omega_{X/k}) \simeq H^1(X, \mathcal{O}_X)^\vee, \quad H^0(X, \mathcal{O}_X) \simeq H^1(X, \omega_{X/k})^\vee.$$

In particular, $\chi(\omega_{X/k}) = -\chi(\mathcal{O}_X)$. Moreover, if X is smooth and geometrically connected, the first isomorphism implies that $p_a(X) = p_g(X)$.

Definition 3.29. Let X be a projective curve over a field k, and let $\mathcal{L}$ be an invertible sheaf on X. We define the *degree of* $\mathcal{L}$ to be the integer

$$\deg_k \mathcal{L} := \chi_k(\mathcal{L}) - \chi_k(\mathcal{O}_X).$$

We omit k in the subscript if there is no ambiguity.

Lemma 3.30. *Let X be a projective curve over a field k.*

(a) *If $\mathcal{L} \simeq \mathcal{O}_X(D)$ for some Cartier divisor D, then $\deg \mathcal{L} = \deg D$.*

(b) *The map $\mathcal{L} \mapsto \deg \mathcal{L}$ induces a group homomorphism $\operatorname{Pic}(X) \to \mathbb{Z}$.*

Proof (a) results from Theorem 3.17. (b) Let K be an infinite field containing k. Then there exist Cartier divisors $D_1, D_2 \in \operatorname{Div}(X_K)$ such that $p^*\mathcal{L}_i \simeq \mathcal{O}_{X_K}(D_i)$, where $p : X_K \to X$ is the projection (Proposition 1.32). Hence

$$\deg_K p^*(\mathcal{L}_1 \otimes \mathcal{L}_2) = \deg_K((D_1)_K + (D_2)_K) = \deg_K p^*\mathcal{L}_1 + \deg_K p^*\mathcal{L}_2.$$

Now $\deg_K p^*\mathcal{L} = \deg_k \mathcal{L}$ for every invertible sheaf $\mathcal{L}$ (Proposition 3.7). Hence $\deg_k$ is an additive function. This proves (b), because $\deg \mathcal{O}_X = 0$.   □

Corollary 3.31. *Let X be an l.c.i. projective curve over k, with arithmetic genus p_a. Then we have*

(a) $\deg \omega_{X/k} = 2(p_a - 1)$;

(b) $\dim_k H^0(X, \omega_{X/k}) = p_a$ *if X is geometrically connected and geometrically reduced.*

Proof (a) This is an immediate consequence of Theorem 3.17 and Remark 3.28. Under the hypothesis of (b), we have $H^0(X, \mathcal{O}_X) = k$ (Corollary 3.3.21). The equality then results from the duality theorem (Remark 3.28).   □

Definition 3.32. Let X be an l.c.i. projective curve over a field k. Any Cartier divisor K on X such that $\mathcal{O}_X(K) \simeq \omega_{X/k}$ is called a *canonical divisor on X*. Such a divisor always exists (see Proposition 1.32). We also denote it by $K_{X/k}$. It is only defined up to linear equivalence.

Remark 3.33. Let X be as above. Let $D \in \operatorname{Div}(X)$. Then the Riemann–Roch theorem states that

$$l(D) - l(K - D) = \deg D + 1 - p_a(X). \tag{3.6}$$

If X is integral and $\deg D > 2p_a(X) - 2 = \deg K$, then

$$l(D) = \deg D + 1 - p_a(X)$$

(Proposition 3.25(b)).

Example 3.34. Let X be an integral l.c.i. projective curve over k, of genus $p_a = 0$ (e.g., $X = \mathbb{P}^1_k$, see Example 3.21). Let $D \in \mathrm{Div}(X)$. We have

$$l(D) = \begin{cases} \deg D + 1 & \text{if } \deg D \geq 0 \\ 0 & \text{otherwise.} \end{cases} \tag{3.7}$$

Indeed, if $\deg D < 0$, then $l(D) = 0$ (Proposition 3.25(b)). By Corollary 3.31, $\deg K = -2$. If $\deg D \geq 0$, $K - D$ is of degree < 0, and therefore $l(K - D) = 0$. Formula (3.6) implies that $l(D) = \deg D + 1$.

When $X = \mathbb{P}^1_k$, we can verify equality (3.7) directly. We have $\mathcal{O}_X(D) \simeq \mathcal{O}_X(n)$ for some $n \in \mathbb{Z}$ (Propositions 2.9, 2.16). Now $\mathcal{O}_X(1) \simeq \mathcal{O}_X(D_0)$ if D_0 is the Cartier divisor associated to a rational point of X. It follows that $n = n \deg \mathcal{O}_X(1) = \deg \mathcal{O}_X(n)$. Equality (3.7) then follows from Lemma 5.3.1.

Example 3.35. Let X be an integral l.c.i. projective curve over k. Let us suppose that X is of arithmetic genus $p_a = 1$. Then

$$\dim_k H^0(X, \omega_{X/k}) = \dim_k H^1(X, \mathcal{O}_X) \geq 1, \quad \deg \omega_{X/k} = 2(p_a - 1) = 0.$$

It follows from Proposition 3.25(c) that $\omega_{X/k} \simeq \mathcal{O}_X$. For any divisor D such that $\deg D > 0$, we then have $l(D) = \deg D$, by Proposition 3.25(b) and formula (3.6). Let us note that conversely, if X is an l.c.i. projective curve over k such that $\omega_{X/k} \simeq \mathcal{O}_X$, then $p_a = 1$, because we have $\deg \omega_{X/k} = 0$.

Exercises

3.1. Let X be a projective curve over a field k, and let k'/k be an extension. We let $p : X_{k'} \to X$ denote the projection morphism. Let $D \in \mathrm{Div}(X)$.

    (a) Show that D is principal if and only if $H^0(X, \mathcal{K}^*) \cap L(D) \neq \emptyset$ and $\deg D = 0$.

    (b) Let us suppose X is integral. Show that D is principal if and only if $p^* D$ is.

3.2. Let X be a normal projective curve over an infinite field k. Let $x_1, \ldots, x_r$ be pairwise distinct closed points of X, and let $n_1, \ldots, n_r \in \mathbb{Z}$.

    (a) Let $D = \sum_i n_i x_i$ (considered as a Cartier divisor). Let $y \in X$ be a closed point distinct from all of the x_i. Show that for n large enough, we have $L(D + ny) \neq \cup_i L(D + ny - x_i)$.

    (b) Show that there exists an $f \in K(X)$ such that $\mathrm{mult}_{x_i}(f) = n_i$ for every $1 \leq i \leq r$. See also the approximation theorem (Lemma 9.1.9).

3.3. Let X be a normal projective curve over a field k. Let $\mathcal{L}$ be a locally free sheaf of rank r on X.

    (a) Let us suppose that $H^0(X, \mathcal{L})$ admits a non-zero global section s. Let $\mathcal{F} = \mathcal{L}/s\mathcal{O}_X$ and $\mathcal{L}' = \mathcal{F}/\mathcal{F}_{\mathrm{tors}}$. Show that $\mathcal{L}'$ is locally free of rank $r - 1$, and that we have an exact sequence

$$0 \to \mathcal{L}'' \to \mathcal{L} \to \mathcal{L}' \to 0,$$

with $\mathcal{L}''$ locally free of rank 1.

(b) Show that we always have an exact sequence as in (a) (consider $\mathcal{L}(n)$ for n large enough). Deduce from this that

$$\chi(\mathcal{L}) = \deg(\det \mathcal{L}) + r\chi(\mathcal{O}_X).$$

3.4. Let X be a projective variety of dimension d over a field k, endowed with a very ample sheaf $\mathcal{O}_X(1)$. Let $\mathcal{F}$ be a coherent sheaf on X. We are going to estimate $\chi(\mathcal{F}(n))$ for $n \in \mathbb{Z}$. Let $D \in \mathrm{Div}(X)$ be effective such that $\mathcal{O}_X(D) \simeq \mathcal{O}_X(1)$ (see Proposition 1.32). Let $Y = V(\mathcal{O}_X(-D))$.

(a) Let $\mathcal{G}$ be the kernel of $\mathcal{F} \otimes \mathcal{O}_X(-D) \to \mathcal{F}$. Show that $\mathcal{O}_X(-D)\mathcal{G} = 0$. Show that for any $n \in \mathbb{Z}$, we have an exact sequence

$$0 \to \mathcal{G}(n) \to \mathcal{F}(n) \to \mathcal{F}(n+1) \to \mathcal{F}|_Y \otimes_{\mathcal{O}_Y} \mathcal{O}_Y(n+1) \to 0,$$

where $\mathcal{O}_Y(1) = \mathcal{O}_X(1)|_Y$ and $\mathcal{O}_Y(m) = \mathcal{O}_Y(1)^{\otimes m}$.

(b) Show by induction on d that there exists a polynomial $P(T) \in \mathbb{Q}[T]$ such that $\chi(\mathcal{F}(n)) = P(n)$ for every $n \in \mathbb{Z}$, with $\deg P(T) \le d$ and $d! \, P(T) \in \mathbb{Z}[T]$. The polynomial $P(T)$, which depends on $\mathcal{F}$ and on the choice of $\mathcal{O}_X(1)$, is called the *Hilbert polynomial* of $\mathcal{F}$.

(c) Show that for any large enough n, we have $P(n) = \dim_k H^0(X, \mathcal{F}(n))$.

(d) If X is a curve over k and $\mathcal{F}$ an invertible sheaf, show that $P(T) = a_1 T + a_0$ with $a_1 = \deg \mathcal{O}_X(1)$.

3.5. Let X be a locally Noetherian scheme of dimension 1. Let U be an open subset of X containing $\mathrm{Ass}(\mathcal{O}_X)$. Let $D \in \mathrm{Div}_+(X)$ be such that $\mathrm{Supp}\, D = X \setminus U$.

(a) Using Proposition 1.15, show that we have a canonical injective homomorphism $\mathcal{O}_X(U) \hookrightarrow H^0(X, \mathcal{K}_X)$.

(b) Let $n \in \mathbb{Z}$. Let $i : U \to X$ be the canonical injection. Show that $\mathcal{O}_X(nD) \to i_*(\mathcal{O}_X(nD)|_U)$ is injective (Lemma 1.9). Deduce from this that $L(nD) \subseteq \mathcal{O}_X(U)$.

(c) Let $f \in \mathcal{O}_X(U)$. Let $x \in \mathrm{Supp}\, D$. Let us write $f_x = a/b \in \mathcal{K}_{X,x}$. Let $\pi \in \mathcal{O}_{X,x}$ be a generator of $\mathcal{O}_X(-D)_x$. Show that there exists an $n_x \ge 1$ such that $\pi^n \in b\mathcal{O}_{X,x}$ for every $n \ge n_x$ (Lemma 2.5.11). Deduce from this that $f_x \in \mathcal{O}_X(nD)_x$.

(d) Show that $\mathcal{O}_X(U) = \cup_{n \ge 1} L(nD)$.

7.4   Algebraic curves

7.4.1   Classification of curves of small genus

Let X be a connected projective curve over a field k. Let $x \in X$ be a regular closed point. For simplicity, we will also denote the Cartier divisor associated to the cycle $[x]$ (see Proposition 2.16) by x.

Proposition 4.1. *Let X be a geometrically integral projective curve over a field k, of arithmetic genus $p_a \leq 0$. Then we have the following properties:*

(a) *The curve X is a smooth conic over k.*

(b) *We have $X \simeq \mathbb{P}_k^1$ if and only if $X(k) \neq \emptyset$.*

(See also Proposition 9.3.16.)

Proof Let X' be the normalization of $X_{\bar{k}}$. Then $H^0(X', \mathcal{O}_{X'}) = \bar{k}$, and hence $p_a(X') \geq 0$. It follows from Proposition 5.4 that $X_{\bar{k}} = X'$. In other words, X is smooth over k.

Let us first show (b). Let us suppose that there exists an $x_1 \in X(k)$. Then $l(x_1) = 2$, by Example 3.34. Hence $X \simeq \mathbb{P}_k^1$ and $\mathcal{O}_X(x_1)$ is very ample, by Corollary 3.24.

Let us now show (a). Let K be a canonical divisor of X, and let $D = -K$. Then $\deg D = 2$ and $l(D) = 3$. We are going to show that $\mathcal{O}_X(D)$ is very ample. This will imply that $\mathcal{O}_X(D)$ induces a closed immersion from X to $\mathbb{P}_k^2$, and the image of X will be a conic, by the genus formula (Example 3.22). To show that $\mathcal{O}_X(D)$ is very ample, we can do a base change and suppose that k is algebraically closed (Exercise 5.1.30). Let $x_1 \in X(k)$. Then $l(D - 2x_1) = 1$, and therefore $D \sim 2x_1$ (Proposition 3.25(b)). Consequently, $\mathcal{O}_X(D) \simeq \mathcal{O}_X(x_1)^{\otimes 2}$. Now we have just seen that $\mathcal{O}_X(x_1)$ is very ample, and hence so is $\mathcal{O}_X(D)$ (Exercise 5.1.27). $\square$

We have seen in the proof of the proposition above that on a smooth projective curve of genus 0, every divisor of degree 1 or 2 is very ample. Therefore every divisor of strictly positive degree is very ample (Exercise 5.1.28(c)). In Proposition 4.4 below, we are going to generalize this type of statement.

Lemma 4.2. *Let X be a projective curve over a field k, and let $D \in \mathrm{Div}_+(X)$ be with support in the regular locus of X. Then $\mathcal{O}_X(D)$ is generated by its global sections if and only if for any $x \in \mathrm{Supp}\, D$, we have $l(D - x) < l(D)$.*

Proof As $D \geq 0$, and hence $1 \in H^0(X, \mathcal{O}_X(D))$, then $\mathcal{O}_X(D)$ is generated by its global sections at every point $x \notin \mathrm{Supp}\, D$. Let $x \in \mathrm{Supp}\, D$. Then

$$\mathcal{O}_X(D - x)_x = \mathfrak{m}_x \mathcal{O}_X(D)_x.$$

If $l(D - x) < l(D)$, there exists an $s \in L(D) \setminus L(D - x)$. We then have $s_x \notin \mathcal{O}_X(D - x)_x$. Hence s_x is a generator of $\mathcal{O}_X(D)_x$, and $\mathcal{O}_X(D)$ is generated by its global sections at x.

Conversely, let us suppose $\mathcal{O}_X(D)$ is generated by its global sections at x. Let $s \in L(D)$ be such that s_x is a generator of $\mathcal{O}_X(D)_x$. Then $s_x \notin L(D - x)_x$, and hence $s \notin L(D - x)$, which implies that $l(D - x) < l(D)$. $\square$

Lemma 4.3. *Let X be a connected smooth projective curve over an algebraically closed field k. Let D be an effective Cartier divisor on X such that for any pair of (not necessarily distinct) points $p, q \in X(k)$, we have*

$$l(D - p - q) < l(D - p) < l(D).$$

Then $\mathcal{O}_X(D)$ is very ample.

Proof It follows from Lemma 4.2 that $\mathcal{O}_X(D)$ is generated by its global sections. Let $\{s_0, \ldots, s_n\}$ be a basis of $L(D)$ and $f : X \to \mathbb{P}_k^n$ be the morphism associated to this basis (Proposition 5.1.31). Let us first show that f is injective. Let p, q be two distinct closed points. Then there exists an $s \in L(D - p) \setminus L(D - p - q)$. We have $s_p \in \mathfrak{m}_p \mathcal{O}_X(D)_p$ while s_q is a generator of $\mathcal{O}_X(D)_q$. Let us write $s = \sum_{0 \le i \le n} \lambda_i s_i$ with $\lambda_i \in k$ and $f(p) = (p_0, \ldots, p_n)$, $f(q) = (q_0, \ldots, q_n)$. Let us fix a basis e of $\mathcal{O}_X(D)_p$. Then $(s_i)_p \in p_i e + \mathfrak{m}_p^2$, and hence $\sum_i \lambda_i p_i = 0$. In the same manner, $\sum_i \lambda_i q_i \ne 0$. This shows that $f(p) \ne f(q)$.

Let us show that the tangent map $T_{f,p}$ is injective for every $p \in X(k)$. This will imply that f is a closed immersion (Exercise 4.4.1). Let us, for example, suppose that s_0 generates $\mathcal{O}_X(D)$ at p. Then $f(p) \in D_+(T_0)$ and the restriction of f to X_{s_0} is given by the homomorphism $\rho : k[T_1/T_0, \ldots, T_n/T_0] \to \mathcal{O}_X(X_{s_0})$, $\rho(T_i/T_0) = s_i/s_0$, where s_i/s_0 denotes the regular function such that $s_i = (s_i/s_0)s_0$ on X_{s_0}. As $l(D - p) > 0$, $D - p$ is equivalent to an effective divisor on X. By Lemma 4.2, $\mathcal{O}_X(D - p)$ is generated by its global sections at p. Thus $\mathfrak{m}_p s_{0,p} = \mathcal{O}_X(D - p)_p = L(D - p) + \mathfrak{m}_p^2 s_{0,p}$. Let $s \in L(D - p)$. Then $s = \sum_{0 \le i \le n} \lambda_i s_i$ with $\lambda_i \in k$, and $\rho(\sum_i \lambda(T_i/T_0)) = s/s_i \in \mathfrak{m}_p$. Hence $\sum_i \lambda_i(T_i/T_0) \in \mathfrak{m}_{f(p)}$. It follows that $\mathfrak{m}_{f(p)}/\mathfrak{m}_{f(p)}^2 \to \mathfrak{m}_p/\mathfrak{m}_p^2$ is surjective, and therefore that $T_{f,p}$ is injective.   □

Proposition 4.4. *Let X be a smooth, geometrically connected, projective curve over a field k of genus g. Let $\mathcal{L}$ be an invertible sheaf on X.*

  (a) *If $\deg \mathcal{L} \ge 2g$, then $\mathcal{L}$ is generated by its global sections.*

  (b) *If $\deg \mathcal{L} \ge 2g + 1$, then $\mathcal{L}$ is very ample.*

Proof We may suppose that k is algebraically closed (Exercises 5.1.29–30). Let D be a Cartier divisor such that $\mathcal{O}_X(D) \simeq \mathcal{L}$. Let us first note that if $\deg D \ge 2g$ (resp. $\deg D \ge 2g + 1$), then

$$l(D - E) = l(D) - \deg E$$

for every effective divisor E such that $\deg E \le 1$ (resp. $\deg E \le 2$). This results from Remark 3.33. In particular, $l(D) \ne 0$. Hence, by linear equivalence, we can reduce to $D \ge 0$. Lemma 4.2 then implies (a). Part (b) is an immediate consequence of Lemma 4.3.   □

Corollary 4.5. *Let X be a smooth, connected, projective curve over k, of genus 1. We suppose that there exists an $o \in X(k)$. Then X is an elliptic curve and $\mathcal{O}_X(3o)$ is very ample.*

Proof Note that $X(k) \ne \emptyset$ implies that X is geometrically connected (Exercise 3.2.11). By the proposition above, $\mathcal{O}_X(3o)$ is very ample. As $l(3o) = 3$, $\mathcal{O}_X(3o)$ induces a closed immersion into $\mathbb{P}_k^2$. The genus formula (Example 3.22) shows that the image of X is necessarily a cubic, that is to say of the form $V_+(F(u, v, w))$ for a homogeneous polynomial F of degree 3.

Let H be a line in $\mathbb{P}_k^2$ such that $H \cap X = \{o\}$ (use Lemma 4.6 below with $m = 1$, $D = 3o$, and $P = \mathbb{P}_k^2$). By an automorphism of $\mathbb{P}_k^2$, we can suppose that

$o = (0, 1, 0)$ and $H = \{w = 0\}$. Then, multiplying u and v by suitable elements of k^* if necessary, F takes the form of Definition 6.1.25. We can also show this assertion directly without using Lemma 4.6, see Exercise 4.1. □

Let D be a Cartier divisor on a projective space P over k. We call the image of $[D]$ by the homomorphism δ defined in Proposition 2.9 the *degree of D*. If $P = \mathbb{P}_k^1$, this definition coincides with that for a divisor on a curve (Remark 3.2).

Lemma 4.6. *Let X be a closed subvariety of dimension ≥ 1 of $P := \mathbb{P}_k^d$. We suppose that X is integral and a complete intersection in P. Let $D \in \mathrm{Div}_+(X)$ be such that $D \simeq i^*\mathcal{O}_P(m)$, where $i : X \to P$ is the canonical injection, and where $m \in \mathbb{Z}$. Then $m \geq 0$, and there exists an effective divisor $H \in \mathrm{Div}_+(P)$ such that $\deg H = m$ and $H|_X = D$. In particular, $(\mathrm{Supp}\, H) \cap X = \mathrm{Supp}\, D$.*

Proof Let H' be a Cartier divisor on P not containing the generic point of X and of degree m. We can then define the restriction $H'|_X$ (Lemma 1.29). There exists a $t \in K(X)^*$ such that $H'|_X + \mathrm{div}(t) = D$. Hence $t \in L(H'|_X)$. Let us consider the homomorphism $L(H') \to L(H'|_X)$ deduced from

$$\phi : \mathcal{O}_P(H') \to i_*i^*\mathcal{O}_P(H') \simeq i_*\mathcal{O}_X(H'|_X)$$

by taking the global sections over P. Up to isomorphism, ϕ is identical to the canonical homomorphism $\mathcal{O}_P(m) \to i_*i^*\mathcal{O}_P(m)$. Now

$$H^0(P, \mathcal{O}_P(m)) \to H^0(P, i_*i^*\mathcal{O}_P(m))$$

is surjective (Exercise 5.3.3(b)), and hence $L(H') \to L(H'|_X)$ is surjective. Let $s \in L(H')$ be an inverse image of t, and $H := H' + \mathrm{div}(s) \geq 0$. As $\deg H = \deg H'$ (Proposition 2.9), we have $m \geq 0$. Moreover, $H|_X = H'|_X + \mathrm{div}(s)|_X = D$. □

Definition 4.7. Let X be a smooth geometrically connected curve over a field k, of genus $g \geq 1$. We say that X is a *hyperelliptic curve* if there exists a finite separable morphism $X \to \mathbb{P}_k^1$ of degree 2. Note that we define here a restricted class of hyperelliptic curves. The general definition is that there exists a finite separable morphism $X \to C$ of degree 2 with C a smooth projective conic.

Lemma 4.8. *Let X be a smooth, geometrically connected, projective curve over a field k, of genus $g \geq 1$. Then X is hyperelliptic if and only if there exists a Cartier divisor D on X such that $l(D) = \deg D = 2$.*

Proof Let us suppose that there exists such a divisor D. As $l(D) \neq 0$, we can reduce to $D \geq 0$, by linear equivalence. Let $x \in \mathrm{Supp}\, D$. If $\deg(D - x) \leq 0$, then $l(D - x) \leq 1$, by Proposition 3.25(c). If $\deg(D - x) = 1$, we have $l(D - x) \leq 1$ by Corollary 3.24, because $g \neq 0$. It follows from Lemma 4.2 that $\mathcal{O}_X(D)$ is generated by its global sections. It therefore defines a morphism $\pi : X \to \mathbb{P}_k^1$ since $l(D) = 2$. This morphism is of degree $2 = \deg D$ by Remark 3.11. It remains to show that π is separable. Let us suppose the contrary. Then $K(X)$ would be a purely inseparable extension of $k(t)$, where t is an variable. Modifying t if necessary, we have $K(X) = k(\sqrt{t})$. Hence X is birational to $\mathbb{P}_k^1$, and consequently of genus 0 (Proposition 3.13).

Conversely, suppose that we have a morphism $\pi : X \to \mathbb{P}^1_k$ of degree 2. Fix a rational point y_0 of $\mathbb{P}^1_k$, considered as a Cartier divisor, and let $D = \pi^* y_0 \in \operatorname{Div}(X)$. Then $\deg D = 2$ (Proposition 3.8). Since $L(D)$ contains $H^0(\mathbb{P}^1_k, \mathcal{O}_{\mathbb{P}^1_k}(y_0))$, we have $l(D) \geq 2$. To show that $l(D) \leq 2$, we can assume k is algebraically closed. Let $x \in \operatorname{Supp} D$; then $l(D) \leq \deg(D - x) + l(x) = 2$ by Proposition 3.25 and Corollary 3.24.   □

Proposition 4.9. *Let X be a smooth, geometrically connected, projective curve. We suppose that X is elliptic or of genus 2. Then X is a hyperelliptic curve.*

Proof Let us first suppose that X is elliptic. Let $o \in X(k)$. Then $\deg(2o) = 2$ and $l(2o) = 2$, by Example 3.35. Hence X is hyperelliptic, by the lemma above. Let us now suppose X is of genus 2. Let $K_{X/k}$ be a canonical divisor of X. Then $\deg K_{X/k} = 2$, and $l(K_{X/k}) = 2$, by Corollary 3.31(b). We once again apply the lemma above.   □

Proposition 4.10. *Let X be a smooth, geometrically connected, projective curve over a field k, of genus $g \geq 1$. Then $\Omega^1_{X/k}$ is generated by its global sections.*

Proof We may assume k is algebraically closed (Exercise 5.1.29). Let K be a canonical divisor, which we can suppose is effective because $l(K) = g > 0$. Let $x \in X(k)$. Then $l(x) = 1$ by Corollary 3.24. It follows by formula (3.6) that

$$l(K - x) = \deg(K - x) + l(x) + 1 - g = g - 1 < l(K),$$

whence the proposition, by virtue of Lemma 4.2.   □

Definition 4.11. Let X be as above. The morphism $X \to \mathbb{P}^{g-1}_k$ defined by $\Omega^1_{X/k}$ (and the choice of a basis of $H^0(X, \Omega^1_{X/k})$) is called the *canonical map*.

Proposition 4.12. *Let X be a smooth, geometrically connected, projective curve of genus $g \geq 2$. Then the canonical map $X \to \mathbb{P}^{g-1}_k$ is a closed immersion if and only if $X_{\bar{k}}$ is not hyperelliptic.*

Proof We may suppose k is algebraically closed. Let K be a canonical divisor on X. Let us suppose that X is not hyperelliptic. Let $E \in \operatorname{Div}_+(X)$ be of degree 2. Lemma 4.8 says that $l(E) = 1$. The Riemann–Roch formula (3.33) then gives

$$l(K - E) = l(E) + \deg(K - E) + 1 - g = g - 2.$$

The condition of Lemma 4.3 is therefore verified with $D = K$ (we have $l(K - x) = g - 1$ for every $x \in X(k)$, by the proposition above). It follows from the same lemma that K is very ample. The converse is an immediate consequence of Proposition 4.34, which we will prove later on.   □

7.4.2   Hurwitz formula

Let $f : X \to Y$ be a finite and *separable morphism* (i.e., $K(Y) \to K(X)$ is a separable extension) of normal projective curves over k. The Hurwitz formula relates the genus of X to that of Y in terms of the ramification indices of f.

Let $A \to B$ be an injective homomorphism of discrete valuation rings. Let t (resp. π) denote a uniformizing parameter for A (resp. of B). Let us recall that the *ramification index* of B over A is the integer $e_{B/A} \geq 1$ such that $tB = \pi^{e_{B/A}} B$ (we put $e_{B/A} = 1$ if $t = 0$). We say that B is *tamely ramified over A* if $B/\pi B$ is separable over A/tA, and if $e_{B/A}$ is prime to char(A/tA) when the latter is non-zero. Let us moreover suppose that $A \to B$ is finite, and that $L := \mathrm{Frac}(B)$ is separable over $K := \mathrm{Frac}(A)$. Let us recall that $\mathrm{Hom}_A(B, A)$ can canonically be identified with the codifferent

$$W_{B/A} = \{\beta \in L \mid \mathrm{Tr}_{L/K}(\beta B) \subseteq A\}$$

(Exercise 6.4.8). We are going to estimate the length of $W_{B/A}/B$ over B.

Proposition 4.13. *Let $A \to B$ be a finite injective homomorphism of discrete valuation rings such that* $\mathrm{Frac}(A) \to \mathrm{Frac}(B)$ *is separable. Then*

$$\mathrm{length}_B(W_{B/A}/B) \geq e_{B/A} - 1.$$

Moreover, equality holds if and only if $A \to B$ is tamely ramified.

Proof   Let $e = e_{B/A}$. Let us write $t = \pi^e u$ with $u \in B^*$. For any $b \in B$, we have

$$\mathrm{Tr}_{L/K}(\pi^{1-e}b) = \mathrm{Tr}_{L/K}(t^{-1}\pi u b) = t^{-1}\,\mathrm{Tr}_{L/K}(\pi u b) \in A.$$

Hence $\pi^{1-e} \in W_{B/A}$, which implies that

$$\mathrm{length}_B(W_{B/A}/B) \geq \mathrm{length}_B(\pi^{1-e}B/B) = e - 1.$$

We have $\mathrm{length}_B(W_{B/A}/B) \geq e$ if and only if $t^{-1} \in W_{B/A}$, which is equivalent to $\mathrm{Tr}_{B/A}(B) \subseteq tA$. Let us set $C = B/tB$. This is equivalent to $\mathrm{Tr}_{C/k}(C) = 0$, where $k = A/tA$. The proposition then results from the lemma below.   □

Lemma 4.14. *Let k be a field, C a finite local k-algebra with residue field k'. Then $\mathrm{Tr}_{C/k}(C) = k$ if and only if k' is separable over k and every nilpotent element ϵ of C verifies $\epsilon^e = 0$ for some integer e prime to char(k).*

Proof   Let k'' be the separable closure of k in k'. Then k'' is a subalgebra of C (apply Proposition 6.2.15 with $S = \mathrm{Spec}\,k$, $X = \mathrm{Spec}\,k''$ étale over S, $Y = \mathrm{Spec}\,C$, and $Y_0 = \mathrm{Spec}\,k'$). As the trace map is transitive and $\mathrm{Tr}_{k''/k}(k'') = k$, we may suppose that k' is purely inseparable of degree m over k. For any $\gamma \in C$, there exists an $\alpha \in k$ such that $\gamma^m - \alpha$ is nilpotent. Let e be its order of nilpotence. Then the polynomial $(T^m - \alpha)^e$ vanishes for $T = \gamma$, with m divisible by char$(k) > 0$. It is then an elementary exercise in linear algebra to see that γ has non-zero trace if and only if $m = 1$ and e is prime to char(k).   □

Definition 4.15. Let $f : X \to Y$ be a finite morphism of normal projective curves over a field k. For any closed point $x \in X$, let e_x denote the ramification index of $\mathcal{O}_{Y,f(x)} \to \mathcal{O}_{X,x}$. We will say that f is *ramified at x* or that *x is a ramification point of f* if f is not étale at x (which is equivalent to $e_x \geq 2$ or $k(x)$ inseparable over $k(f(x))$). The set of ramification points of f is finite (Corollary 4.4.12) and is called the *ramification locus* of f. Its image by f is called the *branch locus* of f. We say that f is *tamely ramified at x* if $\mathcal{O}_{Y,f(x)} \to \mathcal{O}_{X,x}$ is tamely ramified.

Let n be the degree of f. Then Lemma 1.36(c) can be stated as

$$n = \sum_{x \in f^{-1}(y)} e_{x/y} [k(x) : k(y)] \tag{4.8}$$

for any closed point $y \in Y$.

Theorem 4.16. *Let $f : X \to Y$ be a finite morphism of normal projective curves over k. We suppose that f is separable of degree n. Then we have an equality*

$$2p_a(X) - 2 = n(2p_a(Y) - 2) + \sum_x (e'_x - 1)[k(x) : k], \tag{4.9}$$

where the sum takes place over the closed point $x \in X$, e'_x is an integer $\geq e_x$, and $e'_x = e_x$ if and only if f is tamely ramified at x.

Proof Let ω_f be the dualizing sheaf for $f : X \to Y$. We have

$$\omega_{X/k} = \omega_f \otimes f^* \omega_{Y/k}, \tag{4.10}$$

by Lemma 6.4.26(b) and Theorem 6.4.32. In particular, ω_f is an invertible sheaf. By Proposition 3.8 and Corollary 3.31(a), this implies that

$$2p_a(X) - 2 = n(2p_a(Y) - 2) + \deg \omega_f. \tag{4.11}$$

The f separable hypothesis implies that we have an injective canonical homomorphism $\mathcal{O}_X \to \omega_f$ (Exercise 6.4.7(e)), with cokernel with finite support. We have

$$\deg \omega_f = \dim_k H^0(X, \omega_f/\mathcal{O}_X) \tag{4.12}$$

(Definition 3.29 and Lemma 3.16). For any closed point $x \in X$, let us set $e'_x = \mathrm{length}_{\mathcal{O}_{X,x}}(\omega_f/\mathcal{O}_X)_x + 1$. Then we get (4.9).

It remains to estimate e'_x. Let $y = f(x)$, $A = \mathcal{O}_{Y,y}$, $B = (f_* \mathcal{O}_X)_y$. Then $(f_* \omega_f)_y = \mathrm{Hom}_A(B, A)$. Let $\hat{A}$ be the $\mathfrak{m}_y$-adic completion of $\mathcal{O}_{Y,y}$. Then

$$\mathrm{Hom}_A(B, A) \otimes_A \hat{A} = \mathrm{Hom}_{\hat{A}}(B \otimes_A \hat{A}, \hat{A}), \quad B \otimes_A \hat{A} = \oplus_{z \in f^{-1}(y)} \hat{\mathcal{O}}_{X,z}$$

(Exercises 1.2.8 and 4.3.17). Let $\hat{B} = \hat{\mathcal{O}}_{X,x}$. We then have

$$(\omega_f/\mathcal{O}_X)_x = \mathrm{Hom}_{\hat{A}}(\hat{B}, \hat{A})/\hat{B}.$$

The field $\mathrm{Frac}(\hat{B})$ is separable over $\mathrm{Frac}(\hat{A})$, because it is contained in the separable $\mathrm{Frac}(\hat{A})$-algebra $\mathrm{Frac}(\hat{A}) \otimes_{K(Y)} K(X)$. We have $e'_x = \mathrm{length}_{\hat{B}}(W_{\hat{B}/\hat{A}}/\hat{B})$ (Exercise 6.4.8). We conclude with Proposition 4.13 by noting that $\hat{A} \to \hat{B}$ has the same ramification index and the same residue extension as $\mathcal{O}_{Y,y} \to \mathcal{O}_{X,x}$. □

Remark 4.17. If X, Y are smooth and geometrically connected over k, we can express e'_x in another way. We have the exact sequence

$$f^*\Omega^1_{Y/k} \to \Omega^1_{X/k} \to \Omega^1_{X/Y} \to 0,$$

with $f^*\Omega^1_{Y/k} \to \Omega^1_{X/k}$ injective, because it is a homomorphism of invertible sheaves that is injective at the generic point (f separable). Hence, using exact sequence (4.10), we get

$$\Omega^1_{X/Y} \simeq (\omega_f/\mathcal{O}_X) \otimes f^*\Omega^1_{Y/k} \simeq (\omega_f/\mathcal{O}_X) \qquad (4.13)$$

because $\omega_f/\mathcal{O}_X$ has finite support (contained in the ramification locus). It follows that $e'_x = \mathrm{length}_{\mathcal{O}_{X,x}} \Omega^1_{X/Y,x} + 1$.

Remark 4.18. With the hypotheses of the theorem, if f is not tamely ramified at x, we can still compute the index e'_x in some special cases, see [87], III, 6–7, IV, 1.

Corollary 4.19. *Let* $f : X \to Y$ *be a finite separable morphism of normal projective curves over* k, *with* $p_a(X) \geq 0$. *Then* $p_a(X) \geq p_a(Y)$.

Proof By Theorem 4.16, we have $p_a(X) - 1 \geq (\deg f)(p_a(Y) - 1)$. This immediately implies the corollary. $\qquad \square$

Corollary 4.20. *Let* $f : X \to Y$ *be a finite étale morphism of smooth, geometrically connected, projective curves. We suppose that* $g(Y) = 0$. *Then* f *is an isomorphism.*

Proof Indeed, we have $2(g(X) - 1) = -2 \deg f$. As $g(X) \geq 0$, this implies that $\deg f = 1$, and therefore that f is an isomorphism. $\qquad \square$

If $f : X \to Y$ is not necessarily separable, it decomposes into a finite purely inseparable morphism $X \to Z$ (i.e., $K(Z) \to K(X)$ is a purely inseparable extension) followed by a finite separable morphism $Z \to Y$. If $\mathrm{char}(k) = p > 0$, we have an absolute Frobenius $F_k : \mathrm{Spec}\, k \to \mathrm{Spec}\, k$ corresponding to the homomorphism of $\mathbb{F}_p$-algebras $a \mapsto a^p$ (Definition 3.2.21). For any $r \geq 0$, we let $F^r_k : \mathrm{Spec}\, k \to \mathrm{Spec}\, k$ denote the r-uple composition of F_k (F^0_k is the identity), and $X^{(p^r)}$ the variety over k obtained by the base change F^r_k (see Subsection 3.2.4 for the case $r = 1$).

Proposition 4.21. *Let* $f : X \to Y$ *be a finite purely inseparable morphism of normal projective curves over a field* k. *We suppose that* X *is smooth over* k. *Then there exists an* $r \geq 0$ *such that* $Y \simeq X^{(p^r)}$. *In particular,* $p_a(X) = p_a(Y)$.

Proof Let $p = \mathrm{char}(k)$. Then $[K(X) : K(Y)] = p^r$ for some integer $r \geq 0$. We have $K(X) = k(T, \theta)$ with T transcendent over k and θ separable algebraic over $k(T)$ (Proposition 3.2.15). On the other hand, $K(X^{(p^r)}) = k(T^{p^r}, \theta^{p^r})$ and is of index p^r in $K(X)$ (Corollary 3.2.27). Now $k(T^{p^r}, \theta^{p^r}) \subseteq K(Y)$, and we therefore have $K(X^{(p^r)}) = K(Y)$, which shows that $Y \simeq X^{(p^r)}$, by Exercise 4.3.13(a) and Proposition 3.13 (see also Exercise 4.1.19). Finally, as $X^{(p^r)}$ is obtained from X by flat base change, it has the same arithmetic genus as X. $\qquad \square$

Remark 4.22. In the situation of the proposition, if k is perfect, then X and Y are isomorphic as $\mathbb{F}_p$-schemes. Indeed, the Frobenius F is then an isomorphism of $\mathbb{F}_p$-schemes.

As an application of the preceding results, we are going to show Lüroth's theorem:

Corollary 4.23. *Let k be a field, and let L be a subfield of a purely transcendental extension $k(t)$. Then either $L = k$, or L is purely transcendental over k.*

Proof Let us suppose $L \neq k$; then L is of transcendence degree 1 over k. Let Y be the normal projective curve over k such that $K(Y) = L$. We have a finite morphism $f : X = \mathbb{P}^1_k \to Y$ that we can decompose into a finite purely inseparable morphism $X \to Z$, followed by a finite separable morphism $Z \to Y$. By Proposition 4.21, $p_a(Z) = 0$. It follows from Corollary 4.19 that $p_a(Y) = 0$. By Corollary 3.2.14, Y is geometrically integral. As moreover $Y(k) \supseteq f(X(k))$ is non-empty, we have $Y \simeq \mathbb{P}^1_k$ (Proposition 4.1), which means that $L = K(Y)$ is purely transcendental over k. $\qquad\qquad\qquad\qquad\qquad\square$

7.4.3   Hyperelliptic curves

In this subsection, we are going to study hyperelliptic curves (Definition 4.7) in more detail, by 'explicit' methods. Let X be such a curve over a field k, of genus g. Let $f : X \to Y = \mathbb{P}^1_k$ be a finite separable morphism of degree 2. Let us write $Y = U \cup V$ with $U = \operatorname{Spec} k[t]$ and $V = \operatorname{Spec} k[s]$ with $s = 1/t$. Let $U' = f^{-1}(U)$, $V' = f^{-1}(V)$. Hence $X = U' \cup V'$. By formulas (4.13), (4.12), and (4.11),

$$\dim_k H^0(X, \Omega^1_{X/Y}) = 2g + 2. \tag{4.14}$$

What follows is the proof of Proposition 4.24. Let us first suppose that $\operatorname{char}(k) \neq 2$. Then $K(X) = k(t)[y]$ with $y^2 \in k(t)$. Multiplying by an element of $k(t)^*$ if necessary, we may suppose that $y^2 = P(t)$ with $P(t) \in k[t]$ without a square factor. We then easily show that $k[t, Y]/(Y^2 - P(t))$ is normal (e.g., with the method used in Example 4.1.9). It follows that $k[t, y]$ is the integral closure of $\mathcal{O}_Y(U)$ in $K(X)$. Consequently, $\mathcal{O}_X(U') = k[t, y]$. The X smooth hypothesis implies that $P(t)$ has no square factor over $\overline{k}$ (Jacobian criterion); hence $P(t)$ is separable. Let $d = \deg P(t)$, $r = [(d+1)/2]$. Then $(y/t^r)^2$ is a polynomial $P_1(s) \in k[s]$ without a square factor. A similar reasoning shows that $\mathcal{O}_X(V') = k[s, z]$, where $z = t^r y$. Let us now determine d and r. We have

$$\Omega^1_{U'/U} = \mathcal{O}_X(U')dy/(2ydy) \simeq k[t, y]/(y^2 - P(t), y) = k[t]/(P(t)),$$

and $\Omega^1_{V'/V} = k[s]/(P_1(s))$. As $Y = U \cup \{s = 0\}$, we have

$$H^0(X, \Omega^1_{X/Y}) = k[t]/(P(t)) \oplus k[s]\mathfrak{m}/(P_1(s)), \quad \mathfrak{m} = sk[s].$$

The second term is of dimension 0 over k if $P_1(0) \neq 0$ (i.e., $d = 2r$). This also means that f is unramified over $\{s = 0\}$. Otherwise this term is of dimension 1 (i.e., $d = 2r + 1$). Formula (4.14) therefore implies that $2g + 2 = \deg P + \epsilon$, with $\epsilon = 0, 1$. In both cases, $r = g + 1$.

The char$(k) = 2$ case is a bit more complex. The following method is valid in every characteristic, except for the result of the computation of $\Omega^1_{U'/U}$. We have an exact sequence

$$0 \to \mathcal{O}_Y \to f_*\mathcal{O}_X \to \mathcal{L} \to 0 \qquad (4.15)$$

for some coherent sheaf $\mathcal{L}$. As f is flat (Corollary 4.3.10), $f_*\mathcal{O}_X$ is locally free of rank 2 over $\mathcal{O}_Y$. For any point $y \in Y$, exact sequence (4.15) induces an exact sequence

$$0 \to k(y) \to (f_*\mathcal{O}_X) \otimes_{\mathcal{O}_Y} k(y) \to \mathcal{L} \otimes_{\mathcal{O}_Y} k(y) \to 0$$

(the first map on the left is a homomorphism of $k(y)$-algebras, and hence injective). This implies that $\mathcal{L}$ is locally free (Exercise 5.1.15) of rank 1. As $\mathcal{O}_Y(U)$ is principal, $\mathcal{L}|_U$ is free of rank 1, and exact sequence (4.15) splits over U. The same holds over V. In other words, there exist $y \in \mathcal{O}_X(U')$, $z \in \mathcal{O}_X(V')$ such that

$$\mathcal{O}_X(U') = \mathcal{O}_Y(U) \oplus \mathcal{O}_Y(U)y, \quad \mathcal{O}_X(V') = \mathcal{O}_Y(V) \oplus \mathcal{O}_Y(V)z.$$

On $U \cap V$, $\{1, y\}$ and $\{1, z\}$ are two bases of $f_*\mathcal{O}_X(U \cap V)$ over $\mathcal{O}_Y(U \cap V)$. As $\mathcal{O}_Y^*(U \cap V) = k^*t^{\mathbb{Z}}$, multiplying z by an element of k^* if necessary, we have

$$y = R(t) + H(s) + t^r z, \text{ with } r \in \mathbb{Z}, R(t) \in k[t], H(s) \in k[s].$$

We have $r \geq 1$, as $y - R(t)$ would otherwise be a global section of $H^0(X, \mathcal{O}_X) = k$. It follows that $y - R(t) = t^r(H(s)s^r + z)$. Therefore, modifying y and z if necessary, we may suppose that

$$y = t^r z, \quad r \geq 1.$$

We have integral equations

$$y^2 + Q(t)y = P(t), \quad P(t), Q(t) \in k[t],$$

$$z^2 + (Q(t)/t^r)z = (P(t)/t^{2r}).$$

As $z = s^r y$ is integral over $k[s]$, we have $\deg Q(t) \leq r$ and $\deg P(t) \leq 2r + 2$. As far as the ramification is concerned, we have

$$\Omega^1_{U'/U} = \mathcal{O}_X(U')dy/(Q(t)dy) \simeq k[t, y]/(Q(t)) = (k[t]/(Q(t))) \otimes_{\mathcal{O}_Y(U)} \mathcal{O}_X(U').$$

Hence $\dim_k H^0(U', \Omega^1_{U'/U}) = 2 \deg Q(t)$. Similarly, the points lying above the point $\{s = 0\}$ contribute $2(r - \deg Q(t))$ to the dimension of $H^0(X, \Omega^1_{X/Y})$ over k. We therefore have

$$2g + 2 = 2 \deg Q(t) + 2(r - \deg Q(t)) = 2r.$$

Hence $r = g + 1$ and $\deg Q(t) \leq g + 1$. As $P(t)/t^{2r} \in k[s]$, because it is the norm of z over $k[s]$, we have $\deg P(t) \leq 2r = 2g + 2$. The ramification points of f are those of $V(Q(t)) \subset U'$, and also $\{s = 0\}$ if $\deg Q(t) < g + 1$. In that case, $\deg P(t) = 2g + 1$, because otherwise $\{s = 0\}$ would be a singular point of X.

To summarize, we have the following proposition:

Proposition 4.24. *Let* X *be a hyperelliptic curve of genus* g *over a field* k, *with a separable morphism* $f : X \to \mathbb{P}^1_k$ *of degree 2.*

(a) *We have* $K(X) = k(t)[y]$ *with a relation*

$$y^2 + Q(t)y = P(t), \quad P(t), Q(t) \in k[t], \tag{4.16}$$

   with

$$2g + 1 \leq \max\{2 \deg Q(t), \deg P(t)\} \leq 2g + 2.$$

   We can take $Q(t) = 0$ *if* $\mathrm{char}(k) \neq 2$.

(b) *The curve* X *is the union of two affine open schemes*

$$U' = \mathrm{Spec}\, k[t, Y]/(Y^2 + Q(t)Y - P(t)),$$

$$V' = \mathrm{Spec}\, k[s, Z]/(Z^2 + Q_1(s)Z - P_1(s)),$$

   where $Q_1(s) = Q(1/s)s^{g+1}$, $P_1(s) = P(1/s)s^{2g+2}$, *and the two open subschemes glue along* $D(t) \simeq D(s)$ *with relations* $t = 1/s$ *and* $y = t^{g+1}z$.

(c) *The ramification points of* f *are those of* $V(4P(t) + Q(t)^2) \subset U'$, *plus the point* $\{s = 0\} \in V'$ *if* $\deg(4P(t) + Q(t)^2) \leq 2g + 1$.

Remark 4.25. The smoothness of X can be expressed as follows: if $\mathrm{char}(k) \neq 2$, then $4P(t) + Q(t)^2$ is a separable polynomial; if $\mathrm{char}(k) = 2$, then $Q(t)$ and $Q'(t)^2 P(t) + P'(t)^2$ are prime to each other (Jacobian criterion).

Proposition 4.26. *Let* X *be as in the preceding proposition. Then the* k-vector *space* $H^0(X, \Omega^1_{X/k})$ *admits the differentials*

$$\omega_i := \frac{t^i dt}{(2y + Q(t))}, \quad 0 \leq i \leq g - 1$$

as a basis.

Proof By Corollary 6.4.14(b), $\Omega^1_{U'/k}$ is generated by ω_0, and $\Omega^1_{V'/k}$ by $\omega_{g-1} = ds/(2z+Q_1(s))$. We immediately deduce from this that $\omega_i \in H^0(X, \Omega^1_{X/k})$. These differentials are manifestly linearly independent over k. They therefore form a basis, since $\dim_k H^0(X, \Omega^1_{X/k}) = l(K_X) = g$. □

Definition 4.27. Let X be a hyperelliptic curve over a field k. Let us consider a separable morphism $f : X \to \mathbb{P}^1_k$ of degree 2. Let σ be the generator of $\mathrm{Gal}(K(X)/k(t))$. It induces an automorphism of order 2 of X. We also denote this automorphism by σ, and we will call it a *hyperelliptic involution* of X. A point $x \in X(k)$ is ramified for f if and only if $\sigma(x) = x$ (use Proposition 4.24 or Exercise 4.3.19).

Remark 4.28. The automorphism σ acts on $\Omega^1_{X/k}$. At the generic point of X, it acts by $\sigma(h_1 dh_2) = \sigma(h_1)d(\sigma(h_2))$. It follows from the description of Proposition 4.26 that σ acts as $- \mathrm{Id}$ on $H^0(X, \Omega^1_{X/k})$.

Proposition 4.29. *Let X be a hyperelliptic curve over a field k. Let us fix a hyperelliptic involution σ. Let τ be another hyperelliptic involution. Let us suppose that $g \geq 2$, or that τ and σ have a common rational fixed point $x_0 \in X(k)$. Then $\tau = \sigma$. In particular, the hyperelliptic involution is unique if $g \geq 2$.*

Proof If $g \geq 2$, $\tau(\omega_i) = -\omega_i$ for $i = 0, 1$, by Remark 4.28. Now $\tau(\omega_1) = \tau(t\omega_0) = -\tau(t)\omega_0$ in $\Omega^1_{K(X)/k}$, and hence $\tau(t) = t$. This means that τ is an element of $\mathrm{Gal}(K(X)/k(t)) = \{1, \sigma\}$, and hence $\tau = \sigma$.

Let us suppose $g = 1$. Let $f : X \to \mathbb{P}^1_k$ be the degree 2 morphism corresponding to σ. We may assume that $f(x_0)$ is the point $y_0 := \{t = \infty\}$. As $x_0 = f^{-1}(y_0)$, in the second part of the proof of Lemma 4.8 we saw that $L(2x_0) = k + kt$ (let us recall that, by convention, x_0 also denotes the Cartier divisor associated to the cycle $[x_0]$). Consequently, τ leaves $k + kt$ invariant, and therefore induces an automorphism of $k[t]$. If $\tau(t) = t$, the reasoning of the $g \geq 2$ case shows that $\tau = \sigma$. Let us suppose that $\tau(t) \neq t$. We are going to show that this implies an absurdity. As τ is of order 2 over $k[t]$, we have $\tau(t) = \lambda - t$, $\lambda \in k$. If $\mathrm{char}(k) \neq 2$, we may assume that $Q(t) = 0$. Hence

$$-\frac{dt}{2y} = \tau(\omega_0) = \frac{d(\tau(t))}{\tau(2y)} = -\frac{dt}{2\tau(y)}.$$

It follows that $\tau(y) = y$, and therefore $\tau(P(t)) = P(t)$. Consequently, $P(t) \in k[t(\lambda - t)]$ is of even degree. Now, since f is ramified over $t = \infty$, $P(t)$ is of degree 3, a contradiction. The $\mathrm{char}(k) = 2$ case remains. As τ is of order 2, we necessarily have

$$\tau(y) = A(t) + y, \quad A(t) \in k[t], \quad \tau(A(t)) = A(t).$$

A reasoning similar to the above implies that $\tau(Q(t)) = Q(t)$. Hence $\deg Q(t)$ is even and ≤ 1 (Proposition 4.24(c)). It follows that $Q(t) \in k^*$. Applying τ to equation (4.16) of X, we obtain $(Q + A)A = P + \tau(P)$. Now $\deg A$ is even and $\deg(P + \tau(P)) \leq 2$, which implies that $A = 0$ and $\tau(P) = P$. Consequently, $P(t)$ is of degree 4 and does not have any term in t^3. We verify that this contradicts the fact that X is smooth at the point above $\{t = \infty\}$. $\square$

Remark 4.30. Let X be a hyperelliptic curve of genus $g \geq 2$. Then the proposition says that $K(X)$ contains a unique subfield of index 2 isomorphic to $k(t)$. In other words, there is a unique morphism $X \to \mathbb{P}^1_k$ of degree 2 up to automorphisms of X and of $\mathbb{P}^1_k$.

Corollary 4.31. *Let X be a hyperelliptic curve, endowed with a morphism $f : X \to \mathbb{P}^1_k$ of degree 2. Let σ be the hyperelliptic involution of X associated to f. Let τ be an automorphism of X. Let us suppose that $g \geq 2$, or that $g = 1$ and that σ and τ have a common fixed point $x_0 \in X(k)$. Then $\tau\sigma = \sigma\tau$ and τ induces an automorphism $\tilde{\tau}$ of $\mathbb{P}^1_k$ rendering the following diagram commutative:*

$$
\begin{array}{ccc}
X & \xrightarrow{\ \tau\ } & X \\
{\scriptstyle f}\downarrow & & \downarrow{\scriptstyle f} \\
\mathbb{P}^1_k & \xrightarrow{\ \tilde{\tau}\ } & \mathbb{P}^1_k
\end{array}
$$

Proof Let us consider $\tau' := \tau^{-1}\sigma\tau$. This is a hyperelliptic involution of X and we have $\tau'(x_0) = x_0$ if $g = 1$. By the proposition above, $\tau' = \sigma$, whence $\tau\sigma = \sigma\tau$. It follows that τ acts on $K(X)^{\langle\sigma\rangle} = k(t)$. Let $\tilde{\tau}$ be the resulting automorphism of $\mathbb{P}^1_k$. Then $f\tau = \tilde{\tau}f$. □

Definition 4.32. An equation such as (4.16) of Proposition 4.24 is called a *hyperelliptic equation* of X. If, moreover, X is an elliptic curve, a hyperelliptic equation is called an *elliptic equation* of X if $\deg Q(t) \le 1$ and $\deg P(t) \le 3$. The point $\{t = \infty, (y/t^2) = 0\}$ will be called the *origin of X* (relatively to this equation).

Corollary 4.33. *Let X be a hyperelliptic curve of genus g over k. Let*

$$y^2 + Q(t)y = P(t), \quad v^2 + R(u)v = S(u)$$

be two hyperelliptic equations of X.

(a) *Let us suppose that $g \ge 2$. Then there exist*

$$\begin{pmatrix} a & b \\ c & d \end{pmatrix} \in \mathrm{GL}_2(k), e \in k^*, H(t) \in k[t], \deg H \le g+1$$

such that

$$u = \frac{at+b}{ct+d}, \quad v = \frac{H(t)+ey}{(ct+d)^{g+1}}.$$

(b) *Let us suppose that X is elliptic and that the two equations are elliptic with the same origin. Then we have the same conclusion as above, with moreover $c = 0$, $d = 1$, and $\deg H(t) \le 1$.*

Proof The equation $y^2 + Qy = P$ (resp. $v^2 + Rv = S$) induces a hyperelliptic involution σ (resp. τ) of X with $K(X)^{\langle\sigma\rangle} = k(t)$ (resp. $K(X)^{\langle\tau\rangle} = k(v)$). In case (b), the origins of the equations are the fixed points of these involutions. It follows from Proposition 4.29 that $\tau = \sigma$; hence $k(v) = k(t)$. Consequently, $u = (at+b)/(ct+d)$ with $ad - bc \in k^*$ (Exercise 5.1.23). The upper bound for the degrees of R and S (Proposition 4.24) implies that $v' := (ct+d)^{g+1}v$ is integral over $k[t]$. Hence $v' = H(t) + L(t)y$ with $H(t), L(t) \in k[t]$. Let us show that $L(t) \in k^*$. If $c = 0$, this is obvious because $\{1, v\}$ and $\{1, y\}$ are both bases of the same module over $k[t] = k[u]$. Suppose $c \ne 0$. Then we can replace $ct+d$ by t because $k[ct+d] = k[t]$, and $u = a + b/t$ by $1/t$. The result is then contained in Proposition 4.24(b). From the definition of v' we deduce a relation $v'^2 + R_1(t)v' = S_1(t)$ with $\max\{2 \deg R_1(t), \deg S_1(t)\} \le 2g+2$. This implies, after an easy computation, that $\deg H(t) \le g+1$.

To conclude, in case (b) we have moreover $k[t] = k[u]$, and hence $u = at+b$. The inequality $\deg H(t) \le 1$ can be deduced from the hypothesis on the degrees of Q, P, R, S. □

Proposition 4.34. *Let X be a hyperelliptic curve of genus $g \ge 2$ over k. Then the canonical map $X \to \mathbb{P}^{g-1}_k$ decomposes into the canonical morphism $f : X \to \mathbb{P}^1_k$ of degree 2 followed by the $(g-1)$-uple embedding $h : \mathbb{P}^1_k \to \mathbb{P}^{g-1}_k$.*

Proof Let us keep the notation of Proposition 4.24. Let

$$\mathcal{L} =: \left(\frac{dt}{2y + Q(t)} \right)^{-1} \Omega^1_{X/k} \subset \mathcal{K}_X.$$

Then $\{1, t, \ldots, t^{g-1}\}$ is a basis of $H^0(X, \mathcal{L})$ (Proposition 4.26). Hence the canonical map $X \to \mathbb{P}^{g-1}_k$ is defined by $\mathcal{L}$ and the sections $\{1, t, \ldots, t^{g-1}\}$ (Remark 5.1.32). It is clear that $\mathcal{L} \simeq f^*\mathcal{O}_{\mathbb{P}^1_k}(g-1) \simeq f^*(h^*\mathcal{O}_{\mathbb{P}^{g-1}_k}(1))$ (see Exercise 5.1.27), and that the image of a suitable basis of $H^0(\mathbb{P}^{g-1}_k, \mathcal{O}(1))$ in $H^0(X, \mathcal{L})$ is $\{1, t, \ldots, t^{g-1}\}$. This proves that $h \circ f$ coincides with the canonical map. $\square$

7.4.4 Group schemes and Picard varieties

In this subsection, we give, without proof, results on the Jacobian of a smooth projective curve X. In Chapter 10, we will need the structure of the group of n-torsion points $\mathrm{Pic}(X)[n]$ (Corollary 4.41).

Let us first introduce the notion of group scheme. Let S be a scheme. A *group scheme* over S is an S-scheme G endowed with the following morphisms of S-schemes

$$\begin{aligned}
\text{(multiplication)} & \quad m : G \times_S G \to G, \\
\text{(unit section)} & \quad \varepsilon : S \to G, \\
\text{(inverse)} & \quad \mathrm{inv} : G \to G,
\end{aligned}$$

such that we have the commutative diagrams

$$\begin{array}{ccc}
G \times_S G \times_S G & \xrightarrow{m \times \mathrm{Id}_G} & G \times_S G \\
\downarrow{\scriptstyle \mathrm{Id}_G \times m} & & \downarrow{\scriptstyle m} \\
G \times_S G & \xrightarrow{\quad m \quad} & G
\end{array}$$

(associativity),

$$\begin{array}{ccc}
G = G \times_S S & \xrightarrow{\mathrm{Id}_G \times \varepsilon} & G \times_S G \\
& {\scriptstyle \mathrm{Id}_G} \searrow & \downarrow{\scriptstyle m} \\
& & G
\end{array}$$

(right-identity), and

$$\begin{array}{ccc}
G & \xrightarrow{\Delta_{G/S}} G \times_S G \xrightarrow{\mathrm{Id}_G \times \mathrm{inv}} & G \times_S G \\
\downarrow & & \downarrow{\scriptstyle m} \\
S & \xrightarrow{\quad\quad \varepsilon \quad\quad} & G
\end{array}$$

(right-inverse). Let T be an S-scheme. Then the morphisms m and inv canonically induce maps $m(T) : G(T) \times G(T) \to G(T)$, $\mathrm{inv}(T) : G(T) \to G(T)$. The group scheme axioms are equivalent to saying that this data makes $G(T)$ into a group with unit element ε_T for each S-scheme T. It is clear that $G_T \to \operatorname{Spec} T$ canonically admits the structure of a group scheme over T. The fibered product of two group schemes over S is a group scheme over S. We say that G is a *commutative group scheme* if $G(T)$ is commutative for every S-scheme T.

A *subgroup scheme* of G is a (closed) subscheme H of G such that $H(T)$ is a subgroup of $G(T)$ for all S-schemes T. The notion of *morphisms of group schemes* is defined in a natural way. The kernel of a morphism $f : G \to G'$ of group schemes over S is $G \times_{G'} S$, where the second factor is endowed with the structure of G'-scheme via the unit section $\varepsilon : S \to G'$. It is naturally a group scheme. A group scheme G over a field k is called an *algebraic group* over k if G is moreover of finite type over k.

Example 4.35. Let $\mathbb{G}_a = \operatorname{Spec} \mathbb{Z}[T]$. Let $m : \mathbb{G}_a \times_{\operatorname{Spec} \mathbb{Z}} \mathbb{G}_a \to \mathbb{G}_a$ be the morphism corresponding to $\mathbb{Z}[T] \to \mathbb{Z}[T_1, T_2]$, $T \mapsto T_1 + T_2$, $\varepsilon : \operatorname{Spec} \mathbb{Z} \to \mathbb{G}_a$ corresponding to $\mathbb{Z}[T] \to \mathbb{Z}$, $T \mapsto 0$, and inv $: \mathbb{G}_a \to \mathbb{G}_a$ corresponding to $\mathbb{Z}[T] \to \mathbb{Z}[T]$, $T \mapsto -T$. Then it is easy to see that these morphisms endow $\mathbb{G}_a$ with the structure of a group scheme over $\mathbb{Z}$. For any scheme S, we call the group scheme $\mathbb{G}_{a,S} := \mathbb{G}_a \times_{\operatorname{Spec} \mathbb{Z}} S$ over S the *additive group* over S. We verify without difficulty that for any $\mathbb{Z}$-scheme T, $\mathbb{G}_a(T)$ is the additive group $\mathcal{O}_T(T)$.

Example 4.36. Let $\mathbb{G}_m = \operatorname{Spec} \mathbb{Z}[T, 1/T]$. Let $\mathbb{G}_m \times_{\operatorname{Spec} \mathbb{Z}} \mathbb{G}_m \to \mathbb{G}_m$ be the morphism corresponding to $\mathbb{Z}[T, 1/T] \to \mathbb{Z}[T_1, 1/T_1, T_2, 1/T_2]$, $T \mapsto T_1 T_2$. As in the preceding example, we construct the structure of a group scheme on $\mathbb{G}_m$ in such a way that for any $\mathbb{Z}$-scheme T, $\mathbb{G}_m(T)$ is the multiplicative group $\mathcal{O}_T^*(T)$. The group scheme $\mathbb{G}_{m,S} := \mathbb{G}_m \times_{\operatorname{Spec} \mathbb{Z}} S \to S$ is called the *multiplicative group* over S. More generally, let $n \geq 1$, and let

$$\operatorname{GL}_n = \operatorname{Spec} \mathbb{Z}[T_{ij}, 1/\Delta]_{1 \leq i, j \leq n},$$

where Δ is the determinant of the matrix $(T_{ij})_{1 \leq i, j \leq n}$. Then GL_n can be endowed with the structure of a group scheme over $\mathbb{Z}$ such that $\operatorname{GL}_n(T) = \operatorname{GL}_n(\mathcal{O}_T(T))$ is the group of invertible square matrices of order n with coefficients in $\mathcal{O}_T(T)$.

The examples that we have considered are all affine schemes. It is more difficult to construct non-affine examples. Let X be an elliptic curve over a field k together with a point $o \in X(k)$. Then we can show that X has a unique structure of an algebraic group over k such that o is the unit section. The multiplication on $X(k)$ is defined as follows: let $x, y \in X(k)$. Then $l(x + y - o) = 1$ and there exists a unique point $z \in X(k)$ such that $x + y \sim o + z$. By definition, (x, y) will be sent to z by the multiplication. See Proposition 10.1.9.

Definition 4.37. Let k be a field. An *Abelian variety* over k is defined to be an algebraic group that is geometrically integral and proper over k. An Abelian variety is always projective and commutative. See [67], Section 2 and 7.

Let G be a commutative algebraic group over a field k. For any extension k'/k, $G(k')$ is an abstract commutative group. For any $n \in \mathbb{Z}$, we will denote by $G[n]$ the kernel of the multiplication by n morphism $G \to G$. Then $G[n](k')$ is just the kernel of multiplication by n on $G(k')$.

Theorem 4.38. *Let A be an Abelian variety of dimension g over a field k, and let $\bar{k}$ be the algebraic closure of k. Let us fix a non-zero $n \in \mathbb{Z}$.*

(a) *If $(n, \mathrm{char}(k)) = 1$, then $A[n]$ is étale over $\mathrm{Spec}\, k$, and $A[n](\bar{k}) \simeq (\mathbb{Z}/n\mathbb{Z})^{2g}$.*

(b) *If $p = \mathrm{char}(k) > 0$, then there exists an $0 \leq h \leq g$ such that for any $n = p^m$, we have $A[n](\bar{k}) \simeq (\mathbb{Z}/n\mathbb{Z})^h$.*

Proof See [70], Section 6, page 64.   □

A very important tool for the study of algebraic curves is the notion of the Jacobian. Let us fix a smooth, geometrically connected, projective curve over a field k. Let $\mathrm{Pic}^0(X)$ denote the subgroup of $\mathrm{Pic}(X)$ of classes of invertible sheaves of degree 0.

Theorem 4.39. *Let X be a smooth, geometrically connected, projective curve of genus g over a field k. Then there exists an Abelian variety J of dimension g over k such that $J(K) \simeq \mathrm{Pic}^0(X_K)$ for any extension K/k verifying $X(K) \neq \emptyset$. Moreover, the isomorphism is compatible with field extensions.*

Proof See [68], Theorem 1.1. See also Proposition 10.1.9 when X is an elliptic curve.   □

Definition 4.40. The Abelian variety J above is called the *Jacobian of X*.

Corollary 4.41. *Let X be a smooth, connected, projective curve over an algebraically closed field k, of genus g. Let $n \in \mathbb{Z}$ be non-zero.*

(a) *If $(n, \mathrm{char}(k)) = 1$, then $\mathrm{Pic}^0(X)[n] \simeq (\mathbb{Z}/n\mathbb{Z})^{2g}$.*

(b) *If $p = \mathrm{char}(k) > 0$, then there exists an $0 \leq h \leq g$ such that for any $n = p^m$, we have $\mathrm{Pic}^0(X)[n] \simeq (\mathbb{Z}/n\mathbb{Z})^h$.*

Exercises

4.1. Let X be a smooth, geometrically connected, projective curve of genus 1 over a field k. Let $o \in X(k)$ (we suppose that such a point exists). Let us also denote the Cartier divisor associated to the cycle $[o]$ by o.

(a) Let $\{1, t\}$ be a basis of $L(2o)$ and $\{1, t, y\}$ a basis of $L(3o)$. Show that $\{1, t, y, t^2\}$ and $\{1, t, y, t^2, ty\}$ are bases of $L(4o)$ and $L(5o)$, respectively.

(b) Let us consider the elements $1, t, y, t^2, ty, t^3, y^2$ of $L(6o)$. Show that we have a non-trivial linear relation

$$by^2 + a_1 ty + a_3 y = a_0 t^3 + a_2 t^2 + a_4 t + a_6, \quad b, a_i \in k.$$

Deduce from (a) that $a_0 b \neq 0$. Show that by a suitable homothety on t and y, we may suppose that $a_0 = b = 1$.

(c) Show that the closed immersion $X \to \mathbb{P}^2_k$ induced by the basis $\{1, t, y\}$ of $L(3o)$ has for image an elliptic curve in $\mathbb{P}^2_k$.

4.2. Let X be a smooth, geometrically connected, projective curve over a field k, of genus $g = 3$. Show that X is a *plane curve* of degree 4 (i.e., isomorphic to $V_+(F)$, where $F \in k[u, v, w]$ is homogeneous of degree 4) or $X_{\bar{k}}$ is hyperelliptic.

4.3. Show that the converse of Lemma 4.3 is true.

4.4. Let $f : X \to Y$ be a finite separable morphism of degree n between two normal projective curves over a field k. Show that f is étale if and only if $\chi_k(\mathcal{O}_X) = n\chi_k(\mathcal{O}_Y)$.

4.5. Let X be a smooth, geometrically connected, projective curve over a field k. Let $f : X \to \mathbb{P}^1_k$ be a finite separable morphism. Let us fix a parameter t of $\mathbb{P}^1_k$ (i.e., $K(\mathbb{P}^1_k) = k(t)$).

(a) Let us suppose that f is étale outside of the point $\{t = \infty\} \in \mathbb{P}^1_k(k)$ and that f is tamely ramified at every point lying above ∞. Show that f is an isomorphism. Give a counterexample when f is not tamely ramified.

(b) Let us suppose that f is étale outside of the two points $\{t = 0\}$ and $\{t = \infty\}$, and that f is tamely ramified at the points lying above $0, \infty$. Show that $K(X) = k(t)[h]$ with $h^n = \lambda t$, where $n = \deg f$, $\lambda \in k^*$. In particular, if k contains the nth roots of unity, then f is a cyclic covering (i.e., $k(t) \to K(X)$ is a cyclic extension).

4.6. Let X be a smooth, geometrically connected, projective curve over a field k, of genus $g \geq 2$.

(a) Show that $(\Omega^1_{X/k})^{\otimes 3}$ is very ample.

(b) Let $\mathrm{Aut}(X)$ be the group of automorphisms of X. Show that $\mathrm{Aut}(X)$ acts on the k-vector space $V := H^0(X, (\Omega^1_{X/k})^{\otimes 3})$ and induces an injective homomorphism $\mathrm{Aut}(X) \to \mathrm{Aut}_k(V)$.

(c) Show that if k is a finite field, then $\mathrm{Aut}(X)$ is finite. This result stays true over an arbitrary field k. See [34], Corollary V.1.2, for fields of characteristic zero and [83] for any characteristic.

4.7. Let X be a hyperelliptic curve of genus $g \geq 2$, endowed with a separable morphism $f : X \to \mathbb{P}^1_k$ of degree 2. Let $x_0 \in X(k)$. We call x_0 a *Weierstrass point* of X if $l(2x_0) \geq 2$. For example, ramification points of f are Weierstrass points (see the proof of Lemma 4.8). Conversely, we want to show that every Weierstrass point x_0 of X is a ramification point of f.

(a) Let σ denote the hyperelliptic involution of X. Let us suppose that f is unramified at x_0. Then $x'_0 := \sigma(x_0)$ is a point that is distinct from x_0. Let $h \in L(2x_0) \setminus k$. Show that $(h \pm \sigma(h))_\infty = 2[x_0] + 2[x'_0] = 2f^*[f(x_0)]$.

(b) Let us keep the notation of Proposition 4.24. Changing the variable on $\mathbb{P}^1_k$ if necessary, we may assume that $f(x_0) = \{t = \infty\}$. Show that $h = a(t) + b(t)y$ with $a(t), b(t) \in k[t]$ and $b(t) \neq 0$.

(c) By considering the degree of $h - \sigma(h)$, show that we have $g \leq 1$. Conclude.

4.8. Let X, Y be smooth and geometrically connected projective curves over a field k.

(a) Let $f : X \to Y$ be a non-constant morphism. Show that $g(X) \geq g(Y)$.

(b) Let us suppose that $g(X) \geq 2$. Let $f : X \to X$ be a finite morphism. Show that f is an automorphism if it is separable. In general, if $\mathrm{char}(k) = p > 0$, show that f decomposes into a relative Frobenius $F^r_{X/k} : X \to X^{(p^r)}$ followed by an isomorphism $X^{(p^r)} \to X$. These morphisms are morphisms of algebraic varieties over k.

4.9. Let X be a smooth and geometrically connected projective curve over a field k. Let U be a non-empty open subset of X.

(a) Show that the group $\mathrm{Div}(U)$ can be identified with the group of Cartier divisors on X with support contained in U.

(b) Let $\alpha : \mathrm{Pic}^0(X) \to \mathrm{Pic}(U)$ be the restriction of the canonical homomorphism $\mathrm{Pic}(X) \to \mathrm{Pic}(U)$. Let $D_0 \in \mathrm{Div}(U)$ be such that

$$\deg D_0 = \gcd\{\deg D \mid D \in \mathrm{Div}(U)\}.$$

Show that $\mathrm{Coker}(\alpha)$ is generated by the class of $\mathcal{O}_X(D_0)$.

(c) Show that there exists an exact sequence of groups

$$0 \to G \to \mathrm{Pic}^0(X) \to \mathrm{Pic}(U) \to H \to 0$$

where G and H are finitely generated.

(d) Let us suppose that k is algebraically closed. Show that $\mathrm{Pic}^0(X)$ is not finitely generated if $g \geq 1$ (use Corollary 4.41). Deduce from this that $\mathrm{Pic}(U)$ is finitely generated if and only if $g = 0$.

(e) Let $\mathrm{Spec}\, A$ be a smooth and connected affine curve over an algebraically closed field k. Show that $\mathrm{Cl}(A)$ is a finitely generated group if and only if there exists a $P(x) \in k[x]$ such that $A \simeq k[x, 1/P(x)]$.

4.10. Let $P(t) \in k[t]$ be a separable polynomial of even degree ≥ 2 over an algebraically closed field k with $\mathrm{char}(k) \neq 2$. Let us consider the hyperelliptic curve X over k defined by the equation $y^2 = P(t)$. Show that the following properties are equivalent:

(i) There exist $A, B \in k[t]$ such that $A^2 - P(t)B^2 = 1$.

(ii) $k[t, y]^* \neq k^*$.

(iii) Let x_1, x_2 be the points of the support of $(t)_\infty$. Then the divisor $x_1 - x_2 \in \mathrm{Pic}^0(X)$ is an element of finite order.

4.11. Let $E = V_+(F(u, v, w)) \subset \mathbb{P}_k^2$ be an elliptic curve. Let $o = (0, 1, 0)$ and $x \in E(k)$. Let us suppose that there exists a curve $C \subset \mathbb{P}_k^2$ such that $C \cap E = \{x\}$. Show that $x - o \in \mathrm{Pic}^0(X)$ is an element of finite order.

4.12. Let X be a smooth and geometrically connected projective curve over a finite field $k = \mathbb{F}_q$. Let J be its Jacobian.

(a) Show that $J(k)$ is a finite group.

(b) Let $x_1, \ldots, x_n \in X(k)$ be points that are pairwise distinct. Let

$$F := \{f \in K(X)^* \mid \mathrm{Supp}(f) \subseteq \{x_1, \ldots, x_n\}\}.$$

Let $\varphi : F \to \mathbb{Z}^n$ be defined by $f \mapsto (\mathrm{ord}_{x_1}(f), \ldots, \mathrm{ord}_{x_n}(f))$. Show that $\mathrm{Im}\, \varphi$ is of finite index in $\{(r_1, \ldots, r_n) \in \mathbb{Z}^n \mid r_1 + \cdots + r_n = 0\}$. Deduce from this that $F \simeq k^* \oplus \mathbb{Z}^{n-1}$.

4.13. Let C be an affine curve over a field k. We will say that C is an *affine plane curve* if it is isomorphic to a closed subscheme of an open subscheme of $\mathbb{A}_k^2$.

(a) Show that if C is an affine plane curve, then $\omega_{C/k} \simeq \mathcal{O}_C$.

(b) Let X be a smooth, geometrically connected, projective curve over k, of genus $g \geq 2$. Show that there exist at most $(2g - 2)^{2g}$ points $x \in X(k)$ such that $X \setminus \{x\}$ is an affine plane curve.

(c) Let us suppose that X is hyperelliptic over k. Let $x \in X(k)$. Show that the affine curve $X \setminus \{x\}$ is plane if and only if x is a Weierstrass point (see Exercise 4.7).

4.14. Let $G \to S$ be a group scheme. Show that $G \to S$ is separated if and only if the unit section $\varepsilon : S \to G$ is a closed immersion.

4.15. Let G be an algebraic group over a field k, with unit section $o \in G(k)$. Let $x \in G(k)$.

(a) Show that there exists an automorphism $t_x : G \to G$ of G (as an algebraic variety) such that $t_x(o) = x$. Deduce from this that $\mathcal{O}_{G,o} \simeq \mathcal{O}_{G,x}$.

(b) If G is geometrically reduced, show that it is smooth over k.

4.16. Let G be a geometrically reduced algebraic variety over a field k, let $m : G \times_k G \to G$, $\mathrm{inv} : G \to G$, and $\varepsilon : \mathrm{Spec}\, k \to G$ be morphisms of k-schemes. We suppose that these three morphisms induce an abstract group law on $G(\overline{k})$. Using Exercise 3.2.9, show that they induce an algebraic group structure on G.

4.17. Let k be a field and let $U = \mathrm{Spec}\, k[X]$.

(a) Show that the automorphisms of U (as a k-scheme) are defined by $X \mapsto aX + b$ with $a \in k^*$ and $b \in k$.

(b) Let $m : U \times_{\mathrm{Spec}\, k} U \to U$ be an algebraic group law on U. Changing the variables if necessary, we suppose that $X = 0$ is the unit element of the group U. The morphism m corresponds to a homomorphism of k-algebras $k[X] \to K[Y, Z]$. Let $P(Y, Z)$ be the image of X. Show that for any $(y, z) \in k^2$, seen as a rational point of $U \times_{\mathrm{Spec}\, k} U$, we have $m(y, z) = P(y, z)$.

(c) Using (b) on $U_{\overline{k}}$, show that $P(Y, Z) = Y + Z$ (fix one of the variables, say Z, and use the fact that $Y \mapsto P(Y, Z)$ is an automorphism), and that $U \simeq \mathbb{G}_{a,k}$ as algebraic groups over k.

4.18. Let k be a field, $T = \mathrm{Spec}\, k[X, 1/X]$, and $m : T \times_{\mathrm{Spec}\, k} T \to T$ an algebraic group law on T.

(a) Changing the variables if necessary, we suppose that $X = 1$ is the unit element of the group T. The morphism m corresponds to a homomorphism of k-algebras $k[X, 1/X] \to K[Y, 1/Y, Z, 1/Z]$. Let $P(Y, Z)$ be the image of X. Show that there exist $a \in k^*$, $q, r \in \mathbb{Z}$ such that $P(Y, Z) = aY^q Z^r$ (use the fact that $X \in \mathcal{O}_T(T)^*$).

(b) Show that $(y, z) \mapsto ay^q z^r$ defines an (abstract) group law on $\overline{k}^*$ for which 1 is the unit element. Deduce from this that $a = 1$ and $q = r = 1$. Show that $T \simeq \mathbb{G}_{m,k}$ as algebraic groups over k.

7.5   Singular curves, structure of $\mathrm{Pic}^0(X)$

In this section we gather together some results that are rather specific to singular curves. Proposition 5.4 has been used in the proof of Proposition 4.1. A large part of this section is devoted to the structure of the group $\mathrm{Pic}^0(X)$ (Theorem 5.19).

Definition 5.1. Let X be a scheme having only a finite number of irreducible components $X_1, \ldots, X_n$ (endowed with the reduced closed subscheme structure). The disjoint union $X' = \coprod_{1 \le i \le n} X_i'$, where X_i' is the normalization of the integral scheme X_i (Definition 4.1.19), is called the *normalization of* X. By construction, X' is endowed with a surjective integral morphism $\pi : X' \to X$. If X_{red} is the reduced scheme associated to X, then $X_{\mathrm{red}}' = X'$.

Lemma 5.2. Let X be a reduced scheme having only a finite number of irreducible components. Let U be an affine open subset of X.

(a) We have a canonical isomorphism $\mathrm{Frac}(\mathcal{O}_X(U)) \simeq \oplus_{1 \le i \le n} \mathrm{Frac}(\mathcal{O}_{X, \xi_i})$, where the ξ_i are the generic points of X belonging to U.

(b) $\mathcal{O}_{X'}(\pi^{-1}(U))$ is the integral closure of $\mathcal{O}_X(U)$ in $\mathrm{Frac}(\mathcal{O}_X(U))$.

Proof Let us set $A = \mathcal{O}_X(U)$ and let $\mathfrak{p}_i$ be the prime ideal of A corresponding to ξ_i. Then we have a finite injective canonical homomorphism

$$\rho : A \to \oplus_{1 \le i \le n} A/\mathfrak{p}_i,$$

which induces an injection $\rho' : \mathrm{Frac}(A) \rightarrow \oplus_i \mathrm{Frac}(A/\mathfrak{p}_i)$. Let $i_0 \leq n$. There exists an $a \in (\cap_{i \neq i_0} \mathfrak{p}_i) \setminus \mathfrak{p}_{i_0}$. By considering the image of a by ρ', we immediately see that ρ' is surjective, whence (a) (see also Exercises 1.1 and 1.4).

(b) The integral closure A' of A in $\mathrm{Frac}(A)$ is clearly contained (via the injection ρ) in $\oplus_i (A/\mathfrak{p}_i)'$, where $(A/\mathfrak{p}_i)'$ denotes the integral closure in $\mathrm{Frac}(A/\mathfrak{p}_i)$. Now $\oplus_i (A/\mathfrak{p}_i)'$ is integral over $\oplus_i A/\mathfrak{p}_i$, and hence integral over A. It follows that $A' = \oplus_i (A/\mathfrak{p}_i)'$, whence (b). $\qquad\square$

Definition 5.3. Let X be a reduced Noetherian scheme. Let $\xi_1, \ldots, \xi_n$ be the generic points of X. We say that a morphism of finite type $f : Z \rightarrow X$ is a *birational morphism* if Z admits exactly n generic points $\xi'_1, \ldots, \xi'_n$, if $f^{-1}(\xi_i) = \xi'_i$, and if $\mathcal{O}_{X,\xi_i} \rightarrow \mathcal{O}_{Z,\xi'_i}$ is an isomorphism for every i. For example, the normalization morphism $X' \rightarrow X$ is a birational morphism. Using Lemma 5.2, we see that we have a canonical isomorphism $\mathcal{K}_X \rightarrow f_* \mathcal{K}_Z$ as soon as f is birational.

Let X be a reduced curve over a field k, and let $\pi : X' \rightarrow X$ be the normalization morphism. We immediately deduce from Proposition 4.1.27 that π is a finite morphism. We have an exact sequence of coherent sheaves on X:

$$0 \rightarrow \mathcal{O}_X \rightarrow \pi_* \mathcal{O}_{X'} \rightarrow \mathcal{S} \rightarrow 0. \tag{5.17}$$

The support of $\mathcal{S}$ is a closed set not containing any generic point of X; it is therefore a finite set, consisting of the singular points of X. Therefore $\mathcal{S}$ is a skyscraper sheaf. For any $x \in X$, we have $\mathcal{S}_x = \mathcal{O}'_{X,x}/\mathcal{O}_{X,x}$, where $\mathcal{O}'_{X,x}$ is the integral closure of $\mathcal{O}_{X,x}$ in $\mathrm{Frac}(\mathcal{O}_{X,x})$, because normalization commutes with localization. Let

$$\delta_x := \mathrm{length}_{\mathcal{O}_{X,x}} \mathcal{S}_x = [k(x) : k]^{-1} \dim_k \mathcal{S}_x \tag{5.18}$$

(see Exercise 1.6(d) for the second equality). Then $\delta_x = 0$ if and only if x is a normal (hence regular) point of X.

Proposition 5.4. *Let X be a reduced projective curve over a field k. Let $X_1, \ldots, X_n$ be the irreducible components of X. Then*

$$p_a(X) + n - 1 = \sum_{1 \leq i \leq n} p_a(X'_i) + \sum_{x \in X} [k(x) : k] \delta_x,$$

where X'_i is the normalization of X_i.

Proof Exact sequence (5.17) implies that we have the equality $\chi(\pi_* \mathcal{O}_{X'}) = \chi(\mathcal{O}_X) + \chi(\mathcal{S})$ (Lemma 3.16). As π is finite, we have

$$\chi(\pi_* \mathcal{O}_{X'}) = \chi(\mathcal{O}_{X'}) = \sum_{1 \leq i \leq n} \chi(\mathcal{O}_{X'_i}) = n - \sum_{1 \leq i \leq n} p_a(X'_i).$$

Moreover, since $\mathcal{S}$ is a skyscraper sheaf,

$$\chi(\mathcal{S}) = \dim_k H^0(X, \mathcal{S}) = \sum_x \dim_k \mathcal{S}_x = \sum_x [k(x) : k] \delta_x.$$

These equalities imply the proposition. $\qquad\square$

The following proposition in some sense generalizes Proposition 4.4.

Proposition 5.5. *Let X be a projective curve over k. Let $D \in \mathrm{Div}(X)$. Then $\mathcal{O}_X(D)$ is ample if and only if $\deg \mathcal{O}_X(D)|_{X_i} > 0$ for every irreducible component X_i of X.*

Proof By Corollary 5.3.8, we may suppose X is integral. If D is ample, then $\deg D \geq 0$, because otherwise $L(nD) = 0$ and hence $\mathcal{O}_X(nD)$ cannot be generated by its global sections. If $\deg D = 0$, and nD is generated by its global sections, then $\mathcal{O}_X(nD) = \mathcal{O}_X$. Now $\mathcal{O}_X$ is not very ample, because X is not affine (otherwise $X = \mathrm{Spec}\,\mathcal{O}_X(X)$ and would be of dimension 0). Hence $\deg D > 0$.

Conversely, let us suppose that $\deg D > 0$. Let E be an ample divisor. For any $n \geq 1$, we have $l(nD - E) \geq \deg(nD - E) + \chi(\mathcal{O}_X)$. Hence for a sufficiently large n, we have $l(nD - E) \neq 0$. Replacing D if necessary by a divisor that is linearly equivalent to nD, we can suppose that $D \geq E$. Let $\mathcal{F}$ be a coherent sheaf on X. Then $H^1(X, \mathcal{F} \otimes \mathcal{O}_X(nE)) = 0$ for any sufficiently large n (Proposition 5.3.6). We have an exact sequence

$$0 \to \mathcal{O}_X(nE) \to \mathcal{O}_X(nD) \to \mathcal{G}_n \to 0$$

where $\mathcal{G}_n \simeq \mathcal{O}_{nD-nE}$ is a skyscraper sheaf. By tensoring with $\mathcal{F}$, we obtain an exact sequence

$$\mathcal{F} \otimes \mathcal{O}_X(nE) \xrightarrow{\alpha} \mathcal{F} \otimes \mathcal{O}_X(nD) \to \mathcal{F} \otimes \mathcal{G}_n \to 0.$$

As X is of dimension 1 and $\mathcal{F} \otimes \mathcal{G}_n$ is a skyscraper sheaf, the canonical homomorphisms

$$H^1(X, \mathcal{F} \otimes \mathcal{O}_X(nE)) \to H^1(X, \mathrm{Im}\,\alpha), \quad H^1(X, \mathrm{Im}\,\alpha) \to H^1(X, \mathcal{F} \otimes \mathcal{O}_X(nD))$$

are surjective. This implies that $H^1(X, \mathcal{F} \otimes \mathcal{O}_X(nD)) = 0$ and that $\mathcal{O}_X(D)$ is ample (Proposition 5.3.6). See also Exercise 5.3. $\qquad\square$

Definition 5.6. Let X be a locally Noetherian scheme. Let Z be an irreducible component of X. We call the integer $d = \mathrm{length}(\mathcal{O}_{X,\xi})$, where ξ is the generic point of Z, the *multiplicity of Z in X*. We have $d \geq 1$, and $d = 1$ if and only if $\mathcal{O}_{X,\xi}$ is reduced (or, equivalently, if X is reduced on a non-empty open subset containing ξ).

Proposition 5.7. *Let X be a projective curve over a field k, and let $X_1, \ldots, X_n$ be its irreducible components, with respective generic points $\xi_1, \ldots, \xi_n$. We endow each X_i with the reduced subscheme structure. Then for any invertible sheaf $\mathcal{L}$ on X, we have*

$$\deg \mathcal{L} = \sum_{1 \leq i \leq n} d_i \deg(\mathcal{L}|_{X_i}), \tag{5.19}$$

where $d_i = \mathrm{length}(\mathcal{O}_{X,\xi_i})$.

Proof Equality (5.19) is additive on $\mathcal{L}$. Using Proposition 1.32, we may suppose that $\mathcal{L} \simeq \mathcal{O}_X(D)$ for some Cartier divisor D on X. By Lemma 3.6, we can even suppose D effective. Let $x \in X$ be a closed point, and f be a generator of $\mathcal{O}_X(-D)_x$. Let $A = \mathcal{O}_{X,x}$ and let $\mathfrak{p}_1, \ldots, \mathfrak{p}_r$ denote the minimal prime ideals of A. Then it suffices to show the equality

$$\mathrm{length}_A(A/fA) = \sum_{1 \le i \le r} \mathrm{length}_{A_{\mathfrak{p}_i}}(A_{\mathfrak{p}_i})\,\mathrm{length}_{A/\mathfrak{p}_i}(A/(f, \mathfrak{p}_i)).$$

As f is a regular element, this equality can be obtained by taking $M = A$ in the equality

$$e_A(f, M) = \sum_{1 \le i \le r} \mathrm{length}_{A_{\mathfrak{p}_i}}(M_{\mathfrak{p}_i})\,\mathrm{length}_A(A/(f, \mathfrak{p}_i)) \tag{5.20}$$

that we are going to show for every finitely generated A-module M (see Exercise 1.7 for the notation $e_A(f, M)$). As the length over $A_{\mathfrak{p}_i}$ and $e_A(f, \cdot)$ are additive with respect to exact sequences (Lemma 1.23 and Exercise 1.7(c)), with the help of Lemma 1.4, it suffices to show (5.20) for $M = A/\mathfrak{p}$, with $\mathfrak{p}$ prime. If $\mathfrak{p}$ is maximal, then $e_A(f, M) = 0$ and $M_{\mathfrak{p}_i} = 0$; hence (5.20) is true. If $\mathfrak{p}$ is not maximal, it is equal to one of the minimal prime ideals $\mathfrak{p}_i$. As $M_{\mathfrak{p}_j} = A_{\mathfrak{p}_i}$ if $j = i$ and 0 otherwise, equality (5.20) is still true. $\square$

Corollary 5.8. *Let X be a reduced projective curve over a field k. Let $\pi : X' \to X$ be the normalization morphism. Then for any invertible sheaf $\mathcal{L}$ on X, we have $\deg \mathcal{L} = \deg \pi^* \mathcal{L}$.*

Proof This results from Proposition 5.7 and Proposition 3.8. $\square$

Definition 5.9. Let X be a connected projective curve over a field k, with irreducible components $X_1, \ldots, X_n$. We let $\mathrm{Pic}^0(X)$ denote the set of isomorphism classes of invertible sheaves $\mathcal{L}$ such that $\deg \mathcal{L}|_{X_i} = 0$ for every $1 \le i \le n$.

In what follows, we are going to study the structure of the group $\mathrm{Pic}^0(X)$ for a projective curve X over a field k. More specifically, we consider the torsion points (i.e., elements of finite order) of $\mathrm{Pic}^0(X)$. Let G be a commutative group, and $n \in \mathbb{Z}$. We denote the multiplication by n homomorphism by $n_G : G \to G$. From here to the end of the section, *we will suppose k is algebraically closed.*

Definition 5.10. Let G be a commutative group. We will say that G is *unipotent* if there exists an ascending chain of subgroups

$$0 = G_0 \subset G_1 \subset \cdots \subset G_n = G$$

such that G_i/G_{i-1} is isomorphic to the additive group of k for every $1 \le i \le n$. The integer n will be called the *dimension* of G. It is an easy exercise to show that if G is an extension of a unipotent group U_1 by a unipotent group U_2, then G is unipotent of dimension the sum of the dimensions of U_1 and U_2.

It is clear that this definition depends on k. Moreover, the definition of the dimension could depend on the choice of a sequence $(G_i)_i$. We must note that the true definition is that G and the G_i are algebraic groups, and that the homomorphisms that occur are morphisms of algebraic groups. It turns out that the abstract groups that we are going to encounter are in fact groups of rational points of unipotent algebraic groups. In order not to have to develop the theory of algebraic groups, we are forced to do some gymnastics with the definitions. Anyway, we will not really use the dimension of G, but only the fact that $G[n] = 0$ for any n prime to $\mathrm{char}(k)$ (see Corollary 5.23).

Lemma 5.11. *Let X be a connected projective curve over an algebraically closed field k. Let $Z = V(\mathcal{J})$ be a closed subscheme of X defined by a nilpotent sheaf of ideals. Then the homomorphism $R : \mathrm{Pic}(X) \to \mathrm{Pic}(Z)$ induced by the closed immersion $i : Z \to X$ is surjective, with unipotent kernel of dimension $\dim_k H^1(X, \mathcal{O}_X) - \dim_k H^1(Z, \mathcal{O}_Z)$.*

Proof Let $\mathcal{N}_X \subset \mathcal{O}_X$ be the sheaf of nilpotent elements. Let us first suppose that $\mathcal{J}\mathcal{N}_X = 0$. In particular, $\mathcal{J}^2 = 0$. We have an exact sequence of sheaves of groups on X:

$$1 \to 1 + \mathcal{J} \to \mathcal{O}_X^* \to i_*\mathcal{O}_Z^* \to 1.$$

The map $a \to 1+a$ defines an isomorphism of sheaves of groups $\mathcal{J} \simeq 1+\mathcal{J}$. As the first is a quasi-coherent sheaf on X, we have $H^2(X, \mathcal{J}) = 0$ (Proposition 5.2.24). We then obtain an exact sequence of cohomology groups

$$\mathcal{O}_X(X)^* \to \mathcal{O}_Z(Z)^* \to H^1(X, 1 + \mathcal{J}) \to \mathrm{Pic}(X) \xrightarrow{R} \mathrm{Pic}(Z) \to 0$$

(Corollary 5.2.22). Hence R is surjective. Similarly, we have an exact sequence

$$\mathcal{O}_X(X) \to \mathcal{O}_Z(Z) \to H^1(X, \mathcal{J}) \to H^1(X, \mathcal{O}_X) \xrightarrow{S} H^1(Z, \mathcal{O}_Z) \to 0.$$

As $\mathcal{O}_X(X)/\mathcal{N}_X(X)$ is a sub-k-algebra of $\mathcal{O}_{X_{\mathrm{red}}}(X_{\mathrm{red}}) = k$ (Corollary 3.3.21), we have

$$\mathcal{O}_X(X) = k + N_X, \quad \mathcal{O}_Z(Z) = k + N_Z,$$

where $N_X := \mathcal{N}_X(X)$, $N_Z := \mathcal{N}_Z(Z)$ are nilpotent ideals. This implies that

$$\mathcal{O}_X(X)^* = k^*(1 + N_X), \quad \mathcal{O}_Z(Z)^* = k^*(1 + N_Z).$$

Let $\rho : \mathcal{O}_X(X) \to \mathcal{O}_Z(Z)$ be the canonical homomorphism. By directly using the definition of Čech cohomology, we see that the $\mathcal{J}\mathcal{N}_X = 0$ hypothesis entails that $N_Z^2 \subseteq \rho(N_X)$ and that we have the commutative diagram

$$
\begin{array}{ccc}
N_Z & \longrightarrow & H^1(X, \mathcal{J}) \\
\downarrow & & \downarrow \\
1 + N_Z & \longrightarrow & H^1(X, 1 + \mathcal{J})
\end{array}
$$

where the horizontal arrows come from the exact cohomology sequence, and where the vertical arrows are $a \mapsto 1 + a$. This immediately implies that

$$(1 + N_Z)/(1 + \rho(N_X)) \simeq N_Z/\rho(N_X).$$

In other words,

$$\mathrm{Ker}(R) \simeq \mathrm{Ker}(S) = \mathrm{Ker}(H^1(X, \mathcal{O}_X) \to H^1(Z, \mathcal{O}_Z))$$

and $\dim_k \mathrm{Ker}(R) = \dim_k H^1(X, \mathcal{O}_X) - \dim_k H^1(Z, \mathcal{O}_Z)$.

In the general case, we have an ascending chain of closed subschemes

$$Z = X_0 \subseteq X_1 \subseteq \cdots \subseteq X_n = X$$

with $X_i = V(\mathcal{J}\mathcal{N}_X^i)$ (we take n large enough so that $\mathcal{N}_X^n = 0$). Then X_i is defined by a sheaf of ideals $\mathcal{J}_i \subset \mathcal{O}_{X_{i+1}}$ such that $\mathcal{J}_i \mathcal{N}_{X_{i+1}} = 0$. The preceding case immediately implies that $\mathrm{Pic}(X_{i+r}) \to \mathrm{Pic}(X_i)$ is surjective if $r \geq 1$. Its kernel $K_{i+r,i}$ is an extension

$$0 \to K_{i+r,i+1} \to K_{i+r,i} \to K_{i+1,i} \to 0.$$

An induction on r implies that $K_{i+r,i}$ is unipotent and

$$\dim K_{i+r,i} = \dim_k H^1(X_{i+r}, \mathcal{O}_{X_{i+r}}) - \dim_k H^1(X_i, \mathcal{O}_{X_i}),$$

the case $r = 1$ having been proven in the first part. We obtain the lemma by taking $i = 0$ and $r = n$. □

Before studying the Picard group of a reduced curve, we need to examine thoroughly the structure of the singular points. Let X be a reduced curve over an algebraically closed field k, and $\pi : X' \to X$ the normalization morphism. We are going to construct a curve Y lying between X and X', having 'not too complicated' singularities. Let $x \in X$ be a singular point. Let $y_1, \ldots, y_m$ denote the points of $\pi^{-1}(x)$. Let V_x be an affine open neighborhood of x such that x is the only singular point of V_x. Let W_x be the affine curve corresponding to the $\mathcal{O}_X(V_x)$-algebra

$$\{b \in \mathcal{O}_{X'}(\pi^{-1}(V_x)) \mid b(y_1) = \cdots = b(y_m)\}.$$

The morphism $\pi^{-1}(V_x) \to V_x$ decomposes into two finite surjective morphisms $\pi^{-1}(V_x) \to W_x \to V_x$. It follows that $W_x \to V_x$ is an isomorphism over $V_x \setminus \{x\}$. Consequently, the curves W_x glue, as x varies in X_{sing}, to a reduced curve Y such that π decomposes into two finite surjective morphisms $\pi_1 : X' \to Y$ and $\pi_2 : Y \to X$.

Lemma 5.12. *Let us keep the notation and hypotheses above.*

(a) *The morphism* $\pi_2 : Y \to X$ *is a homeomorphism.*

(b) *Let* $y = \pi_2^{-1}(x)$. *There exists an integer* $r \geq 1$ *such that* $\mathfrak{m}_y^r \subseteq \mathfrak{m}_x$. *We then have* $\mathfrak{m}_x^r(\pi_* \mathcal{O}_{X'})_x \subseteq \mathfrak{m}_x$.

(c) *We have exact sequences*

$$0 \to \mathcal{O}_{Y,y}^*/\mathcal{O}_{X,x}^* \to (\pi_* \mathcal{O}_{X'}^*)_x/\mathcal{O}_{X,x}^* \to (\pi_{1*} \mathcal{O}_{X'}^*)_y/\mathcal{O}_{Y,y}^* \to 0, \quad (5.21)$$

$$0 \to \mathcal{O}_{Y,y}/\mathcal{O}_{X,x} \to (\pi_* \mathcal{O}_{X'})_x/\mathcal{O}_{X,x} \to (\pi_{1*} \mathcal{O}_{X'})_y/\mathcal{O}_{Y,y} \to 0, \quad (5.22)$$

isomorphisms

$$(\pi_{1*} \mathcal{O}_{X'}^*)_y/\mathcal{O}_{Y,y}^* \simeq (k^*)^{m-1}, \quad (\pi_{1*} \mathcal{O}_{X'})_y/\mathcal{O}_{Y,y} \simeq k^{m-1},$$

and $U := \mathcal{O}_{Y,y}^*/\mathcal{O}_{X,x}^*$ *is unipotent of dimension* $\dim_k \mathcal{O}_{Y,y}/\mathcal{O}_{X,x}$.

Proof Let $A = \mathcal{O}_X(V_x)$, $B = \mathcal{O}_{X'}(\pi^{-1}(V_x))$, and $C = \mathcal{O}_Y(W_x)$.

(a) Let $\mathfrak{p}$ be the maximal ideal of A corresponding to the point x, $\mathfrak{r}_1, \ldots, \mathfrak{r}_m$ the maximal ideals of B corresponding to the points $y_1, \ldots, y_m$, and $\mathfrak{q} := \cap_{1 \leq i \leq m} \mathfrak{r}_i$. We immediately verify that $\mathfrak{q} \subseteq C$ and that it is the unique maximal ideal of C lying above $\mathfrak{p}$. It follows that π_2 is bijective. As it is moreover proper, it is indeed a homeomorphism.

(b) The ring $C/\mathfrak{p}C$ is local Artinian. Therefore there exists an $N \geq 1$ such that $\mathfrak{q}^N \subseteq \mathfrak{p}C$. The sheaf $(C/\mathfrak{p})^\sim$ is coherent on X, with support $\{x\}$. It follows that $C/\mathfrak{p} = H^0(X, (C/\mathfrak{p})^\sim)$ is an A-module of finite length. Let $\bar{\mathfrak{q}}^n$ be the image of $\mathfrak{q}^n$ in $C/\mathfrak{p}$. Then there exists an $r \geq N$ such that $\bar{\mathfrak{q}}^n = \bar{\mathfrak{q}}^{n+1}$ for every $n \geq r$ (Proposition 1.25). Consequently,

$$\bar{\mathfrak{q}}^r \subseteq \cap_{n \geq r} \bar{\mathfrak{q}}^n \subseteq \cap_{q \geq 1} \mathfrak{p}^q M, \quad M := C/\mathfrak{p}.$$

Let $f \in \mathfrak{q}^r$. By Krull's theorem (Corollary 1.3.13), there exists an $\epsilon \in \mathfrak{p}$ such that $(1 + \epsilon)f \in \mathfrak{p}$. Consequently, $f \in \mathfrak{m}_x \mathcal{O}_{X,x}$. Hence $\mathfrak{m}_y^r \subseteq \mathfrak{m}_x$. As $\mathfrak{q}B \subseteq C$ by definition of C, we have $\mathfrak{q}^r B \subseteq \mathfrak{q}^r$, whence $\mathfrak{m}_x^r(\pi_* \mathcal{O}_{X'})_x \subseteq \mathfrak{m}_x$ by taking the localization.

(c) Complex (5.22) is

$$0 \to C/A \to B/A \to B/C \to 0,$$

which is exact. Similarly, complex (5.21) is exact. The homomorphisms $B^* \to (k^*)^m$, $b \mapsto (b(y_1), \ldots, b(y_m))$; $C^* \to k^*$, $c \mapsto c(y)$ clearly induce an injective homomorphism

$$\rho : B^*/C^* \to (k^*)^m/\Delta(k^{*m}) \simeq (k^*)^{m-1},$$

where $\Delta : k^* \to (k^*)^m$ is the diagonal homomorphism. We have ρ surjective, because $B^* \to (k^*)^m$ is induced by the surjective homomorphism

$$B \to B/\mathfrak{p}B \simeq \oplus_{1\leq i\leq m} B_{\mathfrak{r}_i}/\mathfrak{p}B_{\mathfrak{r}_i} \to \oplus_{1\leq i\leq m} k(y_i) = k^m$$

(see Exercise 2.5.11(c) for the isomorphism in the middle). Hence ρ is an isomorphism. Let us note, moreover, that the same arguments show that $B/C \simeq k^m/\Delta(k) \simeq k^{m-1}$. As we have an exact sequence

$$0 \to C^*/A^* \to B^*/A^* \to B^*/C^* \to 0,$$

and $B^*/A^* = (\pi_* \mathcal{O}_{X'}^*)_x/\mathcal{O}_{X,x}^*$, it suffices to show that $U := C^*/A^*$ verifies the required property. Now $U \simeq (1+\mathfrak{m}_y)/(1+\mathfrak{m}_x)$. In a way similar to the end of the proof of Lemma 5.11, we show that U is unipotent of dimension $\dim_k(\mathfrak{m}_y/\mathfrak{m}_x)$ by using (b). Now we have an isomorphism of k-vector spaces

$$\mathfrak{m}_y/\mathfrak{m}_x \simeq (k+\mathfrak{m}_y)/(k+\mathfrak{m}_x) = \mathcal{O}_{Y,y}/\mathcal{O}_{X,x}.$$

This completes the proof of (c).   □

Definition 5.13. Let X be a reduced curve over an algebraically closed field k. Let $\pi : X' \to X$ be the normalization morphism. We say that a closed point $x \in X$ is an *ordinary multiple point* if $\delta_x = m_x - 1$, where m_x is the number of points of $\pi^{-1}(x)$ and where δ_x is defined by formula (5.18). We will also say that x is an *ordinary m_x-fold* point. If $m_x = 2$, we say that x is an *ordinary double point* or a '*node*'. With the notation of Lemma 5.12 and by virtue of its assertion (c), the point y of Y is an ordinary multiple point. The curve Y is called the *curve with ordinary multiple singularities associated to X*.

If k is not necessarily algebraically closed, we will say that $x \in X$ is an ordinary m-fold point if for every point $x' \in X_{\bar{k}}$ lying above x, we have $X_{\bar{k}}$ reduced in a neighborhood of x' and x' is an ordinary m-fold point.

Example 5.14. Let X be the curve $\operatorname{Spec} k[t,s]/(s^2 - t^2(1+t))$ over a field of characteristic $\operatorname{char}(k) \neq 2$. Then the point $(0,0)$ is an ordinary double point. Indeed, the normalization of X is $\operatorname{Spec} k[u]$ with $u = s/t$ and $u^2 = 1+t$. We easily see that $\delta = 1$ and $m = 2$. We can also step over to the formal completion, and use Exercise 1.3.9. If $\operatorname{char}(k) = 2$, then $(0,0)$ is not an ordinary multiple point because $\delta = m = 1$.

Proposition 5.15. *Let X be a reduced curve over an algebraically closed field k, $x \in X(k)$, $\pi : X' \to X$ the normalization morphism, and $\pi^{-1}(x) = \{y_1, \ldots, y_m\}$. Then the following properties are equivalent:*

(i) *The point x is an ordinary multiple point.*

(ii) *Let V be an affine open neighborhood of x such that $V \cap X_{\mathrm{sing}} = \{x\}$. Then*

$$\mathcal{O}_X(V) = \{f \in \mathcal{O}_{X'}(\pi^{-1}(V)) \mid f(y_1) = \cdots = f(y_m)\}.$$

(iii) *We have an isomorphism*

$$\widehat{\mathcal{O}}_{X,x} \simeq k[[T_1, \ldots, T_m]]/(T_i T_j)_{i \neq j}.$$

(iv) *There exist $n \geq 1$ and an isomorphism*

$$\widehat{\mathcal{O}}_{X,x} \simeq k[[T_1, \ldots, T_n]]/(T_i T_j)_{i \neq j}.$$

Moreover, if these conditions are satisfied, then $\dim T_{X,x} = m$.

Proof Let $X' \xrightarrow{\pi_1} Y \xrightarrow{\pi_2} X$ be the decomposition defined just before Lemma 5.12. Then $\delta_x = m - 1 + \dim_k \mathcal{O}_{Y,y}/\mathcal{O}_{X,x}$ by Lemma 5.12(c). Thus (i) $\Longleftrightarrow \{\pi_2$ is an isomorphism$\} \Longleftrightarrow$ (ii).

(ii) $\Longrightarrow$ (iii). Let $A = \mathcal{O}_{X,x}$, $B = (\pi_*\mathcal{O}_{X'})_x$. Let $\hat{A}$ be the $\mathfrak{m}_x$-adic completion of $\mathcal{O}_{X,x}$, and $\hat{B} = B \otimes_A \hat{A}$. Then we have

$$\hat{B} = \oplus_{1 \leq i \leq m} \widehat{\mathcal{O}}_{X',y_i} = \oplus_{1 \leq i \leq m} k[[t_i]],$$

where t_i is a generator of $\mathfrak{m}_{y_i}$ (Exercise 4.3.17 and Proposition 4.2.27). The surjective homomorphism $\rho : B \to k^m$ defined by $\rho(b) = (b(y_1), \ldots, b(y_m))$ extends to $\hat{\rho} : \hat{B} \to k^m$. We have $A \subseteq \rho^{-1}(\Delta(k))$, where $\Delta : k \to k^m$ is the diagonal map. The hypothesis that x is an ordinary multiple point implies that

$$\dim_k B/A = m - 1 = \dim_k(B/\rho^{-1}(\Delta(k))).$$

So $A = \rho^{-1}(\Delta(k))$. Likewise, $\hat{A} = \hat{\rho}^{-1}(\Delta(k))$. Let us consider the homomorphism

$$\varphi : k[[T_1, \ldots, T_m]]/(T_i T_j)_{i \neq j} \to \oplus_{1 \leq i \leq m} k[[t_i]]$$

which sends the image of T_i in the quotient to t_i. We immediately verify that φ is injective and that its image is precisely $\hat{\rho}^{-1}(\Delta(k))$, whence (iii).

(iv) $\Longrightarrow$ (i). The homomorphism φ above implies that the integral closure $\hat{A}'$ of $\hat{A}$ in $\mathrm{Frac}(\hat{A})$ is isomorphic to $\oplus_{1 \leq i \leq n} k[[t_i]]$, where t_i is the image of T_i in $\hat{A}$. Moreover, we have

$$\hat{B} = B \otimes_A \hat{A} \subseteq \mathrm{Frac}(A) \otimes_A \hat{A} \subseteq \mathrm{Frac}(\hat{A}).$$

As $\hat{B}$ is manifestly integrally closed in $\mathrm{Frac}(\hat{B})$, we have $\hat{A}' = \hat{B}$. In particular, $n = m$ and $\dim_k \hat{B}/\hat{A} = m - 1$. Now $\mathfrak{m}_x^r B \subseteq \mathfrak{m}_x$ (Lemma 5.12(b)) implies that $B/A \to \hat{B}/\hat{A}$ is an isomorphism. Hence $\delta_x = m - 1$, which shows that x is an ordinary m-fold point.

Finally, if x is ordinary, then the tangent space $T_{X,x}$ can be computed with the isomorphism of (iii) and we see that its dimension is equal to m. $\square$

Remark 5.16. The proposition above gives an 'infinitesimal' description of ordinary multiple points. Indeed, let Z be a reduced closed subvariety of $\mathbb{A}_k^m$, union of the m axes, and o the origin of $\mathbb{A}_k^m$. Then $\hat{\mathcal{O}}_{X,x} \simeq \hat{\mathcal{O}}_{Z,o}$ if x is an ordinary m-fold point.

Definition 5.17. Let k be an algebraically closed field. Let $\mathcal{T}_0$ be the set consisting of the trivial group, and let $\mathcal{T}_q$ be the set of groups that are extensions of k^* by an element of $\mathcal{T}_{q-1}$, and $\mathcal{T} = \cup_{q\geq 1}\mathcal{T}_q$. We will call the elements of $\mathcal{T}$ *toric groups*.

As for the definition of unipotent groups, we use for convenience a weakened notion of the true definition of tori (which are algebraic groups over k isomorphic to a power $\mathbb{G}_{m,k}^r$). Let G be a toric group. Then $G \in \mathcal{T}_r$ for some integer $r \geq 0$. We immediately verify that for any non-zero n prime to $\mathrm{char}(k)$, we have $G[n] \simeq \mu_n^r$, where μ_n is the group of nth roots of unity in k. The integer r is called the *dimension* of G.

Lemma 5.18. *Let X be a reduced, connected, projective curve over an algebraically closed field k. Let Y be the (reduced and projective) curve with ordinary multiple singularities associated to X. Then the following properties are true.*

(a) *The canonical homomorphism $\mathrm{Pic}(X) \to \mathrm{Pic}(Y)$ is surjective, with unipotent kernel of dimension $p_a(X) - p_a(Y)$. Moreover, if $p_a(X) = p_a(Y)$, then $Y \to X$ is an isomorphism.*

(b) *The canonical homomorphism $\mathrm{Pic}(Y) \to \mathrm{Pic}(X')$ is surjective, with toric kernel of dimension $t = \mu - c + 1$, where $\mu := \sum_{x\in X(k)}(m_x - 1)$, and where c is the number of irreducible components of X.*

Proof Let $\mathcal{L}$ be an invertible sheaf on X'. Let D' be a Cartier divisor on X' such that $\mathcal{L} \simeq \mathcal{O}_{X'}(D')$. As in the proof of Proposition 1.32, we may suppose that D' is effective and with support disjoint from $\pi^{-1}(X_{\mathrm{sing}})$. Since $\pi_1 : X' \to Y$ is an isomorphism outside of $\pi^{-1}(X_{\mathrm{sing}})$, D' canonically induces a Cartier divisor D on Y with support in $Y \setminus \pi_2^{-1}(X_{\mathrm{sing}})$. We then have $\mathcal{L} \simeq \pi_1^*\mathcal{O}_Y(D)$, and $\mathrm{Pic}(Y) \to \mathrm{Pic}(X')$ is surjective. Likewise, $\mathrm{Pic}(X) \to \mathrm{Pic}(Y)$ is surjective.

Let us consider the exact sequence $0 \to \mathcal{O}_X^* \to \pi_{2*}\mathcal{O}_Y^* \to \mathcal{F} \to 0$, where $\mathcal{F}$ is a skyscraper sheaf. Čech cohomology gives the exact sequence

$$0 \to \mathcal{O}_X(X)^* \to \mathcal{O}_Y(Y)^* \to \mathcal{F}(X) \to \mathrm{Pic}(X) \to \mathrm{Pic}(Y).$$

Now $\pi_2 : Y \to X$ is bijective (Lemma 5.12(a)); hence Y is reduced and connected. It follows that $\mathcal{O}_X(X)^* = \mathcal{O}_Y(Y)^* = k^*$ (Corollary 3.3.21). Consequently, $\mathrm{Ker}(\mathrm{Pic}(X) \to \mathrm{Pic}(Y))$ can be identified with

$$\mathcal{F}(X) = \oplus_{x\in X(k)}\mathcal{O}_{Y,\pi_2^{-1}(x)}^*/\mathcal{O}_{X,x}^*.$$

The latter is a unipotent group of dimension $\sum_{x \in X(k)} \dim_k \mathcal{O}_{Y,\pi_2^{-1}(x)}/\mathcal{O}_{X,x}$ by virtue of Lemma 5.12(c). We have an exact sequence

$$0 \to \mathcal{O}_X \to \pi_{2*}\mathcal{O}_Y \to \mathcal{H} \to 0$$

where $\mathcal{H}$ is a skyscraper sheaf and $\mathcal{H}(X) = \oplus_{x \in X(k)} \mathcal{O}_{Y,\pi_2^{-1}(x)}/\mathcal{O}_{X,x}$. Now the exact cohomology sequence says that $\dim_k \mathcal{H}(X) = p_a(X) - p_a(Y)$. Hence $\mathcal{F}(X)$ is of dimension $p_a(X) - p_a(Y)$, which shows the first part of (a). If $p_a(X) = p_a(Y)$, then $\mathcal{H} = 0$; thus $\pi_2 : Y \to X$ is an isomorphism.

(b) We have an exact sequence $0 \to \mathcal{O}_Y^* \to \pi_{1*}\mathcal{O}_{X'}^* \to \mathcal{G} \to 0$, where $\mathcal{G}$ is a skyscraper sheaf. Taking Čech cohomology, we obtain the exact sequence

$$0 \to k^* \to (k^*)^c \to \mathcal{G}(Y) \to \mathrm{Pic}(Y) \to \mathrm{Pic}(X').$$

The lemma immediately results from Lemma 5.12(c).   □

Theorem 5.19. *Let X be a connected projective curve over an algebraically closed field, with irreducible components $X_1, \dots, X_n$. Let $\pi : X' \to X$ be the normalization morphism, and X_i' the normalization of X_i. Then the following properties are true.*

(a) *The morphism π induces a surjective canonical homomorphism*

$$R : \mathrm{Pic}^0(X) \to \prod_{1 \le i \le n} \mathrm{Pic}^0(X_i').$$

(b) *Let $L = \mathrm{Ker}(R)$. Then L is an extension of a toric group T by a unipotent group U. Let $a = \sum_{1 \le i \le n} g(X_i')$, let t be the integer defined in Lemma 5.18, and $u = \dim_k H^1(X, \mathcal{O}_X) - a - t$. Then T (resp. U) is of dimension t (resp. u).*

Proof (a) Let $\pi_1 : X' \to Y$, $\pi_2 : Y \to X_{\mathrm{red}}$, and $i : X_{\mathrm{red}} \to X$ be the decomposition of $X' \to X$. We have seen previously that the homomorphisms between the Pic associated to these morphisms are surjective. Hence $\mathrm{Pic}(X) \to \mathrm{Pic}(X')$ is surjective. Let $\mathcal{L}$ be an invertible sheaf on X. It follows from Propositions 5.7 and 3.8 that $[\mathcal{L}] \in \mathrm{Pic}^0(X)$ if and only if $[\pi^*\mathcal{L}] \in \mathrm{Pic}^0(X') = \prod_{1 \le i \le n} \mathrm{Pic}^0(X_i')$. Consequently, R is surjective and $L = \mathrm{Ker}(\mathrm{Pic}(X) \to \mathrm{Pic}(X'))$.

(b) Let $K = \mathrm{Ker}(\mathrm{Pic}(X_{\mathrm{red}}) \to \mathrm{Pic}(X'))$ and $U_1 = \mathrm{Ker}(\mathrm{Pic}(X) \to \mathrm{Pic}(X_{\mathrm{red}}))$. There is a canonical exact sequence

$$0 \to U_1 \to L \to K \to 0.$$

By Lemma 5.18, K is an extension of a torus T of dimension t by a unipotent group U_2 of dimension $p_a(X_{\mathrm{red}}) - p_a(Y)$. Let U be the inverse image of U_2 by $L \to K$. Then L is an extension of T by U, and U is an extension of U_2 by U_1. Hence U is unipotent (Lemma 5.11) of dimension the sum of the dimensions:

$$\begin{aligned} \dim U &= (\dim_k H^1(X, \mathcal{O}_X) - p_a(X_{\mathrm{red}})) + (p_a(X_{\mathrm{red}}) - p_a(Y)) \\ &= \dim H^1(X, \mathcal{O}_X) - p_a(Y). \end{aligned} \tag{5.23}$$

Now it can immediately be verified that $p_a(Y) = a + t$, because Y has only ordinary multiple singularities. This completes the proof of the theorem.   □

Remark 5.20. With a little bit of effort, we can show that L is isomorphic to $U \times T$, and that $T \simeq (k^*)^t$. If char(k) $= 0$, then we can also show that $U \simeq k^u$ as groups, using the exponential.

Definition 5.21. The integers $a, t, u \in \mathbb{N}$ of Theorem 5.19 are respectively called the *Abelian, toric, and unipotent ranks* of $\text{Pic}^0(X)$ (or of X). We also denote these ranks by $a(X)$, $t(X)$, and $u(X)$. We have

$$a(X) + t(X) + u(X) = \dim_k H^1(X, \mathcal{O}_X).$$

Remark 5.22. We can show that there exists an algebraic group G, direct product of a torus, and a unipotent algebraic group, such that $G(k) \simeq L$. See [86], V, Section 3.

Corollary 5.23. *Let X be a connected projective curve over an algebraically closed field k. Let n be a non-zero integer prime to char(k). Let a, t be the Abelian and toric ranks of X. Then we have a (non-canonical) isomorphism:*

$$\text{Pic}(X)[n] = \text{Pic}^0(X)[n] \simeq (\mathbb{Z}/n\mathbb{Z})^{t+2a}.$$

Proof We have obviously $\text{Pic}(X)[n] = \text{Pic}^0(X)[n]$. Let us keep the notation of Theorem 5.19. The multiplication by n is an isomorphism on U; we therefore have $L[n] \simeq T[n] \simeq \mu_n^t \simeq (\mathbb{Z}/n\mathbb{Z})^t$. As n_L is surjective, we have an exact sequence

$$0 \to L[n] \to \text{Pic}^0(X)[n] \to \text{Pic}^0(X')[n] \to 0.$$

The corollary then results from Corollary 4.41(a).   □

Corollary 5.24. *Let X be a projective, connected, reduced curve over an algebraically closed field k. Then the unipotent rank of X is zero if and only if the singularities of X are all ordinary.*

Proof This is an immediate consequence of Lemma 5.18(a).   □

Exercises

5.1. Let X be a reduced, connected, projective curve over a field k. Show that $k' := H^0(X, \mathcal{O}_X)$ is a finite field extension of k, and that X naturally has the structure of an algebraic variety over k'. Show that $p_a(X) < 0$ if and only if $k' \neq k$ and $H^1(X, \mathcal{O}_X) = 0$.

5.2. Let X be a projective curve over a field k such that $H^0(X, \mathcal{O}_X) = k$. Show that for any closed curve Z in X (e.g., $Z = X_{\text{red}}$), we have $p_a(Z) \leq p_a(X)$.

5.3. Let X be a proper integral curve over a field k, and let $D \in \text{Div}(X)$ be such that $\deg D > 0$. We are going to give another proof of the fact that

D is ample, without supposing that X is projective (Proposition 5.5). Let $\pi : X' \to X$ be the normalization morphism.

(a) Let $\mathcal{F}$ be a coherent torsion-free sheaf on X. By taking $\mathcal{H}om_{\mathcal{O}_X}(., \mathcal{F})$ of exact sequence (5.17), show that we have an exact sequence

$$0 \to \pi_*\pi^!\mathcal{F} = \mathcal{H}om_{\mathcal{O}_X}(\pi_*\mathcal{O}_{X'}, \mathcal{F}) \to \mathcal{F} \to \mathcal{H} \to 0,$$

where $\mathcal{H}$ is a skyscraper sheaf (see Section 6.4).

(b) Let $\mathcal{G} = \pi^!\mathcal{F}$. Show that it is a coherent sheaf on X' and that there exists an n_0 such that for every $n \geq n_0$, we have

$$H^1(X', \mathcal{G} \otimes_{\mathcal{O}_{X'}} \pi^*\mathcal{O}_X(nD)) = 0.$$

Show that $\pi_*(\mathcal{G} \otimes_{\mathcal{O}_{X'}} \pi^*\mathcal{O}_X(nD)) \simeq (\pi_*\mathcal{G}) \otimes_{\mathcal{O}_X} \mathcal{O}_X(nD)$ (Proposition 5.2.32).

(c) Using the fact that X' is projective (Exercise 4.1.16), show that for every $n \geq n_0$ we have $H^1(X, \mathcal{F} \otimes_{\mathcal{O}_X} \mathcal{O}_X(nD)) = 0$. Conclude.

5.4. Let X be a proper curve over a field k.

(a) Show that there exists an effective Cartier divisor D on X, whose support meets every irreducible component of X.

(b) Show that D is ample (use Corollary 5.3.8 and Exercise 5.3). In particular, X is projective over k.

5.5. Let X be a separated curve over a field k. We suppose that no irreducible component of X is proper. Show that X is affine. (Take the proof of Corollary 5.3.8 and Exercise 5.3 with $\mathcal{L} = \mathcal{O}_X(D) = \mathcal{O}_X$, and use Serre's criterion 5.2.23.)

5.6. Let X be a reduced curve over an algebraically closed field k, and let $x \in X(k)$. Let n_x denote the number of irreducible components of X passing through x. Show that $\delta_x \geq n_x - 1$. Moreover, if equality holds, show that x is an ordinary multiple point.

5.7. Let k be an algebraically closed field, $F(y, z) \in k[y, z]$. Let us write

$$F(y, z) = F_0(y, z) + F_1(y, z) + F_2(y, z) + \dots$$

with F_i homogeneous of degree i. Let us suppose that $F_0 = 0$. Let us consider $X = \mathrm{Spec}\, k[y, z]/(F(y, z))$, and the point $x_0 = (0, 0) \in X$.

(a) Show that X is smooth at x_0 if and only if $F_1 \neq 0$. Let us suppose in what follows that $F_1 = 0$ and $F_2 \neq 0$. Show that by a linear automorphism, we can reduce to $F_2(y, z) = yz, y^2$.

(b) Let us suppose $F_2(y, z) = yz$. By induction, construct two sequences $y_n, z_n, n \geq 1$, such that $y_1 = y, z_1 = z; y_{n+1} - y_n, z_{n+1} - z_n \in (y, z)^{n+1}k[x, y]$; and that

$$F(y, z) \in y_{n+1}z_{n+1} + (y_n, z_n)^{n+3}k[y, z].$$

Deduce from this that $\hat{\mathcal{O}}_{X,x_0} \simeq k[[u, v]]/(uv)$ (so x_0 is an ordinary double point).

(c) Let us suppose that $F_2(y, z) = y^2$. Write

$$F(y, z) = y^2(1 + \varepsilon) + z^{r_1} H_1(z)y + z^{r_2} H_2(z)$$

with $\varepsilon \in (y, z)k[y, z]$, $r_i \geq 2$, and $H_i(0) \neq 0$. Put $r = \min\{r_1, [r_2/2]\}$ and $w = y/z^r \in K(X)$. Show that w is integral over $\mathcal{O}_{X,x_0}$ and that $\delta_{x_0} = r$.

(d) Let F be as in (c) and suppose that $\mathrm{char}(k) \neq 2$. Show that $m_{x_0} = 1$ if $z^{2r_1 - 2r} H_1(z)^2 - 4z^{r_2 - 2r} H_2(z)$ vanishes at 0, and $m_{x_0} = 2$ otherwise.

(e) Let F be as in (c) and suppose that $\mathrm{char}(k) = 2$. Show that $m_{x_0} = 1$ if $r_1 > r$ and $m_{x_0} = 2$ otherwise.

Remark If $F_2 = 0$, one can show, using Exercise 9.2.12(b), that $\delta_{x_0} \geq 3$.

5.8. Let A be an Artinian local algebra over a field k. Let us suppose that the residue field of A is equal to k. Show that $A^* \simeq k^* \times U$, where U is a unipotent group.

5.9. (*Universal property of ordinary multiple points*) Let X be a reduced curve over an algebraically closed field k with a unique singular point $x \in X$. Let $\pi : X' \to X$ be the normalization morphism and $y_1, \ldots, y_m$ the points of $\pi^{-1}(x)$. Let $f : X' \to Z$ be a morphism from X' to an algebraic variety Z over k such that $f(y_1) = \cdots = f(y_m)$. Let us suppose that x is an ordinary multiple point. Show that f decomposes in a unique way as $\pi : X' \to X$ followed by a morphism $X \to Z$.

5.10. Let X be a reduced, connected, projective curve over an algebraically closed field k. Let $X_1, \ldots, X_n$ be closed subschemes of X that are connected curves, pairwise without common irreducible component, and such that $\cup_{1 \leq i \leq n} X_i = X$.

(a) Show that the Abelian ranks of the $\mathrm{Pic}^0(X_i)$ verify

$$a(X) = \sum_{1 \leq i \leq n} a(X_i).$$

(b) Show that if $n = 2$, then $t(X) = t(X_1) + t(X_2) + \mathrm{Card}(X_1 \cap X_2) - 1$. Show that in the general case, we have

$$t(X) \geq \sum_{1 \leq i \leq n} t(X_i).$$

See Exercise 10.3.19 for a more precise relation.

8

Birational geometry of surfaces

In this chapter we introduce some basic tools for studying surfaces. In this book, a *surface* will be a Noetherian integral (in general normal) scheme X of dimension 2, endowed with a projective flat morphism onto a base scheme S that is regular, connected, of dimension 0 or 1. The case when $\dim S = 0$ (hence S is the spectrum of a field) has been abundantly treated in the literature. We will therefore concentrate more particularly on the case when $\dim S = 1$. We can then consider X as a family of curves parameterized by S (more geometric point of view), or as an extension of the generic fiber $X_{K(S)}$ into a scheme over S (more arithmetic point of view). The chapter is organized in the following way. In the first section we study a particularly important class of birational morphisms, the blowing-ups. In fact, birational projective morphisms are predominantly blowing-ups (see Theorem 1.24 for precise conditions). Section 8.2 presents some more commutative algebra, of a higher level than Chapter 1. Certain results are not proven because they extend beyond the scope of this book. At a first reading this section can be skipped. Finally, the last section deals with fibered surfaces, in particular their birational aspect. This section concludes with the resolutions of singularities. By admitting the result in the case of excellent surfaces (Definition 2.35), we obtain a necessary and sufficient condition for the existence of the desingularization for arbitrary fibered surfaces (Theorem 3.50). As a consequence, every fibered surface with smooth generic fiber admits a desingularization (Corollary 3.51).

All schemes considered are locally Noetherian.

8.1 Blowing-ups

The notion of blowing-up is fundamental in the study of birational morphisms of schemes. We will need it to describe birational morphisms between surfaces. We first define blowing-ups as morphisms associated to graded algebraic sheaves

(Definition 1.6). Next we show that these morphisms verify a universal property (Corollary 1.16). When the blowing-up is done along a regular closed subscheme on a regular scheme, Theorem 1.19 gives precise information on the fibers of the blowing-up morphism. Finally, we show that under reasonable hypotheses, any birational projective morphism is a blowing-up (Theorem 1.24).

8.1.1   Definition and elementary properties

Let us start with the local description of blowing-ups. Let A be a Noetherian ring, I an ideal of A. Let us consider the graded A-algebra

$$\tilde{A} = \bigoplus_{d \geq 0} I^d, \quad \text{where} \;\; I^0 := A$$

(it of course depends on I). Let $f_1, \ldots, f_n$ be a system of generators of I. Let $t_i \in I = \tilde{A}_1$ denote the element f_i considered as a homogeneous element of degree 1, not to be confused with the element $f_i \in A = \tilde{A}_0$ of degree 0. We have a surjective homomorphism of graded A-algebras

$$\phi : A[T_1, \ldots, T_n] \rightarrow \tilde{A} \tag{1.1}$$

defined by $\phi(T_i) = t_i$. Hence $\tilde{A}$ is a homogeneous A-algebra. If $P(T)$ is a homogeneous polynomial with coefficients in A, then $P(t_1, \ldots, t_n) = 0$ if and only if $P(f_1, \ldots, f_n) = 0$. In what follows, it is the projective scheme $\text{Proj}\, \tilde{A}$ over $\text{Spec}\, A$ that will interest us.

Definition 1.1. Let $X = \text{Spec}\, A$ be an affine Noetherian scheme, let I be an ideal of A. We let $\tilde{X} = \text{Proj}\, \tilde{A}$, and the canonical morphism $\tilde{X} \rightarrow X$ is called the *blowing-up of X with center (or along) $V(I)$ (or I)*.

Lemma 1.2. *Let A be a Noetherian ring, and let I be an ideal of A.*

(a) *If I is generated by a regular element, then $\tilde{A} \simeq A[T]$. In other words, $\text{Proj}\, \tilde{A} \rightarrow \text{Spec}\, A$ is an isomorphism.*

(b) *We have $\text{Proj}\, \tilde{A} = \emptyset$ if and only if I is nilpotent.*

(c) *The ring $\tilde{A}$ is integral (resp. reduced) if and only if A is integral (resp. reduced).*

(d) *Let B be a flat A-algebra, and let $\tilde{B}$ be the graded B-algebra associated to the ideal IB. Then we have a canonical isomorphism $\tilde{B} \simeq B \otimes_A \tilde{A}$.*

(e) *Let $S_i = T_i/T_1 \in \mathcal{O}(D_+(T_1))$. Then $(\text{Ker}\, \phi)_{(T_1)}$ is equal to*

$$J_1 := \left\{ P(S) \in A[S_2, \ldots, S_n] \mid \exists\, d \geq 0, \; f_1^d P \in (f_1 S_2 - f_2, \ldots, f_1 S_n - f_n) \right\},$$

and we can identify $\tilde{A}_{(t_1)}$ with the sub-A-algebra of A_{f_1} generated by the elements f_i/f_1, $2 \leq i \leq n$.

(f) *Let $J = (f_i T_j - f_j T_i)_{1 \leq i,j \leq n}$. Then $J \subseteq \text{Ker}\, \phi$. Let us suppose that the f_i form a minimal system of generators and that the closed subscheme*

$Z := V_+(J)$ of $\mathbb{P}_A^{n-1}$ is integral; then the closed immersion $\alpha : \operatorname{Proj} \tilde{A} \to Z$ is an isomorphism.

Proof (a) Let f be a generator of I, and let $\phi : A[T] \to \tilde{A}$ be defined as above. It is clear that ϕ is an isomorphism.

(b) We have $\operatorname{Proj} \tilde{A} = \emptyset$ if and only if the t_i are nilpotent (Lemma 2.3.35(c)), which is also equivalent to saying that f_i is nilpotent.

(c) Let us suppose A is integral. Let $a, b \in \tilde{A}$ be such that $ab = 0$ and that $a \neq 0$. Let $a_n, b_m \in A$ be their respective homogeneous components of highest degree; then $a_n b_m = 0$. It follows that $b_m = 0$, and hence $b = 0$. The converse is trivial because A is a subring of $\tilde{A}$. The assertion for reduced rings is shown in the same way.

(d) We canonically have $B \otimes_A I \simeq IB$ (Theorem 1.2.4) and $(IB)^d \simeq B \otimes_A I^d$. Hence

$$\tilde{B} \simeq \oplus_{d \geq 0}(B \otimes_A I^d) \simeq B \otimes_A (\oplus_{d \geq 0} I^d) = B \otimes_A \tilde{A},$$

whence the isomorphism that we were looking for.

(e) Let $P(S) \in A[S_2, \ldots, S_n]$. Multiplying $P(S)$ by a suitable power of f_1 if necessary, we can carry out successive Euclidean divisions and obtain

$$f_1^d P(S) = \sum_{2 \leq i \leq n} Q_i(S)(f_1 S_i - f_i) + a, \quad a \in A. \tag{1.2}$$

If $P(S) \in (\operatorname{Ker} \phi)_{(T_1)}$, then the image of a in $\tilde{A}_{(t_1)}$ is zero. In other words, there exists an $r \geq 0$ such that $at_1^r = 0$. This is equivalent to $af_1^r = 0$. By replacing d by $d+r$, we can suppose that $a = 0$ and we therefore have $P(S) \in J_1$. Conversely, let $P(S) \in J_1$; then $f_1^d P(S) \in (f_1 S_i - f_i)_{1 \leq i \leq n}$ for some $d \geq 0$. Let us write $P(S) = Q(T)/T_1^r$ with $Q(T)$ homogeneous of degree r. There exists an $e \geq 0$ such that $f_1^d T_1^e Q(T)$ belongs to the ideal $(f_i T_j - f_j T_i)_{1 \leq i,j \leq n}$. Hence $f_1^d t_1^e Q(t) = 0$, which implies that $t_1^{d+e} Q(t) = 0$. It follows that $P(S) = (Q(T)T_1^{d+e})/T_1^{r+d+e} \in (\operatorname{Ker} \phi)_{(T_1)}$.

Let us consider the A-algebra homomorphism $\psi : A[S_2, \ldots, S_n] \to A_{f_1}$ which sends S_i to f_i/f_1. Then relation (1.2) immediately implies that $\operatorname{Ker}(\psi) = J_1$. We can therefore identify $\tilde{A}_{(t_1)}$ with its image in A_{f_1}.

(f) It suffices to show that α is an isomorphism on every non-empty principal open set $U_i := D_+(T_i) \cap Z$. By hypothesis, f_i does not belong to the ideal generated by $\{f_j\}_{j \neq i}$. It immediately follows that f_i is non-zero in $\mathcal{O}_Z(U_i)$, and hence not a zero divisor. We then have the result by (e). $\qquad\square$

Example 1.3. Let $A = k[x, y]$ with $y^2 - x^3 = 0$, and $I = xA + yA$. It immediately follows from Lemma 1.2(e) that $\tilde{A}_{(t_1)} = k[u]$, where $u = y/x \in \operatorname{Frac}(A)$. In the same way, we have $\tilde{A}_{(t_2)} = k[v, 1/v]$ with $v = x/y$. Let us note that in this example, $(\operatorname{Ker} \phi)_{(T_1)}$ is not generated by $xS_2 - y$. Indeed, $S_2^2 - x \in (\operatorname{Ker} \phi)_{(T_1)}$ since $x^2(S_2^2 - x) = (xS_2 - y)(xS_2 + y)$, and $S_2^2 - x$ is obviously not a multiple of $xS_2 - y$. Note that the blowing-up morphism $\pi : \operatorname{Proj} \tilde{A} \to \operatorname{Spec} A$ with center $V(I)$ is none other that the normalization of $\operatorname{Spec} A$.

Lemma 1.4. *Let $\widetilde{X} \to X = \operatorname{Spec} A$ be the blowing-up of a Noetherian, integral, affine scheme along a closed subscheme $V(I)$. Let us write $I = (f_1, \ldots, f_n)$, with $f_i \neq 0$ for every i. Then $\widetilde{X}$ is the union of the affine open subschemes $\operatorname{Spec} A_i$, $1 \leq i \leq n$, where A_i is the sub-A-algebra of $\operatorname{Frac}(A)$ generated by the $f_j f_i^{-1} \in \operatorname{Frac}(A)$, $1 \leq j \leq n$.*

Proof This follows from the surjective homomorphism (1.1) and Lemma 1.2(e).

$\square$

Example 1.5. Let k be a field. Let us consider the surface

$$X = \operatorname{Spec} k[S, T, W]/(ST - W^2).$$

Let s, t, w denote the respective images of S, T, W in the quotient. Then the Jacobian criterion shows that X is smooth over k outside of the point x_0 corresponding to the maximal ideal $\mathfrak{m} = (s, t, w)$. Let $\pi : \widetilde{X} \to X$ be the blowing-up with center $\mathfrak{m}$. We are going to show that $\widetilde{X}$ is smooth over k.

Let $A = k[s, t, w] = \mathcal{O}_X(X)$, and let A_1 (resp. A_2; resp. A_3) be the sub-A-algebra of $K(X)$ generated by ts^{-1}, ws^{-1} (resp. st^{-1}, wt^{-1}; resp. tw^{-1}, sw^{-1}). Then $\widetilde{X}$ is covered by the three affine open subschemes $\operatorname{Spec} A_i$, $i = 1, 2, 3$ (Lemma 1.4). In A_1, we have the relations $ts^{-1} = (ws^{-1})^2$, $t = (ts^{-1})s$, and $w = (ws^{-1})s$. It follows that A_1 is the polynomial ring $k[s, ws^{-1}]$. In the same way, we verify that $A_2 = k[t, wt^{-1}]$ and that

$$A_3 = k[w, tw^{-1}, sw^{-1}], \quad (sw^{-1}) \cdot (tw^{-1}) = 1.$$

Hence $\widetilde{X}$ is a union of open subschemes that are isomorphic to open subschemes of $\mathbb{A}_k^2$. In particular, it is smooth over k.

Let us study the fiber $E := \pi^{-1}(x_0)$. We have

$$E \cap \operatorname{Spec} A_1 = V(s) = \operatorname{Spec} k[ws^{-1}],$$

$$E \cap \operatorname{Spec} A_2 = V(t) = \operatorname{Spec} k[wt^{-1}],$$

and

$$E \cap \operatorname{Spec} A_3 = V(w) = \operatorname{Spec} k[tw^{-1}, sw^{-1}]/((sw^{-1}) \cdot (tw^{-1}) - 1).$$

Thus, we see that $E = \mathbb{P}_k^1$ and that $K(E) = k(ws^{-1})$.

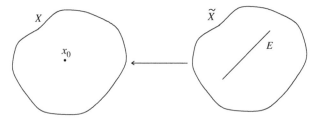

Figure 8. Blowing-up of a surface along a closed point.

We are going to extend the notion of blowing-up to locally Noetherian schemes.

Definition 1.6. Let X be a scheme. A *graded $\mathcal{O}_X$-algebra* $\mathcal{B}$ is a quasi-coherent sheaf of $\mathcal{O}_X$-algebras with a grading $\mathcal{B} = \oplus_{n\geq 0}\mathcal{B}_n$, where the $\mathcal{B}_n$ are quasi-coherent sub-$\mathcal{O}_X$-modules. We say that $\mathcal{B}$ is a *homogeneous $\mathcal{O}_X$-algebra* if, moreover, $\mathcal{B}_1$ is finitely generated, and if $(\mathcal{B}_1)^n = \mathcal{B}_n$ for every $n \geq 1$. For any affine open subset U of X, $\mathcal{B}(U)$ is then a homogeneous $\mathcal{O}_X(U)$-algebra.

Example 1.7. Let X be a scheme. Let $\mathcal{I}$ be a finitely generated quasi-coherent sheaf of ideals over X. Then $\mathcal{B} := \oplus_{n\geq 0}\mathcal{I}^n$ is naturally a homogeneous $\mathcal{O}_X$-algebra.

Lemma 1.8. *Let X be a scheme and $\mathcal{B}$ a graded $\mathcal{O}_X$-algebra. Then there exists a unique X-scheme $f : \operatorname{Proj}\mathcal{B} \to X$ such that for any affine open subscheme U of X, we have an isomorphism of U-schemes $h_U : f^{-1}(U) \simeq \operatorname{Proj}\mathcal{B}(U)$ that is compatible with the restriction to any affine open subscheme $V \subseteq U$.*

Proof The uniqueness results directly from the definition. Let us show existence. Let us first suppose X affine, and let us set $\operatorname{Proj}\mathcal{B} = \operatorname{Proj}\mathcal{B}(X)$. For any affine open subscheme V of X, we have

$$\operatorname{Proj}\mathcal{B}(V) = \operatorname{Proj}(\mathcal{B}(X) \otimes_{\mathcal{O}_X(X)} \mathcal{O}_X(V)),$$

and the right-hand side is canonically isomorphic to $(\operatorname{Proj}\mathcal{B}(X)) \times_X V$ (Proposition 3.1.9). In the general case, let us note that if $\operatorname{Proj}\mathcal{B}$ exists, then $\operatorname{Proj}\mathcal{B}|_W$ exists for any open subscheme W of X, and we have $\operatorname{Proj}\mathcal{B}|_W = \operatorname{Proj}\mathcal{B} \times_X W$. Now we cover X with affine open subschemes X_i. Then the $\operatorname{Proj}\mathcal{B}|_{X_i}$ glue by the existence and uniqueness of the $\operatorname{Proj}\mathcal{B}|_{X_i \cap X_j}$ (Lemma 2.3.33). The resulting scheme is $\operatorname{Proj}\mathcal{B}$. □

Remark 1.9. Let U be an affine open subset of X. Then $\operatorname{Proj}\mathcal{B}(U)$ is endowed with a sheaf $\mathcal{O}(1) = (\mathcal{B}(U)(1))^\sim$ (Example 5.1.19). It is easy to verify that these sheaves glue when the U run through the affine open subsets of X. We will denote the resulting sheaf on $Y := \operatorname{Proj}\mathcal{B}$ by $\mathcal{O}_Y(1)$. This sheaf depends, of course, on $\mathcal{B}$, and not only on the scheme $\operatorname{Proj}\mathcal{B}$. If $\mathcal{B}$ is a homogeneous $\mathcal{O}_X$-algebra, then $\mathcal{O}_Y(1)$ is an invertible sheaf because for any affine open subset U of X, $\mathcal{O}_Y(1)|_{f^{-1}(U)}$ is an invertible sheaf.

Example 1.10. Let $f : P = \mathbb{P}^N_X \to X$ be a projective space over a scheme X. Then $P \simeq \operatorname{Proj}(\oplus_{n\geq 0}f_*\mathcal{O}_P(n))$ by Lemma 5.1.22. And the sheaves $\mathcal{O}_P(1)$ defined in Example 5.1.19 and in the remark above coincide.

Definition 1.11. Let $\mathcal{I}$ be a coherent sheaf of ideals on a locally Noetherian scheme X. The X-scheme $\operatorname{Proj}(\oplus_{n\geq 0}\mathcal{I}^n) \to X$ is called the *blowing-up of X with center* (or *along*) $V(\mathcal{I})$ (or $\mathcal{I}$). This scheme depends on the closed subscheme structure of $V(\mathcal{I})$ and not only on the closed subset $V(\mathcal{I})$. Let us, however, note that we can replace $\mathcal{I}$ by an arbitrary power $\mathcal{I}^n$ without changing the blowing-up (Exercise 2.3.11(a)). See also Exercise 1.6. For simplicity, the scheme $\operatorname{Proj}(\oplus_{n\geq 0}\mathcal{I}^n)$ will also be denoted $\widetilde{X}$. If X is affine, this definition coincides with Definition 1.1.

Let $\pi : Y \to X$ be a morphism of schemes, and let $\mathcal{I}$ be a quasi-coherent sheaf of ideals on X. We have a canonical homomorphism $\pi^*\mathcal{I} \to \pi^*\mathcal{O}_X = \mathcal{O}_Y$. Then the image of $\pi^*\mathcal{I}$ in $\mathcal{O}_Y$ is a quasi-coherent sheaf of ideals on Y. We denote it $(\pi^{-1}\mathcal{I})\mathcal{O}_Y$, or, somewhat abusively, $\mathcal{I}\mathcal{O}_Y$.

Proposition 1.12. *Let X be a locally Noetherian scheme, and let $\mathcal{I}$ be a quasi-coherent sheaf of ideals on X. Let $\pi : \widetilde{X} \to X$ be the blowing-up of X with center $V(\mathcal{I})$. Then the following properties are true.*

(a) *The morphism π is an isomorphism if and only if $\mathcal{I}$ is an invertible sheaf on X.*

(b) *The morphism π is proper.*

(c) *Let $Z \to X$ be a flat morphism with Z locally Noetherian. Let $\widetilde{Z} \to Z$ be the blowing-up of Z with center $\mathcal{I}\mathcal{O}_Z$; then $\widetilde{Z} \simeq \widetilde{X} \times_X Z$.*

(d) *The morphism π induces an isomorphism $\pi^{-1}(X \setminus V(\mathcal{I})) \to X \setminus V(\mathcal{I})$. If X is integral, and if $\mathcal{I} \neq 0$, then $\widetilde{X}$ is integral, and π is a birational morphism.*

(e) *With the notation of Remark 1.9, we have*

$$\mathcal{I}\mathcal{O}_{\widetilde{X}} = \mathcal{O}_{\widetilde{X}}(1).$$

In particular, $\mathcal{I}\mathcal{O}_{\widetilde{X}}$ is an invertible sheaf.

(f) *If X is affine, then $\mathcal{O}_{\widetilde{X}}(1)$ is very ample relative to π (see also Proposition 1.22).*

Proof (a) If $\mathcal{I}$ is invertible, then π is an isomorphism by Lemma 1.2(a). The converse will result from assertion (e), whose proof is independent of (a).

(b) If X is affine, π is projective by construction, in particular π is proper (Theorem 3.3.30). Now properness is a local property on X (Definitions 3.1.21 and 3.3.14), and π is therefore proper in the general case.

(c) results from Lemma 1.2(d).

(d) Let $U = X \setminus V(\mathcal{I})$. Then $\mathcal{I}|_U = \mathcal{O}_U$. Hence $\pi^{-1}(U) \to U$ is an isomorphism by (a) and by applying (c) to the flat morphism $U \to X$.

(e) We can suppose that $X = \operatorname{Spec} A$ and that $\mathcal{I}$ is associated to an ideal $I = (f_1, \ldots, f_n)$ of A. With the notation defined before Lemma 1.2, $\mathcal{I}\mathcal{O}_{\widetilde{X}}|_{D_+(t_i)}$ is generated by f_i for every $1 \leq i \leq n$ because $f_j = f_i \cdot (t_j t_i^{-1})$. This shows that $\mathcal{I}\mathcal{O}_{D_+(t_i)} = \mathcal{O}_{D_+(t_i)}(1)$. Hence $\mathcal{I}\mathcal{O}_{\widetilde{X}} = \mathcal{O}_{\widetilde{X}}(1)$ when X is affine. By construction of $\mathcal{O}_{\widetilde{X}}(1)$, this gives the equality in the general case.

(f) With the notation above, homomorphism (1.1) induces a closed immersion $i : \widetilde{X} \to \mathbb{P}_A^{n-1}$. The proof of (e) shows that $\mathcal{O}_{\widetilde{X}}(1) = i^*\mathcal{O}_{\mathbb{P}_A^{n-1}}(1)$. $\square$

Example 1.13. Let $X = \operatorname{Spec} k[x_1, \ldots, x_m]$ be the affine space of dimension $m \geq 1$ over a field k. Let $\mathcal{I}$ be the sheaf of ideals which defines the origin $o := (0, \ldots, 0)$ (with the reduced scheme structure). Let $\pi : \widetilde{X} \to X$ be the blowing-up with center o. By Proposition 1.12, π is an isomorphism outside of o. Let us study the fiber above o. Let us set $A = \mathcal{O}_X(X)$. It follows from Lemma 1.2(f) that

$$\widetilde{X} = \operatorname{Proj} A[T_1, \dots, T_m]/(x_i T_j - x_j T_i)_{1 \le i, j \le m},$$

because the right-hand side is clearly an integral scheme, as we can verify by taking the affine open covering $\{D_+(T_i)\}_i$. Hence the fiber $\widetilde{X}_o = \operatorname{Proj} k[T_1, \dots, T_m]$ is a projective space of dimension $m - 1$ over k. From a set-theoretical and naive point of view, we 'stretched' the origin until it became a projective space, and we have not touched the rest of the variety. The set $\widetilde{X}(k)$ can be identified with

$$\{(x_1, \dots, x_m, t_1, \dots, t_m) \in k^m \times \mathbb{P}_k^{m-1}(k) \mid x_i t_j = x_j t_i, 1 \le i, j \le m\}.$$

Corollary 1.14. *Let $X, \mathcal{I}$ be as in Proposition 1.12. Let $x \in X$. Then $\widetilde{X} \times_X \operatorname{Spec} A \to \operatorname{Spec} A$, where $A = \mathcal{O}_{X,x}$ or $\widehat{\mathcal{O}}_{X,x}$, is the blowing-up of $\operatorname{Spec} A$ along $V(\mathcal{I}_x)$ or $V(\mathcal{I}_x \widehat{\mathcal{O}}_{X,x})$.*

Proof Indeed, $\operatorname{Spec} A \to X$ is flat. □

8.1.2 Universal property of blowing-up

Proposition 1.15. *Let $f : W \to X$ be a morphism of locally Noetherian schemes. Let $\mathcal{I}$ be a quasi-coherent sheaf of ideals on X, and $\mathcal{J} = (f^{-1}\mathcal{I})\mathcal{O}_W$. Let $\pi : \widetilde{X} \to X$ and $\rho : \widetilde{W} \to W$ denote the blowing-ups of X and of W with respective centers $\mathcal{I}$ and $\mathcal{J}$. Then there exists a unique morphism $\widetilde{f} : \widetilde{W} \to \widetilde{X}$ which makes the following diagram commutative:*

$$
\begin{array}{ccc}
\widetilde{W} & \xrightarrow{\widetilde{f}} & \widetilde{X} \\
\downarrow{\scriptstyle \rho} & & \downarrow{\scriptstyle \pi} \\
W & \xrightarrow{f} & X
\end{array}
$$

Proof The existence of $\widetilde{f}$ is easy. Indeed, let us first suppose that $X = \operatorname{Spec} A$ and $W = \operatorname{Spec} B$ are affine. The canonical homomorphism

$$\oplus_{n \ge 0} I^n \to \oplus_{n \ge 0}(IB)^n, \quad I = \mathcal{I}(X),$$

is a homomorphism of graded algebras, and induces a morphism $\widetilde{W} \to \widetilde{X}$ because $IB = \mathcal{J}(W)$. If W is not affine, we cover it with affine open subschemes W_i. For any affine open subscheme U contained in an intersection $W_i \cap W_j$, we see by construction that the restrictions of $\widetilde{f}_i : \widetilde{W}_i \to \widetilde{X}$ and $\widetilde{f}_j : \widetilde{W}_j \to \widetilde{X}$ coincide on $\widetilde{U}$. Hence the morphisms $\widetilde{f}_i$ glue to a morphism $\widetilde{f} : \widetilde{W} \to \widetilde{X}$. If X is not affine, the same glueing process makes it possible to construct $\widetilde{f}$.

Let us now show the uniqueness of $\tilde{f}$. It suffices to show that there exists a unique morphism $g : \widetilde{W} \to \tilde{X}$ making the following diagram commutative:

We can then suppose that $W = \widetilde{W}$, and therefore that $\mathcal{J}$ is an invertible sheaf. Moreover, as the property is local on X, we can suppose $X = \operatorname{Spec} A$ affine. Let $f_1, \ldots, f_n$ be a system of generators of $I := \mathcal{I}(X)$. Let s_i denote the canonical image of f_i in $\mathcal{J}(W)$. Then $\mathcal{J}$ is an invertible sheaf generated by the sections $s_1, \ldots, s_n$, because $\mathcal{I}$ is generated by the f_i. Let $i : \tilde{X} \to \mathbb{P}_X^{n-1}$ denote the closed immersion induced by homomorphism (1.1). Let $g : W \to \tilde{X}$ be a morphism such that $\pi \circ g = f$. Then we have

$$\mathcal{J} = g^{-1}((\pi^{-1}\mathcal{I})\mathcal{O}_{\tilde{X}})\mathcal{O}_W = g^{-1}(\mathcal{O}_{\tilde{X}}(1))\mathcal{O}_W.$$

We therefore have a surjective homomorphism of invertible sheaves $g^*\mathcal{O}_{\tilde{X}}(1) \to \mathcal{J}$, which is an isomorphism by Exercise 5.1.13. Let $h = i \circ g$; then $h^*\mathcal{O}_{\mathbb{P}_X^{n-1}}(1) \to \mathcal{J}$ is an isomorphism, and it transforms the global sections $T_1, \ldots, T_n$ into the sections $s_1, \ldots, s_n$. There exists a unique morphism h satisfying this condition (Proposition 5.1.31). Now h determines g uniquely because i is a closed immersion, whence the uniqueness of g. □

Corollary 1.16. *Let $\pi : \tilde{X} \to X$ be the blowing-up with center $\mathcal{I}$ of a locally Noetherian scheme. This morphism has the following universal property: for any morphism $f : W \to X$ such that $(f^{-1}\mathcal{I})\mathcal{O}_W$ is an invertible sheaf of ideals on W, there exists a unique morphism $g : W \to \tilde{X}$ making the following diagram commutative:*

Corollary 1.17. *Let us keep the hypotheses of Proposition 1.15. Let us suppose that $W \to X$ is a closed immersion and that the image of W is not contained in the center $V(\mathcal{I})$. Then the morphism $\widetilde{W} \to \tilde{X}$ is also a closed immersion.*

Proof   This results from the construction of $\widetilde{W} \to \tilde{X}$. The hypothesis that W is not contained in $V(\mathcal{I})$ ensures that $\widetilde{W}$ is non-empty. □

Definition 1.18. Let X be a locally Noetherian scheme and $\pi : \tilde{X} \to X$ the blowing-up of X along a closed subscheme $V(\mathcal{I})$. Let W be a closed subscheme of X not contained in $V(\mathcal{I})$. We call the closed subscheme $\widetilde{W} \subseteq \tilde{X}$ the *strict transform* of W. See also Exercise 1.1.

Theorem 1.19. *Let X be a regular locally Noetherian scheme, and $\pi : \widetilde{X} \to X$ be the blowing-up of X along a regular closed subscheme $Y = V(\mathcal{I})$. Then the following properties are true.*

(a) *The scheme $\widetilde{X}$ is regular.*

(b) *For any $x \in Y$, the fiber $\widetilde{X}_x$ is isomorphic to $\mathbb{P}_{k(x)}^{r-1}$, where $r = \dim_x X - \dim_x Y$.*

(c) *Let $Y' = V(\mathcal{I}\mathcal{O}_{\widetilde{X}})$ be the inverse image of Y under π. Then $Y' \to \widetilde{X}$ is a regular immersion. Moreover, if X is affine and if $\mathcal{I}/\mathcal{I}^2$ is free of rank r on Y, then we have*

$$Y' \simeq \mathbb{P}_Y^{r-1}, \quad \omega_{Y'/\widetilde{X}} \simeq \mathcal{O}_{Y'}(-1).$$

Proof (a) We can suppose $X = \operatorname{Spec} A$ is local, with closed point $x \in Y$. As X and Y are regular at x, there exists a system of generators $f_1, \ldots, f_d$ of $\mathfrak{m}_x$ such that $d = \dim \mathcal{O}_{X,x}$ and that $\mathcal{I}_x$ is generated by $f_1, \ldots, f_r$ with $r \leq d$ (Corollary 4.2.15). Let us consider the X-scheme

$$Z = \operatorname{Proj} A[T_1, \ldots, T_r]/(f_i T_j - f_j T_i)_{1 \leq i,j \leq r}.$$

Clearly $Z \to X$ is an isomorphism outside of Y. For any $y \in Y$, the fiber Z_y is isomorphic to $\operatorname{Proj} k(y)[T_1, \ldots, T_r] = \mathbb{P}_{k(y)}^{r-1}$. Hence $Z \to X$ is proper, surjective, with connected fibers. It follows that Z is connected (Exercise 3.3.12). We are going to show that Z is regular (hence integral). It will follow from Lemma 1.2(f) that $\widetilde{X} = Z$ and is therefore regular.

Let $z \in Z$ be a closed point. Then $z \in Z_x$ because $Z \to X$ is proper. Let us, for example, suppose that $z \in D_+(t_1)$ (with the notation of the beginning of the subsection). Then $\mathcal{O}_{Z,z}$ is a localization of $A[S_2, \ldots, S_r]/(f_i - f_1 S_i)_{2 \leq i \leq r}$. As Z_x is regular of dimension $r - 1$ at x, there exist $g_2, \ldots, g_r \in \mathcal{O}_{Z,z}$ such that the maximal ideal $\mathfrak{m}_z$ of $\mathcal{O}_{Z,z}$ is generated by $g_2, \ldots, g_r, f_1, \ldots, f_d$. Now in $\mathcal{O}_{Z,z}$, $f_i = f_1 S_i$ for $2 \leq i \leq r$, and $\mathfrak{m}_z$ is therefore generated by d elements $f_1, g_2, \ldots, g_r, f_{r+1}, \ldots, f_d$. In addition, since z is closed in $\operatorname{Spec} A[S_2, \ldots, S_r]$, we have

$$\dim \mathcal{O}_{Z,z} \geq (d + (r-1)) - (r-1) = d$$

by virtue of Lemma 2.5.16 and Theorem 2.5.15, whence the regularity of Z at the closed points and therefore at every point (Corollary 4.2.17).

(b) We can suppose X is local with closed point x. Then (b) results from the fact that $\widetilde{X} = Z$, with the notation of (a).

(c) Let $\mathcal{J} = \mathcal{I}\mathcal{O}_{\widetilde{X}}$. Then $Y' \to \widetilde{X}$ is a closed immersion that is a local complete intersection of codimension 1 (Proposition 1.12(e)). For the second part of (c), we can suppose $X = \operatorname{Spec} A$ is connected. Let $f_1, \ldots, f_r \in I = \mathcal{I}(X)$ be a family that induces a basis of $\mathcal{I}/\mathcal{I}^2$ over $\mathcal{O}_Y$. Then the X-scheme

$$Z = \operatorname{Proj} A[T_1, \ldots, T_r]/(T_i f_j - T_j f_i)_{1 \leq i,j \leq r} \tag{1.3}$$

is regular and connected by (a), and hence isomorphic to $\widetilde{X}$. It follows that

$$Y' = \widetilde{X} \times_X Y = \mathrm{Proj}(A/I)[T_1, \ldots, T_r] = \mathbb{P}_Y^{r-1}$$

(Proposition 3.1.9). In addition, we have seen that writing Z as in (1.3) induces a closed immersion $i : Z \to \mathbb{P}_X^{r-1}$ such that $\mathcal{J} \simeq i^* \mathcal{O}_{\mathbb{P}_X^{r-1}}(1)$ (Proposition 1.12(e)). Hence $\mathcal{J}/\mathcal{J}^2 \simeq i^* \mathcal{O}_{\mathbb{P}_Y^{r-1}}(1)$. It follows from the definition that

$$\omega_{Y'/\widetilde{X}} = (\mathcal{J}/\mathcal{J}^2)^\vee \simeq \mathcal{O}_{Y'}(-1).$$

$\square$

Example 1.20. Let us once more take $\pi : \widetilde{X} \to X$ from Example 1.5, and let us determine $\omega_{E/\widetilde{X}}$, where $E = \pi^{-1}(x_0) \simeq \mathbb{P}_k^1$. We will keep the notation of that example. Let $\mathcal{J} = \mathfrak{m}\mathcal{O}_{\widetilde{X}}$. Let $U_i = \mathrm{Spec}\, A_i$, $i = 1, 2, 3$. Then

$$\mathcal{J}^\vee|_{U_1} = s^{-1}\mathcal{O}_{U_1}, \quad \mathcal{J}^\vee|_{U_2} = t^{-1}\mathcal{O}_{U_2}, \quad \mathcal{J}^\vee|_{U_3} = w^{-1}\mathcal{O}_{U_3} = s^{-1}\mathcal{O}_{U_3}.$$

Hence $\mathcal{J}^\vee$ is isomorphic to the invertible sheaf $\mathcal{L} := s\mathcal{J}^\vee$ given by

$$\mathcal{L}|_{U_1} = \mathcal{O}_{U_1}, \quad \mathcal{L}|_{U_2} = (wt^{-1})^2 \mathcal{O}_{U_2}, \quad \mathcal{L}|_{U_3} = \mathcal{O}_{U_3}.$$

As $\mathcal{O}_E(E \cap U_2) = k[wt^{-1}]$, these equalities imply that $\mathcal{L}|_E \simeq \mathcal{O}_E(-2)$. Consequently,

$$\omega_{E/\widetilde{X}} = (\mathcal{J}/\mathcal{J}^2)^\vee \simeq \mathcal{L}|_E \simeq \mathcal{O}_E(-2).$$

This does not contradict Theorem 1.19 because x_0 is a singular point of X.

8.1.3   Blowing-ups and birational morphisms

In this subsection we are going to show that projective birational morphisms are predominantly blowing-ups (Theorem 1.24).

Lemma 1.21. *Let X be a scheme. Let $\mathcal{B}$ be a graded $\mathcal{O}_X$-algebra and $\mathcal{N}$ an invertible sheaf on X. Let us consider the graded $\mathcal{O}_X$-algebra $\mathcal{C}$ defined by*

$$\mathcal{C}_n = \mathcal{N}^{\otimes n} \otimes \mathcal{B}_n, \quad n \geq 0.$$

Then we have an isomorphism of X-schemes $\rho : \mathrm{Proj}\,\mathcal{B} \simeq \mathrm{Proj}\,\mathcal{C}$. Moreover, we have $\rho^* \mathcal{O}_{\mathrm{Proj}\,\mathcal{C}}(1) = \mathcal{O}_{\mathrm{Proj}\,\mathcal{B}}(1)$.

Proof The fact that $\mathcal{C}$ is a graded $\mathcal{O}_X$-algebra is clear. Let $\{U_i\}_i$ be a covering of X by affine open subschemes such that for every i, there exists an isomorphism $\varphi_i : \mathcal{N}|_{U_i} \simeq \mathcal{O}_{U_i}$. This isomorphism induces an isomorphism of graded $\mathcal{O}_X(U_i)$-algebras $\psi_i : \mathcal{C}|_{U_i} \simeq \mathcal{B}|_{U_i}$, and therefore an isomorphism of U_i-schemes $\rho_i : \mathrm{Proj}\,\mathcal{B}|_{U_i} \simeq \mathrm{Proj}\,\mathcal{C}|_{U_i}$. If we take another isomorphism $\varphi_i' : \mathcal{N}|_{U_i} \simeq \mathcal{O}_{U_i}$, then we obtain an isomorphism $\rho_i' : \mathrm{Proj}\,\mathcal{B}|_{U_i} \simeq \mathrm{Proj}\,\mathcal{C}|_{U_i}$. The composition $\rho_i^{-1} \circ \rho_i'$ is the automorphism of $\mathrm{Proj}\,\mathcal{B}|_{U_i}$ induced by $\varphi_i' \circ \varphi_i^{-1} : \mathcal{O}_{U_i} \to \mathcal{O}_{U_i}$, which is the multiplication by an $a_i \in \mathcal{O}_X(U_i)^*$. This automorphism induces the identity on the homogeneous elements of degree 0. Consequently, $\rho_i^{-1} \circ \rho_i'$ is the identity morphism on $\mathrm{Proj}\,\mathcal{B}|_{U_i}$. In other words, ρ_i is independent of the choice of an isomorphism $\mathcal{O}_{U_i} \to \mathcal{N}|_{U_i}$. This makes it possible to glue the ρ_i into an isomorphism of X-schemes $\mathrm{Proj}\,\mathcal{B} \to \mathrm{Proj}\,\mathcal{C}$. The last assertion results from the construction of ρ.

$\square$

Proposition 1.22. *Let X be a quasi-projective scheme over an affine Noetherian scheme. Let $\pi : \widetilde{X} \to X$ be the blowing-up of X along a closed subscheme $V(\mathcal{I})$. Then π is a projective morphism. Moreover, $\mathcal{I}\mathcal{O}_{\widetilde{X}}$ is very ample relative to the morphism π.*

Proof Let $\mathcal{N}$ be an ample sheaf on X (see Corollary 5.1.36). We can suppose that $\mathcal{N} \otimes_{\mathcal{O}_X} \mathcal{I}$ is generated by its global sections, if necessary replacing $\mathcal{N}$ by a sufficiently high tensor power. Let us consider the homogeneous $\mathcal{O}_X$-algebra

$$\mathcal{C} = \oplus_{n \geq 0}(\mathcal{N}^{\otimes n} \otimes_{\mathcal{O}_X} \mathcal{I}^n),$$

and the morphism of schemes $f : \mathrm{Proj}\,\mathcal{C} \to X$ associated to it. By hypothesis, the sheaf $\mathcal{C}_1 = \mathcal{N} \otimes_{\mathcal{O}_X} \mathcal{I}$ is generated by its global sections. As X is Noetherian, we deduce from this a surjective homomorphism of $\mathcal{O}_X$-modules

$$\mathcal{O}_X T_0 \oplus \cdots \oplus \mathcal{O}_X T_r = \mathcal{O}_X^{r+1} \to \mathcal{C}_1,$$

which canonically induces a surjective homomorphism of homogeneous $\mathcal{O}_X$-algebras $\mathcal{O}_X[T_0, \ldots, T_r] \to \mathcal{C}$. This homomorphism induces a closed immersion $i : \mathrm{Proj}\,\mathcal{C} \to P := \mathbb{P}_X^r$, and we have $\mathcal{O}_{\mathrm{Proj}\,\mathcal{C}}(1) = i^*\mathcal{O}_P(1)$. Let

$$\rho : \widetilde{X} = \mathrm{Proj}(\oplus_{n \geq 0}\mathcal{I}^n) \to \mathrm{Proj}\,\mathcal{C}$$

be the isomorphism of Lemma 1.21, and let $j : \widetilde{X} \to P$ be the closed immersion $i \circ \rho$. Then

$$\mathcal{I}\mathcal{O}_{\widetilde{X}} = \mathcal{O}_{\widetilde{X}}(1) = \rho^*\mathcal{O}_{\mathrm{Proj}\,\mathcal{C}}(1) = j^*\mathcal{O}_P(1)$$

(see Proposition 1.12(e) for the first equality), which shows that $\mathcal{I}\mathcal{O}_{\widetilde{X}}$ is very ample relative to π. □

Lemma 1.23. *Let X be a Noetherian scheme and $f : Z \to X$ a projective morphism. Let us fix a closed immersion $i : Z \to P := \mathbb{P}_X^N$ and let us set $\mathcal{O}_Z(n) := i^*\mathcal{O}_P(n)$. Then the following properties are true.*

(a) *There exists an $n_0 \geq 1$ such that for any $n \geq n_0$, the canonical homomorphism $(f_*\mathcal{O}_Z(1))^{\otimes n} \to f_*\mathcal{O}_Z(n)$ is surjective.*

(b) *The sheaf*

$$\mathcal{B} := \mathcal{O}_X \oplus (\oplus_{n \geq 1}f_*\mathcal{O}_Z(n))$$

is a graded $\mathcal{O}_X$-algebra and we have an isomorphism of X-schemes $Z \simeq \mathrm{Proj}\,\mathcal{B}$.

Proof (a) Let us consider the exact sequence

$$0 \to \mathcal{I} \to \mathcal{O}_P \to \mathcal{O}_Z \to 0$$

induced by $i : Z \to P$. Let us also denote the structural morphism $P \to X$ by f. Let $n_0 \geq 1$ be such that $R^1 f_*(\mathcal{I}(n)) = 0$ for every $n \geq n_0$ (this exists by

Theorem 5.3.2 and because X is quasi-compact). From the exact sequence

$$0 \to \mathcal{I}(n) \to \mathcal{O}_P(n) \to \mathcal{O}_Z(n) \to 0,$$

we deduce an exact sequence

$$0 \to f_*\mathcal{I}(n) \to f_*\mathcal{O}_P(n) \to f_*\mathcal{O}_Z(n) \to R^1 f_*(\mathcal{I}(n)) = 0$$

for every $n \geq n_0$. As $(f_*\mathcal{O}_P(1))^{\otimes n} \to f_*\mathcal{O}_P(n)$ is surjective for $n \geq 1$, (a) results from the commutative diagram

$$
\begin{array}{ccc}
(f_*\mathcal{O}_P(1))^{\otimes n} & \longrightarrow & f_*\mathcal{O}_P(n) \\
\downarrow & & \downarrow \\
(f_*\mathcal{O}_Z(1))^{\otimes n} & \longrightarrow & f_*\mathcal{O}_Z(n)
\end{array}
$$

(b) We set $\mathcal{B}_n = f_*\mathcal{O}_Z(n)$ if $n \geq 1$ and $\mathcal{B}_0 = \mathcal{O}_X$. We make $\mathcal{B}$ into a graded $\mathcal{O}_X$-algebra via the natural homomorphism

$$f_*\mathcal{O}_Z(n) \times f_*\mathcal{O}_Z(m) \to f_*\mathcal{O}_Z(n+m).$$

By Exercise 2.3.11(a), we can replace $\mathcal{O}_Z(1)$ by a power $\mathcal{O}_Z(d)$, $d \geq 1$, without changing $\operatorname{Proj} \mathcal{B}$. Let n_0 be as in the proof above. By composing $i : Z \to P$ with the n_0-uple embedding $\rho : P \to Q$ (Exercise 5.1.27), we obtain a closed immersion $j : Z \to Q$ into a projective space over X such that $\mathcal{O}_Z(n_0) \simeq j^*\mathcal{O}_Q(1)$. Let $g : Q \to X$ be the structural morphism, and

$$\mathcal{J} = \operatorname{Ker}(\mathcal{O}_Q \to \mathcal{O}_Z) = \rho_*\mathcal{I}.$$

Then $R^1 g_* \mathcal{J}(1) = R^1 f_*(\mathcal{I}(n_0)) = 0$, which, as in (a), implies that we have an exact sequence

$$0 \to \oplus_{n \geq 0} g_* \mathcal{J}(n) \to \oplus_{n \geq 0} g_* \mathcal{O}_Q(n) \to \mathcal{B} \to 0.$$

By Lemma 5.1.29, above any affine open subscheme U of X, the closed subschemes $\operatorname{Proj} \mathcal{B} \times_X U$ and $Z \times_X U$ of $Q \times_X U$ are equal. Therefore $\operatorname{Proj} \mathcal{B}$ and Z are equal in Q.   □

Theorem 1.24. *Let* $f : Z \to X$ *be a projective birational morphism of integral schemes. Suppose that* X *is quasi-projective over an affine Noetherian scheme. Then* f *is the blowing-up morphism of* X *along a closed subscheme.*

Proof   We are going to exploit the isomorphism $\mathcal{O}_{\widetilde{X}}(1) \simeq \mathcal{I}\mathcal{O}_{\widetilde{X}}$ of Proposition 1.22. Let i be a closed immersion of Z into a projective space $P = \mathbb{P}^N_X$, and let us set $\mathcal{L} = i^*\mathcal{O}_P(1)$ (see Definition 5.1.20). By Lemma 1.23, if necessary replacing $\mathcal{L}$ by a positive tensor power (which comes down to changing the immersion of Z into a projective space over X), we have $(f_*\mathcal{L})^{\otimes n} \to f_*(\mathcal{L}^{\otimes n})$ surjective for every $n \geq 1$, and $Z \simeq \mathrm{Proj}\,\mathcal{B}$, where

$$\mathcal{B} = \mathcal{O}_X \oplus \left(\oplus_{n\geq 1} f_*(\mathcal{L}^{\otimes n})\right).$$

As Z is integral, we can identify $\mathcal{L}$ with a subsheaf of $\mathcal{K}_Z$ (Corollary 7.1.19), whence $f_*\mathcal{L} \subseteq f_*\mathcal{K}_Z \simeq \mathcal{K}_X$ because f is birational. Let us set $\mathcal{F} = f_*\mathcal{L}$. We are going to look for an invertible sheaf $\mathcal{N} \subseteq \mathcal{K}_X$ on X such that $\mathcal{N}\mathcal{F} \subseteq \mathcal{O}_X$.

Let $\mathcal{J} = \mathrm{Ann}((\mathcal{F}+\mathcal{O}_X)/\mathcal{O}_X) \subseteq \mathcal{O}_X$. Hence $\mathcal{J}\mathcal{F} \subseteq \mathcal{O}_X$. As X is quasi-projective over an affine scheme, it admits an ample sheaf $\mathcal{M}$ (Corollary 5.1.36). By definition, if necessary replacing $\mathcal{M}$ by a tensor power, $\mathcal{J} \otimes \mathcal{M}$ is generated by its global sections. As $\mathcal{J} \otimes \mathcal{M}$ is a non-zero sheaf, a non-zero global section induces an injective homomorphism $\mathcal{O}_X \hookrightarrow \mathcal{J} \otimes \mathcal{M}$, whence an arrow $\mathcal{M}^\vee \hookrightarrow \mathcal{J}$. In the following commutative diagram of canonical homomorphisms

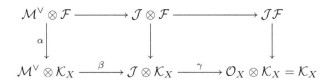

the homomorphisms α, β, γ are injective because $\mathcal{M}^\vee$ and $\mathcal{K}_X$ are flat over $\mathcal{O}_X$. We deduce from this an injective homomorphism $\mathcal{M}^\vee \otimes \mathcal{F} \hookrightarrow \mathcal{K}_X$ whose image $\mathcal{I}$ is contained in $\mathcal{J}\mathcal{F} \subseteq \mathcal{O}_X$. Hence $\mathcal{I}$ is a sheaf of ideals of $\mathcal{O}_X$. Let $\mathcal{N} \subseteq \mathcal{K}_X$ be the image of $\mathcal{M}^\vee$ in $\mathcal{J}$. Then $\mathcal{N}\mathcal{F} = \mathcal{I}$ in $\mathcal{K}_X$.

For any $n \geq 1$, as $\mathcal{L}$ is invertible, we have a canonical isomorphism $\mathcal{L}^{\otimes n} \simeq \mathcal{L}^n \subseteq \mathcal{K}_Z$, and hence an isomorphism $f_*(\mathcal{L}^{\otimes n}) \simeq f_*(\mathcal{L}^n) \subseteq \mathcal{K}_X$. As $\mathcal{F}^{\otimes n} \to f_*(\mathcal{L}^{\otimes n})$ is surjective, by studying the images of the two members in $\mathcal{K}_X$, we obtain

$$\mathcal{I}^n = \mathcal{N}^n\mathcal{F}^n = \mathcal{N}^n f_*(\mathcal{L}^n) \simeq \mathcal{N}^{\otimes n} \otimes f_*(\mathcal{L}^{\otimes n}).$$

The theorem then follows from Lemma 1.21: f is the blowing-up of X along $V(\mathcal{I})$.

□

Remark 1.25. Theorem 1.24 shows that every projective birational morphism is a blowing-up morphism. Hence the notion of blowing-up is somewhat too general. In the theory of desingularization, we are, in general, interested in blowing-ups along a regular center.

8.1.4   Normalization of curves by blowing-up points

As an application of the theory of blowing-ups, we are going to show that we can determine the normalization of an integral projective curve X over a field k by explicit computations. Let us recall that in dimension 1, the notion of normality coincides with that of regularity (Proposition 4.1.12). Let us suppose X is singular. Let $X_1 \to X_0 = X$ be the blowing-up of X_0 along the singular locus (endowed with the reduced scheme structure) of X_0. If X_1 is still singular, we define a blowing-up morphism $X_2 \to X_1$ in the same manner, and so on.

Proposition 1.26. *With the notation above, the sequence*

$$\cdots \to X_n \to X_{n-1} \to \cdots \to X_0 = X$$

is finite. In other words, we desingularize X by a finite number of consecutive blowing-ups with regular centers.

Proof Let $\pi_n : X_n \to X_{n-1}$ be the blowing-up morphism. It is a proper birational morphism of integral curves, and therefore finite. We have an exact sequence of sheaves

$$0 \to \mathcal{O}_{X_{n-1}} \to \pi_{n*}\mathcal{O}_{X_n} \to \mathcal{F}_n \to 0,$$

where $\mathcal{F}_n$ is a skyscraper sheaf whose support is contained in the singular locus of X_{n-1}. We deduce from this a relation between the arithmetic genera (Definition 7.3.19)

$$p_a(X_n) = p_a(X_{n-1}) - \dim_k H^0(X_{n-1}, \mathcal{F}_n).$$

Let $d = [K(X) \cap \overline{k} : k]$. As $H^0(X_n, \mathcal{O}_{X_n}) \subseteq K(X) \cap \overline{k}$, we have $p_a(X_n) \geq 1 - d$. Hence $(p_a(X_n))_n$ is a descending sequence, bounded from below by $1 - d$. Consequently, it is stationary. Let $n \geq 1$ be such that $p_a(X_n) = p_a(X_{n-1})$. Let us show that X_{n-1} is regular. We have $H^0(X_{n-1}, \mathcal{F}_n) = 0$. Now $\mathcal{F}_n$ is a skyscraper sheaf, and hence $\mathcal{F}_n = 0$, which means that $\mathcal{O}_{X_{n-1}} \to \pi_{n*}\mathcal{O}_{X_n}$ is an isomorphism. As π_n is finite (hence affine), it is an isomorphism. It follows from Proposition 1.12(e) that the singular locus of X_{n-1} is defined by a locally invertible ideal, which is impossible. □

Remark 1.27. Let X, Y be integral schemes which are birational. We will say that X *dominates* Y if the birational map $X \dashrightarrow Y$ is defined everywhere. Let X be an integral curve over a field k. Using the fact that $X' \to X$ is finite (Proposition 4.1.27), and that X' dominates all of the X_n, we immediately deduce that the sequence made up of the X_n is stationary, even for non-projective curves.

Remark 1.28. In general, if an integral domain A is given explicitly (e.g., as the quotient of a polynomial ring over a field by an ideal), it is not easy to determine the integral closure of A in $\mathrm{Frac}(A)$. If A is a finitely generated algebra over a field, of Krull dimension 1, then Proposition 1.26 and Lemma 1.4 make it possible to calculate the integral closure of A explicitly.

Exercises

1.1. Let us keep the hypotheses of Definition 1.18. Show that from a set-theoretical point of view, the strict transform of W is the Zariski closure of $\pi^{-1}(W \setminus V(\mathcal{I}))$ in $\widetilde{X}$.

1.2. Let X be a locally Noetherian scheme, $\mathcal{I}$ a coherent sheaf of ideals. Show that for any invertible sheaf $\mathcal{L} \subseteq \mathcal{O}_X$, $\mathcal{L}\mathcal{I}$ defines the same blowing-up morphism as $\mathcal{I}$.

1.3. Let X be a Noetherian scheme, and $\pi : \widetilde{X} \to X$ the blowing-up of X along $V(\mathcal{I})$. Show that if $\mathcal{I}$ is generated by its global sections, then π is a projective morphism.

1.4. Show that Theorem 1.19(b)–(c) remains valid if instead of supposing X, Y regular, we only suppose that $Y \to X$ is a closed immersion that is an l.c.i. (use [65], 15.B, Theorem 27: a regular sequence is quasi-regular).

1.5. (*Arcwise connectedness*) Let X be a connected projective variety over a field k. Let us consider two closed points $x_1, x_2 \in X$. We are going to show that there exists a connected curve in X passing through x_1 and x_2.

(a) Show that we can suppose X is irreducible of dimension ≥ 2.

(b) Let $\pi : \widetilde{X} \to X$ be the blowing-up of X along the reduced closed subvariety $\{x_1, x_2\}$. Show that $\widetilde{X}$ is connected and that $\dim \pi^{-1}(x_i) \geq 1$.

(c) Let D be an ample effective Cartier divisor on $\widetilde{X}$, with support Z. Show that Z meets $\pi^{-1}(x_i)$ for $i = 1, 2$ (Exercise 3.3.4(c)). Show that $\pi(Z)$ is a connected closed subvariety of X (Exercise 7.1.10), passing through x_1, x_2, and that $\dim Z \leq \dim X - 1$.

(d) Conclude by induction on $\dim X$.

1.6. Let us consider the example $X = \operatorname{Spec} k[t, s]$, where k is a field, I the ideal (t, s), and J the ideal (t^2, s). Show that the blowing-ups $X_1 \to X$, $X_2 \to X$ of X with respective centers $\widetilde{I}, \widetilde{J}$ are not isomorphic. Show that there does not even exist any morphism of X-schemes from one to the other.

1.7. Let X be a projective scheme over a Noetherian ring A. Let $\mathcal{L}$ be an invertible sheaf on X generated by global sections $s_0, \ldots, s_n$. Let us consider the morphism $f : X \to \mathbb{P}^n_A$ associated to these sections. We endow $f(X)$ with the structure of scheme-theoretic closure of $f(X)$ (see Exercise 2.3.17).

(a) Show that if $\mathcal{L} \simeq \mathcal{O}_X$, then $f(X)$ is finite over $\operatorname{Spec} A$.

(b) Let us suppose that $f(X)$ is finite over $\operatorname{Spec} A$. Hence $f(X) = \operatorname{Spec} B$ for some finite A-algebra B.

    (1) We can consider X as a scheme over $\operatorname{Spec} B$, with f as structural morphism. Let $g : X \to \mathbb{P}^n_B$ be the morphism associated to the sections $s_0, \ldots, s_n$. Show that $g(X) \to \operatorname{Spec} B$ is an isomorphism

and therefore induces a section $\sigma : \operatorname{Spec} B \rightarrow \mathbb{P}^n_B$ such that $g = f \circ \sigma$.

(2) Show that there exist $b_0, \ldots, b_n \in B$ without common zero such that $D(b_i) = \sigma^{-1}(D_+(T_i))$ and that $\sigma|_{D(b_i)} : D(b_i) \rightarrow D_+(T_i)$ is induced by the homomorphism $T_j/T_i \mapsto b_j/b_i$.

(3) Show that the $s_i b_i \in \mathcal{O}_X(X_{s_i})$ glue to a section $e \in \mathcal{L}(X)$ and that $\mathcal{L} = e\mathcal{O}_X$.

1.8. (*Generalization of* Theorem 1.24) Let X be a reduced quasi-projective scheme over an affine Noetherian scheme Y, let Z be a reduced scheme, and $f : Z \rightarrow X$ a birational morphism (Definition 7.5.3). We are going to show that f is a blowing-up morphism. Let us keep the notation of the proof of Theorem 1.24. It suffices to show that there exists an invertible sheaf $\mathcal{N}$ such that $\mathcal{N} f_* \mathcal{L} \subseteq \mathcal{O}_X$.

(a) Let $\xi_1, \ldots, \xi_n$ be the generic points of X. Show that there exist an affine open subscheme U of X that is the disjoint union of $U_1, \ldots, U_n$ such that $\xi_i \in U_i$, and a section $s \in \mathcal{J}(U)$ such that s is not a zero divisor in $\mathcal{O}_X(U)$.

(b) Let $X \rightarrow \mathbb{P}^m_Y$ be an embedding with $m \geq 1$. Show that there exists a principal open subset $D_+(f)$ of $\mathbb{P}^m_Y$ such that $\xi_i \in D_+(f)$ for every $i \leq n$ and that $D_+(f) \cap X \subseteq U$ (use Proposition 3.3.36).

(c) Show that, if necessary replacing f by a positive power, $f|_X s$ lifts to a global section of $\mathcal{O}_X(d) \otimes \mathcal{J}$, where $d = \deg f$ (Lemma 5.1.25). Show that we have an injective homomorphism $\mathcal{O}_X \rightarrow \mathcal{O}_X(d) \otimes \mathcal{J}$ and conclude.

8.2   Excellent schemes

The aim of this section is to present some topics in commutative algebra, needed for the next section. Certain aspects extend beyond the scope of this book, so not all of the statements will be proven (Propositions 2.11, 2.29, 2.41, and Theorem 2.39). The first part of the section treats catenary schemes. Next we give some rudiments of Cohen–Macaulay schemes, sufficient to show that every regular scheme is universally catenary (Corollary 2.16). Serre's criterion for normality (Theorem 2.23) is also proven. Finally, we present the definition of excellent schemes, as well as some general properties.

8.2.1   Universally catenary schemes and the dimension formula

Definition 2.1. We say that a Noetherian ring A is *catenary* if for any triplet of prime ideals $\mathfrak{q} \subseteq \mathfrak{p} \subseteq \mathfrak{m}$, we have the equality of heights (Subsection 2.5.1)

$$\operatorname{ht}(\mathfrak{m}/\mathfrak{q}) = \operatorname{ht}(\mathfrak{m}/\mathfrak{p}) + \operatorname{ht}(\mathfrak{p}/\mathfrak{q}).$$

We say that a Noetherian ring A is *universally catenary* if every finitely generated A-algebra is catenary. A finitely generated algebra over a universally catenary ring is universally catenary. We say that a locally Noetherian scheme X is *catenary* if its local rings are catenary, and that it is *universally catenary* if $\mathbb{A}_X^n$ is catenary for every $n \geq 0$.

Remark 2.2. Let X be a locally Noetherian scheme. Then X is catenary if and only if for every triplet of irreducible closed subsets $T \subseteq Y \subseteq Z$, we have

$$\operatorname{codim}(T, Z) = \operatorname{codim}(T, Y) + \operatorname{codim}(Y, Z).$$

Lemma 2.3. *Let A be a catenary (resp. universally catenary) Noetherian ring. Then the following properties are true.*

(a) *Let us suppose A is integral. Then for any pair of prime ideals $\mathfrak{p} \subseteq \mathfrak{m}$, we have*
$$\operatorname{ht}(\mathfrak{m}) = \operatorname{ht}(\mathfrak{m}/\mathfrak{p}) + \operatorname{ht}(\mathfrak{p}).$$

(b) *Any localization $S^{-1}A$ and any quotient ring of A is catenary (resp. universally catenary).*

(c) *Any algebraic variety over a field is universally catenary.*

Proof (a) is obvious. (b) The universally catenary case follows immediately from the catenary case. Let us therefore suppose A is catenary. The fact that $S^{-1}A$ is catenary results from the description of the prime ideals of $S^{-1}A$ as those prime ideals $\mathfrak{q}$ of A such that $\mathfrak{q} \cap S = \emptyset$. The property is evident for the quotient rings of A. Finally for (c), it suffices to show that any affine space $\mathbb{A}_k^n$ over a field is catenary, but this follows from Proposition 2.5.23(a). □

Remark 2.4. We will see that a large class of locally Noetherian schemes are universally catenary (Corollary 2.16).

Theorem 2.5 (Dimension formula). *Let X and Y be locally Noetherian integral schemes, $f : X \to Y$ a dominant morphism that is locally of finite type. Then for any $x \in X$, $y = f(x)$, we have*

$$\dim \mathcal{O}_{X,x} + \operatorname{trdeg}_{k(y)} k(x) \leq \dim \mathcal{O}_{Y,y} + \operatorname{trdeg}_{K(Y)} K(X). \qquad (2.4)$$

Moreover, equality holds if Y is universally catenary.

Proof As the property is local on X and Y, we can suppose that $Y = \operatorname{Spec} \mathcal{O}_{Y,y}$ and that X is a closed subscheme of an affine space $Z = \mathbb{A}_Y^n$. We have $\operatorname{trdeg}_{k(y)} k(x) = \dim \overline{\{x\}}$ (Proposition 2.5.19). The proof of the theorem is done by induction on n. If $n = 0$, then f is an isomorphism and there is nothing to prove. Let ξ denote the generic point of Y.

First step: $n = 1$. Then $\operatorname{codim}(X, Z) \leq 1$ because X_ξ is of codimension 0 or 1 in Z_ξ. If $\operatorname{codim}(X, Z) = 0$, then $X = Z$. If x is closed in X_y, then $\dim \mathcal{O}_{X,x} = \dim \mathcal{O}_{Y,y} + 1$ by virtue of Lemma 2.5.16, otherwise x is the generic point of $X_y = \mathbb{A}^1_{k(y)}$, and therefore $\mathrm{trdeg}_{k(y)} k(x) = 1$. In both cases, (2.4) is an equality.

Let us now suppose $\mathrm{codim}(X, Z) = 1$. Then X_ξ is also of codimension 1 in $Z_\xi = \mathbb{A}^1_{k(\xi)}$, and therefore reduced to a closed point of Z_ξ. By what we have just seen,

$$\dim \mathcal{O}_{Z,x} + \mathrm{trdeg}_{k(y)} k(x) = \dim \mathcal{O}_{Y,y} + 1$$

(note that X and Z have the same residue field at x). Let $\mathfrak{p}$ be the prime ideal of $\mathcal{O}(Z)$ defining X, and η be the generic point of X. Then $\mathrm{ht}(\mathfrak{p}\mathcal{O}_{Z,x}) = \dim(\mathcal{O}_{Z,x})_{\mathfrak{p}} = \dim \mathcal{O}_{Z,\eta} = 1$. Hence

$$\dim \mathcal{O}_{Z,x} \geq \dim(\mathcal{O}_{Z,x}/\mathfrak{p}\mathcal{O}_{Z,x}) + \mathrm{ht}(\mathfrak{p}\mathcal{O}_{Z,x}) = \dim \mathcal{O}_{X,x} + 1,$$

and the inequality is an equality if Y is universally catenary (which implies that Z is catenary).

Second step: induction. Let us now suppose that $n \geq 2$ and that the theorem is true for every integral closed subscheme of $\mathbb{A}^{n-1}_Y$. Let $p : Z \to \mathbb{A}^{n-1}_Y$ (resp. $q : Z \to \mathbb{A}^1_Y$) be the projection onto the first $(n-1)$ coordinates (resp. the last coordinate). Let W be the closed subscheme $\overline{p(X)}$ endowed with the reduced (hence integral) scheme structure. Then $(p, q) : X \to W \times \mathbb{A}^1_Y = \mathbb{A}^1_W$ is a closed immersion. The morphism $f : X \to Y$ decomposes into

$$X \to W \to Y, \quad \text{with } W \subseteq \mathbb{A}^{n-1}_Y, X \subseteq \mathbb{A}^1_W.$$

Let $w = p(x)$. It follows from the induction hypothesis and the $n = 1$ case that

$$\dim \mathcal{O}_{W,w} + \mathrm{trdeg}_{k(y)} k(w) \leq \dim \mathcal{O}_{Y,y} + \mathrm{trdeg}_{K(Y)} K(W),$$

$$\dim \mathcal{O}_{X,x} + \mathrm{trdeg}_{k(w)} k(x) \leq \dim \mathcal{O}_{W,w} + \mathrm{trdeg}_{K(W)} K(X).$$

By adding these two inequalities, we obtain (2.4). Moreover, if Y is universally catenary, these two inequalities are equalities because W will also be universally catenary, whence the theorem.   □

Corollary 2.6. *Let $f : X \to Y$ be a finite dominant morphism of locally Noetherian integral schemes. Let us suppose that Y is universally catenary. Then for all $x \in X$, we have $\dim \mathcal{O}_{X,x} = \dim \mathcal{O}_{Y,f(x)}$.*

Corollary 2.7. *Let Y be a locally Noetherian integral scheme. Let $f : X \to Y$ be a proper birational morphism. Then $\dim X = \dim Y$.*

Proof For any $x \in X$, let $y = f(x)$. Then

$$\dim \mathcal{O}_{X,x} \leq \dim \mathcal{O}_{Y,y} + \mathrm{trdeg}_{K(Y)} K(X) - \mathrm{trdeg}_{k(y)} k(x) \leq \dim \mathcal{O}_{Y,y}$$

because $K(Y) \simeq K(X)$, whence $\dim X \leq \dim Y$. The inequality in the other direction results from the fact that f is surjective and closed (Exercise 3.3.3; since the property is local on Y, we can assume it to be Noetherian).   □

Now we can generalize Corollary 4.3.14.

Corollary 2.8. *Let Y be a universally catenary, Noetherian, irreducible scheme. Let $f : X \to Y$ be a flat surjective morphism of finite type. Let us suppose X is equidimensional. Then for any $y \in Y$, X_y is equidimensional of dimension*

$$\dim X_y = \dim X - \dim Y. \tag{2.5}$$

Proof We can suppose X is irreducible. Let $y \in Y$. For any point $x \in X_y$ that is closed in X_y, $k(x)$ is algebraic over $k(y)$. Applying Theorem 2.5 to the morphism $X_{\mathrm{red}} \to Y_{\mathrm{red}}$, we obtain

$$\dim \mathcal{O}_{X,x} = \dim \mathcal{O}_{Y,y} + \dim X_\xi,$$

where ξ is the generic point of Y. By Theorem 4.3.12, $\dim \mathcal{O}_{X_y,x} = \dim X_\xi$ for every closed point x of X_y, which shows that X_y is equidimensional. Finally, by varying x and y in equality (2.5), we obtain $\dim X - \dim Y = \dim X_\xi$. $\quad\square$

8.2.2 Cohen–Macaulay rings

Considering the results of the preceding subsection, it would be interesting to have a large class of universally catenary schemes. In what follows, we are going to show that any scheme that is locally of finite type over a regular locally Noetherian scheme is universally catenary (Corollary 2.16). The proof is relatively elementary except for Proposition 2.11 which we admit. For the proof, we will need the notion of Cohen–Macaulay rings. As a subproduct of this notion, we will also show Serre's criterion for normality.

Definition 2.9. Let A be a ring. Let M be an A-module. We say that an element $a \in A$ is *M-regular* if the map $M \to M$ defined by multiplication by a is injective. A sequence of elements $a_1, \ldots, a_n$ of A is called *M-regular* if a_1 is regular for M and if a_{i+1} is regular for $M/(a_1 M + \cdots + a_i M)$ for every $1 \le i \le n-1$. If I is an ideal of A such that $IM \ne M$, and if the $a_i \in I$, we say that it is an *M-regular sequence in I*. The *I-depth of M*, denoted $\mathrm{depth}_I M$, is the maximal number of elements of an M-regular sequence in I. When A is a Noetherian local ring with maximal ideal $\mathfrak{m}$ and M is finitely generated over A, we also denote the depth $\mathrm{depth}_{\mathfrak{m}} M$ by $\mathrm{depth}\, M$.

Example 2.10. Let A be a Noetherian local ring. Let $\mathfrak{p} \in \mathrm{Spec}\, A$. Then $\mathrm{depth}\,\mathfrak{p} = 0$ if and only if $\mathfrak{p} \in \mathrm{Ass}(A)$. Indeed, $\mathrm{depth}\,\mathfrak{p} = 0$ is equivalent to saying that $\mathfrak{p}$ is contained in the set of zero divisors of A. And the latter is the union of the ideals $\mathfrak{q} \in \mathrm{Ass}(A)$ (Corollary 7.1.3(a)).

Proposition 2.11. *Let $(A, \mathfrak{m})$ be a Noetherian local ring, and let M be a finitely generated A-module of depth $d = \mathrm{depth}\, M$. Then any M-regular sequence in $\mathfrak{m}$ can be completed to an M-regular sequence in $\mathfrak{m}$ with d elements. In particular, if $a \in \mathfrak{m}$ is M-regular, then*

$$\mathrm{depth}(M/aM) = \mathrm{depth}\, M - 1.$$

Proof This comes down to saying that any M-regular sequence in $\mathfrak{m}$ that is maximal for the inclusion has d elements. We will not prove this proposition, which needs the use of homological tools such as the modules $\mathrm{Ext}_A^n(N, M)$. See [65], Chapter 6, (15.C) or [41], $0_{\mathrm{IV}}.16.4.4$. $\quad\square$

Lemma 2.12. *Let M be a finitely generated module over a Noetherian ring A. Let $\operatorname{Supp} M$ denote the support of the coherent sheaf $\widetilde{M}$ on $\operatorname{Spec} A$. Then the following properties are true.*

(a) *Any prime ideal $\mathfrak{q} \in \operatorname{Spec} A$ that is minimal among those containing $\operatorname{Ann}(M)$ belongs to $\operatorname{Ass}(M)$.*

(b) *We have $\operatorname{Supp} M = \cup_{\mathfrak{p} \in \operatorname{Ass}(M)} V(\mathfrak{p})$.*

(c) *Let us suppose A is local and $M \neq 0$. Then we have $\operatorname{depth} M \leq \dim \operatorname{Supp} M$.*

Proof Let $I = \operatorname{Ann} M$; then $\operatorname{Supp} M = V(I)$ (Exercise 5.1.9(d)). The ideal I is contained in every ideal $\mathfrak{p} \in \operatorname{Ass}(M)$ by definition. Thus $\cup_{\mathfrak{p} \in \operatorname{Ass}(M)} V(\mathfrak{p}) \subseteq \operatorname{Supp} M$. As $M_{\mathfrak{q}} \neq 0$, it follows from Lemma 7.1.2 that $\operatorname{Ass}_{A_{\mathfrak{q}}}(M_{\mathfrak{q}})$ is non-empty, and is made up of elements $\mathfrak{p} \in \operatorname{Ass}(M)$ contained in $\mathfrak{q}$. Hence $\mathfrak{p} = \mathfrak{q}$. This proves (a) and the opposite inclusion $\operatorname{Supp} M \subseteq \cup_{\mathfrak{p} \in \operatorname{Ass}(M)} V(\mathfrak{p})$, whence (b).

(c) Let $t \in A$ be an M-regular element in the maximal ideal $\mathfrak{m}$ of A. Then $M/tM \neq 0$ and $\operatorname{Supp}(M/tM) = \operatorname{Supp} M \cap V(tA)$. For any $\mathfrak{p} \in \operatorname{Ass}(M)$, we have $t \notin \mathfrak{p}$, and hence $\dim(V(\mathfrak{p}) \cap V(tA)) \leq \dim V(\mathfrak{p}) - 1$. It follows from (b) that

$$0 \leq \dim \operatorname{Supp}(M/tM) \leq \dim \operatorname{Supp} M - 1.$$

If $t_1, \ldots, t_r$ is an M-regular sequence in $\mathfrak{m}$, then we have

$$0 \leq \dim \operatorname{Supp}(M/(t_1 M + \cdots + t_r M)) \leq \dim \operatorname{Supp} M - r,$$

whence $\operatorname{depth} M \leq \dim \operatorname{Supp} M$. $\qquad\qquad\qquad\qquad\qquad\qquad\qquad\square$

Proposition 2.13. *Let A be a Noetherian local ring and M a non-zero finitely generated A-module. Then we have*

$$\operatorname{depth} M \leq \inf_{\mathfrak{p} \in \operatorname{Ass}(M)} \dim A/\mathfrak{p}.$$

Proof We are going to show the proposition in several steps. Let us fix $\mathfrak{p} \in \operatorname{Ass}(M)$ in (α)–(γ).

(α) There exists an exact sequence of A-modules

$$0 \to K \to M \to L \to 0$$

such that $\operatorname{Ass}(K) = \{\mathfrak{p}\}$ and that $\operatorname{Ass}(L) \subseteq \operatorname{Ass}(M)$. Let us consider the set of submodules N of M such that $\operatorname{Ass}(N) = \{\mathfrak{p}\}$. It is non-empty, and admits a maximal element K. Let $L = M/K$ and $\mathfrak{q} \in \operatorname{Ass}(L)$; then $\{\mathfrak{q}\} = \operatorname{Ass}(F/K)$ for some submodule $F \supseteq K$ of M. We have seen in the proof of Corollary 7.1.5 that $\operatorname{Ass}(F) \subseteq \operatorname{Ass}(F/K) \cup \operatorname{Ass}(K) = \{\mathfrak{p}, \mathfrak{q}\}$. We must have $\mathfrak{q} \in \operatorname{Ass}(F) \subseteq \operatorname{Ass}(M)$ because otherwise $\operatorname{Ass}(F) = \{\mathfrak{p}\}$ and K would not be maximal.

(β) Let us fix an M-regular element $t \in A$. Then $K/tK \to M/tM$ is injective. This comes down to showing that t is regular for L. Let us suppose $tz = 0$ with $z \in L \setminus \{0\}$. We have $\operatorname{Ass}(Az) \neq \emptyset$ (Lemma 7.1.2(a)). Let $\mathfrak{q} \in \operatorname{Ass}(Az)$; then $\mathfrak{q}$ is the annihilator of a submodule of Az. It follows that $t \in \mathfrak{q}$. Now $\mathfrak{q} \in \mathrm{Ass}(M)$ by (α); hence tA is the annihilator of a non-zero submodule of M, a contradiction.

(γ) *Any prime ideal* $\mathfrak{q} \in \mathrm{Spec}\, A$ *that is minimal among those containing* $\mathfrak{p} + tA$ *belongs to* $\mathrm{Ass}(M/tM)$. Indeed, we have

$$\mathrm{Supp}(K/tK) = \mathrm{Supp}\, K \cap V(tA) = V(\mathfrak{p}) \cap V(tA) = V(\mathfrak{p} + tA),$$

where the second equality comes from Lemma 2.12(b). Hence $\mathfrak{q} \in \mathrm{Ass}(K/tK)$ by Lemma 2.12(a). Finally, $\mathfrak{q} \in \mathrm{Ass}(M/tM)$ by virtue of (β).

(δ) *Proof of the proposition by induction on* $d = \mathrm{depth}\, M$. If $d = 0$, then $\mathfrak{m} \subseteq \mathrm{Ann}\, M$, and hence $\mathrm{Ass}(M) \subseteq \{\mathfrak{m}\}$. The inequality is therefore evident. Let us suppose $d \geq 1$ and the proposition is true for every A-module of depth $\leq d-1$. Let t be an M-regular element belonging to the maximal ideal $\mathfrak{m}$ of A. Then

$$\mathrm{depth}(M/tM) = \mathrm{depth}\, M - 1 \tag{2.6}$$

by Proposition 2.11. On the other hand, let $\mathfrak{p} \in \mathrm{Ass}(M)$, and let $\mathfrak{q}$ be as in (γ); then $\mathfrak{q} \in \mathrm{Ass}(M/tM)$. As t is M-regular, we have $t \notin \mathfrak{p}$, and hence

$$\dim A/\mathfrak{p} \geq \dim(A/\mathfrak{q}) + 1 \geq \inf_{\mathfrak{q} \in \mathrm{Ass}(M/tM)} \dim A/\mathfrak{q} + 1 \geq \mathrm{depth}(M/tM) + 1,$$

whence the result by combining with equality (2.6). $\qquad\square$

We say that a Noetherian local ring A is a *Cohen–Macaulay ring* if $\mathrm{depth}\, A = \dim A$ (or if, which is equivalent by Lemma 2.12(c), $\mathrm{depth}\, A \geq \dim A$). A finitely generated A-module M is called Cohen–Macaulay if $\mathrm{depth}\, M = \dim \mathrm{Supp}\, M$.

Example 2.14. Let $(A, \mathfrak{m})$ be a regular, Noetherian, local ring. Then A is Cohen–Macaulay. Indeed, any coordinate system for A forms a regular sequence by Proposition 4.2.11 and Corollary 4.2.15. Hence $\mathrm{depth}\, A \geq \dim A$.

Proposition 2.15. Let $(A, \mathfrak{m})$ be a Cohen–Macaulay Noetherian local ring. Then the following properties are true.

(a) For any regular element $t \in \mathfrak{m}$, A/tA is Cohen–Macaulay.

(b) The scheme $\mathrm{Spec}\, A$ has no embedded points (Definition 7.1.6).

(c) The scheme $\mathrm{Spec}\, A$ is equidimensional.

(d) For any $\mathfrak{p} \in \mathrm{Spec}\, A$, $A_{\mathfrak{p}}$ is Cohen–Macaulay and we have

$$\dim A = \mathrm{ht}(\mathfrak{p}) + \dim A/\mathfrak{p}.$$

(e) The ring A is catenary.

Proof (a) By Proposition 2.11, $\mathrm{depth}(A/tA) = \mathrm{depth}\, A - 1$. By Corollary 7.1.3, t does not belong to any minimal prime ideal of A. It follows from Theorem 2.5.15 that $\dim A/tA = \dim A - 1$. Hence A/tA is Cohen–Macaulay.

(b) By Proposition 2.13, any ideal $\mathfrak{p} \in \mathrm{Ass}(A)$ is a minimal prime ideal. This means that $\mathrm{Spec}\, A$ has no embedded points.

(c) If $\dim A = 0$, there is nothing to prove. Let us suppose $\dim A \neq 0$. Then $\operatorname{depth} A \geq 1$, and hence there exists a regular non-invertible $t \in A$. Let $\mathfrak{p}$ be a minimal prime ideal of A. Then $t \notin \mathfrak{p}$, as we have seen in (a). Let Z be an irreducible component of $V(\mathfrak{p}) \cap V(tA)$. Then $\dim Z = \dim V(\mathfrak{p}) - 1$. Moreover, Z is an irreducible component of $V(tA)$, as we have seen in part (γ) of the proof of Proposition 2.13, using the fact that A/tA is Cohen–Macaulay. Hence, if $V(tA)$ is equidimensional, the same will hold for A. We conclude by induction on $\dim A$.

(d) Let $r = \operatorname{ht}(\mathfrak{p}) = \dim A_\mathfrak{p}$. By induction as above we construct a regular sequence $t_1, \ldots, t_r \in \mathfrak{p}$. As $A \to A_\mathfrak{p}$ is flat, the t_i still form a regular sequence in $A_\mathfrak{p}$ (Lemma 6.3.10). Hence $\operatorname{depth} A_\mathfrak{p} \geq r$, which proves that $A_\mathfrak{p}$ is Cohen–Macaulay. Applying Theorem 2.5.15 successively, we find that $\dim A/(t_1, \ldots, t_r) = \dim A - r$. Now $A/(t_1, \ldots, t_r)$ is Cohen–Macaulay, and hence equidimensional, and $V(\mathfrak{p})$ is an irreducible component of $V(t_1, \ldots, t_r)$. It follows that $\dim V(\mathfrak{p}) = \dim A - r = \dim A - \operatorname{ht}(\mathfrak{p})$.

(e) Let $\mathfrak{p} \subseteq \mathfrak{q}$ be a pair of prime ideals of A. By (d), $A_\mathfrak{q}$ is Cohen–Macaulay and we have

$$\dim A_\mathfrak{q} = \dim A_\mathfrak{q}/\mathfrak{p}A_\mathfrak{q} + \operatorname{ht}(\mathfrak{p}A_\mathfrak{q}).$$

Term by term, this equality translates to

$$\operatorname{ht}(\mathfrak{q}) = \operatorname{ht}(\mathfrak{q}/\mathfrak{p}) + \operatorname{ht}(\mathfrak{p}),$$

which immediately implies that A is catenary. $\qquad\square$

Corollary 2.16. *Any scheme that is locally of finite type over a regular locally Noetherian scheme is universally catenary.*

Proof It suffices to show that every regular Noetherian scheme X is universally catenary. As $\mathbb{A}^n_X$ is regular (Theorem 4.3.36), it suffices to show that every regular Noetherian scheme is catenary. But this is an immediate consequence of Proposition 2.15 and Example 2.14. $\qquad\square$

Definition 2.17. We say that a Noetherian ring is Cohen–Macaulay if its localizations at the prime ideals are Cohen–Macaulay. We say that a locally Noetherian scheme X is Cohen–Macaulay if $\mathcal{O}_{X,x}$ is Cohen–Macaulay for every $x \in X$.

Corollary 2.18. *Let X be a local complete intersection over a regular locally Noetherian scheme. Then X is a Cohen–Macaulay scheme.*

Proof This follows from Example 2.14, Proposition 2.15(a), and Theorem 2.5.15. $\qquad\square$

Properties (R_k) and (S_k), Serre's criterion for normality

Definition 2.19. Let X be a locally Noetherian scheme, and let $k \geq 0$ be an integer. We will say that X *verifies property* (R_k) if X is regular at all of its points of codimension $\leq k$. We will say that X verifies property (S_k) if for any point $x \in X$, we have

$$\operatorname{depth} \mathcal{O}_{X,x} \geq \inf\{k, \dim \mathcal{O}_{X,x}\}.$$

Example 2.20. Let X be a locally Noetherian scheme. Then X always verifies property (S_0). Property (S_1) is equivalent to X having no embedded points (see

Example 2.10). Finally, X verifies (S_k) for every $k \geq 0$ if and only if X is Cohen–Macaulay.

Lemma 2.21. *Let X be a normal locally Noetherian scheme. Then X verifies properties (S_2) and (R_1).*

Proof   We can suppose $X = \operatorname{Spec} A$ affine. Let $\mathfrak{p} \in \operatorname{Spec} A$. If $\operatorname{ht}(\mathfrak{p}) \leq 1$, $A_{\mathfrak{p}}$ is regular (Proposition 4.1.12). Hence X verifies (R_1). Let us suppose $\operatorname{ht}(\mathfrak{p}) \geq 2$. We must show that $\operatorname{depth}(A_{\mathfrak{p}}) \geq 2$. Localizing at $\mathfrak{p}$ if necessary, we can suppose A is local with maximal ideal $\mathfrak{p}$. Let $a \in \mathfrak{p}$ be non-zero. If $\operatorname{depth}(A) \leq 1$, then $\operatorname{depth}(A/aA) = 0$. It follows that $\mathfrak{p} \in \operatorname{Ass}(A/aA)$ (use Corollary 7.1.3). In other words, there exists a $b \in A \setminus aA$ such that $b\mathfrak{p} \subseteq aA$. Let $f = b/a \in \operatorname{Frac}(A)$. Every prime ideal $\mathfrak{q}$ of height 1 is strictly contained in $\mathfrak{p}$ because $\operatorname{ht}(\mathfrak{p}) \geq 2$. Let $c \in \mathfrak{p} \setminus \mathfrak{q}$; then $fc \in (b/a)\mathfrak{p} \subseteq A$, and $f \in A_{\mathfrak{q}}$. By Lemma 4.1.13, we have $f \in A$, hence $b \in aA$, a contradiction. $\square$

Corollary 2.22.   *Any normal locally Noetherian scheme of dimension ≤ 2 is Cohen–Macaulay.*

Theorem 2.23 (Serre criterion). *Let X be a locally Noetherian connected scheme. Then X is normal if and only if it verifies properties (S_2) and (R_1).*

Proof   This theorem is due to Krull in the integral case. We restrict ourselves to this case, leaving the general case to the reader (Exercise 2.8). By the lemma above, the conditions of the theorem are necessary. Let us show that they are sufficient. We can suppose $X = \operatorname{Spec} A$ is affine. Let $f = a/b \in K(X)$ be integral over A. By Proposition 2.11, the scheme $Z := V(b)$ verifies property (S_1). Let $\xi_1, \ldots, \xi_r$ be its generic points. As f is integral over $\mathcal{O}_{X,\xi_i}$, and the latter is normal by hypothesis (R_1), we have $a \in b\mathcal{O}_{X,\xi_i}$. In other words, the image $\bar{a}$ of a in A/bA is zero at the points ξ_i. Now the canonical homomorphism

$$A/bA \hookrightarrow \oplus_{1 \leq i \leq r} \mathcal{O}_{Z,\xi_i}$$

is injective (Lemma 7.1.9). It follows that $\bar{a} = 0$, and hence $f \in A$. $\square$

Corollary 2.24. *Let X be a locally Noetherian connected scheme that is Cohen–Macaulay (e.g., if X is a local complete intersection over a regular locally Noetherian scheme; Corollary 2.18). Then X is normal if and only if it is normal at the points of codimension 1.*

Proof   The fact that X is Cohen–Macaulay implies that it verifies property (S_2). If X is normal at the points of codimension 1, then it also verifies property (R_1). It now suffices to apply the theorem above. $\square$

Corollary 2.25. *Let Y be a normal locally Noetherian scheme, and $f : X \to Y$ a smooth morphism of finite type. Then X is normal.*

Proof   Locally, X is étale over an $\mathbb{A}_Y^n$ (Corollary 6.2.11). As $\mathbb{A}_Y^n$ is normal (Exercise 4.1.5), we only need to consider the case when f is étale. Let $x \in X$ and $y = f(x)$. We have $\dim \mathcal{O}_{Y,y} = \dim \mathcal{O}_{X,x}$ because X_y is of dimension 0 (Theorem 4.3.12). If x is of codimension 1 in X, then $\dim \mathcal{O}_{Y,y} = 1$. Consequently,

$\mathcal{O}_{Y,y}$ is regular and so is $\mathcal{O}_{X,x}$ (Corollary 4.3.24). Hence X verifies (R_1). Let us suppose $\dim \mathcal{O}_{X,x} \geq 2$. Then $\dim \mathcal{O}_{Y,y} \geq 2$ and there exists a regular sequence (a, b) in $\mathfrak{m}_y \mathcal{O}_{Y,y}$. As $\mathcal{O}_{X,x}$ is flat over $\mathcal{O}_{Y,y}$ and $\mathcal{O}_{X,x}/(a)$ is flat over $\mathcal{O}_{Y,y}/(a)$, we deduce from this that a is regular in $\mathcal{O}_{X,x}$, that the image of b in $\mathcal{O}_{X,x}/(a)$ is a regular element, and therefore that (a, b) is a regular sequence in $\mathcal{O}_{X,x}$. Consequently, X verifies (S_2) and is therefore normal.   □

Example 2.26. Let $S = \operatorname{Spec} R$ be an affine Dedekind scheme. Let

$$B = R[x, y]/(F(x, y)), \quad F(x, y) \in R[x, y] \setminus R.$$

We are going to see how to determine whether $X := \operatorname{Spec} B$ is normal or not. As X is Cohen–Macaulay, X is normal if and only if it is normal at its points of codimension 1. These points are either closed points of the generic fiber X_η, or the generic points of the closed fibers of $X \to S$. The fiber X_η is a curve over $K(S)$. Hence its normality is relatively easy to determine, because it is equivalent to regularity (use Corollary 4.2.12).

Let $\xi \in X$ be a generic point of a closed fiber X_s. Let t be a uniformizing parameter for $A := \mathcal{O}_{S,s}$. The point ξ corresponds to the prime ideal generated by t and a polynomial $G(x, y) \in A[x, y]$ whose image $\overline{G}(x, y)$ in $k(s)[x, y]$ is irreducible. We can write

$$F(x, y) = G(x, y)^r H_1(x, y) + t^s H_2(x, y), \quad H_i(x, y) \in A[x, y], r, s \geq 1$$

with $\overline{H}_1(x, y) \notin \overline{G}(x, y)k(s)[x, y]$ and $\overline{H}_2(x, y) \neq 0$. Then X is normal at ξ if and only if either $r = 1$; or $s = 1$ and $\overline{H}_2 \notin \overline{G}(x, y)k(s)[x, y]$ (use Corollary 4.2.12).

With this method, we quickly find that the scheme X of Example 4.1.9 is normal.

Definition 2.27. Let A be a Noetherian ring. We say that A is a *Nagata ring* if for every prime ideal $\mathfrak{p} \in \operatorname{Spec} A$ and for every finite extension L of $\operatorname{Frac}(A/\mathfrak{p})$, the integral closure of $A/\mathfrak{p}$ in L is finite over $A/\mathfrak{p}$.

Example 2.28. (a) Any finitely generated algebra over a field is Nagata (Proposition 4.1.27). (b) Any Dedekind domain of characteristic 0 is Nagata (Propositions 4.1.25 and 4.1.27).

Proposition 2.29. *Let A be a Noetherian ring.*

(a) *If A is local and complete, then it is Nagata.*

(b) *If A is Nagata, then any localization and any finitely generated algebra over A is Nagata.*

(c) *Let us suppose A is integral and Nagata. Then the set of normal points of $\operatorname{Spec} A$ is open in $\operatorname{Spec} A$.*

Proof (a)–(b) See [65], Chapter 12, Theorem 72 and Corollary 2. (c) is a consequence of Proposition 4.1.29.   □

Definition 2.30. We will say that a scheme X is a *Nagata scheme* if it is locally Noetherian and if for every affine open subset U of X, the ring $\mathcal{O}_X(U)$ is Nagata. By the proposition above, the local rings $\mathcal{O}_{X,x}$ are then Nagata rings. Any scheme that is locally of finite type over a field or over a complete discrete valuation ring is Nagata.

Example 2.31 (*A discrete valuation ring that is not Nagata*). Let p be a prime number. The field $\mathbb{F}_p(t)$ as well as its algebraic closure are countable sets. On the other hand, $\mathbb{F}_p[[t]]$ is not. Therefore, there exists an element $s \in \mathbb{F}_p[[t]]$ that is transcendent over $\mathbb{F}_p(t)$. Let us set $K = \mathbb{F}_p(t, s^p) \subset \mathbb{F}_p((t))$. The discrete valuation of $\mathbb{F}_p((t))$ induces a discrete valuation on K. Let $\mathcal{O}_K \subset K$ denote the corresponding discrete valuation ring. Let $L = K[s] \subseteq \mathbb{F}_p((t))$. This is a purely inseparable extension of K of degree p because $s^p \in K$ and $s \notin K$.

Let us show that the integral closure $\mathcal{O}_L$ of $\mathcal{O}_K$ in L is not finite over $\mathcal{O}_K$. Let us suppose the contrary. Then $\mathcal{O}_L$ is free of rank $p = [L : K]$ over $\mathcal{O}_K$. By Exercise 5.3.9, $\operatorname{Spec} \mathcal{O}_L \to \operatorname{Spec} \mathcal{O}_K$ is a homeomorphism. In particular, $\mathcal{O}_L$ is local. Let us note that t is a uniformizing parameter for $\mathcal{O}_K$ and for $\mathcal{O}_L$, and that the residue fields of these rings are equal to $\mathbb{F}_p$. This leads to a contradiction since $\mathcal{O}_L/(t)$ must be free of rank p over $\mathcal{O}_K/(t)$.

Lemma 2.32. *Let X be a reduced Noetherian scheme. Let $X_1, \ldots, X_n$ be its (reduced) irreducible components. Then the normalization morphism $X' \to X$ (Definition 7.5.1) is finite if and only if the normalization morphism $X'_i \to X_i$ is finite for each $i \leq n$.*

Proof We can suppose $X = \operatorname{Spec} A$. Let $\mathfrak{p}_1, \ldots, \mathfrak{p}_n$ be the prime ideals of A corresponding to the irreducible components $X_1, \ldots, X_n$. Then $A \to A'$ decomposes into

$$A \to \oplus_{1 \leq i \leq n} A/\mathfrak{p}_i \to \oplus_{1 \leq i \leq n} (A/\mathfrak{p}_i)'$$

(the $'$ sign denotes the integral closure). See Lemma 7.5.2. Since the first homomorphism is finite, the assertion of the lemma is clear. $\qquad\square$

8.2.3 Excellent schemes

The notion of excellent rings and schemes was introduced by Grothendieck. It essentially answers three types of questions affirmatively:

(A) Do the properties of a local ring A transfer to the completion?

(B) Is the set of points of a scheme X such that $\mathcal{O}_{X,x}$ verifies certain properties (e.g., normal, regular, ...) open in X?

(C) Is the normalization morphism finite?

See [41], IV.7.8.1, for more detail. See also Remark 3.40.

Definition 2.33. Let $(A, \mathfrak{m})$ be a Noetherian local ring. Let $\widehat{A}$ be its completion for the $\mathfrak{m}$-adic topology. We call the fibers of the canonical morphism $\operatorname{Spec} \widehat{A} \to \operatorname{Spec} A$ the *formal fibers* of A. Let us note that in general, this morphism is not of finite type. Let $x \in \operatorname{Spec} A$, the fiber of $\operatorname{Spec} \widehat{A} \to \operatorname{Spec} A$ over x is $\operatorname{Spec}(\widehat{A} \otimes_A$

$k(x)$). It is an affine scheme over $k(x)$. We will say that this fiber is *geometrically regular* if for every finitely generated extension $k'/k(x)$, the scheme $\mathrm{Spec}(\widehat{A} \otimes_A k')$ is regular.

Example 2.34. Let $\mathcal{O}_K$ be a discrete valuation ring with field of fractions K, and residue field k. Then $\mathcal{O}_K$ has two formal fibers, the closed fiber $\mathrm{Spec}\, k \to \mathrm{Spec}\, k$ and the generic fiber $\mathrm{Spec}\, \widehat{K} \to \mathrm{Spec}\, K$. The first is obviously geometrically regular. The second has this property if and only if the extension $\widehat{K}/K$ is separable, that is, if every finitely generated subextension is separable (Definition 6.1.14). Such is the case if, for example, $\mathrm{char}(K) = 0$.

Definition 2.35. Let A be a Noetherian ring. We say that A is *excellent* if it verifies the following three properties:

(i) $\mathrm{Spec}\, A$ is universally catenary.

(ii) For every $\mathfrak{p} \in \mathrm{Spec}\, A$, the formal fibers of $A_\mathfrak{p}$ are geometrically regular.

(iii) For every finitely generated A-algebra B, the set of regular points of $\mathrm{Spec}\, B$ is open in $\mathrm{Spec}\, B$.

Note that conditions (i) and (ii) only relate to the localizations of A at the prime ideals, which is not the case for condition (iii). We say that a locally Noetherian scheme X is *excellent* if there exists an affine covering $\{U_i\}_i$ of X such that $\mathcal{O}_X(U_i)$ is excellent for every i (see also Exercise 2.18).

Example 2.36. The discrete valuation ring in Example 2.31 is not excellent. Indeed, $s \in \widehat{K} = \mathbb{F}_p((t))$, $s \notin K$, and s is inseparable over K.

Notation. Recall that for any locally Noetherian scheme X, $\mathrm{Reg}(X)$ denotes the set of regular points of X.

Lemma 2.37. *Let X be a locally Noetherian scheme. Let us suppose that for every integral closed subscheme Y of X, the set $\mathrm{Reg}(Y)$ contains a non-empty open subset. Then $\mathrm{Reg}(X)$ is open.*

Proof As the property is local, we can suppose X is Noetherian. Let $x \in \mathrm{Reg}(X)$ with $\dim \mathcal{O}_{X,x} = d$. Let us first show that there exists an open neighborhood U of x such that $U \cap \overline{\{x\}} \subseteq \mathrm{Reg}(X)$. As the maximal ideal $\mathfrak{m}_x \mathcal{O}_{X,x}$ is generated by d elements, by restricting X if necessary, we can suppose that X is affine and that the prime ideal $\mathfrak{p} \in \mathrm{Spec}\, \mathcal{O}_X(X)$ corresponding to the point x is generated by d elements. Let $Y = V(\mathfrak{p}) \subseteq X$. By hypothesis, there exists an open subscheme U of X such that $x \in U \cap Y \subseteq \mathrm{Reg}(Y)$. Let $y \in U \cap Y$, let $A = \mathcal{O}_{X,y}$, and $\mathfrak{q} = \mathfrak{p}A$. Then $A_\mathfrak{q} = \mathcal{O}_{X,x}$ is regular of dimension d, $\mathfrak{q}$ is generated by d elements, and $A/\mathfrak{q}$ is also regular. We immediately deduce from this that A itself is regular. Hence $U \cap \overline{\{x\}} = U \cap Y \subseteq \mathrm{Reg}(X)$.

Let F be the Zariski closure of $\mathrm{Sing}(X) := X \setminus \mathrm{Reg}(X)$. Let ξ be a generic point of F. Let us show that $\xi \in \mathrm{Sing}(X)$. Let us suppose the contrary. Then by the above, there exists an open neighborhood V of ξ in X such that $V \cap \overline{\{\xi\}} \subseteq \mathrm{Reg}(X)$. But $V \cap \overline{\{\xi\}}$ meets $\mathrm{Sing}(X)$, a contradiction. We therefore have $\xi \in \mathrm{Sing}(X)$. It follows from Theorem 4.2.16(a) that $\overline{\{\xi\}} \subseteq \mathrm{Sing}(X)$. Hence $\mathrm{Sing}(X) = F$ is closed. $\square$

Corollary 2.38. *Let S be a local Dedekind scheme of dimension 1. Let X be an integral scheme of dimension 2, flat and locally of finite type over S. Then $\mathrm{Reg}(X)$ is open (see also Exercise 2.17).*

Proof We can suppose X is Noetherian. The generic fiber X_η is an integral curve over a field. We therefore have that $\mathrm{Reg}(X_\eta)$ is a non-empty open set (Proposition 4.1.12 and Corollary 4.1.30). Now X_η is open in X, which implies that $\mathrm{Reg}(X)$ contains a non-empty open subset of X. Let Y be a proper closed subscheme of X. Then $\dim Y \le 1$. If $Y \to S$ is dominant, then $\mathrm{Reg}(Y)$ contains the open subset Y_η. If $Y \subseteq X_s$, then Y is an integral curve over a field, and hence $\mathrm{Reg}(Y)$ is open, as we have just seen for X_η. It follows from Lemma 2.37 that $\mathrm{Reg}(X)$ is open in X. □

Theorem 2.39. *We have the following properties concerning excellent schemes.*

(a) *Any complete, Noetherian, local ring (in particular, a field) is excellent.*

(b) *For a Noetherian local ring to be excellent, it suffices that it satisfy conditions (i) and (ii).*

(c) *Let X be an excellent locally Noetherian scheme. Then any scheme that is locally of finite type over X (in particular, any open or closed subscheme of X) is excellent.*

(d) *If X is excellent, then for any affine open subset U of X, $\mathcal{O}_X(U)$ is Nagata. In particular, if X is integral, the normalization morphism $X' \to X$ is finite.*

Proof See [65], 34A, pages 259–260. □

Corollary 2.40.

(a) *Any algebraic variety over a field is excellent.*

(b) *For a regular local ring A to be excellent, it suffices that $\mathrm{Frac}(\hat{A})$ be separable over $\mathrm{Frac}(A)$.*

(c) *Any Dedekind domain A of characteristic 0 is excellent.*

Proof (a) results from Theorem 2.39(a) and (c); (b) is a consequence of Theorem 2.39(b), Corollary 2.16 and [65], (33.C), Theorem 75.

(c) Conditions (i) and (ii) are verified by Corollary 2.16 and Example 2.34. Let us show that A verifies condition (iii). We are going to use Lemma 2.37 and (a). It suffices to show that for every integral scheme X of finite type over A, the set $\mathrm{Reg}(X)$ contains a non-empty open subset. If the image of $X \to \mathrm{Spec}\, A$ is reduced to a point, then X is an algebraic variety over a field, and $\mathrm{Reg}(X)$ is open and non-empty by (a). Let us suppose that $X \to \mathrm{Spec}\, A$ is dominant. Let $K = \mathrm{Frac}(A)$. The function field $K(X)$ is a finite (and *separable* because $\mathrm{char}(K) = 0$) extension of a purely transcendental extension $K(T_1, \ldots, T_d)$. Let B be the integral closure of $A[T_1, \ldots, T_d]$ in $K(X)$. Then $\mathrm{Spec}\, B \to \mathbb{A}_A^d$ is a finite morphism (Proposition 4.1.25), étale at the generic point. It follows from Corollary 4.4.12 that $\mathrm{Spec}\, B \to \mathbb{A}_A^d$ is étale over a non-empty open subset $U \subseteq \mathrm{Spec}\, B$. We have U is regular by virtue of Corollary 4.3.24. The birational map

$X \dashrightarrow U$ induces an isomorphism from a non-empty open subscheme of X to an open subscheme of U, which shows that $\mathrm{Reg}(X)$ contains a non-empty open subset. $\square$

Proposition 2.41. *Let A be an excellent, Noetherian, local ring. Let $\widehat{A}$ be its formal completion.*

(a) *The ring A is normal (resp. is reduced, resp. verifies (S_k), resp. is Cohen–Macaulay) if and only if the same holds for $\widehat{A}$.*

(b) *Let us suppose A is reduced. Then the integral closure A' of A in $\mathrm{Frac}(A)$ is finite over A, and $A' \otimes_A \widehat{A} \simeq (\widehat{A})'$, where the second term is the integral closure of $\widehat{A}$ in its total ring of fractions.*

(c) *With the notation of (b), the irreducible components of $\mathrm{Spec}\, \widehat{A}$ correspond canonically and bijectively to the closed points of $\mathrm{Spec}\, A'$.*

Proof   See [41], IV.7.8.3, (v)–(vii). $\square$

Example 2.42. Let k be a field of $\mathrm{char}(k) \neq 2$. Let A be the local ring of

$$X := \mathrm{Spec}\, k[x, y]/(y^2 - x^2(x + 1))$$

at the point $(0,0)$ (see Example 7.5.14). Then $A' = A[y/x]$ has two maximal ideals, and $\widehat{A} \simeq k[[u, v]]/(uv)$ (Exercise 1.3.10) indeed has two minimal prime ideals, as foreseen by Proposition 2.41(c).

Exercises

2.1. Let $A \subseteq B$ be two Dedekind domains of dimension 1 with $B \subseteq \mathrm{Frac}(A)$.

(a) Show that for any maximal ideal $\mathfrak{q}$ of B, $\mathfrak{p} := \mathfrak{q} \cap A$ is a maximal ideal of A and that $A_{\mathfrak{p}} = B_{\mathfrak{q}}$.

(b) Show that the morphism $\mathrm{Spec}\, B \to \mathrm{Spec}\, A$ induced by the inclusion $A \subseteq B$ is injective. Deduce from this that it is an open immersion if A is a semi-local ring.

(c) Let F be the image of $\mathrm{Spec}\, B \to \mathrm{Spec}\, A$. Show that $B = \cap_{\mathfrak{p} \in F} A_{\mathfrak{p}}$.

2.2. Let Z be a catenary locally Noetherian scheme. Show that if inequality (2.4) of Theorem 2.5 is an equality for every integral closed subscheme Y of Z, then Z is universally catenary.

2.3. Let X be a locally Noetherian integral scheme. Let $\pi : \widetilde{X} \to X$ be the blowing-up of X along a closed subscheme $Z \subset X$.

(a) Let us suppose that π is a finite morphism. For any generic point ξ of Z, show that there exists a generic point η of $\pi^{-1}(Z)$ which maps to ξ. Deduce from this that the irreducible components of Z are of codimension 1 in X.

(b) Show the converse of (a) under the hypothesis that $\dim X = 2$.

2.4. Let $(A, \mathfrak{m})$ be a Noetherian local ring, and let $t \in \mathfrak{m}$ be a regular element. Show that if A/tA is Cohen–Macaulay, then A is Cohen–Macaulay.

2.5. Let A be a Cohen–Macaulay Noetherian ring. Show that every polynomial ring $A[T_1, \ldots, T_n]$ is Cohen–Macaulay. Deduce from this that A is universally catenary.

2.6. Let $(A, \mathfrak{m})$ be a Noetherian local ring. Let M be a non-zero finitely generated A-module. Show that $\operatorname{depth}_A(M) = 0 \Longleftrightarrow \mathfrak{m} \in \operatorname{Ass}_A(M) \Longleftrightarrow$ there exists a submodule of M isomorphic to $A/\mathfrak{m}$.

2.7. Let M be a finitely generated module over a Noetherian ring A. Show that

$$\cup_{x \in M \setminus \{0\}} \operatorname{Ann}(Ax) = \cup_{\mathfrak{p} \in \operatorname{Ass}(M)} \mathfrak{p}.$$

2.8. Let X be a locally Noetherian scheme.

(a) Show that X is reduced if and only if it verifies the properties (S_1) and (R_0). The second property can be interpreted as saying that X is reduced at the generic points.

(b) Let A be a reduced Noetherian ring with minimal ideals $\mathfrak{p}_1, \ldots, \mathfrak{p}_n$. Show that the total ring of fractions $\operatorname{Frac}(A)$ of A (Definition 7.1.11) is equal to $\oplus_{1 \leq i \leq n} \operatorname{Frac}(A/\mathfrak{p}_i)$. Show that $\oplus_{1 \leq i \leq n} A/\mathfrak{p}_i$ is finite, hence integral, over A. Deduce from this that the *integral closure* A' of A in $\operatorname{Frac}(A)$ (i.e., the set of elements of $\operatorname{Frac}(A)$ that are integral over A) is equal to $\oplus_{1 \leq i \leq n} (A/\mathfrak{p}_i)'$, where $(A/\mathfrak{p}_i)'$ is the integral closure of $A/\mathfrak{p}_i$ in $\operatorname{Frac}(A/\mathfrak{p}_i)$.

(c) Show Serre's criterion (Theorem 2.23) without supposing X is integral.

2.9. Let $A \to B$ be a flat homomorphism of Noetherian local rings. Let M be a non-zero A-module. We are going to show that

$$\operatorname{depth}_B(B \otimes_A M) = \operatorname{depth}_A(M) + \operatorname{depth}(C), \qquad (2.7)$$

where C is the local ring $B/\mathfrak{m}_A B$.

(a) Let us suppose that $\operatorname{depth}_A(M) = \operatorname{depth}(C) = 0$. Show, using Exercise 2.6, that $\operatorname{depth}_B(B \otimes_A M) = 0$.

(b) Let us suppose that $\operatorname{depth}_A(M) \neq 0$. Let $x \in \mathfrak{m}_A$ be regular for M. Let us set $A' = A/xA$, $B' = B/xB$. Show that $\operatorname{depth}_{A'}(M/xM) = \operatorname{depth}_A(M) - 1$ and that $\operatorname{depth}_{B'}(B' \otimes_{A'} (M/xM)) = \operatorname{depth}_B(B \otimes_A M) - 1$.

(c) Let us suppose that $\operatorname{depth}(C) \neq 0$. Let $y \in \mathfrak{m}_B$ be an element whose image in C is a regular element. Then $B'' := B/yB$ is flat over A (Lemma 4.3.16). Show that $\operatorname{depth}_{B''}(B'' \otimes_B M) = \operatorname{depth}_B(B \otimes_A M) - 1$ and that $\operatorname{depth}(B''/\mathfrak{m}_A B'') = \operatorname{depth}(C) - 1$.

(d) Show equality (2.7) by induction on its right-hand side.

2.10. Let $A \to B$ be a flat homomorphism of Noetherian local rings. We suppose that B is regular of dimension ≤ 1. Show that there exists an $a \in \mathfrak{m}_A$ such that $\mathfrak{m}_A B = aB$. Deduce from this that $\mathfrak{m}_A = aA$ and that A is regular.
Remark. We can show that A is regular without the hypothesis that $\dim B \leq 1$ ([65], 21.D, Theorem 51 (i)).

2.11. Let $f : X \to Y$ be a surjective flat morphism of locally Noetherian schemes.

(a) Show that if X is normal, then Y is normal.

(b) Let us suppose Y is Cohen–Macaulay. Let $y \in Y$ and $d = \dim X_y$. Show that X_y verifies property (S_k) if and only if X verifies property (S_{d+k}) at every point of X_y. Give a new proof of Lemma 4.1.18.

(c) Show that X is Cohen–Macaulay if and only if Y is Cohen–Macaulay and if the fibers X_y are Cohen–Macaulay.

2.12. Let R be a Dedekind domain in which 2 is invertible. Let $P(x) \in R[x]$. Show that $\operatorname{Spec} R[x, y]/(y^2 - P(x))$ is normal if and only if $P(x)$ is separable (i.e., without multiple root) and if the ideal generated by the coefficients of $P(x)$ is radical.

2.13. Let $(A, \mathfrak{m})$ be a regular, Noetherian, local ring of dimension d, and let M be a finitely generated Cohen–Macaulay A-module such that $\operatorname{Supp} M = \operatorname{Spec} A$. We are going to show that M is free.

(a) Let $k = A/\mathfrak{m}$. Show that there exists an exact sequence

$$0 \to N \to L \to M \to 0$$

of A-modules such that L is free of finite rank and that $L \otimes_A k \to M \otimes_A k$ is an isomorphism (lift a basis of $M \otimes_A k$ over k).

(b) Let $t \in \mathfrak{m}$ be non-zero. Show that t is M-regular (use Exercise 2.7 and Proposition 2.13). Deduce from this that the sequence

$$0 \to N/tN \to L/tL \to M/tM \to 0$$

is exact. Show that M/tM is a Cohen–Macaulay A/tA-module, and that $\operatorname{Supp} M/tM = \operatorname{Spec}(A/tA)$.

(c) Show that $0 \to N \otimes_A k \to L \otimes_A k \to M \otimes_A k \to 0$ is exact. Deduce from this that $N = 0$ and that M is free.

(d) If M is moreover an A-algebra, show that 1 is part of a basis of M (as an A-module).

2.14. Let X be a Noetherian scheme, and let $\pi : \mathbb{A}^1_X \to X$ be the structural morphism.

(a) Show that π admits a section. Deduce from this that π^* is injective.

(b) Let us suppose X is integral. Let $\mathcal{L} \in \operatorname{Pic}(\mathbb{A}^1_X)$. Show that there exists a Cartier divisor D such that $\mathcal{L} \simeq \mathcal{O}(D)$, and that $\operatorname{Supp} D = \pi^{-1}(E)$, where E is a closed subset of X of codimension 1.

(c) Let us suppose X is regular. Show that $\pi^* : \text{Pic}(X) \to \text{Pic}(\mathbb{A}^1_X)$ is an isomorphism. (See also [43], Proposition II.6.6.)

2.15. (*Zariski's purity theorem*) Let Y be a regular locally Noetherian scheme. Let $f : X \to Y$ be a finite surjective morphism.

(a) Let us suppose X is Cohen–Macaulay. Show that for any $y \in Y$, $(f_*\mathcal{O}_X)_y$ is a Cohen–Macaulay $\mathcal{O}_{Y,y}$-module with support $\text{Spec}\,\mathcal{O}_{Y,y}$. Deduce from this that f is flat (use Exercise 2.13).

(b) Show that if $\dim Y = 2$ and X is normal, then f is flat.

(c) Let U be the largest open subscheme of X such that $f|_U : U \to Y$ is étale. We call the closed part $B_f := f(X \setminus U)$ of Y the *branch locus* of f. In the preceding cases, show that if f is generically separable, then either the irreducible components of B_f are of codimension 1 (use Exercise 6.4.7), or $B_f = \emptyset$.

2.16. Let $\mathcal{O}_K$ be a principal ideal domain with field of fractions K. Let $L/K(T)$ be a separable extension of degree 2 and B the integral closure of $A[T]$ in L. Show that $B/A[T]$ is a locally free $A[T]$-module of rank 1 over $A[T]$. Deduce from this that $B = A[T] \oplus b_0 A[T]$ for some $b_0 \in B$ (use Exercise 2.14).

2.17. Let X be a scheme that is locally of finite type over a discrete valuation ring. Show that $\text{Reg}(X)$ is open (use Corollary 2.40(a)).

2.18. Let X be an excellent locally Noetherian scheme. Show that for any affine open subset U of X, $\mathcal{O}_X(U)$ is excellent.

8.3 Fibered surfaces

In this section, we approach the heart of this book, namely relative curves over a Dedekind scheme. We start by studying some elementary properties of the fibers of such a curve. Next, we show that a surface X is essentially determined by its points of codimension 1 (Corollary 3.23). Theorem 3.26 determines the valuations of $K(X)$ with center of codimension 1 in a surface that is birational to X. In Subsection 8.3.3, we explain the process of contraction, which is the inverse of blowing-up. The last subsection is devoted to the statement of the desingularization theorem (Theorem 3.50).

8.3.1 Properties of the fibers

Definition 3.1. Let S be a Dedekind scheme (Definition 4.1.2). We call an integral, projective, flat S-scheme $\pi : X \to S$ of dimension 2 a *fibered surface* over S. The generic point of S will be denoted by η. We call X_η the *generic fiber* of X. A fiber X_s with $s \in S$ closed is called a *closed fiber*. When $\dim S = 1$, X is also called a *projective flat S-curve* (see Lemma 3.3). Note that the flatness of π is equivalent to the surjectivity of π. We will say that X is a *normal* (resp. *regular*) *fibered surface* if X is normal (resp. regular).

A *morphism* (resp. a *rational map*) between fibered surfaces is a morphism (resp. a rational map) that is compatible with the structure of S-schemes.

We can distinguish between two types of fibered surfaces. If $\dim S = 0$, then X is an integral, projective, algebraic surface over a field. It is an 'absolute' surface. We say that this is the 'geometric' case. If $\dim S = 1$, then X is a 'relative curve' over S (see Lemma 3.3). This is the 'arithmetic' case. The terminology of fibered surface really is of interest only when $\dim S = 1$. In this book, we also include the $\dim S = 0$ case when the geometric case is similar to arithmetic case.

Example 3.2. Let $S = \operatorname{Spec} \mathbb{Z}$ and $X = \operatorname{Proj} \mathbb{Z}[x, y, z]/(y^2 z + yz^2 - x^3 + xz^2)$. Let us first show that X is a normal fibered surface. The only point needing verification is that X is normal. First of all, the Jacobian criterion (Theorem 4.2.19) shows that the fibers of $X \to S$ are smooth except at the point corresponding to the prime number $p = 37$. We verify that X_{37} is reduced, and hence X is normal (Lemma 4.1.18). More precisely, the fiber X_{37} is a singular integral curve over $\mathbb{F}_{37}$ with an ordinary double point (Example 7.5.14). See Figure 9.

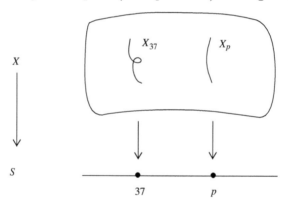

Figure 9. Example 3.2.

In this example, we see that for any prime number p, the fiber X_p is a curve. This is the case in general:

Lemma 3.3. *Let S be a Dedekind scheme of dimension 1, with generic point η. Let $X \to S$ be a fibered (resp. normal fibered) surface. Then X_η is an integral (resp. normal) curve over $K(S)$. For any $s \in S$, X_s is a projective curve over $k(s)$.*

Proof Let ξ be the generic point of X. As $X \to S$ is dominant, $\xi \in X_\eta$. Hence ξ is the generic point of X_η, which shows that X_η is irreducible. In addition, for any $x \in X_\eta$, we have $\mathcal{O}_{X,x} \simeq \mathcal{O}_{X_\eta,x}$ (Exercise 3.7); hence X_η is normal if X is normal. We know that for any closed point $x \in X_s$, we have $\dim \mathcal{O}_{X_s,x} = \dim X_\eta$ (Proposition 4.4.16). It therefore remains to show that $\dim X_\eta = 1$. Let $x \in X$ be a point such that $\dim \mathcal{O}_{X,x} = 2$. Then x is closed and therefore belongs to a closed fiber X_s. It follows that $\dim \mathcal{O}_{X_s,x} = \dim \mathcal{O}_{X,x} - \dim \mathcal{O}_{S,s} = 1$ (Theorem 4.3.12). This proves that $\dim X_\eta = 1$.   □

Let us now study the closed subsets of X.

Proposition 3.4. *Let $\pi : X \to S$ be a fibered surface over a Dedekind scheme of dimension 1.*

(a) *Let x be a closed point of the generic fiber X_η. Then $\overline{\{x\}}$ is an irreducible closed subset of X, finite and surjective to S.*

(b) *Let D be an irreducible closed subset of X. If $\dim D = 1$, then either D is an irreducible component of a closed fiber, or $D = \overline{\{x\}}$, where x is a closed point of X_η.*

(c) *Let x_0 be a closed point of X. Then $\dim \mathcal{O}_{X,x_0} = 2$.*

Proof (a) Let $D = \overline{\{x\}}$. Then D is irreducible because it is the closure of an irreducible subset. Moreover, $\pi(D)$ is a closed subset of S containing the generic point, and hence $\pi(D) = S$. As x is closed in X_η, we have $D \neq X$, and therefore $\dim D \leq 1$. For any $s \in S$, we know that $\dim X_s = 1$ (Lemma 3.3). If $\dim(D \cap X_s) \neq 0$, D would contain an irreducible component of X_s, and would therefore be equal to it, which is impossible. Hence $D \cap X_s$ is finite. It follows that the morphism $\pi|_D : D \to S$ is projective and quasi-finite. It is therefore a finite morphism (Corollary 4.4.7).

(b) The image $\pi(D)$ of D is an irreducible closed subset of S. If it is reduced to a closed point s, then $D \subseteq X_s$. As $\dim D = \dim X_s$ by Lemma 3.3, D is an irreducible component of X_s. If $\pi(D)$ is not reduced to a point, we necessarily have $\pi(D) = S$. Hence D contains a point $x \in X_\eta$. Now $\dim \overline{\{x\}} = 1$ by (a), and therefore we have $D = \overline{\{x\}}$.

(c) As π is proper, $\pi(x_0)$ is a closed point s of S. On the other hand, by Lemma 3.3, $\dim \mathcal{O}_{X_s,x_0} = 1$. It follows that

$$\dim \mathcal{O}_{X,x_0} = \dim \mathcal{O}_{X_s,x_0} + \dim \mathcal{O}_{S,s} = 2$$

(Theorem 4.3.12). □

Definition 3.5. Let $\pi : X \to S$ be a fibered surface over a Dedekind scheme S. Let D be an irreducible Weil divisor (Definition 7.2.4). We say that D is *horizontal* if $\dim S = 1$ and if $\pi|_D : D \to S$ is surjective (hence finite). If $\pi(D)$ is

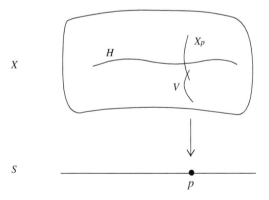

Figure 10. Horizontal divisor H and vertical divisor V.

reduced to a point, we say that D is *vertical*. More generally, an arbitrary Weil divisor will be called *horizontal* (resp. *vertical*) if its components are horizontal (resp. vertical). We will say that a Cartier divisor is *horizontal* (resp. *vertical*) if the associated Weil divisor $[D]$ (Section 7.2) is horizontal (resp. vertical).

By Proposition 3.4, an irreducible horizontal divisor is simply the closure in X of a closed point of X_η.

Corollary 3.6. *Let $\pi : X \to S$ be a fibered surface over a Dedekind scheme S of dimension 1. Let $s \in S$. Then the following properties are true.*

(a) *The fiber X_s is a projective curve over $k(s)$, and we have the equality of arithmetic genera $p_a(X_s) = p_a(X_\eta)$.*

(b) *If X_η is geometrically connected (e.g., if $\mathcal{O}_S \simeq \pi_* \mathcal{O}_X$, see Corollary 5.3.17), then the same holds for X_s.*

(c) *If X_η is geometrically integral, then the canonical homomorphism $\mathcal{O}_S \to \pi_* \mathcal{O}_X$ is an isomorphism.*

(d) *Let us suppose that X is a regular scheme. Then the morphisms $X \to S$ and $X_s \to \operatorname{Spec} k(s)$ are local complete intersections, and we have the relation $\omega_{X_s/k(s)} = \omega_{X/S}|_{X_s}$ between the dualizing sheaves.*

Proof   (a) is contained in Lemma 3.3 and Proposition 5.3.28.

(b) Let $s \in S$ be a closed point. Localizing at s if necessary, we may suppose that $S = \operatorname{Spec} A$, where A is local and s is the closed point of S. Let us first show that X_s is connected. Let $B = \mathcal{O}_X(X)$ and $L = \mathcal{O}_X(X_\eta)$. Then X is canonically a projective scheme over B. It follows from Corollary 5.3.17 that the fibers of $X \to \operatorname{Spec} B$ are geometrically connected. As X_η is geometrically connected, L is a finite purely inseparable extension of $K(S)$ (Corollary 3.3.21). It follows that $\operatorname{Spec} B \to \operatorname{Spec} A$ is bijective (Proposition 3.2.7). Consequently X_s is connected.

Let k' be a finite simple extension of $k(s)$. There exists a discrete valuation ring A' that is finite over A, with residue field k'. Then $X_{A'}$ is flat and projective over A', with geometrically connected generic fiber. It follows from the above that $X_{k'}$ is connected. As every finite extension of $k(s)$ decomposes into a sequence of simple extensions, we see that X_s is geometrically connected. See also Exercise 5.3.9.

To show (c), we can suppose S is affine. We must show that $\mathcal{O}_X(X) = \mathcal{O}_S(S)$. As $\mathcal{O}_X(X)$ is integral over $\mathcal{O}_S(S)$ (Proposition 3.3.18) and is contained in $\mathcal{O}_X(X_\eta)$, and $\mathcal{O}_X(X_\eta) = K(S)$ since X_η is geometrically integral (Corollary 3.3.21), we indeed have the desired equality. Finally, (d) follows from Example 6.3.18 and Theorem 6.4.9(b).   □

Remark 3.7. In the situation of Corollary 3.6(c), we have $H^0(X_\eta, \mathcal{O}_{X_\eta}) = K(S)$. But in general, $H^0(X_s, \mathcal{O}_{X_s}) \neq k(s)$. This is a problem of cohomological flatness (see Theorem 5.3.20 and Exercise 5.3.14).

Let $X \to S$ be as in Corollary 3.6. If its generic fiber is not geometrically connected, we can reduce to this by suitably replacing S:

Proposition 3.8. *Let* $\pi : X \to S$ *be a normal fibered surface and* η *the generic point of* S. *Let* $\rho : S' \to S$ *be the normalization of* S *in* $H^0(X_\eta, \mathcal{O}_{X_\eta})$. *Then we have the following properties.*

(a) *The morphism* ρ *is finite and flat,* $X \to S$ *factors into* $\pi' : X \to S'$ *followed by* ρ, *and* π' *makes* X *into a normal fibered surface over* S', *with generic fiber equal to* X_η.

(b) *The morphism* $\pi' : X \to S'$ *has geometrically connected fibers.*

(c) *Let* $s \in S$ *be such that* $\dim \mathcal{O}_{S,s} = 1$, *and* $s_1, \ldots, s_n$ *be the points of* $\rho^{-1}(s)$. *Then* X_s *is the disjoint union of* X_i, $1 \le i \le n$, *where* X_i *is a scheme of finite type over* S' *such that* X_{s_i} *is a closed subscheme of* X_i *defined by a sheaf of nilpotent ideals.*

Proof (a) The sheaf $\pi_*\mathcal{O}_X$ is a coherent sheaf of algebras on S. Let $\rho :$ $S' \to S$ be the S-scheme $\operatorname{Spec} \pi_*\mathcal{O}_X \to S$ (Exercise 5.1.17). Then π factors into $X \to S' \to S$. As X is normal, so is S'. We have $\rho_*\mathcal{O}_{S'} = \pi_*\mathcal{O}_X$ coherent and torsion-free (hence flat) over $\mathcal{O}_S$, which means that ρ is finite and flat. As $K(S') = H^0(X_\eta, \mathcal{O}_{X_\eta})$ and S' is normal and finite over S, it is the normalization of S in $H^0(X_\eta, \mathcal{O}_{X_\eta})$ (Definition 4.1.24). Finally, $X \to S'$ is projective (Corollary 3.3.32(e)) and clearly flat. Therefore $X \to S'$ is a normal fibered surface.

Let $s \in S$. Then we have

$$X_s = X \times_S \operatorname{Spec} k(s) = X \times_{S'} (S' \times_S \operatorname{Spec} k(s)). \qquad (3.8)$$

As $S' \times_S \operatorname{Spec} K(S) = \operatorname{Spec} K(S')$, taking for s the generic point of S, we obtain

$$X \times_{S'} \operatorname{Spec} K(S') = X \times_S \operatorname{Spec} K(S) = X_\eta,$$

which proves (a). Let s be as in (c). Then $S' \times_S \operatorname{Spec} k(s)$ is the disjoint union of the schemes $\operatorname{Spec}(\mathcal{O}_{S',s_i}/(t))$, where t is a uniformizing parameter for $\mathcal{O}_{S,s}$. Equality (3.8) implies that X_s is the disjoint union of the $X_i :=$ $X \times_{S'} \operatorname{Spec}(\mathcal{O}_{S',s_i}/(t))$, $1 \le i \le n$. It is clear that X_{s_i} is a closed subscheme of X_i and that they have the same underlying topological space, whence (c). Finally, (b) is simply a consequence of Corollary 5.3.17. $\qquad \square$

Lemma 3.9. *Let* $\pi : X \to S$ *be a normal fibered surface with* $\dim S = 1$. *Let us fix a closed point* $s \in S$. *Let* $\Gamma_1, \ldots, \Gamma_r$ *be the irreducible components of* X_s. *Then the following properties are true.*

(a) *We have the following equality of Weil divisors in* X:

$$X_s = \sum_{1 \le i \le r} d_i \Gamma_i,$$

where d_i *is the multiplicity of* Γ_i *in* X_s (*Definition 7.5.6*). *If* X *is regular, we have the equality of Cartier divisors* $X_s = \pi^*s$.

(b) *Let $t \in \mathcal{O}_{S,s}$ be a uniformizing parameter, and let ν_i be the normalized valuation of $K(X)$ associated to the discrete valuation ring $\mathcal{O}_{X,\xi_i}$, where ξ_i is the generic point of Γ_i. Then $d_i = \nu_i(t)$.*

Proof (a) The first equality follows from the definition of the 1-cocycle X_s (Definition 7.2.12). The second equality is obvious. (b) Let t_i be a uniformizing parameter for $\mathcal{O}_{X,\xi_i}$. Then $t = t_i^{e_i} u_i$ with $e_i = \nu_{\Gamma_i}(t)$ and $u_i \in \mathcal{O}_{X,\xi_i}^*$. We have

$$d_i = \text{length}\, \mathcal{O}_{X_s,\xi_i} = \text{length}\, \mathcal{O}_{X,\xi_i}/(t) = \text{length}\, \mathcal{O}_{X,\xi_i}/(t_i^{e_i}) = e_i.$$

This completes the proof.   □

Example 3.10. Let us fix integers $n, q > 0$. Let Y be the scheme $\text{Proj}\, \mathbb{Z}[u, v]$, with function field $\mathbb{Q}(u/v)$. Let X be the normalization of Y in the extension $\mathbb{Q}(u/v)[w]$, $w^2 = n(1 + (u/v)^q)$. Let us suppose n is without a square factor. Then it is easy to see that X is the union of the affine open subschemes

$$\text{Spec}\, \mathbb{Z}[u/v, w]/(w^2 - n(1 + (u/v)^q)), \quad \text{Spec}\, \mathbb{Z}[v/u, z]/(z^2 - n(v/u)^\alpha(1 + (v/u)^q)),$$

where $z = w(u/v)^{-[(q+1)/2]}$ and $\alpha = q - 2[q/2]$. Indeed, let Z be the union of these two schemes. It suffices to show that $Z_{\mathbb{Q}}$ is normal, and that every closed fiber is irreducible and contains a regular (hence normal) point of Z, which implies that Z is normal at the points of codimension 1. We then apply Serre's criterion (Corollary 2.24) to show that Z is normal. For any prime number $p > 2$, we can verify that the fiber X_p (which is irreducible) is of multiplicity equal to 2 if $p \mid n$, and to 1 otherwise. In $p = 2$, the multiplicity is 2 if $2 \mid n$ or if q is even, and 1 otherwise.

Proposition 3.11. *Let $\pi : X \to S$ be a fibered surface over a Dedekind scheme S. We suppose that the generic fiber X_η is smooth. Then there exists a non-empty open subset V of S such that $\pi^{-1}(V) \to V$ is smooth. In other words, X_s is smooth over $k(s)$ except maybe for a finite number of s.*

Proof The set X_{sm} of points where π is smooth is open (Corollary 6.2.12). Consequently, $X \setminus X_{\text{sm}}$, and therefore $\pi(X \setminus X_{\text{sm}})$, is a closed subset. Let

$$V = S \setminus \pi(X \setminus X_{\text{sm}}).$$

This is a non-empty (since it contains η) open subset of S. Its complement in S is finite because $\dim S \leq 1$. By construction, $\pi^{-1}(V) \to V$ is smooth.   □

Example 3.12. Let $P(x) = x^3 + a_1 x^2 + a_2 x + a_3 \in \mathbb{Z}[x]$ be an irreducible separable polynomial over $\mathbb{Q}$. Let $X = \text{Proj}\, \mathbb{Z}[x, y, z]/(y^2 z + x^3 + a_1 x^2 z + a_2 x z^2 + a_3 z^3)$. Then $X \to \text{Spec}\, \mathbb{Z}$ is smooth over the open subset $\text{Spec}\, \mathbb{Z} \setminus V(2\Delta)$ of $\text{Spec}\, \mathbb{Z}$, where $\Delta \in \mathbb{Z}$ is the discriminant of $P(x)$. This can be seen using the Jacobian criterion (4.2.19).

Remark 3.13. Let $S = \text{Spec}\, \mathbb{Z}$. Let $X \to S$ be as in Corollary 3.6. Then $X \to S$ is never smooth if $g(X_\eta) \neq 0$. That is to say, there always exists a point $s \in S$ such that X_s is not smooth (J.-M. Fontaine [35]).

Definition 3.14. We will call a regular fibered surface $X \to S$ over a Dedekind scheme S of dimension 1 an *arithmetic surface*.

Example 3.15. Let $n \in \mathbb{N} \setminus \{0\}$ be without a square factor. Let

$$X = \operatorname{Proj} \mathbb{Z}[x, y, z]/(xy - nz^2).$$

Using Corollary 4.2.12, we easily verify that X is regular. It is therefore an arithmetic surface over $\mathbb{Z}$. Its generic fiber is isomorphic to $\mathbb{P}^1_{\mathbb{Q}}$. Let p be a prime number. The fiber X_p of X over the point $p\mathbb{Z}$ is given by

$$X_p = \operatorname{Proj} \mathbb{F}_p[x, y, z]/(xy - \overline{n}z^2),$$

where $\overline{n}$ is the image of n in $\mathbb{F}_p$. We see that X_p is smooth (even isomorphic to $\mathbb{P}^1_{\mathbb{F}_p}$) if p does not divide n. Otherwise X_p is reduced and is the union of two projective lines meeting at a unique point.

Theorem 3.16 (Lichtenbaum [56], Theorem 2.8). *Let $S = \operatorname{Spec} A$ be an affine Dedekind scheme. Let $\pi : X \to S$ be a proper flat morphism with fibers of dimension 1. Let us suppose that X is regular. Then π is projective.*

Proof (See also Remark 9.3.5.) We can suppose that $\dim S = 1$ (Exercise 4.1.16) and X is connected. Let us first recall that the X regular hypothesis implies that every Weil divisor is a Cartier divisor (Proposition 7.2.16). The proof of the theorem consists of finding an effective horizontal Cartier divisor D which meets all of the irreducible components of all of the fibers X_s, $s \in S$.

Let D_0 be a non-zero, effective, horizontal Cartier divisor (such a divisor exists, it suffices to take the closure of a closed point of X_η). We know that X_η is projective over $K(S)$ (Exercise 4.1.16) and $\mathcal{O}_X(D_0)|_{X_\eta}$ is ample (Proposition 7.5.5). By Proposition 5.1.37, there is a non-empty open subset U of S such that $\mathcal{O}_X(D_0)|_{\pi^{-1}(U)}$ is ample. Consequently, for any closed point $s \in U$, $\mathcal{O}_X(D_0)|_{X_s}$ is ample and D_0 meets all of the irreducible components of the fibers X_s (Proposition 7.5.5). Let $s_1, \dots, s_r$ be the points of $S \setminus U$. For each s_i, there exists a horizontal effective Weil (hence Cartier) divisor D_i on X which meets all of the irreducible components of X_{s_i} (see Lemma 3.35). Let us set $D = D_0 + D_1 + \cdots + D_n$. Then D is an effective Cartier divisor which meets all of the irreducible components of the fibers X_s, $s \in S$. Since X_s is a projective curve (Exercise 7.5.4), $\mathcal{O}_X(D)|_{X_s}$ is ample (Proposition 7.5.5). It follows from Corollary 5.3.24 that $\mathcal{O}_X(D)$ is ample. So $X \to \operatorname{Spec} A$ is projective. $\square$

8.3.2 Valuations and birational classes of fibered surfaces

We study the points of codimension 1 in a fibered surface X and their relation to the valuations of the function field $K(X)$. Let S be a scheme, X a proper integral scheme over S. For any non-trivial valuation ν of the field $K(X)$ (Definition 3.3.22), we let $\mathcal{O}_\nu$ denote the valuation ring of ν. Then $\operatorname{Spec} \mathcal{O}_\nu$ is made up of two points: the generic point ξ and the closed point. Hence $\{\xi\}$ is an open subset of $\operatorname{Spec}\mathcal{O}_\nu$. Moreover, $K(X) = \operatorname{Frac}(\mathcal{O}_\nu) = k(\xi)$ is the residue field of the generic point of X. The composition of the canonical morphisms $\operatorname{Spec} K(X) \to X$, $X \to S$ therefore induces a rational map $\operatorname{Spec}\mathcal{O}_\nu \dashrightarrow S$.

Definition 3.17. Let $X \to S$ be as above. We say that a valuation ν of $K(X)$ is an *S-valuation* if the rational map $\operatorname{Spec}\mathcal{O}_\nu \dashrightarrow S$ is a morphism. This comes down to saying that there exists an $s \in S$ such that $\mathcal{O}_\nu$ dominates $\mathcal{O}_{S,s}$ (that is to say that, by identifying $K(S)$ to a subfield of $K(X)$, we have $\mathcal{O}_{S,s} \subseteq \mathcal{O}_\nu$ and $\mathfrak{m}_s \subseteq \mathfrak{m}_\nu$). If $\mathcal{O}_{S,s}$ is a discrete valuation ring, then ν is a valuation that extends that of $\mathcal{O}_{S,s}$. If s is the generic point of S, then ν is constant on $K(S)^*$.

We will say that ν has *center* $x \in X$ if there exists a morphism of S-schemes $f : \operatorname{Spec}\mathcal{O}_\nu \to \operatorname{Spec}\mathcal{O}_{X,x}$ which makes the following diagram commutative

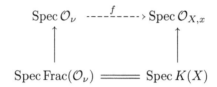

By the valuative criterion of properness (Theorem 3.3.25), if $X \to S$ is proper, then every S-valuation has a center in X. Finally, if $S = \operatorname{Spec}\mathbb{Z}$, we will not specify the base scheme.

The S-valuations of $K(X)$ play an important role in the algebraic geometry of the Zariski school. In some sense, the set of these valuations corresponds to a 'maximal' model of $K(X)$ over S ([97], VI, Section 17). In this book, we will essentially be interested in discrete valuations. This will suffice most of the time.

Definition 3.18. Let X be an integral scheme. We will say that a valuation ν of $K(X)$ is *of the first kind in X* if it admits a center in X that is a normal point of codimension 1. A valuation of the first kind is necessarily discrete.

Remark 3.19. Let us suppose X is normal. For any closed irreducible subset $\Gamma \subset X$ of codimension 1, the local ring $\mathcal{O}_{X,\xi}$ at the generic point ξ of Γ is a discrete valuation ring, and the corresponding (normalized) valuation, denoted by ν_Γ or ν_ξ, is of the first kind in X because its center in X is ξ. Conversely, if ν is of the first kind in X, then the closure $\Gamma = \overline{\{\xi\}}$ of the center ξ of ν is a closed irreducible subset of codimension 1. Thus, we see that there is a canonical bijective correspondence between the S-valuations of $K(X)$ of the first kind in X and the irreducible Weil divisors on X.

Let us suppose that S is a Dedekind scheme of dimension 1 and that $X \to S$ is a normal fibered surface. For any irreducible vertical divisor $\Gamma \subseteq X_s$, the associated valuation ν_Γ dominates the discrete valuation ring $\mathcal{O}_{S,s}$. On the other hand, if Γ is horizontal, ν_Γ is constant on $K(S)^*$.

Theorem 3.20. *Let S be a Dedekind scheme. Let $f : X \dashrightarrow Y$ be a birational map of normal fibered surfaces over S. Let us suppose that for every $\xi \in Y$ of codimension 1, the center of ν_ξ in X is of codimension 1. Then f is a morphism.*

Proof   Let Z be the graph of f. We must show that the projection $p_1 : Z \to X$ is an isomorphism. As Z is a closed subscheme of $X \times_S Y$, it is proper over S. Hence p_1 is a proper birational morphism. Let $x \in X$ be such that p_1 is not an isomorphism above x. By virtue of Proposition 4.4.3(b), x is a point of codimension ≥ 2, and therefore closed. Moreover, Z_x has no isolated points, and hence $\dim Z_x = 1$.

Let $s \in S$ be the image of x in S, z a generic point of Z_x, and ξ the image of z under the projection $p_2 : Z \to Y$. As Z_x is contained in $\{x\} \times_S Y = \mathrm{Spec}\, k(x) \times_{\mathrm{Spec}\, k(s)} Y_s$, which is finite over Y_s, $\overline{\{\xi\}}$ is of dimension 1 (and therefore of codimension 1 in Y). Applying Proposition 4.4.3(b) to p_2, we see that $p_2 : Z \to Y$ is an isomorphism above a neighborhood of ξ; hence $\mathcal{O}_{Z,z} = \mathcal{O}_{Y,\xi}$ dominate $\mathcal{O}_{X,x}$. This means that ν_ξ has center x in X, which contradicts the hypothesis of the theorem since x is of codimension 2.   □

Definition 3.21.   Let $f : X \dashrightarrow Y$ be a birational map of normal fibered surface over S. Let $Z \subset X \times_S Y$ be the graph of f, and $p_1 : Z \to X$, $p_2 : Z \to Y$ the projections. For any closed point $x \in X$, we call the closed subset $p_2(p_1^{-1}(x))$ of Y the *total transform of x by f*. If f is defined at x, then the total transform of x is none other than the point $f(x)$. In the general case, we still let $f(x)$ denote the total transform of x by f.

With this definition, Theorem 3.20 can also be stated as follows: f is defined everywhere if for every closed point $x \in X$, the total transform of x is a closed point of Y. The proof of Theorem 3.20 in fact shows the following, more precise, result.

Proposition 3.22.   *Let $f : X \dashrightarrow Y$ be a birational map as in Theorem 3.20. Let us suppose that f is not defined at a point $x \in X$. Then x is closed, and the total transform $f(x)$ is connected, of dimension 1. Moreover, if ξ is the generic point of $f(x)$, then f^{-1} is defined at ξ and we have $f^{-1}(\xi) = x$.*

Corollary 3.23.   *Let $f : X \dashrightarrow Y$ be a birational map of normal fibered surfaces over S. We identify $K(X)$ with $K(Y)$ via f. Then f induces an isomorphism if and only if every discrete valuation of the first kind in X is also of the first kind in Y, and vice versa.*

Remark 3.24.   This corollary is false in dimension ≥ 3. See [69], III.9, p. 291, Example 0.

Remark 3.25.   Let X, Y be two fibered surfaces over S with $\dim S = 1$. We suppose that $K(X) = K(Y)$. We therefore have a birational map $f : X \dashrightarrow Y$. As X_η and Y_η are normal projective curves over $K(S)$, f is defined at every point of X_η (Proposition 4.1.16). Let $s \in S$ be a closed point. Then f is defined at the generic points of X_s. By virtue of the corollary above, f induces an isomorphism if and only if f establishes a bijection between the generic points of X_s and those of Y_s for every $s \in S$; that is to say, if X_s and Y_s have, to some extent, birationally the same irreducible components.

Theorem 3.26.   *Let X be a Noetherian integral scheme. Let ν be a non-trivial valuation of $K(X)$, with residue field $k(\nu)$. We suppose that ν has a center $x \in X$.*

(a) *We have*

$$\text{trdeg}_{k(x)} k(\nu) \leq \dim \mathcal{O}_{X,x} - 1. \tag{3.9}$$

(b) *(Zariski) Let us suppose that X is universally catenary and Nagata (Definitions 2.1 and 2.30). Then (3.9) is an equality if and only if there exists a proper birational morphism $f : Y \to X$ such that ν is a valuation of the first kind in Y (see also Exercise 3.14).*

Proof Let $f_1, \ldots, f_n \in \mathcal{O}_\nu$ be such that their images in $k(\nu)$ form an algebraically independent family over $k(x)$. Let us show the following assertion:

(α) *There exists a blowing-up morphism $X' \to X$ such that the center x' of ν in X' verifies $\text{trdeg}_{k(x)} k(x') \geq n$.*

Let $a_0, a_1, \ldots, a_n \in \mathfrak{m}_x \mathcal{O}_{X,x}$ be such that $f_i = a_i/a_0$ for every $1 \leq i \leq n$. Let U be an affine open neighborhood of x such that $a_i \in \mathcal{O}_X(U)$ for every $i \geq 0$. Let $U' \to U$ be the blowing-up of U along the closed subscheme $V(I)$, where $I = (a_0, a_1, \ldots, a_n)$. As $U' \to U$ is proper, ν admits a center $x' \in U'$ (Corollary 3.3.26). The ideal $I\mathcal{O}_{U',x'}$ is principal, and hence generated by an a_r. If $r \neq 0$, we have $a_0/a_r \in \mathcal{O}_{U',x'} \cap \mathcal{O}_\nu^* = \mathcal{O}_{U',x'}^*$. Hence $I\mathcal{O}_{U',x'}$ is still generated by a_0. It follows that $f_i = a_i/a_0 \in \mathcal{O}_{U',x'}$. In particular, $\text{trdeg}_{k(x)} k(x') \geq n$. It now suffices to show that $U' \to U$ extends to a blowing-up of X. Let us consider the scheme-theoretic closure Z of $V(I)$ in X (Exercise 2.3.17(e)). Let $\pi : X' \to X$ be the blowing-up of X along Z. Then $U' \to U$ coincides with $\pi^{-1}(U) \to U$.

Let us show (a). Let n be an integer $\leq \text{trdeg}_{k(x)} k(\nu)$. Let $X' \to X$ be as in (α). By Theorem 2.5, we have

$$\dim \mathcal{O}_{X',x'} + \text{trdeg}_{k(x)} k(x') \leq \dim \mathcal{O}_{X,x}.$$

As x' is not the generic point of X' (because ν would otherwise be the trivial valuation), we have $n \leq \dim \mathcal{O}_{X,x} - 1$, which proves (a).

(b) Let us suppose $\text{trdeg}_{k(x)} k(\nu) = \dim \mathcal{O}_{X,x} - 1$. Let $f_1, \ldots, f_n \in \mathcal{O}_\nu$ be such that their images in $k(\nu)$ form a transcendence basis of $k(\nu)$ over $k(x)$. Let $X' \to X$ be as in (α). Let $Y \to X'$ be the normalization. This is a finite morphism because X' is Nagata (Proposition 2.29(b)). Hence ν admits a center $y \in Y$ which is a point lying above x'. Theorem 2.5 implies that

$$\dim \mathcal{O}_{Y,y} + \text{trdeg}_{k(x)} k(y) \leq \dim \mathcal{O}_{X,x} = 1 + \text{trdeg}_{k(x)} k(\nu)$$
$$= 1 + \text{trdeg}_{k(x)} k(y).$$

Consequently, $\dim \mathcal{O}_{Y,y} = 1$, and it is therefore a discrete valuation ring (Proposition 4.1.12), whence $\mathcal{O}_\nu = \mathcal{O}_{Y,y}$ (Lemma 3.3.24). This means that ν is of the first kind in Y. The converse can be deduced in the same way with Theorem 2.5 and the hypothesis that X is universally catenary. $\square$

8.3.3   Contraction

Let $X \to S$ be a normal fibered surface over a Dedekind scheme of dimension 1. Let $E_1, \ldots, E_n$ be irreducible vertical divisors on X. The aim of this subsection is to construct, when that is possible, a fibered surface $Y \to S$ and a birational projective morphism $f : X \to Y$ such that $f(E_i)$ is reduced to a point for each $i \leq n$, and that f is an isomorphism outside of $\cup_i E_i$ (in other words, $\cup_i E_i$ is the exceptional locus of f, Definition 7.2.21). The principal result of this subsection (Theorem 3.36) comes from [15], Section 6.7.

Definition 3.27. Let $X \to S$ be a normal fibered surface. Let $\mathcal{E}$ be a set of integral (projective) vertical curves on X. A normal fibered surface $Y \to S$ together with a projective birational morphism $f : X \to Y$ such that for every integral vertical curve E on X, the set $f(E)$ is a point if and only if $E \in \mathcal{E}$ is called a *contraction* (or a *contraction morphism*) of the $E \in \mathcal{E}$.

Proposition 3.28. *Let $X \to S$ be a normal fibered surface, and let $\mathcal{E}$ be a set of integral vertical curves on X. If a contraction $f : X \to Y$ of the $E \in \mathcal{E}$ exists, then it is unique up to unique isomorphism.*

Proof Let $f' : X \to Y'$ be another contraction. Then there exists a unique birational map $g : Y \dashrightarrow Y'$ such that $f' = g \circ f$. It follows from Corollary 3.23 that g is in fact an isomorphism.   □

Lemma 3.29. *Let X be a scheme that is locally of finite type over a Noetherian ring A. Let $\mathcal{L}$ be an invertible sheaf on X generated by global sections $s_0, \ldots, s_n$. Let us consider the morphism $f : X \to \mathbb{P}^n_A$ associated to these sections. Let $s \in \operatorname{Spec} A$, and let Z be a connected closed subscheme of X_s that is projective over $k(s)$. Then $f(Z)$ is reduced to a point if and only if $\mathcal{L}|_Z \simeq \mathcal{O}_Z$ (see also Exercise 1.7).*

Proof Let $t_0, \ldots, t_n$ be the respective images of $s_0, \ldots, s_n$ in $H^0(Z, \mathcal{L}|_Z)$. Then $\mathcal{L}|_Z$ is generated by $t_0, \ldots, t_n$, and the restriction of f to Z is the morphism associated to the sections $t_0, \ldots, t_n$. We can therefore suppose that $X = Z$ and that X is a connected projective variety over a field k. We use the notation of Proposition 5.1.31. Let us recall that

$$X_{s_i} := \{x \in X \mid \mathcal{L}_x = s_{i,x} \mathcal{O}_{X,x}\}.$$

The construction of f shows that $f^{-1}(D_+(T_i)) = X_{s_i}$.

Let us suppose $f(X) = \{y\}$. We have, for example, $y \in D_+(T_0)$. This implies that $X = X_{s_0}$. Hence $\mathcal{L} = s_0 \mathcal{O}_X \simeq \mathcal{O}_X$. Conversely, let us suppose that $\mathcal{L} = e\mathcal{O}_X$. Let $B = H^0(X, \mathcal{O}_X)$ and $Y = \operatorname{Spec} B$. Then Y is finite over $\operatorname{Spec} k$ (Theorem 5.3.2). We have $s_i = eb_i$ for some $b_i \in B$. The fact that the s_i generate $\mathcal{L}$ implies that the b_i generate the unit ideal of B. In other words, the b_i generate the sheaf $\mathcal{O}_Y$ on Y. Let $g : Y \to \mathbb{P}^n_k$ be the associated morphism. Then f decomposes into the canonical morphism $X \to Y$ followed by g. As Y is finite over $\operatorname{Spec} k$, it follows that $f(X)$ is finite and discrete. Now X is connected, and hence $f(X)$ is reduced to a point.   □

Proposition 3.30. *Let $X \to S$ be a normal fibered surface with $\dim S = 1$. Let $\mathcal{E}$ be a set of integral (projective) vertical curves on X. Then the following conditions are equivalent.*

(i) *The contraction $g : X \to Y$ of the $E \in \mathcal{E}$ exists.*

(ii) *There exists a Cartier divisor D on X such that $\deg(D|_{X_\eta}) > 0$, that $\mathcal{O}_X(D)$ is generated by its global sections, and that for any integral vertical curve E, we have $\mathcal{O}_X(D)|_E \simeq \mathcal{O}_E$ if and only if $E \in \mathcal{E}$.*

Proof (i) $\Longrightarrow$ (ii) We are going to present D as a Cartier divisor such that $\mathcal{O}_X(D) \simeq g^*\mathcal{L}$ for some ample sheaf $\mathcal{L}$ on Y. The birational morphism $X \to Y$ is an isomorphism outside of a proper closed subset of X. This closed subset, being of dimension at most 1, has only a finite number of points of codimension 1. Now these points are in a bijective correspondence with the $E \in \mathcal{E}$. It follows that $\mathcal{E}$, and therefore $g(\mathcal{E})$, is finite.

We embed Y in a projective space $\mathbb{P}^N_S$ and we project this space to $P := \mathbb{P}^N_{\mathcal{O}_S(S)}$. There exists a hypersurface $V_+(F)$ of P which does not meet the images of $g(\mathcal{E})$ and the generic point of Y (Proposition 3.3.36(a)). Let d be the degree of F. Then $F \in H^0(P, \mathcal{O}_P(d))$. Let D_0 be the Cartier divisor associated to F (Exercise 7.1.13). Then $\mathcal{O}_P(D_0) \simeq \mathcal{O}_P(d)$ and $\operatorname{Supp} D_0 = V_+(F)$. As $S \to \operatorname{Spec} \mathcal{O}_S(S)$ is dominant, so is $p : \mathbb{P}^N_S \to P$. Let $D_1 = p^*D_0$ (Definition 7.1.34). Then D_1 is an effective Cartier divisor, and we have

$$Y \not\subseteq \operatorname{Supp} D_1, \quad \operatorname{Supp} D_1 \cap (\cup_{E \in \mathcal{E}} g(E)) = \emptyset, \quad \mathcal{O}_{\mathbb{P}^N_S}(D_1) \simeq \mathcal{O}_{\mathbb{P}^N_S}(d).$$

The restriction $D_2 := D_1|_Y$ is an effective Cartier divisor whose support does not meet $g(\mathcal{E})$, and such that $\mathcal{O}_Y(D_2)$ is very ample. Replacing D_2 by a multiple if necessary, we can suppose that $\mathcal{O}_Y(D_2)$ is generated by its global sections.

Let $D = g^*D_2$. Then $D > 0$, $\mathcal{O}_X(D) \simeq g^*\mathcal{O}_Y(D_2)$ is generated by its global sections. For any $E \in \mathcal{E}$, we have $\operatorname{Supp} D \cap E = \emptyset$, and hence $\mathcal{O}_X(D)|_E \simeq \mathcal{O}_E$. If E is an integral vertical curve that does not belong to $\mathcal{E}$, then $g(E)$ is a vertical curve on Y, and g induces a finite birational morphism $h : E \to g(E)$. In addition, we have $\mathcal{O}_X(D)|_E \simeq h^*(\mathcal{O}_Y(D_2)|_{g(E)})$, which implies that $\mathcal{O}_X(D)|_E$ is ample (Exercise 5.2.16). Hence $\mathcal{O}_X(D)|_E \not\simeq \mathcal{O}_E$.

(ii) $\Longrightarrow$ (i) As X_η is an integral projective curve over a field, $\mathcal{O}_X(D)|_{X_\eta}$ is an ample divisor (Proposition 7.5.5). Replacing D by a positive multiple (which does not change the set $\mathcal{E}$), we can suppose that $\mathcal{O}_X(D)|_{X_\eta}$ is very ample.

Let us first construct the contraction in the case when $S = \operatorname{Spec} A$ is affine. Let $f : X \to \mathbb{P}^n_A$ be a morphism associated to sections $s_0, \ldots, s_n \in H^0(X, \mathcal{O}_X(D))$ which generate $\mathcal{O}_X(D)$. Let Z be the closed subset $f(X) \subseteq \mathbb{P}^n_A$ endowed with the reduced (hence integral) scheme structure. Then f induces a dominant morphism $X \to Z$ that we still denote by f. This is an isomorphism on the generic fiber because $\mathcal{O}_X(D)|_{X_\eta}$ is very ample. In particular, $f : X \to Z$ is birational. As this morphism is projective (Corollary 3.3.32(e)), $f_*\mathcal{O}_X$ is a coherent sheaf on Z (Corollary 5.3.5). Moreover, it is a sheaf of integrally closed algebras because X is normal. Consequently, the normalization morphism $\pi : Y \to Z$ coincides with the canonical morphism $\operatorname{Spec} f_*\mathcal{O}_X \to Z$ (Exercise 5.1.17) and is a finite morphism. As X is normal, f factors into $g : X \to Y$ followed by $\pi : Y \to Z$.

Let E be an integral vertical curve on X. By Lemma 3.29, $f(E)$ is a point if and only if $\mathcal{O}_X(D)|_E \simeq \mathcal{O}_E$, that is, if $E \in \mathcal{E}$. As π is a finite morphism, $g(E)$ is reduced to a point if and only if $f(E)$ is a point, which shows the proposition in the S affine case. The general case follows from the affine case and the uniqueness of the contraction (Proposition 3.28). □

Remark 3.31. One can verify that the proof of implication (i) $\implies$ (ii) does not use the dim $S = 1$ hypothesis. Moreover, the construction of the divisor D shows that $E \cap \operatorname{Supp} D = \emptyset$ if and only if $E \in \mathcal{E}$.

Lemma 3.32. Let X be a fibered surface over an affine Dedekind scheme $S = \operatorname{Spec} A$ of dimension dim $S = 1$. Let D be an effective horizontal Cartier divisor on X, and $\mathcal{L} = \mathcal{O}_X(D)$. Then there exists an $m_0 \geq 1$ such that $\mathcal{L}^{\otimes m}$ is generated by its global sections for every $m \geq m_0$.

Proof We will identify $\mathcal{L}^{\otimes n}$ with $\mathcal{L}^n = \mathcal{O}_X(nD)$. We can suppose $D \neq 0$. The restriction $\mathcal{O}_X(D)|_{X_\eta}$ is an ample sheaf (Proposition 7.5.5). Hence there exists an n_0 such that

$$H^1(X, \mathcal{L}^n) \otimes_A \operatorname{Frac}(A) = H^1(X_\eta, \mathcal{L}^n|_{X_\eta}) = 0$$

for every $n \geq n_0$ (Corollary 5.2.27 and Proposition 5.3.6). In other words, the A-module $H^1(X, \mathcal{L}^n)$ is torsion. As it is finitely generated, it is also of finite length over A. We endow D with the closed subscheme structure $V(\mathcal{O}_X(-D))$. By tensoring the exact sequence

$$0 \to \mathcal{O}_X(-D) \to \mathcal{O}_X \to \mathcal{O}_D \to 0$$

by $\mathcal{L}^{n+1}$ and taking the cohomology, we obtain an exact sequence

$$H^0(X, \mathcal{L}^{n+1}) \to H^0(D, \mathcal{L}^{n+1}|_D) \to H^1(X, \mathcal{L}^n)$$
$$\to H^1(X, \mathcal{L}^{n+1}) \to H^1(D, \mathcal{L}^{n+1}|_D) = 0$$

because D is finite over $\operatorname{Spec} A$, and hence affine. Consequently, the sequence of the lengths $(\operatorname{length} H^1(X, \mathcal{L}^n))_{n \geq n_0}$ is descending, and hence stationary starting at some rank $m_0 - 1$. The homomorphism

$$H^0(X, \mathcal{L}^m) \to H^0(D, \mathcal{L}^m|_D)$$

is then surjective for every $m \geq m_0$, because $H^1(X, \mathcal{L}^{m-1}) \to H^1(X, \mathcal{L}^m)$ is an isomorphism (Lemma 7.1.23). As $\mathcal{L}^m|_D$ is generated by its global sections because D is affine (Theorem 5.1.7), Nakayama's lemma implies that the canonical homomorphism

$$H^0(X, \mathcal{L}^m) \otimes_{\mathcal{O}_X(X)} \mathcal{O}_{X,x} \to \mathcal{L}_x^m \qquad (3.10)$$

is surjective for every $x \in \operatorname{Supp} D$. If $x \notin \operatorname{Supp} D$, then $\mathcal{L}_x^m = \mathcal{O}_{X,x}$ and homomorphism (3.10) is still surjective because $1 \in \mathcal{O}_X(X) \subseteq \mathcal{L}^m(X)$. Consequently, $\mathcal{L}^m$ is generated by its global sections. □

Proposition 3.30 and Lemma 3.32 show that for a contraction morphism to exist, it suffices that there exist an effective Cartier divisor that meets the closed fibers in a suitable way.

Definition 3.33. Let A be a Noetherian local ring. We say that A is *Henselian* if every finite A-algebra is a direct sum of local A-algebras.

Example 3.34. If $(A, \mathfrak{m})$ is a Noetherian local ring, complete for the $\mathfrak{m}$-adic topology, then A is Henselian (Exercise 4.3.17). Thus every Noetherian local ring A is dominated by a Henselian local ring. There exists a 'smallest' Henselian local ring A^h that dominates A (see [80], VIII or [15], Section 2.3). This is called the *Henselization of A*. It can be seen as a direct limit of local rings that are localizations of étale A-algebras. If A is a discrete valuation ring, then A^h is equal to $\hat{A} \cap \mathrm{Frac}(A)^{\mathrm{alg}}$.

Lemma 3.35. *Let $\pi : X \to S$ be a fibered surface over a Dedekind scheme of dimension 1. Let us fix closed points $s \in S$ and $x \in X_s$. Then the following properties are true.*

(a) *There exists an irreducible horizontal Weil divisor Δ containing x.*

(b) *Let us suppose that S is local and that $\mathcal{O}_{S,s}$ is Henselian. Then there exists an effective horizontal Cartier divisor D such that $X_s \cap \mathrm{Supp}\, D = \{x\}$.*

Proof (a) Let $\mathfrak{m}_x$ (resp. $\mathfrak{m}_s$) be the maximal ideal of $\mathcal{O}_{X,x}$ (resp. of $\mathcal{O}_{S,s}$). Let $\mathfrak{p}_1, \dots, \mathfrak{p}_n$ be the prime ideals of $\mathcal{O}_{X,x}$ that are minimal among those containing $\mathfrak{m}_s \mathcal{O}_{X,x}$. As $\dim \mathcal{O}_{X,x} = 2$, Theorem 2.5.12 implies that $\mathfrak{m}_x \neq \mathfrak{p}_i$. There therefore exists an

$$f \in \mathfrak{m}_x \setminus (\cup_i \mathfrak{p}_i)$$

(see [65], 1.B). Thus f is a regular element and $V(f)$ does not contain any irreducible component of $\mathrm{Spec}\, \mathcal{O}_{X_s,x}$. There exist an affine open subset $W \ni x$ and a regular element $g \in \mathcal{O}_X(W)$ such that $g_x = f$, and that $V(g) \cap W_s = \{x\}$. Let Δ be the closure in X of an irreducible component of $V(g)$ passing through x. Then Δ is an irreducible horizontal divisor passing through x.

(b) Let $E := V(g)$ and let W be as in (a). Then E can be seen as an effective Cartier divisor on W. Moreover, $E \to S$ is quasi-projective and quasi-finite because $\dim E_s < \dim X_s = 1$. By Corollary 4.4.8, there exists an affine open neighborhood U of x such that $E \cap U$ is an open subscheme of a scheme Z that is finite over S. Let D be the connected component of Z containing x. By the hypothesis on $\mathcal{O}_{S,s}$, Z is a disjoint union of local schemes. Hence D is local, and consequently $D \subset E \cap V(g) \subset X$. Finally, as D is closed in Z, it is finite over S, and hence closed in X (Exercise 3.3.22, Theorem 3.3.30, and Proposition 3.3.16(e)). It is therefore a Cartier divisor, as we want.   □

Theorem 3.36 ([15], Proposition 6.7/4). *Let $X \to S$ be a normal fibered surface over the spectrum of a Henselian discrete valuation ring. Then for any proper subset $\mathcal{E}$ of the set of irreducible components of X_s, the contraction morphism of the $E \in \mathcal{E}$ exists.*

Proof Let $Z_1, \dots, Z_n$ be the irreducible components of X_s that do not belong to $\mathcal{E}$. By Lemma 3.35(b), there exists an effective horizontal Cartier divisor D_i such that $X_s \cap \mathrm{Supp}\, D_i$ is a point of Z_i that belongs to none of the components of $\mathcal{E}$. Let $D = \sum_{1 \le i \le n} D_i$. By virtue of Lemma 3.32, if necessary replacing D by a multiple, we can even suppose that $\mathcal{O}_X(D)$ is generated by its global sections. The theorem then results from Proposition 3.30. Indeed, $\mathcal{O}_X(D)|_E \simeq \mathcal{O}_E$ if $E \in \mathcal{E}$ since $E \cap \operatorname{Supp} D = \emptyset$; if $E \notin \mathcal{E}$, then E is some Z_i and $\deg \mathcal{O}_X(D)|_E \ge \deg \mathcal{O}_X(D_i)|_{Z_i} > 0$, and hence $\mathcal{O}_X(D)|_E \not\simeq \mathcal{O}_E$. □

Remark 3.37. Further on, we will see the possibility of contracting certain vertical divisors on an arbitrary basis S (Subsection 9.4.1).

Remark 3.38. If we leave out the Henselian hypothesis, Theorem 3.36 is no longer true in general. See [15], Lemma 6.7/6.

8.3.4 Desingularization

Given the good properties that regular Noetherian schemes enjoy, it is very important for many problems to be able to reduce to the case of regular schemes. One way to do this is to 'resolve singularities'. That is the object of the beautiful theory of desingularization.

Definition 3.39. Let X be a reduced locally Noetherian scheme. A proper birational (Definition 7.5.3) morphism $\pi : Z \to X$ with Z regular is called a *desingularization of X* (or a *resolution of singularities of X*). If π is an isomorphism above every regular point of X, we say that it is a *desingularization in the strong sense*.

Remark 3.40. Let X be a universally catenary, locally Noetherian scheme. If every integral scheme Y that is finite over X admits a desingularization, then X is excellent. See [41], IV.7.9.5. Thus we see the importance of the hypothesis of excellence in questions of desingularization.

Example 3.41. Let X be a reduced curve over a field k; the normalization $X' \to X$ (Definition 7.5.1) is a desingularization. More generally, let X be an excellent, reduced, Noetherian scheme of dimension 1; then the normalization $X' \to X$ is a desingularization (Theorem 2.39(d)). Hence the problem of the existence of desingularizations essentially concerns schemes in higher dimensions.

Theorem 3.42 (Hironaka, [46]). *Let X be a reduced algebraic variety over a field of characteristic 0, or more generally a reduce scheme that is locally of finite type over an excellent, reduced, locally Noetherian, scheme of characteristic 0 (i.e., char $k(x) = 0$ for every $x \in X$). Then X admits a desingularization in the strong sense.*

In arbitrary characteristics, this theorem is an open problem. We refer the reader to the book [45] for more bibliographical references to the theory of resolution of singularities. In the absence of a desingularization theorem, it sometimes turns out that a weaker version suffices for the applications. This is the notion of alteration introduced by A.J. de Jong. An *alteration* of a locally Noetherian integral scheme X is a morphism $Y \to X$ that can be decomposed into a proper birational morphism $Y \to W$ and a finite surjective morphism $W \to X$.

Theorem 3.43 (de Jong, [49]). *Let X be a separated integral scheme of finite type over a complete discrete valuation ring (that can be a field). Then there exists an alteration $Y \to X$ with Y regular.*

Theorems 3.42 and 3.43 will not be used in this book. Let us return to the problem of desingularization. Let X be an excellent, reduced, Noetherian scheme of dimension 2. Let us consider the following sequence of proper birational morphisms:

$$\cdots \to X_{n+1} \to X_n \to \cdots \to X_1 \to X, \qquad (3.11)$$

where $X_1 \to X$ is the normalization of X, and for every $i \geq 1$, $X_{i+1} \to X_i$ is the composition of the blowing-up $X_i' \to X_i$ of the singular locus $\mathrm{Sing}(X_i) := X_i \setminus \mathrm{Reg}(X_i)$ (which is closed because X_i is excellent) endowed with the reduced scheme structure, and of the normalization $X_{i+1} \to X_i'$. The sequence stops at n when X_n is regular.

Theorem 3.44 (Lipman, [58], [57]). *Let X be an excellent, reduced, Noetherian scheme of dimension 2. Then the sequence above is finite. In particular, X admits a desingularization in the strong sense.*

For surfaces over a field, this theorem was proven independently by Zariski and Abhyankar. See references in [45]. It is impossible to give a proof here for these theorems, which extend widely beyond the scope of this book. For Theorem 3.44, see also the presentation of the proof by M. Artin [6].

Corollary 3.45. *Let S be an excellent Dedekind scheme. Let $X \to S$ be a fibered surface. Then X admits a desingularization in the strong sense.*

Remark 3.46. This assertion was proven by Abhyankar [2] under a slightly different hypothesis: S is a Nagata scheme (Definition 2.27) and the residue fields of S at the closed points are perfect.

Lemma 3.47. *Let $f : X \to Y$ be a finite birational morphism of reduced Noetherian schemes.*

(a) *Let us suppose that there exists an invertible sheaf of ideals $\mathcal{L} \subseteq \mathcal{O}_Y$ such that f is an isomorphism above $Y \setminus V(\mathcal{L})$. Then f is the blowing-up morphism of Y along a closed subscheme $V(\mathcal{I})$ whose support is the set $V(\mathcal{L})$.*

(b) *Let us suppose Y is quasi-projective over an affine Noetherian scheme. Then f is a projective morphism.*

Proof (a) The quotient $f_*\mathcal{O}_X/\mathcal{O}_Y$ is a coherent sheaf on Y with support in $V(\mathcal{L})$. As Y is Noetherian, there exists an $r \geq 1$ such that $\mathcal{L}^r f_*\mathcal{O}_X \subseteq \mathcal{O}_Y$ (because $\mathcal{L}$ is contained in $\sqrt{\mathrm{Ann}(f_*\mathcal{O}_X/\mathcal{O}_Y)}$, see Exercise 5.1.9). Let $\mathcal{I} = \mathcal{L}^{r+1} f_*\mathcal{O}_X$. We have

$$\mathcal{L}^{r+1} \subseteq \mathcal{I} = \mathcal{L} \cdot \mathcal{L}^r f_*\mathcal{O}_X \subseteq \mathcal{L}.$$

Let T be a variable. As f is finite, we have

$$X = \mathrm{Spec}\, f_*\mathcal{O}_X = \mathrm{Proj}(\oplus_{n \geq 0}\mathcal{B}_n), \quad \mathcal{B}_n = f_*\mathcal{O}_X \cdot T^n.$$

By Lemma 1.21, we have

$$X \simeq \mathrm{Proj}\,(\oplus_{n\geq 0}\mathcal{C}_n) \simeq \mathrm{Proj}\,\oplus_{n\geq 0}\mathcal{I}^n, \quad \mathcal{C}_n = \mathcal{B}_n \otimes_{\mathcal{O}_Y} (\mathcal{L}^{r+1})^{\otimes n}.$$

(b) Let $\mathcal{N}$ be an ample sheaf on Y such that $f_*\mathcal{O}_X \otimes_{\mathcal{O}_Y} \mathcal{N}$ is generated by its global sections (Corollary 5.1.36). Then we have a surjective homomorphism

$$\mathcal{O}_Y^{r+1} \to (f_*\mathcal{O}_X \cdot T) \otimes_{\mathcal{O}_Y} \mathcal{N}.$$

We deduce from this a closed immersion $X \to \mathbb{P}_Y^r$ as at the beginning of the proof of Proposition 1.22, using Lemma 1.21. □

In what follows we are going to consider the statement of Corollary 3.45 when S is not necessarily excellent. Let us first consider the local case. Let R be an arbitrary discrete valuation ring, let $\widehat{R}$ denote the completion of R, and t a uniformizing parameter for R.

Lemma 3.48. *Let Y be a flat locally Noetherian R-scheme. Let us consider the blowing-up $f : W \to \widehat{Y} := Y \times_{\mathrm{Spec}\,R} \mathrm{Spec}\,\widehat{R}$ along a closed subscheme $V(\mathcal{I})$ with support contained in $V(t)$. Then there exists a blowing-up $g : X \to Y$ along a closed subscheme $V(\mathcal{I}_0)$ with support contained in $V(t)$ such that f is obtained from g by the base change $\mathrm{Spec}\,\widehat{R} \to \mathrm{Spec}\,R$.*

Proof Let $\pi : \widehat{Y} \to Y$ be the canonical morphism. As Y is flat over R, the canonical homomorphism $\mathcal{O}_Y \to \pi_*\mathcal{O}_{\widehat{Y}} = \mathcal{O}_Y \otimes_R \widehat{R}$ is injective. Let $\mathcal{I}_0 = \pi_*\mathcal{I} \cap \mathcal{O}_Y$. We immediately deduce from the hypothesis $V(\mathcal{I}) \subseteq V(t)$ that $\mathcal{I} = \pi^*\mathcal{I}_0$. As $R \to \widehat{R}$ is flat, the lemma results from Proposition 1.12(c). □

Lemma 3.49. *Let $X \to \mathrm{Spec}\,R$ be a fibered surface. Let $\widehat{K} = \mathrm{Frac}(\widehat{R})$, $\widehat{X} = X \times_{\mathrm{Spec}\,R} \mathrm{Spec}\,\widehat{R}$, and $p : \widehat{X} \to X$ be the projection morphism. Then the following properties are true.*

(a) *The morphism p induces an isomorphism between the special fibers of $\widehat{X}$ and of X.*

(b) *For any $x \in X_s$, we have an isomorphism*

$$\widehat{\mathcal{O}}_{\widehat{X}, p^{-1}(x)} \simeq \widehat{\mathcal{O}}_{X, x}$$

of the completions of the local rings.

(c) *Let us suppose X_K is regular and that X admits a desingularization. Then $X_{\widehat{K}}$ is regular.*

(d) *Let us suppose $X_{\widehat{K}}$ is regular. Then X admits a desingularization in the strong sense. More precisely, sequence (3.11) is defined (i.e., the normalization morphisms are finite and the singular loci are closed) and is finite.*

Proof (a) results immediately from the fact that t is a uniformizing parameter for $\widehat{R}$ and that $\widehat{R}/t\widehat{R} = R/tR$.

(b) We deduce from (a) that $\mathcal{O}_{\widehat{X},p^{-1}(x)} = \mathcal{O}_{X,x} \otimes_R \widehat{R}$. Let $\mathfrak{m}_x$ be the maximal ideal of $\mathcal{O}_{X,x}$. For any $n \geq 1$, we have

$$(\mathcal{O}_{X,x} \otimes_R \widehat{R})/(\mathfrak{m}_x^n) = (\mathcal{O}_{X,x}/\mathfrak{m}_x^n) \otimes_R \widehat{R} = (\mathcal{O}_{X,x}/\mathfrak{m}_x^n) \otimes_{R/t^n R} (R/t^n R) \otimes_R \widehat{R}$$
$$= \mathcal{O}_{X,x}/\mathfrak{m}_x^n,$$

because $\widehat{R}/t^n \widehat{R} = R/t^n R$. This shows (taking $n = 1$) that $\mathfrak{m}_x$ generates the maximal ideal of $\mathcal{O}_{\widehat{X},p^{-1}(x)}$, and implies the desired isomorphism by taking the projective limit as n varies.

(c) Let $Z \to X$ be a desingularization. Let us set $\widehat{Z} = Z \times_{\mathrm{Spec}\,R} \mathrm{Spec}\,\widehat{R}$, and let $z \in Z_{\widehat{K}}$. As $\widehat{Z}$ is proper over $\mathrm{Spec}\,\widehat{R}$, $\overline{\{z\}}$ contains a point of $\widehat{Z}_s$. Since the latter is contained in $\mathrm{Reg}(\widehat{Z})$ by (b), z is a regular point by Theorem 4.2.16(a). By the X_K regular hypothesis, we obtain $X_K = Z_K$, and hence $X_{\widehat{K}} = Z_{\widehat{K}}$ is regular.

(d) For any affine open subset U of X, we have

$$\mathcal{O}_X(U) \otimes_R \widehat{R} \subseteq K(X) \otimes_R \widehat{R} = K(X) \otimes_K \widehat{K}.$$

As $X_{\widehat{K}}$ is reduced, the last term is reduced. Hence $\widehat{X}$ is reduced, projective over $\widehat{R}$, and of dimension 2 since its fibers are of dimension 1.

Let us consider the finite sequence (3.11) associated to $\widehat{X}$ (Theorem 3.44). As the generic fiber of $\widehat{X}$ is regular by hypothesis, the normalization $\widehat{X}_1 \to \widehat{X}$ is the blowing-up along a closed subscheme with support in $V(t\mathcal{O}_{\widehat{X}})$ (Lemma 3.47(a)). It follows from Lemma 3.48 that this morphism comes from a blowing-up morphism $f : X_1 \to X$ over R. The exceptional locus of f is in the special fiber of X_1. It follows from (a) that $X_1 \to X$ is quasi-finite and projective (Proposition 1.22), and therefore finite. Now X_1 is normal (Exercise 4.1.6), and hence f is the normalization morphism of X.

The singular locus $\mathrm{Sing}(X_1)$ of X_1 is finite and closed (Corollary 2.38), and its inverse image in $\widehat{X}_1$ is the singular locus thereof, by virtue of (b). Moreover, by endowing the singular loci with the reduced scheme structure, we have

$$\mathrm{Sing}(X_1) = \mathrm{Sing}(X_1) \times_{\mathrm{Spec}\,R} \mathrm{Spec}\,\widehat{R} = \mathrm{Sing}(\widehat{X}_1),$$

where the first equality comes from the fact that $\mathrm{Sing}(X_1)$ is contained in the special fiber, and the second comes from the fact that the middle term is reduced (since it is equal to the term on the left). We therefore see that the sequence (3.11) associated to $\widehat{X}$ is obtained from the one associated to X by the base change $\mathrm{Spec}\,\widehat{R} \to \mathrm{Spec}\,R$. This concludes the proof of the lemma, since if $\widehat{X}_n$ is regular, then so is X_n. $\qquad\square$

Theorem 3.50. *Let S be a Dedekind scheme of dimension 1, and let $\pi : X \to S$ be a fibered surface with regular generic fiber. Then the following conditions are equivalent.*

(i) *The scheme X admits a desingularization.*

(ii) *The set* $\mathrm{Reg}(X)$ *is open, and the curve* $X \times_S \mathrm{Spec}\,\mathrm{Frac}(\widehat{\mathcal{O}}_{S,s})$ *is regular for every closed point* $s \in S$.

(iii) *The set* $\mathrm{Sing}(X)$ *is contained in a finite union of closed fibers* $X_{s_1}, \ldots, X_{s_r}$, *and the curve* $X \times_S \mathrm{Spec}\,\mathrm{Frac}(\widehat{\mathcal{O}}_{S,s_i})$ *is regular for every* $i \le r$.

Moreover, if one of these conditions is satisfied, then X admits a desingularization $Z \to X$ in the strong sense. More precisely, sequence (3.11) is defined and is finite. Finally, $Z \to X$ is a projective morphism if S is affine.

Proof (i) $\Longrightarrow$ (ii). Let $Z \to X$ be a desingularization. As it is a birational morphism, X contains a dense regular open subset. Consequently, the singular locus $\mathrm{Sing}(X)$ is contained in a finite number of closed fibers, which implies that $\mathrm{Reg}(X)$ is open by using Corollary 2.38. The second property is nothing more than Lemma 3.49(c).

(iii) $\Longrightarrow$ (ii) The reasoning above shows that $\mathrm{Reg}(X)$ is open. For any $s \in S$ such that $X_s \subset \mathrm{Reg}(X)$, we have $X \times_S \mathrm{Spec}\,\mathrm{Frac}(\widehat{\mathcal{O}}_{S,s})$ is regular by virtue of Lemma 3.49(c), whence (ii).

As (ii) trivially implies (iii), it remains to show that (ii) implies (i). Let $f : X_1 \to X$ be the normalization morphism. For any $s \in S$, the base change $X_1 \times_S \mathrm{Spec}\,\mathcal{O}_{S,s} \to X \times_S \mathrm{Spec}\,\mathcal{O}_{S,s}$ is the normalization morphism, which is finite by Lemma 3.49(d). Hence f is a finite morphism. As X_1 is birational to X, we deduce by the same reasoning as above that $\mathrm{Reg}(X_1)$ is open. Hence $\mathrm{Sing}(X_1)$ is closed and finite (because it does not contain any point of codimension 1). Consequently, sequence (3.11) is well defined. It is finite by once more applying Lemma 3.49(d) to the finite number of fibers contained in $\mathrm{Sing}(X)$ (let us note that the normalization and blowing-up commute with the base change $\mathrm{Spec}\,\mathcal{O}_{S,s} \to S$). Finally, the projectivity of $Z \to X$ when S is affine results from Lemma 3.47(b), Proposition 1.22, and because the composition of projective morphisms is projective (Corollary 3.3.32(b)). $\qquad\square$

Corollary 3.51. *Let $X \to S$ be a fibered surface. Let us suppose that $\dim S = 1$ and that X has a smooth generic fiber. Then $X \to S$ verifies the conditions of Theorem 3.50. In particular, X admits a desingularization in the strong sense.*

Proof By Proposition 3.11, $X \to S$ is smooth above a non-empty open subscheme V of S. Hence X_V is regular. This shows that condition (iii) of Theorem 3.50 is satisfied. $\qquad\square$

Remark 3.52. Since desingularization is a problem of local nature, it is clear that Theorem 3.50 and Corollary 3.51 are true for any integral flat (not necessarily proper) curve over S.

Let us conclude with two examples of resolutions of singularities.

Example 3.53. Let R be a discrete valuation ring with uniformizing parameter t and residue field k. Let $a \in R$ be a non-zero element, and let us consider

$$X = \mathrm{Spec}\,R[x,y]/(xy - a).$$

We are going to see how to resolve the possible singularities of X using the sequence (3.11). The morphism $X \to \operatorname{Spec} R$ is smooth everywhere if $a \in R^*$ and is smooth only outside of the closed point $x = y = 0$ of the special fiber otherwise. In particular, X is regular outside of this point. If $\nu(a) = 1$, then X is regular (Corollary 4.2.12).

Let us suppose $e = \nu(a) \geq 2$. Let us write $a = t^e u$ with $u \in R^*$. The scheme X is normal (Lemma 4.1.18). Let $X_1 \to X$ be the blowing-up of the reduced singular point (that corresponds to the maximal ideal (x, y, t)). Let us consider the following elements of $K(X)$:

$$x_1 = x/t, y_1 = y/t, y_2 = y/x, t_2 = t/x, t_3 = t/y.$$

Using Lemma 1.4, we see that X_1 is a union of three open subschemes $\operatorname{Spec} A_i$, $i = 1, 2, 3$, with

$$A_1 = R[x, y, x_1, y_1] = R[x_1, y_1], \quad \text{with } x_1 y_1 = t^{e-2} u,$$

$$A_2 = R[x, y, y_2, t_2] = R[x, t_2], \quad \text{with } xt_2 = t,$$

because $y_2 = t_2^{e-2} x^e u$. And by symmetry,

$$A_3 = R[y, t_3], \quad \text{with } yt_3 = t.$$

If $e = 2$, then X_1 is regular and its special fiber is of the form

where the horizontal component is a projective line over the residue field k and the two oblique components are affine lines over k (a cross means that we have removed a point from a projective curve). If $e \geq 3$, then $\operatorname{Spec} A_2$ and $\operatorname{Spec} A_3$ are regular. The open subscheme $\operatorname{Spec} A_1$ is regular if $e = 3$ and singular otherwise. The special fiber of X_1 is of the form

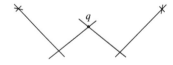

where q is the singular point of X_1 if $e \geq 4$. As $\operatorname{Spec} A_1$ is of a form similar to X, with e replaced by $e - 2$, its blowing-up along the point q is computed in the same way as $X_1 \to X$. Thus we see that after $[e/2]$ successive blowing-ups of singular points, we end up with a desingularization $f : Z \to X$ of X. If $e \geq 2$, the inverse image of the singular point of X under f is a chain of $e-1$ projective lines over k:

Figure 11. Special fiber of Z with $e+1$ irreducible components.

In the preceding example, the desingularization consists uniquely of blowing up isolated singular points. But this is not the case in general, even if we start out with a normal surface.

Example 3.54. Let R be as in the example above. We suppose char $k \neq 2, 3$. Let

$$X = \operatorname{Spec} R[x, y], \quad y^2 = t(x^3 + t^3).$$

Then X has a smooth generic fiber, it is normal (use Example 2.26), and its unique singular point q corresponds to the maximal ideal (x, y, t). Let $X_1 \to X$ be the blowing-up of X along the point q. Then X_1 contains an open subscheme

$$U_1 = \operatorname{Spec} R[x_1, y_1], \quad y_1^2 = t^2(x_1^3 + 1),$$

where $x_1 = x/t$ and $y_1 = y/t$. The normalization of U_1 is

$$U_1' = \operatorname{Spec} R[x_1, z_1], \quad z_1^2 = x_1^3 + 1,$$

where $z_1 = y_1/t$. Hence X_1 is not normal. We can verify that the normalization $X_1' \to X_1$ resolves the singularities of X_1, and that the special fiber of X_1' is the union of an affine line (strict transform of the special fiber X_k of X) of multiplicity 2 in X_k (Definition 7.5.6), and of an elliptic curve.

$$2\, \widetilde{X}_k$$

elliptic

Exercises

3.1. Let $S = \operatorname{Spec} \mathbb{Z}$, $X = \operatorname{Proj} \mathbb{Z}[T_0, T_1]$. Let us consider the point $x \in D_+(T_0) \subset \mathbb{P}^1_{\mathbb{Q}}$ corresponding to the ideal $(t_1^2 - 2)\mathbb{Q}[t_1]$ where $t_1 = T_0^{-1}T_1$. Determine $\overline{\{x\}}$.

3.2. Let $X = \mathbb{P}^1_{\mathbb{Z}}$, and let $f : X_{\mathbb{Q}} \to X_{\mathbb{Q}}$ be an automorphism corresponding to a matrix of $\operatorname{PGL}_2(\mathbb{Q})$. Determine the domain of definition of the rational map $X \dashrightarrow X$ induced by f.

3.3. Let R be a discrete valuation ring, with uniformizing parameter t, $S = \operatorname{Spec} R$, and X a normal, flat, Noetherian scheme over S. We are going to study the set N of points $x \in X_s$ such that $\mathcal{O}_{X_s,x}$ is not reduced.

(a) Let Γ be an irreducible component of X_s. Let ν_Γ be the normalized valuation of $K(X)$ associated to the generic point of Γ. Show that the multiplicity of Γ in X_s (Definition 7.5.6) is equal to $\nu_\Gamma(t)$.

(b) Show that if $\nu_\Gamma(t) \geq 2$, then $\Gamma \subseteq N$.

(c) Let x be a closed point of X_s. We suppose that the irreducible components of X_s containing x are of multiplicity 1.

   (1) Let $a \in \mathcal{O}_{X,x}$ be such that $a^n = tb$ with $n \geq 2$ and $b \in \mathcal{O}_{X,x}$. Let $f = a/t \in K(X)^*$. Show that for any irreducible component Γ of X_s passing through x, we have $\nu_\Gamma(f) \geq 0$.

   (2) Show that $x \notin \operatorname{Supp}(f)_\infty$ and that $f \in \mathcal{O}_{X,x}$. Deduce from this that $x \notin N$.

(d) Conclude that N is the union of the irreducible components of X_s of multiplicity at least 2.

(e) Give a quick proof using Corollary 2.22 and Proposition 2.11.

(f) Let k be a field and $n \geq 2$. Study the example $\operatorname{Spec} k[u,v]/(u^n, uv)$, and show that property (d) above is not true for this curve.

3.4. Give explicit equations of a fibered surface $X \to \operatorname{Spec} \mathbb{Z}$ such that $X_{\mathbb{Q}} \simeq \mathbb{P}^1_{\mathbb{Q}}$ and that X_p is an irreducible curve of multiplicity 2, for some prime number p.

3.5. Let $f : X \to S$ be a fibered surface over a Dedekind scheme of dimension 1, with generic fiber $X_{K(S)} \simeq \mathbb{P}^1_{K(S)}$. Let $s \in S$ be such that X_s is geometrically integral.

(a) Suppose S is local. Show that $X_s \simeq \mathbb{P}^1_{k(s)}$ using the fact that $p_a(X_s) = 0$ and that $X_s(k(s)) \neq \emptyset$.

(b) Let D be the Cartier divisor corresponding to a section of $X \to S$. Show that, under the S local assumption, $\mathcal{O}_X(D)$ is very ample.

(c) Show that there exists an open neighborhood V of s such that $f^{-1}(V) \simeq \mathbb{P}^1_V$.

3.6. Let $\pi : X \to S$ be a smooth arithmetic surface with geometrically connected generic fiber of genus g.

(a) Let us suppose $g = 0$. Show that $\pi_*(\omega^\vee_{X/S})$ is locally free of rank 2 over S. Show that $\omega^\vee_{X/S}$ is very ample if S is affine. Deduce from this that we have an isomorphism of S-schemes $X \simeq \operatorname{Proj} \mathcal{B}$, where $\mathcal{B}$ is the graded $\mathcal{O}_S$-algebra

$$\mathcal{B} := \oplus_{n \geq 0} (\pi_*(\omega^\vee_{X/S}))^{\otimes n}.$$

(b) Let us suppose $g \geq 1$. Show that $\pi_* \omega_{X/S}$ is locally free of rank g over S. Show that $\omega_{X/S}$ is generated by its global sections if S is affine.

Let us moreover suppose that no geometric fiber $X_{\bar{s}}$ is hyperelliptic (in particular, $g \geq 3$). Show that $\omega_{X/S}$ is very ample if S is affine, and that in general, $X \simeq \operatorname{Proj}\mathcal{B}$ with

$$\mathcal{B} := \oplus_{n \geq 0}(\pi_*(\omega_{X/S}))^{\otimes n}.$$

(c) Let us suppose that $g \geq 2$ and that the generic fiber X_η is hyperelliptic, a double covering of $\mathbb{P}^1_{K(S)}$. Let σ be the hyperelliptic involution of X_η. Show that σ extends in a unique way to an involution on X, and that $X/\langle\sigma\rangle$ is a smooth arithmetic surface with generic fiber isomorphic to $\mathbb{P}^1_{K(S)}$. Deduce from this that X_s is a hyperelliptic curve for every $s \in S$.

3.7. Let $f : X \to Y$ be a morphism of schemes. Let $y \in Y$ and

$$p : \ X' = X \times_Y \operatorname{Spec}\mathcal{O}_{Y,y} \to X$$

be the projection morphism.

(a) Show that p induces an isomorphism of the fibers $X'_y \simeq X_y$.

(b) For any $x \in X_y$, show that $\mathcal{O}_{X',p^{-1}(x)} \simeq \mathcal{O}_{X,x}$.

3.8. Let X, Y be normal fibered surfaces over a Dedekind scheme S with $\dim S = 1$. Let $f : X \to Y$ be a birational S-morphism.

(a) Show that f is surjective.

(b) Let us suppose that for any $s \in S$, the curves X_s and Y_s have the same number of irreducible components. Show that f is an isomorphism.

(c) Deduce from this that every birational morphism $X \to X$ is an automorphism.

(d) Let $f : X \to X$ be a non-constant morphism. Show that it is finite.

3.9. Let $f : X \to Y$ be a finite morphism of normal fibered surfaces over S with $\dim S = 1$. Let ξ be a generic point of a closed fiber X_s, and ξ' its image in Y_s.

(a) Show that the multiplicity of $\overline{\{\xi'\}}$ in Y_s divides that of $\overline{\{\xi\}}$ in X_s.

(b) Show that $k(\xi') \subseteq k(\xi)$. Deduce from this that if $\overline{\{\xi\}}$ is geometrically reduced (resp. geometrically integral), then so is $\overline{\{\xi'\}}$.

3.10. Let $X \to S$ be an arithmetic surface.

(a) Let $S' \to S$ be an étale morphism with S' connected. Show that S' is a Dedekind scheme, and that $X \times_S S' \to S'$ is an arithmetic surface.

(b) Give an example where (a) is false if $S' \to S$ is not étale.

3.11. Let $\pi : X \to S$ be an arithmetic surface, $s \in S$ a closed point. Show that for any $x \in X_s$, we have $\dim_{k(x)} T_{X_s,x} \leq 2$.

See [95] for a converse (by imposing additional conditions of global nature).

3.12. Let $\pi : X \to S$ be an arithmetic surface, $x \in X$ a closed point, and $s = \pi(x)$. Let t be a uniformizing parameter for $\mathcal{O}_{S,s}$. Show that x is a regular point of X_s if and only if $t \notin \mathfrak{m}_x^2 \mathcal{O}_{X,x}$.

3.13. Let $\pi : X \to S$ be a fibered surface with dim $S = 1$. Let $X \to S' \to S$ be the decomposition of π as in Proposition 3.8. Show that if X_s is reduced, then $S' \to S$ is étale above s. Show that if $X \to S$ is smooth, then the same holds for $X \to S'$.

3.14. We are going to show a more explicit version of Theorem 3.26(b). Let X be a Nagata, Noetherian, integral scheme. Let ν be a valuation of $K(X)$ with center $x \in X$. Let us suppose that $k(\nu)/k(x)$ is of finite type and that

$$\operatorname{trdeg}_{k(x)} k(\nu) = \dim \mathcal{O}_{X,x} - 1.$$

If ν is not of the first kind in X, we consider the blowing-up $X_1 \to X$ with center $\overline{\{x\}}$ (reduced subscheme). As $X_1 \to X$ is proper, ν has a center x_1 in X_1. Once more, we blow X_1 up along $\overline{\{x_1\}}$. We thus construct a sequence

$$\cdots \to X_n \to \cdots \to X_1 \to X_0 = X,$$

where $X_{n+1} \to X_n$ is the blowing-up of X_n along $\overline{\{x_n\}}$, with x_n the center of ν in X_n. We want to show that ν is of the first kind in X_n for some n.

(a) Let us set $x_0 = x$. The local rings $\mathcal{O}_{x_n} := \mathcal{O}_{X_n,x_n}$ form an ascending sequence of subrings of $\mathcal{O}_\nu$. Let $f \in \mathcal{O}_\nu$. For every $n \geq 0$, let us set

$$r_n = \min\{\nu(b_n) \mid b_n \in \mathcal{O}_{x_n},\ b_n f \in \mathcal{O}_{x_n}\}.$$

Show that the r_n form a strictly descending sequence whenever $r_n > 0$. Deduce from this that $\cup_{n \geq 0} \mathcal{O}_{x_n} = \mathcal{O}_\nu$.

(b) Show that there exists an $r \geq 0$ such that $\dim \mathcal{O}_{x_n} = 1$ for every $n \geq r$.

(c) Let $\mathcal{O}'_{x_n}$ be the integral closure of $\mathcal{O}_{x_n}$. Show that the Spec $\mathcal{O}'_{x_n}$, $n \geq r$, form a descending sequence of open and finite subschemes of Spec $\mathcal{O}'_{x_0}$ (Exercise 3.1). They are therefore equal from some rank $q \geq r$ on.

(d) Show that the quotients $\mathcal{O}_{x_n}/\mathcal{O}_{x_q}$, $n \geq q$, are of finite length, and that they form an ascending sequence that is stationary from some rank m on.

(e) Show that $\mathcal{O}_{x_m} = \mathcal{O}_\nu$.

3.15. Let X, Y be two fibered surfaces over a Dedekind scheme S of dimension 1. Let us suppose that X, Y are smooth over S, that there exists an isomorphism $f_\eta : X_\eta \to Y_\eta$ of the generic fibers, and that $p_a(X_\eta) \geq 1$. We want to show that f_η extends uniquely to an isomorphism $f : X \to Y$. See also Example 9.3.15.

(a) Show the uniqueness of f if it exists.

(b) Show that we can suppose that X, Y have connected fibers over S (use Proposition 3.8). Show that Y_s is then irreducible, and that $p_a(Y_s) = p_a(X_\eta)$.

(c) We identify X_η and Y_η via the isomorphism f_η. Let us suppose that the birational map $f_\eta : X \dashrightarrow Y$ is not a morphism. Let ν be the valuation of $K(Y)$ corresponding to the generic point of Y_s. Using the method of Exercise 3.14(a)–(b), show that there exists a sequence of blowing-ups with regular closed centers $X_n \to X_{n-1} \to \cdots \to X_0 = X$ such that ν is centered at a point of codimension 1 of X_n.

(d) Using Theorem 1.19, show that Y_s is a rational curve. Deduce from this a contradiction with (b).

3.16. Let $f : X \to Y$ be a dominant morphism of finite type between locally Noetherian integral schemes. Let ν be a discrete valuation of $K(X)$, and let ν' denote the valuation $\nu|_{K(Y)}$ of $K(Y)$.

(a) Show that $\operatorname{trdeg}_{k(\nu')} k(\nu) \leq \operatorname{trdeg}_{K(Y)} K(X)$.

(b) Let us suppose that ν is of the first kind, that it admits a center $x \in X$, and that Y is universally catenary and Nagata. Show that ν' is of the first kind in some blowing-up of Y.

3.17. Let A be a Dedekind domain of dimension 1. Let $f : X \to Y$ be a proper birational morphism of normal fibered surfaces over A. Let us suppose that Y is regular.

(a) Show that the exceptional locus $E \subset X$ of f is made up of a finite number of vertical divisors $E_1, \ldots, E_n$ (Theorem 7.2.22).

(b) Show that there exists a Cartier divisor $D > 0$ on Y such that $Y \setminus \operatorname{Supp} D$ is affine and contains $\cup_i f(E_i)$.

(c) Using the divisor f^*D and Proposition 3.30, show that f is the contraction morphism of the E_i. Deduce from this that f is a projective morphism.

3.18. Let X be a locally Noetherian integral scheme. Let F be a connected subset of X and $f : X \to Y$ a projective birational morphism with Y normal and locally Noetherian, $f(F) = y_0$, and $X \setminus F \simeq Y \setminus \{y_0\}$ (so f contracts F to a point). Let $g : X \to Z$ be a dominant morphism with Z integral and $g(F) = z_0$.

(a) Show that $f_*\mathcal{O}_X = \mathcal{O}_Y$ and that via the isomorphism $K(Y) \simeq K(X)$, we have $\mathcal{O}_{Y,y_0} = \cap_{x \in F} \mathcal{O}_{X,x}$.

(b) Let us identify $K(Z)$ with a subfield of $K(X)$ via g. Show that $\mathcal{O}_{Z,z_0} \subseteq \mathcal{O}_{X,x}$ for every $x \in F$. Deduce from this that $\mathcal{O}_{Z,z_0} \subseteq \mathcal{O}_{Y,y_0}$ and that the rational map $h : Y \dashrightarrow Z$ is a morphism with $g = h \circ f$.

3.19. Let $f : X \to S$ be a fibered surface over a Dedekind scheme of dimension 1. Let $s_1, \ldots, s_n$ be closed points of S and $\{E_{ij}\}_{i,j}$ irreducible vertical divisors of X with $E_{ij} \subset X_{s_i}$. Show that the contraction of the E_{ij} exists if and only if for $i \le n$, the contraction of the E_{ij} exists in the fibered surface $X \times_S \operatorname{Spec} \mathcal{O}_{S,s_i} \to \operatorname{Spec} \mathcal{O}_{S,s_i}$.

3.20. Let X be a proper normal surface over a field k. Let $E_1, \ldots, E_n$ be prime divisors on X. Let us suppose that there exists an effective Cartier divisor D on X such that $\mathcal{O}_X(D)|_{E_i} \simeq \mathcal{O}_{E_i}$ for every $i \le n$ and that for any prime divisor Γ different from the E_i, we have $\deg \mathcal{O}_X(D)|_\Gamma > 0$. Show that Lemma 3.32 is true for (X, D) and that there exists a contraction morphism of the divisors $E_1, \ldots, E_n$.

3.21. Let $\mathcal{O}_K$ be a discrete valuation ring that is Nagata. Let X be a fibered surface over $\mathcal{O}_K$. Let us consider the $\mathcal{O}_K$-valuations of $K(X)$ of the first kind $\nu_1, \ldots, \nu_r$.

(a) Show that there exists a birational morphism $X_1 \to X$ such that the ν_i have centers of codimension 1 in X_1.

(b) Let us suppose that $\mathcal{O}_K$ is, moreover, Henselian. Show that there exists a fibered surface X_2, birational to X, such that the valuations induced by the generic points of the special fiber of X_2 are exactly the ν_i, $1 \le i \le r$.

3.22. Let X be an excellent, Noetherian, integral scheme of dimension 1. Show that the desingularization of X can be accomplished by a sequence of blowing-ups with regular centers as in Proposition 1.26

3.23. Let $X = \operatorname{Proj} \mathbb{Z}[x, y, z]/(3x^3 + 4y^3 + 5z^3)$.

(a) Show that for any prime number $p \ne 2, 3, 5$, the fiber X_p is an elliptic curve over $\mathbb{F}_p$.

(b) Show that X_5 is contained in the regular locus $\operatorname{Reg}(X)$ and that X_5 is the union of a projective line over $\mathbb{F}_5$ and a singular irreducible conic over $\mathbb{F}_5$.

(c) Show that X_2 contains a unique singular point q_2 of X, and that after two blowing-ups of singular points, one resolves the singularity q_2. See Figure 12.

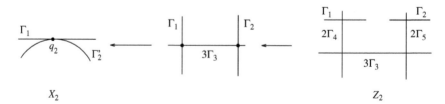

Figure 12. Resolution of singularities of $X \times_{\operatorname{Spec} \mathbb{Z}} \operatorname{Spec} \mathbb{Z}_2$.

The components $\Gamma_1, \Gamma_3, \Gamma_4$ are projective lines over $\mathbb{F}_2$, Γ_2' is a singular irreducible conic over $\mathbb{F}_2$, and Γ_2, Γ_5 are projective lines over $\mathbb{F}_4$.

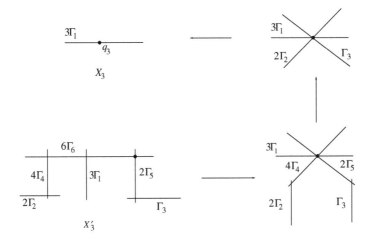

Figure 13. Partial resolution of singularities of $X \times_{\operatorname{Spec}\mathbb{Z}} \operatorname{Spec}\mathbb{Z}_3$.

(d) Show that X_3 contains a unique singular point q_3 of X, and that after three blowing-ups of singular points, one obtains a surface X' with only one singular point contained in X_3'. Using techniques we will develop in the next chapters, it is possible to show that after some more blowing-ups, one obtains a desingularization $Z \to X$ with Z_3 of type II* (see Subsection 10.2.1, Figure 41).

3.24. Let R be a discrete valuation ring. Let $X \to \operatorname{Spec} R$ be a normal fibered surface with smooth generic fiber. Show that $X \times_R \widehat{R}$ is normal.

3.25. Let X be a reduced locally Noetherian scheme. Show that if X admits a desingularization in the strong sense, then the set $\operatorname{Reg}(X)$ of regular points of X is open.

3.26. Let $f : X \to S$ be a fibered surface over a Dedekind scheme of dimension 1, with smooth generic fiber. Let L be a finite extension of $K(X)$. Show that the normalization of X in L (Definition 4.1.24) is a finite morphism (reduce to the case when S is local and proceed as for Lemma 3.49).

3.27. Let $X \to S$ be a fibered surface as in Theorem 3.50 with S affine. Let $\pi_n : X_n \to X$ be the last element of the sequence (3.11) associated to X (hence X_n is regular while X_{n-1} is singular).

(a) Let $f : Z \to X$ be a proper birational morphism with Z regular. Show that f factors in a unique way into $g : Z \to X_n$ and $\pi_n : X_n \to X$. Deduce from this that f is projective (use Exercise 3.17).

(b) Show that there exists a birational projective morphism $X' \to X_n$, made up of successive blowing-ups with finite regular centers, that factors into $X' \to Z \to X_n$.

3.28. Let K be a non-excellent discrete valuation field. Let L/K be a finite inseparable extension contained in the completion $\widehat{K}$ of K.

  (a) Show that the integral closure $\mathcal{O}_L$ of $\mathcal{O}_K$ in L is never finite over $\mathcal{O}_K$.

  (b) Let $C = \mathbb{P}^1_L$, considered as an algebraic curve over K. Show that there exists a scheme X, integral, projective, and flat over $\mathcal{O}_K$, such that $X_K \simeq C$.

  (c) Show that there does not exist any alteration $Y \to X$ with Y regular.

9

Regular surfaces

This chapter is devoted to regular fibered surfaces and presents several fundamental theorems. Section 9.1 develops intersection theory, starting from the local definition. In the case of arithmetic surfaces, we show that the intersection matrix associated to a closed fiber is negative semi-definite (Theorem 1.23). We also show relations between numerical invariants of the generic fiber and those of the closed fibers (Proposition 1.35 and Theorem 1.37). The intersection theory presented here is rather elementary, even though it is well adapted to the case of regular surfaces. We invite the reader to study the subject more thoroughly with the book of Fulton [37] that treats intersection theory in a general setting.

Section 9.2 studies the relation between morphisms of regular fibered surfaces and intersection numbers. First, we classify birational morphisms (factorization theorem, 2.2). Next, we show the projection formula (Theorem 2.12) for dominant projective morphisms. We then use these results to show a strong version of the embedded resolution of curves in a regular surface (Theorem 2.26). Finally, we describe the local structure of arithmetic surfaces $X \to S$ whose closed fibers are normal crossings divisors (Proposition 2.34).

In Section 9.3, we start by proving Castelnuovo's criterion (Theorem 3.8), which characterizes integral vertical curves that can be contracted to a regular point. This criterion makes it possible to prove the existence of relatively minimal models in a birational equivalence class of a fixed fibered surface (Proposition 3.19), and also the existence of the minimal model for 'sufficiently general' surfaces (Theorem 3.21 and also Exercise 3.2). In the same spirit, we introduce the notion (and show the existence) of minimal desingularization and minimal desingularization with normal crossings (Propositions 3.32 and 3.36). The last section is more technical. There we show the numerical Artin contraction criterion (Theorem 4.7). Combined with Theorem 4.15, this generalizes Castelnuovo's criterion. Using this technique, we show the existence of the canonical model (Proposition 4.20). To conclude, we study Weierstrass models of elliptic curves in relation to the discriminant. We show that a Weierstrass model is minimal if and only if its minimal desingularization is the minimal regular model (Theorem 4.35).

Even if our principal interest is the study of arithmetic surfaces, we systematically include the geometric case (dim $S = 0$) when it does not weigh down the proof. The exceptions essentially concern Subsections 9.3.2, 9.3.3 (minimal surfaces), 9.4.3 (canonical model), and 9.4.4, whose subject is specific to arithmetic surfaces.

9.1   Intersection theory on a regular surface

In this section, we define the intersection number of two divisors (of which at least one is vertical) on a regular fibered surface. This provides a tool of foremost importance for the study of these surfaces. The intersection number is first defined locally for two effective divisors with no common component. It is then extended to the general case by bilinearity.

9.1.1   Local intersection

We fix a regular, Noetherian, connected scheme X of dimension 2. Let us recall that the Cartier divisors on X can be identified with the Weil divisors on X (Proposition 7.2.16). For any Cartier divisor D, we let $\mathcal{O}_X(D)$ denote the invertible sheaf associated to D. If D is effective, then $\mathcal{O}_X(-D)$ is a sheaf of ideals of $\mathcal{O}_X$. Consequently, D is naturally endowed with the closed subscheme structure $V(\mathcal{O}_X(-D))$ of X.

Let D and E be two effective divisors on X with no common irreducible component. Let $x \in X$ be a closed point. As $\operatorname{Supp} D \cap \operatorname{Supp} E = \{x\}$ or $\emptyset$ in a neighborhood of x, we have $\sqrt{\mathcal{O}_X(-D)_x + \mathcal{O}_X(-E)_x} \supseteq \mathfrak{m}_x\mathcal{O}_{X,x}$. Hence $\mathcal{O}_{X,x}/(\mathcal{O}_X(-D)_x + \mathcal{O}_X(-E)_x)$ is an Artinian ring, and, consequently, of finite length (Proposition 7.1.25).

Definition 1.1. Let D and E be two effective divisors on X, with no common irreducible component. Let $x \in X$ be a closed point. We call the integer

$$i_x(D, E) = \operatorname{length}_{\mathcal{O}_{X,x}} \mathcal{O}_{X,x}/(\mathcal{O}_X(-D)_x + \mathcal{O}_X(-E)_x)$$

the *intersection number* or *intersection multiplicity* of D and E at x. It is a non-negative integer, and $i_x(D, E) = 0$ if and only if $x \notin \operatorname{Supp} D \cap \operatorname{Supp} E$.

Example 1.2. Let $X = \operatorname{Spec} k[u, v]$ be the affine plane over a field k. Let $x \in X$ be the point corresponding to the maximal ideal (u, v), let $D = V(u)$ and $E = V(u + v^r)$ for some $r \geq 1$. Then $i_x(D, E) = \operatorname{length}(k[u, v]/(u, u + v^r)) = r$.

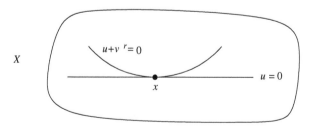

Figure 14. Two divisors which meet with multiplicity $r \geq 2$.

Example 1.3. Let $X = \operatorname{Spec} \mathbb{Z}[1/6, u, v]/(v^2 + u^3 + p)$, where p is a prime number different from 2 and 3. This is a regular scheme. Let $D = (u)$ and $E = (p)$. Then D and E meet at a unique point x corresponding to the ideal (p, u, v). We have $i_x(D, E) = 2$.

Lemma 1.4. *Let X be a regular, Noetherian, connected scheme of dimension 2, $x \in X$ a closed point. For any triplet of effective Cartier divisors D, E, F that, pairwise, have no common component, the following properties are true.*

(a) $i_x(D, E) = i_x(E, D)$.
(b) *Let $j : E \to X$ be the closed immersion. Then the divisor $D|_E := j^*D$ (see Lemma 7.1.29(b)) is an effective Cartier divisor on E, and we have the isomorphism and equality*

$$\mathcal{O}_E(D|_E) \simeq \mathcal{O}_X(D)|_E, \quad i_x(D, E) = \operatorname{mult}_x(D|_E)$$

if $x \in E$ (Definition 7.1.27).
(c) $i_x(D + F, E) = i_x(D, E) + i_x(F, E)$.

Proof (a) The symmetry of $i_x(\cdot, \cdot)$ follows from the definition.

(b) By Corollary 8.2.18 and Proposition 8.2.15(b), the scheme D has no embedded point. Therefore, $D|_E$ is well defined, and its properties are contained in Lemma 7.1.29. (We can also see this by using the fact that the local rings of X are unique factorization domains.) As

$$\mathcal{O}_{E,x}/\mathcal{O}_E(-D|_E)_x = \mathcal{O}_{X,x}/(\mathcal{O}_X(-D)_x + \mathcal{O}_X(-E)_x),$$

we have

$$\operatorname{mult}_x(D|_E) = \operatorname{length}_{\mathcal{O}_{E,x}} \mathcal{O}_{E,x}/\mathcal{O}_E(-D|_E)_x = i_x(D, E).$$

(c) We have $(D + F)|_E = D|_E + F|_E$. It suffices to apply (b) and the additivity of the function mult_x (Definition 7.1.27). $\qquad\square$

Corollary 1.5. *Let D, E be two divisors on X without common component. We can write $D = D_1 - D_2$, $E = E_1 - E_2$, with D_i, E_j effective and pairwise without common component. Let us set*

$$i_x(D, E) := i_x(D_1, E_1) - i_x(D_1, E_2) - i_x(D_2, E_1) + i_x(D_2, E_2).$$

Then this number is independent of the choice of the D_i and E_j. Moreover, $i_x(\cdot, \cdot)$ is symmetric bilinear in its definition domain (that is, the equalities of Lemma 1.4(a), (c) are true for Cartier divisors D, E, F which are pairwise without common component).

Proof This immediately follows from Lemma 1.4. $\qquad\square$

Definition 1.6. Let Y be a regular Noetherian scheme, and let D be an effective Cartier divisor on Y. We say that D has *normal crossings* at a point $y \in Y$ if there exist a system of parameters $f_1, \ldots, f_n$ of Y at y, an integer $0 \le m \le n$, and integers $r_1, \ldots, r_m \ge 1$ such that $\mathcal{O}_Y(-D)_y$ is generated by $f_1^{r_1} \cdots f_m^{r_m}$. We say that D has normal crossings if it has normal crossings at every point $y \in Y$. We say that the prime divisors $D_1, \ldots, D_\ell$ meet *transversally* at $y \in Y$ if they are pairwise distinct and if the divisor $D_1 + \cdots + D_\ell$ has normal crossings at y.

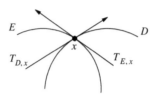

Figure 15. Two divisors that meet transversally.

Remark 1.7. Under the hypotheses of Definition 1.6, we sometimes say that D has *strictly normal crossings*, keeping the term normal crossings for the more general case that there exists an étale morphism $\pi : Z \to Y$ such that $\pi^* D$ has normal crossings in the sense of Definition 1.6.

Let D be an effective Cartier divisor on X. The closed immersion $D \to X$ induces an injective linear map on the tangent spaces $T_{D,x} \to T_{X,x}$ for every $x \in X$. We will identify $T_{D,x}$ with its image in $T_{X,x}$.

Proposition 1.8. *Let X be a regular, Noetherian, connected scheme of dimension 2. Let D, E be two distinct prime divisors, and $x \in \operatorname{Supp} D$. The following properties are true.*

(a) *The divisor D has normal crossings at x if and only if the scheme D is regular at x.*

(b) *Let us moreover suppose that $x \in \operatorname{Supp} D \cap \operatorname{Supp} E$. Then the following properties are equivalent:*

  (i) *D and E meet transversally at x.*

  (ii) *The maximal ideal $\mathfrak{m}_x$ of $\mathcal{O}_{X,x}$ is generated by $\mathcal{O}_X(-D)_x$ and $\mathcal{O}_X(-E)_x$.*

  (iii) *$i_x(D, E) = 1$.*

  (iv) *The schemes D and E are regular at x, and we have $T_{D,x} \oplus T_{E,x} = T_{X,x}$.*

Proof (a) follows from Corollary 4.2.15.

(b) The equivalence (i) $\Longleftrightarrow$ (ii) follows directly from the definition. The equivalence (ii) $\Longleftrightarrow$ (iii) follows from the fact that a non-zero module has length 1 if and only if it is a field.

Let f, g be respective generators of $\mathcal{O}_X(-D)_x$ and $\mathcal{O}_X(-E)_x$. Let $d_x f$, $d_x g$ be their respective differentials at x. Then $T_{D,x} = (d_x f)^\perp \subset T_{X,x}$ and

$T_{E,x} = (d_x g)^\perp$. Let us suppose that (ii) is verified. Then $\mathfrak{m}_x = (f, g)$. Hence $T_{D,x} + T_{E,x} = T_{X,x}$. In addition, the maximal ideal of $\mathcal{O}_{D,x}$ is generated by the image of g in $\mathcal{O}_{D,x}$, and hence $\dim_{k(x)} T_{D,x} \leq 1$. This implies that D is regular at x. The same holds for E. A dimension argument now shows that the spaces $T_{D,x}$ and $T_{E,x}$ form a direct sum. Conversely, let us suppose that we have (iv). Then $d_x f$ and $d_x g$ generate $T_{X,x}^\vee$, and hence $(f, g) = \mathfrak{m}_x$ by Nakayama. $\qquad\square$

We conclude this subsection with the 'moving lemma', which can be useful in computing the global intersection number of two divisors on a regular fibered surface.

Lemma 1.9. *Let K be a field and let $\nu_1, \ldots, \nu_r$ be non-trivial discrete valuations of K that are pairwise non-equivalent. The following properties are true.*

(a) *For any $m \geq 0$, there exists an $h \in K$ such that $\nu_1(h - 1) \geq m$ and $\nu_i(h) \geq m$ for every $i \geq 2$.*

(b) *(Approximation theorem) For any i, we suppose given an element $f_i \in K$ and an integer $n_i \in \mathbb{Z}$. Then there exists an $f \in K$ such that $\nu_i(f - f_i) = n_i$ for every i.*

Proof (a) Let us show by induction on r that there exists an $a \in K$ such that $\nu_1(a) < 0$ and $\nu_i(a) > 0$ if $i \geq 2$. Let us first note that for two distinct discrete valuations ν and ν', there exists an $e \in K$ such that $\nu(e) \leq 0$ and $\nu'(e) > 0$. Let us first suppose $r = 2$. Let $b, c \in K$ be such that $\nu_1(b) \leq 0$, $\nu_2(b) > 0$, and $\nu_1(c) > 0$, $\nu_2(c) \leq 0$. Then $a = b/c$ is appropriate. Let us suppose $r \geq 3$. Let $e \in K$ be such that $\nu_1(e) < 0$ and $\nu_r(e) > 0$. By the induction hypothesis, there exists an $f \in K$ such that $\nu_1(f) < 0$ and $\nu_i(f) > 0$ if $2 \leq i \leq r - 1$. If $\nu_r(f) \geq 0$, then $a = ef^n$ is appropriate for n large enough. If, on the contrary, $\nu_r(f) < 0$, then for any $n > 0$, we have

$$\nu_i(ef^n/(1 + f^n)) = \begin{cases} \nu_i(e) + n\nu_i(f) & \text{if} \quad 2 \leq i \leq r - 1 \\ \nu_i(e) & \text{if} \quad i = 1, r. \end{cases}$$

Hence $a = ef^n/(1 + f^n)$ is appropriate for n large enough. Consequently, for any $r \geq 2$, there exists an $a \in K$ such that $\nu_1(a) < 0$ and $\nu_i(a) > 0$ for $2 \leq i \leq r$.

Let $a \in K$ be as above, and let $m \geq 0$. Taking $h = a^m/(1 + a^m)$, we have the desired properties.

(b) Let $g_i \in K$ be such that $\nu_i(g_i - f_i) = n_i$. We fix an integer $m \geq n_i - \nu_i(g_j) + 1$ for every i, j. By (a), there exist $h_i \in K$ such that $\nu_i(h_i - 1) \geq m$ and $\nu_j(h_i) \geq m$ if $j \neq i$. Let us set

$$f = h_1 g_1 + h_2 g_2 + \cdots + h_r g_r.$$

Then we have $\nu_i(f - g_i) \geq n_i + 1$. Hence $\nu_i(f - f_i) = \nu_i(f - g_i + g_i - f_i) = n_i$. $\quad\square$

Corollary 1.10 ('Moving lemma'). *Let X be a normal, Noetherian, connected, separated scheme. Let D and E be two Weil divisors on X. Then there exists an $f \in K(X)$ such that $(f) + D$ and E have no common component.*

Proof Let $\nu_1, \ldots, \nu_r$ be the discrete valuations of $K(X)$ induced by the generic points ξ_i of the irreducible components Γ_i of $\operatorname{Supp} E$. Let us show that these valuations are pairwise distinct. Let us, for example, suppose that $\nu_1 = \nu_2$. Then $\mathcal{O}_{X,\xi_1} = \mathcal{O}_{X,\xi_2} =: A$. The canonical morphisms $\phi_i : \operatorname{Spec} A = \operatorname{Spec} \mathcal{O}_{X,\xi_i} \to X$, $i = 1, 2$, coincide at the generic point of $\operatorname{Spec} A$. Hence they coincide on a non-empty open subset of $\operatorname{Spec} A$. By Proposition 3.3.11 (since X is presumed separated), we have $\phi_1 = \phi_2$; hence $\xi_1 = \xi_2$, which is impossible.

Let us write $D = \sum_{1 \le i \le r} n_i \Gamma_i + D'$, where D' is a divisor with no common component with E, and $n_i \in \mathbb{Z}$. By Lemma 1.9(b), there exists an $f \in K(X)$ such that $\nu_i(f) = -n_i$ for every $i \le r$. It follows that $\operatorname{mult}_{\xi_i}(D + (f)) = 0$ for every $i \le r$. Hence $D + (f)$ and E do not have any common component. $\square$

On a regular scheme, we have a finer version of the moving lemma. Let us recall that the support $\operatorname{Supp} D$ of a Cartier divisor D on a Noetherian scheme X is the set of $x \in X$ such that $\mathcal{O}_X(D)_x \ne \mathcal{O}_{X,x}$ as subgroups of $\mathcal{K}_{X,x}$. If X is regular, we have $\operatorname{Supp} D = \operatorname{Supp}[D]$ (Exercise 7.2.4(b)).

Proposition 1.11. *Let X be a regular, connected, quasi-projective scheme over an affine Noetherian scheme. Let D be a Cartier divisor on X, and let $x_1, \ldots, x_m \in X$. Then there exists a $D' \sim D$ such that $x_i \notin \operatorname{Supp} D'$ for every $i \le m$.*

Proof We may suppose that the x_i are pairwise distinct. Using the equivalence between Cartier divisors and Weil divisors on X (Propositions 7.2.14(c) and 7.2.16), we can assume that D is effective. As the x_i belong to the same affine open subscheme (Proposition 3.3.36), we can suppose that X is affine. Moreover, by replacing x_i by a point belonging to $\overline{\{x_i\}}$, we can suppose that x_i is closed. Let us show the proposition by induction on m. Let us set $x = x_m$. As X is affine, there exists a $\pi \in H^0(X, \mathcal{O}_X(-D))$ such that π_x generates $\mathcal{O}_X(-D)_x$. We have $\mathcal{O}_X(D + \operatorname{div}(\pi))_x = \mathcal{O}_{X,x}$; hence $x \notin \operatorname{Supp}(D + \operatorname{div}(\pi))$ and the proposition is true for $m = 1$. Let us suppose $m \ge 2$ and the proposition true for $m - 1$. By replacing D by a linearly equivalent Cartier divisor, we can suppose that $x_i \notin \operatorname{Supp} D$ for every $i \le m - 1$, and that $x \in \operatorname{Supp} D$.

Let I (resp. $\mathfrak{m}_i$) be the radical ideal of $H^0(X, \mathcal{O}_X)$ corresponding to the closed subset $\operatorname{Supp} D$ (resp. to the point x_i). Then there exists an integer $r \ge 1$ such that $I^r \subseteq H^0(X, \mathcal{O}_X(-D))$. For any $i \le m - 1$, we have $x_i \notin \operatorname{Supp} D \cup \{x_1, \ldots, x_{i-1}, x_{i+1}, \ldots, x_m\}$. There therefore exist $g_i \in I \cap \cap_{j \ne i, m} \mathfrak{m}_j$ such that $g_i \notin \mathfrak{m}_i$. Let us set

$$f = \pi + \sum_{1 \le i \le m-1} a_i g_i^{r+1},$$

where the $a_i \in H^0(X, \mathcal{O}_X)$ are chosen such that $\pi + a_i g_i^{r+1} \notin \mathfrak{m}_i$ for each $i \le m-1$ (this is possible because $g_i \notin \mathfrak{m}_i$). We then have $f \notin \mathfrak{m}_i$ for every $i \le m - 1$. Let us set $D' = D + \operatorname{div}(f) \ge 0$. For any $i \le m - 1$, $x_i \notin \operatorname{Supp} D'$ because $\mathcal{O}_X(D')_{x_i} = \mathcal{O}_X(D)_{x_i} = \mathcal{O}_{X,x_i}$. Finally, as $(g_i^{r+1})_x \in I \mathcal{O}_X(-D)_x \subseteq \pi_x \mathfrak{m}_x \mathcal{O}_{X,x}$, we have $f_x \mathcal{O}_{X,x} = \mathcal{O}_X(-D)_x$. Hence $\mathcal{O}_X(-D')_x = \mathcal{O}_{X,x}$ and $x \notin \operatorname{Supp} D'$. $\square$

9.1.2   Intersection on a fibered surface

Let us fix for the entire subsection a regular fibered surface $\pi : X \to S$ over a Dedekind scheme. The fact that S is not necessarily 'compact' (i.e., a proper algebraic variety over a field) makes it in general impossible to define the intersection of two arbitrary divisors on X. However, this obstacle can easily be circumvented when one of the divisors in question is vertical. This is what we are going to do in this subsection. Let us recall that on X and S, there is an equivalence between the Weil divisors and the Cartier divisors (Proposition 7.2.16). Let $s \in S$ be a closed point. The set $\mathrm{Div}_s(X)$ of Cartier divisors on X with support in X_s forms a subgroup of $\mathrm{Div}(X)$ (we have $\mathrm{Div}_s(X) = \mathrm{Div}(X)$ if $\dim S = 0$). If $\dim S = 1$, then $\mathrm{Div}_s(X)$ has the irreducible components of X_s as a basis.

Let E be an effective Cartier divisor on X such that $E \leq X_s$ (the condition is empty if $\dim S = 0$). Then the scheme E is a closed subscheme of X_s, and it is therefore a projective curve over $k(s)$.

Theorem 1.12. *Let $X \to S$ be a regular fibered surface. Let $s \in S$ be a closed point. Then there exists a unique bilinear map (of $\mathbb{Z}$-modules)*

$$i_s : \mathrm{Div}(X) \times \mathrm{Div}_s(X) \to \mathbb{Z}$$

which verifies the following properties:

(a) *If $D \in \mathrm{Div}(X)$ and $E \in \mathrm{Div}_s(X)$ have no common component, then*

$$i_s(D, E) = \sum_x i_x(D, E)[k(x) : k(s)],$$

where x runs through the closed points of X_s.

(b) *The restriction of i_s to $\mathrm{Div}_s(X) \times \mathrm{Div}_s(X)$ is symmetric.*

(c) *$i_s(D, E) = i_s(D', E)$ if $D \sim D'$.*

(d) *If $0 < E \leq X_s$, then*

$$i_s(D, E) = \deg_{k(s)} \mathcal{O}_X(D)|_E$$

(Definition 7.3.29).

Proof   The uniqueness of i_s follows from its bilinearity and from property (d) since $\mathrm{Div}_s(X)$ is generated by the prime divisors with support in X_s (see also Exercise 1.4). Let us now construct i_s. Let $(D, E) \in \mathrm{Div}(X) \times \mathrm{Div}_s(X)$. Let us write $E = \sum_{1 \leq i \leq r} n_i \Gamma_i$ with $n_i \in \mathbb{Z}$, and where the Γ_i are the irreducible components of X_s. Let us set

$$i_s(D, E) = \sum_{1 \leq i \leq r} n_i \deg_{k(s)} \mathcal{O}_X(D)|_{\Gamma_i}. \tag{1.1}$$

This is a map that is bilinear in D because the function deg is a group homomorphism. And it is clearly bilinear in E. It remains to show that i_s verifies properties (a)–(d).

(a) By the bilinearity of i_s and of the i_x (Corollary 1.5), we can suppose that D and E are two distinct prime divisors. By Lemma 1.4, we have $\mathcal{O}_X(D)|_E = \mathcal{O}_E(D|_E)$, $[D|_E] = \sum_x i_x(D,E)[x]$, and therefore

$$\deg_{k(s)} \mathcal{O}_X(D)|_E = \deg_{k(s)}[D|_E] = \sum_x i_x(D,E)[k(x):k(s)].$$

(b) Let us write $D = \sum_j m_j \Gamma_j$ and $E = \sum_i n_i \Gamma_i$. Then

$$i_s(D,E) = \sum_{i,j} n_i m_j i_s(\Gamma_i, \Gamma_j).$$

We have $i_s(\Gamma_i, \Gamma_j) = i_s(\Gamma_j, \Gamma_i)$. Indeed, there is nothing to prove if $\Gamma_i = \Gamma_j$. Otherwise, we apply (a) and Lemma 1.4(a).

(c) By bilinearity, we can suppose E is prime. If $D \sim D'$, then $\mathcal{O}_X(D)|_E \simeq \mathcal{O}_X(D')|_E$, whence $\deg_{k(s)} \mathcal{O}_X(D)|_E = \deg_{k(s)} \mathcal{O}_X(D')|_E$ (Lemma 7.3.30(b)).

(d) Let us write $E = \sum_{1 \le i \le r} n_i \Gamma_i$ with $n_i \ge 0$ and the Γ_i prime. Let ξ_i be the generic point of Γ_i. Then

$$\mathcal{O}_{E,\xi_i} = \mathcal{O}_{X,\xi_i}/\mathcal{O}_X(-E)_{\xi_i} = \mathcal{O}_{X,\xi_i}/\mathcal{O}_X(-n_i\Gamma_i)_{\xi_i}.$$

As in the proof of Lemma 8.3.9(b), we have $\operatorname{length} \mathcal{O}_{E,\xi_i} = n_i$. By Proposition 7.5.7, we have

$$\deg_{k(s)} \mathcal{O}_X(D)|_E = \sum_i n_i \deg_{k(s)} \mathcal{O}_X(D)|_{\Gamma_i} = i_s(D,E),$$

whence (d).   □

Remark 1.13. Let $(D,E) \in \operatorname{Div}(X) \times \operatorname{Div}_s(X)$. Let $A = \mathcal{O}_{S,s}$. Then the intersection number $i_s(D,E)$ can be computed after the base change $\operatorname{Spec} A \to S$. Indeed, we can suppose that E is a prime divisor. The closed immersion $i : E \to X$ then decomposes into the closed immersion $j : E \to Y := X_A$ followed by the projection $p : Y \to X$. We have

$$\mathcal{O}_X(D)|_E = i^* \mathcal{O}_X(D) = j^* \mathcal{O}_Y(p^*D),$$

and hence $i_s(D,E) = \deg_{k(s)} \mathcal{O}_X(D)|_E = \deg_{k(s)} \mathcal{O}_Y(p^*D)|_E$.

Remark 1.14. If $E \in \operatorname{Div}_s(X)$ is a prime divisor (therefore an integral curve over $k(s)$), then

$$i_s(D,E) = \deg_{k(s)} f^*(\mathcal{O}_X(D)),$$

where f is the composition of the normalization morphism $E' \to E$ and the closed immersion $E \to X$. This follows from Proposition 7.3.8. This identity is sometimes useful to define $i_s(D,E)$.

Definition 1.15. Let $\pi : X \to S$ be a regular fibered surface and $s \in S$ a closed point. For any $D \in \mathrm{Div}(X)$ and any $E \in \mathrm{Div}_s(X)$, we call the integer $i_s(D, E)$ the *intersection number of D and E*. The number $i_s(E, E)$ is called the *self-intersection* of E and denoted E^2. In general, if E is a vertical divisor, we denote by $D \cdot E$ the 0-cycle

$$D \cdot E := \sum_{s \in S} i_s(D, E)[s]$$

on S, where it is implied that if the support of $D \cdot E$ is concentrated in a point s, then we identify $D \cdot E$ with the number $i_s(D, E)$. In general, the context eliminates any possible ambiguity.

Remark 1.16. If $\dim S = 0$, then $\mathrm{Div}_s(X) = \mathrm{Div}(X)$, and i_s induces a symmetric bilinear form $\mathrm{Div}(X) \times \mathrm{Div}(X) \to \mathbb{Z}$. By Theorem 1.12, this factors through a bilinear form

$$\mathrm{Pic}(X) \times \mathrm{Pic}(X) \to \mathbb{Z}.$$

Let $\mathrm{Num}(X)$ be the quotient of $\mathrm{Pic}(X)$ by the subgroup of divisors D that are numerically equivalent to 0 (i.e., $D \cdot E = 0$ for every E). One can show that $\mathrm{Num}(X)$ is free and finitely generated over $\mathbb{Z}$, and that the bilinear form has signature $(1, -1, \ldots, -1)$ on $\mathrm{Num}(X) \otimes_{\mathbb{Z}} \mathbb{R}$. This is Hodge's index theorem (see for example [43], Theorem V.1.9).

Remark 1.17. Let D, F be two Cartier divisors on X with no common component. Then we can define the 0-cycle $D \cdot F$ on S by setting

$$D \cdot F = \sum_{s \in S} \left(\sum_{x \in X_s} i_x(D, F)[k(x) : k(s)] \right) [s].$$

This operation is bilinear and symmetric in its domain of definition (Corollary 1.5), but is not compatible with the linear equivalence relation (see Remark 1.33). See also Remark 2.13.

Remark 1.18. Let $X \to S$ be a not necessarily regular fibered surface. Let D, E be divisors on X with E vertical with support in X_s and D Cartier; then we can still define the intersection number $i_s(D, E)$ by once more taking formula (1.1) of the proof of Theorem 1.12. If E is contained in $\mathrm{Reg}(X)$ and is such that $0 < E \le X_s$, then E^2 is defined, and is equal to $\deg_{k(s)} \mathcal{O}_X(E)|_E$. See [57], Section 13. Intersection theory on a normal surface is described for instance in [71], II (b).

Example 1.19. Let us suppose $S = \mathrm{Spec}\, k$ with k a field. Let D and E be two prime divisors on X, with no common component. Then

$$D \cdot E \ge \text{number of intersection points of } D \text{ and } E,$$

with equality if and only if D and E meet transversally at points that are rational over k. In this case, the intersection number coincides with the intuition. As a particular case, if $X = \mathbb{P}_k^2$, then two distinct lines H_1, H_2 on X meet transversally at a rational point. Hence $H_1 \cdot H_2 = 1$.

Corollary 1.20 (Bézout identity). *Let $X = \mathbb{P}^2_k$ be the projective plane over a field k. Let $F, G \in k[x, y, z]$ be two coprime homogeneous polynomials of respective degrees m and n. Then we have*

$$V_+(F) \cdot V_+(G) = mn.$$

In particular, the set $V_+(F) \cap V_+(G)$ contains at most mn points.

Proof We have seen (Proposition 7.2.9) that $V_+(F) \sim mH_1$ and $V_+(G) \sim nH_2$, where H_1, H_2 are arbitrary planes on X. We can therefore take $H_1 \neq H_2$. By Theorem 1.12 and the example above, we have $V_+(F) \cdot V_+(G) = mnH_1 \cdot H_2 = mn$.   □

Proposition 1.21. *Let $\pi : X \to S$ be an arithmetic surface (Definition 8.3.14), and let $s \in S$ be a closed point. The following properties are true.*

(a) *For any $E \in \mathrm{Div}_s(X)$, we have $E \cdot X_s = 0$.*

(b) *Let $\Gamma_1, \ldots, \Gamma_r$ be the irreducible components of X_s, of respective multiplicities $d_1, \ldots, d_r$. Then for any $i \leq r$, we have*

$$\Gamma_i^2 = -\frac{1}{d_i} \sum_{j \neq i} d_j \, \Gamma_j \cdot \Gamma_i.$$

Proof (a) By virtue of Remark 1.13, we can suppose S is local, and hence s is a principal divisor. It follows that $X_s = \pi^* s$ (Lemma 8.3.9) is also a principal divisor, whence $E \cdot X_s = X_s \cdot E = 0$ by Theorem 1.12(c). Let $i \leq r$. Then

$$0 = \Gamma_i \cdot X_s = \sum_{1 \leq j \leq r} d_j \, \Gamma_i \cdot \Gamma_j = d_i \Gamma_i^2 + \sum_{j \neq i} d_j \, \Gamma_j \cdot \Gamma_i,$$

which implies equality (b).   □

Example 1.22. Let $\mathcal{O}_K$ be a discrete valuation ring with uniformizing parameter t, field of fractions K, and residue field k such that char $k \neq 2$. Let us write $\mathbb{P}^1_{\mathcal{O}_K} = \mathrm{Spec}\, \mathcal{O}_K[x] \cup \mathrm{Spec}\, \mathcal{O}_K[1/x]$. Let X be the integral closure of $\mathbb{P}^1_{\mathcal{O}_K}$ in the field $K[y]/(y^2 - x^6 + t)$. Then X is the union of the open subschemes

$$U = \mathrm{Spec}\, \mathcal{O}_K[x, y]/(y^2 - x^6 + t), \quad W = \mathrm{Spec}\, \mathcal{O}_K[v, z]/(z^2 - 1 + tv^6)$$

(we identify v with $1/x$ and z with y/x^3). One can immediately verify that U and W are regular. Thus X is an arithmetic surface over $S = \mathrm{Spec}\, \mathcal{O}_K$. The curve U_s, where s is the closed point of S, has two irreducible components $V(y - x^3)$ and $V(y + x^3)$. Let Γ_1, Γ_2 be their respective closures in X. They are the irreducible components of X_s. They intersect at a unique point $p \in U$ corresponding to the maximal ideal (t, x, y). We have

$$\Gamma_1 \cdot \Gamma_2 = \mathrm{length}\, \mathcal{O}_{X,p}/(y - x^3, y + x^3) = \mathrm{length}\, k[x, y]/(x^3, y) = 3.$$

Therefore the appearance of X_s is represented by Figure 16. It follows from Proposition 1.21 that $\Gamma_1^2 = \Gamma_2^2 = -3$.

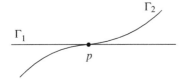

Figure 16. Two divisors that meet with multiplicity 3.

Let us keep the hypotheses of Proposition 1.21. Let us denote the real vector space $\mathrm{Div}_s(X) \otimes_{\mathbb{Z}} \mathbb{R}$ by $\mathrm{Div}_s(X)_{\mathbb{R}}$. It is of finite dimension and has the images of the irreducible components of X_s as a basis. By tensoring with $\mathbb{R}$, the map i_s induces a symmetric bilinear form $\langle \cdot, \cdot \rangle_s : \mathrm{Div}_s(X)_{\mathbb{R}} \times \mathrm{Div}_s(X)_{\mathbb{R}} \to \mathbb{R}$.

Theorem 1.23. *Let $X \to S$ be an arithmetic surface and $s \in S$ a closed point. Then the bilinear form $\langle \cdot, \cdot \rangle_s$ on $\mathrm{Div}_s(X)_{\mathbb{R}}$ is negative semi-definite. Moreover, if X_s is connected (e.g., if the generic fiber is geometrically connected), then $\langle v, v \rangle_s = 0$ if and only if $v \in X_s \mathbb{R}$.*

Proof Let $\Gamma_1, \ldots, \Gamma_r$ be the irreducible components of X_s and d_i the multiplicity of Γ_i in X_s. We can suppose $r \geq 2$. Let us set $a_{ij} = \Gamma_i \cdot \Gamma_j$ and $b_{ij} = d_i d_j a_{ij}$. We have $b_{ij} \geq 0$ if $i \neq j$, and for any $i \leq r$, $\sum_j b_{ij} = X_s \cdot (d_i \Gamma_i) = 0$ by Proposition 1.21. Likewise, $\sum_i b_{ij} = 0$ by symmetry. Let $v = \sum_i x_i \Gamma_i \in \mathrm{Div}_s(X)_{\mathbb{R}}$. Let us set $y_i = x_i / d_i$. We then have

$$\langle v, v \rangle_s = \sum_{1 \leq i,j \leq r} a_{ij} x_i x_j = \sum_{1 \leq i,j \leq r} b_{ij} y_i y_j = -\sum_{1 \leq i,j \leq r,\ i<j} b_{ij}(y_i - y_j)^2 \leq 0.$$

This shows that $\langle \cdot, \cdot \rangle_s$ is negative semi-definite. If $\langle v, v \rangle_s = 0$, then $y_i = y_j$ if $b_{ij} \neq 0$, that is to say, if $\Gamma_i \cap \Gamma_j \neq \emptyset$. Consequently, the y_i are equal for the Γ_i belonging to the same connected component of X_s. It follows that $y_i = y_j$ for every i, j if X_s is connected. This implies that $v = X_s y_1$. Conversely, every element of $X_s \mathbb{R}$ is isotropic by Proposition 1.21(a).   □

Corollary 1.24. *Let $\pi : X \to S$ be an arithmetic surface over the spectrum of a discrete valuation ring. Let $\Gamma_1, \ldots, \Gamma_r$ be the irreducible components of X_s, of respective multiplicities $d_1, \ldots, d_r$. Let us suppose that the d_i have no common divisor and that $k(s)$ is perfect. Then $X \to S$ is cohomologically flat (Exercise 5.3.14). See also Exercise 1.14.*

Proof Let us first suppose that $\mathcal{O}_S = \pi_* \mathcal{O}_X$. Hence X_s is geometrically connected (Corollary 8.3.6). Let $Z_0 = (X_s)_{\mathrm{red}}$. Then $H^0(Z_0, \mathcal{O}_{Z_0})$ is reduced. Let Z be a divisor such that $Z_0 \leq Z \leq X_s$ and that $H^0(Z, \mathcal{O}_Z)$ is reduced. Let us suppose that $Z < X_s$. We are going to construct a divisor $Z < Z' \leq X_s$ such that $H^0(Z', \mathcal{O}_{Z'})$ is reduced. The hypothesis on the d_i implies that $X_s - Z \notin X_s \mathbb{R}$. It follows from Theorem 1.23 that $(X_s - Z)^2 < 0$. There therefore exists an irreducible component Γ of X_s such that $(X_s - Z) \cdot \Gamma < 0$. This implies that

$X_s - Z \geq \Gamma$. Let us consider $Z' = Z + \Gamma$. Then $Z < Z' \leq X_s$. We have the exact sequence

$$0 \to \mathcal{O}_X(-Z)/\mathcal{O}_X(-Z') \to \mathcal{O}_X/\mathcal{O}_X(-Z') \to \mathcal{O}_X/\mathcal{O}_X(-Z) \to 0$$

which is, term by term, isomorphic to the exact sequence

$$0 \to \mathcal{O}_X(-Z)|_\Gamma \to \mathcal{O}_{Z'} \to \mathcal{O}_Z \to 0. \tag{1.2}$$

As $(-Z) \cdot \Gamma < 0$, we have $H^0(\Gamma, \mathcal{O}_X(-Z)|_\Gamma) = 0$, whence an exact sequence

$$0 \to H^0(Z', \mathcal{O}_{Z'}) \to H^0(Z, \mathcal{O}_Z)$$

which implies that $H^0(Z', \mathcal{O}_{Z'})$ is reduced. By repeating this process a finite number of times, we deduce that $H^0(X_s, \mathcal{O}_{X_s})$ is reduced. As X_s, and therefore $\operatorname{Spec} H^0(X_s, \mathcal{O}_{X_s})$, is geometrically connected over $k(s)$, $H^0(X_s, \mathcal{O}_{X_s})$ is a purely inseparable field extension of $k(s)$, and hence equal to $k(s)$. Consequently, $X \to S$ is cohomologically flat (Corollary 5.3.22).

Let us now show the general case. Let $\pi' : X \to S'$, $S' \to S$ be the decomposition of $X \to S$ as in Proposition 8.3.8, with $\mathcal{O}_{S'} = \pi'_* \mathcal{O}_X$. Let us first show that $X \to S'$ is cohomologically flat. Let $s' \in S'$ be a closed point and let $X' = X \times_{S'} \operatorname{Spec} \mathcal{O}_{S',s'}$. Any irreducible component Γ' of $X'_{s'}$ dominates an irreducible component Γ of X_s. Let ξ, ξ' be the respective generic points of Γ, Γ'. Then $\mathcal{O}_{X,\xi} = \mathcal{O}_{X',\xi'}$. Therefore the multiplicity of Γ' in $X'_{s'}$, which is also the ramification index of $\mathcal{O}_{S',s'} \to \mathcal{O}_{X',\xi'}$, divides the multiplicity of Γ in X_s. By the first case, $X' \to \operatorname{Spec} \mathcal{O}_{S',s'}$ is cohomologically flat. Therefore so is $X \to S'$ (Exercise 5.3.14(a)). Consequently, $X \to S'$ is cohomologically flat.

Let t be a uniformizing parameter for $\mathcal{O}_{S,s}$. Let

$$Y = X \times_{S'} \operatorname{Spec}(\mathcal{O}(S')/(t)).$$

Then $H^0(X, \mathcal{O}_X) \otimes_{\mathcal{O}_{S,s}} k(s) = H^0(X, \mathcal{O}_X) \otimes_{\mathcal{O}(S')} \mathcal{O}(S')/(t) \simeq H^0(Y, \mathcal{O}_Y)$. Now $Y = X_s$. Hence $X \to S$ is cohomologically flat. $\qquad\square$

Remark 1.25. If $k(s)$ is not perfect, we have the same result if we suppose that the gcd d' of the $d_i e_i$, where e_i is the geometric multiplicity of Γ_i (Exercise 1.9), is equal to 1. More generally, it suffices to suppose that d' is prime to $\operatorname{char}(k(s))$ ([81], Théorème 7.2.1).

Let $f : X \to Y$ be a birational morphism of fibered surfaces over S with X regular and Y normal. Let $y \in Y$ be a closed point such that $\dim X_y \geq 1$. By Corollary 5.3.16, X_y is a connected curve over $k(y)$. Let $\Gamma_1, \ldots, \Gamma_r$ denote the irreducible components of X_y. These are vertical divisors contained in X_s, where s is the image of y in S. We are going to show an analogue of Theorem 1.23 with the bilinear form defined by the intersection numbers $i_s(\Gamma_i, \Gamma_j)$.

Lemma 1.26. *With the notation above, there exists an effective divisor $D = \sum_{1 \le i \le r} d_i \Gamma_i$ with $d_i > 0$, $D \cdot \Gamma_i \le 0$, and $D^2 < 0$.*

Proof Let $C > 0$ be an effective divisor on Y whose support contains y. We have a decomposition

$$f^*C = \tilde{C} + D,$$

where $\tilde{C}$ is an effective divisor whose support does not have any irreducible component contained in X_y, and where $D = \sum_i d_i \Gamma_i$ with $d_i \ge 0$. There exists an affine open neighborhood V of y such that $\mathcal{O}_Y(C)|_V = a^{-1}\mathcal{O}_V$ with an $a \in \mathcal{O}_Y(V)$. It follows that $\mathcal{O}_X(f^*C)|_{f^{-1}(V)}$ is free. Consequently, $f^*C \cdot \Gamma_i = 0$ for every $i \le r$. On the other hand, $\tilde{C}$ and Γ_i have no common component, and hence $\tilde{C} \cdot \Gamma_i \ge 0$. Thus

$$D \cdot \Gamma_i = f^*C \cdot \Gamma_i - \tilde{C} \cdot \Gamma_i \le 0.$$

Let ν_i be the discrete valuation of $K(X)$ associated to the generic point of Γ_i; then $d_i = \nu_i(a) > 0$ because $a_y \in \mathfrak{m}_y \mathcal{O}_{Y,y}$. It remains to show that $D^2 < 0$. From a set-theoretical point of view, we have $f^*C = f^{-1}(C) = \tilde{C} \cup X_y$; hence $f(\tilde{C})$ is a closed subset of C containing $C \setminus \{y\}$. This implies that $\tilde{C} \to C$ is surjective. Therefore $\tilde{C}$ meets X_y. Let Γ_{i_0} be such that $\Gamma_{i_0} \cap \tilde{C} \ne \emptyset$. Then the computations above show that $D \cdot \Gamma_{i_0} < 0$, which implies that $D^2 \le d_{i_0} D \cdot \Gamma_{i_0} < 0$. □

Theorem 1.27. *Let $f : X \to Y$ be a birational morphism of fibered surfaces over S with X regular and Y normal. Let $y \in Y$ be a closed point such that $\dim X_y \ge 1$. Let $\Gamma_1, \ldots, \Gamma_r$ denote the irreducible components of X_y. Let $x_1, \ldots, x_r \in \mathbb{R}$. Then we have*

$$\sum_{1 \le i \le r} x_i x_j \Gamma_i \cdot \Gamma_j \le 0,$$

the equality taking place only if $x_1 = \cdots = x_r = 0$.

Proof The proof is similar to that of Theorem 1.23, with the divisor D defined as in Lemma 1.26 instead of X_s. With the notation of the proof of Theorem 1.23, we have $b_{ij} \ge 0$ if $i \ne j$, $\sum_j b_{ij} = D \cdot (d_i \Gamma_i) \le 0$, and

$$\sum_{i,j} b_{ij} y_i y_j = \sum_i \left(\sum_j b_{ij} \right) y_i^2 - \sum_{i<j} b_{ij}(y_i - y_j)^2 \le 0. \tag{1.3}$$

If the inequality is an equality, then $\sum_i d_i(D \cdot \Gamma_i)y_i^2 = 0$. The $D^2 < 0$ hypothesis implies that $y_i = 0$ for at least one $i \le r$. Using (1.3) and the connectedness of X_y, we deduce from this that all of the y_i are zero. □

Remark 1.28. Theorem 1.27 remains true if we replace Y by a normal Noetherian scheme of dimension 2, f by a projective birational morphism with X regular, and $i_s(\Gamma_i, \Gamma_j)$ by $i_y(\Gamma_i, \Gamma_j) := \deg_{k(y)} \mathcal{O}_X(\Gamma_i)|_{\Gamma_j}$. The proof is the same.

9.1.3   Intersection with a horizontal divisor, adjunction formula

Proposition 1.21 and Theorems 1.23, 1.27 give information concerning the intersection of two vertical divisors. In this subsection, let us study the situation where one of the divisors is horizontal. On account of Proposition 8.3.4, a horizontal prime divisor is just the closure in X of a closed point of the generic fiber. In all of this subsection, $X \to S$ will be an arithmetic surface.

Lemma 1.29. *Let $\pi : X \to S$ be an arithmetic surface. Let D be a horizontal prime divisor and V a vertical divisor on X. Let $h = \pi|_D : D \to S$. Then the Weil divisor associated to the Cartier divisor $V|_D$ (Lemma 7.1.29) verifies the identity*

$$h_*[V|_D] = \sum_{s \in S} i_s(V, D)[s].$$

Proof   This is a consequence of Lemma 1.4(b) and of Definition 7.2.17.   □

Proposition 1.30. *Let $\pi : X \to S$ be an arithmetic surface. Let η be the generic point of S and $s \in S$ a closed point. Then for any closed point $P \in X_\eta$, we have*

$$\overline{\{P\}} \cdot X_s = [K(P) : K(S)],$$

where $\overline{\{P\}}$ is the Zariski closure of $\{P\}$ in X, endowed with the reduced closed subscheme structure.

Proof   Let us also denote the Cartier divisor corresponding to the point s by s, and let us denote the divisor $\overline{\{P\}}$ by D. Let $i : D \to X$ be the canonical closed immersion and $h : D \to S$ the finite surjective morphism $\pi \circ i$. As $X_s = \pi^* s$ (Lemma 8.3.9), we have $X_s|_D = i^*(\pi^* s) = h^* s$. By Theorem 7.2.18, we have

$$h_*[X_s|_D] = h_*[h^* s] = d[s], \quad \text{where } d = [K(D) : K(S)].$$

It follows from Lemma 1.29 that $D \cdot X_s = d$. As $K(D) = K(P)$, the proposition is proven.   □

Remark 1.31. Let $D \in \mathrm{Div}(X)$. We can write $D = D_1 + V_1 + V_2$, where D_1 is a combination of horizontal prime divisors, $V_1 \in \mathrm{Div}_s(X)$, and V_2 is vertical, with support in $\pi^{-1}(S \setminus \{s\})$. We then have $D \cdot X_s = D_1 \cdot X_s$. The divisor D induces, in a natural way, a divisor D_η on X_η, and we have $D_\eta = (D_1)_\eta$. It follows from Proposition 1.30 that

$$D \cdot X_s = \deg_{K(S)}(D_1)_\eta.$$

Corollary 1.32. *Let $X \to S$ be an arithmetic surface, $P \in X_\eta$ a K-rational point, and $D = \overline{\{P\}}$. Then $X_s \cap D$ is reduced to a point $p \in X_s(k(s))$, and X_s is smooth at p. In particular, p belongs to a single irreducible component of X_s, which is moreover of multiplicity 1 in X_s.*

Proof Let $\Gamma_1, \ldots, \Gamma_m$ be the irreducible components of X_s passing through p, with respective multiplicities $d_1, \ldots, d_m$ in X_s. Then

$$\sum_i d_i \, D \cdot \Gamma_i = D \cdot X_s = 1$$

by Proposition 1.30. It follows that $m = d_1 = D \cdot \Gamma_1 = 1$. Hence D and Γ_1 meet at a unique point p, which is moreover rational over $k(s)$ (Theorem 1.12(a)). We have X_s regular at p by virtue of Proposition 1.8. It is in fact smooth at p since $p \in X_s(k(s))$ (Proposition 4.3.30). See also Exercise 4.3.25(c).   □

Remark 1.33. Let $S = \operatorname{Spec}\mathbb{Z}$. Then any closed point $s \in S$ is a principal divisor. Hence X_s is a principal divisor. However, Proposition 1.30 implies that $D \cdot X_s \neq 0$ if $D > 0$ and D is horizontal. This shows that we cannot have a reasonable intersection theory (i.e., compatible with the linear equivalence of divisors) for two arbitrary divisors on X (as in the $\dim S = 0$ case). This comes from the fact that the base S is not 'compact'. One way to overcome this problem is to 'compactify' $\operatorname{Spec}\mathbb{Z}$. That is Arakelov theory, see [54] for an introduction. See also [37] for an intersection theory in arbitrary dimension.

Let $\pi : X \to S$ be a regular fibered surface. As π is a local complete intersection (Example 6.3.18), we have a canonical sheaf $\omega_{X/S}$ on X that is an invertible sheaf (Definition 6.4.7) and an r-dualizing sheaf (Theorem 6.4.32), where $r = 1$ is the relative dimension of $X \to S$.

Definition 1.34. Let $X \to S$ be a regular fibered surface. We call any Cartier divisor $K_{X/S}$ on X such that $\mathcal{O}_X(K_{X/S}) \simeq \omega_{X/S}$ a *canonical divisor* on X (relative to S). Such a divisor exists because X is integral (Corollary 7.1.19).

In what follows, we are going to study the intersection number of the canonical divisor with a vertical divisor E. Let us note that for any vertical divisor E, by virtue of Theorem 1.12(c), the intersection number $K_{X/S} \cdot E$ depends uniquely on $\omega_{X/S}$ and not on a choice of a representative $K_{X/S}$. Let C be a projective curve over a field k, and let us recall that, by definition, the *arithmetic genus* $p_a(C)$ is $1 - \chi_k(\mathcal{O}_C)$. If k is a finite extension of a field k', then C is also a projective curve over k'. The arithmetic genus of C is not, in general, the same over k' as over k. On a fibered surface $X \to S$, a curve C contained in a fiber X_s is considered as a curve over $k(s)$ (if $C \leq X_s$), unless explicitly mentioned otherwise.

Proposition 1.35. *Let $X \to S$ be an arithmetic surface, $s \in S$ a closed point, and $\Gamma_1, \ldots, \Gamma_r$ the irreducible components of X_s, with respective multiplicities $d_1, \ldots, d_r$. Let $K_{X/S}$ be a canonical divisor on X. Then we have*

$$2p_a(X_\eta) - 2 = K_{X/S} \cdot X_s = \sum_{1 \leq i \leq r} d_i \, K_{X/S} \cdot \Gamma_i.$$

Proof   By Corollary 8.3.6(d), we have $\omega_{X/S}|_{X_s} \simeq \omega_{X_s/k(s)}$. It follows that

$$\sum_i d_i\, K_{X/S} \cdot \Gamma_i = K_{X/S} \cdot X_s = \deg_{k(s)} \mathcal{O}_X(K_{X/S})|_{X_s} = \deg_{k(s)} \omega_{X_s/k(s)}.$$

In addition, we have

$$\deg_{k(s)} \omega_{X_s/k(s)} = -2\chi_{k(s)}(\mathcal{O}_{X_s}) = -2\chi_{k(\eta)}(\mathcal{O}_{X_\eta}) = 2p_a(X_\eta) - 2$$

(Corollary 7.3.31 and Proposition 5.3.28), which completes the proof.   □

Lemma 1.36. *Let X be a locally Noetherian scheme. Let D be an effective Cartier divisor on X. Then $\omega_{D/X} = \mathcal{O}_X(D)|_D$.*

Proof   By definition, the closed subscheme $D \subset X$ is defined by the sheaf of ideals $\mathcal{I} = \mathcal{O}_X(-D)$. Therefore the conormal sheaf $\mathcal{C}_{D/X}$ (Definition 6.3.7) verifies

$$\mathcal{C}_{D/X} = \mathcal{I}/\mathcal{I}^2 = \mathcal{I} \otimes_{\mathcal{O}_X} \mathcal{O}_X/\mathcal{I} = \mathcal{O}_X(-D) \otimes_{\mathcal{O}_X} \mathcal{O}_D = \mathcal{O}_X(-D)|_D,$$

whence

$$\omega_{D/X} = (\mathcal{O}_X(-D)|_D)^\vee = \mathcal{O}_X(D)|_D.$$

□

The following theorem generalizes Proposition 1.35.

Theorem 1.37 (Adjunction formula). *Let $X \to S$ be a regular fibered surface, $s \in S$ a closed point, and $E \in \mathrm{Div}_s(X)$ such that $0 < E \le X_s$ (the second inequality is an empty condition if $\dim S = 0$). Then we have*

$$\omega_{E/k(s)} \simeq (\mathcal{O}_X(E) \otimes \omega_{X/S})|_E,$$

and if $K_{X/S}$ is a canonical divisor,

$$p_a(E) = 1 + \frac{1}{2}(E^2 + K_{X/S} \cdot E).$$

Proof   We have

$$\omega_{E/S} = \omega_{E/X} \otimes_{\mathcal{O}_E} \omega_{X/S}|_E = \mathcal{O}_X(E)|_E \otimes_{\mathcal{O}_E} \omega_{X/S}|_E$$

by Theorem 6.4.9(a) and Lemma 1.36. The hypothesis on E implies that $E \to S$ factors into $f : E \to \mathrm{Spec}\, k(s)$ followed by $\mathrm{Spec}\, k(s) \to S$, so the same theorem and lemma show that

$$\omega_{E/S} = \omega_{E/k(s)} \otimes_{\mathcal{O}_E} f^* \omega_{\mathrm{Spec}\, k(s)/S} \simeq \omega_{E/k(s)},$$

which shows the first isomorphism. The equality for $p_a(E)$ results from this isomorphism and from the equality $\chi(\mathcal{O}_E) = -(\deg \omega_{E/k(s)})/2$ (Corollary 7.3.31).   □

Example 1.38. Let us consider Example 1.22. We have already seen that $\Gamma_i^2 = -3$. On the other hand, it is easy to see that $d_i = 1$. Combining Proposition 1.35 and the adjunction formula, we find $p_a(\Gamma_i) = 0$ (which can very easily be verified) and $K_{X/S} \cdot \Gamma_i = 1$.

Exercises

1.1. Let X be a quasi-projective variety over an infinite field. Let $x_1, \ldots, x_n \in X$ of which none is the generic point of X. Let us consider a Cartier divisor D on X. Drawing inspiration from the proof of Proposition 7.1.32, show (without using Proposition 1.11) that there exists a Cartier divisor $D' \sim D$ such that $x_i \notin \operatorname{Supp} D'$ for every i.

1.2. Let X be an irreducible quasi-projective algebraic variety over an infinite field. Let $D_1, \ldots, D_n$, where $n = \dim X$, be Cartier divisors on X. Show that there exist Cartier divisors $D_i' \sim D_i$ such that

$$\dim \cap_{1 \leq i \leq r} \operatorname{Supp} D_i' = n - r$$

for every $r \leq n$. This generalizes the 'moving lemma' for projective algebraic varieties.

1.3. Let X be a regular Noetherian scheme and $D = \sum_{1 \leq i \leq m} n_i \Gamma_i$ an effective Cartier divisor on X. We suppose the Γ_i is prime and pairwise distinct, and $n_i > 0$. Show that D has normal crossings at $x \in X$ if and only if for every $i \leq m$, the scheme Γ_i is regular at x, and if the sub-$k(x)$-vector spaces $T_{\Gamma_i,x}$ of $T_{X,x}$, $1 \leq i \leq m$, form a direct sum.

1.4. Let $X \to S$ be a regular fibered surface. Show that the bilinear map i_s of Theorem 1.12 is uniquely determined by properties (a) and (c).

1.5. Let E, D be two distinct smooth projective curves contained in $\mathbb{P}_k^2$, where k is a field. Let $x \in E \cap D$ be a rational point. Show that $i_x(D, E) \geq 2$ if and only if the respective tangents of E and D at x, considered as lines in $\mathbb{P}_k^2$, coincide.

1.6. Let $X \to S$ be an arithmetic surface and X_s a closed fiber. Let $C_1, \ldots, C_m$ denote the connected components of X_s. Let $V \in \operatorname{Div}_s(X)$ be a vertical divisor with support in X_s. Show that the following properties are equivalent:

  (i) $V \cdot D = 0$ for every $D \in \operatorname{Div}_s(X)$;

  (ii) $V^2 = 0$;

  (iii) $V \in \oplus_{1 \leq i \leq m} \mathbb{Q} C_i$.

1.7. Let $X \to S$ be an arithmetic surface, η the generic point of S, and $s \in S$ a closed point. Let us suppose that X_s is connected. Let us consider the $\mathbb{Q}$-vector space $V = (\operatorname{Div}_s(X) \otimes_{\mathbb{Z}} \mathbb{Q})/X_s\mathbb{Q}$.

(a) Show that i_s induces a symmetric negative definite bilinear form on V.

(b) Let $D \in \operatorname{Div}(X)$ be such that $\deg D_\eta = 0$. Show that $E \mapsto \langle D, E \rangle_s$ induces a linear form $\rho_D : V \to \mathbb{Q}$.

(c) Show that there exist an $n \in \mathbb{N}$ and $E_0 \in \operatorname{Div}_s(X)$ such that $\langle nD - E_0, E \rangle_s = 0$ for every $E \in \operatorname{Div}_s(X)$.

1.8. Let $X \to S$ be an arithmetic surface, $s \in S$ a closed point, $d_1, \ldots, d_r$ the multiplicities of the irreducible components of X_s, and $d = \gcd_i \{d_i\}$. Show that d divides $p_a(X_\eta) - 1$ (apply the adjunction formula to $E := d^{-1}X_s$).

1.9. Let $X \to S$ be an arithmetic surface over a local scheme S of dimension 1. Let Γ be an irreducible component of the closed fiber X_s. Let $k = k(s)$, k^s the separable closure of k, and $\bar{\xi}$ a generic point of $\Gamma_{\bar{k}}$. Let us set

$$r_\Gamma = [k(\Gamma) \cap k^s : k], \quad e_\Gamma := \operatorname{length} \mathcal{O}_{\Gamma_{\bar{k}}, \bar{\xi}}.$$

We know that r_Γ is the number of irreducible components of $\Gamma_{\bar{k}}$ (Exercise 3.2.12). The integer e_Γ is called the *geometric multiplicity of* Γ. It is clear that $e_\Gamma = 1$ if and only if Γ is geometrically reduced.

(a) Let F be a field extension of k. Let $k \to k'$ and $k' \to k''$ be finite purely inseparable extensions. Let F' (resp. F'') be the residue field of $F \otimes_k k'$ (resp. of $F' \otimes_{k'} k''$). Show that

$$\operatorname{length}(F \otimes_k k') = [k' : k]/[F' : F]$$

(use Exercise 7.1.6(d)) and that

$$\operatorname{length}(F \otimes_k k'') = \operatorname{length}(F \otimes_k k') \operatorname{length}(F' \otimes_{k'} k'')$$

(b) Let $\mathcal{L} \in \operatorname{Pic}(\Gamma)$. We want to show that $\deg_k \mathcal{L}$ is a multiple of $r_\Gamma e_\Gamma$.

    (1) Show that one can suppose k is separably closed, and hence $r_\Gamma = 1$.

    (2) Let k' be a finite extension of k. Let $\Gamma' = (\Gamma_{k'})_{\operatorname{red}}$ and $f : \Gamma' \to \Gamma$ be the canonical morphism. Show that

$$\deg_k \mathcal{L} = (\operatorname{length} \mathcal{O}_{\Gamma_{k'}, \xi'}) \deg_{k'} f^*\mathcal{L},$$

where ξ' is the generic point of $\Gamma_{k'}$ (use Propositions 7.3.7-8 and (a)), and that

$$e_\Gamma = (\operatorname{length} \mathcal{O}_{\Gamma_{k'}, \xi'}) e_{\Gamma'}.$$

    (3) Show that there exists a k' as above such that Γ' is geometrically reduced. Deduce from this that e_Γ divides $\deg_k \mathcal{L}$. See also [15], Corollary 9.1/8.

(c) Let us keep the notation of Exercise 1.8. Let us set

$$d'' = \gcd_{1 \le i \le n} \{d_i r_{\Gamma_i} e_{\Gamma_i}\}.$$

Show that d'' divides $2p_a(X_\eta) - 2$.

(d) Let I be *the index* of X_η, that is to say the gcd of the $[K(P) : K(\eta)]$ when P runs through the closed points of X_η. Show that d'' divides I.

Remark. If $K(\eta)$ is an finite extension of the field $\mathbb{Q}_p$ of p-adic numbers, we can show that $d'' = I$ ([23], Théorème 3.1).

1.10. Let k be a field. Let $C \subset \mathbb{P}_k^2$ be a curve of degree n. Show, using the adjunction formula, that $p_a(C) = (n-1)(n-2)/2$ (see also Example 7.3.22).

1.11. Let $X \to S$ be a regular fibered surface. Let D be a vertical divisor on X with support in a fiber X_s. Show that $D^2 + K_{X/S} \cdot D \in 2\mathbb{Z}$. We can thus define the arithmetic genus virtually as $p_a(D) := 1 + (D^2 + K_{X/S} \cdot D)/2$.

1.12. Let $\pi : X \to S$ be an arithmetic surface with smooth and geometrically connected generic fiber X_η. Let $\mathcal{L} \in \operatorname{Pic}(X)$ be such that $\mathcal{L}|_{X_\eta} \simeq \mathcal{O}_{X_\eta}$ and that $\deg \mathcal{L}|_{X_s} = 0$ for every $s \in S$.

    (a) Show that there exists a vertical Cartier divisor V such that $\mathcal{L} \simeq \mathcal{O}_X(V)$.

    (b) For any closed point $s \in S$, let d_s denote the gcd of the multiplicities of the irreducible components of X_s. Let d be the lcm of the d_s. Show that d is finite and that dV is a sum of closed fibers. Deduce from this that $\mathcal{L}^{\otimes d} \simeq \pi^*\mathcal{M}$ for some $\mathcal{M} \in \operatorname{Pic}(S)$.

    (c) Let us suppose that π admits a section $\sigma : S \to X$ and that $\sigma^*(\mathcal{L}^{\otimes d}) \simeq \mathcal{O}_S$. Show that $\mathcal{L}^{\otimes d} \simeq \mathcal{O}_X$.

1.13. Let $\pi : X \to S$ be an arithmetic surface such that $\pi_*\mathcal{O}_X = \mathcal{O}_S$ and that π is cohomologically flat (Exercise 5.3.14).

    (a) Let $\mathcal{L} \in \operatorname{Pic}(X)$ be such that $\mathcal{L}|_{X_s} \simeq \mathcal{O}_{X_s}$ for every $s \in S$. Show that $\pi_*\mathcal{L}$ is an invertible sheaf on S (use Theorem 5.3.20(b)). Show that the canonical homomorphism $\pi^*\pi_*\mathcal{L} \to \mathcal{L}$ is surjective by considering its restriction to the fibers X_s. Deduce from this that $\mathcal{L} \simeq \pi^*\pi_*\mathcal{L}$.

    (b) Let us suppose that π admits a section $\sigma : S \to X$. Let $\mathcal{L}_1, \mathcal{L}_2 \in \operatorname{Pic}(X)$. Show that $\mathcal{L}_1 \simeq \mathcal{L}_2$ if and only if $\mathcal{L}_1|_{X_s} \simeq \mathcal{L}_2|_{X_s}$ for every $s \in S$ and if $\sigma^*\mathcal{L}_1 \simeq \sigma^*\mathcal{L}_2$.

1.14. Let $\pi : X \to S$ be an arithmetic surface. Let $\Gamma_1, \ldots, \Gamma_n$ be the irreducible components of a connected fiber X_s, of respective multiplicities $d_1, \ldots, d_n$, and let $d = \gcd\{d_i\}_i$. Let us fix an $m \le n$ and let us set $Z_0 = \sum_{1 \le i \le m} \Gamma_i$, $Z = \sum_{1 \le i \le m} d_i d^{-1} \Gamma_i$.

    (a) Show that there exists a chain of divisors $Z_0 < Z_1 < \cdots < Z_r < Z_{r+1} = Z$ such that $Z_{j+1} - Z_j$ is a prime divisor Δ_j and that $Z_j \cdot \Delta_j > 0$.

    (b) Show that the canonical homomorphism $H^0(Z, \mathcal{O}_Z) \to H^0(Z_0, \mathcal{O}_{Z_0})$ is injective.

1.15. Let C, E be two proper smooth curves over a field k, and $f : C \to E$ a finite morphism. Let us set $X = C \times_{\operatorname{Spec} k} E$. Let us consider the graph $\Gamma_f \subseteq X$ of f endowed with the reduced closed subscheme structure.

(a) Let $p_1 : X \to C$ and $p_2 : X \to E$ denote the projections. Then p_1 induces an isomorphism $\varphi : \Gamma_f \simeq C$. Show that $\omega_{X/k} \simeq p_1^* \omega_{C/k} \otimes p_2^* \omega_{E/k}$ and that $\omega_{X/k}|_{\Gamma_f} \simeq \varphi^* \omega_{C/k} \otimes \varphi^* f^* \omega_{E/k}$.

(b) Show that

$$\deg_k \omega_{X/k}|_{\Gamma_f} = 2g(C) - 2 + (\deg f)(2g(E) - 2).$$

Deduce from this that $\Gamma_f^2 = (\deg f)(2 - 2g(E))$.

(c) Let us henceforth suppose that $C = E$. Let $\Delta \subset X$ denote the diagonal. Show that $\Delta^2 = 2 - 2g(C)$.

(d) Let us suppose $f \neq \mathrm{Id}_C$. Let $x \in X(k) \cap \Delta \cap \Gamma_f$, let $y = p_1(x)$, and let t be a uniformizing parameter for $\mathcal{O}_{C,y}$. Show that

$$i_x(\Gamma_f, \Delta) = \text{length } \mathcal{O}_{C,y}/(\sigma(t) - t),$$

where σ is the automorphism of $\mathcal{O}_{C,y}$ induced by f.

(e) Let us take a finite field $k = \mathbb{F}_{p^r}$ of characteristic $p > 0$, and let $f : C \to C$ be the Frobenius F_C^r (see Subsection 3.2.4). Show that the divisors Γ_f, Δ meet transversally and that $\Gamma_f \cap \Delta \subseteq X(k)$. Deduce from this that the cardinal N of $C(k)$ is given by $N = \Gamma_f \cdot \Delta$.

Remark. The computations above make it possible to show the inequality (*Riemann hypothesis for curves over a finite field*)

$$|N - (q + 1)| \leq 2g(C)\sqrt{q}.$$

See [43], Exercises V.1.9–1.10.

9.2   Intersection and morphisms

We study the behavior of intersection numbers with respect to morphisms of regular fibered surfaces. The first result of the section (factorization theorem) describes projective birational morphisms of regular fibered surfaces as being made up of blowing-ups of closed points. Next, Theorem 2.12 give rules that allow us to compute intersection numbers of divisors which are inverse images or direct images of dominant morphisms. Finally, given a Cartier divisor, we show how to make it into a divisor with normal crossings by a sequence of blowing-ups of points (Theorem 2.26, embedded resolution). The local structure of arithmetic surfaces $X \to S$ with normal crossings (Definition 2.29) is described in Proposition 2.34.

9.2.1   Factorization theorem

We are going to determine the birational morphisms $f : X \to Y$ of regular fibered surfaces over S. Such a morphism is necessarily projective (Corollary 3.3.32(e)).

In what follows, when we talk about the blowing-up $\pi : \widetilde{Y} \to Y$ of a regular scheme Y along a closed point y, the closed subscheme $\{y\}$ will always be endowed with the reduced scheme structure. Such a blowing-up on a regular scheme of dimension 2 also has names such as *monoidal transformation*, σ-*process*, *dilatation*, etc. These blowing-ups form a fundamental example of morphisms of regular fibered surfaces, as Theorem 2.2 will show.

Lemma 2.1. *Let $f : X \to Y$ be a birational morphism of regular fibered surfaces over S. Let $y \in Y$ be a closed point such that $\dim X_y \geq 1$. Then f factors into $X \xrightarrow{g} \widetilde{Y} \xrightarrow{\pi} Y$, where π is the blowing-up of Y with center y.*

Proof Let us consider the birational map $g = \pi^{-1} \circ f : X \dashrightarrow \widetilde{Y}$, and let us suppose that it is not defined at a point $x \in X$. By Proposition 8.3.22, the total transform $g(x)$ is a prime divisor E on $\widetilde{Y}$, g^{-1} is defined at the generic point ξ of E, and sends ξ to x. As π is an isomorphism outside of y, we have $x \in f^{-1}(y)$ and $E = \pi^{-1}(y)$ (by Theorem 8.1.19(b), $\pi^{-1}(y)$ is irreducible). By identifying the function fields $K(X)$ and $K(Y)$, we therefore have the relations of domination of local rings

$$\mathcal{O}_{Y,y} \subseteq \mathcal{O}_{X,x} \subseteq \mathcal{O}_{\widetilde{Y},\xi}.$$

The closed subset X_y is pure of codimension 1 (use Proposition 4.4.2 or Theorem 7.2.22). Let D be the sum, as divisor, of the irreducible components of X_y. Let α be a generator of $\mathcal{O}_X(-D)_x$. Then

$$\mathfrak{m}_y \mathcal{O}_{X,x} \subseteq \sqrt{\mathfrak{m}_y \mathcal{O}_{X,x}} = \alpha \mathcal{O}_{X,x}.$$

We have $\alpha^{-1} \mathfrak{m}_y \mathcal{O}_{X,x} = \mathcal{O}_{X,x}$ because otherwise

$$\mathfrak{m}_y \mathcal{O}_{X,x} \subseteq \alpha \mathfrak{m}_x \mathcal{O}_{X,x} \subseteq \mathfrak{m}_x^2 \mathcal{O}_{X,x} \subseteq \mathfrak{m}_\xi^2 \mathcal{O}_{\widetilde{Y},\xi},$$

and hence $\mathfrak{m}_y \mathcal{O}_{\widetilde{Y},\xi} \subseteq \mathfrak{m}_\xi^2 \mathcal{O}_{\widetilde{Y},\xi}$, which is impossible since $\widetilde{Y}_y$ is reduced by Theorem 8.1.19(b). Consequently, $\mathfrak{m}_y \mathcal{O}_{X,x} = \alpha \mathcal{O}_{X,x}$ is principal. This is in contradiction with the universal property of the blowing-up $\widetilde{Y} \to Y$ (Corollary 8.1.16). Hence $g : X \dashrightarrow \widetilde{Y}$ is a morphism. $\square$

Theorem 2.2 (Factorization theorem). *Let S be a Dedekind scheme, and let $f : X \to Y$ be a birational morphism of regular fibered surfaces over S. Then f is made up of a finite sequence of blowing-ups along closed points.*

Proof Let $\mathcal{E} \subset X$ be the exceptional locus of f (Definition 7.2.21). We can suppose $\mathcal{E} \neq \emptyset$. Let $y \in f(\mathcal{E})$. Then $\dim X_y \geq 1$ by Corollary 4.4.3(b). Let $\pi : \widetilde{Y} \to Y$ be the blowing-up of Y with center y. By Lemma 2.1, f factors into $g : X \to \widetilde{Y}$ and $\pi : \widetilde{Y} \to Y$. Let $\mathcal{E}' \subset X$ be the exceptional locus of g. Let Γ be an element of $\mathcal{E}$ whose image in $\widetilde{Y}$ is $\pi^{-1}(y)$; then $\Gamma \notin \mathcal{E}'$. In other words, $\mathcal{E}'$ is strictly contained in $\mathcal{E}$. We can therefore conclude the proof of the theorem by induction on the number of irreducible components of $\mathcal{E}$. $\square$

Corollary 2.3. *Let $f : X \to Y$ be as in Theorem 2.2. Let us suppose that the exceptional locus of f is irreducible. Then f is the blowing-up of Y along a closed point y.*

Proof   With each blowing-up of a closed point, the number of divisors in the exceptional locus grows with one. This immediately implies the corollary.   □

Remark 2.4.   Theorem 2.2 is false for regular schemes (even smooth algebraic varieties over $\mathbb{C}$) of dimension ≥ 3. That is to say that a projective birational morphism does not necessarily decompose into a finite sequence of blowing-ups with regular centers. However, there exists an analogue if, in the decomposition, we admit birational maps which are inverses of blowing-ups with regular centers. Cf. [4].

Proposition 2.5.   *Let Y be a regular fibered surface, and let $f : X \to Y$ be the blowing-up of Y along a closed point $y \in Y_s$. Let E denote the scheme X_y. Then $E \simeq \mathbb{P}^1_{k(y)}$, $k(y) = H^0(E, \mathcal{O}_E)$, and we have*

$$\mathcal{O}_X(E)|_E \simeq \mathcal{O}_E(-1), \quad E^2 = -[k(y) : k(s)].$$

Proof   The fact that $E \simeq \mathbb{P}^1_{k(y)}$ is in Theorem 8.1.19(b). This immediately implies that $H^0(E, \mathcal{O}_E) = k(y)$. On the other hand, the same theorem says that $\omega_{E/X} \simeq \mathcal{O}_E(-1)$. Applying Lemma 1.36, we obtain $\mathcal{O}_X(E)|_E \simeq \mathcal{O}_E(-1)$ and

$$E^2 = \deg_{k(s)} \mathcal{O}_E(-1) = [k(y) : k(s)] \deg_{k(y)} \mathcal{O}_E(-1) = -[k(y) : k(s)].$$

See also Exercise 2.3.   □

Remark 2.6.   Let $f : X \to Y$ be a projective birational morphism of regular Noetherian schemes. Let $\mathcal{E} \subseteq X$ be the exceptional locus of f. When Y is as in Theorem 2.2, any prime divisor E contained in $\mathcal{E}$ is a rational curve by this same theorem and Proposition 2.5. More generally, if Y is a regular (and excellent) Noetherian scheme of arbitrary dimension, a theorem of Abhyankar ([3], Proposition 3) states that E is birationally ruled above a scheme over $f(E)$; that is, $k(E)$ is a purely transcendental extension of a finitely generated field L over $k(f(E))$, and $\operatorname{trdeg}_L k(E) \geq 1$. See Exercise 2.13.

Theorem 2.7 (Elimination of points of indeterminacy). *Let $X \to S$ be a regular fibered surface. Let $\varphi : X \dashrightarrow Z$ be a rational map from X to a projective S-scheme Z. Then there exist a projective birational morphism $f : \tilde{X} \to X$ made up of a finite sequence of blowing-ups of closed points*

$$\tilde{X} = X_n \to X_{n-1} \to \cdots \to X_0 = X,$$

and a morphism $g : \tilde{X} \to Z$ making the following diagram commutative:

Proof Let $\Gamma \subseteq X \times_S Z$ be the graph of φ. This is an S-scheme and the projection $p : \Gamma \to X$ is projective birational. Let us show that Γ admits a desingularization. This is Theorem 8.3.44 if $\dim S = 0$. Let us therefore suppose that $\dim S = 1$. Let $K = K(S)$. Then $p_K : \Gamma_K \to X_K$ is proper birational with X_K normal, and hence p_K is an isomorphism. We can therefore apply Theorem 8.3.50. Let $g : \widetilde{X} \to \Gamma$ be a desingularization morphism, and $f : \widetilde{X} \to X$ the composition $p \circ g$. It now suffices to apply Theorem 2.2 to the morphism f. □

9.2.2 Projection formula

Definition 2.8. Let X, Y be Noetherian schemes, and let $f : X \to Y$ be a proper morphism. For any prime cycle Z on X (Section 7.2), we set $W = f(Z)$ and

$$f_* Z = \begin{cases} [K(Z) : K(W)]W & \text{if } K(Z) \text{ is finite over } K(W) \\ 0 & \text{otherwise.} \end{cases}$$

By linearity, we define a homomorphism f_* from the group of cycles on X to the group of cycles on Y. This generalizes Definition 7.2.17. It is clear that the construction of f_* is compatible with the composition of morphisms.

Remark 2.9. We can interpret the intersection of two divisors in terms of direct images of cycles. Let C, D be two Cartier divisors on a regular fibered surface $X \to S$, of which at least one is vertical. Let us suppose that C is effective and has no common component with $\operatorname{Supp} D$. Let $h : C \to S$ denote the morphism induced by $X \to S$. Then

$$C \cdot D = h_*[D|_C].$$

This generalizes Lemma 1.29. See also Remark 2.13.

Lemma 2.10. *Let* $f : X \to Y$ *be a projective birational morphism of Noetherian integral schemes such that* $\mathcal{O}_Y \to f_* \mathcal{O}_X$ *is an isomorphism. Then the following properties are true.*

(a) *There exists an open subset V of Y such that $f^{-1}(V) \to V$ is an isomorphism, and that $\operatorname{codim}(Y \setminus V, Y) \geq 2$.*

(b) *Let Z be a cycle of codimension 1 in X. Then $f_* Z$ is a cycle of codimension 1 in Y.*

Proof (a) Let V be the open subset of Y defined by Proposition 4.4.2. Let y be a point of codimension 1 in Y. We have to prove that $y \in V$. Let x be a generic point of X_y. Then $\dim \mathcal{O}_{X,x} \neq 0$ because f is dominant. By Theorem 8.2.5, we have $\operatorname{trdeg}_{k(y)} k(x) = 0$, hence X_y is finite and $y \in V$.

(b) We can suppose that Z is a prime cycle. Let x be the generic point of Z and $y = f(x)$. If $y \in V$, then y has codimension $\dim \mathcal{O}_{Y,y} = \dim \mathcal{O}_{X,x} = 1$, and $f_* Z = \overline{\{y\}}$ is a cycle of codimension 1. Let us suppose $y \notin V$. Let F be the irreducible component of X_y containing x. Then $\dim F \geq 1$ because X_y has no isolated point by definition of V. Hence $\operatorname{trdeg}_{k(y)} k(x) = \dim F \geq 1$, and $f_* Z = 0$. □

We can now generalize Theorem 7.2.18. See also [37], Proposition 1.4(b).

Proposition 2.11. Let $f : X \to Y$ be a surjective projective morphism of Noetherian integral schemes. We suppose that $[K(X) : K(Y)] = n$ is finite. Then for any Cartier divisor D on Y, we have

$$f_*[f^*D] = n[D].$$

Proof   The morphism f factors into a projective birational morphism $X \to Y' := \operatorname{Spec} f_* \mathcal{O}_X$ followed by a finite surjective morphism $Y' \to Y$ (Exercise 5.3.11). It suffices to show that the proposition is true for each of these morphisms. Finite morphism are dealt with in Theorem 7.2.18. We can therefore suppose that f is birational and $f_* \mathcal{O}_X = \mathcal{O}_Y$. Then $f_*[f^*D] - [D]$ is a cycle of codimension 1 (Lemma 2.10(b)), and its restriction to some open subset $V \subseteq Y$ with $\operatorname{codim}(Y \setminus V, Y) \geq 2$ is zero (Lemma 2.10(a)). Hence $f_*[f^*D] - [D] = 0$. $\square$

Theorem 2.12. Let $f : X \to Y$ be a dominant morphism of regular fibered surfaces over S. Let C (resp. D) be a divisor on X (resp. on Y). Then the following properties are true.

(a) For any divisor E on X such that $f(\operatorname{Supp} E)$ is finite, we have $E \cdot f^*D = 0$.

(b) Let us suppose that C or D is vertical. Then

$$C \cdot f^*D = f_*C \cdot D \quad \text{(Projection formula)}, \qquad (2.4)$$

where f_*C is the Cartier divisor on Y such that $[f_*C] = f_*[C]$.

(c) The extension $K(X)/K(Y)$ is finite. Let F be a vertical divisor on Y. Then f^*F is vertical and we have

$$f^*F \cdot f^*D = [K(X) : K(Y)]F \cdot D.$$

Proof   (a) We can suppose that E is a vertical prime divisor. Let $y = f(E)$. Then $\mathcal{O}_Y(D)$ is free on an open neighborhood V of y. It follows that $\mathcal{O}_X(f^*D) = f^*\mathcal{O}_Y(D)$ is free on $f^{-1}(V) \supset E$, and hence $\mathcal{O}_X(f^*D)|_E \simeq \mathcal{O}_E$, which, in particular, implies that $f^*D \cdot E = 0$.

(b) We can suppose that C is a prime divisor. If $f(C)$ is a point, then $f_*[C] = 0$ and equality (2.4) is true by (a). Let us therefore suppose that $\dim f(C) = 1$. Using the moving lemma 1.10, we can suppose that $\operatorname{Supp} D$ does not contain $f(C)$. Hence $\operatorname{Supp} f^*D \subseteq f^{-1}(\operatorname{Supp} D)$ does not contain C. Let $\pi : X \to S$, $\pi' : Y \to S$ be the structural morphisms. With the notation of Remark 2.13,

$$C \cdot f^*D = \pi_*([C].f^*D) = \pi'_*f_*([C].f^*D) = \pi'_*(f_*[C].D) = f_*C \cdot D.$$

(c) The morphism of the generic fibers $X_\eta \to Y_\eta$ is a dominant morphism of algebraic curves over $K(S)$. This implies that $K(Y) \to K(X)$ is finite. We have $f_*[f^*F] = [K(X) : K(Y)][F]$ by Proposition 2.11. It now suffices to apply (b) to the pair (f^*F, D). $\square$

Remark 2.13. We have a more precise version of Part (b) of Theorem 2.12. Let us first define intersection cycles on an integral Noetherian scheme X of dimension 2. Let $Z \in Z^1(X)$ be a cycle of codimension 1 on X and let D be a Cartier divisor on X such that $\operatorname{Supp} D$ does not contain any irreducible component of Z. We denote by $Z_0(X)$ the subgroup of $Z(X)$ of 0-cycles on X. We define $Z.D \in Z_0(X)$ in the following way. Write $Z = \sum_i n_i Z_i$ with Z_i irreducible. Then $D|_{Z_i}$ is a Cartier divisor on Z_i (Lemma 7.1.29). We let

$$Z.D := \sum_i n_i [D|_{Z_i}] \in Z_0(X)$$

where $[D|_{Z_i}]$ is the 0-cycle on Z_i (hence on X) associated to $D|_{Z_i}$ (Definition 7.2.12). The 0-cycle $Z.D$ is obviously additive in Z and in D. In the case when Z is the cycle $[C]$ associated to some effective Cartier divisor C, we have

$$[C].D = [D|_C].$$

Indeed, if $[C] = \sum_i n_i Z_i$ with Z_i irreducible of generic point ξ_i, then $n_i = \operatorname{mult}_{\xi_i}(C) = \operatorname{length} \mathcal{O}_{C,\xi_i}$ is equal to the multiplicity of Z_i in the scheme C. In the course of the proof of Proposition 7.5.7, we saw that $\operatorname{mult}_x(D|_C) = \sum_i n_i \operatorname{mult}_x(D|_{Z_i})$. Hence the equality $[D|_C] = \sum_i n_i [D|_{Z_i}] = [C].D$.

If $\pi : X \to S$ is a regular fibered surface and $Z = [C]$ for some Cartier divisor C on X, then we have clearly

$$\pi_*([C].D) = C \cdot D, \quad i_s(C,D) = \sum_{x \in X_s} \operatorname{mult}_x([C].D)[k(x) : k(s)]$$

the second equality being true only if C or D has support contained in X_s.

Let $f : X \to Y$ be a projective surjective morphism of integral Noetherian schemes of dimension 2. Let $Z \in Z^1(X)$, $D \in \operatorname{Div}(Y)$. Suppose that $f^{-1}(\operatorname{Supp} D)$ does not contain any irreducible component of Z. Then we have in $Z_0(Y)$:

$$f_*(Z.f^*D) = (f_*Z).D \quad \text{(Projection formula)}.$$

To prove this formula, we can suppose that Z is irreducible. Let $V = f(Z)$. If V is reduced to a single point, then we see easily that both sides of the above formula vanish. Suppose now that V is one-dimensional. Then, similarly to the proof of Lemma 7.3.10, $g = f|_Z : Z \to V$ is a surjective finite morphism. By Proposition 7.1.38, we have

$$f_*(Z.f^*D) = g_*[(f^*D)|_Z] = g_*[g^*(D|_V)] = [k(Z) : k(V)]V.D = (f_*Z).D.$$

Example 2.14. Let $f : X \to Y$ be a finite dominant morphism of regular Noetherian schemes of dimension 2, of degree $n = [K(X) : K(Y)]$. Let C and D be distinct prime divisors on Y. Let us suppose that $C' := f^{-1}(C)$ and $D' := f^{-1}(D)$ are irreducible. We have $f_*C' = [K(C') : K(C)]C$. Let e_C denote the ramification index of $\mathcal{O}_{Y,\xi} \to \mathcal{O}_{X,\xi'}$, where ξ (resp. ξ') is the generic point of C (resp. of C'). Let us define e_D in a similar way. Then $f^*C = e_C C'$ (Exercise 7.2.3(b)), and $n = e_D[K(D') : K(D)]$ (Lemma 7.1.36(c)). The projection formula described in Remark 2.13 then gives

$$n i_y(C, D) = e_C e_D \sum_{x \in X_y} i_x(C', D')[k(x) : k(y)].$$

Base change

Let $X \to S$ be a regular fibered surface. Let $\lambda : S' \to S$ be a dominant morphism of Dedekind schemes with $\dim S' = \dim S$. For example, S' can be $\operatorname{Spec} \mathcal{O}_{S,s}$ or $\operatorname{Spec} \widehat{\mathcal{O}}_{S,s}$ for a closed point $s \in S$, or a scheme that is finite and surjective over S. Let us suppose that $X \times_S S'$ is integral and admits a desingularization $X'' \to X \times_S S'$. In general, $X'' \to S$ is not projective, nor even of finite type. We therefore cannot use Theorem 2.12 directly with the morphism $f : X'' \to X$ (composition of $X'' \to X \times_S S'$ and the projection $X \times_S S' \to X$), except when $S' \to S$ is finite. For any $s' \in S'$, if $s = \lambda(s')$, then

$$e_{s'/s} := \operatorname{length}_{\mathcal{O}_{S',s'}}(\mathcal{O}_{S',s'}/\mathfrak{m}_s \mathcal{O}_{S',s'})$$

is finite because $\mathcal{O}_{S',s'}/\mathfrak{m}_s \mathcal{O}_{S',s'}$ is of dimension 0.

Proposition 2.15. *Let $X \to S$ be a regular fibered surface. Let $\lambda : S' \to S$ be a dominant morphism of Dedekind schemes with $\dim S' = \dim S$. Let us suppose that $X' := X \times_S S'$ is integral and admits a desingularization $g : X'' \to X'$. Let $f : X'' \to X$ be the natural morphism. Let C, D be Cartier divisors on X, of which at least one is vertical. Then for any closed point $s \in S$ and $s' \in \lambda^{-1}(s)$, we have*

$$i_{s'}(f^*C, f^*D) = e_{s'/s} i_s(C, D). \tag{2.5}$$

Proof We can suppose that S, S' are local. Using the moving lemma 1.10, and by bilinearity, we reduce to the case when C and D are two distinct prime divisors. Let $p : X' \to X$ denote the projection morphism and let $C' = p^*C$, $D' = p^*D$. We are in position to apply the projection formula of Remark 2.13 to the birational morphism g. We then get

$$g_*([f^*C].f^*D) = g_*([g^*C'].g^*D') = g_*[g^*C'].D' = [C'].D' = [D'|_{C'}]$$

Let π, π', π'' be respectively the structural morphisms of X, X' and X''. Then $\pi''_*([f^*C].f^*D) = \pi'_*[D'|_{C'}]$. Let $\mathcal{J} = \mathcal{O}_X(-C) + \mathcal{O}_X(-D)$. Let $x \in \operatorname{Supp} C$. Then $\operatorname{mult}_x([D|_C]) = \operatorname{length}(\mathcal{O}_{C,x}/\mathcal{J}\mathcal{O}_{C,x}) = [k(x) : k(s)]^{-1} \operatorname{length}_{\mathcal{O}_s}(\mathcal{O}_X/\mathcal{J})_x$ (Lemma 7.1.36(a)). Therefore

$$i_s(C, D) = \operatorname{mult}_s(\pi_*[D|_C]) = \operatorname{length}_{\mathcal{O}_s} H^0(X, \mathcal{O}_X/\mathcal{J})$$

because $\mathcal{O}_X/\mathcal{J}$ is a skyscraper sheaf on X. Similarly,

$$i_{s'}(f^*C, f^*D) = \operatorname{mult}_{s'}(\pi'_*[D'|_{C'}]) = \operatorname{length}_{\mathcal{O}_{s'}} H^0(X', \mathcal{O}_{X'}/(\mathcal{J})).$$

As $H^0(X', \mathcal{O}_{X'}/(\mathcal{J})) = H^0(X, \mathcal{O}_X/\mathcal{J}) \otimes_{\mathcal{O}_S} \mathcal{O}_{S'}$, we have $i_{s'}(f^*C, f^*D) = e_{s'/s} i_s(C, D)$ by Exercise 7.1.8(b).   □

Corollary 2.16. *Let us keep the hypotheses and notation of Proposition 2.15. Let us moreover suppose that $S' \to S$ is étale, or that $S' = \operatorname{Spec} \widehat{\mathcal{O}}_{S,s}$ (for a closed point $s \in S$), or that $\dim S = 0$. Then we have*

$$i_{s'}(f^*C, f^*D) = i_s(C, D).$$

Proof Under the additional hypotheses, we have $e_{s'/s} = 1$. □

9.2.3 Birational morphisms and Picard groups

Let us now study the behavior of Cartier divisors with respect to birational morphisms of regular fibered surfaces.

Lemma 2.17. *Let T be a normal locally Noetherian scheme. Let $F \subset T$ be a closed subset such that $\operatorname{codim}(F, T) \geq 2$. Let $U = T \setminus F$. Then we have the following properties.*

(a) *For any invertible sheaf $\mathcal{L}$ on T, the restriction $H^0(T, \mathcal{L}) \to H^0(U, \mathcal{L})$ is an isomorphism.*

(b) *The restriction homomorphism $\operatorname{Pic}(T) \to \operatorname{Pic}(U)$ is injective.*

Proof (a) We cover T with open subschemes U_i such that $\mathcal{L}|_{U_i}$ is free for every i. Then $H^0(U_i, \mathcal{L}) \to H^0(U \cap U_i, \mathcal{L})$ is an isomorphism by Theorem 4.1.14. Hence $H^0(T, \mathcal{L}) \to H^0(U, \mathcal{L})$ is an isomorphism.

(b) Let $\mathcal{L}$ be an invertible sheaf on T such that $\mathcal{L}|_U \simeq \mathcal{O}_U$. Let e be a generator of $\mathcal{L}|_U$. We can suppose $e \in H^0(T, \mathcal{L})$ by (a). For any affine open subscheme V of one of the U_i, we have $H^0(V, \mathcal{L}) = e\mathcal{O}_T(V)$ by the same arguments as in (a). Hence $\mathcal{L} = e\mathcal{O}_T$. □

Proposition 2.18. *Let $X \to S$ be a regular fibered surface. Let $\pi : \widetilde{X} \to X$ be the blowing-up of X along a point $x \in X$. Let $E = \pi^{-1}(x) \simeq \mathbb{P}^1_{k(x)}$ be the exceptional locus of π. Then $\operatorname{Pic}(E)$ is free over $\mathbb{Z}$, generated by $\mathcal{O}_{\widetilde{X}}(E)|_E$, and we have a split exact sequence*

$$1 \to \operatorname{Pic}(X) \to \operatorname{Pic}(\widetilde{X}) \to \operatorname{Pic}(E) \to 1, \tag{2.6}$$

where the two homomorphisms in the middle are respectively induced by π and by the closed immersion $E \to \widetilde{X}$.

Proof We have $\mathcal{O}_{\widetilde{X}}(E)|_E \simeq \mathcal{O}_E(-1)$ by Proposition 2.5; it is therefore a basis of $\operatorname{Pic}(E)$ over $\mathbb{Z}$ (Proposition 7.2.9). Let $\mathcal{L}$ be an invertible sheaf on X. Then $\mathcal{L}$ is free on an open neighborhood $V \ni x$. It follows that $\pi^*\mathcal{L}|_{\pi^{-1}(V)}$ is free. In particular, $\pi^*\mathcal{L}|_E \simeq \mathcal{O}_E$. Hence (2.6) is a complex. The injectivity of $\operatorname{Pic}(X) \to \operatorname{Pic}(\widetilde{X})$ comes from the injective canonical homomorphism $\operatorname{Pic}(X) \to \operatorname{Pic}(X \setminus \{x\})$ (Lemma 2.17) and the isomorphism $X \setminus \{x\} \simeq \widetilde{X} \setminus E$. The homomorphism $\operatorname{Pic}(\widetilde{X}) \to \operatorname{Pic}(E)$ is surjective because it sends $\mathcal{O}_{\widetilde{X}}(E)$ onto a basis of $\operatorname{Pic}(E)$, and it admits a section because $\operatorname{Pic}(E)$ is free. It remains to show that (2.6) is exact in the middle. Let $\mathcal{F}$ be an invertible sheaf on $\widetilde{X}$ such that $\mathcal{F}|_E \simeq \mathcal{O}_E$.

By Lemma 2.17, there exists an $\mathcal{L} \in \operatorname{Pic}(X)$ such that $\mathcal{F}|_{\widetilde{X} \setminus E} = \mathcal{L}|_{X \setminus \{x\}}$. Hence $\mathcal{F} \otimes (\pi^* \mathcal{L})^\vee \simeq \mathcal{O}_X(rE)$ for some $r \in \mathbb{Z}$. Taking the degrees of the restrictions to E, we obtain $rE^2 = 0$. Hence $r = 0$ and $\mathcal{F} \simeq \pi^* \mathcal{L}$, which proves the exactness of sequence (2.6). $\qquad\square$

Definition 2.19. Let X be a regular Noetherian scheme, $x \in X$ a point, and D an effective Cartier divisor on X. We let $\mathfrak{m}_x$ denote the maximal ideal of $\mathcal{O}_{X,x}$. We call the greatest integer $n \geq 0$ such that $\mathcal{O}_X(-D)_x \subseteq \mathfrak{m}_x^n$ the *multiplicity of D at x*. This number is finite because $\cap_{i \geq 0} \mathfrak{m}_x^i = 0$ by Krull's theorem (Corollary 1.3.13). We denote this integer by $\mu_x(D)$.

Remark 2.20. We have $\mu_x(D) \geq 1$ if and only if $x \in \operatorname{Supp} D$, and $\mu_x(D) = 1$ if and only if $x \in \operatorname{Supp} D$ and if the scheme D is regular at x (Corollary 4.2.12).

Example 2.21. Let us suppose $d = \dim \mathcal{O}_{X,x} \geq 1$. Let f be a generator of the ideal $\mathcal{O}_X(-D)_x$, and $u_1, \ldots, u_d$ generators of $\mathfrak{m}_x$. We can write

$$f = P(u_1, \ldots, u_d) + Q, \quad \text{with } P(u_1, \ldots, u_d) \in \mathcal{O}_{X,x}^*[u_1, \ldots, u_d], \ Q \in \mathfrak{m}_x^{\mu+1}$$

and P homogeneous of degree μ. We then have $\mu_x(D) = \mu$. This is a consequence of Exercise 4.2.13.

Definition 2.22. Let $\pi : \widetilde{X} \to X$ be a projective birational morphism of regular, Noetherian, integral schemes. Let D be an effective Cartier divisor on X. We define the *strict transform of D in $\widetilde{X}$* to be the effective Cartier divisor $\widetilde{D}$ on $\widetilde{X}$ such that $\mathcal{O}_{\widetilde{X}}(-\widetilde{D})$ defines the scheme-theoretic closure of $\pi^{-1}(D \setminus F)$ in $\widetilde{X}$, where F is the image under π of the exceptional locus of π. If we write $D = \sum_i d_i \Gamma_i$ with the Γ_i prime, then $\widetilde{D} = \sum_i d_i \widetilde{\Gamma}_i$ and $\widetilde{\Gamma}_i$ is the strict transform of Γ_i defined in 8.1.18.

Proposition 2.23. *Let X be a regular fibered surface, and let $\pi : \widetilde{X} \to X$ be the blowing-up of X along a closed point x. Let $E \subset \widetilde{X}$ denote the exceptional locus of π. Then for any effective Cartier divisor D on X, we have*

$$\pi^* D = \widetilde{D} + \mu_x(D)E.$$

Proof The difference $\pi^* D - \widetilde{D}$ is a Cartier divisor with support in E, and hence $\pi^* D = \widetilde{D} + rE$ for some integer $r \in \mathbb{Z}$. Let U be an open affine neighborhood of x, sufficiently small so that the maximal ideal $\mathfrak{m}$ of $\mathcal{O}_X(U)$ defining x is generated by two elements u, v and that $\mathcal{O}_U(-D|_U)$ is generated by an element $f \in \mathcal{O}_X(U)$. Let $A = \mathcal{O}_X(U)$. In the proof of Theorem 8.1.19(c), we have seen that $\pi^{-1}(U)$ is the union of open subschemes $W = \operatorname{Spec} A[w]$ and $\operatorname{Spec} A[w']$ with $w = v/u, w' = u/v \in K(X)$, and that $\mathcal{O}_W(-E|_W) = (u)$. Let $\mu = \mu_x$ and let $f = P(u,v) + Q$ be a presentation of f as in Example 2.21. As $Q \in \mathfrak{m}^{\mu+1} = (u,v)^{\mu+1}$, we have

$$f = u^\mu P(1, w) + u^{\mu+1} g, \quad g \in \mathcal{O}_{\widetilde{X}}(W)$$

in $\mathcal{O}_{\widetilde{X}}(W)$. The quotient $\mathcal{O}_{\widetilde{X}}(W)/(u)$ is a polynomial ring $k(x)[w]$. Hence

$$P(1,w) \notin u\mathcal{O}_{\widetilde{X}}(W) = \mathcal{O}_{\widetilde{X}}(-E)(W),$$

which implies that $\mathrm{mult}_\xi(\pi^*D) = \mathrm{mult}_\xi(f) = \mu$, where ξ is the generic point of E. The proposition is therefore proven. Incidentally, we see that $\widetilde{D} \cap W$ is the Cartier divisor defined by $P(1,w) + ug \in \mathcal{O}_{\widetilde{X}}(W)$. □

Proposition 2.24. *Let $\pi : \widetilde{X} \to X$ be as in Proposition 2.23. Then we have an isomorphism*

$$\omega_{\widetilde{X}/S} = \pi^*\omega_{X/S} \otimes \mathcal{O}_{\widetilde{X}}(E).$$

Proof As $\widetilde{X} \setminus E \to X \setminus \{x\}$ is an isomorphism, the sheaves $\omega_{\widetilde{X}/S}$ and $\pi^*\omega_{X/S}$ are identical on $\widetilde{X} \setminus E$. We therefore have

$$\omega_{\widetilde{X}/S} = \pi^*\omega_{X/S} \otimes \mathcal{O}_{\widetilde{X}}(rE)$$

for some integer $r \in \mathbb{Z}$. It follows that

$$\omega_{\widetilde{X}/S}|_E = \pi^*\omega_{X/S}|_E \otimes \mathcal{O}_{\widetilde{X}}(rE)|_E.$$

The computation of the degrees of these sheaves over $k(x)$ using the adjunction formula (Theorem 1.37) and Theorem 2.12(a) then gives $r = 1$. □

Corollary 2.25. *Let $f : X \to Y$ be a birational morphism of regular fibered surfaces over S. Then the following properties are true.*

(a) *Let ξ be the generic point of X. The restriction*

$$H^0(X, \omega_{X/S}) \to (\omega_{X/S})_\xi = \Omega^1_{K(X)/K}$$

is injective.

(b) *We have*

$$H^0(Y, \omega_{Y/S}) = H^0(X, \omega_{X/S})$$

as subgroups of $\Omega^1_{K(X)/K}$.

Proof (a) follows from the fact that X is integral and that $\omega_{X/S}$ is locally free.
  (b) We can suppose that $f : X \to Y$ is the blowing-up of a point $y \in Y$, by virtue of the factorization theorem. By Proposition 2.24, $f^*\omega_{Y/S} \subseteq \omega_{X/S}$, and hence $H^0(Y, \omega_{Y/S}) \subseteq H^0(X, \omega_{X/S})$. In addition,

$$H^0(X, \omega_{X/S}) \subseteq H^0(X \setminus f^{-1}(y), \omega_{X/S}) = H^0(Y \setminus \{y\}, \omega_{Y/S}) = H^0(Y, \omega_{Y/S})$$

(Lemma 2.17(a) for the last equality), whence the desired equality. □

9.2.4   Embedded resolutions

The aim of this subsection is to show the following theorem:

Theorem 2.26 (Embedded resolution). *Let S be a Dedekind scheme and $X \to S$ a regular fibered surface. Let us fix an effective Cartier divisor D on X. Let us suppose that the scheme D is excellent. Then there exists a projective birational morphism $f : X' \to X$ with X' regular, such that f^*D is a divisor with normal crossings.*

Remark 2.27. If D is an integral curve, then the strict transform $\widetilde{D}$ of D in X' is an irreducible component of f^*D. The theorem therefore implies that $\widetilde{D}$ is a regular curve (Exercise 1.3). This is the reason why this theorem is also called the *Theorem of the embedded resolution of curves in surfaces.*

Remark 2.28. The D excellent hypothesis is satisfied if S is excellent (Theorem 8.2.39(c)) or if D is vertical. Indeed, in the second case, D is a scheme of finite type over an Artinian ring, and we can apply Theorem 8.2.39(a). Let us note that the D excellent hypothesis is necessary in the theorem. Let us consider the following counterexample. Let K be a non-excellent discrete valuation field (Example 8.2.31) of characteristic p. Let $a \in \mathcal{O}_K$ be such that $a \in \widehat{K}^p \setminus K^p$, and let D be the zeros divisor of $\mathrm{div}(T^p - a)$ on $X = \mathbb{P}^1_{\mathcal{O}_K}$. Then D is integral, isomorphic to $\mathrm{Spec}\,\mathcal{O}_K[T]/(T^p - a)$. If Theorem 2.26 were true for the pair (D, X), then the normalization of D would be finite over D, which is not the case by Exercise 8.3.28.

Definition 2.29. Let $X \to S$ be an arithmetic surface. For simplicity, we will say that $X \to S$ has *normal crossings* if for every closed point $s \in S$, the divisor X_s on X has normal crossings.

Corollary 2.30. *Let $X \to S$ be an arithmetic surface that has only a finite number of singular fibers (e.g., if its generic fiber is smooth; see Proposition 8.3.11). Then there exists a projective birational morphism $X' \to X$ such that $X' \to S$ is an arithmetic surface with normal crossings.*

Proof   We apply Theorem 2.26 with D the sum of the singular fibers.   □

Lemma 2.31. *Let $X \to S$ and D be as in Theorem 2.26. Let $\pi : \widetilde{X} \to X$ be the blowing-up of a closed point $x \in \mathrm{Supp}\,D$. Let E denote the exceptional divisor $\pi^{-1}(x)$ (see Definition 3.1). Let us suppose that D has normal crossings at x. Then π^*D has normal crossings at the points $x' \in E$. Moreover, E meets the irreducible components of the strict transform $\widetilde{D}$ in at most two points, and these intersection points are rational over $k(x)$. (See Lemma 3.35 for the converse.)*

Proof   Let us first suppose that x belongs to a single irreducible component Γ of D. Let $\widetilde{\Gamma}$ be the strict transform of Γ in $\widetilde{X}$. We have

$$\pi^*\Gamma = \widetilde{\Gamma} + E, \quad E \cdot \pi^*\Gamma = E \cdot \widetilde{\Gamma} + E^2 = 0$$

(Proposition 2.23 and Theorem 2.12(a)). It follows that

$$\deg_{k(x)} \mathcal{O}_{\widetilde{X}}(\widetilde{\Gamma})|_E = -\deg_{k(x)} \mathcal{O}_{\widetilde{X}}(E)|_E = 1$$

(Theorem 8.1.19(b)). This shows that $\widetilde{\Gamma}$ and E intersect transversally at a unique point that is rational over $k(x)$. As $E \simeq \mathbb{P}^1_{k(x)}$ is regular, π^*D has normal crossings at the points of E.

Let us suppose that two irreducible components Γ_1, Γ_2 of D pass through x. Let $\widetilde{\Gamma}_i$ be the strict transform of Γ_i in $\widetilde{X}$. In a way similar to what was discussed above, we show that the $\widetilde{\Gamma}_i$ meet E transversally at points that are rational over $k(x)$, and that $\widetilde{\Gamma}_1 \cap \widetilde{\Gamma}_2 = \emptyset$. □

Lemma 2.32. *Let us take the hypotheses of Theorem 2.26. If D is reduced, then there exists a morphism $f : X' \to X$ made up of a finite sequence of blowing-ups of closed points such that the irreducible components of f^*D are regular.*

Proof Let x be a singular point of D. Let $\pi : \widetilde{X} \to X$ be the blowing-up of X with center x. By Proposition 2.23, we have $\pi^*D = \widetilde{D} + \mu_x(D)E$ with $E \simeq \mathbb{P}^1_{k(x)}$. Hence the possibly singular irreducible components of π^*D are those of $\widetilde{D}$. The restriction $\pi|_{\widetilde{D}} : \widetilde{D} \to D$ is the blowing-up of D with center x (Corollary 8.1.17). As the normalization of D is finite over D by the D excellent hypothesis, the method of Proposition 8.1.26 (see Remark 8.1.27) shows that after a finite number of blowing-ups, the strict transform of D becomes a regular scheme, whence the lemma. □

Proof of Theorem 2.26. By Lemma 2.32 above, we can suppose that the irreducible components of D are regular. Let $x \in \operatorname{Supp} D$ be a closed point belonging to at least two irreducible components of D. Let $\pi : \widetilde{X} \to X$ be the blowing-up of X with center x and $E = \pi^{-1}(x)$. Let $D = \sum_{1 \le i \le r} d_i \Gamma_i$ be the decomposition of D as a sum of prime divisors. By virtue of Proposition 2.23, if $x \in \Gamma_i$, we have $\pi^*\Gamma_i = \widetilde{\Gamma}_i + E$ since Γ_i is regular. As in the proof of Lemma 2.31, we see that $\widetilde{\Gamma}_i \cdot E = -E^2$ and that $\widetilde{\Gamma}_i$ and E meet transversally (at a unique point that is rational over $k(x)$). For any $i \le r$, we have $\pi_*(\widetilde{\Gamma}_i) = \Gamma_i$. Hence we have the following relation in the group of 0-cycles on S:

$$\Gamma_i \cdot \Gamma_j = \widetilde{\Gamma}_i \cdot \pi^*\Gamma_j = \widetilde{\Gamma}_i \cdot \widetilde{\Gamma}_j + \widetilde{\Gamma}_i \cdot E$$

(Theorem 2.12(b)). It follows that if $i \ne j$ and if $x \in \Gamma_i$, then

$$0 \le \widetilde{\Gamma}_i \cdot \widetilde{\Gamma}_j = \Gamma_i \cdot \Gamma_j - [k(x) : k(s)][s] < \Gamma_i \cdot \Gamma_j, \qquad (2.7)$$

where s is the image of x in S. Let us also note that $\widetilde{\Gamma}_i$ is regular because it is finite birational to Γ_i, and hence isomorphic to Γ_i. Thus, after a finite number of blowing-ups, we can suppose that the irreducible components Γ_i of D are regular and that they meet pairwise transversally and at at most one point.

Let $x \in \operatorname{Supp} D$ be an intersection point of the irreducible components of D, and $\pi : \widetilde{X} \to X$ the blowing-up of X with center x. Then inequality (2.7) shows that the $\widetilde{\Gamma}_i$, for $\Gamma_i \ni x$, are pairwise disjoint. As the divisor $E = \pi^{-1}(x)$ meets all of the $\widetilde{\Gamma}_i$ transversally at at most one point, by successively blowing-up the intersection points that are not transversal to D, we see that the inverse image of D is a divisor with normal crossings. □

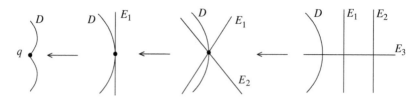

Figure 17. Embedded resolution of a cuspidal singularity.

Example 2.33. Let $\mathcal{O}_K$ be a discrete valuation ring, with uniformizing parameter t. Let $X = \operatorname{Proj} \mathcal{O}_K[x, y, z]/(yz^2 + x^3 + tz^3)$. Then X is smooth over $S = \operatorname{Spec} \mathcal{O}_K$ outside of the point $q \in D_+(z)$ with coordinates $x = y = t = 0$ and $z = 1$. Let us set $D = X_s$. Then the special fibers of the blowing-ups needed to make D into a normal crossings divisor are described in Figure 17, where the darkened points are the centers of the blowing-ups. For simplicity, the strict transform of a divisor is denoted by the same letter. Using Proposition 2.23, we see that in the final special fiber, the divisors D, E_1, E_2, E_3 are of respective multiplicities 1, 2, 3, and 6.

The local structure of arithmetic surfaces with normal crossings is relatively simple to describe.

Proposition 2.34. *Let $X \to S$ be an arithmetic surface with normal crossings. Let $x \in X_s$ be a closed point. Let us fix a uniformizing parameter t of $\mathcal{O}_{S,s}$. The following properties are true.*

(a) *There exist a flat scheme of finite type Z over S of relative dimension 2, a closed point $z \in Z_s$ such that Z_s is regular at z, and a surjective homomorphism $\mathcal{O}_{Z,z} \to \mathcal{O}_{X,x}$ whose kernel is generated by an element $F \in \mathcal{O}_{Z,z}$. The form of F can be specified as follows:*

(b) *If a single irreducible component Γ_1 of X_s, of multiplicity d_1, passes through x, then $F = u^{d_1} - ta$, where u is part of a system of parameters of $\mathcal{O}_{Z,z}$, and where $a \in \mathcal{O}_{Z,z}^*$. Moreover, if Γ_1 is smooth at x, then Z is smooth at z.*

(c) *If two irreducible components Γ_1, Γ_2 of X_s, of respective multiplicities d_1, d_2, pass through x, then $F = u^{d_1} v^{d_2} - ta$, where $\{t, u, v\}$ is a system of parameters of $\mathcal{O}_{Z,z}$ and $a \in \mathcal{O}_{Z,z}^*$.*

Proof (a) As the property is of local nature in X, we can replace X by an affine open neighborhood of x, and embed X in a smooth affine scheme $Y \to S$ (e.g., an affine space over S). Let $n = \dim_x Y_s$ and let y be the image of x in Y. Let $\mathfrak{m}_y$ denote the maximal ideal of $\mathcal{O}_{Y,y}$, $\overline{\mathfrak{m}}_y$ (resp. $\overline{\mathfrak{m}}_x$) the maximal ideal of $\mathcal{O}_{Y_s,y}$ (resp. of $\mathcal{O}_{X_s,x}$), and I the ideal of $\mathcal{O}_Y(Y)$ defining the closed subscheme $X \subset Y$. As $\dim_{k(x)} T_{X_s,x} \leq \dim_{k(x)} T_{X,x} = 2$, there exist $f_3, \ldots, f_n \in I$ whose images in $\overline{\mathfrak{m}}_y/\overline{\mathfrak{m}}_y^2$ are free over $k(y)$ and belong to the kernel of $\overline{\mathfrak{m}}_y/\overline{\mathfrak{m}}_y^2 \to \overline{\mathfrak{m}}_x/\overline{\mathfrak{m}}_x^2$. Let Z be the closed subscheme $V(f_3, \ldots, f_n) \subseteq Y$, and $z = y$. Replacing Z by an open neighborhood of z if necessary, it is easy to see that it verifies the desired properties.

Let us show (c). Let α, β be respective generators of $\mathcal{O}_X(-\Gamma_1)_x$ and $\mathcal{O}_X(-\Gamma_2)_x$, and $u, v \in \mathcal{O}_{Z,z}$ respective preimages of α and β. The hypothesis that X has normal crossings implies that $(\alpha, \beta) = \mathfrak{m}_x \mathcal{O}_{X,x}$. As

$$\dim_{k(x)} T_{X_s, x} = 2 = \dim_{k(z)} T_{Z_s, z},$$

the maximal ideal $\mathfrak{m}_z \mathcal{O}_{Z_s, z}$ is generated by the images of u and v. It follows that $\mathfrak{m}_z \mathcal{O}_{Z,z} = (t, u, v)$. The equality of divisors $\operatorname{div}(t) = d_1 \Gamma_1 + d_2 \Gamma_2$ implies the equality of ideals $(t) = (\alpha^{d_1} \beta^{d_2})$ in $\mathcal{O}_{X,x}$. Consequently, there exists an $a \in \mathcal{O}_{Z,z}^*$ such that $u^{d_1} v^{d_2} - ta \in F\mathcal{O}_{Z,z}$. We therefore have a surjective homomorphism

$$\mathcal{O}_{Z,z}/(u^{d_1} v^{d_2} - ta) \to \mathcal{O}_{X,x}.$$

Since the terms on both sides are regular (hence integral) local rings of dimension 2, this homorphism is therefore an isomorphism. Consequently, $F = u^{d_1} v^{d_2} - ta$, up to a factor in $\mathcal{O}_{Z,z}^*$.

Let us show (b). The proof of the first part is similar to that of (c). Let us suppose that Γ_1 is smooth at x. Let $\overline{Z}$ (resp. $\overline{\Gamma}_1$) denote the base change of Z (resp. of Γ_1) to $\overline{k}$. Let $\overline{z}$ be a point of $\overline{Z}$ lying over z. Then it induces a point $\overline{x}$ of $\overline{\Gamma}_1$ lying over x. Since $\mathcal{O}_{Z,z}/(u) = \mathcal{O}_{\Gamma_1, x}$, by tensoring with $\overline{k}$ and then localizing, we get $\mathcal{O}_{\overline{Z}, \overline{z}}/(u) = \mathcal{O}_{\overline{\Gamma}_1, \overline{x}}$. As the right-hand side is regular, this immediately implies that $\mathcal{O}_{\overline{Z}, \overline{z}}$ also is, and hence Z is smooth at z. □

Remark 2.35. If $k(x)$ is separable over $k(s)$, then Z is smooth at z (Proposition 4.3.30). If $k(x) = k(s)$, then we have an isomorphism

$$\hat{\mathcal{O}}_{X,x} \simeq \hat{\mathcal{O}}_K[[u, v]]/(F(u, v)),$$

with $F(u, v) = u^{d_1} - ta$ or $F(u, v) = u^{d_1} v^{d_2} - ta$ according to whether X_s is irreducible or not at x, and $a \in \mathcal{O}_K^* + (u, v)$. This follows from the fact that $\hat{\mathcal{O}}_{Z,z} \simeq \hat{\mathcal{O}}_K[[u, v]]$ (Exercise 6.2.1).

Remark 2.36. Let $S' \to S$ be a finite morphism of Dedekind schemes. The local structure of X described in Proposition 2.34 makes it possible to find the desingularization of $X \times_S S'$ when $S' \to S$ is tamely ramified. We will return to this in the next chapter (Proposition 10.4.6).

In higher dimension, Theorem 2.26 generalizes as such if the residue fields of S are of characteristic 0 (Hironaka, [46]). In the general case, we only have a version with alteration morphisms (see their definition before Theorem 8.3.43).

Theorem 2.37 (de Jong, [49], Theorems 6.5, 8.2). *Let $X \to S$ be an integral projective scheme over the spectrum of a complete discrete valuation ring. Let Z be a closed subset of X. Then there exists an alteration $f : X_1 \to X$ such that X_1 is regular, and that $f^{-1}(Z)$ is the support of an effective Cartier divisor with normal crossings.*

Exercises

2.1. Let Y be a regular locally Noetherian scheme of dimension 2, and let $f : X \to Y$ be a proper birational morphism with X regular.

(a) Show that f is made up of a sequence of blowing-ups along reduced closed points.

(b) Show that if Y is quasi-projective over an affine Noetherian scheme, then f is projective.

2.2. Let $X \to S$ be an arithmetic surface, $E \subseteq X_s$ an exceptional divisor (see Definition 3.1). We suppose E is of multiplicity 1 in X_s. Show that E meets a unique other irreducible component Γ of X_s, that E and Γ meet transversally at a unique point p that is rational over $H^0(E, \mathcal{O}_E)$, and that Γ is also of multiplicity 1 in X_s.

2.3. Using Proposition 2.23, give another proof of Proposition 2.5.

2.4. Let $f : X \to Y$ be a birational morphism of regular fibered surfaces over S. Let $E_1, \ldots, E_n$ be the irreducible components of the exceptional locus of f.

(a) Show that $\mathrm{Pic}(X) \simeq \mathrm{Pic}(Y) \oplus \mathbb{Z}^n$.

(b) Show that

$$\omega_{X/S} \simeq f^*\omega_{Y/S} \otimes_{\mathcal{O}_X} \mathcal{O}_X(D),$$

where D is a linear combination of the E_i with strictly positive coefficients.

2.5. Let D, E be two effective Cartier divisors on a regular Noetherian scheme X of dimension 2. Show that for any closed point $x \in X$, we have

$$\mu_x(D + E) = \mu_x(D) + \mu_x(E).$$

(Use Example 2.21.)

2.6. Let $X \to S$ be an arithmetic surface. Let $D, F \in X(S)$ be two distinct sections and $X_{s_1}, \ldots, X_{s_n}$ the fibers where they meet. Show that there exists a projective birational morphism $f : X' \to X$ with X' regular such that $f^*(D + F + \sum_i X_{s_i})$ has normal crossings, and that the exceptional locus of f contains exactly $\sum_x i_x(D, F)$ irreducible components.

2.7. Let $X \to S$ be an arithmetic surface, $x \in X_s$ a closed point of X, and $\Gamma_1, \ldots, \Gamma_r$ the irreducible components of X_s passing through x, of respective multiplicities $d_1, \ldots, d_r$.

(a) Show that there exist a flat scheme $Z \to S$ of finite type and of relative dimension 2, a closed point $z \in Z_s$ that is regular in Z_s, and an isomorphism

$$\mathcal{O}_{Z,z}/(u_1^{d_1} \cdots u_r^{d_r} - ta) \simeq \mathcal{O}_{X,x},$$

where $u_i \in \mathfrak{m}_z \mathcal{O}_{Z,z}$ is such that its image in $\mathcal{O}_{X,x}$ is a generator of $\mathcal{O}_X(-\Gamma_i)_x$, t is a uniformizing parameter $\mathcal{O}_{S,s}$, and $a \in \mathcal{O}_{Z,z}^*$.

(b) Let $\mu_i = \mu_x(\Gamma_i)$. Show that we can choose the u_i in such a way that $u_i \in \mathfrak{m}_z^{\mu_i} \setminus \mathfrak{m}_z^{\mu_i+1}$ for every $i \leq r$.

2.8. Let $\mathcal{O}_K$ be a discrete valuation ring with residue field k, and $X \to$ Spec $\mathcal{O}_K$ an arithmetic surface. Let Γ, Δ be irreducible components of X_k.

(a) Let k' be a finite Galois extension of k containing $k(\Gamma) \cap k^s$. Show that $\Gamma_{k'}$ has $r := r_\Gamma$ irreducible components $\Gamma_1, \ldots, \Gamma_r$, and that the Γ_i are isomorphic as algebraic varieties over k.

(b) Show that there exists a finite étale extension $\mathcal{O}_{K'}/\mathcal{O}_K$ of discrete valuation rings such that k' is the residue field of $\mathcal{O}_{K'}$. Let $p : X_{\mathcal{O}_{K'}} \to X$ be the projection morphism. Show that $p_*\Gamma_i = r^{-1}[k' : k]\Gamma$.

(c) Let $\Delta_1, \ldots, \Delta_q$ be the irreducible components of $\Delta_{k'}$. Denoting the respective intersections on X and on $X_{\mathcal{O}_{K'}}$ by i_k and $i_{k'}$, show that for any $i \leq r$, we have

$$i_k(\Delta, \Gamma) = \sum_{1 \leq j \leq q} r_\Gamma i_{k'}(\Delta_j, \Gamma_i).$$

(Use the projection formula with the morphism p.) Show that Δ_j has the same geometric multiplicity e_{Δ_j} (Exercise 1.9) as Δ. Deduce from this that $i_k(\Delta, \Gamma)$ is divisible by each of the integers $r_\Gamma e_\Gamma$, $r_\Gamma e_\Delta$, $r_\Delta e_\Gamma$, and $r_\Delta e_\Delta$.

(d) Show that in general, neither $r_\Gamma r_\Delta$ nor $e_\Gamma e_\Delta$ divides $i_k(\Delta, \Gamma)$.

2.9. Let $X \to S$ be an arithmetic surface and $s \in S$ a closed point. Let us write $X_s = \sum_{1 \leq i \leq n} d_i \Gamma_i$. Let us set

$$d = \gcd_i\{d_i\}, \quad d' = \gcd_i\{r_i d_i\}, \quad d'' = \gcd_i\{r_i e_i d_i\},$$

where $r_i = r_{\Gamma_i}$ and e_i is the geometric multiplicity of Γ_i (see Exercise 1.9).

(a) Let $\pi : \tilde{X} \to X$ be the blowing-up of a closed point $x \in X_s$ and let us denote the exceptional divisor (Definition 3.1) by $E \subset \tilde{X}_s$. Show that

$$\tilde{X}_s = \sum_i d_i \tilde{\Gamma}_i + \left(\sum_i \mu_x(\Gamma_i) d_i \right) E, \quad r_{\tilde{\Gamma}_i} = r_i, \quad e_{\tilde{\Gamma}_i} = e_i,$$

and that

$$r_E = [k(x) \cap k(s)^{\text{sep}} : k(s)], \quad r_E e_E = [k(x) : k(s)].$$

(b) With the notation of (a), show that $\tilde{\Gamma}_i \cdot E = \mu_x(\Gamma_i)[k(x) : k(s)]$. Deduce from this the divisibility relations

$$r_i e_i \mid \mu_x(\Gamma_i) r_E e_E, \quad r_i \mid \mu_x(\Gamma_i) r_E$$

(use Exercises 1.9(b) and 2.8(c)).

(c) Let $X' \to S$ be an arithmetic surface that is birational to X. Show that the integers d, d', d'' associated to X'_s are the same as those associated to X_s.

(d) Show that

$$\widetilde{\Gamma}_i \cdot E = \mu_x(\Gamma_i)[k(x) : k(s)], \quad \widetilde{\Gamma}_i \cdot \widetilde{\Gamma}_j = \Gamma_i \cdot \Gamma_j - \mu_x(\Gamma_i)\mu_x(\Gamma_j)[k(x) : k(s)].$$

2.10. Let $X \to S$ be an arithmetic surface. Let $P \in X_\eta$ be a closed point. Let $\overline{\{P\}}$ denote the Zariski closure of $\{P\}$ in X. Let $s \in S$ be a closed point and $\Gamma_1, \ldots, \Gamma_n$ the irreducible components of X_s. Show that

$$[K(P) : K(S)] \geq \sum_{x \in \overline{\{P\}} \cap X_s} \mu_x(X_s)[k(x) : k(s)],$$

and that for any $x \in X_s$,

$$\mu_x(X_s) = \sum_{1 \leq i \leq n} d_i \mu_x(\Gamma_i),$$

where d_i is the multiplicity of Γ_i in X_s (use Proposition 1.30 and Exercise 2.9(a)).

2.11. Let $\mathcal{O}_K$ be a Henselian discrete valuation ring (Definition 8.3.33), and $X \to \operatorname{Spec} \mathcal{O}_K$ be a fibered surface.

(a) Show that for any closed point $P \in X_\eta$, the set $\overline{\{P\}} \cap X_s$ is reduced to a point.

(b) Let us suppose X is regular. Let $x \in X_s$ be a closed point, and let us once more take the notation of Exercise 2.7. Show that if $\operatorname{Card} k(x) \geq \sum_{1 \leq i \leq r} \mu_i$ (e.g., if $\operatorname{Card} k(x) \geq \mu_x(X_s)$), then there exists a $u \in \mathfrak{m}_z \setminus \mathfrak{m}_z^2$ such that u divides none of the u_i modulo t.

(c) Under the hypotheses of (b), let us consider the image $\tilde{u} \in \mathcal{O}_{X,x}$ of u, its zeros divisor $\operatorname{div}(\tilde{u})_0$ on X, and an irreducible component D of $\operatorname{div}(\tilde{u})_0$ passing through x. Show that $i_x(D, X_s) \leq \sum_i d_i \mu_i$. Let $P \in X_\eta$ be the point D_K. Show that

$$[K(P) : K] = \mu_x(X_s)[k(x) : k(s)].$$

See Exercise 2.10 for the computation of the second term.

(d) Let $\mathcal{O}_K = \mathbb{Z}_p$ be the ring of p-adic integers. Let us consider the scheme

$$X = \operatorname{Proj} \mathcal{O}_K[u, v, w]/(uv(u^{p-1} - v^{p-1}) + pw^{p+1})$$

and the point $x \in X_s$ with homogeneous coordinates $(0, 0, 1)$. Show that X is an arithmetic surface over $\mathcal{O}_K$, that $\mu_x(X_s) = p + 1$, and that for any point $P \in X_\eta$ such that $x \in \overline{\{P\}}$, we have $[K(P) : K] \geq 2p$ (consider the blowing-up of X with center x). This shows that in (c) the condition on $\operatorname{Card} k(x)$ is necessary.

2.12. Let $X \to S$ be a regular fibered surface, $\Gamma \subset X_s$ a vertical prime divisor, $\widetilde{X} \to X$ the blowing-up of a closed point $x \in X_s$, and $\widetilde{\Gamma}$ the strict
transform of Γ in $\widetilde{X}$.

(a) Let $K_{X/S}$ (resp. $K_{\widetilde{X}/S}$) be a canonical divisor of X (resp. of $\widetilde{X}$). Let
$E \subset \widetilde{X}$ be the exceptional divisor. Show that

$$K_{\widetilde{X}/S} \cdot \widetilde{\Gamma} = K_{X/S} \cdot \Gamma - \mu E^2, \quad \widetilde{\Gamma}^2 = \Gamma^2 + \mu^2 E^2,$$

where $\mu = \mu_x(\Gamma)$.

(b) Show the equality

$$p_a(\widetilde{\Gamma}) = p_a(\Gamma) - \frac{(\mu^2 - \mu)}{2} [k(x) : k(s)].$$

2.13. We are going to show Abhyankar's theorem stated in Remark 2.6. Let
$f : X \to Y$ be a birational morphism of finite type of Noetherian schemes
with X normal, and Y regular and excellent. Let E be a prime divisor
contained in the exceptional locus of f (see Theorem 7.2.22). Let $F = f(E)$.

(a) Let ξ denote the generic point of E and y_0 that of F. Let $Y_1 \to Y$
be the blowing-up of Y with center F. Show that $X \to Y$ factors
through a birational map $f_1 : X \dashrightarrow Y_1$ defined at ξ. Let $y_1 = f_1(\xi)$.
Let $Y_2 \to Y_1$ be the blowing-up of Y_1 with center $\overline{\{y_1\}}$. Show that
f_1 factors through a birational map $f_2 : X \dashrightarrow Y_2$. We thus define a
sequence of birational maps $f_n : X \dashrightarrow Y_n$.

(b) Show, using the method of Exercise 8.3.14, that there exists a greatest
$r \geq 1$ such that $\dim \mathcal{O}_{Y_r, y_r} \geq 2$.

(c) Show that, if necessary replacing Y by an open neighborhood of y_0,
the schemes $\overline{\{y_n\}}$ and Y_{n+1} are regular for every $n \geq 0$ (use the Y
excellent hypothesis and Theorem 8.1.19).

(d) Show that $f_{r+1} : X \dashrightarrow Y_{r+1}$ is an isomorphism above an open
neighborhood of y_{r+1}.

(e) Show that E is birational to $\mathbb{P}^d_{F_r}$, where $F_r = \overline{\{y_r\}}$, and that $d \geq 1$.
This is *Abhyankar's theorem*.

9.3 Minimal surfaces

Contrarily to proper algebraic curves over a field, any regular fibered surface
$X \to S$ has an infinity of regular fibered surfaces that are birational but not
isomorphic to it. The aim of the theory of minimal surfaces it to determine
a canonical element in each birational equivalence class, when possible. If we
consider the class of the regular fibered surfaces that are birationally equivalent

to a given surface, it contains a minimal element if $X \to S$ is sufficiently general (Theorem 3.21 and Exercise 3.2). We also consider the class of the regular fibered surfaces that dominate and are birational to a given surface, and the class of the arithmetic surfaces with normal crossings that are birational to a given surface (Subsection 9.3.4).

9.3.1   Exceptional divisors and Castelnuovo's criterion

Definition 3.1. Let $X \to S$ be a regular fibered surface. A prime divisor E on X is called an *exceptional divisor* (or (-1)-*curve*) if there exist a regular fibered surface $Y \to S$ and a morphism $f : X \to Y$ of S-schemes such that $f(E)$ is reduced to a point, and that $f : X \setminus E \to Y \setminus f(E)$ is an isomorphism. In other words, an exceptional divisor is an integral curve that can be contracted (Subsection 8.3.3) to a regular point. Let us note that as $f(E)$ is a closed point, its image in S is also a closed point. Hence E is a vertical divisor.

Remark 3.2. The contraction $f : X \to Y$ of E is also the blowing-up of Y along the closed point $f(E)$. This follows from Corollary 2.3.

The principal result of this subsection is Castelnuovo's theorem that characterizes exceptional divisors (Theorem 3.8). Its proof uses the theorem on formal functions (for Lemma 3.6(b)). From here to the end of the subsection, we fix a regular fibered surface $X \to S$ and a vertical prime divisor E contained in a closed fiber X_s. We will suppose that $E \simeq \mathbb{P}^1_{k'}$ for some (necessarily finite) extension k' of $k(s)$, and that $E^2 < 0$.

Lemma 3.3. *Let H be an effective Cartier divisor on X with $H^1(X, \mathcal{O}_X(H)) = 0$. Then the following properties are true.*

(a) *We have $H^1(X, \mathcal{O}_X(H + iE)) = 0$ for every $0 \le i \le r := H \cdot E/(-E^2)$.*

(b) *Let us moreover suppose that $\mathcal{O}_X(H)$ is generated by its global sections and that r is an integer. Then $\mathcal{O}_X(H + rE)|_E \simeq \mathcal{O}_E$ and $\mathcal{O}_X(H + rE)$ is generated by its global sections.*

Proof (a) We are going to proceed by induction on i. The property is true for $i = 0$ by hypothesis. Let us suppose it is true at rank i with $0 \le i \le r - 1$. We have

$$(H + (i+1)E) \cdot E = (r - (i+1))(-E^2) \ge 0;$$

hence $\mathcal{O}_X(H + (i+1)E)|_E \simeq \mathcal{O}_E(a)$ with $a \ge 0$. It follows from Lemma 5.3.1 that $H^1(E, \mathcal{O}_X(H + (i+1)E)|_E) = 0$. In addition, we have an exact sequence

$$0 \to \mathcal{O}_X(H + iE) \to \mathcal{O}_X(H + (i+1)E) \to \mathcal{O}_X(H + (i+1)E)|_E \to 0. \quad (3.8)$$

Taking the cohomology, we then obtain an exact sequence

$$0 = H^1(X, \mathcal{O}_X(H + iE)) \to H^1(X, \mathcal{O}_X(H + (i+1)E)) \to 0.$$

Hence $H^1(X, \mathcal{O}_X(H + (i+1)E)) = 0$, which shows (a).

(b) We have $(H+rE)\cdot E = 0$; hence $\mathcal{O}_X(H+rE)|_E \simeq \mathcal{O}_E$ (Example 7.3.34). In particular, $\mathcal{O}_X(H+rE)|_E$ is generated by its global sections. Exact sequence (3.8) with $i = r-1$ implies that we have the exact sequence

$$H^0(X, \mathcal{O}_X(H+rE)) \to H^0(E, \mathcal{O}_X(H+rE)|_E) \to H^1(X, \mathcal{O}_X(H+(r-1)E)) = 0.$$

Nakayama's lemma then shows that $\mathcal{O}_X(H + rE)$ is generated by its global sections at the points $x \in E$. The hypothesis on H implies that $\mathcal{O}_X(H + rE)$ is generated by its global sections at the points $x \notin E$, whence (b). □

Lemma 3.4. *There exists a contraction morphism $f : X \to Y$ of E (but Y is not necessarily regular).*

Proof We can suppose S is affine. Let $\mathcal{L}$ be an ample sheaf on X. Then $\mathcal{L}$ is generated by its global sections (Proposition 5.1.31(a)). In particular, $H^0(X, \mathcal{L}) \neq 0$. There therefore exists an effective Cartier divisor H_0 such that $\mathcal{L} \simeq \mathcal{O}_X(H_0)$. Moreover, if necessary replacing H_0 by a multiple, we can suppose that $H^1(X, \mathcal{O}_X(nH_0)) = 0$ for every $n \geq 1$ (Theorem 5.3.2(b)). For any vertical prime divisor Γ on X, $\mathcal{O}_X(H_0)|_\Gamma$ is ample; hence $H_0 \cdot \Gamma > 0$ (Proposition 7.5.5). Let $m = -E^2 > 0$, $r = H_0 \cdot E > 0$, and let us set $D = mH_0 + rE$. For any vertical prime divisor $\Gamma \neq E$, we have $D \cdot \Gamma \geq mH_0 \cdot \Gamma > 0$; hence $\mathcal{O}_X(D)|_\Gamma \not\simeq \mathcal{O}_\Gamma$. Applying Lemma 3.3 with $H = mH_0$, we see that $\mathcal{O}_X(D)$ is generated by its global sections and that $\mathcal{O}_X(D)|_E \simeq \mathcal{O}_E$. By Proposition 8.3.30 (and Exercise 8.3.20 if $\dim S = 0$), the contraction $f : X \to Y$ of E exists. □

Remark 3.5 (*Regular proper algebraic surfaces*). Let X be a regular proper algebraic surface over a field k. Then X is projective over k. Indeed, using Chow's lemma (Remark 3.3.34) and Theorem 8.3.44, there exists a projective birational morphism $f : Z \to X$ with Z regular and projective over k. Next, Theorem 2.2 can be applied to f because its proof is based on Theorem 8.3.20, which is valid for normal schemes of dimension 2 that are proper over a Dedekind scheme S. We can therefore decompose $Z \to X$ into a sequence of blowing-ups of closed points $Z = X_n \to \cdots \to X_0 = X$. We then reduce to the case when $f : Z \to X$ is the blowing-up of a closed point $x \in X$. Let $E = f^{-1}(x)$. Proposition 2.5 and Lemma 3.4 imply that there exists a contraction morphism $Z \to X'$ of E. Theorem 8.3.20 (the version for proper schemes) implies that $X \simeq X'$, and hence X is projective over k.

Let us note that the same proof works if X is a regular scheme of dimension 2 that is proper and flat over an affine Dedekind scheme S of dimension 1, provided that we have the resolution of singularities (hence, for example, if S is excellent or if $X \to S$ has a smooth generic fiber, cf. Subsection 8.3.4). This gives another proof of Theorem 8.3.16 in that case.

Let $f : X \to Y$ be the contraction defined in Lemma 3.4. We have $f_*\mathcal{O}_X = \mathcal{O}_Y$ by Corollary 4.4.3(a). Let $y = f(E)$. We are going to determine the tangent space $T_{Y,y}$. For the proof of the following lemma, we need the theorem on formal functions (Remark 5.3.12).

Lemma 3.6. *Let V be an affine open neighborhood of y, $U = f^{-1}(V)$, $\mathcal{J} = \mathcal{O}_X(-E)|_U$. Let $n \geq 0$. Then the following properties are true.*

(a) For any $m \geq n+1$, we have $H^1(U, \mathcal{J}^n/\mathcal{J}^m) = 0$.

(b) We have $H^1(U, \mathcal{J}^n) = 0$.

(c) The sheaf $\mathcal{J}^n$ is generated by its global sections.

(d) We have $H^0(U, \mathcal{J}^n) = \mathfrak{m}^n$.

Proof (a) Let $d = -\deg_{k'} \mathcal{O}_X(E)|_E = -E^2/[k' : k(s)]$. Then $\mathcal{O}_X(-E)|_E \simeq \mathcal{O}_E(d)$ because $\deg : \text{Pic}(E) \to \mathbb{Z}$ is an isomorphism. For any integer $k \geq 0$, we have

$$\mathcal{J}^k/\mathcal{J}^{k+1} = \mathcal{J}^k \otimes_{\mathcal{O}_X} \mathcal{O}_E = \mathcal{O}_X(-kE)|_E \simeq \mathcal{O}_E(kd).$$

It follows that $H^1(U, \mathcal{J}^k/\mathcal{J}^{k+1}) \simeq H^1(E, \mathcal{O}_E(kd)) = 0$ (Lemma 5.3.1(c)). We have an exact sequence

$$0 \to \mathcal{J}^m/\mathcal{J}^{m+1} \to \mathcal{J}^n/\mathcal{J}^{m+1} \to \mathcal{J}^n/\mathcal{J}^m \to 0,$$

whence the exact sequence

$$0 = H^1(X, \mathcal{J}^m/\mathcal{J}^{m+1}) \to H^1(X, \mathcal{J}^n/\mathcal{J}^{m+1}) \to H^1(X, \mathcal{J}^n/\mathcal{J}^m).$$

This implies that $H^1(X, \mathcal{J}^n/\mathcal{J}^m)$ is zero by induction on $m \geq n+1$.

(b) Let $A = \mathcal{O}_Y(V)$ and let $\mathfrak{m}$ denote the maximal ideal of A corresponding to the point $y \in V$. By the theorem on formal functions (Remark 5.3.12), we have an isomorphism

$$H^1(U, \mathcal{J}^n) \otimes_A \widehat{A} \simeq \varprojlim_k H^1(U, \mathcal{J}^n/\mathfrak{m}^k \mathcal{J}^n),$$

where $\widehat{A}$ is the $\mathfrak{m}$-adic completion of A. As $\sqrt{\mathfrak{m}\mathcal{O}_U} = \mathcal{J}$, there exists an $r \geq 1$ such that $\mathcal{J}^r \subseteq \mathfrak{m} \subseteq \mathcal{J}$. It follows that

$$H^1(U, \mathcal{J}^n) \otimes_A \widehat{A} = \varprojlim_{k \geq 1} H^1(U, \mathcal{J}^n/\mathfrak{m}^k \mathcal{J}^n) = \varprojlim_{m \geq n} H^1(U, \mathcal{J}^n/\mathcal{J}^m) = 0.$$

Now $H^1(U, \mathcal{J}^n)$ is a finitely generated A-module with support in $\{y\}$ since $U \to V$ is projective and is an isomorphism outside of y; hence $H^1(U, \mathcal{J}^n) = H^1(U, \mathcal{J}^n) \otimes_A A_{\mathfrak{m}}$. It follows that $H^1(U, \mathcal{J}^n) = 0$ by the faithful flatness of $A_{\mathfrak{m}} \to \widehat{A}$.

(c) To show that $\mathcal{J}^n$ is generated by its global sections, it suffices to consider the case when $n = 1$. It is clear that $\mathcal{J}$ is generated by $H^0(U, \mathcal{J})$ at the points of $U \setminus E$. The exact sequence

$$0 \to \mathcal{J}^2 \to \mathcal{J} \to \mathcal{J}|_E \to 0$$

implies that $H^0(U, \mathcal{J}) \to H^0(E, \mathcal{J}|_E)$ is surjective since $H^1(U, \mathcal{J}^2) = 0$. As $\mathcal{J}|_E \simeq \mathcal{O}_E(d)$ is generated by its global sections, Nakayama's lemma implies that $\mathcal{J}$ is generated by its global sections at the points $x \in E$.

(d) Let $h \in \mathfrak{m}$. Then $h \in \mathcal{O}_X(U)$ and is zero on E; hence $h \in H^0(U, \mathcal{J})$. Conversely, if $h \in H^0(U, \mathcal{J})$, then $h \in \mathcal{O}_X(U) = \mathcal{O}_Y(V)$ and $h \in \mathfrak{m}$ because otherwise, it would be invertible on a neighborhood of E, whence $H^0(U, \mathcal{J}) = \mathfrak{m}$. To show that $H^0(U, \mathcal{J}^n) = \mathfrak{m}^n$, it suffices to show that if $\mathcal{F}, \mathcal{G}$ are two coherent sheaves on U, generated by their respective global sections, then the canonical homomorphism

$$\varphi : \; H^0(U, \mathcal{F}) \otimes_A H^0(U, \mathcal{G}) \to H^0(U, \mathcal{F} \otimes_{\mathcal{O}_U} \mathcal{G})$$

is surjective. The proof that follows is taken from Lipman, [57], Section 7.2. As $H^0(U, \mathcal{F})$ and $H^0(U, \mathcal{G})$ are finitely generated over A because $U \to V$ is projective, we have exact sequences

$$0 \to \mathcal{K}_1 \to \mathcal{O}_U^p \xrightarrow{\alpha} \mathcal{F} \to 0, \quad 0 \to \mathcal{K}_2 \to \mathcal{O}_U^q \xrightarrow{\beta} \mathcal{G} \to 0$$

with $p, q \in \mathbb{N}$, whence an exact sequence

$$H^0(U, \mathcal{O}_U^p) \to H^0(U, \mathcal{F}) \to H^1(U, \mathcal{K}_1) \to H^1(U, \mathcal{O}_U) = 0.$$

We can choose α such that the homomorphism on the left is surjective. It follows that $H^1(U, \mathcal{K}_1) = 0$. Likewise, $H^1(U, \mathcal{K}_2) = 0$ for a suitable choice of β. The tensor product $\gamma := \alpha \otimes \beta$ induces an exact sequence

$$0 \to \operatorname{Ker} \gamma \to \mathcal{O}_U^p \otimes \mathcal{O}_U^q \to \mathcal{F} \otimes \mathcal{G} \to 0.$$

As $\operatorname{Ker} \gamma$ is a quotient of $(\mathcal{K}_1 \otimes \mathcal{O}_U^q) \oplus (\mathcal{K}_2 \otimes \mathcal{O}_U^p)$ (Exercise 1.1.3), and $U \to V$ has fibers of dimension ≤ 1, we have an exact sequence

$$0 = H^1(U, (\mathcal{K}_1 \otimes \mathcal{O}_U^q) \oplus (\mathcal{K}_2 \otimes \mathcal{O}_U^p)) \to H^1(U, \operatorname{Ker} \gamma) \to 0$$

(Proposition 5.2.34). Hence $H^1(U, \operatorname{Ker} \gamma) = 0$ and

$$H^0(U, \mathcal{O}_U^p \otimes \mathcal{O}_U^q) \to H^0(U, \mathcal{F} \otimes \mathcal{G})$$

is surjective. The commutative diagram

$$
\begin{array}{ccc}
H^0(U, \mathcal{O}_U^p) \otimes H^0(U, \mathcal{O}_U^q) & \longrightarrow & H^0(U, \mathcal{F}) \otimes H^0(U, \mathcal{G}) \\
\| & & \downarrow{\varphi} \\
H^0(U, \mathcal{O}_U^p \otimes \mathcal{O}_U^q) & \longrightarrow & H^0(U, \mathcal{F} \otimes \mathcal{G})
\end{array}
$$

shows that φ is surjective. $\qquad\square$

Theorem 3.7. Let $X \to S$ be a regular fibered surface. Let $E \subseteq X_s$ be a vertical prime divisor. We suppose that $E \simeq \mathbb{P}^1_{k'}$ for some extension $k'/k(s)$, and that $E^2 < 0$. Let us set

$$d = -E^2/[k' : k(s)] = \deg_{k'} \mathcal{O}_X(-E)|_E = \deg_{k'} \mathcal{N}_{E/X}$$

(Definition 6.3.7). Then the following properties are true.

(a) There exists a contraction morphism $f : X \to Y$ of E.

(b) Let $y = f(E)$, and let $\mathfrak{m}_y$ be the maximal ideal of $\mathcal{O}_{Y,y}$. Then we have $k(y) = k'$ and there exists an isomorphism of $k(y)$-algebras

$$\oplus_{n \geq 0} \mathfrak{m}_y^n/\mathfrak{m}_y^{n+1} \simeq \oplus_{n \geq 0} H^0(E, \mathcal{O}_E(dn)).$$

(c) Let $T_{Y,y}$ be the Zariski tangent space of Y at y. Then

$$\dim_{k(y)} T_{Y,y} = d + 1.$$

Proof (a) is Lemma 3.4. (b) With the notation of Lemma 3.6, we have

$$\mathfrak{m}^n/\mathfrak{m}^{n+1} = H^0(U, \mathcal{J}^n)/H^0(U, \mathcal{J}^{n+1}) \simeq H^0(U, (\mathcal{J}^n/\mathcal{J}^{n+1}))$$

for every $n \geq 0$, because $H^1(U, \mathcal{J}^n) = 0$. Now the term on the right is equal to $H^0(E, \mathcal{O}_X(-nE)|_E) \simeq H^0(E, \mathcal{O}_E(nd))$. Taking $n = 0$, we obtain $k(y) = H^0(E, \mathcal{O}_E) = k'$. Finally, taking $n = 1$, we have

$$\dim_{k(y)} T_{Y,y} = \dim_{k(y)} H^0(E, \mathcal{O}_E(d)) = d + 1,$$

which shows (c).   □

Theorem 3.8 (Castelnuovo's criterion). Let $X \to S$ be a regular fibered surface. Let $E \subset X_s$ be a vertical prime divisor. Let us set $k' = H^0(E, \mathcal{O}_E)$. Then E is an exceptional divisor if and only if $E \simeq \mathbb{P}^1_{k'}$ and if $E^2 = -[k' : k(s)]$.

Proof This is just Proposition 2.5 and Theorem 3.7(c).   □

Remark 3.9. The proof of Castelnuovo's criterion in the case of arithmetic surfaces was given independently by Lichtenbaum [56], 3.10, and Shafarevich [90], Lecture 6, pages 102–114. See also the presentation of [20], Section 6. For the geometric case (dim $S = 0$), see for example [43], V.5.7. It seems that, at least in the arithmetic case, the theorem on formal functions is indispensable. Later on, we will see a theorem of Artin–Lipman (Theorem 4.15) that generalizes Theorem 3.7.

With the canonical divisor $K_{X/S}$ (Definition 1.34), we can give another characterization of exceptional divisors.

Proposition 3.10. *Let $\pi : X \to S$ be a regular fibered surface. Let $K_{X/S}$ be a canonical divisor and $E \subset X_s$ a vertical prime divisor on X. Then the following properties are true.*

(a) *The divisor E is exceptional if and only if $K_{X/S} \cdot E < 0$ and $E^2 < 0$. Moreover, we then have $K_{X/S} \cdot E = E^2$.*

(b) *Let us suppose that $H^0(X, \omega_{X/S}^{\otimes q}) \neq 0$ for some $q \geq 1$ if $\dim S = 0$, and $p_a(X_\eta) \geq 1$ if $\dim S = 1$. Then E is an exceptional divisor if and only if $K_{X/S} \cdot E < 0$.*

Proof (a) Let us set $k' = H^0(E, \mathcal{O}_E)$. Let us first suppose that E is an exceptional divisor. Let $m = [k' : k(s)]$. Then $E^2 = -m$ and $\chi_{k(s)}(\mathcal{O}_E) = m$ by Proposition 2.5. Moreover, the adjunction formula (Theorem 1.37) implies that

$$K_{X/S} \cdot E = -E^2 - 2\chi_{k(s)}(\mathcal{O}_E) = -m < 0.$$

Let us show the converse. By computing the degrees over k' in the adjunction formula, we obtain

$$\deg_{k'} \omega_{X/S}|_E + \deg_{k'} \mathcal{O}_X(E)|_E = -2\chi_{k'}(\mathcal{O}_E) = -2 + 2\dim_{k'} H^1(E, \mathcal{O}_E),$$

with the two terms on the left strictly negative. It follows that

$$H^1(E, \mathcal{O}_E) = 0, \quad \deg_{k'} \mathcal{O}_X(-E)|_E = 1.$$

Hence $p_a(E) \leq 0$ and $\deg_{k'} : \operatorname{Pic}(E) \to \mathbb{Z}$ is surjective. By Proposition 3.16(b) that we will show in the next subsection, E is then isomorphic to $\mathbb{P}^1_{k'}$.

(b) In the $\dim S = 1$ case, we can suppose S is affine because the property is local on S. We then have

$$H^0(X, \omega_{X/S}) \otimes_{\mathcal{O}_S(S)} K(S) = H^0(X_\eta, \omega_{X_\eta/K(S)}) \simeq H^1(X_\eta, \mathcal{O}_{X_\eta}) \neq 0.$$

Hence $H^0(X, \omega_{X/S}) \neq 0$ and there exists a canonical divisor $K_{X/S} \geq 0$. If $\dim S = 0$, there exists an effective divisor $K' \sim qK_{X/S}$. Let us set $q = 1$ if $\dim S = 1$. Let us suppose $K_{X/S} \cdot E < 0$. Then $K' \cdot E < 0$. We then have $K' = aE + D$ with $a \geq 1$ and $D \geq 0$ with no common component with E. We have $aE^2 = K' \cdot E - D \cdot E < 0$, and hence $E^2 < 0$. It follows from (a) that E is an exceptional divisor. $\square$

The following proposition is an analogue of Lemma 2.1, except that we turn the blowing-up process around.

Proposition 3.11. *Let $f : X \to Y$ be a birational morphism of fibered surfaces over S with X regular and Y normal, and let E be an exceptional divisor on X contained in the exceptional locus of f. Let us consider the contraction $\pi : X \to X'$ of E. Then f factors in a unique way into $X \to X' \to Y$.*

Proof This is an immediate consequence of Theorem 8.3.20. $\square$

9.3.2   Relatively minimal surfaces

Definition 3.12. We say that a regular fibered surface $X \to S$ is *relatively minimal* if it does not contain any exceptional divisor. By Theorem 2.2, this is equivalent to saying that every birational morphism of regular fibered surfaces $X \to Y$ is an isomorphism. We say that $X \to S$ is *minimal* if every birational map of regular fibered S-surfaces $Y \dashrightarrow X$ is a birational morphism. A minimal surface is, of course, relatively minimal. See Corollary 3.24 for the converse.

Proposition 3.13. *Let $X \to S$ be a minimal arithmetic surface. Let η be the generic point of S. Then the canonical map $\mathrm{Aut}_S(X) \to \mathrm{Aut}_{X_\eta}(X_\eta)$ is bijective. In other words, any automorphism of X_η extends in a unique way to an automorphism of X.*

Proof Let $\sigma : X_\eta \to X_\eta$ be an automorphism. It then induces a birational morphism $\sigma' : X \dashrightarrow X$. By hypothesis, σ' is a morphism. Applying the same reasoning to σ^{-1}, we see that σ' is an automorphism.   □

Definition 3.14. Let $X \to S$ be a normal fibered surface. We call a regular fibered surface $Y \to S$ together with a birational map $Y \dashrightarrow X$ a *(regular) model of X over S*. Let us note that if $\dim S = 1$, then $Y_\eta \dashrightarrow X_\eta$ is a birational map of projective normal curves; it is therefore an isomorphism.

A *morphism* of two (regular) models Y, Z of X is a morphism of fibered S-surfaces $Y \to Z$ that is compatible with the birational maps $Y \dashrightarrow X$, $Z \dashrightarrow X$. A (regular) model Y of X is called the *minimal (regular) model* (resp. a *relatively minimal (regular) model*) of X if it is minimal (resp. relatively minimal) as a regular fibered surface over S. If it exists, the minimal model is unique.

Example 3.15. Let $X \to S$ be a smooth arithmetic surface. Then $X \to S$ is relatively minimal. Indeed, any vertical divisor V on X verifies $V^2 = 0$ (Proposition 1.21(a)), and hence V is not exceptional. If moreover $g(X_\eta) \geq 1$, then $X \to S$ is minimal (see Corollary 3.24).

In this and the following subsection, we will only be interested in arithmetic surfaces. The case of smooth projective surfaces over a field is more difficult, and the statements are a little bit different. See Exercise 3.2 and the references that are given there. In the following proposition, we gather together some information on integral curves of arithmetic genus $p_a \leq 0$.

Proposition 3.16. *Let C be an integral, l.c.i., projective curve over a field K such that $p_a(C) \leq 0$. Let us set $L = H^0(C, \mathcal{O}_C)$. Then C is also a projective curve over L. Moreover, the following properties are true.*

(a) *The curve C is a conic over L. It contains at most one singular point, which will then be rational over L.*

(b) *We have $\mathrm{Pic}^0(C) = 1$. The image of the injective homomorphism $\deg_L : \mathrm{Pic}(C) \to \mathbb{Z}$ is equal to $e\mathbb{Z}$, where e is the minimum of the degrees $[k(x) : L]$ of the regular closed points $x \in C$. Moreover, $e = 1$ if and only if $C \simeq \mathbb{P}^1_L$.*

(c) *Let us suppose C is regular and non-smooth over L. Then there exists a finite morphism $C \to \mathbb{P}_L^1$ that is purely inseparable of degree 2.*

Proof The $p_a(C) \le 0$ hypothesis is equivalent to $H^1(C, \mathcal{O}_C) = 0$. The restriction $H^0(C, \mathcal{O}_C) \to \mathcal{O}_C(U)$ for every open subset U of C makes $\mathcal{O}_C$ into a sheaf of L-algebras. Hence C is a curve over L. Let us henceforth suppose that $L = K$. Then $p_a(C) = 0$.

(a) Let us consider the dualizing sheaf $\omega_{C/K}$. By Example 7.3.34, we know that $\deg_K \omega_{C/K}^\vee = 2$ and $\dim_K H^0(C, \omega_{C/K}^\vee) = 3$. We are going to show that $\omega_{C/K}^\vee$ is very ample. Let $D > 0$ be a Cartier divisor such that $\mathcal{O}_C(D) \simeq \omega_{C/K}^\vee$. Let us first show that $\mathcal{O}_C(D)$ is generated by its global sections. For this, it suffices to show that for every $x \in \operatorname{Supp} D$, the homomorphism $H^0(C, \mathcal{O}_C(D)) \to \mathcal{O}_C(D) \otimes k(x)$ is surjective. Let $\mathcal{M}$ denote the sheaf of ideals of $\mathcal{O}_C$ corresponding to the reduced closed subscheme $\{x\}$. We then have an exact sequence

$$0 \to \mathcal{M}\mathcal{O}_C(D) \to \mathcal{O}_C(D) \to \mathcal{O}_C(D) \otimes k(x) \to 0.$$

It therefore suffices to show that $H^1(C, \mathcal{M}\mathcal{O}_C(D)) = 0$. We have an exact sequence

$$0 \to \mathcal{O}_C(-D) \to \mathcal{M} \to \mathcal{F} \to 0,$$

where the support of $\mathcal{F}$ is in $\{x\}$. Tensoring this sequence by $\mathcal{O}_C(D)$ and taking its cohomology, we obtain an exact sequence

$$H^1(C, \mathcal{O}_C) \to H^1(C, \mathcal{M}\mathcal{O}_C(D)) \to 0,$$

whence $H^1(C, \mathcal{M}\mathcal{O}_C(D)) = 0$ and $\mathcal{O}_C(D)$ is generated by its global sections.

Let $\varphi : C \to \mathbb{P}_K^2$ be the morphism associated to a basis of $H^0(C, \mathcal{O}_C(D))$. Let us show that φ is a closed immersion. Let us denote the global sections of $\mathcal{O}_C(nD)$ by $L(nD)$. We have

$$\mathcal{O}_C(C \setminus \operatorname{Supp} D) = \cup_{n \ge 0} L(nD)$$

(Exercise 7.3.5). Taking once more the proof of Lemma 3.6(d) (we use the fact that $H^1(C, \mathcal{O}_C) = 0$), we prove that the canonical homomorphism $L(D)^n \to L(nD)$ is surjective for every $n \ge 1$. This immediately implies that the restriction of φ to $C \setminus \operatorname{Supp} D$ is a closed immersion. Let $x \in \operatorname{Supp} D$. There exists an $f \in L(D)$ such that $f_x \mathcal{O}_{C,x} = \mathcal{O}_C(D)_x$. Hence $x \notin \operatorname{Supp}(D+(f))$. Now the morphism φ is also the one associated to $D+(f)$ for a suitable basis (Remark 5.1.32); what we have just seen shows that φ is a closed immersion on $C \setminus \operatorname{Supp}(D+(f))$, which contains x. Consequently, φ is a closed immersion. Its image is a plane curve Q with $H^1(Q, \mathcal{O}_Q) = 0$; it is therefore a conic (Example 7.3.22). The assertion concerning C_{Sing} is an easy consequence of the classification of conics done in Exercise 4.3.22. We can also use the exact sequence

$$0 \to \mathcal{O}_C \to \pi_* \mathcal{O}_{C'} \to \mathcal{G} \to 0,$$

where $\pi : C' \to C$ is the normalization morphism, and the fact that there exists a Cartier divisor $\pi^* D$ of degree 2 over K.

(b) Let $\mathcal{L} \in \operatorname{Pic}^0(C)$. By Example 7.3.34, we have $H^0(C, \mathcal{L}) \neq 0$, and hence $\mathcal{L} \simeq \mathcal{O}_C$ by virtue of Proposition 7.3.25(c), whence $\operatorname{Pic}^0(C) = 1$. Hence the homomorphism $\deg_K : \operatorname{Pic}(C) \to \mathbb{Z}$ is injective, and its image $\delta\mathbb{Z}$ contains $e\mathbb{Z}$. The effective divisor D in the proof of (a) verifies $\deg_K D = 2$. Hence $\delta = 1$ or 2. If $\delta = 1$, there exists an effective divisor D_0 of degree 1. Its support is then a regular rational point x_0 (because the maximal ideal $\mathfrak{m}_{x_0} \mathcal{O}_{C, x_0}$ is equal to $\mathcal{O}_C(-D_0)_{x_0}$). We therefore have $\delta = e$. Let us suppose $\delta = 2$. If the support of D is reduced to a point x of degree 2 over K, then x is regular (same reasoning as above), and hence $e = 2 = \delta$. Otherwise, $\operatorname{Supp} D$ is a singular point x. Replacing D by a linearly equivalent effective divisor D', we can suppose that $\operatorname{Supp} D'$ does not contain x, and therefore contains a regular point since C has at most one singular point, whence once more $e = 2 = \delta$. Finally, if $e = 1$, C contains a regular rational point. Hence $C \simeq \mathbb{P}^1_K$ by Exercise 4.3.22(b).

(c) Since C is a conic, by Exercise 4.3.22 we have $\operatorname{char}(K) = 2$, and there exist $a, b, c \in K$ such that $C = V_+(ax^2 + by^2 + cz^2)$, with, for example, $a \neq 0$. The projection onto (y, z) then induces a purely inseparable morphism $C \to \mathbb{P}^1_K$ of degree 2.   □

Lemma 3.17. *Let $X \to S$ be an arithmetic surface. Then there exist only a finite number of fibers of $X \to S$ containing exceptional divisors.*

Proof Let us first note that if $X \to S$ has a smooth generic fiber, then by removing a finite number of closed fibers, $X \to S$ is smooth (Proposition 8.3.11), and hence relatively minimal (Example 3.15). The general case is slightly more complex, especially when the generic fiber is a conic. We can suppose $S = \operatorname{Spec} \mathcal{O}_K$ is affine. Let us first treat the case when $H^0(X, \omega_{X/S}) \neq 0$. The set $\mathcal{B}$ of points $x \in X$ where $\omega_{X/S}$ is not generated by its global sections (Definition 5.1.2) is a closed subset, which does not contain the generic point, by the hypothesis that $H^0(X, \omega_{X/S}) \neq 0$. Let E be an exceptional divisor. By Proposition 3.10, $\deg \omega_{X/S}|_E < 0$. It follows that $H^0(E, \omega_{X/S}|_E) = 0$. It immediately follows from this that $E \subseteq \mathcal{B}$. As $\mathcal{B}$ is a proper closed subset of X, the lemma is proven.

Let us now suppose $H^0(X, \omega_{X/S}) = 0$. Let $X \to S' \to S$ be the decomposition as in Proposition 8.3.8. It suffices to show that the fibers of $X \to S'$ are irreducible, except for a finite number of them. By hypothesis, $H^0(X_K, \omega_{X_K/K}) = H^0(X, \omega_{X/S}) \otimes_{\mathcal{O}_K} K = 0$. By duality (Remark 6.4.21), we have $H^1(X_K, \mathcal{O}_{X_K}) = 0$. Let $L = K(S')$. By Proposition 3.16(c), X_K is smooth over L or purely inseparable over $\mathbb{P}^1_L$. The smooth case was seen at the beginning of the proof. Let us therefore suppose that there exists a finite purely inseparable morphism $\pi_L : X_K \to \mathbb{P}^1_L$. Replacing S by a dense open subscheme if necessary, π_L extends to a finite purely inseparable morphism $\pi : X \to \mathbb{P}^1_{\mathcal{O}_L}$. In particular, it is a homeomorphism (Exercise 5.3.9(a)). Therefore the fibers of $X \to \operatorname{Spec} \mathcal{O}_L$ are irreducible.   □

Remark 3.18. We can also show Lemma 3.17 with the help of the following result: *Let $f : X \to Y$ be a morphism of finite type of locally Noetherian irreducible schemes. Let us suppose that the generic fiber X_η is non-empty and* geometrically irreducible. Then X_y is (geometrically) irreducible for every point y of a dense open subscheme of Y ([41], Proposition IV.9.7.8).

Proposition 3.19. *Let $f : X \to S$ be an arithmetic surface. Then there exists a birational morphism $X \to Y$ of arithmetic surfaces over S, with Y relatively minimal.*

Proof Let $X_0 = X \to X_1 \to \dots \to X_n \to \dots$ be a sequence of contractions of exceptional divisors. We must show that the sequence is necessarily finite. Let B_n be the (finite) set of points $s \in S$ such that $(X_n)_s$ contains an exceptional divisor. Then $B_{n+1} \subseteq B_n$. Moreover, the total number of irreducible components contained in the fibers $(X_n)_s$, $s \in B_n$, decreases strictly with n. Therefore the sequence is finite. $\square$

9.3.3 Existence of the minimal regular model

We are going to show the existence of minimal models (Definition 3.14) for arithmetic surfaces whose generic fibers have arithmetic genus $p_a \geq 1$. We will also show that the minimal model is compatible with étale base change (Proposition 3.28).

Lemma 3.20. *Let $Y \to S$ be a normal fibered surface. Let us suppose that Y admits two regular models X_1, X_2 without relation of domination between them. Then there exist a regular model Z of Y that dominates the X_i, and an exceptional divisor E_1 on Z contained in the exceptional locus of $Z \to X_1$ (in other words, the image of E_1 in X_1 is a point) such that*

(i) *either the image of E_1 in X_2 is still an exceptional divisor;*

(ii) *or there exist another exceptional divisor E_2 on Z, and an integer $\mu \geq 1$ with $(E_1 + \mu E_2)^2 \geq 0$.*

Moreover, if the X_i dominate Y, then the E_i are contained in the exceptional locus of $Z \to Y$.

Proof Let Γ be the graph of the birational map $X_1 \dashrightarrow X_2$. Then we have a birational morphism $\Gamma \to X_1$, and we have $\Gamma_\eta \simeq (X_1)_\eta$ if $\dim S = 1$. Hence Γ admits a desingularization $Z \to \Gamma$ (Theorem 8.3.50). Let us consider the birational morphisms $f_1 : Z \to X_1$ and $f_2 : Z \to X_2$. Let E_1 be an exceptional divisor on Z contained in the exceptional locus of f_1. If E_1 is contained in the exceptional locus of f_2, then the f_i factor into $Z \to Z'$ and $g_i : Z' \to X_i$, where $Z \to Z'$ is the contraction of E_1 (Proposition 3.11). By the hypothesis that X_1, X_2 have no relation of domination, g_i is not an isomorphism. Through a finite sequence of such contractions, we can reduce to the case when E_1 is not contained in the exceptional locus of f_2. If E_1 does not meet the exceptional locus of f_2, then $f_2(E_1)$ is an exceptional divisor on X_2 and we are in case (i). Let us suppose that E_1 meets the exceptional locus of f_2. By successive contractions of the exceptional divisors on Z contained in the exceptional locus of f_2 and that do not meet E_1, we end up finding an exceptional divisor E_2 on Z different from E_1 and such that $E_1 \cap E_2 \neq \emptyset$. Let us show that we are in case (ii).

Let us consider the contraction $\pi : Z \to Z'$ of E_2. Let us set $D_1 = \pi(E_1)$. Then there exists a $\mu \geq 1$ such that $\pi^* D_1 = E_1 + \mu E_2$ (Proposition 2.23). Let s be the image of the E_i in S. By Theorem 2.12(a), we have $E_2 \cdot \pi^* D_1 = 0$ and therefore

$$\begin{aligned} \pi^* D_1 \cdot \pi^* D_1 &= E_1 \cdot \pi^* D_1 = E_1^2 + \mu E_1 \cdot E_2 \\ &= [k' : k(s)](-1 + \mu \deg_{k'} \mathcal{O}_Z(E_2)|_{E_1}) \geq 0, \end{aligned} \tag{3.9}$$

where $k' = H^0(E_1, \mathcal{O}_{E_1})$. Hence $(E_1 + \mu E_2)^2 \geq 0$. Finally, by construction, E_i is in the exceptional locus of $Z \to X_i$; hence if the X_i dominate Y, then the E_i are in the exceptional locus of $Z \to Y$. □

Theorem 3.21. *Let* $X \to S$ *be an arithmetic surface with generic fiber of genus* $p_a(X_\eta) \geq 1$. *Then* X *admits a unique minimal model over* S, *up to unique isomorphism.*

Proof The uniqueness of a minimal model (up to unique isomorphism) follows from the definition. We already know that X admits relatively minimal models (Proposition 3.19). The existence of the minimal model is equivalent to saying that two relatively minimal models X_1, X_2 of X are isomorphic. Let us suppose that this is not the case. Let $D = E_1 + \mu E_2$ be the divisor on a regular model Z defined in Lemma 3.20 (case (i) of the lemma cannot occur since X_2 is relatively minimal). This is a vertical divisor contained in a fiber Z_s. As $D^2 \geq 0$, it follows from Theorem 1.23 that $D = rZ_s$ for some rational number $r > 0$. By Proposition 1.35 and Proposition 3.10, we have

$$2p_a(X_\eta) - 2 = 2p_a(Z_\eta) - 2 = K_{Z/S} \cdot Z_s = (E_1 \cdot K_{Z/S} + \mu E_2 \cdot K_{Z/S})/r < 0,$$

whence $p_a(X_\eta) \leq 0$, a contradiction. □

Remark 3.22. This theorem was first proven by Lichtenbaum and Shafarevich. See the references in Remark 3.9.

Remark 3.23. Theorem 3.21 is false without the $p_a(X_\eta) \geq 1$ hypothesis. Indeed, and let us take $X_1 = \mathbb{P}^1_S$, let X be the blowing-up of X_1 with center a closed point $x \in X_1(k(s))$. In X_s, the strict transform E of $(X_1)_s$ is an exceptional divisor. Let $X \to X_2$ be the contraction of E. Then the models X_1 and X_2 of X are relatively minimal, but not isomorphic as models of X (more precisely, the birational map $X_1 \dashrightarrow X_2$ induced by the identity on the generic fiber does not extend to a morphism because the generic points of the fibers $(X_1)_s$ and $(X_2)_s$ induce distinct valuations in $K(X)$, even if, abstractly, we have $X_1 \simeq X_2 \simeq \mathbb{P}^1_S$). See also Exercise 3.1.

Corollary 3.24. *Let* $X \to S$ *be a relatively minimal arithmetic surface, with generic fiber* X_η *verifying* $p_a(X_\eta) \geq 1$. *Then* X *is minimal.*

Proof This is an immediate consequence of Theorem 3.21 and of the definition of a relatively minimal surface. □

Definition 3.25. Let D be a divisor on a regular fibered surface $X \to S$. We say that D is *numerically effective* if $D \cdot C \geq 0$ for every vertical prime divisor C. For example, an ample divisor is numerically effective by Proposition 7.5.5 and the fact that the restriction of an ample divisor to a closed subscheme remains ample.

Corollary 3.26. *Let $X \to S$ be an arithmetic surface with $p_a(X_\eta) \geq 1$. Let $K_{X/S}$ be a canonical divisor. Then $X \to S$ is minimal if and only if $K_{X/S}$ is numerically effective.*

Proof Indeed, by Proposition 3.10(b), $K_{X/S}$ is numerically effective if and only if X is relatively minimal, which, in turn, is equivalent to X being minimal by Corollary 3.24. □

Corollary 3.27. *Let $\pi : X \to S$ be a minimal arithmetic surface whose generic fiber is an elliptic curve. Then $\pi_* \omega_{X/S}$ is an invertible sheaf on S, and the canonical homomorphism $\pi^* \pi_* \omega_{X/S} \to \omega_{X/S}$ is an isomorphism.*

Proof We have $\omega_{X/S}|_{X_\eta} \simeq \mathcal{O}_{X_\eta}$ (Example 7.3.35). Hence there exists a vertical divisor V on X such that $\omega_{X/S} \simeq \mathcal{O}_X(V)$. Let $s \in S$ be a closed point. We have

$$V \cdot X_s = 2p_a(X_\eta) - 2 = 0,$$

and $V \cdot \Gamma \geq 0$ for every irreducible component Γ of X_s, since V is numerically effective. It follows that $V \cdot \Gamma = 0$. As X_s is connected (Corollary 8.3.6(b)), we have $V \in d_s X_s + V'$ with $d_s \in \mathbb{Q}$ and $\operatorname{Supp} V' \cap X_s = \emptyset$ (Theorem 1.23). Now the gcd of the multiplicities of the irreducible components of X_s is equal to 1 (Corollary 1.32), and we therefore have $d_s \in \mathbb{Z}$. Consequently, $V = \sum_{s \in S} d_s X_s$ (finite sum). Let $\mathcal{L} = \mathcal{O}_S(\sum_s d_s s) \in \operatorname{Pic}(S)$. Then $\mathcal{O}_X(V) = \pi^* \mathcal{L}$, and $\pi_* \mathcal{O}_X(V) = \mathcal{L}$. The corollary can immediately be deduced from this. □

Proposition 3.28. *Let $X \to S$ be an arithmetic surface such that $p_a(X_\eta) \geq 1$. Let $S' \to S$ be a morphism. Let us suppose that $S' \to S$ is étale surjective, or that S is the spectrum of a discrete valuation ring R and that $S' = \operatorname{Spec} \widehat{R}$. Then $X \to S$ is minimal if and only if $X \times_S S' \to S'$ is minimal.*

Proof The fibered surface $X' := X \times_S S'$ over S' is regular by Corollary 4.3.24 for the $S' \to S$ étale case, and by Lemma 8.3.49(a)–(b) for the completion case. If X is not minimal (hence not relatively minimal by Corollary 3.24), we contract an exceptional divisor E by a morphism $X \to Z$. Then $X' \to Z \times_S S'$ is the contraction of $E \times_S S'$ (which is a divisor on S' because $S' \to S$ is surjective) with $Z \times_S S'$ regular. Consequently, X' is not relatively minimal.

Let us show the converse. Let us suppose that X' contains an exceptional divisor E'. Let $p : X' \to X$ be the projection morphism, $E = p(E')$, and $q = p|_{E'}$. Then $\omega_{X'/S'} = p^* \omega_{X/S}$ (Theorem 6.4.9(b)) and $q : E' \to E$ is a finite surjective morphism of integral projective curves over $k(s)$, where s is the image of E in S. It follows from Proposition 7.3.8 that

$$[K(E') : K(E)] \deg_{k(s)} \omega_{X/S}|_E = \deg_{k(s)} \omega_{X'/S'}|_{E'} < 0.$$

Hence E is an exceptional divisor by virtue of Proposition 3.10(b). See also another proof in Exercise 3.6. □

Remark 3.29. Proposition 3.28 is not true if we replace minimal by relatively minimal and omit the condition $p_a(X_\eta) \geq 1$. See Exercise 3.3.

Corollary 3.30. *Let $X \to S$ be as in* Proposition 3.28. *Let $T \to S$ be a smooth morphism. Then for any point $\xi \in T$ of codimension 1, $X \times_S \operatorname{Spec} \mathcal{O}_{T,\xi}$ is a minimal arithmetic surface over* $\operatorname{Spec} \mathcal{O}_{T,\xi}$.

Proof Let us first note that $X \times_S T$ is smooth over X, and hence regular (Theorem 4.3.36). Therefore $X \times_S \operatorname{Spec} \mathcal{O}_{T,\xi}$ is regular. Let s be the image of ξ in S. If s is the generic point, then the special fiber of $X \times_S \operatorname{Spec} \mathcal{O}_{T,\xi}$ is regular. The minimality follows immediately in this case. Let us therefore suppose s is closed. By Corollary 6.2.11, if necessary reducing T, we can suppose that T is étale over a scheme $\mathbb{A}_S^n$. The image ξ' of ξ in $\mathbb{A}_S^n$ is the generic point of $\mathbb{A}_{k(s)}^n$. The irreducible components of $X \times_S \operatorname{Spec} k(\xi')$ are those of X_s obtained by base change. As the degree is invariant under change of base field (use Theorem 7.3.17 because the Euler–Poincaré characteristic is invariant under extension of the base field), we see, as in the proof of Proposition 3.28, that $X \times_S \operatorname{Spec} \mathcal{O}_{\mathbb{A}_S^n,\xi'}$ is minimal. To conclude, it suffices to apply Proposition 3.28 to the étale morphism $\operatorname{Spec} \mathcal{O}_{T,\xi} \to \operatorname{Spec} \mathcal{O}_{\mathbb{A}_S^n,\xi'}$.   □

9.3.4 Minimal desingularization and minimal embedded resolution

We are going to use the preceding methods to show the existence of models that are minimal for certain types of criteria. Let $Y \to S$ be a normal fibered surface. If Y admits a desingularization, it is never unique.

Definition 3.31. Let Y be a normal Noetherian scheme. We call a desingularization morphism $Z \to Y$ such that every other desingularization morphism $Z' \to Y$ factors uniquely through $Z' \to Z \to Y$ a *minimal desingularization of* Y. By definition, if a minimal desingularization exists, then it is unique up to unique isomorphism. If Y is already regular, then it is its own minimal desingularization.

Proposition 3.32. *Let $Y \to S$ be a normal fibered surface. If Y admits a desingularization, then it admits a minimal desingularization. More precisely, if $X \to Y$ is a desingularization such that no exceptional divisor of X is contained in the exceptional locus of $X \to Y$, then it is a minimal desingularization.*

Proof Let $X_1 \to Y$ be a desingularization. By successive contractions of exceptional divisors contained in the exceptional locus of $X_1 \to Y$, we can suppose that X_1 no longer contains an exceptional divisor contained in the exceptional locus of $X_1 \to Y$. It remains to show that if $X_2 \to Y$ is another desingularization with this property on the exceptional divisors, then we have an isomorphism of Y-schemes $X_2 \simeq X_1$. Let us suppose the contrary. Then X_1 does not dominate X_2 and vice versa. Let Z and the divisor $D = E_1 + \mu E_2$ on Z be as defined in Lemma 3.20 (case (i) of the lemma cannot occur because the image of E_1 in X_2 would then be an exceptional divisor contained in the exceptional locus of

$X_2 \to Y$). The support of D is contained in the exceptional locus of $Z \to Y$ and we have $D^2 \geq 0$. This is impossible by Theorem 1.27. $\qquad\square$

Example 3.33. Let R be a non-trivial discrete valuation ring with residue field k. Let $a \in R$ with $\nu(a) \geq 2$. Let us consider the normal fibered surface

$$Y = \operatorname{Proj} R[u, v, w]/(uv - aw^2)$$

over R. It contains the affine open subscheme $\operatorname{Spec} R[x,y]/(xy - a)$ of Example 8.3.53. Let q be the closed point $x = y = 0$ of the special fiber of Y. It is the unique singular point of Y. Then the desingularization $f : X \to Y$ of q done in Example 8.3.53 is minimal. Indeed, with the help of Proposition 1.21, we see that every irreducible component Γ of the exceptional locus of f verifies $\Gamma^2 = -2$ and $\Gamma \simeq \mathbb{P}_k^1$, and therefore Γ is not an exceptional divisor. We note that the integer $\nu(a) - 1$ is equal to the number of irreducible components of $f^{-1}(q)$.

Remark 3.34. Let $Y \to S$ be a fibered surface admitting a desingularization; hence the sequence (3.11) of Subsection 8.3.4 relative to Y is finite. In the example above, the first regular scheme Y_n in this sequence is the minimal desingularization of X. But this is not the case in general. See Exercise 3.7. On the other hand, if we know beforehand that the singular points of Y are of a certain type (rational singularities), then Y_n is the minimal desingularization. See Exercise 4.7.

We have seen that, in general, an arithmetic surface $X \to S$ admits a regular model with normal crossings that dominates it (Corollary 2.30). We are going to show that there exists one, X', that is *minimal*; that is to say that for any regular model Y of X, if Y has normal crossings and dominates X, then Y dominates X'.

Lemma 3.35. *Let $X' \to S$ be an arithmetic surface with normal crossings. Let $E \subset X_s'$ be an exceptional divisor and $\pi : X' \to X$ the contraction of E. Then X has normal crossings if and only if E meets the other irreducible components of X_s' in at most two points, and if these points are rational over $k' := H^0(E, \mathcal{O}_E)$.*

Proof We will use the computations of Subsection 9.2.4 freely. Let us suppose that E verifies the conditions of the lemma. Let $x \in X_s$ be the image of E. Then there exist at most two irreducible components of X_s passing through x. Let Γ be one of these components. Then

$$-\mu_x(\Gamma)E^2 = E \cdot \widetilde{\Gamma} = [k' : k(s)].$$

It follows that $\mu_x(\Gamma) = 1$, and hence Γ is regular at x. If there exist two irreducible components Γ_1, Γ_2 of X_s passing through x, then their strict transforms $\widetilde{\Gamma}_1$ and $\widetilde{\Gamma}_2$ do not meet in E, and we have

$$\Gamma_1 \cdot \Gamma_2 = \pi^*\Gamma_1 \cdot \pi^*\Gamma_2 = (\widetilde{\Gamma}_1 + E) \cdot (\widetilde{\Gamma}_2 + E) = \widetilde{\Gamma}_1 \cdot \widetilde{\Gamma}_2 + [k' : k(s)].$$

As $k' = k(x)$, we have $i_x(\Gamma_1, \Gamma_2) = 1$. Hence X has normal crossings at x. The converse is nothing more than Lemma 2.31. $\qquad\square$

Proposition 3.36. *Let* $X \to S$ *be an arithmetic surface having only a finite number of singular fibers (e.g., if X_η is smooth, see Proposition 8.3.11). Then the following properties are true.*

(a) *There exists a regular model X' of X that dominates X, has normal crossings, and is minimal for this property.*

(b) *If $p_a(X_\eta) \geq 1$, then there exists a regular model of X that has normal crossings, and is minimal for this property.*

Proof Let us show (a). Assertion (b) follows by applying (a) to the minimal model of X. Let X' be a regular model with normal crossings which dominates X. By successive contractions of exceptional divisors as in Lemma 3.35 and that are contained in the exceptional locus of $X' \to X$, we can suppose that X' no longer contains any of these exceptional divisors. It remains to show that X' is then minimal.

Let $X_i \to X$, $i = 1, 2$, be two regular models as above. Then they are dominated by a regular model Z with normal crossings. Let E be an exceptional divisor on Z contained in the exceptional locus of $Z \to X_1$. Then E verifies the conditions of Lemma 3.35. If E does not meet the exceptional locus of $Z \to X_2$, then its image in X_2 is an exceptional divisor verifying the conditions of Lemma 3.35, which contradicts the hypothesis on X_2. Consequently, E meets the exceptional locus of $Z \to X_2$. We arrive at a contradiction by successive contractions as in the proof of Proposition 3.32.   □

Remark 3.37. In Proposition 3.36, we can replace X by a normal fibered surface that admits a desingularization with normal crossings. The proof is the same.

Exercises

3.1. Let $X \to S$ be an arithmetic surface. We suppose that X_η is a conic over $K(S)$.

(a) Show that X is relatively minimal if and only if X_s is irreducible for every $s \in S$ (use Proposition 3.10). Show that X_s irreducible implies X_s integral (use Exercise 1.8).

(b) Let us suppose that $X_s \simeq \mathbb{P}^1_{k(s)}$. Show that X is not minimal. Show that the hypothesis on X_s is satisfied if and only if X_s is integral and if X_η has a point that is rational over $\mathrm{Frac}\,\widehat{\mathcal{O}}_{S,s}$ (use Corollary 6.2.13).

(c) Let us suppose X is relatively minimal but not minimal. Show that there exist an $s \in S$ and a regular model Z that dominates X such that $Z_s = E_1 + E_2$ with the E_i exceptional divisors. Show that $-2 = K_{Z/S} \cdot E_1 + K_{Z/S} \cdot E_2$. Deduce from this that $E_i \simeq \mathbb{P}^1_{k(s)}$ and that $X(\mathrm{Frac}\,\widehat{\mathcal{O}}_{S,s}) \neq \emptyset$.

(d) Conclude that X admits a minimal model if and only if for every closed point $s \in S$, we have $X(\mathrm{Frac}\,\widehat{\mathcal{O}}_{S,s}) = \emptyset$.

3.2. Let $X \to \operatorname{Spec} k$ be a regular, connected, projective surface over a field k that is relatively minimal (otherwise, we can show that after a finite number of successive contractions of exceptional divisors, X becomes relatively minimal; see [43], Theorem V.5.8, or [90], pages 62–65). Let us suppose that

$$H^0(X, \omega_{X/k}^{\otimes q}) \neq 0 \quad \text{for some } q \geq 1. \tag{3.10}$$

We are going to show that $X \to \operatorname{Spec} k$ is minimal. Let us suppose the contrary.

(a) Show that there exist a regular model Z of X and two distinct exceptional divisors E_1, E_2 on Z such that $E_1 \cap E_2 \neq \emptyset$ (draw inspiration from the proof of Theorem 3.21).

(b) Show that there exists a canonical divisor $K_{Z/k}$ such that $qK_{Z/k} \geq 0$ (use Corollary 2.25). Let us write

$$qK_{Z/k} = a_1 E_1 + a_2 E_2 + D$$

with $a_1, a_2 \geq 0$ and $D \geq 0$ with support containing neither E_1 nor E_2.

(c) Use the equality $K_{Z/k} \cdot E_1 = E_1^2$ (Proposition 3.10(a)) to show that

$$(a_1 - q)(-E_1^2) \geq a_2 E_1 \cdot E_2 \quad \text{and} \quad (a_1 - q) \geq a_2.$$

Conclude with a contradiction by inverting the roles of E_1 and E_2.

Remark. Condition (3.10) can be considered as the equivalent of the $p_a(X_\eta) \geq 1$ condition for an arithmetic surface. Let us suppose k is algebraically closed. We can show that (3.10) is not satisfied if and only if X is *rational* (i.e., birational to $\mathbb{P}_k^2$) or *ruled* (i.e., isomorphic to $C \times_k \mathbb{P}_k^1$ for some smooth projective curve C over k). See [43], Theorem V.6.1, and [90], Lecture 7, pages 140–150.

(d) Show that rational surfaces and ruled surfaces are not minimal.

3.3. Let $\mathcal{O}_K$ be a discrete valuation ring, with uniformizing parameter t, and residue field k. We suppose that $\operatorname{char}(k) \neq 2$ and that $k \neq k^2$. Let $a \in \mathcal{O}_K$ be such that its image in k is not a square. Let us consider

$$X = \operatorname{Proj} \mathcal{O}_K[u, v, w]/(u^2 - av^2 + tw^2).$$

(a) Show that X is a relatively minimal arithmetic surface over $\mathcal{O}_K$.

(b) Let $\mathcal{O}_{K'} = \mathcal{O}_K[T]/(T^2 - a)$. Show that $\mathcal{O}_{K'}$ is étale over $\mathcal{O}_K$, but that $X_{\mathcal{O}_{K'}}$ is not relatively minimal over $\mathcal{O}_{K'}$.

3.4. Let $Y \to S$ be a normal fibered surface admitting a desingularization. Let $f : X \to Y$ be the minimal desingularization. Show that every automorphism σ of Y canonically induces an automorphism σ' of X, and that the correspondence $\sigma \mapsto \sigma'$ induces an isomorphism from $\mathrm{Aut}_S(Y)$ to the group

$$\{\tau \in \mathrm{Aut}_S(X) \mid \tau(f^{-1}(\mathrm{Sing}(Y))) = f^{-1}(\mathrm{Sing}(Y))\}.$$

3.5. Let S be a normal (hence integral) Noetherian scheme. Let $S' \to S$ be a surjective étale morphism. We want to show that there exists an integral étale S-scheme $T \to S$ that is finite and Galois (i.e., $G = \mathrm{Gal}(K(T)/K(S))$ acts on T and $T/G \simeq S$), and that $T \to S$ factors into $T \to S' \to S$.

(a) Let $K = K(S)$. Let N/K be the Galois closure of $K(S')$ over K and $\sigma \in \mathrm{Gal}(N/K)$. Show that the correspondence

$$V \mapsto \sigma(\mathcal{O}_{S'}(V)) \subset N, \quad V \text{ affine open subsets of } S'$$

induces a sheaf of $\mathcal{O}_S$-algebras and defines an étale S-scheme that we will denote by $(S')^\sigma$.

(b) Let S'' be the fibered product of the $(S')^\sigma$ over S, σ running through the group $\mathrm{Gal}(N/K)$. Show that S'' (endowed with the projection onto S') satisfies the requested properties for T, except that S'' may not be connected.

(c) Show that there exists a connected component T of S'' that maps surjectively onto S', and that this T has the desired properties.

3.6. Let $X \to S$ be a minimal arithmetic surface with $p_a(X_\eta) \geq 1$. Let $S' \to S$ be a surjective étale morphism. We are going to give another proof of the fact that $X_{S'}$ is minimal (Proposition 3.28).

(a) Let $T \to S' \to S$ be as in Exercise 3.5. Let $Z \to T$ be the minimal model of X_T. Show that the group $G := \mathrm{Gal}(K(T)/K(S))$ acts on Z and that Z/G is a normal fibered surface over S, dominated by X and birational to X.

(b) Show that $Z/G \times_S T$ is normal (use Corollary 8.2.25). Deduce from this that the canonical morphism $Z \to Z/G \times_S T$ is an isomorphism, and that Z/G is regular.

(c) Show that X_T is minimal. Deduce from this that $X_{S'}$ is minimal.

3.7. Let $\mathcal{O}_K$ be a discrete valuation ring, with uniformizing parameter t and residue characteristic $\neq 2, 3$. Let $X = \mathrm{Proj}\, \mathcal{O}_K[u, v, w]/(u^2 w + v^3 + t^6 w^3)$. Determine the minimal desingularization of X.

3.8. Let $X \to S$ be a minimal arithmetic surface with $p_a(X_\eta) = 1$. Show that there exists a vertical divisor $V \in \oplus_s \mathbb{Q} X_s$ such that $\omega_{X/S} \simeq \mathcal{O}_X(V)$.

3.9. Let $\pi : X \to S$ be an arithmetic surface with geometrically integral generic fiber. Let U be an open subset of X such that $U \supset X_\eta$ and that $U \cap X_s \neq \emptyset$ for every $s \in S$.

(a) Show that $\mathcal{O}_X(X) = \mathcal{O}_X(U)$.

(b) Let us suppose that X_η is an elliptic curve. Show that we have $\omega_{X/S}(X) = \omega_{X/S}(U)$ (use Corollary 3.27).

3.10. Let X be a regular, connected, quasi-projective scheme over an affine Dedekind scheme S, with geometrically irreducible generic fiber. Let $x_1, \ldots, x_m \in X$ and let D be a Cartier divisor on X.

(a) Let S_0 be the set of closed points $s \in S$ such that X_s is not irreducible. This is a finite set (Remark 3.18). Show that there exists a Cartier divisor D', linearly equivalent to D, and whose support does not contain any of the points x_i nor any of the irreducible components of the X_s, $s \in S_0$ (use Proposition 1.11).

(b) Show that $D' = H + V$, where H is a horizontal Cartier divisor (i.e., its support does not contain any irreducible component of a fiber), and where V is vertical (i.e., sum of a finite number of irreducible components of closed fibers).

(c) ([21], Appendix) Let us suppose that $\mathrm{Pic}(S)$ is a torsion group (i.e., every element is of finite order). Show that there exists a $q \geq 1$ such that $qD \sim qH$. In other words, qD is linearly equivalent to a horizontal divisor whose support does not contain any of the points $x_1, \ldots, x_m$.

9.4 Applications to contraction; canonical model

Let $X \to S$ be a minimal regular fibered surface. Then the canonical divisor $K_{X/S}$ is in general numerically effective (Corollary 4.26 and Exercise 4.2). If we want it to be ample, we must remove the vertical prime divisors C such that $K_{X/S} \cdot C = 0$. In this section, we will show that it is possible to contract every effective vertical divisor C such that $K_{X/S} \cdot C = 0$ (Corollary 4.7). The first two subsections are devoted to the existence theorem for contractions and to the computation of tangent spaces (Theorems 4.2 and 4.15). The content of these two subsections has for the most part been taken from articles [8], [9], and [57], with some ad hoc adaptations.

In the third subsection we define the notion of canonical model. Let $X \to S$ be an arithmetic surface; its fiber over a closed point s can be very complex. The canonical model, when the generic fiber of X is of arithmetic genus ≥ 2, is obtained by the contraction of all of the vertical prime divisors C such that $K_{X/S} \cdot C = 0$. Even if we lose the regularity of the surface, we gain in the simplicity of the closed fibers. For example, the number of irreducible components of a fiber is then bounded from above by a number that only depends on the arithmetic genus of the generic fiber (Proposition 4.24). Moreover, we 'do not modify' the dualizing sheaf (Corollary 4.18), but make it ample (Proposition 4.20).

When the generic fiber of $X \to S$ is an elliptic curve, the canonical model does not exist. But the minimal Weierstrass model plays a similar role. As we will explain in the last subsection. The objects introduced in this subsection, minimal discriminant, dualizing sheaf of the minimal Weierstrass model, are very useful in the arithmetic study of elliptic curves.

9.4.1   Artin's contractibility criterion

Definition 4.1. Let $X \to S$ be a regular fibered surface, and let Z be an effective vertical divisor, with irreducible components $\Gamma_1, \ldots, \Gamma_r$. We set

$$\mathrm{Pic}^0(Z) = \{ \mathcal{L} \in \mathrm{Pic}(Z) \mid \deg \mathcal{L}|_{\Gamma_i} = 0, \ i = 1, \ldots, r \}$$

(see also Definition 7.5.9). Let us recall that we have a canonical isomorphism $\mathrm{Pic}(Z) \simeq H^1(Z, \mathcal{O}_Z^*)$ (Exercise 5.2.7(c)).

Theorem 4.2 (Artin [8], Theorem 2.3). *Let $X \to S$ be a regular fibered surface. Let us consider a reduced effective vertical divisor $\Delta = \sum_{1 \le i \le r} \Gamma_i$ contained in a closed fiber X_s. Let us suppose that:*

(a) *the intersection matrix $(\Gamma_i \cdot \Gamma_j)_{1 \le i,j \le r}$ is negative definite;*

(b) *for any divisor $Z > 0$ with support in Δ, we have $\mathrm{Pic}^0(Z) = 1$.*

Then there exists a contraction morphism $f : X \to Y$ of the $\Gamma_1, \ldots, \Gamma_r$.

Proof Let M be the free $\mathbb{Z}$-module generated by the Γ_i. Let us consider the linear map

$$\varphi : M \to M^\vee := \mathrm{Hom}_{\mathbb{Z}}(M, \mathbb{Z}), \quad V \mapsto \sum_{1 \le i \le r} (V \cdot \Gamma_i) \Gamma_i^*,$$

where $\{\Gamma_i^*\}_i$ is the dual basis. This is the canonical map induced by the intersection form $M \times M \to \mathbb{Z}$. By condition (a), φ is injective. Consequently, $M^\vee$ has the same rank as $\varphi(M)$. There therefore exists an integer $m \ge 1$ such that $mM^\vee \subseteq \varphi(M)$. For any divisor H on X, let $\rho_H \in M^\vee$ denote the linear form defined by $\Gamma_i \mapsto H \cdot \Gamma_i$. Then $m\rho_H \in \varphi(M)$. Let us note that $\rho_V = \varphi(V)$ if $V \in M$.

Let us fix an ample effective divisor H on X. Replacing H by a sufficiently large multiple if necessary, we can suppose that $\mathcal{O}_X(H)$ is generated by its global sections, that $H^1(X, \mathcal{O}_X(H)) = 0$, and from the above, that there exists a cycle $Z \in M$ such that $\rho_H = \varphi(-Z)$. Let us show that $Z > 0$. For any vertical prime divisor Γ, $\mathcal{O}_X(H)|_\Gamma$ is ample (this immediately results from Definition 5.1.33), and hence $H \cdot \Gamma > 0$ (Proposition 7.5.5). Let us write Z as a difference $Z = Z_0 - Z_\infty$ of two effective divisors with no common component. If $Z_\infty > 0$, then

$$-Z_0 \cdot Z_\infty + Z_\infty^2 = -Z \cdot Z_\infty = H \cdot Z_\infty > 0.$$

This is impossible because $Z_0 \cdot Z_\infty \ge 0$, and $Z_\infty^2 < 0$ by hypothesis (a). Consequently, $Z_\infty = 0$ and $Z > 0$. Let us set $D = H + Z$. We are going to show that D fulfills condition (ii) of Proposition 8.3.30 (or that of Exercise 8.3.20 if $\dim S = 0$) for the set $\mathcal{E} = \{\Gamma_i\}_{1 \le i \le r}$. This will prove the theorem.

Let Γ be a vertical prime divisor. If $\Gamma \notin M$, then $\Gamma \cdot D \geq \Gamma \cdot H > 0$. Let us suppose that Γ is equal to one of the Γ_i; then $D \cdot \Gamma = \rho_H(\Gamma) + \varphi(Z)(\Gamma) = 0$, and hence $\mathcal{O}_X(D)|_\Gamma \simeq \mathcal{O}_\Gamma$ since $\mathrm{Pic}^0(\Gamma) = 1$ by hypothesis. If $\dim S = 1$, then $\deg D_\eta = \deg H_\eta > 0$. It remains to show that $\mathcal{O}_X(D)$ is generated by its global sections at the points $x \in X$. This follows from the hypothesis on H if $x \notin \mathrm{Supp}\, Z$. We have $\mathcal{O}_X(D)|_Z \in \mathrm{Pic}^0(Z)$; hence $\mathcal{O}_X(D)|_Z \simeq \mathcal{O}_Z$. In particular, $\mathcal{O}_X(D)|_Z$ is generated by its global sections. Let us consider the exact sequence

$$0 \to \mathcal{O}_X(H) \to \mathcal{O}_X(D) \to \mathcal{O}_X(D)|_Z \to 0.$$

The $H^1(X, \mathcal{O}_X(H)) = 0$ hypothesis then implies that we have an exact sequence

$$H^0(X, \mathcal{O}_X(D)) \to H^0(Z, \mathcal{O}_X(D)|_Z) \to 0. \tag{4.11}$$

By Nakayama's lemma, this implies that $\mathcal{O}_X(D)$ is generated by its global sections at the points $x \in \mathrm{Supp}\, Z$. □

Remark 4.3. It immediately follows from Theorem 1.27 that condition (a) of Theorem 4.2 is a necessary condition. If $\dim S = 1$, this condition is equivalent to saying that the union of the Γ_i does not contain any connected component of X_s (Exercise 1.6). However, (b) is not a necessary condition. For example, under the hypotheses of Theorem 8.3.36, the contraction exists under condition (a) alone. To our knowledge, there do not exist necessary and sufficient numerical conditions for the existence of the contraction of Δ.

In the proof of Theorem 4.2, we defined a homomorphism $\rho : \mathrm{Pic}(X) \to M^\vee$, $D \mapsto \rho_D$. Let us study its kernel.

Lemma 4.4. *Let us keep the hypotheses of Theorem 4.2. Let K be a Cartier divisor on X such that $K \cdot \Gamma_i = 0$ for every $i \leq r$. Then there exists a Cartier divisor K', linearly equivalent to K and such that $\mathrm{Supp}\, K' \cap \Delta = \emptyset$. In other words, there exists an open neighborhood W of Δ such that $\mathcal{O}_X(K)|_W \simeq \mathcal{O}_X|_W$.*

Proof ([8], Corollary 2.6) Let us keep the notation of the proof of Theorem 4.2. We can choose H such that $H^1(X, \mathcal{O}_X(H')) = 0$, where $H' = H + K$. Let $f \in H^0(X, \mathcal{O}_X(D))$ be such that its image in $\mathcal{O}_X(D)|_Z$ is a basis of the latter over $\mathcal{O}_Z$ (such an f exists because $\mathcal{O}_X(D)|_Z \simeq \mathcal{O}_Z$ and we have exact sequence (4.11)). Let $D_1 = D + \mathrm{div}(f)$. Then $(\mathrm{Supp}\, D_1) \cap \Delta = \emptyset$. We have $\rho_{H'} = \varphi(-Z)$ by the hypothesis on K. The same reasoning implies that there exists a $D'_1 \sim D' := H' + Z$ such that $\mathrm{Supp}\, D'_1 \cap \Delta = \emptyset$. It now suffices to take $K' = D'_1 - D_1 \sim K$. □

Let us study condition (b) of Theorem 4.2. Let $X \to S$ be a regular fibered surface. Let Z be an effective vertical divisor with support contained in the closed fiber X_s. We set

$$p_a(Z) = 1 + \frac{1}{2}(Z^2 + K_{X/S} \cdot Z) \in \mathbb{Z}. \tag{4.12}$$

(See Exercise 1.11.) If $\dim S = 0$ or if $\dim S = 1$ and $Z \leq X_s$, such that Z is a projective curve over $k(s)$, then by Theorem 1.37, $p_a(Z)$ is the arithmetic genus of Z (Definition 7.3.19).

Lemma 4.5. *Let $X \to S$ be a regular fibered surface. Let Z be an effective vertical divisor with support in a closed fiber X_s. Let us consider the following conditions.*

(i) *For any divisor Z' such that $0 < Z' \le Z$, we have $p_a(Z') \le 0$.*
(ii) $H^1(Z, \mathcal{O}_Z) = 0$.
(iii) $\mathrm{Pic}^0(Z) = 1$.

Then (i) *implies* (ii) *and* (iii).

Proof Let us write $Z = \sum_i a_i \Gamma_i$, where the Γ_i are the irreducible components of Z. We are going to show the lemma by induction on $a := \sum_i a_i$. The case $a = 1$ (hence Z integral) results from Proposition 3.16. Let us therefore suppose $a \ge 2$. The $p_a(Z) \le 0$ hypothesis is equivalent to saying that $(Z + K_{X/S}) \cdot Z < 0$. There therefore exists an irreducible component Γ of Z such that

$$(Z + K_{X/S}) \cdot \Gamma < 0. \tag{4.13}$$

Let us set $D = Z - \Gamma$. As the divisor D verifies the conditions of the lemma, by the induction hypothesis, we have $H^1(D, \mathcal{O}_D) = 0$ and $\mathrm{Pic}^0(D) = 1$. We have an exact sequence of sheaves on X

$$0 \to \mathcal{I} := \mathcal{O}_X(-D)/\mathcal{O}_X(-Z) \to \mathcal{O}_Z \to \mathcal{O}_D \to 0 \tag{4.14}$$

(because $\mathcal{O}_Z = \mathcal{O}_X/\mathcal{O}_X(-Z)$ and $\mathcal{O}_D = \mathcal{O}_X/\mathcal{O}_X(-D)$) with $\mathcal{I} \simeq \mathcal{O}_X(-D)|_\Gamma$. By duality (Remark 6.4.21) and the adjunction formula (Theorem 1.37), we have

$$\begin{aligned} H^1(\Gamma, \mathcal{O}_X(-D)|_\Gamma) &\simeq H^0(\Gamma, \mathcal{O}_X(K_{X/S} + \Gamma + D)|_\Gamma) \\ &\simeq H^0(\Gamma, \mathcal{O}_X(Z + K_{X/S})|_\Gamma). \end{aligned} \tag{4.15}$$

As Γ is integral, inequality (4.13) implies the vanishing of the terms of (4.15). The cohomology sequence of (4.14) therefore implies that $H^1(Z, \mathcal{O}_Z) = 0$. It remains to determine $\mathrm{Pic}^0(Z)$.

Let us distinguish between two cases. Let us first suppose that $2\Gamma \le Z$. Then $2D \ge Z$, and hence $\mathcal{I}^2 = 0$. As in the proof of Lemma 7.5.11, we then have an exact sequence

$$0 \to 1 + \mathcal{I} \to \mathcal{O}_Z^* \to \mathcal{O}_D^* \to 1$$

with $1 + \mathcal{I} \simeq \mathcal{I}$ as sheaves of groups. Hence $H^1(Z, 1 + \mathcal{I}) = 1$. The cohomology sequence of the exact sequence above then implies that the homomorphism $\mathrm{Pic}(Z) \to \mathrm{Pic}(D)$ is injective, and hence $\mathrm{Pic}^0(Z) \to \mathrm{Pic}^0(D)$ is injective, whence the triviality of $\mathrm{Pic}^0(Z)$. The case when Γ is not contained in the support of D remains. If $D \cap \Gamma = \emptyset$, we have $\mathrm{Pic}^0(Z) = \mathrm{Pic}^0(D) \oplus \mathrm{Pic}^0(\Gamma) = 1$. Let us therefore suppose $D \cap \Gamma \ne \emptyset$. Let $k_1 = H^0(\Gamma, \mathcal{O}_\Gamma)$. We have $p_a(\Gamma) = 1 - n$, where $n = [k_1 : k(s)]$. The $(Z + K_{X/S}) \cdot \Gamma < 0$ condition implies that $0 < D \cdot \Gamma < 2 - 2p_a(\Gamma) = 2n$. As Γ is a curve over k_1, we have $D \cdot \Gamma \in n\mathbb{N}^*$. It follows that $D \cdot \Gamma = n$. In other words, D and Γ meet transversally at a point x

that is rational over k_1 (and D is reduced at x). We have an exact sequence of sheaves of groups

$$1 \to \mathcal{O}_Z^* \to \mathcal{O}_D^* \oplus \mathcal{O}_\Gamma^* \xrightarrow{\alpha} \mathcal{S} \to 1, \tag{4.16}$$

where $\mathcal{S}$ is the skyscraper sheaf with support in $\{x\}$ and such that $\mathcal{S}_x = k(x)^* = k_1^*$, and where $\alpha(f,g) = f(x)g(x)^{-1}$. This immediately implies that the homomorphism

$$\mathrm{Pic}(Z) \to \mathrm{Pic}(D) \oplus \mathrm{Pic}(\Gamma)$$

induced by the exact sequence (4.16) is injective. This proves that $\mathrm{Pic}^0(Z) = 1$. □

Remark 4.6. For any $0 < Z' \le Z$, we have a surjective canonical homomorphism $H^1(Z, \mathcal{O}_Z) \to H^1(Z', \mathcal{O}_{Z'})$. Hence if (ii) is true, then $H^1(Z', \mathcal{O}_{Z'}) = 0$. If $Z \le X_s$ (e.g., if $\dim S = 0$), this implies that

$$p_a(Z') = 1 - \dim_{k(s)} H^0(Z', \mathcal{O}_{Z'}) \le 0.$$

Hence (i) is equivalent to (ii) under this hypothesis. If $k(s)$ is moreover algebraically closed, the three conditions are equivalent. See Exercises 4.3–4.

Corollary 4.7. *Let $X \to S$ be a regular fibered surface. Let $\Gamma_1, \ldots, \Gamma_r$ be vertical prime divisors contained in a closed fiber X_s such that*

$$K_{X/S} \cdot \Gamma_i = 0, \quad i = 1, \ldots, r,$$

and that the intersection matrix $(\Gamma_i \cdot \Gamma_j)_{1 \le i,j \le r}$ is negative definite. Then $p_a(Z) \le 0$ for every $Z > 0$ with support in $\cup_i \Gamma_i$. In particular, there exists a contraction morphism $f : X \to Y$ of the Γ_i.

Proof Let $Z = \sum_i a_i \Gamma_i > 0$. Then

$$2p_a(Z) - 2 = Z^2 + K_{X/S} \cdot Z = Z^2 < 0.$$

Hence $p_a(Z) \le 0$. We can therefore apply Lemma 4.5 and Theorem 4.2. □

Proposition 4.8. *Let $\pi : X \to S$ be an arithmetic surface, $s \in S$ a closed point, Γ an irreducible component of X_s, and $k' = H^0(\Gamma, \mathcal{O}_\Gamma)$. Then we have*

(i) $K_{X/S} \cdot \Gamma = 0$

if and only if one of the following conditions is satisfied:

(ii) $H^1(\Gamma, \mathcal{O}_\Gamma) = 0$ *and* $\Gamma^2 = -2[k' : k(s)]$.
(iii) Γ *is a conic over* k' *and* $\deg_{k'} \mathcal{O}_X(\Gamma)|_\Gamma = -2$.
(iv) $p_a(X_\eta) = 1$ *and* Γ *is a connected component of* X_s.

Moreover, conditions (ii) and (iii) are equivalent.

Proof The equivalence of (ii) and (iii) results from Proposition 3.16 and the equality

$$\Gamma^2 = \deg_{k(s)} \mathcal{O}_X(\Gamma)|_\Gamma = [k' : k(s)] \deg_{k'} \mathcal{O}_X(\Gamma)|_\Gamma.$$

The adjunction formula gives

$$\Gamma^2 + K_{X/S} \cdot \Gamma = 2p_a(\Gamma) - 2 = 2[k' : k(s)](-1 + \dim_{k'} H^1(\Gamma, \mathcal{O}_\Gamma)).$$

Let us suppose $\Gamma^2 < 0$. Then (i) is clearly equivalent to (ii). Let us therefore consider the $\Gamma^2 = 0$ case. Then Γ is a connected component of X_s (Exercise 1.6). If X_s is connected, then $X_s = d\Gamma$ for some $d \geq 1$. The equality

$$2p_a(X_\eta) - 2 = dK_{X/S} \cdot \Gamma$$

(Proposition 1.35) implies the equivalence (i) $\iff$ (iv). Let us show the general case. Let $\pi' : X \to S'$, $S' \to S$ be the decomposition of π as in Proposition 8.3.8. Let $s' = \pi'(\Gamma)$. We have $\omega_{X/S} = \omega_{X/S'} \otimes \pi'^* \omega_{S'/S}$ (Theorem 6.4.9(a)), and hence

$$\deg_{k(s)} \omega_{X/S}|_\Gamma = \deg_{k(s)} \omega_{X/S'}|_\Gamma = [k(s') : k(s)] \deg_{k(s')} \omega_{X/S'}|_\Gamma.$$

As the fiber $X_{s'}$ is connected (Theorem 5.3.15), the discussion above implies that (i) is equivalent to $p_a(X_{\eta'}) = 1$, where η' is the generic point of S'. Now $p_a(X_\eta) = 1$ if and only if

$$\dim_{k(\eta)} H^0(X_\eta, \mathcal{O}_{X_\eta}) = \dim_{k(\eta)} H^1(X_\eta, \mathcal{O}_{X_\eta}),$$

which is independent of the base field $K(S)$. This completes the proof. $\square$

Definition 4.9. The vertical prime divisors Γ verifying condition (ii) of the proposition and smooth over k' are sometimes called (-2)-*curves*.

Remark 4.10. Let $\Delta = \sum_i \Gamma_i$ be a connected vertical divisor such that $(\Gamma_i \cdot \Gamma_j)_{i,j}$ is negative definite and that for any divisor $Z > 0$ with support in Δ, we have $p_a(Z) \leq 0$. Then the configuration of Δ is relatively rigid. We can classify all of the configurations coming from such Δ. See [57], Section 24. If, moreover, $K_{X/S} \cdot \Gamma_i = 0$ for every i, then the list is even more restricted. We will return to this in the next chapter (Proposition 10.1.53).

9.4.2   Determination of the tangent spaces

Let $X \to S$ be a regular fibered surface and $f : X \to Y$ the contraction morphism defined in Theorem 4.2. We are going to determine the tangent space of Y at the singular points (Theorem 4.15).

Lemma 4.11. Let Z be an effective vertical divisor on a regular fibered surface $X \to S$. Let us suppose that $H^1(Z, \mathcal{O}_Z) = 0$ and that $\text{Pic}^0(Z) = 1$. Let $\mathcal{L} \in \text{Pic}(Z)$ be such that $\deg \mathcal{L}|_\Gamma \geq 0$ for every irreducible component Γ of Z. Then $H^1(Z, \mathcal{L}) = 0$ and $\mathcal{L}$ is generated by its global sections.

Proof For any irreducible component Γ of Z, the surjective homomorphism $\mathcal{O}_Z \to \mathcal{O}_\Gamma$ implies that $H^1(\Gamma, \mathcal{O}_\Gamma) = 0$. Let us first suppose that each component Γ of Z contains a regular point x of degree $e(\Gamma)$ over $H^0(\Gamma, \mathcal{O}_\Gamma)$ (Proposition 3.16(b)) so that $\text{Pic}(\Gamma) = \mathcal{O}_\Gamma(x)\mathbb{Z}$, and that, moreover, x does not belong to any other component of Z. This is always possible if $k(s)$ is infinite (taking into account the fact that Γ is a conic over $H^0(\Gamma, \mathcal{O}_\Gamma)$). We are going to show that $\mathcal{L}$ is represented by an effective divisor $D_\mathcal{L}$.

The point $x \in \Gamma$ corresponds to an effective Cartier divisor on Γ. Let f be a generator of $\mathcal{O}_X(-\Gamma)_x$ and $g \in \mathcal{O}_{X,x}$ be a preimage of the generator of $\mathfrak{m}_x \mathcal{O}_{\Gamma,x}$. Then the image of g in $\mathcal{O}_{X,x}/(f^a) = \mathcal{O}_{Z,x}$, where a is the multiplicity of Z in Γ, is not a zero divisor (we use the fact that $\mathcal{O}_{X,x}$ is factorial); it therefore defines an effective Cartier divisor D_Γ on Z whose support is the point x. Moreover, $\mathcal{O}_Z(D_\Gamma)|_\Gamma = \mathcal{O}_\Gamma(x)$ and the restriction of $\mathcal{O}_Z(D_\Gamma)$ to any other irreducible component is trivial. Let $\Gamma_1, \ldots, \Gamma_r$ be the irreducible components of Z and let $D_i \in \text{Div}(Z)$ denote the Cartier divisor D_{Γ_i} associated to Γ_i as above. For any $\mathcal{L} \in \text{Pic}(Z)$, let us set

$$D_\mathcal{L} := \sum_{1 \le i \le r} (\deg \mathcal{L}|_{\Gamma_i})(\deg \mathcal{O}_Z(D_i)|_{\Gamma_i})^{-1} D_i.$$

Let us note that the coefficients are integers because $\mathcal{O}_Z(D_i)|_{\Gamma_i}$ is a generator of $\text{Pic}(\Gamma_i)$. We have $\deg(\mathcal{L} \otimes \mathcal{O}_Z(-D_\mathcal{L}))|_{\Gamma_j} = 0$ for every $j \le r$, which means that $\mathcal{L} \otimes \mathcal{O}_Z(-D_\mathcal{L}) \in \text{Pic}^0(Z)$, whence $\mathcal{L} \simeq \mathcal{O}_Z(D_\mathcal{L})$ with $D_\mathcal{L} \ge 0$. We have an exact sequence

$$0 \to \mathcal{O}_Z \to \mathcal{L} \to \mathcal{L}|_{D_\mathcal{L}} \to 0.$$

As $H^1(Z, \mathcal{O}_Z) = 0$ by hypothesis and $\mathcal{L}|_{D_\mathcal{L}}$ is a skyscraper sheaf, we obtain an exact sequence

$$H^0(Z, \mathcal{L}) \to H^0(Z, \mathcal{L}|_{D_\mathcal{L}}) \to 0 \to H^1(Z, \mathcal{L}) \to 0.$$

Hence $H^1(Z, \mathcal{L}) = 0$ and $\mathcal{L}$ is generated by its global sections at the points of $\text{Supp}\, D_\mathcal{L}$. As $D_\mathcal{L} \ge 0$, $\mathcal{L}$ is also generated by its global sections outside of $\text{Supp}\, D_\mathcal{L}$, which proves the lemma.

The case that $k(s)$ is finite remains. It is easy to see that there exists a finite separable extension $k'/k(s)$ such that the irreducible components of $Z_{k'}$ verify the properties stated at the beginning of the proof. There exists a finite étale morphism $S' \to S$ with S' local with closed point s', $\mathcal{O}_{S,s} \subseteq \mathcal{O}_{S',s'}$, and $k(s') = k'$. Then $Z \times_S S'$ is an effective divisor on the arithmetic surface $X \times_S S' \to S'$. We leave it to the reader to verify the lemma with the help of the result on $Z \times_S S'$. $\square$

Lemma 4.12. Let $\Gamma_1, \ldots, \Gamma_r$ be vertical prime divisors on a regular fibered surface such that $(\Gamma_i \cdot \Gamma_j)_{i,j}$ is negative definite. Let $\Delta = \sum_{1 \le i \le r} \Gamma_i$. Then the following properties are true.

(a) *There exists a smallest effective divisor $Z = \sum_i a_i \Gamma_i$ such that $Z \geq \Delta$ and that $Z \cdot \Gamma_i \leq 0$ for every $i \leq r$.*

(b) *The ring $H^0(Z, \mathcal{O}_Z)$ is a field if Δ is connected.*

Proof (a) It is clear that we can suppose Δ is connected (the divisor Z will be the sum of the Z associated to each connected component of Δ). As the intersection matrix $(\Gamma_i \cdot \Gamma_j)_{i,j}$ is negative definite, there exists a non-zero $Z \in \oplus_i \Gamma_i \mathbb{Z}$ such that $Z \cdot \Gamma_i \leq 0$ for every $i \leq r$. In a way similar to the proof of Theorem 4.2, we deduce from this that Z is effective. We even have $Z \geq \Delta$. Indeed, in the opposite case, there exists an $i_0 \leq r$ such that Γ_{i_0} is not an irreducible component of Z but meets an irreducible component of Z (because Δ is supposed connected). Now this implies that $Z \cdot \Gamma_{i_0} > 0$, which is contrary to the hypothesis on Z. Let $Z' = \sum_i a_i' \Gamma_i \geq \Delta$ be another divisor such that $Z' \cdot \Gamma_i \leq 0$ for every i. Let us show that

$$Z'' := \sum_i \min\{a_i, a_i'\}\Gamma_i$$

verifies the same property. Let $i \leq r$. Let us, for example, suppose that $a_i \leq a_i'$. Then

$$Z'' \cdot \Gamma_i = a_i \Gamma_i^2 + \sum_{j \neq i} \min\{a_j, a_j'\}\Gamma_j \cdot \Gamma_i \leq a_i \Gamma_i^2 + \sum_{j \neq i} a_j \Gamma_j \cdot \Gamma_i = Z \cdot \Gamma_i \leq 0.$$

The existence of the smallest divisor Z is now clear.

(b) Let us first show that $H^0(Z, \mathcal{O}_Z)$ is a subring of the field $k' = H^0(\Delta, \mathcal{O}_\Delta)$. Let Δ' be a divisor such that $\Delta \leq \Delta' \leq Z$ and $H^0(\Delta', \mathcal{O}_{\Delta'}) \subseteq k'$ which is maximal for this property. Let us show that $\Delta' = Z$. Let us suppose the contrary. Then $\Delta' < Z$. There therefore exists an irreducible component Γ of Δ such that $\Delta' \cdot \Gamma > 0$. We have

$$Z - \Delta' \geq 0, \quad (Z - \Delta') \cdot \Gamma = Z \cdot \Gamma - \Delta' \cdot \Gamma < 0,$$

and hence $Z - \Delta' \geq \Gamma$. In other words, $\Delta' + \Gamma \leq Z$. We have an exact sequence

$$0 \to \mathcal{O}_X(-\Delta')|_\Gamma \to \mathcal{O}_{\Delta'+\Gamma} \to \mathcal{O}_{\Delta'} \to 0.$$

As $\deg \mathcal{O}_X(-\Delta')|_\Gamma < 0$, we have $H^0(\Gamma, \mathcal{O}_X(-\Delta')|_\Gamma) = 0$, and hence

$$H^0(\Delta' + \Gamma, \mathcal{O}_{\Delta'+\Gamma}) \subseteq H^0(\Delta', \mathcal{O}_{\Delta'}) \subseteq k',$$

which contradicts the maximality hypothesis on Δ', whence $Z = \Delta'$ and hence $H^0(Z, \mathcal{O}_Z) \subseteq k'$. Let $s \in S$ be the image of Δ. Then $H^0(Z, \mathcal{O}_Z)$ is a finite $\mathcal{O}_{S,s}$-algebra. As k' is finite over $k(s)$, this implies that $H^0(Z, \mathcal{O}_Z)$ is a field. $\square$

Definition 4.13. The divisor Z of Lemma 4.12(a) is called the *fundamental divisor* for the Γ_i (or for $\Delta = \sum_i \Gamma_i$).

Let $\Delta = \sum_{1 \le i \le r} \Gamma_i$ be a connected divisor verifying the hypotheses of Theorem 4.2. Let Z be its fundamental divisor. We suppose that $H^1(Z, \mathcal{O}_Z) = 0$. Let $y = f(\Delta)$. We are going to compute the tangent space $T_{Y,y}$ in the manner of Lemma 3.6, with the fundamental divisor Z playing the role of E. Let V be an affine open neighborhood of y, $\mathfrak{m}$ the maximal ideal of $\mathcal{O}_Y(V)$ corresponding to y, $U = f^{-1}(V)$, and $\mathcal{J} = \mathcal{O}_X(-Z)|_U$.

Lemma 4.14. *Let $n \ge 0$. Then the following properties are true.*

(a) *For any $m \ge n+1$, we have $H^1(U, \mathcal{J}^n/\mathcal{J}^m) = 0$.*

(b) *We have $H^1(U, \mathcal{J}^n) = 0$.*

(c) *The sheaf $\mathcal{J}^n$ is generated by its global sections.*

(d) *We have $H^0(U, \mathcal{J}^n) = \mathfrak{m}^n$ and $H^0(Z, \mathcal{O}_Z) = k(y)$.*

Proof We have $\mathcal{J}^n/\mathcal{J}^m \simeq \mathcal{O}_X(-nZ)|_{(m-n)Z}$. For any $i \le r$, we have $(-nZ) \cdot \Gamma_i \ge 0$ by hypothesis. It follows from Lemma 4.11 that

$$H^1(U, \mathcal{J}^n/\mathcal{J}^m) = H^1(Z, \mathcal{J}^n/\mathcal{J}^m) = H^1(Z, \mathcal{O}_X(-nZ)|_{(m-n)Z}) = 0,$$

and that $\mathcal{J}|_Z = \mathcal{O}_X(-Z)|_Z$ is generated by its global sections. Properties (b) and (c) can then be proven in the same way as in Lemma 3.6. Let us now show (d) for $n = 1$. We have $H^0(U, \mathcal{O}_U) = \mathcal{O}_Y(V)$ and therefore $H^0(U, \mathcal{J}) \subseteq \mathfrak{m}$. From the exact sequence

$$0 \to \mathcal{J} \to \mathcal{O}_U \to \mathcal{O}_Z \to 0,$$

we deduce the exact sequence

$$0 \to H^0(U, \mathcal{J}) \to \mathcal{O}_Y(V) \to H^0(Z, \mathcal{O}_Z) \to H^1(U, \mathcal{J}) = 0.$$

Now $H^0(Z, \mathcal{O}_Z)$ is a field by Lemma 4.12(b); it follows that $H^0(U, \mathcal{J})$ is a maximal ideal, and therefore that $H^0(U, \mathcal{J}) = \mathfrak{m}$. The exact sequence above shows that $H^0(Z, \mathcal{O}_Z) = k(y)$. For $n \ge 2$, the proof is the same as that of Lemma 3.6(d). □

Theorem 4.15 (Artin [9], Theorem 4; Lipman [58], Theorem 27.1). *Let $X \to S$ be a regular fibered surface. Let $\Delta = \sum_{1 \le i \le r} \Gamma_i$ be a reduced connected vertical divisor contained in a closed fiber X_s. We suppose that*

(a) *the intersection matrix $(\Gamma_i \cdot \Gamma_j)_{1 \le i,j \le r}$ is negative definite;*

(b) *for any divisor $Z' > 0$ with support in Δ, we have*

$$\mathrm{Pic}^0(Z') = 1, \quad H^1(Z', \mathcal{O}_{Z'}) = 0.$$

Let $f : X \to Y$ be the contraction of the irreducible components of Δ (Theorem 4.2). Let Z be the fundamental divisor for Δ, $y = f(\Delta)$, and $\mathfrak{m}_y$ the maximal ideal of $\mathcal{O}_{Y,y}$. Then $k(y) = H^0(Z, \mathcal{O}_Z)$ and for every $n \ge 1$,

$$\dim_{k(y)} \mathfrak{m}_y^n/\mathfrak{m}_y^{n+1} = n(-Z^2)/[k(y) : k(s)] + 1.$$

In particular,

$$\dim_{k(y)} T_{Y,y} = (-Z^2)/[k(y) : k(s)] + 1.$$

Proof Let $\mathcal{J}$ be as defined before Lemma 4.14. Then the exact sequence

$$0 \to \mathcal{J}^{n+1} \to \mathcal{J}^n \to \mathcal{J}^n/\mathcal{J}^{n+1} = \mathcal{O}_X(-nZ)|_Z \to 0$$

induces the exact sequence

$$0 \to \mathfrak{m}_y^{n+1} \to \mathfrak{m}_y^n \to H^0(Z, \mathcal{O}_X(-nZ)|_Z) \to 0$$

(Lemma 4.14(d) and (b)). It follows that

$$\dim_{k(y)} \mathfrak{m}_y^n/\mathfrak{m}_y^{n+1} = \dim_{k(y)} H^0(Z, \mathcal{O}_X(-nZ)|_Z).$$

For any $i \le r$, we have $(-nZ) \cdot \Gamma_i \ge 0$ by definition of Z, and hence, by Lemma 4.11, $H^1(Z, \mathcal{O}_X(-nZ)|_Z)$ is zero. Let us consider Z as a curve over $k(y) = H^0(Z, \mathcal{O}_Z)$ (Lemma 4.14(d)). We have

$$\chi_{k(y)}(\mathcal{O}_X(-nZ)) = \deg_{k(y)}(\mathcal{O}(-nZ)|_Z) + \chi_{k(y)}(\mathcal{O}_Z) = n(-Z^2)/[k(y):k(s)] + 1,$$

where the first equality comes from Theorem 7.3.17 and the second results from Lemma 4.5, whence the theorem.   □

Remark 4.16. Let E be a vertical prime divisor verifying the hypotheses of Theorem 3.7. Then $\Delta := E$ verifies the hypotheses of Theorem 4.15 and the fundamental divisor Z is none other than E itself. We then see that Theorem 3.7 is a particular case of Theorems 4.2 and 4.15. However, we prefer to present its proof separately in order not to weigh down the proof of Castelnuovo's theorem.

Remark 4.17. If Δ is not connected in Theorem 4.15, we decompose it as a sum of connected components Δ_i, each with a fundamental divisor Z_i. Let y_i be the image of Δ_i in Y and s_i the image of y_i in S; then we have

$$\dim_{k(y)} T_{Y,y_i} = (-Z_i^2)/[k(y_i):k(s_i)] + 1.$$

The proof is the same.

9.4.3   Canonical models

Corollary 4.18. *Let* $X \to S$ *be a regular fibered surface. Let* $\Gamma_1, \ldots, \Gamma_r$ *be vertical prime divisors on* X *such that* $K_{X/S} \cdot \Gamma_i = 0$ *for every* $i \le r$ *and that the intersection matrix* $(\Gamma_i \cdot \Gamma_j)_{i,j}$ *is negative definite. Let* $f : X \to Y$ *be the contraction morphism of the* Γ_i *(Corollary 4.7). Then the following properties are true.*

(a) *For any* $y \in Y$, *we have* $\dim_{k(y)} T_{Y,y} \le 3$, *and* Y *is an l.c.i. over* S.

(b) *For any* $q \in \mathbb{Z}$, *we have*

$$f_*(\omega_{X/S}^{\otimes q}) = \omega_{Y/S}^{\otimes q}, \quad f^*(\omega_{Y/S}^{\otimes q}) = \omega_{X/S}^{\otimes q}. \tag{4.17}$$

Proof Let us first note that the Γ_i satisfy the hypotheses of Theorem 4.15 by Corollary 4.7 and Lemma 4.5.

(a) We can suppose that $\cup_{1 \leq i \leq r} \Gamma_i$ are connected and that y is their image in Y. Let Z be the fundamental divisor of the Γ_i. We have seen that Z could be considered as a curve over $k(y)$, where $y = f(Z)$ (Lemma 4.14(d)). The adjunction formula says that

$$2p_a(Z) - 2 = Z^2/[k(y):k(s)] + K_{X/S} \cdot Z/[k(y):k(s)] = Z^2/[k(y):k(s)],$$

the arithmetic genus $p_a(Z)$ being computed over $k(y)$. Now, on $k(y)$ we have

$$p_a(Z) = \dim_{k(y)} H^1(Z, \mathcal{O}_Z) = 0.$$

Hence $Z^2 = -2[k(y):k(s)]$. It follows from Theorem 4.15 that $\dim_{k(y)} T_{Y,y} = 3$. By a reasoning similar to that used at the beginning of the proof of Proposition 2.34, we can embed an open neighborhood V of y in a regular scheme W of dimension 3. As $\dim V = 2$, V is a hypersurface in W, and is therefore a complete intersection in W. It follows that Y is an l.c.i. over S.

(b) Let $\Delta = \sum_i \Gamma_i$. Let F denote the finite closed set $f(\Delta)$. Then $X \setminus \Delta \to Y \setminus F$ is an isomorphism, by definition. Hence

$$\omega_{X/S}|_{X \setminus \Delta} = f^*(\omega_{Y/S}|_{Y \setminus F}), \quad f_*(\omega_{X/S}|_{X \setminus \Delta}) = \omega_{Y/S}|_{Y \setminus F}. \tag{4.18}$$

In addition, there exists an open neighborhood W of Δ on which $\omega_{X/S}$ is free (Lemma 4.4). There exists an open neighborhood V of F such that $U := f^{-1}(V) \subseteq W$ and that $\omega_{Y/S}|_V$ is free. Hence

$$\omega_{X/S}|_U = e\mathcal{O}_U, \quad \omega_{Y/S}|_V = \delta\mathcal{O}_V. \tag{4.19}$$

Consequently, there exists an $a \in \mathcal{O}_V(V \setminus F)^* = \mathcal{O}_V(V)^*$ (Theorem 4.1.14) such that $e|_{U \setminus \Delta} = af^*(\delta|_{V \setminus F})$. Replacing δ by $a\delta$, we can suppose that $e = f^*\delta$. The identities (4.18) and (4.19) then imply property (b). $\qquad\square$

Example 4.19. Let $X \to S$ be a minimal arithmetic surface with $p_a(X_\eta) = 1$ and connected fibers (e.g., if X_η is an elliptic curve over $K(S)$). Let $s \in S$ be a closed point, and $X_s = \sum_{1 \leq i \leq m} d_i \Gamma_i$. Then Proposition 1.35 implies that

$$0 = 2p_a(X_\eta) - 2 = \sum_{1 \leq i \leq m} d_i K_{X/S} \cdot \Gamma_i.$$

It follows from Corollary 3.26 that $K_{X/S} \cdot \Gamma_i = 0$ for every i. Hence any proper subset of the set of irreducible components of X_s can be contracted to a point whose tangent space will be of dimension 3.

Proposition 4.20. *Let $X \to S$ be a minimal arithmetic surface with $p_a(X_\eta) \geq 2$ and S affine. Let $\mathcal{E}$ be the set of vertical prime divisors Γ such that $\deg \omega_{X/S}|_\Gamma = 0$. Then the following properties are true.*

(a) *The set $\mathcal{E}$ is finite and there exists a contraction morphism $f : X \to Y$ of the $\Gamma \in \mathcal{E}$ (the S affine hypothesis is useless here).*

(b) *The sheaf $\omega_{Y/S}$ is ample.*

(c) *There exists an $m \geq 1$ such that $\omega_{X/S}^{\otimes m}$ is generated by its global sections.*

(d) *Let $\varphi : X \to \mathbb{P}_S^N$ be the morphism associated to a generating system of the $\mathcal{O}_S(S)$-module $H^0(X, \omega_{X/S}^{\otimes m})$. Then the morphism $\varphi : X \to \varphi(X)$ coincides with f.*

Proof   (a) As $\omega_{X/S}$ is ample over X_η (see Proposition 7.5.5), by Proposition 5.1.37(b), there exists an open subscheme V of S such that $\omega_{X_V/V}$ is ample. The set $\mathcal{E}$ is contained in the set of irreducible components of the X_s, $s \in S \setminus V$. Hence $\mathcal{E}$ is finite. As for Proposition 4.8, we see that the intersection matrix of the components of $\mathcal{E}$ is negative definite. There therefore exists a contraction morphism $f : X \to Y$ of the $\Gamma \in \mathcal{E}$ by virtue of Corollary 4.7.

(b) Let Γ be an irreducible component of Y_s and Γ' its strict transform in X. Then the restriction $h : \Gamma' \to \Gamma$ of f is a finite birational morphism. As

$$\omega_{X/S}|_{\Gamma'} \simeq (\pi^* \omega_{Y/S})|_{\Gamma'} = h^*(\omega_{Y/S}|_\Gamma), \quad \deg \omega_{Y/S}|_\Gamma = \deg \omega_{X/S}|_{\Gamma'}$$

(Proposition 7.3.8), we have $\deg \omega_{Y/S}|_\Gamma > 0$ because $\Gamma' \notin \mathcal{E}$. Hence $\omega_{Y/S}|_{Y_s}$ is ample (Proposition 7.5.5). By virtue of Corollary 5.3.24, $\omega_{Y/S}$ is ample.

(c) By (b), there exists an $m \geq 1$ such that $\omega_{Y/S}^{\otimes m}$ is generated by its global sections. Hence $\omega_{X/S}^{\otimes m} \simeq f^*(\omega_{Y/S}^{\otimes m})$ is generated by its global sections.

(d) It follows from (c) and Lemma 4.4 that the sheaf $\omega_{X/S}^{\otimes m}$ verifies the hypotheses of Proposition 8.3.30(b) for the set $\mathcal{E}$. The equality between f and φ results from the construction of f in the proof of that proposition. $\square$

Definition 4.21. Let $X \to S$ be a minimal arithmetic surface with $p_a(X_\eta) \geq 2$. Let $f : X \to Y$ be the contraction of the vertical prime divisors Γ such that $K_{X/S} \cdot \Gamma = 0$. The surface $Y \to S$ is called the *canonical model* of X. It is singular as soon as there exists at least one contracted component.

Remark 4.22. See [12] for a study of the canonical model of surfaces 'of general type' over $\mathbb{C}$.

Example 4.23. Let $\mathcal{O}_K$ be a discrete valuation ring, with uniformizing parameter t and residue field k of char$(k) \neq 2, 3$. Let us fix $n \geq 1$. Let us consider the scheme X_0 over $\mathcal{O}_K$, normalization of $\mathbb{P}_{\mathcal{O}_K}^1 = \operatorname{Spec} \mathcal{O}_K[x] \cup \operatorname{Spec} \mathcal{O}_K[1/x]$ in

$$K(X_0) = K(x)[y]/(y^2 - (x^2 + t^n)(x^3 + 1)).$$

Then X_0 is the union of the affine open subschemes

$$U = \operatorname{Spec} \mathcal{O}_K[x, y]/(y^2 - (x^2 + t^n)(x^3 + 1)),$$

$$V = \operatorname{Spec} \mathcal{O}_K[x_1, y_1]/(y_1^2 - x_1(1 + t^n x_1^2)(1 + x_1^3)), \quad x_1 = 1/x, y_1 = y/x^3.$$

The open subscheme V is smooth over $\mathcal{O}_K$, and U contains a unique singular point p corresponding to the maximal ideal (t, x, y). Let us determine the desingularization X of X_0 obtained by blowing-up and normalizing as in Subsection 8.3.4. Let $z \in \widehat{\mathcal{O}}_{U,p}$ be a square root of $1 + x^3$ (Exercise 1.3.9). We see that

$$\widehat{\mathcal{O}}_{U,p} \simeq \widehat{\mathcal{O}}_K[[x, v]]/((v - x)(v + x) - t^n), \quad v = y/z.$$

By Example 8.3.53, X_p is made up of a chain of n projective lines $\Gamma_1, \ldots, \Gamma_n$ over k, with self-intersection $\Gamma_i^2 = -2$. We easily deduce from this that X_k is the union of an elliptic curve E and of the Γ_i.

Figure 18. Contraction of (-2)-curves.

The model X is minimal by Castelnuovo's criterion. The Γ_i are the irreducible components of X_k with zero intersection with $K_{X/\mathcal{O}_K}$. Hence the canonical model of X is none other than X_0 itself.

Let $X \to S$ be a minimal arithmetic surface. Then a closed fiber X_s can have as many irreducible components as we want, even if we fix the genus $p_a(X)$ and S, which is not the case for the canonical model.

Proposition 4.24. *Let $Y \to S$ be the canonical model of a minimal arithmetic surface. Let $s \in S$ be a closed point and n the number of irreducible components of Y_s. Then $n \leq 2p_a(X_\eta) - 2$.*

Proof   The curve Y_s is an l.c.i. over $k(s)$ since Y is an l.c.i over S (Corollaries 4.18 and 6.3.24). By Theorem 6.4.9(b), we have $\omega_{Y/S}|_{Y_s} \simeq \omega_{Y_s/k(s)}$. It follows that

$$\deg_{k(s)} \omega_{Y_s/k(s)} = -2\chi_{k(s)}(\mathcal{O}_{Y_s}) = -2\chi_{k(\eta)}(\mathcal{O}_{Y_\eta}) = 2p_a(X_\eta) - 2$$

(Corollary 7.3.31 and Proposition 5.3.28). Let $F_1, \ldots, F_n$ be the irreducible components of Y_s, with respective multiplicities $d_1, \ldots, d_n$. As $\omega_{Y/S}$ is ample, we have $\deg \omega_{Y/S}|_{F_i} > 0$. By virtue of Proposition 7.5.7, we then have

$$\deg_{k(s)} \omega_{Y/S}|_{Y_s} = \sum_{1 \leq i \leq n} d_i \deg_{k(s)}(\omega_{Y/S}|_{F_i}) \geq \sum_{1 \leq i \leq n} d_i \geq n. \qquad (4.20)$$

This completes the proof. $\square$

9.4.4   Weierstrass models and regular models of elliptic curves

Let $S = \operatorname{Spec} A$ be an affine Dedekind scheme of dimension 1. In this subsection, we fix an elliptic curve E over $K = K(S)$, endowed with a privileged rational point $o \in E(K)$. By Definition 6.1.25, E admits a homogeneous equation (*Weierstrass equation*)

$$v^2 z + (a_1 u + a_3 z)vz = u^3 + a_2 u^2 z + a_4 u z^2 + a_6 z^3, \tag{4.21}$$

with o corresponding to the point $(0, 1, 0)$ and such that its discriminant is non-zero (see formula (4.27)). The equation

$$y^2 + (a_1 x + a_3)y = x^3 + a_2 x^2 + a_4 x + a_6 \tag{4.22}$$

of the affine open subscheme $D_+(z)$ is called an *affine Weierstrass equation of E*. If the $a_i \in A$, we will say that (4.21) is an *integral equation* of E (in fact of the pair (E, o)). We then associate the S-scheme

$$W = \operatorname{Proj} A[u, v, z]/(v^2 z + (a_1 u + a_3 z)vz - (u^3 + a_2 u^2 z + a_4 u z^2 + a_6 z^3))$$

to it. Given an integral equation (4.21) of E, we call the surface $W \to S$ the *Weierstrass model* of E over S associated to (4.21).

Remark 4.25. Two integral equations of E can induce isomorphic Weierstrass models (as S-schemes). But we do not know how to go from one equation to the other by a 'fractional linear transformation'. See Exercise 4.12.

Proposition 4.26. *Let* $\pi : W \to S$ *be the Weierstrass model above. Then this is a normal fibered surface that verifies the following properties.*

(a) *The morphism* π *is smooth at the points of* $\overline{\{o\}}$.

(b) *For any* $s \in S$, W_s *is geometrically integral.*

(c) *The morphism* π *is an l.c.i. Let us set* $x = u/z$, $y = v/z \in K(W)$, *and*

$$\omega := \frac{dx}{2y + (a_1 x + a_3)} \in \Omega^1_{K(W)/K}. \tag{4.23}$$

Then $\omega_{W/S} = \omega \mathcal{O}_W$. *In particular,* $\pi_* \omega_{W/S} = \omega \mathcal{O}_S$ *is free on* S.

Proof It is easy to see that $\mathcal{O}_W$ is torsion-free over $\mathcal{O}_S$. Hence $W \to S$ is flat (Corollary 1.2.14). Its normality results from (b) by virtue of Lemma 4.1.18 or 8.2.11. Let us show (a). For any $s \in S$, we have

$$W_s = \operatorname{Proj} k(s)[u, v, z]/(v^2 z + (\bar{a}_1 u + \bar{a}_3 z)vz - (u^3 + \bar{a}_2 u^2 z + \bar{a}_4 u z^2 + \bar{a}_6 z^3)),$$

where $\bar{a}$ denotes the image of $a \in A$ in $k(s)$ (Proposition 3.1.9). Hence $\overline{\{o\}} \cap W_s$ consists of the point $o_s \in W_s(k(s))$ with coordinates $(0, 1, 0)$. We see that W_s is regular at o_s by the Jacobian criterion.

(b) Let us note that W_s is a cubic over $k(s)$. Its intersection with the line $z = 0$ is reduced to a smooth point o_s. As this line meets all of the irreducible components of W_s (Exercise 3.3.4(c)), we deduce from this that W_s is irreducible. If it were not integral, then it would be of the form $3L$, with L a line, which would imply that W_s does not have any smooth point. As this argument is valid over every extension of $k(s)$, W_s is geometrically integral.

(c) As, by construction, W is a global complete intersection over S, it is a fortiori an l.c.i. The determination of $\omega_{W/S}$ is similar to the case of elliptic curves over a field (Proposition 6.1.26). Let U be the open subscheme $D_+(z)$ of W. We have

$$\mathcal{O}_W(U) = A[x, y]/(y^2 + (a_1 x + a_3)y - (x^3 + a_2 x^2 + a_4 x + a_6)),$$

and therefore $\omega_{U/S} = \omega \mathcal{O}_U$ by Corollary 6.4.14. Let V be the open subset $D_+(v)$ of W. Let us set $t = u/v$, $w = z/v$. Then $\mathcal{O}_W(V) = A[t, w]/(F(t, w))$, where

$$F(t, w) = w + (a_1 t + a_3 w)w - (t^3 + a_2 t^2 w + a_4 t w^2 + a_6 w^3)).$$

Hence $\omega_{V/S}$ is generated by the rational differential

$$\omega' := \frac{dw}{\partial F/\partial t}.$$

Now, using the relations

$$t = \frac{x}{y}, \quad w = \frac{1}{y}, \quad dw = -\frac{1}{y^2} dy, \quad dy = (-a_1 y + 3x^2 + 2a_2 x + a_4)\omega,$$

in $K(W)$ and in $\Omega^1_{K(W)/K}$, we find by direct computations that $\omega' = \omega$. As $W = U \cup V$, we find that $\omega_{W/S} = \omega \mathcal{O}_W$. Finally, we have $\pi_* \mathcal{O}_W = \mathcal{O}_S$ by Corollary 8.3.6(c). □

Remark 4.27. In what follows, we are going to show that the conditions of Proposition 4.26 characterize Weierstrass models (Proposition 4.30).

Lemma 4.28. *Let C be an integral projective curve over a field k. Let us suppose that $p_a(C) = 1$, and that C contains a smooth point p that is rational over k. Then we have the following properties:*

(a) *For $n = 0$ or 1, we have $H^0(C, \mathcal{O}_C(np)) = k$, while $H^1(C, \mathcal{O}_C) \simeq k$ and $H^1(C, \mathcal{O}_C(p)) = 0$.*

(b) *For any $n \geq 2$, we have $\dim_k H^0(C, \mathcal{O}_C(np)) = n$, $H^1(C, \mathcal{O}_C(np)) = 0$, and $\mathcal{O}_C(np)$ is generated by its global sections.*

Proof (a) Let $L(np) = H^0(C, \mathcal{O}_C(np))$. We have $L(0) \subseteq K(C) \cap \bar{k} \subseteq \mathcal{O}_{C,p}$ because C is normal at p. The last inclusion induces an inclusion in $k(p) = k$. Hence $L(0) = k$. We have $\dim_k H^1(C, \mathcal{O}_C) = 1$ because $p_a(C) = 1$. Let $f \in L(p)$. Let us show that $f \in k$. Let us suppose the contrary. Then we have $\mathrm{div}(f) = q - p$ for a regular rational point $q \neq p$. We can once more take the construction before

Lemma 7.3.10 and define a finite morphism $\pi : C \to \mathbb{P}^1_k$ such that $\pi^*\infty = p$, where ∞ is a rational point of $\mathbb{P}^1_k$. Proposition 7.3.8 implies that π is birational. It is therefore an isomorphism, which is contrary to the $p_a(C) = 1$ hypothesis. Consequently, $L(p) = k$. We have $H^1(C, \mathcal{O}_C(p)) = 0$ because $\chi_k(\mathcal{O}_C(p)) = 1$ by the Riemann–Roch theorem 7.3.17.

(b) We have an exact sequence

$$0 \to \mathcal{O}_C((n-1)p) \to \mathcal{O}_C(np) \to \mathcal{O}_C(np)|_p \to 0,$$

which induces the exact cohomology sequence

$$0 \to L((n-1)p) \to L(np) \to V \to H^1(C, \mathcal{O}_C((n-1)p)) \to H^1(C, \mathcal{O}_C(np)) \to 0,$$

with $\dim_k V = 1$. We then determine the dimensions of the $H^i(C, \mathcal{O}_C(np))$ by induction on $n \geq 2$. We use (a) to start the induction at $n = 2$. Finally, $\mathcal{O}_C(np)$ is generated by its global sections by Lemma 7.4.2. □

Lemma 4.29. *Let* $\pi : W \to S = \operatorname{Spec} A$ *be a fibered surface such that* $W_\eta = E$, *that* $O := \overline{\{o\}}$ *is contained in the smooth locus of* W, *and that* W_s *is integral for every* $s \in S$. *Let us consider* O *as a Cartier divisor on* W. *Then the following properties are true.*

(a) *For any* $n \geq 2$, $\mathcal{O}_W(nO)$ *is generated by its global sections.*

(b) *The sheaf* $\mathcal{L} = R^1\pi_*\mathcal{O}_W$ *is invertible on* S. *Let us suppose that it is free. For any* $n \geq 2$, *there exists an exact sequence*

$$0 \to \pi_*\mathcal{O}_W((n-1)O) \to \pi_*\mathcal{O}_W(nO) \to \mathcal{L}^{\otimes n} \to 0, \qquad (4.24)$$

$\pi_*\mathcal{O}_W(n)$ *is free of rank* n, *and the canonical homomorphism*

$$\bigoplus_{2a+3b\leq n} (\pi_*\mathcal{O}_W(2O))^{\otimes a} \otimes (\pi_*\mathcal{O}_W(3O))^{\otimes b} \to \pi_*\mathcal{O}_W(nO)$$

is surjective.

Proof Let us first note that for any $s \in S$, we have $p_a(W_s) = p_a(W_\eta) = 1$ (Proposition 5.3.28). Therefore the curve W_s over $k(s)$ verifies the hypotheses of Lemma 4.28. Hence for any $n \geq 0$, the dimensions of the $H^i(W_s, \mathcal{O}_W(nO)_s)$ over $k(s)$ depend only on n and on i. By Theorem 5.3.20, the $R^i\pi_*\mathcal{O}_W(nO)$ are locally free on S, and we have

$$(R^i\pi_*\mathcal{O}_W(nO))_s \otimes_{\mathcal{O}_{S,s}} k(s) = H^i(W_s, \mathcal{O}_W(nO)|_{W_s}). \qquad (4.25)$$

Moreover, as S is affine and of dimension 1, the term on the left is equal to $H^i(W, \mathcal{O}_W(nO)) \otimes_A k(s)$. Property (a) is then a consequence of Lemma 4.28(b) and of Nakayama's lemma. This also implies that $\mathcal{L}$ is locally free of rank 1, and the $\pi_*\mathcal{O}_W(n)$ locally free of rank n.

Let $\sigma : S \to W$ denote the section corresponding to O, and $i : O \to W$ the canonical closed immersion. For any $n \geq 1$, we have the exact sequence

$$0 \to \mathcal{O}_W((n-1)O) \to \mathcal{O}_W(nO) \to i_* i^* \mathcal{O}_W(nO) \to 0. \qquad (4.26)$$

By the computations of Lemma 4.28 and (4.25), we have $R^1 \pi_* \mathcal{O}_W((n-1)O) = 0$ if $n \geq 2$. Taking π_* in the exact sequence above, we obtain an exact sequence over S for every $n \geq 2$:

$$0 \to \pi_* \mathcal{O}_W((n-1)O) \to \pi_* \mathcal{O}_W(nO) \to \mathcal{L}_n \to 0,$$

where $\mathcal{L}_n$ is the invertible sheaf

$$\mathcal{L}_n = \pi_* i_* (i^* \mathcal{O}_W(nO)) = \sigma^*(\mathcal{O}_W(nO)) = (\sigma^* \mathcal{O}_W(O))^{\otimes n} = \mathcal{L}_1^{\otimes n}.$$

Taking π_* of sequence (4.26) with $n = 1$, we obtain an exact sequence

$$0 \to \mathcal{O}_S \to \pi_* \mathcal{O}_W(O) \to \mathcal{L}_1 \to R^1 \pi_* \mathcal{O}_W \to 0.$$

By Lemma 4.28(a) and the isomorphism (4.25), we have $(\pi_* \mathcal{O}_W(O)/\mathcal{O}_S) \otimes k(s) = 0$ for every $s \in S$. Hence, by Nakayama, we obtain $\pi_* \mathcal{O}_W(O) = \mathcal{O}_S$. It follows that $\mathcal{L}_1 = \mathcal{L}$, which shows exact sequence (4.24). The rest of the lemma can be shown by using this exact sequence and by induction on $n \geq 2$. $\qquad \square$

Proposition 4.30. *Let $\pi : W \to S = \operatorname{Spec} A$ be a fibered surface such that $W_\eta = E$, that $O := \overline{\{o\}}$ is contained in the smooth locus of W, and that W_s is integral for every $s \in S$. Then the following properties are true.*

(a) *The scheme $W \to S$ is an l.c.i. and $\pi_* \omega_{W/S}$ is an invertible sheaf.*

(b) *Let us suppose that $\pi_* \omega_{W/S}$ is free over S. Then $W \to S$ is the Weierstrass model of E associated to an integral equation.*

Proof Let us first suppose that $R^1 \pi_* \mathcal{O}_W$ is free on S. Let us set $L(nO) = H^0(W, \mathcal{O}_W(nO))$ for every $n \geq 0$. By Lemma 4.29(b), there exist $x \in L(2O)$, $y \in L(3O)$ such that $\{1, x\}$ is a basis of $L(2O)$ over A and that $\{1, x, y\}$ is a basis of $L(3O)$ over A. Moreover, the images of x^3 and y^2 (as elements of $L(6O) \subset K(W)$) in $L(6O)/L(5O) \simeq H^1(W, \mathcal{O}_W)^{\otimes 6}$ are both bases. There therefore exists an $\alpha \in A^*$ such that

$$y^2 - \alpha x^3 \in L(5O) = A + Ax + Ax^2 + Axy + Ay.$$

Replacing y by $\alpha^{-1} y$ and x by $\alpha^{-1} x$, we can suppose $\alpha = 1$. There therefore exist $a_i \in A$ such that

$$y^2 + (a_1 x + a_3)y = x^3 + a_2 x^2 + a_4 x + a_6.$$

This implies that the morphism $\varphi : W \to \mathbb{P}_A^2$ associated to the basis $\{1, x, y\}$ of $L(3O)$ sends $W \setminus O$ (and therefore W) into the cubic W' over A with equation

$$v^2 z + (a_1 uz + a_3 z^2)v = u^3 + a_2 u^2 z + a_4 uz^2 + a_6 z^3.$$

We know by Proposition 7.4.4 that φ is a closed immersion at the generic fiber. As in Proposition 4.26, we see that W' is normal, with integral fibers. Hence

$\varphi: W \to W'$ is birational, and quasi-finite because W_s is irreducible, and hence finite. Consequently, φ is an isomorphism.

In the general case, $R^1\pi_*\mathcal{O}_W$ is locally free (Lemma 4.29(b)). It follows from the particular case above that $W \to S$ is an l.c.i. By duality (Section 6.4, formula (4.12)), we have a canonical isomorphism $\pi_*\omega_{W/S} \simeq (R^1\pi_*\mathcal{O}_W)^\vee$. Hence $\pi_*\omega_{W/S}$ is invertible. If it is free, then $R^1\pi_*\mathcal{O}_W$ is also free, which concludes the proof of the proposition.   □

Discriminant and Weierstrass models

To equation (4.21), we associate its discriminant Δ. If char$(K) \neq 2$, by definition

$$\Delta = 2^{-4}\,\mathrm{disc}\left(4(x^3 + a_2x^2 + a_4x + a_6) + (a_1x + a_3)^2\right). \tag{4.27}$$

An explicit computation (see, for example, [91], Section III.1) shows that Δ is a polynomial

$$\Delta \in \mathbb{Z}[a_1, a_2, a_3, a_4, a_6]. \tag{4.28}$$

If char$(K) = 2$, then the formula for the discriminant is obtained by reduction of the polynomial above modulo 2. The merit of the discriminant is that its vanishing characterizes the smoothness of E (see [91], loc. cit.). In addition, relation (4.28) shows that if the $a_i \in A$, then $\Delta \in A$. Let W be the Weierstrass model associated to an integral equation (4.21). We call the element $\Delta \in A$ modulo A^* the *discriminant of W*. We denote it Δ_W. This discriminant does not depend on the choice of integral equation (Corollary 4.32).

Lemma 4.31. *Let us consider an equation (4.21) of (E, o). Let Δ be its discriminant, $\omega = dx/(2y + (a_1x + a_3)) \in H^0(E, \omega_{E/K})$ (formula (4.23)). Then the element*

$$\Delta\omega^{\otimes 12} \in (H^0(E, \omega_{E/K}))^{\otimes 12}$$

is independent of the choice of equation.

Proof Let (4.21)′ be another (affine) equation of E with variables x', y'. By Corollary 7.4.33(b), we have

$$x' = ax + b, \quad y' = ey + cx + d, \quad \text{with } a, b, c, d \in K, e \in K^*.$$

As y'^2 and x'^3 have the same coefficient in the affine equation, we have $a^3 = e^2$. In other words, $a = \alpha^2$ and $e = \alpha^3$ by taking $\alpha = e/a$. A direct computation shows that

$$\omega' = \alpha^{-1}\omega, \quad \Delta' = \alpha^{12}\Delta,$$

where ω' and Δ' are the differential and discriminant associated to an equation (4.21)′, whence $\Delta'\omega'^{\otimes 12} = \Delta\omega^{\otimes 12}$.   □

Corollary 4.32. *If (4.21) and (4.21)′ are two integral equations defining the same Weierstrass model W, then $\Delta' \in \Delta A^*$.*

Proof With the notation above, we have $\omega' \in \omega A^*$ since these differentials are bases of $\omega_{W/S}$. It follows from Lemma 4.31 that $\Delta' \in \Delta A^*$.   □

Definition 4.33. For any $s \in S$, let ν_s denote the valuation of $K(S)$ associated to the ring $\mathcal{O}_{S,s}$. Let W be a Weierstrass model of E over S, with discriminant $\Delta_W \in A/A^*$. We will say that W is *minimal at s* if $\nu_s(\Delta_W)$ is the smallest among the valuations of the discriminants of the integral equations of E. The integer $\nu_s(\Delta_W)$ is then called the *minimal discriminant* of E at s. We will say that W is *minimal* if it is minimal at every $s \in S$. A Weierstrass model that is minimal at some s always exists in an evident manner. On the other hand, a global minimal Weierstrass model does not always exist.

Starting with an equation (4.21) of E, we obtain an integral equation by a suitable change of variables. The Weierstrass model associated to this equation admits a desingularization $\widetilde{W}$ since E is smooth (Corollary 8.3.51).

Definition 4.34. Let E be an elliptic curve over K. We call the minimal arithmetic surface $X \to S$ with generic fiber isomorphic to E the *minimal regular model of E*. Such a model exists by applying Proposition 3.19 and Corollary 3.24 to the surface $\widetilde{W} \to S$ above. It is unique by the definition of minimality. Moreover, X is independent of the choice of W and of o.

The minimal discriminant and minimal regular model are important objects for the study of the arithmetic of E. We invite the reader to consult [92], IV. We are now going to link the minimal Weierstrass model to the minimal regular model.

Theorem 4.35. *Let S be an affine Dedekind scheme of dimension 1, E an elliptic curve over $K = K(S)$ with a given rational point $o \in E(K)$, and $\rho : X \to S$ the minimal regular model of E over S.*

(a) *The set $\mathcal{E}$ of vertical prime divisors Γ on X such that $\Gamma \cap \overline{\{o\}} = \emptyset$ is finite, and there exists a contraction morphism $f : X \to W$ of the divisors belonging to $\mathcal{E}$. Moreover, $W \to S$ is an l.c.i. and we have $f_*\omega_{X/S} = \omega_{W/S}$ and $\omega_{X/S} = f^*\omega_{W/S}$.*

(b) *Let us suppose that the invertible sheaf $\rho_*\omega_{X/S}$ is free on S (e.g., $\mathrm{Pic}(S){=}0$). Then $\pi : W \to S$ is the Weierstrass model over S associated to an integral equation, and we have*

$$\omega_{X/S} = \omega\mathcal{O}_X, \quad \omega_{W/S} = \omega\mathcal{O}_W, \tag{4.29}$$

where ω is the differential associated to an integral equation of W defined in Proposition 4.26(c). In particular,

$$H^0(X, \omega_{X/S}) = H^0(W, \omega_{W/S}) = \omega\mathcal{O}_S(S). \tag{4.30}$$

(c) *Under the hypotheses of (b), W is a (global) minimal Weierstrass model.*

(d) *Let us suppose that E admits a minimal Weierstrass model W'. Then $\rho_*\omega_{X/S}$ is free on S and $W' \simeq W$. In other words, the minimal Weierstrass model is unique.*

Proof (a) If there exists a component of $\mathcal{E}$ that is contained in X_s, then X_s is not irreducible. It follows from Corollary 8.3.6(b) and Proposition 8.3.11 that $\mathcal{E}$ is finite. As X is regular, $\overline{\{o\}}$ is an effective Cartier divisor. The existence of the contraction $f : X \to W$ results from Lemma 8.3.32 and Proposition 8.3.30 (see also Example 4.19). For any $s \in S$, $\overline{\{o\}}$ meets only one irreducible component of X_s that is moreover of multiplicity 1 (Corollary 1.32). Hence W_s is irreducible, of multiplicity 1. As W is normal, this implies that W_s is integral (Exercise 8.3.3(d) or (e)). By Proposition 4.30, locally on S, W is a Weierstrass model. It follows that $W \to S$ is an l.c.i.

The equality $f_*\omega_{X/S} = \omega_{W/S}$ is a direct consequence of Corollary 4.18 because the contracted components have a zero intersection with $K_{X/S}$ (Example 4.19). But we are going to give another, more direct, proof. To show this equality, we can suppose S is local. By Propositions 4.30 and 4.26, $\omega_{W/S} = \omega \mathcal{O}_W$. Likewise, by Corollary 3.27, $\omega_{X/S} = \omega_0 \mathcal{O}_X$ is free. In $\Omega^1_{K(X)/K}$, we have $\omega_0 = \omega h$ with $h \in K(X)^*$. The divisor $\mathrm{div}(h)$ on X has support in the exceptional locus $\mathcal{E}$ of f and we have $\mathrm{div}(h) \cdot \mathrm{div}(h) = 0$. Theorem 1.27 then implies that $\mathrm{div}(h) = 0$, and hence $h \in \mathcal{O}_X(X)^* = A^*$. This implies that $f_*\omega_{X/S} = \omega_{W/S}$. Finally, $\omega_{X/S} = f^*f_*\omega_{X/S} = f^*\omega_{W/S}$ by Corollary 3.27.

(b) We know by Corollary 3.27 that $\rho_*\omega_{X/S}$ is invertible. Let us suppose it is free. It follows from (a) that $\pi_*\omega_{W/S} = \rho_*\omega_{X/S}$ is free. By Proposition 4.30, W is a Weierstrass model of E over S. Equalities (4.29) follow from Corollary 3.27, Proposition 4.26, and (a). We have $\rho_*\mathcal{O}_X = \pi_*\mathcal{O}_W = \mathcal{O}_S$ by Corollary 8.3.6(c), whence (4.30).

(c) Let W' be a Weierstrass model, let $g : X' \to W'$ be a minimal desingularization, and F the exceptional locus of g. Then we have the following relations in $\Omega^1_{K(E)/K}$:

$$H^0(X', \omega_{X'/S}) \subseteq H^0(X' \setminus F, \omega_{X'/S}) = H^0(W' \setminus g(F), \omega_{W'/S}) = H^0(W', \omega_{W'/S}).$$

The last equality comes from Lemma 2.17(a). By definition, we have a birational morphism $X' \to X$. By (b) and Corollary 2.25, we have

$$H^0(W, \omega_{W/S}) = H^0(X, \omega_{X/S}) = H^0(X', \omega_{X'/S}) \subseteq H^0(W', \omega_{W'/S}).$$

The equality $\Delta_W \omega_W^{\otimes 12} = \Delta_{W'} \omega_{W'}^{\otimes 12}$ (Lemma 4.31) then implies that $\Delta_{W'} \in \Delta_W A$. Hence W is minimal.

(d) Let us first suppose that $\rho_*\omega_{X/S}$ is free. Let us keep the notation of the proof of (c). The reasoning in (c) shows that $\omega_{X'/S} = \omega'\mathcal{O}_X$, where $\omega' \in \Omega^1_{K(X)/K}$ is a basis of $\omega_{W'/S}$. Comparing $\omega_{X'/S}$ with the pull-backs of $\omega_{X/S}$ and of $\omega_{W'/S}$, we see that there exists a divisor D on X' such that $\omega_{X'/S} = \omega'\mathcal{O}_{X'}(D)$, and that $\mathrm{Supp}\, D$ is both equal to the union of the exceptional divisors of X' (Exercise 2.4) and contained in the exceptional locus of $X' \to W'$. Now, by hypothesis, the exceptional locus of $X' \to W'$ does not contain any exceptional divisor. Hence $X' = X$ and X dominates W'. Let O' be the closure of $o \in E(K)$ in W'. Then O' coincides with $g(\overline{\{o\}})$ at the generic fiber. They are therefore equal. Let $s \in S$ and let Γ_0 be the irreducible component of X_s passing through $\overline{\{o\}} \cap X_s$. If

$g(\Gamma_0)$ is a point, as X is minimal, it follows from the factorization theorem 2.2 that this point is singular in W', and hence singular in W'_s. Now this is also the point $O' \cap W'_s$, a contradiction with the hypothesis that O' is contained in the smooth locus of W'. Consequently, $g(\Gamma_0)$ is equal to W'_s. Hence $X \to W'$ is obtained by contraction of the same vertical divisors as W, whence $W' = W$ (Proposition 8.3.28).

In the general case, we have already seen that $\rho_* \omega_{X/S}$ is locally free of rank 1. Hence, locally on S, W' coincides with W, which implies that $W = W'$. The fact that $\rho_* \omega_{X/S}$ is free then comes from the equality $\rho_* \omega_{X/S} = \pi_* \omega_{W/S}$ in (a). □

Remark 4.36. The $\rho_* \omega_{X/S}$ free condition is equivalent to $\omega_{X/S}$ free since $\omega_{X/S} = \rho^* \rho_* \omega_{X/S}$ by Corollary 3.27 and $\pi_* \mathcal{O}_X = \mathcal{O}_S$.

Corollary 4.37. *Let E be an elliptic curve over K. Let us suppose that E admits a minimal Weierstrass model W over $S = \operatorname{Spec} A$. Then the minimal desingularization of W is isomorphic to the minimal regular model X of E over S. Moreover, the differential ω associated to W (formula (4.23)) is also a basis of $\omega_{X/S}$.*

Corollary 4.38. *Let $W_0 \to S$ be a Weierstrass model of E over $S = \operatorname{Spec} A$. Let us suppose that W_0 is dominated by the minimal regular model X of E. Then W_0 is the minimal Weierstrass model of E over S.*

Proof In fact, $X \to W_0$ coincides with the contraction morphism of Theorem 4.35(a) (Example 4.19). So the corollary is a consequence of Theorem 4.35(c). □

Remark 4.39. Let us keep the notation of Theorem 4.35. Let U be an open subscheme of X such that $E \subset U$ and that $U \cap X_s \neq \emptyset$ for every $s \in S$. Then we have

$$H^0(U, \omega_{X/S}) = H^0(X, \omega_{X/S})$$

(Exercise 3.9). This is particularly interesting when we take for U the *Néron model* of E (this is the greatest open subscheme of X that is smooth over S, see Subsection 10.2.2).

Exercises

4.1. Let $X \to S$ be a regular fibered surface. Let Δ be a reduced vertical divisor contained in the fiber X_s. Let us suppose that the intersection matrix of the irreducible components of Δ is negative definite. Here is how we can find its fundamental divisor Z algorithmically. If there exists a component Γ_1 of Δ such that $\Delta \cdot \Gamma_1 > 0$, let us set $\Delta_1 = \Delta + \Gamma_1$ (otherwise $Z = \Delta$). If there exists a component Γ_2 of Δ such that $\Delta_1 \cdot \Gamma_2 > 0$, then let us set $\Delta_2 = \Delta_1 + \Gamma_2$ (otherwise $Z = \Delta_1$). We thus construct an ascending sequence of divisors Δ_i. Show that there exists an $n \geq 1$ such that $Z = \Delta_n$.

4.2. Let us keep the notation of Exercise 4.1 and let us suppose $\dim S = 1$. Show that the fundamental divisor Z verifies $Z < X_s$ (write $X_s = Z_1 + Z_2$ with Z_j effective and Z_2 with no common component with Δ, and show that $Z \le Z_1$).

4.3. Let $X \to S$ be a regular fibered surface, and let Z be an effective vertical divisor contained in a closed fiber X_s. We suppose that $k(s)$ is algebraically closed. We want to show that $H^1(Z, \mathcal{O}_Z) = 0$ is equivalent to $\mathrm{Pic}^0(Z) = 1$.

  (a) Show the equivalence for Z integral.

  (b) Let us suppose Z is reduced and reducible. Let us write $Z = Z_1 + Z_2$ with the Z_i with no common component. Show that we have an exact sequence

  $$0 \to \mathcal{O}_Z \to \mathcal{O}_{Z_1} \oplus \mathcal{O}_{Z_2} \to \mathcal{G} \to 0,$$

  where $\mathcal{G}$ is a skyscraper sheaf with support in $Z_1 \cap Z_2$. Show that

  $$\mathrm{Pic}(Z) \simeq \mathrm{Pic}(Z_1) \oplus \mathrm{Pic}(Z_2)$$

  and

  $$H^1(Z, \mathcal{O}_Z) \simeq H^1(Z_1, \mathcal{O}_{Z_1}) \oplus H^1(Z_2, \mathcal{O}_{Z_2}).$$

  (c) Let Γ be an irreducible component of Z such that $2\Gamma \le Z$. Use the method of Lemma 7.5.11 to show that $\mathrm{Pic}^0(Z) \simeq \mathrm{Pic}^0(Z - \Gamma)$ if and only if $H^1(Z, \mathcal{O}_Z) \simeq H^1(Z - \Gamma, \mathcal{O}_{Z-\Gamma})$. Note that we cannot use the results of Lemma 7.5.11 directly because the divisors here are not necessarily curves over a field.

  (d) Show that $H^1(Z, \mathcal{O}_Z) = 0 \iff \mathrm{Pic}^0(Z) = 1$.

4.4. Let $X \to S$ and Δ be as in Exercise 4.1. Let Z be the fundamental divisor of Δ. We suppose that $H^1(Z, \mathcal{O}_Z) = 0$.

  (a) Show that $p_a(Z') \le 0$ for every $0 < Z' \le Z$ (use Exercise 4.2 if $\dim S = 1$). Deduce from this that $\mathrm{Pic}^0(Z) = 1$.

  (b) Show that for any $n \ge 1$, we have $H^1(nZ, \mathcal{O}_{nZ}) = 0$ and $\mathrm{Pic}^0(nZ) = 1$.

  (c) Let $Z' > 0$ be with support in $\mathrm{Supp}\, Z$. Show that we have $H^1(Z', \mathcal{O}_{Z'}) = 0$ and $\mathrm{Pic}^0(Z') = 1$.

4.5. Let $X \to S$ be an arithmetic surface with $p_a(X_\eta) \le 0$. Let $\Delta = \sum_i \Gamma_i$ be a reduced vertical divisor that does not contain any connected component of X_s for any $s \in S$. Show that there exists a contraction morphism of the components Γ_i (use Exercises 4.2 and 4.4). This result has also been shown using methods from p-adic analysis (see [64], Theorem 0.1, [78], Proposition 4).

4.6. Show that in Proposition 4.24, we can replace n by the number of irreducible components of $Y_{\overline{k(s)}}$.

4.7. Let $f : X \to Y$ be the contraction of Δ as in Theorem 4.15. Let Z be the fundamental divisor of Δ, and $y = f(\Delta)$.

   (a) Let $\mathcal{M} \subset \mathcal{O}_Y$ be the sheaf of ideals corresponding to $y \in Y$. Show that $\mathcal{M}\mathcal{O}_X = \mathcal{O}_X(-Z)$.

   (b) Let $Y_1 \to Y$ be the normalization of the blowing-up of Y with center y. Show that $X \to Y$ factors into $X \to Y_1 \to Y$, and that $X \to Y_1$ is the contraction of a set of vertical prime divisors verifying the conditions of Theorem 4.15.

   (c) Show that the sequence (3.11) of Subsection 8.3.4 associated to Y ends with X.

   (d) Let us suppose $\dim S = 1$. Show that $T_{Y_s,y} = T_{Y,y}$ if and only if $2Z \le X_s$.

4.8. Let $X \to S$ be an arithmetic surface such that for some $s \in S$, X_s is a union of two elliptic curves meeting transversally at a point that is rational over $k(s)$. Show that $\omega_{X/S}$ is not generated by its global sections. Find the smallest integer n such that $\omega_{X/S}^{\otimes n}$ is generated by its global sections.

4.9. Let $\mathcal{O}_K$ be a discrete valuation ring with residue field k and field of fractions K.

   (a) Let k'/k be a finite extension. Show that there exists a closed point $x \in \mathbb{P}_k^1$ such that $k(x) = k'$ if and only if k' is simple over k (i.e., generated by a single element).

   (b) Let k'/k be simple. Show that there exists a normal fibered surface $X \to \operatorname{Spec} \mathcal{O}_K$ such that $X_K \simeq \mathbb{P}_K^1$, X_k is integral, $H^1(X_k, \mathcal{O}_{X_k}) = 0$, and that the normalization of X_k is isomorphic to $\mathbb{P}_{k'}^1$. (Blow up $\mathbb{P}_{\mathcal{O}_K}^1$ along the point x as in (a) and contract the non-exceptional component.)

   (c) Show that X_k is not an l.c.i. if $[k' : k] \ge 3$.

4.10. Let C be a projective curve over a field k and $p \in C(k)$ as in Lemma 4.28. We want to show that $\mathcal{O}_C(3p)$ is very ample and induces an isomorphism from C to a cubic on $\mathbb{P}_k^2$.

   (a) Show that C is geometrically integral.

   (b) Show that we can reduce to the case when k is algebraically closed and when C is singular.

   (c) Let $\pi : C' \to C$ be the normalization of C. Show that $C' \simeq \mathbb{P}_k^1$. Let $B = \mathcal{O}_C(C \setminus \{p\})$. Show that $\mathcal{O}_{C'}(C' \setminus \pi^{-1}(p)) = k[T]$ and that $k[T] = B \oplus kT$.

   (d) Show that there exist $a, b \in k$ such that $x := T^2 + aT \in B$ and $y := T^3 + bT \in B$. Show that $B = k[x, y]$ and that we have the relation

$$y^2 + (3ax + a^3 + ab)y = x^3 + 2bx^2 + (b^2 + a^2b)x.$$

   (e) Deduce from this that C is a cubic.

4.11. Let A be a Dedekind ring of dimension 1, with field of fractions K, and E an elliptic curve over K with a rational point $o \in E(K)$. Let σ be the hyperelliptic involution of E such that $\sigma(o) = o$ (Proposition 7.4.29).

(a) Show that on equation (4.21), σ is defined by

$$(u, v, z) \to (u, -v - a_1 u - a_3 z, z).$$

(b) Let W be a Weierstrass model of E over S. Show that the action of σ on E extends to W and that the quotient scheme $W/\langle\sigma\rangle$ (Exercises 2.3.21 and 3.3.23) is isomorphic to $\mathbb{P}_S^1$ ($= \operatorname{Proj} A[u, z]$ if W is associated to equation (4.21)).

4.12. Let (E, o) be an elliptic curve over K as in Exercise 4.11. Let

$$y^2 + (a_1 x + a_3)y = x^3 + a_2 x^2 + a_4 x + a_6,$$

$$y'^2 + (b_1 x' + b_3)y' = x'^3 + a_2 x'^2 + a_4 x' + a_6$$

be two integral equations of (E, o) that induce isomorphic Weierstrass models. Show that there exist $a \in A^*, b, c, d \in A$ such that

$$x' = a^2 x + b, \quad y' = a^3 y + cx + d.$$

(Use Exercise 4.11.)

4.13. Let E be an elliptic curve over the field of fractions K of a Dedekind ring A of dimension 1. Let $o, o' \in E(K)$. Show that for any closed $s \in \operatorname{Spec} A$, the minimal discriminant of (E, o) at s is the same as that of (E, o') at s. (Use Theorem 4.35.)

4.14. Let $E \to \operatorname{Spec} K$ be as in Exercise 4.11. Let us consider the divisor

$$\mathcal{D} := \sum_{s \in S} \nu_s(\Delta_s)[s]$$

on S, where $\nu_s(\Delta_s)$ is the minimal discriminant of E at s.

(a) Let $\rho : X \to S$ be the minimal regular model of E over S. Show that $\mathcal{O}_S(\mathcal{D}) \simeq (\rho_* \omega_{X/S})^{\otimes 12}$.

(b) Show that if $\mathcal{D}$ is principal and if $\operatorname{Pic}(S)$ does not have any non-trivial point of order dividing 6, then E admits a global minimal Weierstrass model.

4.15. Let E be an elliptic curve over K as in Exercise 4.11. Let $W \to S$ be a Weierstrass model of E over S. Let us suppose that any $s \in S$ such that W_s is singular is principal (i.e., the maximal ideal of A defining the point s is principal). Show that E admits a minimal Weierstrass model over S.

4.16. Let $X \to S$ be a minimal arithmetic surface, with generic fiber X_η smooth of genus $g(X_\eta) = 1$.

(a) Show that $\omega_{X_\eta/K} \simeq \mathcal{O}_{X_\eta}$. Deduce from this that the canonical divisor $K_{X/S}$ is linearly equivalent to an effective vertical divisor V.

(b) Let us fix a closed point $s \in S$, let d denote the gcd of the multiplicities of the irreducible components of X_s, and V_s the connected component of V that meets X_s. Show that $V \cdot \Gamma = 0$ for any vertical prime divisor Γ. Deduce from this that $V_s \sim r(d^{-1}X_s)$ for some integer $0 \le r \le d - 1$.

(c) Show that $\omega_{X/S} \simeq \mathcal{O}_X$ if $X_\eta(K) \neq \emptyset$.

10

Reduction of algebraic curves

For a long time, we have known that the arithmetic of an algebraic variety V defined over a number field K is only well understood if we take into account its behavior modulo the finite and infinite places of K. The theory of reduction of varieties is concerned with the finite places. Crudely speaking, we extend V to a scheme $\mathcal{V}$ over the ring of integers $\mathcal{O}_K$ of K, while trying to preserve as many good properties of V as possible. The reduction of V modulo a maximal ideal $\mathfrak{p}$ is the fiber of $\mathcal{V}$ over the point of $\operatorname{Spec} \mathcal{O}_K$ corresponding to $\mathfrak{p}$. Today, the theory of reduction is very well understood for algebraic curves. This chapter attempts to present some essential aspects of this theory. The first section contains generalities on reduction. The second section treats the case of elliptic curves in detail (reduction of the minimal regular model, Néron model, and potential semi-stable reduction). The theory of stable reduction is broached in Section 10.3. Finally, the last section is devoted to the proof of the fundamental theorem of Deligne–Mumford (4.3) which stipulates that a smooth projective curve of genus ≥ 2 always admits a stable reduction after a suitable base change.

In all of this chapter, the considered fibered surfaces will all be defined over a Dedekind scheme of dimension 1. They are therefore relative curves.

10.1 Models and reductions

In this section we study general aspects of reduction. We begin by defining the different types of models of algebraic curves, accompanying this with some examples presented in a relatively detailed manner. Next we define the notion of reduction and show some general properties of good reduction. If $\mathcal{X} \to S$ is a surjective projective morphism onto the spectrum of a Henselian discrete valuation ring, we have a reduction map that sends a closed point of $\mathcal{X}_\eta$ to a closed point of $\mathcal{X}_s$. This map is described concretely in Corollary 1.34; its fibers are described in Proposition 1.40.

We conclude the section with a combinatorial study of the special fibers of minimal regular models. We show, in particular, a finiteness theorem (Proposition 1.57) which stipulates that for a fixed genus, there essentially exist only a finite number of possible configurations for these special fibers.

10.1.1   Models of algebraic curves

Definition 1.1. Let S be a Dedekind scheme of dimension 1, with function field K. Let C be a normal, connected, projective curve over K. We call a normal fibered surface $\mathcal{C} \to S$ together with an isomorphism $f : \mathcal{C}_\eta \simeq C$ a *model of C over S*. We will say that and a *regular model of C* if $\mathcal{C}$ is regular. More generally, we will say that a model $(\mathcal{C}, f)$ verifies a property (P) if $\mathcal{C} \to S$ verifies (P). The property (P) can, for example, be the fact of being smooth, minimal regular, or regular with normal crossings, etc. A *morphism* $\mathcal{C} \to \mathcal{C}'$ of two models of C is a morphism of S-schemes that is compatible with the isomorphisms $\mathcal{C}_\eta \simeq C$, $\mathcal{C}'_\eta \simeq C$.

Example 1.2. If C is an elliptic curve over K, then the Weierstrass models of C over K (Subsection 9.4.4) are models of C over S.

Remark 1.3. In practice, when we talk of a model of C over S, we only mention the scheme $\mathcal{C} \to S$ with generic fiber isomorphic to C. It is only useful to specify the choice of the isomorphism $\mathcal{C}_\eta \simeq C$ in rare situation. For example, if $C = \mathbb{P}^1_K$, the examples considered in Remark 9.3.23 show that C admits smooth models over S that are not isomorphic among themselves as models of C over S, but are isomorphic to $\mathbb{P}^1_S$.

Example 1.4. Let C be a normal projective curve over K, defined by homogeneous polynomials $F_1, \ldots, F_m \in K[T_0, \ldots, T_n]$. Let us suppose that $S = \operatorname{Spec} A$. Multiplying the F_i by elements of $A \setminus \{0\}$ if necessary, we can make the F_i have coefficients in A. If, by chance, the scheme $\mathcal{C}_0 := \operatorname{Proj} A[T_0, \ldots, T_n]/(F_1, \ldots, F_m)$ is normal, then it is a model of C over S because its generic fiber is isomorphic to C (Proposition 3.1.9).

Example 1.5. Let $q \geq 1$ be a square-free integer. Let C be the projective curve over $\mathbb{Q}$ defined by the equation

$$x^q + y^q + z^q = 0.$$

By the Jacobian criterion (Theorem 4.2.19), we easily see that C is smooth over $\mathbb{Q}$. Let $\mathcal{C}$ be the closed subscheme of $\mathbb{P}^2_\mathbb{Z}$ defined by the same equation. Let us show that $\mathcal{C}$ is normal, and it will consequently be a model of C over $\mathbb{Z}$. The Jacobian criterion shows that $\mathcal{C} \to \operatorname{Spec} \mathbb{Z}$ is smooth outside of the primes p that divide q. Let p be a prime factor of q. The integer $r = q/p$ is prime to p by hypothesis. We have

$$\mathcal{C}_p = \operatorname{Proj} \mathbb{F}_p[x, y, z]/(x^r + y^r + z^r)^p.$$

We deduce from this that $\mathcal{C}_p$ is irreducible and that $(\mathcal{C}_p)_{red}$ is the closed subvariety $V_+(x^r + y^r + z^r)$ over $\mathbb{F}_p$. As $\mathcal{C}$ is a complete intersection, and is regular at the generic fiber, to show the normality of C it suffices to show its normality at the generic point of C_p. We can therefore restrict ourselves to the affine open subscheme $U := \operatorname{Spec} \mathbb{Z}[x, y]/(x^q + y^q + 1)$ of C (to simplify, we write x, y instead of $x/z, y/z$). The prime ideal corresponding to C_p is generated by $x^r + y^r + 1$. Let us suppose that $p \geq 3$ (the $p = 2$ case can be treated in a similar way). We have

$$(T + S)^p = T^p + S^p + p(T + S)F(T, S), \quad F(T, S) \in \mathbb{Z}[T, S]$$

with F homogeneous and $F \notin p\mathbb{Z}[T, S]$. Hence

$$x^q + y^q + 1 = (x^r + y^r + 1)^p - p\left((x^r + y^r)F(x^r, y^r) + (x^r + y^r + 1)F(x^r + y^r, 1)\right).$$

The polynomial $F(x^r, y^r)$ modulo p is homogeneous and non-zero, and hence non-divisible by $x^r + y^r + 1$. The reasoning of Example 8.2.26 shows that C is normal. Moreover, the intersection of the singular locus of U with U_p is defined by the ideal $(p, x^r + y^r + 1, F(x^r, y^r))$.

Remark 1.6. The construction of a model of C over S is a way to extend C to a scheme that is surjective onto S. In a similar way, we can define the notion of *model* for any algebraic variety V over K. It is a scheme $\mathcal{V}$, surjective and flat over S, endowed with an isomorphism $\mathcal{V}_\eta \simeq V$. We would of course also like to preserve as many of the properties of V as possible. For normal projective curves, we ask that the models be normal and projective. These exist in very general situations, for example if S is affine and if we suppose S is excellent or C smooth. On the other hand, if we take smooth projective curves, we cannot always preserve the smoothness (see Exercise 1.5). If A is an Abelian variety over K (Definition 7.4.37), it is not always possible to extend A to an Abelian scheme, that is to say, to a proper group scheme over S. But we can always extend A to a scheme over S while preserving the smoothness (an Abelian variety is smooth) and the group structure. See Definition 2.7.

Remark 1.7. Let C be a smooth, connected, projective curve over K. We have two numerical invariants at our disposal, the arithmetic genus $p_a(C)$ and the geometric genus $g(C) = \dim_K H^1(C, \mathcal{O}_C)$. These two integers are equal if C is geometrically connected. In the general case, if $n = 1, 2$, the conditions $p_a(C) \geq n$ and $g(C) \geq n$ are equivalent. We will say *genus of C* for the geometric genus. The arithmetic genus suits the theory of reduction better because of its invariance in the fibers.

Proposition 1.8. *Let us suppose S is affine. Let C be a smooth projective curve of genus g over K. Then C admits a relatively minimal regular model (resp. a regular model with normal crossings) over S. If, moreover, $g \geq 1$, then C admits a unique minimal regular model $C_{\min}$, and a unique minimal regular model with normal crossings.*

Proof Let C_0 be as in Example 1.4. Let $\mathcal{C}$ be the Zariski closure of C in C_0, endowed with the reduced (hence integral) closed subscheme structure. Then $\mathcal{C} \to S$ is a fibered surface with generic fiber isomorphic to C. By desingularization of $\mathcal{C}$ (Corollary 8.3.51), we obtain a regular model of C over S. The remainder of the properties are just applications of Proposition 9.3.19, of Corollary 9.2.30, of Theorem 9.3.21, and of Proposition 9.3.36(b).   □

Remark 1.9. In Proposition 1.8, the S affine hypothesis is needed to assure the existence of a projective scheme over S with generic fiber isomorphic to C. Otherwise there always exists a regular, flat, proper scheme over S, with generic fiber isomorphic to C, which, taking into account Theorem 8.3.16, is not very different from a fibered surface (which is projective by definition).

Definition 1.10. Let C be a smooth projective curve over K, of genus $g \geq 2$. Let us suppose that C admits a minimal regular model $\mathcal{C}_{\min}$ over S (which is the case if S is affine). We call the canonical model $\mathcal{C}_{\mathrm{can}}$ of the minimal surface $\mathcal{C}_{\min}$ (Definition 9.4.21) the *canonical model of C over S*.

Remark 1.11. Given a smooth projective curve C over K, there exist a plethora of normal models of C over S. They are not all interesting. In general, we are mostly interested in minimal regular models, regular models with normal crossings, or canonical models. If E is a (pointed) elliptic curve over K, the interesting models are the minimal regular model and the minimal Weierstrass model, which is the analogue of the canonical model of curves of genus ≥ 2. If we consider E as an Abelian variety (7.4.37), there is in addition the Néron model to consider (Definition 2.7).

Let C be a curve over K defined by homogeneous equations. The following diagram sums up the way to produce the minimal regular model and the canonical model of C over S. The arrows are morphisms of schemes. If C is an elliptic curve with a point $o \in C(K)$, we must replace $\mathcal{C}_{\mathrm{can}}$ by the minimal Weierstrass model and the morphism $\mathcal{C}_{\min} \to \mathcal{C}_{\mathrm{can}}$ by the contraction of the (-2)-curves that do not meet the section $\overline{\{o\}}$.

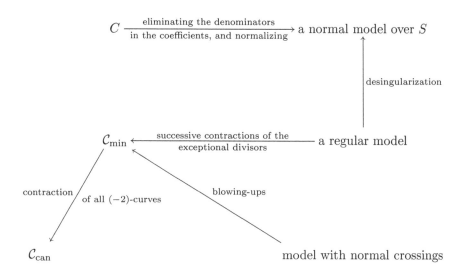

Example 1.12. Let C be the curve over $\mathbb{Q}$ defined by the homogeneous equation

$$x^4 + y^4 + z^4 = 0. \tag{1.1}$$

We are going to determine $\mathcal{C}_{\min}$ and $\mathcal{C}_{\mathrm{can}}$. By taking the same equation over $\mathbb{Z}$, we obtain a projective scheme $\mathcal{C}$ over $\mathbb{Z}$. By the Jacobian criterion, we see that $\mathcal{C}$ is smooth over $\mathbb{Z}$ outside of $p = 2$. Let us show that $\mathcal{C}$ is normal. Let us, for example, consider the affine open subscheme $U = D_+(z)$ of $\mathcal{C}$ defined by the equation $x^4 + y^4 + 1 = 0$. The left-hand side can be written as $F^4 + 2G$ for some polynomials F, G. The singular points of U contained in the fiber U_2 correspond to the zeros of G modulo 2. For the purpose of computations, we should have F and G as simple as possible. Setting $v = y + 1$ and $u = x - v$, the equation becomes

$$u^4 + 2((v^2 - v + 1)^2 + 3v^2u^2 + 2v^3u + 2vu^3) = 0. \qquad (1.2)$$

A reasoning similar to that of Example 1.5 shows that U is normal. More precisely, U_2 is an affine line parameterized by v, of multiplicity 4. The singular point q of U corresponds to the ideal $(2, u, v^2 - v + 1)$. In particular, the point $x = 0, y = z = 1$ of the fiber U_2 is regular in U. By symmetry, we see that the point $x = y = 1, z = 0$ of $\mathcal{C}_2 \setminus U_2$ is regular in $\mathcal{C}$. Let $\widetilde{U} \to U$ be the blowing-up of U with center q. Using Lemma 8.1.4, we can show, in a manner similar to Examples 8.1.5 and 8.3.53, that $\widetilde{U}$ is a desingularization of U. The scheme $\widetilde{U}$ consists of three affine pieces. The first one is $\operatorname{Spec} A_1$, with A_1 the sub-$\mathbb{Z}$-algebra $\mathbb{Z}[u, v, u_1, v_1]$ of $K(\mathcal{C})$ where $u_1 = u/2$, $v_1 = (v^2 - v + 1)/2$. By substituting these relations in (1.2), we obtain

$$2u_1^4 + v_1^2 + 3v^2u_1^2 + v^3u_1 + 4vu_1^3 = 0.$$

Hence modulo 2, we have $v^2 - v + 1 = 0$ and $v_1^2 + 3v^2u_1^2 + v^3u_1 = 0$. Therefore, the fiber of $\operatorname{Spec} A_1$ over 2 is a smooth affine conic over $\mathbb{F}_2[v]/(v^2 - v + 1) = \mathbb{F}_4$. The second affine piece of $\widetilde{U}$ is $\operatorname{Spec} A_2$, with $A_2 = \mathbb{Z}[u, v, t_1, s_1]$, where $t_1 = 2/u$ and $s_1 = (v^2 - v + 1)/u$. We have the relation

$$u + t_1 s_1^2 + 3v^2 t_1 + v^3 t_1^2 + vu^2 t_1^2 = 0.$$

The fiber of $\operatorname{Spec} A_2$ over 2 is the union of an affine line over $\mathbb{F}_4$ of multiplicity 2 (defined by $t_1 = 0$), and a smooth affine conic over $\mathbb{F}_4$ of multiplicity 1, defined by $s_1^2 + 3v^2 + v^3 t_1 = 0$ (note that the image of v in $A_2/2A_2$ is a generator of $\mathbb{F}_4$ over $\mathbb{F}_2$). The function field of this conic is generated by s_1 over $\mathbb{F}_4$. Since $s_1 = v_1/u_1$, we see that this conic will glue with $\operatorname{Spec} A_1/2A_1$ to give a projective smooth conic Γ_1. Furthermore, the two irreducible components of $\operatorname{Spec} A_2$ meet each other in $t_1 = u = 0$, $s_1 = v$, with multiplicity 2 over $\mathbb{F}_4$. We leave the computation of the third affine piece of $\widetilde{U}$ to the reader. Let $\widetilde{\mathcal{C}}$ be the blowing-up of $\mathcal{C}$ with center q. In Figure 19, we represent the fibers over 2 of $\mathcal{C}$, $\widetilde{\mathcal{C}}$, $\mathcal{C}_{\min}$, the minimal model with normal crossings $\mathcal{C}_{\mathrm{nc}}$, and of $\mathcal{C}_{\mathrm{can}}$. We have $\Gamma_0 \simeq \mathbb{P}^1_{\mathbb{F}_2}$, Γ_1 is a smooth conic over $\mathbb{F}_4$, and $\Gamma_2 \simeq \mathbb{P}^1_{\mathbb{F}_4}$. All intersection points are rational over $\mathbb{F}_4$. In $\widetilde{\mathcal{C}}$, we have $\Gamma_0 \cdot \Gamma_2 = 2$, $\Gamma_2 \cdot \Gamma_1 = 4$ (the intersection numbers are computed over $\mathbb{F}_2$; and we must take into account the fact that the intersection points are rational over $\mathbb{F}_4$). Hence $\Gamma_0^2 = -1$ (Proposition 9.1.21). Thus Γ_0 is an exceptional divisor (Theorem 9.3.8), and it is contracted to a point $q_1 \in \mathcal{C}_{\min}$ that

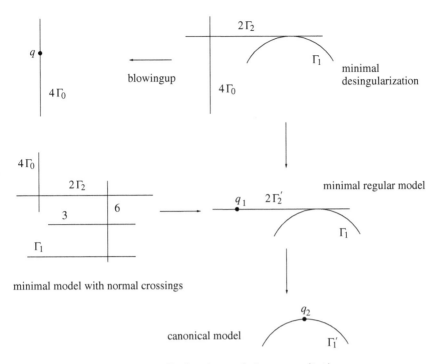

Figure 19. Reductions of the curve (1.1).

is rational over $\mathbb{F}_2$ because Γ_0 contains a rational point over $\mathbb{F}_2$. After contraction of Γ_0, Γ_2' is a singular conic (since it is birational to Γ_2 and contains a rational point over $\mathbb{F}_2$). Therefore, no component is exceptional and we get the minimal regular model $\mathcal{C}_{\min}$. In this model, the adjunction formula (Proposition 9.1.35) implies that $\Gamma_2' \cdot K_{\mathcal{C}_{\min}/\mathbb{Z}} = 0$. Hence the fiber of $\mathcal{C}_{\mathrm{can}}$ over 2 consists of only one irreducible component Γ_1'. Finally, the minimal model with normal crossings is found using the computations of the proof of Theorem 9.2.26.

Remark 1.13. Let $\mathcal{O}_L$ be a discrete valuation ring, étale over $\mathbb{Z}_{2\mathbb{Z}}$, and such that its residue field contains $\mathbb{F}_4$. By Proposition 1.17 below, the minimal regular model (resp. canonical model) of C_L over $\operatorname{Spec} \mathcal{O}_L$ is obtained from $\mathcal{C}_{\min}$ (resp. $\mathcal{C}_{\mathrm{can}}$) by base change. The special fibers of these models are described in Figure 20. On the other hand, we can contract Γ_0 in the special fiber of $\mathcal{C}_{\mathrm{nc}} \times \operatorname{Spec} \mathcal{O}_L$ and we still have a model with normal crossings. Therefore, $\mathcal{C}_{\mathrm{nc}}$ does not commute with étale base change. However, one can contract the exceptional divisor Γ_0 in $\mathcal{C}_{\mathrm{nc}}$ and get a regular model $\mathcal{C}_{\mathrm{nc}}'$. The minimal model of C_L with normal crossings is obtained from $\mathcal{C}_{\mathrm{nc}}'$ by base change.

Example 1.14. Let us consider an example that is very close to the curve (1.1), but whose models are different. Let C be the projective curve over $\mathbb{Q}$ defined by the equation

$$x^4 + y^4 - z^4 = 0. \qquad (1.3)$$

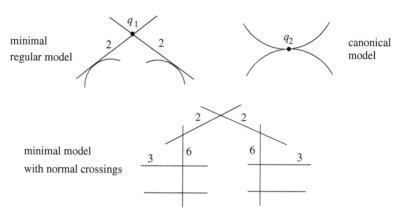

regular model

canonical
model

minimal model
with normal crossings

Figure 20. Reductions of the curve (1.1) after étale extension.

Let $\mathcal{C}$ be the projective scheme over $\mathbb{Z}$ defined by the same equation. Then $\mathcal{C}$ is a model of C over $\mathbb{Z}$, smooth outside of $p = 2$. Let us consider the affine open subscheme $U = D_+(z)$ of $\mathcal{C}$; it has equation $x^4 + y^4 - 1 = 0$. Let us set $u = x + y - 1$ and $v = y - 1$. Then we have

$$u^4 - 4vu^3 + 6v^2u^2 - 4v^3u + 2v(v+1)(v^2 + v + 2) = 0.$$

The singular points of $\mathcal{C}$ contained in U_2 are therefore q_1, corresponding to $p = 2$, $(u, v) = (0, 0)$, and q_2, corresponding to $p = 2$, $(u, v) = (0, 1)$. We verify that the point of $\mathcal{C}_2 \setminus U_2$ is regular in $\mathcal{C}$. Hence $\mathcal{C}$ is normal, with fiber over 2 irreducible of multiplicity 4 because $u^4 = 2F$ with $F \in \mathcal{O}_\mathcal{C}(U)$ and not identically zero on U_2. A direct computation shows that the blowing-up of U with center q_1 is a resolution of singularities of U above q_1. Now we have an automorphism $\sigma : x \mapsto y$, $\sigma : y \mapsto x$ of U which sends q_1 to q_2, and therefore we know that the blowing-up of U with center q_2 will have the same appearance as the blowing-up with center q_1. In Figure 21, we represent the fibers over 2 of the minimal desingularization of $\mathcal{C}$, of the minimal regular model, of the canonical model, and of the minimal regular model with normal crossings. All of the irreducible components are isomorphic to $\mathbb{P}^1_{\mathbb{F}_2}$. The integers in the figure denote the multiplicity of the components. The non-marked components are of multiplicity 1.

Remark 1.15. In the examples above, we see that given a curve C, it is not always easy to determine its minimal regular model explicitly. For examples of the determination of the minimal regular model and of applications, see [63] for the curve $x^p + y^p = z^p$, where p is a prime number. The model is computed over the ring of integers of $\mathbb{Q}(\xi_p)$, where ξ_p is a primitive pth root of unity.

Proposition 1.16. *Let S be a Dedekind scheme of dimension 1. Let C be a smooth projective curve over $K = K(S)$ of genus $g \geq 1$ (resp. $g \geq 2$) admitting a minimal regular model $\mathcal{C}_{\min}$ (resp. a canonical model $\mathcal{C}_{\can}$) over S. Then any automorphism of C extends in a unique way to an isomorphism σ of $\mathcal{C}_{\min}$ (resp. of $\mathcal{C}_{\can}$).*

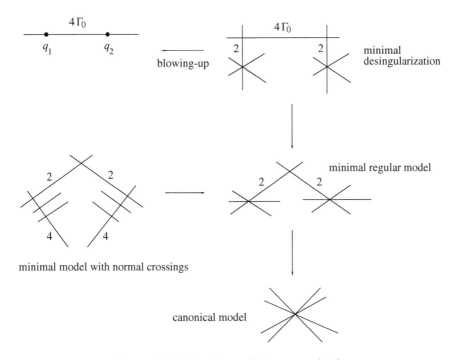

Figure 21. Reductions of the curve (1.3).

Proof   For the minimal regular model, it is a direct translation of Proposition 9.3.13. Let Γ be a vertical prime divisor of $\mathcal{C}_{\min}$ such that $K_{\mathcal{C}_{\min}/S} \cdot \Gamma = 0$. As $\sigma(\Gamma)^2 = \Gamma^2$ (Theorem 9.2.12(c)), Proposition 9.4.8 implies that $K_{\mathcal{C}_{\min}/S} \cdot \sigma(\Gamma) = 0$ (we can use the fact that $\sigma^* K_{\mathcal{C}_{\min}/S} \sim K_{\mathcal{C}_{\min}/S}$ by the uniqueness of the dualizing sheaf). Hence σ acts on the set of divisors contracted by the morphism $\mathcal{C}_{\min} \to \mathcal{C}_{\mathrm{can}}$. By the uniqueness of the contraction (Proposition 8.3.28), σ induces an automorphism of $\mathcal{C}_{\mathrm{can}}$. $\square$

Proposition 1.17. *Let C, S be as in Proposition 1.16. Let S' be a Dedekind scheme of dimension 1 that is étale over S or equal to $\operatorname{Spec} \widehat{\mathcal{O}}_{S,s}$ for a closed point $s \in S$. Let $K' = K(S')$. Let us suppose C is of genus $g \geq 1$ (resp. $g \geq 2$) and admitting a minimal regular (resp. canonical) model $\mathcal{C}$ over S. Then $\mathcal{C} \times_S S'$ is the minimal regular (resp. canonical) model of $C_{K'}$ over S'.*

Proof   If $\mathcal{C}$ is the minimal regular model, then $\mathcal{C} \times_S S'$ is minimal by Proposition 9.3.28 (the $S' \to S$ surjective hypothesis is only used in that proposition for the validity of the converse). Let us now consider the canonical model. We can suppose S and S' are affine. By construction of the canonical model, the minimal desingularization of $\mathcal{C}$ is the minimal regular model $\mathcal{C}_{\min}$. Let $p : \mathcal{C}'_{\min} := \mathcal{C}_{\min} \times_S S' \to \mathcal{C}_{\min}$ be the projection morphism. Then

$$H^0(\mathcal{C}'_{\min}, \omega^{\otimes m}_{\mathcal{C}'_{\min}/S'}) = H^0(\mathcal{C}_{\min}, \omega^{\otimes m}_{\mathcal{C}_{\min}/S}) \otimes_{\mathcal{O}(S)} \mathcal{O}(S')$$

for every $m \in \mathbb{Z}$ because $\mathcal{O}(S')$ is flat over $\mathcal{O}(S)$ and $\omega_{\mathcal{C}'_{\min}/S'} = p^*\omega_{\mathcal{C}_{\min}/S}$. The fact that $\mathcal{C}_{\mathrm{can}} \times_S S'$ is the canonical model of $C_{K'}$ then results immediately from Proposition 9.4.20(d). $\qquad\square$

10.1.2   Reduction

In all of this section, S will be a Dedekind scheme of dimension 1, and we will set $K = K(S)$.

Definition 1.18. Let C be a normal projective curve over K. Let us fix a closed point $s \in S$. We call the fiber $\mathcal{C}_s$ of a model $\mathcal{C}$ of C a *reduction of C at s*. If S is the spectrum of a Dedekind ring A, and if $\mathfrak{p}$ is the maximal ideal of A corresponding to s, we also call $\mathcal{C}_s$ a *reduction of C modulo $\mathfrak{p}$*. It is clear that the notion of reduction depends strongly on the choice of a model $\mathcal{C}$.

Definition 1.19. Let C be as above. We will say that C has *good reduction at $s \in S$* if it admits a smooth model over $\operatorname{Spec} \mathcal{O}_{S,s}$. This implies that C is smooth over K. If C does not have good reduction at s, we will say that C has *bad reduction at s*. Note that if C has a non-smooth model over $\operatorname{Spec} \mathcal{O}_{S,s}$, this does not necessarily imply that C has bad reduction at s. We will say that C has *good reduction over S* if it has good reduction at every $s \in S$.

Example 1.20. Let p be a prime number $\neq 3$. Then the curve

$$C = \operatorname{Proj} \mathbb{Q}[x, y, z]/(x^3 + y^3 + p^3 z^3)$$

admits an evident model $\mathcal{C}$ over $\mathbb{Z}$ by taking the same equation over $\mathbb{Z}$. The reduction $\mathcal{C}_p$ is a singular curve. Meanwhile, $\operatorname{Proj} \mathbb{Z}[x, y, w]/(x^3 + y^3 + w^3)$, where $w = pz$, is also a model of C over $\mathbb{Z}$, but is smooth over p. Hence C has good reduction at p. In this somewhat artificial example, the good reduction of C is easy to reveal. In the general case, it is sometimes necessary to know the minimal regular model.

Proposition 1.21. *Let S be a Dedekind scheme of dimension 1, C a smooth projective curve over $K = K(S)$ of genus $g \geq 1$.*

(a) *The curve C has good reduction at $s \in S$ except perhaps for a finite number of s.*

(b) *Let us suppose S is affine. Then C has good reduction over S if and only if the minimal regular model $\mathcal{C}_{\min}$ of C over S is smooth. Moreover, this implies that $\mathcal{C}_{\min}$ is the unique smooth model of C over S.*

(c) *('Étale' base change) Let $S' \to S$ be as in Proposition 1.17. Let $s' \in S'$ and let s be its image in S. Then $C_{K'}$ has good reduction at s' if and only if C has good reduction at s.*

Proof (a) The curve $C \to \operatorname{Spec} K$ extends to a projective scheme $\mathcal{C} \to U$ over a non-empty open subscheme U of S (see the beginning of Example 1.4). By taking the irreducible component of $\mathcal{C}$ containing C and endowing it with the reduced scheme structure, we can suppose that $\mathcal{C}$ is integral, and hence flat over U. By

Proposition 8.3.11, there exists a non-empty open subscheme V of U such that $\mathcal{C}_V \to V$ is smooth. Hence C has good reduction over V. As S has dimension 1, $S \setminus V$ is finite, whence (a).

(b) Let us suppose that C has good reduction over S. Let $\mathcal{C}_{\min}$ be the minimal regular model of C over S (which exists by Proposition 1.8). Let $s \in S$. Let $\mathcal{X} \to \operatorname{Spec} \mathcal{O}_{S,s}$ be a smooth model of C. Then $\mathcal{C}_{\min} \times_S \operatorname{Spec} \mathcal{O}_{S,s} \simeq \mathcal{X}$ (Example 9.3.15), which implies that $\mathcal{C}_{\min}$ is smooth. The converse is trivial. Any smooth model of C over S is relatively minimal (Example 9.3.15), and hence isomorphic to $\mathcal{C}_{\min}$. Finally, (c) results from (b) and from Proposition 9.3.28.   □

Example 1.22. The curves of Examples 1.12 and 1.14 have good reduction at every $p \neq 2$, and have bad reduction at $p = 2$. This results from Proposition 1.21(b) and from the knowledge of the minimal regular models.

Corollary 1.23. *Let E be an elliptic curve over $K = K(S)$. Let $s \in S$, let W be the minimal Weierstrass model of E over $\operatorname{Spec} \mathcal{O}_{S,s}$, and Δ the discriminant of W (Subsection 9.4.4). Then the following properties are equivalent:*

(i) *E has good reduction at s;*

(ii) *W_s is smooth over $k(s)$;*

(iii) *$\Delta \in \mathcal{O}_{S,s}^*$.*

Proof Let us suppose that E has good reduction. Let $\mathcal{X}$ be the minimal regular model of E over $\operatorname{Spec} \mathcal{O}_{S,s}$. Then $\mathcal{X}_s$ is smooth by Proposition 1.21(b). Now Theorem 9.4.35(a) implies that $\mathcal{X} = W$, whence W_s is smooth. The converse is true by definition. Finally, the equivalence (ii) $\Longleftrightarrow$ (iii) is a property of the discriminant (cf. [91], Section III.1).   □

Lemma 1.24. *Let C be a smooth, connected, projective curve over K, of arithmetic genus $p_a(C) \geq 2$. Let $\mathcal{C}$ be a model of C over S. Let us suppose that $\mathcal{C}_s$ contains an irreducible component Γ whose normalization Γ' is smooth of arithmetic genus $p_a(\Gamma') \geq p_a(C)$. Then C has good reduction at s. Moreover, the reduction at s of the smooth model of C over $\operatorname{Spec} \mathcal{O}_{S,s}$ is isomorphic to Γ'.*

Proof We can suppose S is local. By desingularization (Corollary 8.3.51), and by replacing Γ by its strict transform (which does not modify its normalization), we can reduce to $\mathcal{C}$ regular. Let $f : \mathcal{C} \to \mathcal{X} := \mathcal{C}_{\min}$ be the canonical morphism to the minimal regular model of C. As Γ is not a rational curve, it follows from Castelnuovo's criterion (Theorem 9.3.8) that Γ cannot be contracted to a point in $\mathcal{X}$. In other words, $f(\Gamma)$ is a curve Γ_0 on $\mathcal{X}$ that is birational to Γ. Let $\pi : \Gamma' \to \Gamma_0$ be the normalization morphism. The exact sequence

$$0 \to \mathcal{O}_{\Gamma_0} \to \pi_* \mathcal{O}_{\Gamma'} \to \mathcal{S} \to 0$$

with a skyscraper sheaf $\mathcal{S}$ implies that

$$p_a(\Gamma_0) = p_a(\Gamma') + \dim_{k(s)} H^0(\Gamma_0, \mathcal{S}) \geq p_a(\Gamma') \geq p_a(C) \qquad (1.4)$$

(see the proof of Proposition 7.5.4). Let $\Gamma_1, \ldots, \Gamma_r$ be the irreducible components of $\mathcal{X}_s$ other than Γ_0. We have $K_{\mathcal{X}/S} \cdot \Gamma_i \geq 0$ for every $i \geq 0$ (Proposition 9.3.10).

Using the adjunction formulas (Proposition 9.1.35 and Theorem 9.1.37), we obtain

$$2p_a(C) - 2 = d_0(2p_a(\Gamma_0) - 2 - \Gamma_0^2) + \sum_{i \geq 1} d_i K_{\mathcal{X}/S} \cdot \Gamma_i,$$

where d_i is the multiplicity of Γ_i in $\mathcal{X}_s$. Taking into account relation (1.4) and the fact that $\Gamma_0^2 \leq 0$ (Theorem 9.1.23), this implies that $K_{\mathcal{X}/S} \cdot \Gamma_i = 0$ if $i \geq 1$, and

$$\Gamma_0^2 = 0, \quad d_0 = 1, \quad p_a(\Gamma_0) = p_a(C), \quad H^0(\Gamma_0, \mathcal{S}) = 0.$$

Hence $\Gamma' \to \Gamma_0$ is an isomorphism and Γ_0 is smooth over $k(s)$. Let us show that $\mathcal{X}_s$ is connected. Let us suppose the contrary. Let F be a connected component of $\mathcal{X}_s$ that does not contain Γ_0. Then $K_{\mathcal{X}/S} \cdot \Gamma_i = 0$ for every $\Gamma_i \subseteq F$. A reasoning similar to that of Proposition 9.4.8 then implies $p_a(C) = p_a(\mathcal{X}_\eta) = 1$, which is contrary to our hypothesis. Hence $\mathcal{X}_s$ is connected and $\mathcal{X}_s = \Gamma_0$ (Theorem 9.1.23) is smooth. Consequently, C has good reduction at s.   □

Corollary 1.25. *Let C be a smooth, connected, projective curve over K of genus $g \geq 2$, $s \in S$ a closed point. Let W^0 be a quasi-projective scheme over S, such that W_η^0 is an open subscheme of C, that W_s^0 is smooth over $k(s)$, and that $K(W_s^0)$ is the function field of a smooth projective curve of genus g. Then C has good reduction at s and, if necessary replacing S by an open neighborhood of s, W^0 is an open subscheme of the smooth model of C over S.*

Proof We can suppose S is local. By hypothesis, W^0 is an open subscheme of a projective scheme $\mathcal{C}$ over S. We can suppose $\mathcal{C}$ is integral. As W^0 is smooth over S, it is contained in the open subset of regular points of $\mathcal{C}$. By Corollary 8.3.51, $\mathcal{C}$ admits a desingularization in the strong sense. We can therefore suppose $\mathcal{C}$ is regular. Let $\mathcal{C} \to \mathcal{C}_{\min}$ be the canonical morphism to the minimal regular model. Then $\mathcal{C}_{\min}$ is smooth over S by Lemma 1.24. Moreover, the reasoning used in the proof of that lemma shows that $W^0 \to \mathcal{C}_{\min}$ is quasi-finite and birational; it is therefore an open immersion (Corollary 4.4.8).   □

Example 1.26. Let $S = \operatorname{Spec} \mathcal{O}_K$, where $\mathcal{O}_K$ is a discrete valuation ring, with field of fractions K and residue field k of $\operatorname{char}(k) \neq 2$ (see Exercise 1.9 for the general case). Let C be the hyperelliptic curve of genus $g \geq 1$ over K (see Subsection 7.4.3) defined by an affine equation

$$y^2 = P(x), \quad P(x) \in K[x],$$

with $P(x)$ separable. Then C has good reduction if and only if after a suitable change of variables (Corollary 7.4.33) we can define C by an equation as above with $P(x) \in \mathcal{O}_K[x]$ such that its image in $k[x]$ is separable of degree $2g + 1$ or $2g + 2$. Indeed, it is a sufficient condition by considering the affine scheme

$$W^0 = \operatorname{Spec} \mathcal{O}_K[x, y]/(y^2 - P(x))$$

and applying Corollary 1.25. Let us show that it is also a necessary condition.

Let us suppose that C has good reduction; hence $\mathcal{C}_{\min}$ is smooth over S. Let σ be the hyperelliptic involution on C. It acts on $\mathcal{C}_{\min}$ (Proposition 1.16) and the quotient $\mathcal{D} := \mathcal{C}_{\min}/\langle\sigma\rangle$ is a model over S of $D := C/\langle\sigma\rangle \simeq \mathbb{P}^1_K$. As we have a dominant morphism $\mathcal{C}_{\min} \to \mathcal{D}$, the special fiber $\mathcal{D}_s$ is necessarily geometrically integral because $(\mathcal{C}_{\min})_s$ is. It follows that $\mathcal{D} \simeq \mathbb{P}^1_S$ (Exercise 8.3.5). Let $V^0 = \operatorname{Spec} \mathcal{O}_K[v]$ be an open subscheme of $\mathcal{D}$ isomorphic to $\mathbb{A}^1_S$ and W^0 its inverse image in $\mathcal{C}_{\min}$. Then there exists a $z \in \mathcal{O}_{\mathcal{C}_{\min}}(W^0)$ such that

$$\mathcal{O}_{\mathcal{C}_{\min}}(W^0) = \mathcal{O}_K[v] \oplus z\mathcal{O}_K[v]$$

(Exercise 8.2.16). Hence there exist $F(v), G(v) \in \mathcal{O}_K[v]$ such that $z^2 + G(v)z = F(v)$. As 2 is invertible in $\mathcal{O}_K$ by hypothesis, we can replace z by $z + G(v)/2$ and reduce to $G = 0$. We have a surjective homomorphism

$$\mathcal{O}_K[v, Z]/(Z^2 - F(v)) \to \mathcal{O}_{\mathcal{C}_{\min}}(W^0), \quad Z \mapsto z.$$

Since the left-hand term is clearly integral of dimension 2, it is an isomorphism. The W^0_s smooth hypothesis implies that the image $\bar{F}(v)$ of $F(v)$ in $k[v]$ is a separable polynomial. The reasoning used before Proposition 7.4.24 implies that $\bar{F}(v)$ is of degree $2g+1$ or $2g+2$. Finally, Corollary 7.4.33 shows that the equation $z^2 = F(v)$ can be obtained from $y^2 = P(x)$ by a suitable change of variables.

By Proposition 1.21(c), a curve that has bad reduction will have bad reduction after any étale base change. However, if we admit ramified base change, the situation is different.

Definition 1.27. Let C be a smooth projective curve over $K = K(S)$. We will say that C has *potential good reduction at* $s \in S$ if there exist a morphism $S' \to S$ from a Dedekind scheme S' to S and a point $s' \in S'$ lying above s such that $C_{K(S')}$ has good reduction at s'. If C has good reduction at s, then it has potential good reduction at s.

Example 1.28. Let us consider the hyperelliptic curve C over $\mathbb{Q}_2$ defined by the equation

$$y^2 = x^4 - 1.$$

We are going to show that C has potential good reduction and give an explicit smooth model over an explicit extension of $\mathbb{Q}$. We leave it to the reader to verify that C has bad reduction over $\mathbb{Z}_2$. Let us set

$$x = 1 + x_1^{-1}, \quad y = y_1 x_1^{-2}.$$

Then we have $y_1^2 = 4x_1^3 + 6x_1^2 + 4x_1 + 1$. Let us set

$$x_1 = v + \alpha, \quad y_1 = 2z + (\beta v + \gamma)$$

with scalars $\alpha, \beta, \gamma \in \overline{\mathbb{Q}}_2$ that we are going to choose so that

$$z^2 + (\beta v + \gamma)z = v^3.$$

For this it suffices to solve the system

$$\begin{cases} 4(1 + 4\alpha + 6\alpha^2 + 4\alpha^3)(6 + 12\alpha) = (4 + 12\alpha + 12\alpha^2)^2 \\ \gamma^2 = 1 + 4\alpha + 6\alpha^2 + 4\alpha^3 \\ \beta^2 = 6 + 12\alpha. \end{cases}$$

A direct computation shows that we have a solution in $\overline{\mathbb{Q}}_2$ with $|\alpha| = |2|^{-1/4}$, $|\beta| = |2|^{1/2}$, and $|\gamma - 1| = |2|^{1/4}$ (see formula (1.6) for a reminder of the definition of $|.|$). Let $L = \mathbb{Q}_2(\alpha, \beta, \gamma)$, and let $W^0 = \operatorname{Spec} \mathcal{O}_L[v, z]/(z^2 + (\beta v + \gamma)z - v^3)$. Then W^0 is smooth over $\mathcal{O}_L$, and its special fiber is an open subscheme of an elliptic curve. Hence C_L has good reduction.

Let us conclude with a sufficient condition for bad reduction.

Proposition 1.29. *Let C be a smooth, geometrically connected, projective curve over $K = K(S)$. Let us fix $s \in S$. Let W^0 be a quasi-projective scheme over S such that W_η^0 is an open subscheme of C, that W_s^0 is integral, and that $K(W_s^0)$ is the function field of a normal curve Γ of arithmetic genus $1 \le p_a(\Gamma) < p_a(C)$. Then C has bad reduction. If, moreover, $\Gamma \times_{\operatorname{Spec} k(s)} \operatorname{Spec} \overline{k(s)}$ is integral and non-rational, then C does not have potential good reduction.*

Proof We can suppose S is local. We know by Lemma 4.1.18 that W^0 is a normal scheme. The set of regular points $\operatorname{Reg}(W^0)$ is an open subset of W^0 (Corollary 8.2.38) that contains the generic point of W^0. Hence, if necessary restricting W_s^0, we can suppose W_s^0 and W^0 regular. As in the proof of Corollary 1.25, W^0 is then isomorphic to an open subscheme of a regular model $\mathcal{C}$ of C over S. Let Γ be the closure of W_s^0 in $\mathcal{C}_s$, and $f : \mathcal{C} \to \mathcal{C}_{\min}$ the canonical morphism to the minimal regular model of C. By Castelnuovo's criterion, Γ is not an exceptional divisor, and hence $\Gamma \to f(\Gamma)$ is a birational morphism. If C has good reduction, then $\mathcal{C}_{\min}$ is smooth (Proposition 1.21(b)) and its special fiber is smooth and connected (Corollary 8.3.6(b)). Hence $\Gamma \to f(\Gamma) = (\mathcal{C}_{\min})_s$ is an isomorphism, which is contrary to the $p_a(\Gamma) < p_a(C)$ hypothesis. Hence C does not have good reduction.

Let us now suppose $\Gamma \times \operatorname{Spec} \overline{k(s)}$ is integral and non-rational. Let S' be a Dedekind scheme of dimension 1 that dominates S, and $s' \in S'$ a closed point lying above s. Then $K(W_s^0 \times \operatorname{Spec} k(s'))$ is the function field of the normalization Γ' of $\Gamma \times \operatorname{Spec} k(s')$, with

$$1 \le p_a(\Gamma') \le p_a(\Gamma \times \operatorname{Spec} k(s')) = p_a(\Gamma) < p_a(C) = p_a(C_{K(S')}),$$

where the first inequality comes from the $\Gamma \times \operatorname{Spec} \overline{k(s)}$ non-rational hypothesis. The preceding reasoning shows that $C_{K(S')}$ has bad reduction at s'. Hence C does not have potential good reduction. $\qquad\square$

Example 1.30. Let $\mathcal{O}_K$ be a discrete valuation ring with uniformizing parameter t and residue field k of characteristic $\operatorname{char}(k) \neq 2$. Let $g \ge 2$. Let us consider the hyperelliptic curve C of genus g over K given by the equation

$$y^2 = (x^3 + ax + b)(x^{2g-1} + tcx + td), \quad a, b, c, d \in \mathcal{O}_K,$$

with $(4a^3 + 27b^2)b \in \mathcal{O}_K^*$. Let

$$W^0 = \operatorname{Spec} \mathcal{O}_K[x, y, 1/x]/(y^2 - (x^3 + ax + b)(x^{2g-1} + tcx + td)).$$

Then $W_k^0 = \operatorname{Spec} k[x, z, 1/x]/(z^2 - (x^3 + \tilde{a}x + \tilde{b})x)$ with $z = y/x^{g-1}$ and where $\tilde{a}, \tilde{b}$ are the respective images of a, b in k. It is easy to see that W_k^0 is an open subscheme of a projective hyperelliptic curve of genus 1 over k. Hence C has bad reduction, and this over every extension of $\mathcal{O}_K$.

10.1.3 Reduction map

This subsection is not specific to curves. Let S be a Dedekind scheme of dimension 1. Let $\mathcal{X} \to S$ be a surjective proper morphism, with generic fiber X, and $s \in S$ a closed point. We would like to define a map $X \to \mathcal{X}_s$. Let us suppose that $S = \operatorname{Spec} \mathcal{O}_{S,s}$ and that $\mathcal{O}_{S,s}$ is Henselian (e.g., complete). For any closed point $x \in X$, the Zariski closure $\overline{\{x\}}$ is an irreducible finite scheme over S; it is therefore a local scheme. Its closed point is the point of $\overline{\{x\}} \cap \mathcal{X}_s$.

Definition 1.31. Let S be the spectrum of a Henselian discrete valuation ring $\mathcal{O}_K$. Let $\mathcal{X}$ be a proper scheme over S, with generic fiber X. Let X^0 denote the set of closed points of X. The map $r : X^0 \to \mathcal{X}_s$, which to every closed point $x \in X$ associates the point $\overline{\{x\}} \cap \mathcal{X}_s$, is called the *reduction map of X*. We also say that x *reduces* or *specializes* to $r(x)$. For fixed X, the map r depends, of course, on the choice of $\mathcal{X}$. We will denote this map by $r_{\mathcal{X}}$ if we want to specify the choice of an $\mathcal{X}$.

Let $\mathcal{O}_K$ be a Henselian discrete valuation ring, with field of fractions K and residue field k. Then the integral closure $\mathcal{O}_{\overline{K}}$ of $\mathcal{O}_K$ in the algebraic closure $\overline{K}$ is a valuation ring that dominates $\mathcal{O}_K$, and that has residue field $\overline{k}$. There exists a valuation $\nu_{\overline{K}} : \overline{K} \to \mathbb{Q} \cup \{\infty\}$ that extends that of K, defined in the following manner: if $\alpha \in \overline{K}$, we set

$$\nu_{\overline{K}}(\alpha) = [K(\alpha) : K]^{-1} \nu_K(N_{K[\alpha]/K}(\alpha)) \tag{1.5}$$

(see [55], Section XII.2). It is sometimes more convenient to use the *absolute value* instead of the valuation. Let us fix a real number $0 < \theta < 1$. For any $\alpha \in \overline{K}$, we set

$$|\alpha| = \theta^{\nu_{\overline{K}}(\alpha)}. \tag{1.6}$$

The choice of θ is not canonical. In general, if $\mathcal{O}_K$ dominates $\mathbb{Z}_p$, we take $\theta = p^{-\nu_K(p)}$. For any $a \in \mathcal{O}_{\overline{K}}$, let $\tilde{a}$ denote its image in the residue field $\overline{k}$.

Lemma 1.32. *Let $\mathcal{O}_K$ be a Henselian discrete valuation ring. Let x be a closed point of the generic fiber of $\mathcal{X} = \operatorname{Proj} \mathcal{O}_K[T_0, \ldots, T_n]$, with homogeneous coordinates $(x_0, \ldots, x_n) \in \mathcal{O}_{\overline{K}}^{n+1}$ with at least one $x_{i_0} \in \mathcal{O}_{\overline{K}}^*$. Then $r_{\mathcal{X}}(x) = (\tilde{x}_0, \ldots, \tilde{x}_n)$.*

Proof We can suppose $i_0 = 0$ and $x_0 = 1$. Then $\tilde{x} := r_{\mathcal{X}}(x) \in D_+(T_0)$. Let Z be the reduced closed subscheme of $\mathcal{X}$ with support $\overline{\{x\}}$. Then $D_+(T_0)$ is an open neighborhood of $\tilde{x}$ and therefore contains the generic point x of Z. Let t_i be the image of T_i/T_0 in $\mathcal{O}_{Z,\tilde{x}}$. Then x_i is the image of t_i in $k(x)$, and the homogeneous coordinates x'_i of $\tilde{x}$ are the images of the t_i in $k(\tilde{x})$. As Z is finite over $\mathcal{O}_K$ (because it is quasi-finite and proper over $\mathcal{O}_K$), we have $\mathcal{O}_{Z,\tilde{x}}$ finite over $\mathcal{O}_K$. Hence $\mathcal{O}_{\overline{K}}$ is a local ring that dominates $\mathcal{O}_{Z,\tilde{x}}$. This implies that $x'_i = \tilde{x}_i$. We can sum up the situation in the following commutative diagram:

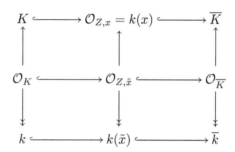

Remark **1.33.** Let x be a closed point of $\operatorname{Proj} K[T_0, \ldots, T_n]$. As $\mathcal{O}_{\overline{K}}$ is a valuation ring, we can always find homogeneous coordinates $(x_0, \ldots, x_n)$ for x so that $x_i \in \mathcal{O}_{\overline{K}}$ and that at least one of them is invertible in $\mathcal{O}_{\overline{K}}$.

Corollary 1.34. *Let C be a normal projective curve over the field of fractions of a Henselian discrete valuation ring $\mathcal{O}_K$, and let $\mathcal{C} \subseteq \operatorname{Proj} \mathcal{O}_K[T_0, \ldots, T_n]$ be a model of C. Let $x = (x_0, \ldots, x_n)$ be a closed point of C, with homogeneous coordinates $x_i \in \mathcal{O}_{\overline{K}}$ of which at least one belongs to $\mathcal{O}^*_{\overline{K}}$. Then $r(x) = (\tilde{x}_0, \ldots, \tilde{x}_n)$.*

Proof Indeed, when we have fixed a closed immersion of $\mathcal{C}$ in a larger proper scheme $\mathcal{X}$, it immediately follows from the definition that $r_{\mathcal{C}}(x) = r_{\mathcal{X}}(x)$. It now suffices to apply the preceding lemma. Let us note that the proof of the corollary is valid in arbitrary dimension.   □

Remark **1.35.** If $\mathcal{O}_K$ is not Henselian, we can define the reduction of the rational points $x \in X(K)$ because $\overline{\{x\}} \cap \mathcal{X}_s$ is reduced to one point. The results above remain true when replacing $\mathcal{O}_{\overline{K}}$ by $\mathcal{O}_K$ itself.

Proposition 1.36. *Let S be a Dedekind scheme of dimension 1, $\mathcal{X} \to S$ a dominant morphism of finite type with $\mathcal{X}$ irreducible. Let $\tilde{x} \in \mathcal{X}_s$ be a closed point of a closed fiber. Then there exists a closed point x of $\mathcal{X}_\eta$ such that $\tilde{x} \in \overline{\{x\}}$.*

Proof As the question is of topological nature, we can suppose $\mathcal{X}$ is integral. Hence $\mathcal{X}$ is flat over S. We are going to use induction on $d = \dim \mathcal{O}_{\mathcal{X},\tilde{x}}$. If $\dim \mathcal{O}_{\mathcal{X},\tilde{x}} = 1$, then $\dim \mathcal{O}_{\mathcal{X}_s,\tilde{x}} = 0$ (Theorem 4.3.12), and therefore $\dim \mathcal{X}_\eta = 0$ (Lemma 4.4.13). This implies that $\mathcal{X}_\eta$ is reduced to a point x. We necessarily have $\tilde{x} \in \overline{\{x\}}$ because x is the generic point of $\mathcal{X}$.

Let us suppose that $d \geq 2$. Similarly to the proof of Lemma 8.3.35(a), we construct an integral closed subscheme $\mathcal{Y}$ of $\mathcal{X}$ such that $\tilde{x} \in \mathcal{Y}$, $\mathcal{Y}$ is not contained in $\mathcal{X}_s$, and $\mathcal{Y}_s$ does not contain any irreducible component of $\mathcal{X}_s$. We then have $\mathcal{Y}$ flat of finite type over S and

$$\dim \mathcal{O}_{\mathcal{Y},\tilde{x}} = \dim \mathcal{O}_{\mathcal{Y}_s,\tilde{x}} + 1 < \dim \mathcal{O}_{\mathcal{X}_s,\tilde{x}} + 1 = \dim \mathcal{O}_{\mathcal{X},\tilde{x}}.$$

By the induction hypothesis, there exists a closed point x of $\mathcal{Y}_\eta$ such that $\tilde{x} \in \overline{\{x\}}$. As $\mathcal{Y}$ is closed in $\mathcal{X}$, the point x verifies the required property. □

Remark 1.37. See [41], IV.14.5.3 for a more general statement.

Corollary 1.38. *Let S be the spectrum of a Henselian discrete valuation ring $\mathcal{O}_K$. Let $\mathcal{X}$ be an irreducible scheme, proper and dominant over S, with generic fiber X. Then the reduction map $r_\mathcal{X} : X^0 \to \mathcal{X}_s$ is surjective onto the set of closed points.*

Definition 1.39. Let S be the spectrum of a complete discrete valuation ring. Let $\mathcal{X}$ be a scheme that is proper and surjective over S with generic fiber X, and $\tilde{x} \in \mathcal{X}_s$ a closed point. We will call the set $r_\mathcal{X}^{-1}(\tilde{x}) \subseteq X^0$ the *formal fiber* of $\mathcal{X}$ over $\tilde{x}$. We denote this set by $X_+(\tilde{x})$ (dropping the reference to x in the notation even though it depends on the model $\mathcal{X}$).

Let L be an extension of K. The set of morphisms $f \in X(L)$ with image in $X_+(\tilde{x})$ will be denoted $X_+(\tilde{x})(L)$. See Proposition 3.2.18(b). Let $\mathcal{O}_L$ denote the integral closure of $\mathcal{O}_K$ in L. If L is finite over K, then $\mathcal{O}_L$ is a discrete valuation ring that is finite over $\mathcal{O}_K$ (Proposition 8.2.29(a) or, more elementarily, [55], Section XII.6), and hence also complete (Corollary 1.3.14).

Proposition 1.40. *Let S be the spectrum of a complete discrete valuation ring $\mathcal{O}_K$. Let $\mathcal{X} \to S$ be a surjective proper morphism, and $\tilde{x} \in \mathcal{X}_s$ a closed point. Let L be a finite extension of K. Then the following properties are true.*

(a) *There exist canonical bijections*

$$X_+(\tilde{x})(L) \simeq \mathrm{Mor}_S(\mathrm{Spec}\,\mathcal{O}_L, \mathrm{Spec}\,\mathcal{O}_{\mathcal{X},\tilde{x}}) \simeq \mathrm{Mor}_S(\mathrm{Spec}\,\mathcal{O}_L, \mathrm{Spec}\,\widehat{\mathcal{O}}_{\mathcal{X},\tilde{x}}).$$

(b) *Let $f_1,\dots,f_q \in \mathcal{O}_{\mathcal{X},\tilde{x}}$ be such that their images in $\mathcal{O}_{\mathcal{X}_s,\tilde{x}}$ generate its maximal ideal. Let us suppose that $\tilde{x}$ is rational over $k(s)$. Then the homomorphism*

$$\varphi : \mathcal{O}_K[[T_1,\dots,T_q]] \to \widehat{\mathcal{O}}_{\mathcal{X},\tilde{x}}, \quad T_i \mapsto f_i$$

is surjective and induces a bijection

$$X_+(\tilde{x})(L) \simeq \{(t_1,\dots,t_q) \in \mathfrak{m}_L^q \mid F(t_1,\dots,t_q) = 0, \forall F \in \mathrm{Ker}\,\varphi\}, \quad (1.7)$$

where $\mathfrak{m}_L$ is the maximal ideal of $\mathcal{O}_L$, and $\mathfrak{m}_L^q = \mathfrak{m}_L \times \cdots \times \mathfrak{m}_L$ (q times). In particular, if $\mathcal{X} \to S$ is smooth at $\tilde{x}$ and if $q = \dim_{\tilde{x}} \mathcal{X}_s$, then φ induces a bijection

$$X_+(\tilde{x})(L) \simeq \mathfrak{m}_L^q. \quad (1.8)$$

(c) Let U be an affine open neighborhood of $\tilde{x}$ in $\mathcal{X}$. Let us suppose that U is a closed subscheme of $\mathbb{A}^n_{\mathcal{O}_K}$ and that $\tilde{x}$ corresponds to the origin of $\mathbb{A}^n_{k(s)}$. Then we have a bijection

$$X_+(\tilde{x})(L) \simeq U(L) \cap \mathfrak{m}^n_L, \tag{1.9}$$

where the second member is considered as a subset of $\mathbb{A}^n_{\mathcal{O}_K}(L) = L^n$.

Proof (a) Let $f : \operatorname{Spec} L \to X$ be an element of $X_+(\tilde{x})(L)$, $x \in X_+(\tilde{x})$ the image of the point of $\operatorname{Spec} L$ under f, and $Z = \overline{\{x\}}$ the Zariski closure in $\mathcal{X}$ endowed with the structure of reduced (hence integral) closed subscheme. As Z is local, we have $Z = \operatorname{Spec} \mathcal{O}_{Z,\tilde{x}}$. The closed immersion $Z \to \mathcal{X}$ induces a surjective homomorphism $\mathcal{O}_{\mathcal{X},\tilde{x}} \to \mathcal{O}_{Z,\tilde{x}}$ and therefore a closed immersion $i : Z \to \operatorname{Spec} \mathcal{O}_{\mathcal{X},\tilde{x}}$. As $\mathcal{O}_{Z,\tilde{x}} \subseteq L$ is a local ring dominated by $\mathcal{O}_L$, f extends to a morphism $\operatorname{Spec} \mathcal{O}_L \to Z$. We obtain a morphism $\operatorname{Spec} \mathcal{O}_L \to \operatorname{Spec} \mathcal{O}_{\mathcal{X},\tilde{x}}$ by composition with i. Conversely, let $g : \operatorname{Spec} \mathcal{O}_L \to \operatorname{Spec} \mathcal{O}_{\mathcal{X},\tilde{x}}$ be a morphism. By composition with the canonical morphism $\operatorname{Spec} \mathcal{O}_{\mathcal{X},\tilde{x}} \to \mathcal{X}$, we obtain a morphism $h : \operatorname{Spec} \mathcal{O}_L \to \mathcal{X}$. The image of this morphism is an irreducible closed subscheme Z of relative dimension 0 over $\operatorname{Spec} \mathcal{O}_K$ because $\operatorname{Spec} \mathcal{O}_L \to \operatorname{Spec} \mathcal{O}_K$ is finite. By definition, its generic point x belongs to $r^{-1}(Z \cap \mathcal{X}_s) = r^{-1}(\tilde{x})$. Hence $h_K \in X_+(\tilde{x})(L)$. We easily verify that these two maps that we have just defined are mutual inverses. The second isomorphism results from the universal property of the formal completion and from the fact that $\mathcal{O}_L$ is complete.

(b) For any $m \geq 1$, we have $\mathcal{O}_{\mathcal{X},\tilde{x}} \subseteq \mathcal{O}_K[f_1, \ldots, f_q] + (f_1, \ldots, f_q)^m$. Hence the homomorphism

$$\mathcal{O}_K[T_1, \ldots, T_q]/(T_1, \ldots, T_q)^m \to \mathcal{O}_{\mathcal{X},\tilde{x}}/\mathfrak{m}^m_{\tilde{x}}$$

induced by φ is surjective, whence the surjectivity of φ, by using Lemma 1.3.1. Let $\widehat{\mathcal{X}} = \operatorname{Spec} \mathcal{O}_K[[T_1, \ldots, T_q]]/\operatorname{Ker} \varphi$. By (a), $X_+(\tilde{x})(L)$ can be identified with the set $\operatorname{Mor}_S(\operatorname{Spec} \mathcal{O}_L, \widehat{\mathcal{X}})$. Bijection (1.7) is then an immediate consequence of Lemma 1.41 below. When $\mathcal{X}_s$ is smooth at $\tilde{x}$ and $q = \dim_{\tilde{x}} \mathcal{X}_s$, $\widehat{\mathcal{O}}_{\mathcal{X},\tilde{x}}$ is a regular (hence integral) local ring of dimension $q + 1 = \dim \mathcal{O}_K[[T_1, \ldots, T_q]]$ (Lemma 4.2.26). It follows that φ is an isomorphism, whence (1.8).

(c) As $\mathcal{O}_{\mathcal{X},\tilde{x}} = \mathcal{O}_{U,\tilde{x}}$, and by (a), the set $X_+(\tilde{x})(L)$ can be identified with $\operatorname{Mor}_S(\operatorname{Spec} \mathcal{O}_L, \widehat{\mathcal{O}}_{U,\tilde{x}})$. Then (1.9) results from Lemma 1.41 and from Proposition 3.2.18(d). We can also see this bijection with the help of Lemma 1.32. □

Lemma 1.41. Let $A \to B$ be a homomorphism of complete local rings. Let $q \geq 1$, let I be an ideal of $A[[T_1, \ldots, T_q]]$, and t_i the image of T_i in the quotient by I. Let θ be the map

$$\operatorname{Hom}_A(A[[T_1, \ldots, T_q]]/I, \; B) \to \mathfrak{m}^q_B, \quad \varphi \mapsto (\varphi(t_1), \ldots, \varphi(t_q)),$$

where Hom_A means continuous homomorphisms of local A-algebras, and where $\mathfrak{m}_B$ is the maximal ideal of B. Then θ is injective. Its image is equal to the set

$$Z_B(I) := \{(a_1, \ldots, a_q) \in \mathfrak{m}^q_B \mid F(a_1, \ldots, a_q) = 0, \quad \forall F \in I\}.$$

Proof Let us first suppose $I = 0$. As φ is a local homomorphism, $\varphi(T_i)$ indeed belongs to $\mathfrak{m}_B$. The map θ is bijective because φ is completely determined by the $\varphi(T_i) \in \mathfrak{m}_B$. In the general case, a continuous homomorphism $A[[T_1,\ldots,T_q]]/I \to B$ corresponds to a continuous homomorphism $\varphi : A[[T_1,\ldots,T_q]] \to B$ such that $I \subseteq \operatorname{Ker}\varphi$. This last condition comes down to saying that $F(\varphi(t_1),\ldots,\varphi(t_q)) = 0$ for every $F \in I$, because the continuity of φ implies that $\varphi(F(T_1,\ldots,T_q)) = F(\varphi(t_1),\ldots,\varphi(t_q))$, whence the lemma. $\square$

Remark 1.42. The formal fiber $X_+(\tilde{x})$ does not have the structure of a scheme. However, it can be endowed with the structure of a *rigid analytic space* over K. This structure reflects certain properties of $\operatorname{Spec}\mathcal{O}_{X,\tilde{x}}$. Cf. [13] and [36] for an introduction to the theory of rigid analytic spaces.

Example 1.43. Let $\mathcal{O}_K$ be a complete discrete valuation ring with uniformizing parameter t and residue characteristic (i.e., characteristic of the residue field) different from 2. Let $n \geq 1$ and

$$U = \operatorname{Spec}\mathcal{O}_K[u,v]/(u^2 - (v^2 + t^n)P(v)), \quad P(v) \in \mathcal{O}_K[v], P(0) = 1,$$

and let $\mathcal{X}$ be a projective scheme over $\operatorname{Spec}\mathcal{O}_K$ containing U as an open subscheme. Let $\tilde{x}$ be the point of U corresponding to the maximal ideal (t, u, v). In the ring of formal power series $\mathcal{O}_K[[v]]$, $P(v) \in \mathcal{O}_K[[v]]^*$ is a square (Exercise 1.3.9). Hence if we set $u' = u/\sqrt{P(v)}$, and $u_1 = u' - v$, $v_1 = u' + v$, then

$$\widehat{\mathcal{O}}_{U,\tilde{x}} \simeq \mathcal{O}_K[[u_1,v_1]]/(u_1 v_1 - t^n).$$

For any finite extension L/K, we have bijections

$$X_+(\tilde{x})(L) \simeq \{(a,b) \in \mathfrak{m}_L^2 \mid ab = t^n\} \simeq \{a \in L \mid |t|^n < |a| < 1\}.$$

10.1.4 Graphs

Let $X \to S$ be an arithmetic surface. The study of a singular closed fiber X_s of X is sometimes of combinatorial nature. It is then more convenient to represent X_s by its dual graph (Definition 1.48). We are going to compare invariants associated to X_s to those of the dual graph (Proposition 1.51) and classify the dual graphs of curves that are contracted in closed points in the canonical model (Proposition 1.53). Finally, we show that for fixed $p_a(X_\eta)$, there exist only a finite number of possible types for the dual graph of X_s (Proposition 1.57).

Definition 1.44. A (finite) *graph* G consists of a finite set V whose elements are called *vertices*, and for every pair of vertices v_1, v_2, a finite (possibly empty) set whose elements are called the *edges* (joining v_1 and v_2). We say that v_1, v_2 are the *end vertices* of these edges. In general, we represent a graph as in Figure 22, where the circles are the vertices, and the segments or curves between two circles are the edges. A *subgraph* of G consists of a subset V' of the vertices of G and of a subset of edges with end vertices belonging to V'. A *path* between two vertices v_1, v_n is a set of vertices $v_1, v_2, \ldots, v_n$ that are pairwise distinct, with v_i, v_{i+1} joined by an edge e_i. The integer $n - 1$ is called the *length* of the path. In Figure 22, the vertices $v_1, v_2, \ldots, v_6$ (with the edges $e_1, \ldots, e_5$) form a path of length 5. We say that G is *connected* if two arbitrary vertices of G are joined by a path. In general, a *connected component* is a maximal connected subgraph (for the inclusion).

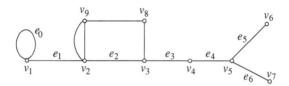

Figure 22. Example of a graph.

Definition 1.45. Let G be a graph. A *circuit* in G is a path $v_1, \ldots, v_n$ to which we add an edge e joining the vertices v_1, v_n, the edge e being different from those already appearing in the path. For example, in the graph of Figure 22, there are three circuits passing through v_2 and v_9, and one circuit passing through v_1. A connected graph is a *tree* if it does not contain any circuit. This comes down to saying that there is exactly one path between two vertices of G. The graph represented by Figure 22 is not a tree. On the other hand, the subgraph made up of the vertices $v_1, \ldots, v_7$ and of the edges $e_1, \ldots, e_6$ is a tree.

We can 'measure' the complexity of a graph with its Betti number.

Definition 1.46. Let G be a graph with ν vertices and ε edges. We call the integer $\beta(G) = \varepsilon - \nu + 1$ the (first) *Betti number* of G. The Betti number of Figure 22 is equal to $11 - 9 + 1 = 3$.

Lemma 1.47. *Let G be a connected graph. Then $\beta(G) \geq 0$. Equality holds if and only if G is a tree.*

Proof Let v_0 be a vertex of G, $e_1, \ldots, e_m$ the edges of which one end vertex is v_0, G' the subgraph of G obtained by deleting v_0 and $e_1, \ldots, e_m$, and $G_1, \ldots, G_r$ the connected components of G'. As G is connected, for any component G_i, there exists at least one edge of G that joins v_0 to G_i, and hence $m \geq r$. Now it is easy to see that

$$\beta(G) = \sum_{1 \leq i \leq r} \beta(G_i) + m - r.$$

Hence induction on the number of vertices of G implies that $\beta(G) \geq 0$. Finally, it is easy to see that G is a tree if and only if the G_i are trees and if $r = m$. The equality above implies, once more by induction on the number of vertices of G, the equivalence between '$\beta(G) = 0$' and 'G is a tree'.   □

Definition 1.48. Let $C > 0$ be a vertical divisor contained in a closed fiber X_s of a regular fibered surface $X \to S$. Let $\Gamma_1, \ldots, \Gamma_n$ be its irreducible components. We associate a graph G to C in the following manner. The vertices of G are the irreducible components $\Gamma_1, \ldots, \Gamma_n$, and there are $\Gamma_i \cdot \Gamma_j$ edges between Γ_i and Γ_j if $i \neq j$. This graph is called the *dual graph* of C.

Be careful: the definition above is not the unique way to associate a graph to C. According to our needs, we may want to define a different graph, for example to attach a loop to a vertex Γ_i if Γ_i contains an ordinary double point (a *loop* is an edge whose end vertices coincide). See Definition 3.17.

Example 1.49. Let G' be the graph of Figure 22 from which we delete the vertices $v_4, \ldots, v_7$ and the edges $e_0, e_3, \ldots, e_6$. Then it is the dual graph of the curves of Figure 23.

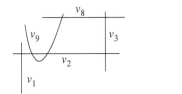 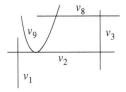

Figure 23. Two curves having the same dual graph.

Lemma 1.50. *Let* $(A, \mathfrak{m})$ *be a regular local (hence factorial) ring of dimension 2, let be* $f_1, \ldots, f_n \in \mathfrak{m}$ *pairwise relatively prime irreducible elements. Then the canonical homomorphism*

$$\varphi : A/(f_1 f_2 \cdots f_n) \to \oplus_{1 \leq i \leq n} A/(f_i)$$

is injective, and its cokernel is of length $\sum_{i<j} \mathrm{length}_A \, A/(f_i, f_j)$.

Proof The injectivity of φ comes from the fact that A is factorial. For any $1 \leq m \leq n$, let $g_m = f_1 f_2 \cdots f_m$ and let K_m denote the cokernel of φ_m : $A/(g_m) \to \oplus_{1 \leq i \leq m} A/(f_i)$. We have a canonical exact sequence

$$0 \to A/(g_n) \to A/(g_{n-1}) \oplus A/(f_n) \to A/(g_{n-1}, f_n) \to 0.$$

The middle term canonically injects into $(\oplus_{1 \leq i \leq n-1} A/(f_i)) \oplus A/(f_n)$, with cokernel isomorphic to K_{n-1}. We deduce from this an exact sequence

$$0 \to A/(g_{n-1}, f_n) \to K_n \to K_{n-1} \to 0.$$

Applying Lemma 7.1.26 to $A/(f_n)$, we have

$$\mathrm{length}_A \, A/(g_{n-1}, f_n) = \sum_{1 \leq i \leq n-1} \mathrm{length}_A \, A/(f_i, f_n).$$

We immediately obtain the assertion on $\mathrm{length}_A \, K_n$ by induction on n. $\square$

Proposition 1.51. *Let* C *be a divisor as in Definition 1.48 and such that* $C \leq X_s$. *Let* G *be the dual graph of* C. *Then the following properties are true.*

(a) *The curve* C *is connected if and only if* G *is connected.*

(b) *Let us suppose* C *is reduced with irreducible components* $\Gamma_1, \ldots, \Gamma_n$. *Then* $\beta(G) = p_a(C) - \sum_{1 \leq i \leq n} p_a(\Gamma_i)$.

(c) *Let us suppose k is algebraically closed and C is connected. Let t, u be the toric and unipotent ranks of C (Definition 7.5.21). Then we have $\beta(G) \leq t + u$.*

Proof (a) A path in G corresponds to a sequence of irreducible components $\Gamma_1, \ldots, \Gamma_r$ of C such that $\Gamma_i \cap \Gamma_{i+1} \neq \emptyset$. We immediately deduce from this the equivalence between the connectedness of C and that of G.

(b) The divisor C is a curve over $k(s)$. The cokernel $\mathcal{F}$ of the canonical homomorphism

$$0 \to \mathcal{O}_X/\mathcal{O}_X(-C) \to \oplus_{1 \leq i \leq n} \mathcal{O}_X/\mathcal{O}_X(-\Gamma_i)$$

is a skyscraper sheaf with support in the intersection points of C, and we have $\dim_{k(s)} \mathcal{F}(X) = \sum_{i<j} \Gamma_i \cdot \Gamma_j$ by Lemma 1.50 above. It follows that

$$\sum_{1 \leq i \leq n} \chi_{k(s)}(\mathcal{O}_{\Gamma_i}) = \chi_{k(s)}(\mathcal{O}_C) + \sum_{1 \leq i < j \leq n} \Gamma_i \cdot \Gamma_j,$$

and therefore

$$\beta(G) = \sum_{i<j} \Gamma_i \cdot \Gamma_j - n + 1 = p_a(C) - \sum_{1 \leq i \leq n} p_a(\Gamma_i).$$

(c) Let $\Gamma_i' \to \Gamma_i$ be the normalization of Γ_i and $C' \to C_{\mathrm{red}}$ that of C_{red}. Then by definition

$$t + u = \dim_k H^1(C, \mathcal{O}_C) - \sum_{1 \leq i \leq n} p_a(\Gamma_i') \geq \dim_k H^1(C_{\mathrm{red}}, \mathcal{O}_{C_{\mathrm{red}}}) - \sum_{1 \leq i \leq n} p_a(\Gamma_i').$$

It follows from (b) that $t + u \geq \beta(G')$, where G' is the dual graph of C_{red}. Now $G' = G$ by definition of the dual graph, whence the result. $\qquad\qquad\square$

Until the end of the subsection, we fix a discrete valuation ring $\mathcal{O}_K$ with algebraically closed residue field k and an arithmetic surface $X \to S = \operatorname{Spec} \mathcal{O}_K$. We will suppose that the generic fiber X_η is geometrically connected, so that the special fiber X_s is also connected (Corollary 8.3.6(b)).

Lemma 1.52. *Let $R > 0$ be a reduced, connected, vertical divisor on X. We suppose that the irreducible components Γ of R all verify $K_{X/S} \cdot \Gamma = 0$, and that R does not contain all of the irreducible components of X_s. Let C be a divisor such that $0 < C \leq R$ and that is connected. Then the following properties are true.*

(a) *We have $p_a(C) = 0$ and $C^2 = -2$.*

(b) *Let D be an effective vertical divisor on X such that $\operatorname{Supp} C \cup \operatorname{Supp} D$ is strictly contained in X_s. Then $C \cdot D < -2D^2$.*

(c) *The divisor R has normal crossings and its dual graph is a tree.*

Proof   (a) The C connected and reduced hypothesis implies that $H^0(C, \mathcal{O}_C) = k$ since k is algebraically closed. It follows that $p_a(C) \geq 0$. We have $C \cdot K_{X/S} = 0$. Hence the adjunction formula implies that

$$0 > C^2 = 2p_a(C) - 2 \geq -2,$$

whence $p_a(C) = 0$ and $C^2 = -2$.

(b) Let p, q be integers with $q > 0$. We have $(pC+qD)^2 < 0$ (Theorem 9.1.23). Let $r = p/q$; then $2r^2 - 2(C \cdot D)r - D^2 > 0$. Since this is true for every rational number r, we have $(C \cdot D)^2 + 2D^2 < 0$, whence (b).

(c) Let $\Gamma_1, \dots, \Gamma_m$ be irreducible components of R that meet at a common point. Let us set $C = \sum_{2 \leq i \leq m} \Gamma_i$. Then C is connected and it follows from (b) that

$$(m - 1)^2 \leq (C \cdot \Gamma_1)^2 < -2\Gamma_1^2 = 4.$$

Hence $m = 2$ and $\Gamma_1 \cdot \Gamma_2 = 1$. Consequently, R is a divisor with normal crossings. As $p_a(R) = 0$ by (a), it follows from Proposition 1.51(b) and Lemma 1.47 that the dual graph of R is a tree.   □

Proposition 1.53. *Let $R > 0$ be a reduced, connected, vertical divisor on X. We suppose that the irreducible components Γ of R all verify $K_{X/S} \cdot \Gamma = 0$, and that R does not contain all of the irreducible components of X_s. Let G be the dual graph of R. Then G is of one of the forms of Figure 24, where the indices indicate the number of vertices in the graph (with $N \geq 4$ for the graph D_N).*

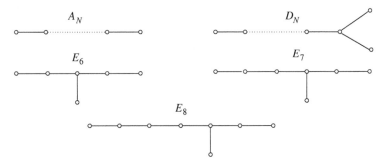

Figure 24. Classification of dual graphs.

Proof   We will use the fact that the intersection form is negative definite on the $\mathbb{Q}$-vector space whose basis is made up of the irreducible components of R (Theorem 9.1.23). By Lemma 1.52, we know that R has normal crossings. Let us first show that G has at most one *node*, that is to say a vertex that is an end vertex of at least three edges. Let us suppose the contrary; then G contains a subgraph of the form $\tilde{D}_N$ with $N+1$ vertices ($N \geq 5$). Let C be the divisor that is the sum of the irreducible components corresponding to the vertices of $\tilde{D}_N$, with multiplicities those noted next to the vertices. Then a direct computation shows that $C^2 = 0$, a contradiction. Hence G has at most one node.

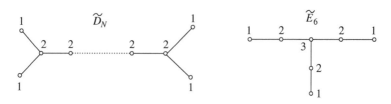

If G does not have any node, then it is of the form A_N. Let us now suppose that it has a node v. By the same method as above (consider $\widetilde{D}_4$), we see that there exist exactly three edges leaving v. Each edge extends to a maximal path (for the inclusion) leaving v, which we call a *branch* with origin v. Let us show that there exists at least one branch of length 1 with origin v. Indeed, in the opposite case, G contains a subgraph of the form $\widetilde{E}_6$. By considering the sum of the divisors appearing in $\widetilde{E}_6$ with the indicated multiplicities, we find a divisor C with support in R and with $C^2 = 0$, which is impossible. If there exist two branches with origin v of length 1, then G is of the form D_N.

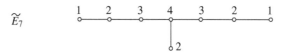

Let us now suppose that there exist two branches of length ≥ 2 with origin v. By a similar reasoning with the graph $\widetilde{E}_7$, we see that one of them must be of length exactly 2. If the third is of length 2, 3, or 4, then G is of the form E_6, E_7, or E_8.

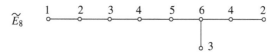

By considering the graph $\widetilde{E}_8$, we see that this third branch cannot be of length ≥ 5, so our list is complete. We note that the graphs $\widetilde{D}_N$ ($N \geq 4$), $\widetilde{E}_6$, $\widetilde{E}_7$, and $\widetilde{E}_8$ correspond to the types I^*_{N-4}, IV^*, III^*, and II^* of reduction of elliptic curves (see the next section).   □

Remark 1.54. The graphs of Figure 24 are called *Dynkin diagrams*. They are related to the theory of Coxeter groups and root systems. See [47]. Let $f : X \to Y$ be the contraction of the irreducible components of R and let $y = f(R)$. We know that, locally at y, Y is a hypersurface in a regular scheme of dimension 3 (Corollary 9.4.18(a)). The classification of the dual graphs of the Rs can also be accomplished by explicitly computing the minimal desingularization at y. This was done by Du Val [30] in the complex case and by Lipman [57] in the general case. Following this course, we can exhibit, for each Dynkin diagram as above, a divisor R in an arithmetic surface with the corresponding dual graph.

Let $n, d_1, \ldots, d_n, k_1, \ldots, k_n$ be integers with $n, d_i \geq 1$, and $k_i \in \mathbb{Z}$. Let M be a symmetric matrix $M = (m_{ij})_{i,j} \in M_{n \times n}(\mathbb{Z})$ such that $m_{ij} \geq 0$ if $i \neq j$. Let

$$T = (d_1, \ldots, d_n, k_1, \ldots, k_n, M),$$

defined up to permutations of the set $\{1, \ldots, n\}$. To T, we associate a graph $G(T)$ whose vertices are indexed by $1, \ldots, n$. Two distinct vertices i and j are joined by exactly m_{ij} edges, and there is no edge between i and itself. We call d_i the *multiplicity* of the vertex i in $G(T)$.

Definition 1.55. We call T a *type* if, moreover, the following conditions are verified:

(α) The graph $G(T)$ is connected.

(β) The bilinear form $\langle , \rangle$ on $\mathbb{Q}^n$ associated to M is negative semi-definite and the isotropic subspace (the vectors $v \in \mathbb{Q}^n$ such that $\langle v, v \rangle = 0$) is generated by $(d_1, \ldots, d_n)$.

(γ) For any $i \leq n$, $p_i := 1 + (m_{ii} + k_i)/2$ is an integer ≥ 0.

We will say that T is *minimal* if $k_i \geq 0$ for all i. We call the rational number $g(T) = 1 + (\sum_{1 \leq i \leq n} d_i k_i)/2$ the *genus* of T. It is in fact an integer.

Let us note that condition (β) implies that $(d_1, \ldots, d_n)$ is orthogonal to every $\mathbb{Q}^n$, and therefore

$$d_i m_{ii} = - \sum_{j \neq i} m_{ij} d_j. \qquad (1.10)$$

Example 1.56. Let X be an arithmetic surface as before Lemma 1.52. We moreover suppose that $p_a(X_\eta) \geq 1$. Let $X_s = \sum_{1 \leq i \leq n} d_i \Gamma_i$ be the decomposition of X_s as a Cartier divisor. Then

$$(d_1, \ldots, d_n, k_1, \ldots, k_n, M),$$

where $k_i = K_{X/S} \cdot \Gamma_i$, and where M is the intersection matrix $(\Gamma_i \cdot \Gamma_j)_{1 \leq i,j \leq n}$, is a type of genus $p_a(X_\eta)$. If X is minimal, then the associated type is minimal by Proposition 9.3.10.

Let $d, N \geq 1$. We will say that a path with N vertices in $G(T)$ is of the form $_d A_N$ if its vertices v_i are of multiplicity d, if $k_i = 0$, and if the path only meets the rest of the graph at its end vertices (in fact, this automatically follows from the equalities $m_{ii} = -2$ and (1.10)).

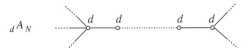

We will say that a graph G is a *union of subgraphs* $G_1, \ldots, G_r$ if each vertex and each edge of G belongs to one of the graphs G_i.

Proposition 1.57 ([11], Theorem 1.6). *Let $g \geq 2$ be an integer. Then there exists a constant $c(g)$ depending only on g such that for any minimal type T of genus g, we have:*

(i) *the d_i are bounded from above by $c(g)$;*

(ii) *the graph $G(T)$ is a union of a subgraph which has at most $c(g)$ vertices and edges, and of at most $c(g)$ paths of the form $_d A_N$.*

Proof For simplicity, we will say that a subgraph of G is *bounded from above* by a constant c if its number of vertices, its number of edges, and the multiplicities of its vertices are all bounded from above by c.

Let $v_1, \ldots, v_n$ denote the vertices of G, V_0 the set of v_i such that $k_i > 0$. The equality $2g - 2 = \sum_{1 \le i \le n} d_i k_i$ implies that $d_i \le 2g - 2$ if $v_i \in V_0$, and that V_0 has cardinal $\le 2g - 2$. Moreover, if $v_i \in V_0$, we have

$$\sum_{j \ne i} d_j m_{ij} = d_i(-m_{ii}) \le d_i(k_i + 2) = O(g).$$

Hence the number of edges leaving a vertex $\in V_0$ is $O(g)$, and for any j, the multiplicity d_j is $O(g)$ if $m_{ij} \ne 0$ for some $v_i \in V_0$. We define a subgraph G_0 of G as follows: the vertices of G_0 are the elements of V_0 and those that are joined to them by an edge; the edges of G_0 are those of G of which at least one end vertex belongs to V_0. Then G_0 is bounded from above by $O(g^3)$. Let V_1 be the set of vertices of G that do not belong to V_0, let G_1 be the subgraph of G of which V_1 is the set of vertices, and of which the edges are those of G with end vertices contained in V_1. Then G is the union of the subgraphs G_0 and G_1.

Let us study G_1. Let $H_1, \ldots, H_r$ be its connected components. As G is connected, each H_i contains a vertex of G_0. It follows that $r = O(g^3)$ (this is obviously not an optimal upper bound, but we are not concerned by that question). We can therefore restrict ourselves to a connected component H of G_1. As $g \ge 2$, we have $V_0 \ne \emptyset$ and therefore the bilinear form M restricted to the subvector space generated by the vertices of H is negative definite. For any pair of vertices v_i, v_j of H, we have $k_i = 0$; hence $m_{ii} = -2$ and $m_{ij} = 0$ or 1(using the fact that $\langle v_i + v_j, v_i + v_j \rangle < 0$ for the bilinear form defined by M). As the reasoning of Proposition 1.53 is still valid, we see that H is of one of the forms A_N, D_N, or E_q ($q = 6, 7, 8$). Let v_1 be a vertex of H and of G_0; then $d_1 = O(g^3)$. Let v_2 be a vertex of H joined to v_1 by an edge. As $2d_1 = d_2 + \sum_{i \ne 1,2} d_i m_{i1}$ (formula (1.10)), we have $d_2 \le 2d_1$. Thus we see that $d_i = O(2^N g^3)$ for every $v_i \in H$ if H has N vertices. It follows that H is bounded from above by $O(g^3)$ if it is of type E_q.

Let us therefore suppose that H is of the form D_{N+2} with $N \ge 2$ (the case of the A_N can be treated in the same manner, but more simply). Let $v_1, v_2, \ldots, v_{N+2}$ be the vertices of H with $m_{i,i+1} = 1$ for every $1 \le i \le N$, $m_{N,N+2} = 1$, and $m_{N+1,N+2} = 0$.

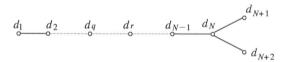

Let $2 \le r \le N$. If $d_r > d_{r-1}$, as $2d_{r-1} \ge d_r + d_{r-2}$ by formula (1.10), we have $d_{r-1} > d_{r-2}$, and hence $d_r > d_{r-1} > \cdots > d_1$. Likewise, if $d_r < d_{r-1}$, then we have $d_N < d_{N-1} < \cdots < d_{r-1}$. There therefore exist $1 \le q \le r \le N$ such that

$$d_1 < \cdots < d_q = d_{q+1} = \cdots = d_r > d_{r+1} > \cdots > d_N.$$

If $r < N$, the inequality $2d_r > d_{r+1} + d_{r-1}$ implies that v_r is a vertex of G_0; hence $d_r = O(g^3)$ and therefore $q = O(g^3)$, $N - r = O(g^3)$. Moreover, we always have $d_{N+1}, d_{N+2} \le 2d_N$. Hence H is the union of a ${}_dA_{r-q+1}$ with $d = d_r$ and of two subgraphs bounded from above by $O(g^3)$. The same reasoning holds if $q > 1$. There therefore remains the case $q = 1$ and $r = N$. As H contains a vertex of G_0, and $d_N \le 2d_{N+1}$, we see as before that $d_i = O(g^3)$ for every $1 \le i \le N+2$, and that H is the union of ${}_{d_1}A_N$ and of two subgraphs bounded from above by $O(g^3)$. $\square$

Exercises

1.1. Show that Proposition 1.8 is true when we replace the S affine hypothesis by S a normal algebraic curve over a field.

1.2. Let S be an affine Dedekind scheme of dimension 1. Let C be a geometrically reduced normal curve over $K(S)$. Show that C admits a model over S.

1.3. Let $S' \to S$ be a morphism of Dedekind schemes of dimension 1. Let C be a projective curve over $K(S)$ with good reduction over S. Show that $C_{K(S')}$ has good reduction over S'.

1.4. Let C be the projective curve over $\mathbb{Q}$ defined by the equation $x^3 y + y^3 z + z^3 x = 0$ (it is the *Klein curve*). Show that the same equation over $\mathbb{Z}$ defines a normal model of C over $\mathbb{Z}$. Show that C has good reduction at every $p \ne 7$.

1.5. Let $\mathcal{O}_K$ be a complete discrete valuation ring, with separably closed residue field. Let C be a normal projective curve over K. Show that the following properties are equivalent:

  (i) $C(K) \ne \emptyset$,

  (ii) C admits a model $\mathcal{C}$ over $\operatorname{Spec} \mathcal{O}_K$ whose special fiber $\mathcal{C}_s$ contains an irreducible component of multiplicity 1 that is geometrically reduced.

1.6. Let S be a Dedekind scheme of dimension 1. Let C be a smooth conic over $K(S)$. Show that C has good reduction at $s \in S$ if and only if there exists a discrete valuation ring $\mathcal{O}_L$, étale over $\mathcal{O}_{S,s}$, such that $C(L) \ne \emptyset$.

1.7. Let C be a smooth projective curve over K as in Lemma 1.24 but with $p_a(C) = 1$ and C geometrically connected. Let X be the minimal regular model of C over $\operatorname{Spec} \mathcal{O}_{S,s}$. Show that $X_s = d\Gamma$, where Γ is a smooth projective curve over $k(s)$ of arithmetic genus $p_a(\Gamma) = 1$, and where $d \ge 1$. Show that $d = 1$ if $C(K) \ne \emptyset$.

1.8. Let C be a normal projective curve over a discrete valuation field K, and let $\mathcal{C}$ be a regular model of C over $\mathcal{O}_K$. For any closed point $\tilde{x} \in \mathcal{C}_s$, and for any $x \in C_+(\tilde{x})$, show that

$$[k(x) : K] \geq \sum_{1 \leq i \leq m} \mu_{\tilde{x}}(\Gamma_i)d_i,$$

where $\Gamma_1, \ldots, \Gamma_m$ are the irreducible components of C_s passing through $\tilde{x}$, $\mu_{\tilde{x}}(\Gamma_i)$ is the multiplicity of Γ_i at $\tilde{x}$, and d_i is the multiplicity of Γ_i in C_s.

1.9. Let us keep the hypotheses and notation of Example 1.26, but without the char$(k) \neq 2$ hypothesis.

(a) Let us suppose that C admits a hyperelliptic equation $y^2 + Q(x)y = P(x)$ with $P, Q \in \mathcal{O}_K[x]$, which in $k[x, y]$ is the equation of a smooth hyperelliptic curve of genus g. Show that C has good reduction.

(b) Conversely, let us suppose that C has good reduction. Let us consider the finite morphism $W^0 \to V^0 = \operatorname{Spec} \mathcal{O}_K[v]$ and the decomposition $\mathcal{O}_{C_{\min}}(W^0) = \mathcal{O}_K[v] \oplus z\mathcal{O}_K[v]$. We therefore have a relation $z^2 + G(v)z = F(v)$ so that

$$\mathcal{O}_{C_{\min}}(W^0) = \mathcal{O}_K[v, Z]/(Z^2 + G(v)Z - F(v)). \qquad (1.11)$$

We want to show that it is possible to choose z so that

$$\deg G(v) \leq g + 1, \quad \deg F(v) \leq 2g + 2. \qquad (1.12)$$

We know that W_K^0 is defined by an equation $z_0^2 + G_0(v)z_0 = F_0(v)$ with $F_0(v), G_0(v) \in K[v]$ and $\deg G_0(v) \leq g + 1$, $\deg F_0(v) \leq 2g + 2$ (see the proof of Proposition 7.4.24). Show that we can suppose $F_0, G_0 \in \mathcal{O}_K[v]$ and hence $z_0 \in \mathcal{O}_{C_{\min}}(W^0)$. Deduce from this that there exist a non-zero $a \in \mathcal{O}_K$ and an $H(v) \in \mathcal{O}_K[v]$ such that $z_0 = az + H(v)$. Show that if $a \in \mathcal{O}_K^*$, then (1.11) and (1.12) are satisfied when replacing z by z_0. Let us therefore suppose $\nu(a) > 0$. Show that the image of $H(v)$ in $k[v]$ is of degree $\leq g + 1$. Let us write $H(v) = H_0(v) + H_1(v)$ with $\deg H_0(v) \leq g + 1$ and $H_1(v) \in t\mathcal{O}_K[v]$, where t is a uniformizing parameter for $\mathcal{O}_K$.

(c) Show that $(z_0 - H_0(v))/t \in \mathcal{O}_{C_{\min}}(W^0)$ and that it enjoys the same properties as z_0. Conclude by induction on $\nu(a)$.

1.10. Let C be a hyperelliptic curve over a discrete valuation field K of residue characteristic different from 2. Let $y^2 = P(x)$ be an affine equation for C and $\alpha_1, \ldots, \alpha_n$ be its roots in $\overline{K}$. Let L be a finite extension of K containing the α_i, and ν_L be a valuation of L that extends that of K. Show that C has potential good reduction if and only if $\nu_L(\alpha_i - \alpha_j)$, $i \neq j$, is independent of the pair (i, j) (use Example 1.26).

1.11. Let $\mathcal{O}_K$ be a discrete valuation ring of residue characteristic $\neq 2$, with uniformizing parameter t. Let C be the hyperelliptic curve over K with equation

$$y^2 = x^6 + atx^4 + bt^2x^2 + t^3,$$

with $a, b \in \mathcal{O}_K$ and such that $-4a^3 + a^2b^2 + 18ab - 4b^3 - 27$ (the discriminant of the polynomial $T^3 + aT^2 + bT + 1$) is invertible in $\mathcal{O}_K$.

(a) Show that the special fiber of $\mathcal{C}_{\min}$ is the union of an elliptic curve of multiplicity 2 and of two projective lines of multiplicity 1.

(b) Show that C has bad reduction over $\mathcal{O}_K$, but that it has good reduction over $\mathcal{O}_L$, where $L = K[\sqrt{t}]$.

1.12. Let S be a Dedekind scheme of dimension 1 with $\mathrm{char}(K(S)) = 0$. Let $\pi : \mathcal{X} \to S$ be a flat morphism of finite type.

(a) Show that there exist smooth schemes $\mathcal{X}_i \to V_i$, $1 \le i \le n$, with V_i an open subscheme of S, $\mathcal{X}_i$ admitting an immersion $f_i : \mathcal{X}_i \to \mathcal{X}$, and such that $\cup_{1 \le i \le n} f_i(V_i)$ contains $\pi^{-1}(V)$ for some non-empty open subset V of S.

(b) Show that there exists a non-empty open subset V of S, containing at least one closed point of S, such that for any $s \in V$ and for any $\tilde{x} \in X_s(k(s))$, there exists a closed point $x \in \mathcal{X}_{K_s}(K_s)$, where $K_s := \mathrm{Frac}(\widehat{\mathcal{O}}_{S,s})$), which reduces to $\tilde{x}$.

1.13. Let $X \to T$ be a flat proper morphism to a regular locally Noetherian scheme T. Let us suppose T is irreducible with generic point ξ. Let $x \in X_\xi$ be a closed point and $\overline{\{x\}}$ its Zariski closure in X, endowed with the structure of reduced closed subscheme.

(a) Show that $\overline{\{x\}}$ is finite over T.

(b) Show that if x is rational over $k(\xi)$, then $\overline{\{x\}} \to T$ is an isomorphism. Thus we can define a reduction map $X_\xi(k(\xi)) \to X_t(k(t))$ for every point $t \in T$.

(c) Show that (a) and (b) are false if we leave out the $X \to T$ flat hypothesis.

1.14. Let S be the spectrum of a complete discrete valuation ring. Let $\mathcal{X} \to S$ be a morphism of finite type, and $\tilde{x}$ a closed point of $\mathcal{X}_s$.

(a) Let us suppose $\mathcal{X}$ is affine. Let $i : \mathcal{X} \to \mathcal{Y}$ be an open immersion into a projective scheme $\mathcal{Y}$ over S. Show that $Y_+(\tilde{x}) \subseteq X^0$. Let us set $X_+(\tilde{x}) = Y_+(\tilde{x})$. Show that $X_+(\tilde{x})$ is independent of the choice of $\mathcal{Y}$.

(b) Let U and V be affine open neighborhoods of $\tilde{x}$. Show that $U_+(\tilde{x}) = V_+(\tilde{x})$.

(c) Let us set $X_+(\tilde{x}) = U_+(\tilde{x})$ for an arbitrary affine open neighborhood U of $\tilde{x}$. This set is called the *formal fiber* of $\mathcal{X}$ over $\tilde{x}$. Show that Proposition 1.40 is true for $X_+(\tilde{x})$.

1.15. Let S be the spectrum of a complete discrete valuation ring $\mathcal{O}_K$. Let $f : \mathcal{X} \to \mathcal{Y}$ be a morphism of schemes of finite type over S. Let us fix a closed point $\tilde{x} \in \mathcal{X}_s$ and let us set $\tilde{y} = f(\tilde{x})$.

(a) Show that f canonically induces a map $f_{\tilde{x}} : X_+(\tilde{x}) \to Y_+(\tilde{y})$.

(b) Let us suppose $\mathcal{X}$, $\mathcal{Y}$ are smooth over S. Let $d = \dim_{\tilde{x}} \mathcal{X}_s$ and $e = \dim_{\tilde{y}} \mathcal{Y}_s$. Show that there exist formal power series

$F_1, \ldots, F_e \in \mathcal{O}_K[[T_1, \ldots, T_d]]$ such that for any finite extension L/K, via isomorphism (1.8) of Proposition 1.40, the map $\mathfrak{m}_L^d \to \mathfrak{m}_L^e$ induced by $f_{\tilde{x}}$ is defined by

$$(t_1, \ldots, t_d) \mapsto (F_1(t_1, \ldots, t_d), \ldots, F_e(t_1, \ldots, t_d))$$

with the F_i are independent of L.

(c) Let $\mathcal{G}$ be a smooth group scheme over S, and $\tilde{o}$ be the unit element of $\mathcal{G}_s$ and $d = \dim_{\tilde{o}} \mathcal{G}_s$. Show that there exist $F_1, \ldots, F_d \in \mathcal{O}_K[[T_1, \ldots, T_d, V_1, \ldots, V_d]]$ such that for any finite subextension L of $\overline{K}$, the map $\mathfrak{m}_L^d \times \mathfrak{m}_L^d \to \mathfrak{m}_L^d$ defined by

$$(t_1, \ldots, t_d, v_1, \ldots, v_d) \mapsto (F_1(t_1, \ldots, t_d, v_1, \ldots, v_d), \ldots,$$
$$F_d(t_1, \ldots, t_d, v_1, \ldots, v_d))$$

induces a group law on $\mathfrak{m}_L^d \simeq (\mathcal{G}_K)_+(L)$ that coincides with the one induced by $\mathcal{G}_K(L)$. This law is commutative if $\mathcal{G}$ is a commutative group scheme. We call

$$\mathfrak{m}_{\overline{K}}^d := \varinjlim_L \mathfrak{m}_L^d,$$

where the direct limit is taken over the set of finite subextensions of $\overline{K}$, endowed with the group law as above, the *formal group* of $\mathcal{G}$ at $\tilde{o}$.

1.16. Let S be the spectrum of a discrete valuation ring and $K = K(S)$. Let $X \to S$ be an arithmetic surface with generic fiber isomorphic to $\mathbb{P}_K^1$ and containing a unique exceptional divisor E.

(a) Using Exercise 9.3.1, show that any relatively minimal regular model of $\mathbb{P}_K^1$ over S is smooth over S, and that any regular model of $\mathbb{P}_K^1$ over S has normal crossings.

(b) Show that E meets at most two other irreducible components, and that the multiplicity of E is the sum of the multiplicities of the irreducible components it intersects.

(c) Show that the dual graph of X_s is of the form where the top left vertex represents E, the multiplicities of the vertices are decreasing (resp. strictly decreasing) from left to right (resp. from top to bottom), and two of the multiplicities are equal to 1.

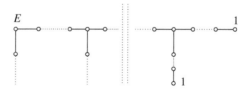

1.17. Let $\mathcal{O}_K$ be a complete discrete valuation ring with algebraically closed residue field k. Let $\mathcal{X}$ be a (normal) model of $X := \mathbb{P}_K^1$ over $\mathcal{O}_K$.

(a) Show that the irreducible components of $(\mathcal{X}_s)_{\mathrm{red}}$ are isomorphic to $\mathbb{P}^1_k$, and that the intersection points of $(\mathcal{X}_s)_{\mathrm{red}}$ are ordinary double points (use Theorem 7.5.19).

(b) Let $\tilde{x} \in \mathcal{X}_s$ be a point belonging to a unique irreducible component Γ_1 of $\mathcal{X}_s$, of multiplicity d_1. Let $\mathcal{X} \to \mathcal{Y}$ be the contraction of the irreducible components of $\mathcal{X}_s$ that are different from Γ_1 (Exercise 9.4.5). Let $\mathcal{Z} \to \mathcal{Y}$ be the minimal desingularization. Using Exercise 1.16 with the model $\mathcal{Y}$, show that $\mathcal{X}$ is regular at $\tilde{x}$ if and only if for every $x \in X_+(\tilde{x})$, we have $[k(x) : K] \geq d_1$.

(c) Let us suppose that $\tilde{x} \in \mathcal{X}_s$ is an intersection point belonging to $\Gamma_1 \cap \Gamma_2$. Let d_1, d_2 be the multiplicities of these components. Show that if $\tilde{x}$ is a regular point of $\mathcal{X}$, then for any $x \in X_+(\tilde{x})$, we have $[k(x) : K] \geq d_1 + d_2$.

(d) Let $\tilde{x}$ be as in (c) and let us suppose that $\tilde{x}$ is singular in $\mathcal{X}$. We want to show that there exists an $x \in X_+(\tilde{x})$ with $[k(x) : K] \leq \min\{d_1, d_2\}$. Let $f : \mathcal{X} \to \mathcal{Y}$ be the contraction of the components of $\mathcal{X}_s$ that are different from Γ_1, Γ_2, and let $g : \mathcal{Z} \to \mathcal{Y}$ be the minimal desingularization. Show that there exists an irreducible component Γ of $\mathcal{Z}_s$, different from the strict transforms of Γ_1, Γ_2, which is of multiplicity lesser than or equal to $\min\{d_1, d_2\}$, and such that $g(\Gamma) = f(\tilde{x})$. Deduce from this the existence of an $x \in X_+(\tilde{x})$ as desired.

1.18. Let G be a connected graph.

(a) Let G_1 be a subgraph of G. Show, by induction on the number of vertices of G_1, that $\beta(G) \geq \beta(G_1)$.

(b) Let $G_1, \ldots, G_m$ be subgraphs of G. We suppose that the G_i are pairwise without common vertex. Show that $\beta(G) \geq \sum_{1 \leq i \leq m} \beta(G_i)$.

1.19. Let us once again take the hypotheses and notation of Exercise 7.5.10. Let G be the graph defined as follows (following an idea communicated by Moret-Bailly): the vertices of G are the connected curves $X_1, \ldots, X_n$ and the intersection points $X_i \cap X_j$, $i \neq j$. Each intersection point x and each $X_i \ni x$ gives an edge that joins x to X_i. Show that G is a connected graph, and, by a direct computation, that we have

$$t(X) = \sum_{1 \leq i \leq n} t(X_i) + \beta(G).$$

10.2 Reduction of elliptic curves

We study, in the specific case of elliptic curves E, the notions of models and of reduction introduced in Section 10.1. First, we classify the different types of reduction of the minimal regular model. It is essentially a consequence of the classification of graphs (Proposition 1.53), but we are going to give a finer classification, notably when the residue field is not algebraically closed. Then, we will involve ourselves in the Néron model of E. We show that this model exists and is isomorphic to the smooth locus of the minimal regular model (Theorem 2.14). We will also discuss the group of components of the Néron model and the filtration induced by the rational points of E. The last subsection makes the criterion for potential good reduction explicit in terms of the modular invariant j (Proposition 2.33).

In all of this section, S will be a Dedekind scheme of dimension 1, E an elliptic curve over $K = K(S)$ together with a privileged rational point $o \in E(K)$.

10.2.1   Reduction of the minimal regular model

Let us suppose, in this subsection, that S is the spectrum of a discrete valuation ring $\mathcal{O}_K$ with residue field k. We will suppose k *perfect* or $\mathrm{char}(k) \neq 2, 3$ (see also Remark 2.3). We let $\mathcal{E}$ denote the minimal regular model of E over S, and W the minimal Weierstrass model of E over S. Then W is obtained by contracting certain vertical divisors on $\mathcal{E}$ (Theorem 9.4.35; be careful of the difference in notation). By Corollary 9.3.27, the canonical divisor $K_{\mathcal{E}/S}$ is trivial.

Let $\Gamma_1, \ldots, \Gamma_n$ be the irreducible components of the special fiber $\mathcal{E}_s$, of respective multiplicities $d_1, \ldots, d_n$. By Proposition 9.4.8, if $n \geq 2$, then Γ_i is a conic over the field $k_i := H^0(\Gamma_i, \mathcal{O}_{\Gamma_i})$, and we have $\Gamma_i^2 = -2[k_i : k]$. A conic C over a field k' that is not smooth cannot be geometrically irreducible because it would be geometrically integral (if $\mathrm{char}(k) \neq 2$, use Exercise 4.3.22; if $\mathrm{char}(k) = 2$ use the k perfect hypothesis), and therefore smooth (Proposition 7.4.1). After a suitable quadratic extension, C becomes the union of two projective lines meeting at one point.

Dual graph of $\mathcal{E}_s$

Let us suppose k is algebraically closed. Let G be the dual graph of $\mathcal{E}_s$ (Definition 1.48), with vertices $v_1, \ldots, v_n$ corresponding respectively to the components $\Gamma_1, \ldots, \Gamma_n$. Let us suppose that Γ_1 is the irreducible component that meets $\{o\}$. If $n = 1$, then G is a vertex v_1 without edge. Let us suppose $n \geq 2$. Then, for any proper subset I of $\{1, \ldots, n\}$, $\sum_{i \in I} \Gamma_i$ is a divisor with normal crossings (Lemma 1.52). Let G_1 be the subgraph of G whose vertices are $v_2, \ldots, v_n$ and whose edges are those of G joining two vertices of G_1. Proposition 1.53 classifies all of the possibilities for G_1. Moreover, the equality

$$2d_i = \sum_{1 \leq j \leq n,\, j \neq i} d_j \Gamma_i \cdot \Gamma_j \tag{2.13}$$

(Proposition 9.1.21) for $i = 1$ implies that G_1 has at most two connected components because $d_1 = 1$. Taking into account the fact that the classification of Proposition 1.52 applies to the subgraphs made up of any proper subset of $\Gamma_1, \ldots, \Gamma_n$, we easily deduce from this that G is of the form $\widetilde{A}_n$ (Figure 25 with $n \geq 2$ vertices), or one of the graphs (whose vertices have the indicated multiplicities) $\widetilde{D}_{n-1}$ ($n \geq 5$), $\widetilde{E}_{n-1}$ ($n = 7, 8, 9$) that we saw in the proof of Proposition 1.53. The multiplicities of the vertices are determined by the equalities (2.13) and $d_1 = 1$.

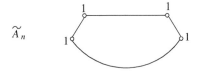

$$\widetilde{A}_n$$

Figure 25. Graph of type $\widetilde{A}_n$.

Let us return to the study of the curve $\mathcal{E}_s$ itself.

Lemma 2.1. *Let us once again suppose k is perfect or char$(k) \neq 2,3$. Let $\Gamma := W_s$ be the special fiber of the minimal Weierstrass model of E over S. Then Γ is a geometrically irreducible cubic. Let Γ^0 denote the smooth locus of Γ. The curve Γ verifies one of the following properties:*

(1) *Γ is smooth over k and is an elliptic curve over k;*

(2) *Γ admits a unique singular point p, and p is rational over k. Let $\pi : \Gamma' \to \Gamma$ be the normalization of Γ. Then $\Gamma' \simeq \mathbb{P}^1_k$. Moreover, we have one of the following three possibilities:*

(2.1) *$\pi^{-1}(p)$ is made up of two k-rational points, and $\Gamma^0 \simeq \mathbb{A}^1_k \setminus \{0\}$;*

(2.2) *$\pi^{-1}(p)$ is a point q and $k(q)$ is separable of degree 2 over k;*

(2.3) *$\pi^{-1}(p)$ is a rational point and $\Gamma^0 \simeq \mathbb{A}^1_k$.*

Proof Let Δ be the minimal discriminant of E. If $\Delta \in \mathcal{O}_K^*$, then W is smooth. As Γ contains a rational point $\overline{\{o\}} \cap \Gamma$, it is an elliptic curve (Corollary 7.4.5). Let us henceforth suppose Δ is non-invertible. Hence Γ is not smooth. Let us suppose k is perfect (we leave it to the reader to verify the case char$(k) \neq 2,3$ by writing an elliptic equation with coefficients $a_1 = a_3 = 0$). Then Γ is singular, by Corollary 4.3.33. Let us consider the exact sequence of sheaves

$$0 \to \mathcal{O}_\Gamma \to \pi_* \mathcal{O}_{\Gamma'} \to \mathcal{S} \to 0$$

with a skyscraper sheaf $\mathcal{S}$ whose support is the singular locus of Γ. The long exact cohomology sequence is

$$0 \to k \to k \to \mathcal{S}(\Gamma) \to H^1(\Gamma, \mathcal{O}_\Gamma) \to H^1(\Gamma', \mathcal{O}_{\Gamma'}) \to 0,$$

the first two terms being equal to k because Γ' contains a rational point $\overline{\{o\}} \cap \Gamma$. As $p_a(\Gamma) = p_a(E) = 1$, and $\mathcal{S} \neq 0$, we obtain $H^1(\Gamma', \mathcal{O}_{\Gamma'}) = 0$. Hence $\Gamma' \simeq \mathbb{P}^1_k$ (Propositions 7.4.1 and 9.4.26(b)), and $\dim_k \mathcal{S}(\Gamma) = 1$; hence Γ contains exactly one singular point p that is rational over k.

The set $\pi^{-1}(p)$ is the support of the finite scheme Γ'_p. The latter is the spectrum of $A := (\pi_* \mathcal{O}_{\Gamma'})_p \otimes k(p)$. As $(\pi_* \mathcal{O}_{\Gamma'})_p / \mathcal{O}_{\Gamma,p} = \mathcal{S}(\Gamma)$ is of dimension 1 over k, we deduce from this that A is of dimension 2 over k (as a k-vector space). Hence Spec A is two rational points, or a quadratic point, or one rational point (and then A is not reduced), which proves the lemma. □

Definition 2.2. We will say that E has *split multiplicative reduction, non-split multiplicative reduction*, or *additive reduction* according to whether we are in case (2.1), (2.2), or (2.3) of Lemma 2.1.

We are now going to study the classification of the $\mathcal{E}_s$ following the notation of Kodaira. But as k is not algebraically closed, we will add indices in the symbols to indicate the case of non-split multiplicative reduction and the case of irreducible components that are not geometrically irreducible. We say that E is of type I_0 if it has good reduction. We say that E is of type I_1 (resp. $I_{1,2}$) if $\mathcal{E}_s$ is irreducible and if E has split (resp. non-split) multiplicative reduction. We say that E is of type II if $\mathcal{E}_s$ is irreducible and if E has additive reduction.

Figure 26. Reduction of type I_1.   Figure 27. Reduction of type II.

In what follows, we will suppose that $n \geq 2$. Let us recall that $\Gamma_1 \supset \overline{\{o\}} \cap \mathcal{E}_s$. Let $r_i = [H^0(\Gamma_i, \mathcal{O}_{\Gamma_i}) : k] = [k(\Gamma_i) \cap \overline{k} : k]$. We have

$$2r_i d_i = \sum_{1 \leq j \leq n,\, j \neq i} d_j \Gamma_i \cdot \Gamma_j \qquad (2.14)$$

(Proposition 9.1.21). It is essentially this equality that will allow us to list the different possibilities for the configuration of $\mathcal{E}_s$. Let us note that r_i divides $\Gamma_i \cdot \Gamma_j$ (Exercise 9.2.8). Also note that if $p \in \Gamma_i \cap \Gamma_j$ is an intersection point, then we have the inequalities

$$[k(p) : k] \geq r_i \geq (d_j/2d_i)\Gamma_i \cdot \Gamma_j. \qquad (2.15)$$

As Γ_1 contains a rational point, we have $d_1 = r_1 = 1$. Hence

$$2 = \sum_{i \geq 2} d_i \Gamma_1 \cdot \Gamma_i, \qquad (2.16)$$

and Γ_1 meets the other components in at most two points.

Let us suppose that Γ_1 meets the other components in two distinct points. The equality above immediately implies that these points are rational over k, and that $\mathcal{E}_s$ has normal crossings at these points. We then see that $\mathcal{E}_s$ is as in Figure 28, with $\Gamma_i \simeq \mathbb{P}^1_k$ and where two consecutive components meet transversally at a k-rational point. We say that $\mathcal{E}_s$ is of type I_n. As W_s is isomorphic to Γ_1, we see that E has split multiplicative reduction.

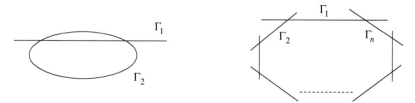

Figure 28. Reduction of type I_n, $n \geq 2$.

Let us suppose that Γ_1 meets the other irreducible components at a point p of degree $[k(p) : k] = 2$. It follows from equality (2.16) that $d_2 = \Gamma_1 \cdot \Gamma_2 = 1$ (this implies, in particular, that Γ_1, Γ_2 are regular at p, by Proposition 9.1.8(b)), and

Γ_2 is a conic over the field $k' := H^0(\Gamma_2, \mathcal{O}_{\Gamma_2}) \subseteq k(p)$. If $k' = k$, then Γ_2 does not meet any component other than Γ_1 because $\Gamma_2^2 = -2$. If $k' = k(p)$, then $\Gamma_2 \simeq \mathbb{P}^1_{k(p)}$ and we have $1 = \sum_{i \geq 3} d_i(2^{-1}\Gamma_i \cdot \Gamma_2)$. Hence Γ_2 meets a component Γ_3 at a point $p_2 \neq p$ that is regular in Γ_3 and with residue field $k(p_2) = k(p)$. We continue the discussion with Γ_3, and we find that $\mathcal{E}_s$ is as in Figure 29, with $\Gamma_i \simeq \mathbb{P}^1_{k(p)}$ for $2 \leq i \leq n-1$, Γ_n a conic over k, and with intersection points having residue field $k(p)$. We will say that $\mathcal{E}_s$ is of type $I_{2n-1,2}$ if Γ_n is singular, and of type $I_{2n-2,2}$ otherwise (hence Γ_n smooth). Over $\bar{k}$, these types become respectively I_{2n-1} and I_{2n-2}. The reduction of E is non-split multiplicative.

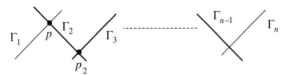

Figure 29. Reduction of type $I_{2n-2,2}$ or $I_{2n-1,2}$.

From now on, Γ_1 meets the other components at a unique rational point $p \in \mathcal{E}_s(k)$. The curve E therefore has additive reduction. If Γ_1 meets two other components, then $\mathcal{E}_s$ is represented by Figure 30, with $\Gamma_i \simeq \mathbb{P}^1_k$, and where the components meet pairwise transversally at p. We say that $\mathcal{E}_s$ is of type IV.

After the preceding cases, there remains the case when Γ_1 meets only one other component Γ_2. We therefore have $2 = d_2\Gamma_1 \cdot \Gamma_2$. If $\Gamma_1 \cdot \Gamma_2 = 2$, then $d_2 = 1$ and Γ_2 is a conic over k. Either Γ_2 is singular, and $\mathcal{E}_s$ is as in Figure 31 (we will say that $\mathcal{E}_s$ is of type IV$_2$; over $\bar{k}$, this type becomes IV). Or Γ_2 is regular; then $\Gamma_2 \simeq \mathbb{P}^1_k$ because $\Gamma_1 \cap \Gamma_2$ is rational over k by hypothesis, and $\mathcal{E}_s$ is then as in Figure 32. In this case we say that $\mathcal{E}_s$ is of type III.

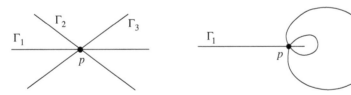

Figure 30. Reduction of type IV.   Figure 31. Reduction of type IV$_2$.

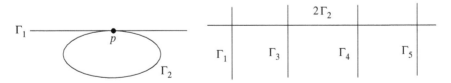

Figure 32. Reduction of type III.   Figure 33. Reduction of type I_0^*.

From now on, let us suppose that $d_2 = 2$, and hence $\Gamma_1 \cdot \Gamma_2 = 1$. It follows that $\Gamma_2 \simeq \mathbb{P}^1_k$. We have $3 = \sum_{i \geq 3} d_i\Gamma_2 \cdot \Gamma_i$. If the components that meet Γ_2 are of multiplicity 1, then either $\mathcal{E}_s$ is as in Figure 33 with $\Gamma_i \simeq \mathbb{P}^1_k$ and intersection points that are rational over k; or $\mathcal{E}_s$ is as in Figure 34, with $\Gamma_i \simeq \mathbb{P}^1_k$ if $i \leq 3$

and $\Gamma_4 \simeq \mathbb{P}^1_{k(p_2)}$ with $[k(p_2) : k] = 2$, and where $\Gamma_2 \cap \Gamma_3$ is a k-rational point; or $\mathcal{E}_s$ is as in Figure 35 with $[k(p_3) : k] = 3$ and $\Gamma_3 \simeq \mathbb{P}^1_{k(p_3)}$. In these three cases, we will say, respectively, that $\mathcal{E}_s$ is of type I_0^*, $I_{0,2}^*$, $I_{0,3}^*$. Over $\bar{k}$, they are all of type I_0^*.

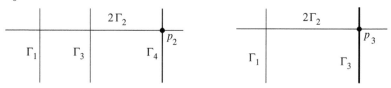

Figure 34. Reduction of type $I_{0,2}^*$.   Figure 35. Reduction of type $I_{0,3}^*$.

We proceed to the case when one of the components that meet Γ_2 is of multiplicity 2. Then $\mathcal{E}_s$ is either as in Figure 36, with $\Gamma_i \simeq \mathbb{P}^1_k$, and the intersection points rational over k (we then say that $\mathcal{E}_s$ is of type I_{n-5}^*); or $\mathcal{E}_s$ is as in Figure 37, with $\Gamma_i \simeq \mathbb{P}^1_k$ if $i \leq n-1$, and $\Gamma_n \simeq \mathbb{P}^1_{k(p_2)}$ with $k(p_2)$ quadratic over k. We will then say that $\mathcal{E}_s$ is of type $I_{n-4,2}^*$ (type I_{n-4}^* over $\bar{k}$). In the figures that follow, the integers indicate the multiplicities of the irreducible components. The non-marked components are of multiplicity 1.

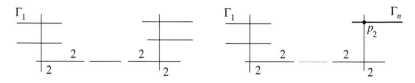

Figure 36. Type I_{n-5}^*.   Figure 37. Type $I_{n-4,2}^*$.

From now on, let us suppose that Γ_2 meets Γ_3 with $d_3 = 3$; then $\Gamma_3 \simeq \mathbb{P}^1_k$ and the intersection point $\Gamma_2 \cap \Gamma_3$ is rational over k. We have $4 = \sum_{i \geq 4} d_i \Gamma_3 \cdot \Gamma_i$, and $d_i \geq 2$ if $\Gamma_i \cap \Gamma_2 \neq \emptyset$.

If the components that meet Γ_3 are of multiplicity 2, then either $\mathcal{E}_s$ is represented by Figure 38, with $\Gamma_i \simeq \mathbb{P}^1_k$ and the intersection points rational over k; or Γ_3 meets a single component Γ_4 distinct from Γ_2, and $\mathcal{E}_s$ is represented by Figure 39, where the intersection point $p_2 \in \Gamma_3 \cap \Gamma_4$ verifies $[k(p_2) : k] = 2$. Moreover, $\Gamma_i \simeq \mathbb{P}^1_{k(p_2)}$ if $i = 4$ or 5, and the intersection point of Γ_4 and Γ_5 is rational over $k(p_2)$. In these two cases, we will respectively say that $\mathcal{E}_s$ is of type IV^* or IV_2^* (type IV^* over $\bar{k}$).

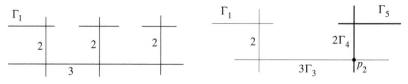

Figure 38. Type IV^*.   Figure 39. Type IV_2^*.

From now on, let us suppose that Γ_3 meets Γ_4 and that $d_4 = 4$. By an analogous reasoning, we show that $\mathcal{E}_s$ is either as in Figure 40 and we say that

$\mathcal{E}_s$ is of type III*, or as in Figure 41, and we say that $\mathcal{E}_s$ is of type II*. In both cases we have $\Gamma_i \simeq \mathbb{P}^1_k$ and the intersection points are rational over k. The classification ends here.

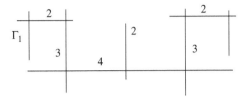

Figure 40. Reduction of type III*.

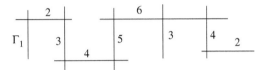

Figure 41. Reduction of type II*.

Remark 2.3. If k is not perfect, the classification above still gives all of the possibilities for $\mathcal{E}_s$, because the method is purely combinatorial. However, the irreducible components are not necessarily as we have described them. For example, a component of $\mathcal{E}_s$ can be of multiplicity 1 without being geometrically reduced.

Remark 2.4. Starting with an elliptic equation for E with coefficients in $\mathcal{O}_K$, there exists an algorithm, called the *Tate algorithm* [93], that determines whether the Weierstrass model associated to the equation is minimal, and that also determines the reduction type of E. See also [92], IV.9, which contains a more detailed proof of this algorithm.

Remark 2.5. The classification of $\mathcal{E}_s$ was first given by Kodaira [52], whose notation we adopted here. It was also done by Néron [75], chap. III, in a more arithmetic setting.

Remark 2.6. For curves of genus 2, the classification of the possible special fibers of the minimal regular model is given in [76] and [74]. The number of possible types then rises to approximatively 120.

10.2.2 Néron models of elliptic curves

Let S be a Dedekind scheme of dimension 1, with function field $K = K(S)$. Let (E, o) be an elliptic curve over K (where $o \in E(K)$ is a fixed rational point), and $\mathcal{E}$ its minimal regular model over S (whose existence is assured if S is affine, see Proposition 2.8). Let $\mathcal{N}$ denote the open subscheme of $\mathcal{E}$ made up of the points that are smooth over S (Corollary 6.2.12). We are going to show that E is an Abelian variety, and that $\mathcal{N}$ is the Néron model of E over S.

Definition 2.7. Let S be a Dedekind scheme of dimension 1, with function field $K = K(S)$. Let A be an Abelian variety over K. We define the *Néron model* *of A over S* to be a scheme $\mathcal{A} \to S$ which is smooth, separated, and of finite type, with generic fiber isomorphic to A, and that verifies the following universal property: for any smooth scheme X over S, the canonical map

$$\mathrm{Mor}_S(X, \mathcal{A}) \to \mathrm{Mor}_K(X_K, A)$$

is bijective.

The universal property implies that $\mathcal{A}$ is unique up to isomorphism and that the algebraic group structure of A extends in a unique way to the structure of a group scheme on $\mathcal{A} \to S$ (take $X = \mathcal{A} \times_S \mathcal{A}$). The notion of Néron model was invented by A. Néron [75], who also showed the existence of such a model. The Néron model is a very powerful tool for the study of the arithmetic properties of Abelian varieties. Cf. [15] for a modern treatment of the subject, or [7] for a rapid overview.

Algebraic group law on E

Let us first show that (E, o) can be endowed with the structure of an algebraic group (and hence of an Abelian variety since E is proper) such that o is the unit element. In Lemma 2.8 and Proposition 2.9, the field K is arbitrary. For any extension K'/K, the elements of $E(K')$ are seen both as points of $E_{K'}$ that are rational over K', and as K-morphisms $\mathrm{Spec}\, K' \to E$ (isomorphism (1.2) of Remark 3.1.6).

Lemma 2.8. *Let E be an elliptic curve over a field K, endowed with a rational point $o \in E(K)$.*

(a) *For any (not necessarily algebraic) field extension $K \to K'$, and for any $(x, y) \in E(K') \times E(K')$, there exists a unique point $m_{K'}(x, y) \in E(K')$ such that*

$$m_{K'}(x, y) + o \sim x + y$$

as Cartier divisors on $E_{K'}$. The map $m_{K'}$ makes $E(K')$ into a commutative group, with unit element o. Moreover, if $K' \subseteq K''$, then $m_{K'}$ is the restriction of $m_{K''}$.

(b) *Let $x \in E(K)$. Then there exists an automorphism of K-schemes $t_x : E \to E$, called the translation by x, such that for any extension K' of K, the map $E(K') \to E(K')$ induced by t_x is the translation by x. Moreover, by considering x as a point x' of $E(K')$, $t_{x'}$ is obtained from t_x by base change.*

Proof (a) By Riemann–Roch, $H^0(E_{K'}, \mathcal{O}_{E_{K'}}(x + y - o)) \neq 0$ (Remark 7.3.33). Hence $x + y - o$ is linearly equivalent to an effective divisor D of degree 1 over K'. Consequently, D is a rational point z. In addition, this point is unique by Corollary 7.3.12. The rest of the properties can be verified immediately.

(b) Let ξ be the generic point of E and $L = K(E)$. Let $i : \mathrm{Spec}\, L \to E$ be the canonical morphism. By Proposition 7.3.13(b), the canonical map

$$\theta : \mathrm{Mor}_K(E, E) \to \mathrm{Mor}_K(\mathrm{Spec}\, L, E) = E(L), \qquad f \mapsto f \circ i$$

is bijective. Let $\xi' = m_L(x, \xi) \in E(L)$ and let $t_x : E \to E$ be the morphism $\theta^{-1}(\xi')$. Then $t_x(\xi) = \xi'$ (we identify ξ with the morphism i). We therefore have a linear equivalence of Cartier divisors $x+\xi \sim o+\xi'$ on E_L. Let $j : E_L \to E \times_K E$ be the morphism $\mathrm{Id}_E \times i$. Then $j(\xi) \in \Delta_E$, the diagonal of $E \times_K E$, and $j(\xi') \in \Gamma'_{t_x} := (\mathrm{Id}_E \times t_x)^{-1}(\Delta_E)$ (this is the graph of t_x with permuted coordinates). Let us consider the Cartier divisor

$$F := \{x\} \times E + \Delta_E - (\{o\} \times E + \Gamma'_{t_x})$$

on $X := E \times_K E$. It follows that $j^*F \sim 0$. Let us consider X as a fibered surface over E via the second projection $q : X \to E$. We have just seen that the restriction of F to the generic fiber X_ξ of X is trivial. Let $f \in K(X) = K(X_\xi)$ be such that $j^*F = j^* \mathrm{div}(f)$. Then the support of $F - \mathrm{div}(f)$ does not meet X_ξ and is therefore contained in a finite union of closed fibers. Since $X \to E$ is smooth, each closed fiber is a prime divisor, hence $F - \mathrm{div}(f)$ is a sum of closed fibers. In other word, we have

$$\{x\} \times E + \Delta_E - (\{o\} \times E + \Gamma'_{t_x}) \sim q^*D$$

for some Cartier divisor D on E. Let $y : \mathrm{Spec}\, K' \to E$ be an element of $E(K')$. Then $y^*\mathcal{O}_E(D) \simeq \mathcal{O}_{\mathrm{Spec}\, K'}$. By taking the inverse image of the relation above in $E \times_K \mathrm{Spec}\, K' = X \times_E \mathrm{Spec}\, K'$, we obtain the relation $x + y - (o + t_x(y)) \sim 0$ on $E_{K'}$. Hence

$$t_x(y) = m_{K'}(x,y), \quad \text{for every } y \in E(K'). \tag{2.17}$$

If $z \in E(K)$, then this relation with $K' = \overline{K}$ and Exercise 3.2.9 imply that $t_x \circ t_z = t_{m_K(x,z)}$. By taking for $z \in E(K)$ the inverse of x for the group law defined by m_K, we obtain $t_x \circ t_z = t_o = \mathrm{Id}_E$. Hence t_x is an automorphism. Similarly, relation (2.17) implies that $t_{x'}$ is obtained from t_x by base change. $\square$

Proposition 2.9. *Let E be an elliptic curve over a field K, endowed with a rational point $o \in E(K)$. Then E has the structure of an Abelian variety over K such that o is the unit element and that for any extension K' of K, the group law on $E(K')$ defined in Lemma 2.8(a) is induced by the algebraic group structure on E.*

Proof Let us keep the notation of the proof of Lemma 2.8. Let $t_\xi : E_L \to E_L$ be the automorphism associated to the point $\xi \in E(L) = E_L(L)$. The second projection $q : E \times_K E \to E$ endows $E \times_K E$ with the structure of a minimal fibered surface over E. Hence t_ξ extends to an automorphism $t : E \times_K E \to E \times_K E$ of E-schemes (Proposition 9.3.13). Let $m : E \times_K E \to E$ be the morphism $p \circ t$, where $p : E \times_K E \to E$ is the first projection. Let us show that m verifies the stated properties. Let $x \in E(K)$. We have

$$t(x, \xi) = (t_\xi(x), \xi) = (m_L(x, \xi), \xi) = (t_x(\xi), \xi) \in \Gamma'_{t_x},$$

where $\Gamma'_{t_x} := \{(y, z) \in E \times_K E \mid y = t_x(z)\}$. Hence $t(\{x\} \times E) \subseteq \Gamma'_{t_x}$. For any $y \in E$, we then have $t(x, y) = (t_x(y), y)$. It follows from equality (2.17) that

$$m(x, y) = t_x(y) = m_K(x, y), \quad \text{if } x, y \in E(K).$$

Let $\mathrm{inv}_E : E \to E$ be the morphism $p \circ t^{-1} \circ (o, \mathrm{Id}_E)$. For any $y \in E(K)$, we have $\mathrm{inv}_E(y) = t_y^{-1}(o)$, and hence $m(y, \mathrm{inv}_E(y)) = t_y(\mathrm{inv}_E(y)) = o$.

Let K' be an extension of K. As E is geometrically integral, $K' \otimes_K L$ is the function field of $E_{K'}$ (Corollary 3.2.14(c)), and the generic point of $E_{K'}$ is the inverse image of ξ under the projection $E_{K'} \to E$. Hence the construction of m is compatible with the base change $\operatorname{Spec} K' \to \operatorname{Spec} K$, which implies that the properties above concerning m and inv_E are valid over K'. Hence m, inv_E induce the group law $m_{K'}$ of Lemma 2.8(a) on $E(K')$. Consequently, by Exercise 7.4.16, m indeed defines an algebraic group law on E. $\qquad\square$

Remark 2.10. With an explicit Weierstrass equation of E, it is possible to write the group law on E explicitly. Cf. [91], III.2–3. The method is based on the fact that if $x, y \in E(K)$ and if z is the inverse of $m(x,y)$ for the group law on $E(K)$, then $x + y + z \sim 3o$ as Cartier divisors on E.

Remark 2.11. The group law on E is commutative, that is to say that we have a commutative diagram

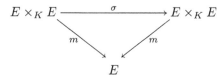

where σ is the permutation of the coordinates. Indeed, since the (abstract) group $E(\overline{K})$ is commutative by construction, it suffices to apply Exercise 3.2.9. As a trivial consequence, if $x, y \in E(K)$, then we have $t_x \circ t_y = t_y \circ t_x$.

Néron model

Lemma 2.12. *Let S be a Dedekind scheme of dimension 1, with function field $K = K(S)$. Let (E, o) be an elliptic curve over K, $\mathcal{E}$ its minimal regular model over S, and $\mathcal{N}$ the open subscheme of the smooth points of $\mathcal{E}$ over S. Then the following properties are true.*

(a) *The canonical maps $\mathcal{N}(S) \to \mathcal{E}(S) \to E(K)$ are bijective.*

(b) *For any section $x \in \mathcal{E}(S)$, the translation $t_{x_K} : E \to E$ associated to $x_K \in E(K)$ extends to an automorphism $t_x : \mathcal{E} \to \mathcal{E}$.*

(c) *Let $m : E \times_K E \to E$ be the algebraic group law on E defined in Proposition 2.9. Then the automorphism $t = (m, q) : E \times_K E \to E \times_K E$, where $q : E \times E \to E$ is the second projection, extends to an automorphism $t : \mathcal{E} \times_S \mathcal{N} \to \mathcal{E} \times_S \mathcal{N}$.*

(d) *Let $p : \mathcal{N} \times_S \mathcal{N} \to \mathcal{N}$ be the first projection. Then t induces an automorphism $\tau : \mathcal{N} \times_S \mathcal{N} \to \mathcal{N} \times_S \mathcal{N}$ and $p \circ \tau$ defines a smooth group scheme structure on $\mathcal{N} \to S$.*

Proof (a) Let $\varepsilon : \operatorname{Spec} K \to E$ be a section. By Corollary 3.3.26, ε extends to a morphism $\operatorname{Spec} \mathcal{O}_{S,s} \to \mathcal{E}$ for any $s \in S$. The latter extends to a morphism $\varepsilon_V : V \to \mathcal{E}$ for an open neighborhood V of s (Exercise 3.2.4). As $\mathcal{E}$ is separated, the ε_V (when s runs through the closed points of S) glue to a morphism $S \to \mathcal{E}$ (Proposition 3.3.11). Hence $\mathcal{E}(S) \to E(K)$ is surjective, and hence bijective because $\mathcal{E}$ is separated over S. As $\mathcal{E}$ is regular, the sections $\mathcal{E}(S)$ have their image in the smooth locus (Corollary 9.1.32), whence the equality $\mathcal{N}(S) = \mathcal{E}(S)$. Property (b) results from Proposition 9.3.13.

(c) Let us consider the automorphism t as a birational map $t : \mathcal{E} \times_S \mathcal{N} \dashrightarrow \mathcal{E} \times_S \mathcal{N}$. We must first show that t is defined everywhere. For this, we can suppose that S is local. By the descent of the definition domain by faithfully flat morphisms (Exercise 5.2.14), and because the formation of $\mathcal{E}$ and of $\mathcal{N}$ commute with étale base change and with completion (Proposition 9.3.28), we can suppose that S is the spectrum of a complete discrete valuation ring with separably closed residue field.

Abusively, we can imagine the first coordinate of $t(x, y)$ as the translation by y applied to x. The proof that follows is based on this interpretation. Let λ be a generic point of $\mathcal{N}_s$ and $T = \operatorname{Spec} \mathcal{O}_{\mathcal{N},\lambda}$. Then T is regular, local of dimension 1. The scheme $\mathcal{E} \times_S \mathcal{N}$ is smooth over $\mathcal{E}$, and is therefore regular (Theorem 4.3.36). Consequently, $\mathcal{E} \times_S T$ is also regular. It is therefore a minimal arithmetic surface over T (Corollary 9.3.30). By Proposition 9.3.13, t extends to an automorphism of T-schemes $\mathcal{E} \times_S T \to \mathcal{E} \times_S T$. As a birational map, t is therefore defined at the points of $\mathcal{E} \times_S U$ for some open neighborhood U of λ with $E \subset U$. We are going to show that t is defined everywhere by performing translations on U. Let $x \in \mathcal{N}(S)$ be a section. Let us consider the automorphism

$$t' = (t_x \times t_x) \circ t \circ (\operatorname{Id}_{\mathcal{E}} \times t_x^{-1}) : \ \mathcal{E} \times_S t_x(U) \to \mathcal{E} \times_S t_x(U).$$

Then t and t' coincide on $\operatorname{Spec} K$ because E is a commutative group (Remark 2.11); hence they coincide on $\mathcal{E} \times_S (U \cap t_x(U))$. Consequently, t is defined on $\mathcal{E} \times_S t_x(U)$. Thus the domain of definition of t contains $\mathcal{E} \times_S (\cup_x t_x(U))$, where x runs through the set $\mathcal{N}(S)$. Let us show that $\cup_x t_x(U) = \mathcal{N}$. *Let us suppose $k(s)$ is perfect* (hence algebraically closed). Let us fix a closed (hence rational) point $y_s \in U_s$. Let $z_s \in \mathcal{N}_s$ be closed. Then y_s, z_s lift respectively to sections $y, z \in \mathcal{N}(S)$ (Corollary 6.2.13). Let $x = t_y^{-1}(z)$. Then $t_x(y) = t_y(x) = z$. Therefore, $z_s = t_x(y_s) \in t_x(U)$. See Exercise 2.5 for the general case (when $k(s)$ is not necessarily perfect).

Finally, the automorphism $\operatorname{inv}_E : E \to E$ extends to an automorphism of $\mathcal{E}$ (by the uniqueness of the minimal regular model) that induces an automorphism $\operatorname{inv}_{\mathcal{N}} : \mathcal{N} \to \mathcal{N}$. It is easy to verify that $(\operatorname{Id}_{\mathcal{E}} \times \operatorname{inv}_{\mathcal{N}}) \circ t \circ (\operatorname{Id}_{\mathcal{E}} \times \operatorname{inv}_{\mathcal{N}})$ is the inverse of t. Hence t is an automorphism.

(d) The image of $\mathcal{N} \times_S \mathcal{N}$ under t is the smooth locus of $\mathcal{E} \times_S \mathcal{N}$, which is equal to $\mathcal{N} \times_S \mathcal{N}$. By construction, at the generic fiber, $p \circ \tau$ and $\operatorname{inv}_{\mathcal{N}}$ induce the algebraic group structure m, inv_E on E. To verify that $p \circ \tau$ and $\operatorname{inv}_{\mathcal{N}}$ induce the structure of a group scheme over S (with $\{o\}$ as unit element), we must verify that the diagrams in the definition of group schemes are commutative. But they are commutative at the generic fiber, and hence commutative over S (Proposition 3.3.11). $\qquad\square$

Remark 2.13. Here we have used an ad hoc method to prove the existence of a group scheme structure on $\mathcal{N}$. For Abelian varieties of arbitrary dimension, the right notion to use is that of the normal law, which is a group law defined birationally on $\mathcal{E}$. See [7], 1.12 and 1.15.

Theorem 2.14. *Let S be a Dedekind scheme of dimension 1, with function field $K = K(S)$. Let E be an elliptic curve over K with minimal regular model $\mathcal{E}$ over S. Then the open subscheme $\mathcal{N}$ of smooth points of $\mathcal{E}$ is the Néron model of E over S.*

Proof By Lemma 2.12, $\mathcal{N}$ is a smooth group scheme, separated of finite type over S (because contained in $\mathcal{E}$), with generic fiber isomorphic to E. It remains to show the universal property. Let X be a smooth scheme over S, and let $f : X_\eta \to E$ be a morphism considered as a rational map $X \dashrightarrow \mathcal{N}$. Let $\xi \in X$ be a point of codimension 1, and $T = \operatorname{Spec} \mathcal{O}_{X,\xi}$. As we have seen in the proof of Lemma 2.12(c), $\mathcal{E} \times_S T \to T$ is a minimal arithmetic surface; its smooth locus is $\mathcal{N} \times_S T$. Hence $\mathcal{N}(T) = \mathcal{N}_T(T) \to E_{K(T)}(K(T)) = E(K(X))$ is bijective. Therefore f is defined at ξ. The following theorem of Weil implies that f is defined on X. $\square$

Theorem 2.15 (Weil). *Let $G \to S$ be a group scheme that is smooth and separated over a normal Noetherian scheme S, let X be a smooth scheme over S, and $f : X \dashrightarrow G$ a rational map. If f is defined at the points of codimension 1 and at the generic points of all fibers of $X \to S$, then f is defined everywhere.*

Proof See [15], 4.4, Theorem 1, or [7], Proposition 1.3. The idea is to consider the rational map

$$F : X \times_S X \dashrightarrow G, \quad (x,y) \mapsto f(x)f(y)^{-1}$$

(that is to say $m \circ (\operatorname{Id}_G \times \operatorname{inv}_G) \circ (f \times f)$) and to show that the domain of definition of F is the product of the domain of definition of f with itself, and that it contains the diagonal Δ_X. $\square$

Example 2.16. Let E be an elliptic curve over $\mathbb{Q}$ defined by an affine equation

$$y^2 + y = x^3 + 1.$$

Let W be the Weierstrass model over $\mathbb{Z}$ corresponding to this equation. Then W is smooth over $\mathbb{Z}$ outside of $p = 3, 5$. The fiber W_5 contains a unique singular point q_5 defined by $p = 5, x = 0, y = 2$. This point is regular in W (Kodaira type II, see Figure 27). The fiber W_3 contains a unique singular point at $p = 3$, $x = 1$, and $y = 1$. This point is also singular in W. Let $\mathcal{E}$ be the blowing-up of W with center at this point. Then $\mathcal{E}$ is regular, and its fiber over $p = 3$ is the union of two projective lines over $\mathbb{F}_3$ meeting at a rational point q_3, with multiplicity 2 (Kodaira type III, see Figure 32). The Néron model is then equal to $\mathcal{E} \setminus \{q_3, q_5\}$. The fiber $\mathcal{N}_p$ is connected for $p \neq 3$, while $\mathcal{N}_3$ is made up of two connected components.

Summary
Let us suppose S is affine. Let W be the minimal Weierstrass model of E over S. The relations between the models W, $\mathcal{E}$, and $\mathcal{N}$ are the following: $\mathcal{E}$ is the minimal desingularization of W; $\mathcal{N}$ is the open subscheme of the points of $\mathcal{E}$ that

are smooth over S. Moreover, these schemes are birational to each other and we have

$$H^0(\mathcal{E}, \omega_{\mathcal{E}/S}) = H^0(\mathcal{N}, \omega_{\mathcal{N}/S}) = H^0(W, \omega_{W/S})$$

as subgroups of $H^0(E, \omega_{E/K})$. See Remark 9.4.39.

Component group, filtration

Lemma 2.17. *Let k be a field, A a k-algebra that is a subalgebra of a finitely generated k-algebra B. Then there exists a largest subalgebra A^{et} of A that is finite and étale over k. Moreover, the points of $\operatorname{Spec} A^{et}$ correspond to the connected components of $\operatorname{Spec} A$.*

Proof An algebra is finite étale over k if and only if it is a finite direct sum of finite separable extensions of k. Let A_1, A_2 be two finite étale subalgebras of A. Then $A_1 \otimes_k A_2$ is étale over k. The compositum $A_1 A_2$ is a quotient of $A_1 \otimes_k A_2$ and is therefore étale over k. We can therefore take for A^{et} the compositum of the finite étale subalgebras of A. It remains to show that A^{et} is finite over k. It is clear that $A^{et} \subseteq B^{et}$, and that $A^{et} = (A/\sqrt{0})^{et}$. We can therefore reduce to the case when A is integral, finitely generated over k. It follows that $A^{et} \subseteq \bar{k} \cap \operatorname{Frac}(A)$ is an algebraic field that is finitely generated over k; it is therefore finite over k.

By definition, an idempotent element $e \in A$ verifies the equation $e^2 - e = 0$. Hence $e \in A^{et}$. Consequently, A and A^{et} have the same idempotent elements, and their connected components correspond bijectively (Exercise 2.4.6). But $\operatorname{Spec} A^{et}$ is finite over $\operatorname{Spec} k$; its connected components coincide with its points, which concludes the proof. □

Proposition 2.18. *Let X be an algebraic variety over a field k.*

(a) *There exist a unique scheme $\pi_0(X)$, finite étale over k, and a morphism $f : X \to \pi_0(X)$ verifying the following universal property: any k-morphism $X \to Z$ of X to a finite étale k-scheme Z factors in a unique way as*

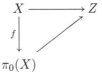

(b) *The morphism $f : X \to \pi_0(X)$ is surjective. For any $x \in X$, the fiber $X_{f(x)}$ is the connected component of x in X. In particular, the points of $\pi_0(X)$ correspond bijectively to the connected components of X.*

(c) *Let L/k be an extension. Then we have a canonical isomorphism $\pi_0(X_L) \simeq \pi_0(X) \times_k \operatorname{Spec} L$, the π_0 on the left being that of algebraic varieties over L.*

(d) *Let Y be another algebraic variety over k. Then we have a canonical isomorphism $\pi_0(X \times_k Y) \simeq \pi_0(X) \times_k \pi_0(Y)$.*

Proof Let $X_1, \ldots, X_n$ be affine open subschemes that cover X. Then $\mathcal{O}_X(X)$ is a subalgebra of $\oplus_{1 \le i \le n} \mathcal{O}_X(X_i)$. We can therefore define $\mathcal{O}_X(X)^{et}$. Let us note that X has the same connected components as $\operatorname{Spec} \mathcal{O}_X(X)$ (Exercise 2.4.6). Let us set $\pi_0(X) = \operatorname{Spec} \mathcal{O}_X(X)^{et}$. It is then easy to verify properties (a) and (b). The proof of properties (c) and (d) is slightly longer. See [94], Section 6.5, or [28], I, Section 4, n° 6. $\square$

Definition 2.19. Let X be an algebraic variety over a field k. The finite étale k-scheme $\pi_0(X)$ is called the *scheme of connected components of X*.

Definition 2.20. Let G be an algebraic group over a field k. The connected component G^0 of G containing the unit element of G is called the *identity component of G*, and $\pi_0(G)$ is called the *group of components* of G.

Corollary 2.21. *Let k be a field.*

(a) *Let X be an algebraic variety over k. Then the set of rational points $\pi_0(X)(k)$ corresponds to the connected components of X that are geometrically connected. A connected component containing a rational point is geometrically connected.*

(b) *Let G be an algebraic group over k; then G^0 is an open algebraic subgroup of G. The scheme $\pi_0(G)$ is a finite étale algebraic group over k.*

Proof   (a) We can suppose X is connected. Hence $\pi_0(X) = \operatorname{Spec} k'$ for some finite separable extension k'/k. By Proposition 2.18(c), we have

$$\pi_0(X_{\overline{k}}) \simeq \pi_0(X) \times_k \operatorname{Spec} \overline{k} = \operatorname{Spec}(k' \otimes_k \overline{k}).$$

Hence X is geometrically connected if and only if $\operatorname{Spec}(k' \otimes_k \overline{k})$ is reduced to a point. As k' is separable over k, this is equivalent to $k' = k$. If C is a connected component of X with $C(k) \neq \emptyset$, then its image under $X \to \pi_0(X)$ is a rational point, which implies that C is geometrically connected (we can use Exercise 3.2.11).

(b) Let o be the unit element of G. As G^0 contains $o \in G(k)$, its image $\pi_0(G^0)$ is also a rational point. It follows that $\pi_0(G^0 \times_k G^0) \simeq \pi_0(G^0) \times_k \pi_0(G^0)$ is a rational point. The group law $m : G \times_k G \to G$ therefore maps $G^0 \times_k G^0$ onto a connected subset of G containing o. We deduce from this that m induces a morphism $G^0 \times_k G^0 \to G^0$. This immediately implies that G is an algebraic subgroup of G. By the universal property of π_0, m induces a morphism

$$\pi_0(G) \times_k \pi_0(G) \simeq \pi_0(G \times_k G) \to \pi_0(G).$$

Likewise, inv_G induces a morphism $\pi_0(G) \to \pi_0(G)$. It is easy to verify that this induces an algebraic group law on $\pi_0(G)$ whose unit element corresponds to the identity component of G. $\square$

From here to the end of the subsection, S will be the spectrum of a discrete valuation ring $\mathcal{O}_K$ with residue field k. As usual, the closed point of S is denoted by s. We will suppose that k is perfect or that $\operatorname{char}(k) \neq 2, 3$. Let us return to the Néron model $\mathcal{N}$ of E over S.

Remark 2.22 (*Structure of the algebraic group $\mathcal{N}_s^0$*). If E has good reduction, then $\mathcal{N}_s^0 = \mathcal{N}_s$ is an elliptic curve over k (Lemma 2.1). If E has split multiplicative (resp. additive) reduction, then $\mathcal{N}_s^0$ is isomorphic to the algebraic group $\mathbb{G}_{m,k}$ (resp. $\mathbb{G}_{a,k}$) by the uniqueness of the algebraic group structure on $\mathbb{A}_k^1 \setminus \{0\}$ (resp. $\mathbb{A}_k^1$), see Exercises 7.4.17–18. If E has non-split multiplicative reduction, $\mathcal{N}_s^0$ is an algebraic group over k that is isomorphic to $\mathbb{G}_m$ over a quadratic extension of k. We say that it is a *non-split torus*. See Exercise 2.7 for more details.

Reduction type	I_0	$I_n, n \geq 1$	$I_{n,2}, n \geq 1$	The rest
$\mathcal{N}_s^0$	elliptic	$\mathbb{G}_{m,k}$	non-split torus	$\mathbb{G}_{a,k}$

Remark 2.23. With the classification of $\mathcal{E}_s$ (Subsection 10.2.1), we can determine the set $\mathcal{N}_s$ explicitly. If $\mathcal{E}_s$ is irreducible, then $\mathcal{N}_s = \mathcal{N}_s^0$. Let us suppose $\mathcal{E}_s$ is non-irreducible. Then $\mathcal{N}_s$ is the complement in $\mathcal{E}_s$ of the irreducible components of multiplicity ≥ 2, of the intersection points of $\mathcal{E}_s$, and if $\mathcal{E}_s$ is of type $I_{2n-1,2}$ (Figure 29), of the singular point of Γ_n.

Remark 2.24. Let U be a connected component of $\mathcal{N}_s$. Let Γ be the irreducible component of $\mathcal{E}_s$ containing U. Then U is geometrically connected if and only if it is geometrically integral, since U is smooth. This is also equivalent to Γ geometrically integral (Corollary 3.2.14(c)). Let Φ_E denote the group of components $\pi_0(\mathcal{N}_s)$ of $\mathcal{N}_s$. It follows from Corollary 2.21(a) that the rational points $\Phi_E(k)$ correspond to the components Γ that are geometrically integral and of multiplicity 1 in $\mathcal{E}_s$. We can read these components in the classification of the $\mathcal{E}_s$. The components of multiplicity 1 that are not geometrically irreducible are: the $\mathbb{P}_{k'}^1$ with $[k' : k] \geq 2$; the component Γ_n if E is of type $I_{2n-1,2}$ (Figure 29); and the component Γ_2 if E is of type IV_2 (Figure 31). As for the set Φ_E, it simply corresponds to the irreducible components of $\mathcal{E}_s$ of multiplicity 1. Later on, in Section 10.4, we will see a method to determine the group structure on $\Phi_E(\bar{k})$ (Exercise 4.7(b)). We then have the following table ($n \geq 1$):

Reduction type	I_n	III, III*	IV, IV*	I_{2n}^*	I_{2n+1}^*	The rest
$\Phi_E(\bar{k})$	$\mathbb{Z}/n\mathbb{Z}$	$\mathbb{Z}/2\mathbb{Z}$	$\mathbb{Z}/3\mathbb{Z}$	$(\mathbb{Z}/2\mathbb{Z})^2$	$\mathbb{Z}/4\mathbb{Z}$	0

We deduce from this the structure of $\Phi_E(k)$ ($n \geq 2$):

Reduction type	$I_{2n-2,2}$	$I_{2n-3,2}$	IV_2, IV_2^*	$I_{n-2,2}^*$	$I_{0,3}^*$	The rest
$\Phi_E(k)$	$\mathbb{Z}/2\mathbb{Z}$	0	0	$\mathbb{Z}/2\mathbb{Z}$	0	$\Phi_E(\bar{k})$

The universal property of the Néron model implies that the canonical map $\mathcal{N}(S) \to E(K)$, induced by the base change $\operatorname{Spec} K \to S$, is bijective. By composing its inverse with the canonical map $\mathcal{N}(S) \to \mathcal{N}_s(k)$, we obtain a reduction map $r : E(K) \to \mathcal{N}_s(k)$.

Lemma 2.25. *The map* $r : E(K) \to \mathcal{N}_s(k)$ *above is a group homomorphism.*

Proof Let $T \to S$ be a morphism of schemes. Then we have a commutative diagram

$$
\begin{array}{ccc}
\mathcal{N} \times_S \mathcal{N} & \longrightarrow & \mathcal{N} \\
\uparrow & & \uparrow \\
\\
\mathcal{N}_T \times_T \mathcal{N}_T & \longrightarrow & \mathcal{N}_T
\end{array}
$$

where the horizontal arrows are the group scheme law and where the vertical arrows are the canonical projections. We immediately deduce from this that the canonical map $\mathcal{N}(S) \to \mathcal{N}(T)$ is a group homomorphism. We obtain the lemma by applying this remark to $T = \operatorname{Spec} K$ and $T = \operatorname{Spec} k$.   □

Let us fix $s \in S$ and let $E^0(K) = r^{-1}(\mathcal{N}_s^0(k))$, $E^1(K) = r^{-1}(\tilde{o})$, where $\tilde{o}$ is the unit element of $\mathcal{N}_s$. These are subgroups of $E(K)$, by the lemma above.

Proposition 2.26 (Filtration of $E(K)$). *Let us suppose that S is the spectrum of a complete discrete valuation ring $\mathcal{O}_K$ with finite residue field k (e.g., $K = \mathbb{Q}_p$). Then we have a bijection $E^1(K) \simeq \mathfrak{m}_K$, where $\mathfrak{m}_K$ is the maximal ideal of $\mathcal{O}_K$, and group isomorphisms*

$$ E^0(K)/E^1(K) \simeq \mathcal{N}_s^0(k), \quad E(K)/E^0(K) \simeq \Phi_E(k). \tag{2.18} $$

Proof The k finite hypothesis is only useful to show the surjectivity of the last of the three homomorphisms. The first bijection is a specific case of Proposition 1.40(b). The two homomorphisms of (2.18) are injective by definition. By Corollary 6.2.13, the homomorphism $r : E(K) \to \mathcal{N}_s(k)$ is surjective. Hence $E^0(K) \to \mathcal{N}_s^0(k)$ is surjective. It remains to show that $\mathcal{N}_s(k) \to \Phi_E(k)$ is surjective. This is equivalent to saying that if U is a connected component of $\mathcal{N}_s$ that is moreover geometrically connected, then U contains a rational point.

We can suppose that E does not have good reduction. Let U be a geometrically connected component. Let Γ be the irreducible component of $\mathcal{E}_s$ containing U. Then Γ is a geometrically integral conic (see the beginning of Remark 2.24). It is smooth by Proposition 7.4.1. It follows from the k finite hypothesis that Γ contains a rational point (Exercise 4.3.23) and therefore $\Gamma \simeq \mathbb{P}_k^1$ (Proposition 7.4.1). Now the classification of the $\mathcal{E}_s$ shows that Γ meets the other irreducible components at at most two points. Hence $\Gamma \setminus U$ contains at most two points (this also results from the fact that $U_{\bar{k}} \simeq \mathcal{N}_{\bar{k}}^0$ as a variety, and that the latter is the complement of one or two points in $\mathbb{P}^1$), which implies that $U(k) \neq \emptyset$.   □

10.2.3   Potential semi-stable reduction

We have already seen the notion of potential good reduction (Definition 1.27). Let us introduce the notion of potential multiplicative reduction.

Definition 2.27. Let S be a Dedekind scheme of dimension 1, and let E be an elliptic curve over $K = K(S)$. We say that E has *potential multiplicative reduction* at a closed point $s \in S$ if there exists a discrete valuation ring $\mathcal{O}_L$ that dominates $\mathcal{O}_{S,s}$ such that the reduction of E_L is multiplicative (Definition 2.2). We say that E has *potential semi-stable reduction* if it has potential good or potential multiplicative reduction. We will show in Proposition 2.33 that E always has potential semi-stable reduction.

Lemma 2.28. *Let C be a smooth projective curve over K of genus $g \geq 1$, let $\mathcal{C}$ be a model of C over S, and Γ an irreducible component of $\mathcal{C}_s$ of multiplicity 1. Let us suppose that there exists an $x \in \Gamma$ whose inverse image in the normalization of Γ contains at least two points. Then the birational map $\mathcal{C} \dashrightarrow \mathcal{C}_{\min}$ to the minimal regular model of C over S is an isomorphism over a non-empty open subset of Γ.*

Proof Let $\pi : \mathcal{C}_1 \to \mathcal{C}$ be the minimal desingularization of $\mathcal{C}$ (Proposition 9.3.32 and Corollary 8.3.51), $\rho_1 : \mathcal{C}_1 \to \mathcal{C}_{\min}$ the canonical morphism, and Γ_1 the strict transform of Γ in $\mathcal{C}_1$. We must show that $\rho_1(\Gamma_1)$ is not reduced to a point.

Let us suppose the contrary. Then ρ_1 factors into a morphism $\rho_2 : \mathcal{C}_1 \to \mathcal{C}_2$ followed by $\mathcal{C}_2 \to \mathcal{C}_{\min}$ which is the blowing-up of the point $\rho_1(\Gamma_1)$ (Lemma 9.2.1). Let us set $\Gamma_2 = \rho_2(\Gamma_1)$. Then Γ_2 has multiplicity 1 in $(\mathcal{C}_2)_s$, and we have $\Gamma_2 \simeq \mathbb{P}^1_{k'}$ for some extension k' of $k(s)$ with $\Gamma_2^2 = -[k' : k(s)]$ (Proposition 9.2.5). The birational morphism $\Gamma_1 \to \Gamma_2$ is therefore an isomorphism. By hypothesis, $\pi^{-1}(x) \cap \Gamma_1$ contains at least two points. Zariski's connectedness principle (Corollary 5.3.16) says that $\pi^{-1}(x)$ is connected. Hence $\rho_2(\pi^{-1}(x))$ is a closed connected subset of $\mathcal{C}_2$ that meets Γ_2 at at most two points; it is therefore of dimension 1. Let D be the effective divisor on $\mathcal{C}_2$ associated to $\rho_2(\pi^{-1}(x))$ (endowed with the reduced scheme structure). Then we have

$$1 = -\Gamma_2^2 [k' : k(s)]^{-1} \geq D \cdot \Gamma_2 [k' : k(s)]^{-1} = \deg_{\Gamma_2} D|_{\Gamma_2} \geq 2$$

(the first inequality results from Proposition 9.1.21(b)), which is absurd. $\square$

Corollary 2.29. *Let W be a Weierstrass model of E. Let us suppose that there exists a point of W_s whose inverse image in the normalization of W'_s contains at least two points. Then W is minimal and E has split multiplicative reduction at s.*

Proof It results from Lemma 2.28 and from Theorem 8.3.20 that W is dominated by the minimal regular model $\mathcal{E}$, and hence W is minimal (Corollary 9.4.38). As W_s is singular, the proof of Lemma 2.1 (which does not depend on the hypothesis on the residue field $k(s)$ once we know that W_s is singular) shows that E has split multiplicative reduction at s. $\square$

Proposition 2.30. *Let E be an elliptic curve over $K = K(S)$.*

(a) *If E has split multiplicative reduction at s, then for any discrete valuation ring $\mathcal{O}_L$ dominating $\mathcal{O}_{S,s}$, E_L has split multiplicative reduction.*

(b) *The curve E cannot simultaneously have potential multiplicative reduction and potential good reduction at the same point $s \in S$.*

Proof (a) Let W be the minimal Weierstrass model of E over $\mathcal{O}_{S,s}$. Then $W \times_{\mathrm{Spec}\,\mathcal{O}_{S,s}} \mathrm{Spec}\,\mathcal{O}_L$ is a Weierstrass model of E_L that verifies the conditions of Corollary 2.29. It is therefore minimal, and E_L has split multiplicative reduction.

(b) We can suppose S is local and $\mathcal{O}_{S,s}$ complete. Let $\mathcal{O}_{L_1}$ (resp. $\mathcal{O}_{L_2}$) be a finite extension of $\mathcal{O}_{S,s}$ such that E_{L_1} (resp. E_{L_2}) has good reduction (resp. multiplicative reduction). Enlarging $\mathcal{O}_{L_2}$ if necessary, we can suppose that E_{L_2} has split multiplicative reduction. There exists a discrete valuation ring $\mathcal{O}_L$ that dominates $\mathcal{O}_{L_1}$ and $\mathcal{O}_{L_2}$. Then E_L has both good reduction and multiplicative reduction, which is impossible by the uniqueness of the minimal regular model. $\qquad\square$

In the remainder of the subsection, we are going to determine whether E has potential good reduction (resp. potential multiplicative reduction) in terms of the modular invariant j of E. Let E be an elliptic curve over K defined by an affine Weierstrass equation

$$y^2 + (a_1 x + a_3)y = x^3 + a_2 x^2 + a_4 x + a_6. \tag{2.19}$$

To this equation, we associate the discriminant Δ (see Subsection 9.4.4) and the following invariants:

$$b_2 = a_1^2 + 4a_2, \quad b_4 = 2a_4 + a_1 a_3, \quad c_4 = b_2^2 - 24b_4, \quad j = c_4^3/\Delta.$$

It is well known that $j \in K$ is independent of the choice of the equation ([91], Section III.1). This scalar is called the *modular invariant* or *j-invariant* of E. We also denote it by $j(E)$. By construction, it is clear that $j(E)$ remains the same after extension of the base field.

Lemma 2.31. *Let E be an elliptic curve over a field K.*

(a) *If $\mathrm{char}(K) \neq 2$, then there exists a finite extension L/K such that E admits an equation of the form*

$$v^2 = u(u-1)(u-\lambda), \quad \lambda \in L. \tag{2.20}$$

(b) *If $\mathrm{char}(K) \neq 3$, then there exists a finite extension L/K such that E admits an equation of the form*

$$v^2 + (au+b)v = u^3, \quad a, b \in L. \tag{2.21}$$

Proof Let us first fix an equation (2.19) of E.

(a) We can take $a_1 = a_3 = 0$. Let L be an extension of K containing the roots of $x^3 + a_2 x^2 + a_4 x + a_6$. There exist $\alpha, \beta \in L$ such that the map $x \mapsto \alpha x + \beta$ transforms these roots into $0, 1, \lambda$ for some $\lambda \in L$. Enlarging L if necessary, we have $\alpha^3 = \gamma^2$ for some $\gamma \in L$. Let us set $u = \alpha x + \beta$ and $v = \gamma y$. Then we have an equation (2.20) for E_L with $\lambda \in L$.

(b) Since $\mathrm{char}(K) \neq 3$, we can eliminate a_2 in equation (2.19) by a translation on x. Let $\alpha, \beta, \gamma \in \overline{K}$ (considered as variables) and let us set $x = u + \alpha$, $y = v + \beta u + \gamma$. We then obtain

$$v^2 + ((2\beta + a_1)u + 2\gamma + \alpha a_1 + a_3)v = u^3 + A_2 u^2 + A_4 u + A_6$$

with

$$A_2 = \beta^2 + a_1\beta - 3\alpha$$
$$A_4 = (2\gamma + \alpha a_1 + a_3)\beta + (a_1\gamma - 3\alpha^2 - a_4)$$
$$A_6 = \gamma^2 + (\alpha a_1 + a_3)\gamma - (\alpha^3 + a_4\alpha + a_6).$$

We want to solve the system

$$A_2 = A_4 = A_6 = 0. \tag{2.22}$$

The first two equations are equivalent to

$$\alpha = (\beta^2 + a_1\beta)/3$$
$$(2\beta + a_1)\gamma = -(\alpha a_1 + a_3)\beta + (3\alpha^2 + a_4).$$

If there exists a solution of the system $A_2 = A_4 = 0$ with $2\beta + a_1 = 0$, then system (2.22) clearly has a solution because the first two equations do not impose any constraint on γ. In the opposite case, by substituting these relations in A_6 and by multiplying by $(2\beta + a_1)^2$, we obtain a polynomial equation of degree 8 in β, with dominant coefficient $1/27$. Hence (2.22) always admits a solution (α, β, γ) in $\overline{K}$. Let L be a finite extension of K containing α, β, γ; then E_L has an equation of the form (2.21). □

Remark 2.32. In the $\mathrm{char}(K) \neq 3$ case, the existence of an equation of the form (2.21) over an extension L/K is connected to the existence of 3-torsion points of $E(L)$. Indeed, if $P \in E(L)[3]$ (i.e., $3P = o$ for the group law on $E(L)$), then Lemma 7.4.6 shows that there exists a line H in $\mathbb{P}_L^2$ such that $H \cap E = \{P\}$. It follows that $i_P(H, E_L) = 3$. By a suitable change of variables, we can suppose that P corresponds to the point $\{u = v = 0\}$ and that the tangent at P is the line $\{y = 0\}$. This implies that $a_6 = a_4 = a_2 = 0$. Conversely, if we have an equation of the form (2.21) over L, then the point $(0,0)$ is an inflexion point, and hence a point of order 3 in $E(L)$.

Proposition 2.33. *Let S be a Dedekind scheme of dimension 1. Let E be an elliptic curve over $K = K(S)$ and $s \in S$ a closed point. Then the following properties are true.*

(a) *If $j(E) \in \mathcal{O}_{S,s}$, then E has potential good reduction at s.*

(b) *If $j(E) \notin \mathcal{O}_{S,s}$, then E has potential multiplicative reduction at s.*

(c) *The implications above are equivalences.*

Proof Property (c) results from (a), (b), and from Proposition 2.30(b). Let us therefore show (a) and (b). We can suppose S is local and $\mathcal{O}_{S,s}$ complete. Let

L be a finite extension of K as in Lemma 2.31, let B be the integral closure of $\mathcal{O}_{S,s}$ in L, and $\mathcal{O}_L$ be the localization of B at a maximal ideal. It is consequently a discrete valuation ring that dominates $\mathcal{O}_K$ (Proposition 4.1.31).

Let us first treat the case that $\mathrm{char}(k(s)) \neq 2$, and therefore $\mathrm{char}(K) \neq 2$. Multiplying u and v if necessary by an invertible element of a quadratic extension of L, we can suppose that $\lambda \in \mathcal{O}_L$ in equation (2.20) of E_L. A direct computation shows that

$$j(E) = 2^8 \frac{(\lambda^2 - \lambda + 1)^3}{\lambda^2(\lambda - 1)^2}.$$

Equation (2.20) induces a Weierstrass model W of E_L over $\mathrm{Spec}\,\mathcal{O}_L$, with discriminant $\Delta = 2^4\lambda^2(\lambda - 1)^2 \in L^*$. If $j(E) \in \mathcal{O}_{S,s}$, then we immediately verify that $\lambda, \lambda - 1 \in \mathcal{O}_L^*$. Hence $\Delta \in \mathcal{O}_L^*$ and E_L has good reduction over $\mathrm{Spec}\,\mathcal{O}_L$. If $j(E) \notin \mathcal{O}_{S,s}$, then λ or $\lambda - 1$ belongs to the maximal ideal of $\mathcal{O}_L$. Let us suppose that it is λ. Then the point $\{u = v = 0\}$ of W_s is singular, and there are two points $v/u = \pm\sqrt{-1}$ in the normalization of W_s that lie above this point. Consequently, E_L has multiplicative reduction (Corollary 2.29).

Let us now suppose $\mathrm{char}(k(s)) \neq 3$. Let us consider equation (2.21) of E_L. Enlarging L again and multiplying u and v by an invertible element if necessary, we can suppose that $a, b \in \mathcal{O}_L$, and that at least one of the two is invertible. Let W be the Weierstrass model of E_L over $\mathcal{O}_L$ associated to equation (2.21). Let Δ be its discriminant. A direct computation shows that

$$\Delta = b^3(a^3 - 27b), \quad j(E) = a^3(a^3 - 24b)^3/\Delta.$$

If $j(E) \in \mathcal{O}_{S,s}$, we easily deduce from this that $\Delta \in \mathcal{O}_L^*$, and hence E_L has good reduction. Let us suppose $j(E) \notin \mathcal{O}_{S,s}$. We have two possibilities. The first is that b belongs to the maximal ideal of $\mathcal{O}_L$. Then $a \in \mathcal{O}_L^*$, W_s contains a singular point x defined by $u = v = 0$, and there exist two points $v/u = 0$ or $-a$ lying above x in the normalization of W_s. Hence E_L has multiplicative reduction, by Corollary 2.29. The second possibility is that $b \in \mathcal{O}_L^*$ and $a^3 - 27b$ belongs to the maximal ideal of $\mathcal{O}_L$. In this case, let us set $c = a/3$, $u = u_1 - c^2$, $v = v_1 + c^3$. Then W_s contains an affine open subscheme with equation

$$v_1^2 + 3\tilde{c}u_1v_1 = u_1^3 - 3\tilde{c}^2u_1^2,$$

where $\tilde{c}$ is the image of c in the residue field of $\mathcal{O}_L$. Enlarging L if necessary, we can suppose that the polynomial $T^2 + 3T + 3$ has its two (distinct) roots in the residue field of $\mathcal{O}_L$. We then see that in the normalization of W_s, there exist two points above the singular point $\{u_1 = v_1 = 0\}$, which are $\{u_1 = v_1 - (\tilde{c}u_1)\alpha = 0\}$, with α root of $T^2 + 3T + 3$. Hence E_L has multiplicative reduction. $\square$

Remark 2.34. For curves of higher genus, there does not exist any criterion for potential good reduction as simple as that of Proposition 2.33. For curves of genus 2, an analogue exists using Igusa invariants [48] (see also [59], Théorème 1). For an Abelian variety A, we have a criterion for good reduction, called the *criterion of Néron–Ogg–Shafarevich*, and a criterion for potential good reduction in terms of the torsion points of A ([88], Theorems 1–2).

Exercises

2.1. Let $\mathcal{E}$ be the minimal regular model of an elliptic curve as in Subsection 10.2.1. Let Γ be an irreducible component of $\mathcal{E}_s$ that is a singular conic over k. Show that for any $x \in E(K)$, we have $\overline{\{x\}} \cap \Gamma = \emptyset$.

2.2. Let E be an elliptic curve over a discrete valuation field as in Subsection 10.2.1. Let $\mathcal{E}'$ be the regular model of E with normal crossings.

(a) Let us suppose that the reduction of E is of type II or III. Show that $\mathcal{E}'_s$ is as in Figure 42 (use Exercise 9.2.9(d)). The integers are the multiplicities of the components.

Figure 42. Models with normal crossings for types II and III.

(b) Let us suppose that the reduction of E is of type IV or IV$_2$. Show that $\mathcal{E}'_s$ is represented by Figure 43, the figure on the left corresponding to type IV and the one on the right to type IV$_2$. The irreducible components are projective lines over k, except Γ, which is a projective line over a quadratic extension of k.

Figure 43. Models with normal crossings for types IV, IV$_2$.

(c) If the reduction of E is of type I$_{2n-1,2}$ (Figure 29 with Γ_n singular), show that $\mathcal{E}'_s$ can be obtained by replacing Γ_n by $\Gamma'_n \simeq \mathbb{P}^1_{k(p)}$, and by adding a line $\Gamma_{n+1} \simeq \mathbb{P}^1_k$ of multiplicity 2, the intersection point $\Gamma'_n \cap \Gamma_{n+1}$ having residue field $k(p)$.

Figure 44. Model with normal crossings for type I$_{2n-1,2}$.

(d) Show that outside of the preceding cases, $\mathcal{E}$ already has normal crossings.

2.3. Let E be an elliptic curve over a discrete valuation field K. By refining the computations of Lemma 2.31 and of Proposition 2.33, show that there exists a separable extension L/K of degree at most 24 such that E_L has good reduction or multiplicative reduction.

2.4. Let (E, o) be an elliptic curve over a field K. For any extension K'/K, let us consider the map

$$f_{K'} : E(K') \to \mathrm{Pic}^0(E_{K'}), \qquad x \mapsto \mathcal{O}_{E_{K'}}(x - o).$$

Show that $f_{K'}$ is a group isomorphism and that its construction is compatible with base change. Deduce from this that E is isomorphic to its Jacobian.

2.5. Let $t : \mathcal{E} \times_S \mathcal{N} \dashrightarrow \mathcal{E} \times_S \mathcal{N}$ be the birational map of Lemma 2.12(c), with S local, $\mathcal{O}_{S,s}$ complete, and $k(s)$ separably closed.

(a) Let $x \in \mathcal{N}(S)$ be a section. Show that $t|_{\mathcal{E} \times_S x}$ can be identified with the translation t_x on $\mathcal{E}$.

(b) Let $x, x' \in \mathcal{N}(S)$ be such that $x_s \neq x'_s$. Let us consider their graphs Γ_{t_x} and $\Gamma_{t_{x'}}$ in $\mathcal{N} \times_S \mathcal{N}$. Let us suppose that $\Gamma_{t_x} \cap \Gamma_{t_{x'}} \neq \emptyset$. Show that it is a closed subset of $\mathcal{N} \times_S \mathcal{N}$, of codimension ≥ 2 (use the regularity of $\mathcal{N} \times_S \mathcal{N}$), and that the first projection $p : \mathcal{N} \times_S \mathcal{N} \to \mathcal{N}$ induces an isomorphism from $\Gamma_{t_x} \cap \Gamma_{t_{x'}}$ onto an irreducible component of $\mathcal{N}_s$. Deduce from this that $\Gamma_{t_x} \cap \Gamma_{t_{x'}}$ contains a rational point of $\mathcal{N}_s$ (Proposition 3.2.20), and conclude with a contradiction.

(c) Let U_s be a dense open subset of $\mathcal{N}_s$. Show that the union of the $t_x(U_s)$, where x runs through the set $\mathcal{N}(S)$, is equal to $\mathcal{N}_s$. Using the method of the proof of Lemma 2.12(c), deduce from this that t is defined everywhere even when $k(s)$ is not perfect.

(d) Let us suppose that $k(s)$ is perfect and that the reduction of E is of type I_1 or II. Show that the translation t_x fixes the singular point of $\mathcal{E}_s$. In particular, (b) is false if we consider the graphs of the translations in $\mathcal{E} \times_S \mathcal{E}$.

2.6. Let A be an Abelian variety over the function field K of a Dedekind scheme S of dimension 1. Let $\mathcal{A} \to S$ be its Néron model (of which we admit the existence). Let $\mathcal{A}^0 := \cup_{s \in S} \mathcal{A}_s^0$ be the union of the identity components of the fibers $\mathcal{A}_s$ (including the generic fiber).

(a) Show that there exist a non-empty open subscheme V of S and a smooth projective scheme $\mathcal{A}' \to V$ with generic fiber isomorphic to A.

(b) Show that there exist a non-empty open subscheme U of V such that $\mathcal{A}'_U \simeq \mathcal{A}_U$.

(c) Deduce from this that $\mathcal{A}_U \to U$ is smooth projective, with geometrically connected fibers, and that $\mathcal{A}^0$ is an open subset and open group subscheme in $\mathcal{A}$. We call it the *identity component* of $\mathcal{A}$.

2.7. Let T be an algebraic group over a field k. We say that T is a *torus of dimension* 1 if there exists a finite extension k'/k such that $T_{k'} \simeq \mathbb{G}_{m;k'}$. Let us suppose that $\dim T = 1$ and that T is not isomorphic to $\mathbb{G}_{m,k}$.

(a) Show that $T \simeq \mathbb{P}_k^1 \setminus \{q\}$ as algebraic varieties, where q is a closed point such that $k(q)$ is separable of degree 2 over k. Show that there exists an algebraic group isomorphism $f : T_{k(q)} \to \mathbb{G}_{m,k(q)}$ (use Exercise 7.4.18).

(b) Let $\sigma \in \mathrm{Gal}(k(q)/k)$ be a generator. It canonically induces k-automorphisms of $T_{k(q)}$ and of $\mathbb{G}_{m,k(q)}$ that we indifferently denote by σ. Let us set $g = \sigma^{-1} f \sigma f^{-1} : \mathbb{G}_{m,k(q)} \to \mathbb{G}_{m,k(q)}$. Show that g is a $k(q)$-automorphism, and that $g \neq \mathrm{Id}$ (otherwise f would be defined over k).

(c) Show that g is the automorphism $\mathrm{inv} : z \to z^{-1}$ on $\mathbb{G}_m$. Deduce from this that f induces an isomorphism of abstract groups

$$T(k) \simeq \{z \in k(q)^* \mid \sigma(z)z = 1\}.$$

(d) Let θ be a generator of $k(q)$ over k and $\theta^2 + a\theta + b = 0$ its characteristic polynomial over k. Show that $T \simeq \mathrm{Spec}\, k[X,Y]/(X^2 + aXY + bY^2 - 1)$ as an algebraic variety.

2.8. Give a new proof of the fact that the curve of Example 1.28 has potential good reduction at 2.

2.9. Give an example of an elliptic curve over $\mathbb{Q}$ that has potential good reduction at every prime number p. Show that no elliptic curve can have potential multiplicative reduction at every prime number.

10.3 Stable reduction of algebraic curves

When we are interested in the classification of smooth projective curves over a field k, we are led to study the families of such curves. And, inevitably, families appear that degenerate, that is to say that at least one member of the family becomes a singular curve. This is the phenomenon of bad reduction cf. Section 10.1. Among the singular curves, the simplest ones are those whose singularities are ordinary double points (Definition 7.5.13). These curves (with moreover a combinatorial condition on the graph, see Definition 3.1 below) are called stable curves. The introduction of these curves was first motivated by the study of the moduli space of smooth curves (cf. the fundamental article [26]). But it turns out that the notion is also very important in arithmetic geometry. It allows us to have a better understanding of the different reductions of an algebraic curve. The profound nature of the stable reduction resides in the representation of the Galois group of the base field in the Tate module of the Jacobian of the curve. This aspect will unfortunately not be developed here. It appears furtively in Proposition 3.42. The reader can refer to [15], [26], [1].

10.3.1 Stable curves

In this subsection, the base scheme will, most of the time, be the spectrum of a field.

Definition 3.1. Let C be an algebraic curve over an algebraically closed field k. We say that C is *semi-stable* if it is reduced, and if its singular points are ordinary double points (Definition 7.5.13). We say that C is *stable* if, moreover, the following conditions are verified:

(1) C is connected and projective, of arithmetic genus $p_a(C) \geq 2$.

(2) Let Γ be an irreducible component of C that is isomorphic to $\mathbb{P}^1_k$. Then it intersects the other irreducible components at least three points.

Definition 3.2. We say that a curve C over a field k is *semi-stable* (resp. *stable*) if its extension $C_{\overline{k}}$ to the algebraic closure $\overline{k}$ of k is a semi-stable (resp. stable) curve over $\overline{k}$.

Example 3.3. A smooth curve over a field k is semi-stable. It is stable if it is moreover geometrically connected and of genus ≥ 2. In Figure 45, we have represented curves that are unions of projective lines over k (except Γ, which is a rational curve with an ordinary double point) that meet at ordinary double points. The curve on the left is semi-stable but not stable, while the one on the right is stable.

Figure 45. Semi-stable curve and stable curve.

Example 3.4. Let k be a field with char$(k) \neq 2$, $p(x) \in k[x]$ a polynomial whose roots in $\overline{k}$ are of order at most 2. Let U be the affine plane curve $y^2 = p(x)$. Then U is semi-stable. Indeed, we can suppose k is algebraically closed. Let $u \in U$ be a closed point with coordinates (a, b). The Jacobian criterion shows that U is smooth at u if $p(a) \neq 0$ or if a is a simple root of $p(x)$. Let us suppose that a is a double root. Then $y^2 = (x - a)^2 q(x)$ with $q(a) \neq 0$. Exercise 1.3.9 implies that $q(x) = h(x)^2$ with $h(x) \in k[[x]]^*$. We easily deduce from this that

$$\widehat{\mathcal{O}}_{U,u} \simeq k[[x - a, y]]/(y - (x - a)h(x))(y + (x - a)h(x)) = k[[z, v]]/(zv)$$

with $z = y - (x - a)h(x)$, $v = y + (x - a)h(x)$. This shows that u is an ordinary double point. We can also see this by computing the normalization of U. Indeed, by writing $p(x) = p_1(x)p_2(x)^2$ with $p_1(x)$ separable, it is easy to see that the integral closure of $\mathcal{O}_U(U)$ is $k[x, w]/(w^2 - p_1(x))$, with $w = y/p_2(x)$. Note that if $p_1(x) \notin k$, then U is irreducible, and that if $p_1(x) \in k$, then U has two irreducible components.

Example 3.5. Let us keep the hypotheses of Example 3.4 and let us suppose, moreover, that $p(x)$ is of degree $d \geq 5$. Let V be the affine curve $z^2 = p(1/t)t^r$, where $r = d$ if d is even and $r = d+1$ otherwise. Then U and V can be glued along Spec $k[x, y, 1/x]/(y^2 - p(x))$ via the identification $x = 1/t$ and $y = t^{-r/2}z$. We then obtain a projective curve D over k (because D is, in an evident manner, a degree 2-covering of $\mathbb{P}_k^1$) that is geometrically connected (because U is) and semi-stable, by Example 3.4 (the points of $D \setminus U$ are smooth). Using Proposition 7.5.4, we find that the arithmetic genus of D is $(r-2)/2 \geq 2$. Consequently, D is a stable curve over k.

Example 3.6 (*Stable curves of genus 2*). Let k be an algebraically closed field. Using Proposition 7.5.4, we easily see that there exist seven possible stable curves of genus 2 over k, as represented in Figure 46. The integers indicate the genus of the normalization of the component, and the non-marked components are rational.

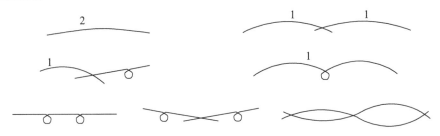

Figure 46. List of the stable curves of genus 2.

Proposition 3.7. *Let C be a semi-stable curve over a field k.*

(a) *If C is regular, then it is smooth over k.*

(b) *Let us suppose C is singular at a point x. Let $\pi : C' \to C$ be the normalization morphism. Then for $y \in \pi^{-1}(x)$, $k(x)$ and $k(y)$ are separable over k.*

(c) *Let $x \in C$ be a singular point and let us suppose that the points of $\pi^{-1}(x)$ are rational over k; then $\pi^{-1}(x)$ contains exactly two points y_1, y_2. Let V be an affine open neighborhood of x such that $V_{\text{sing}} = \{x\}$. Then we have*

$$\mathcal{O}_C(V) = \{f \in \mathcal{O}_{C'}(\pi^{-1}(V)) \mid f(y_1) = f(y_2)\}, \quad \widehat{\mathcal{O}}_{C,x} \simeq k[[u,v]]/(uv).$$

(d) *The curve C is an l.c.i. over k.*

Proof (a) We can suppose $C = \operatorname{Spec} A$ affine, regular, and integral. We already know that C_{k^s} is regular over the separable closure k^s of k (Corollary 4.3.24). We can therefore suppose that k is separably closed. By hypothesis, C is geometrically reduced. Let B be the integral closure of A in $K(C) \otimes_k \bar{k}$. Then for any $b \in B$, there exists a power $q \geq 1$ of $\operatorname{char}(k)$ such that $b^q \in K(C)$. As b^q is integral over A, we have $b^q \in A$. Consequently, if $\mathfrak{p}$ is a prime ideal of A, then $\sqrt{\mathfrak{p}B}$ is a prime ideal of B. In other words, as $C_{\bar{k}} \to C$ is a homeomorphism (Proposition 3.2.7), the normalization morphism $(C_{\bar{k}})' \to C_{\bar{k}}$ is bijective. As $C_{\bar{k}}$ is semi-stable, and in particular its singular points are double points, $(C_{\bar{k}})' \to C_{\bar{k}}$ is an isomorphism.

(b) We can suppose k is separably closed, $C = \operatorname{Spec} A$ affine, and that x is the unique singular point of C. Let $\bar{x}$ denote the point of $C_{\bar{k}}$ lying above

x. By (a), $\overline{x}$ is the unique singular point of $C_{\overline{k}}$. Let $\pi : \operatorname{Spec} B \to \operatorname{Spec} A$ and $\operatorname{Spec} D \to \operatorname{Spec}(A \otimes_k \overline{k})$ be the normalization morphisms. Then $A \otimes_k \overline{k} \subseteq B \otimes_k \overline{k} \subseteq D$, and $\dim_{\overline{k}} D/(A \otimes_k \overline{k}) = 1$ because $C_{\overline{k}}$ has a unique singular point that is moreover an ordinary double point. As $A \neq B$, we deduce from this that $D = B \otimes_k \overline{k}$, and therefore that $\dim_k B/A = 1$. In addition, $\operatorname{Spec} D \to \operatorname{Spec} B$ is a homeomorphism, and hence $\pi^{-1}(x)$ contains exactly two points y_1, y_2. Let $\mathfrak{m}$ be the maximal ideal of A corresponding to the point x. We have $B/A \simeq (B/\mathfrak{m})/k(x)$ and surjective homomorphisms of k-vector spaces

$$(B/\mathfrak{m})/k(x) \to (B/\mathfrak{m}B)/k(x) \to (k(y_1) \oplus k(y_2))/k(x).$$

This immediately implies that $k(x) = k(y_1) = k(y_2) = k$.

(c) As the points of $\pi^{-1}(x)$ are rational over k, x is also rational over k. The projection morphisms $C'_{\overline{k}} \to C'$ and $C_{\overline{k}} \to C$ are therefore bijections above the points of $\pi^{-1}(x)$ and x, whence $\operatorname{Card} \pi^{-1}(x) = 2$. The other properties can be shown as in Proposition 7.5.15.

(d) Let $x \in C$. Let us show that $\dim_{k(x)} \Omega^1_{C,x} \otimes_{\mathcal{O}_{C,x}} k(x) \le 2$. Let us first suppose that k is algebraically closed. Let $x \in C$ be a closed point. The tangent space $T_{C,x}$ to C at x coincides with that of $\operatorname{Spec} \widehat{\mathcal{O}}_{C,x}$. It follows from (c) that $\dim_k T_{C,x} \le 2$. Hence $\dim_{k(x)} \Omega^1_{C,x} \otimes k(x) \le 2$ (Lemma 6.2.1). In the general case, let $\overline{k}$ be the algebraic closure of k and $\overline{x} \in C_{\overline{k}}$ a point lying above x. Then we have $\Omega^1_{C_{\overline{k}}, \overline{x}} = \Omega^1_{C,x} \otimes_{\mathcal{O}_{C,x}} \mathcal{O}_{C_{\overline{k}}, \overline{x}}$ (Propositions 5.1.14(a) and 6.1.24(a)). It follows that

$$\dim_{k(x)} \Omega^1_{C,x} \otimes_{\mathcal{O}_{C,x}} k(x) = \dim_{k(\overline{x})} \Omega^1_{C_{\overline{k}}, \overline{x}} \otimes_{\mathcal{O}_{C_{\overline{k}}, \overline{x}}} k(\overline{x}) \le 2.$$

By Lemma 6.2.4, C is locally a closed (hence principal) subscheme of a smooth surface over k. Consequently C is an l.c.i. over k.

Let us note that a faster proof consists of saying that the property of being an l.c.i. is invariant under faithfully flat morphisms (Remark 6.3.23), and it therefore suffices to show that in the case when k is algebraically closed, $\operatorname{Spec} \widehat{\mathcal{O}}_{C,x}$ is an l.c.i., which follows from (c). But the proof above gives a more precise result: C is locally a curve on a smooth surface over k.   □

Definition 3.8. Let C be a semi-stable curve over a field k, let $\pi : C' \to C$ be the normalization morphism, and $x \in C$ a singular point. We will say that x is a *split ordinary double point* (or simply that x is *split*) if the points of $\pi^{-1}(x)$ are all rational over k. This implies that x is rational over k. Proposition 3.7(b) says that every singular point becomes split over a finite separable extension of k.

Example 3.9. If C is the special fiber of the minimal Weierstrass model of an elliptic curve E and if E has multiplicative reduction, then C is semi-stable. Its singular point is a split ordinary double point if and only if E has split multiplicative reduction.

Example 3.10. Let $p(x) \in k[x]$ be as in Example 3.4. We can write $p(x) = p_1(x) p_2(x)^2$ with p_1, p_2 separable (in fact, $p_2(x) = \gcd\{p(x), p'(x)\}$). Then the curve $\operatorname{Spec} k[x, y]/(y^2 - p(x))$ has split ordinary double points if and only if $p_2(x)$ has all of its roots in k and if $p_1(a) \in k^2$ for every root a of $p_2(x)$.

Let x be a point on a curve C over a field. Let Z_1, Z_2 be two distinct irreducible components passing through x. In what follows, we will say that Z_1, Z_2 meet *transversally* at x if we have the relation $T_{C,x} = T_{Z_1,x} \oplus T_{Z_2,x}$ of the tangent spaces.

Lemma 3.11. *Let C be a semi-stable curve over a field k, $x \in C$ a split ordinary double point such that at least two irreducible components of C pass through x. Then x belongs to exactly two irreducible components Z_1, Z_2. Moreover, they are smooth at x and meet transversally at x.*

Proof Let $\mathcal{O}'_{C,x}$ be the integral closure of $\mathcal{O}_{C,x}$, $\mathfrak{p}_1, \ldots, \mathfrak{p}_n$ $(n \geq 2)$ the minimal prime ideals of $\mathcal{O}_{C,x}$. We have the inclusions

$$\mathcal{O}_{C,x} \subsetneq \oplus_{1 \leq i \leq n} \mathcal{O}_{C,x}/\mathfrak{p}_i \subseteq \mathcal{O}'_{C,x}.$$

As $\dim_{k(x)} \mathcal{O}'_{C,x}/\mathcal{O}_{C,x} = 1$ by Proposition 3.7(c), the second inclusion is an equality. Let y_1, y_2 be the points of the normalization $C' \to C$ lying above x. Then $\mathcal{O}'_{C,x} = \mathcal{O}_{C',y_1} \oplus \mathcal{O}_{C',y_2}$, which implies that $n = 2$, and that $\mathcal{O}_{C,x}/\mathfrak{p}_i = \mathcal{O}_{C',y_i}$ (up to reordering). It follows that the Z_i are regular, and hence smooth, at x (since x is rational over k). Moreover, $\widehat{\mathcal{O}}_{C,x}/(\mathfrak{p}_i) \simeq \widehat{\mathcal{O}}_{C',y_i}$ is integral, and hence the $\mathfrak{p}_i$ generate minimal prime ideals in $\widehat{\mathcal{O}}_{C,x} \simeq k[[u,v]]/(uv)$. It follows that $\widehat{\mathcal{O}}_{C,x}/(\mathfrak{p}_1, \mathfrak{p}_2) \simeq k[[u,v]]/(uv, u, v) = k$, which implies that $\mathfrak{p}_1 + \mathfrak{p}_2$ is equal to the maximal ideal of $\mathcal{O}_{C,x}$, and hence $T_{C,x} = T_{Z_1,x} + T_{Z_2,x}$. It is a direct sum because $\dim_{k(x)} T_{C,x} > 1$. $\square$

Lemma 3.12. *Let C be a semi-stable curve over a field k. Let $\pi : C' \to C$ be the normalization morphism.*

(a) *We have an exact sequence*

$$0 \to \pi_* \omega_{C'/k} \to \omega_{C/k} \to \mathcal{F} \to 0, \tag{3.23}$$

where $\mathcal{F}$ is a skyscraper sheaf with support equal to C_{sing} and such that $\mathcal{F}_x = k(x)$ for every $x \in C_{\text{sing}}$.

(b) *Let D be the divisor $\sum_{y \in \pi^{-1}(C_{\text{sing}})} [y]$ on C'. Then*

$$\pi^* \omega_{C/k} \simeq \omega_{C'/k}(D). \tag{3.24}$$

Proof As π is an isomorphism above the regular open subscheme $\text{Reg}(C)$ of C, $\pi^* \omega_{C/k} \simeq \omega_{C'/k}$ on $\pi^{-1}(\text{Reg}(C))$. Hence the lemma is of local nature around the singular points. We can suppose $C = \text{Spec } A$, $C' = \text{Spec } B$, and that C has only one singular point x. By Lemma 6.4.26(b) and Exercise 6.4.9(a), we have

$$\omega_{B/k} = (\omega_{A/k} \otimes_A B) \otimes_B I = \omega_{A/k} \otimes_A I,$$

where I is the ideal $\{a \in A \mid aB \subseteq A\}$ of A and of B. Let $\mathfrak{m}$ be the maximal ideal of A corresponding to the point x. Then $I \subseteq \mathfrak{m}$. Let k'/k be a finite extension such that $\pi^{-1}(x) = \{y_1, y_2\} \subset C'(k')$. If $f \in \mathfrak{m}B$, then $f(y_1) = f(y_2) = 0$. It follows from Proposition 3.7(c) that $\mathfrak{m}B \subseteq A \otimes_k k'$. In other words, $(\mathfrak{m}(B/A)) \otimes_k k' = 0$.

Hence $\mathfrak{m}(B/A) = 0$ and $\mathfrak{m} = I$, which proves that

$$\omega_{A/k}/\omega_{B/k} \simeq \omega_{A/k} \otimes_A A/I \simeq A/\mathfrak{m} = k(x),$$

whence exact sequence (3.23). As $\mathfrak{m}B \subseteq A$, we have $\mathfrak{m}B = \mathfrak{m}$. Hence $B/I = B/\mathfrak{m}$. It remains to show that $B/\mathfrak{m} = \oplus_{y \in \pi^{-1}(x)} k(y)$. As $\operatorname{Spec} B/\mathfrak{m} = C'_x = \pi^{-1}(x)$, it suffices to show that C'_x is reduced, or even that it is geometrically reduced. We can therefore suppose k is algebraically closed. Now in that case, $\dim_k B/A = 1$ (Lemma 7.5.12(c)), and hence $\dim_k B/\mathfrak{m} = 2 = \operatorname{Card} \pi^{-1}(x)$. This implies that $B/\mathfrak{m} = \oplus_{y \in \pi^{-1}(x)} k(y)$, and therefore that $\mathfrak{m} = \mathcal{O}_{C'}(-D)$. ◻

Corollary 3.13. *Let C be a stable curve over a field k. Then the dualizing sheaf $\omega_{C/k}$ is ample.*

Proof We can suppose k is algebraically closed because the formation of $\omega_{C/k}$ commutes with the base change $k \to \bar{k}$ (Theorem 6.4.9(b)); likewise, the ampleness of a sheaf is invariant under this base change (Exercise 5.1.29).

Let Γ be an irreducible component of C and $\rho : \Gamma' \to \Gamma$ the normalization morphism. By restricting isomorphism (3.24) to Γ', we obtain

$$\rho^*(\omega_C|_\Gamma) \simeq \omega_{\Gamma'/k}(E),$$

where E is the divisor $\sum_{y \in \pi^{-1}(C_{\text{sing}}) \cap \Gamma'} [y]$ on Γ'. This implies that

$$\deg \omega_C|_\Gamma = \deg \rho^*(\omega_C|_\Gamma) = 2p_a(\Gamma') - 2 + \operatorname{Card}(\pi^{-1}(C_{\text{sing}}) \cap \Gamma')$$
$$= 2p_a(\Gamma) - 2 + \operatorname{Card}(\overline{\{C \setminus \Gamma\}} \cap \Gamma).$$

(Proposition 7.5.4). If $p_a(\Gamma) = 0$, then, by hypothesis, $\overline{\{C \setminus \Gamma\}} \cap \Gamma$ contains at least three points, and hence $\deg \omega_C|_\Gamma > 0$. If $p_a(\Gamma) = 1$, then $\overline{\{C \setminus \Gamma\}} \cap \Gamma$ contains at least one point, because otherwise $C = \Gamma$ and C would be of genus 1. Hence $\deg \omega_C|_\Gamma > 0$. Finally, if $p_a(\Gamma) \geq 2$, then the same inequality is trivially true. Consequently, $\deg \omega_C|_\Gamma > 0$ for every irreducible component Γ, which proves that $\omega_{C/k}$ is ample (Proposition 7.5.5). ◻

Definition 3.14. Let $f : X \to S$ be a morphism of finite type to a scheme S. We say that f is *semi-stable*, or that X is a *semi-stable curve over S*, if f is flat and if for any $s \in S$, the fiber X_s is a semi-stable curve over $k(s)$. We say that f is *stable of genus $g \geq 2$*, or that X is a *stable curve over S of genus $g \geq 2$*, if f is proper, flat, with stable fibers of arithmetic genus g.

Proposition 3.15. *Let $f : X \to S$ be a semi-stable curve over a scheme S.*

  (a) *Let $S' \to S$ be a morphism. Then $X \times_S S' \to S'$ is semi-stable. If f is stable, then the same holds for $X \times_S S' \to S'$.*

  (b) *If S is locally Noetherian, then $X \to S$ is an l.c.i.*

  (c) *If S is a Dedekind scheme and if the generic fiber of $X \to S$ is normal, then X is normal.*

Proof (a) results from the fact that flat (resp. proper) morphisms remain flat (resp. proper) after base change (Propositions 4.3.3 and 3.3.16), and from the definition of semi-stable and stable curves. (b) is a consequence of Corollary 6.3.24 and of Proposition 3.7(d). Finally, (c) results from Lemma 4.1.18 because the fibers of f are reduced. ☐

Remark 3.16. Let us suppose $X \to S$ is stable and S affine. Then $\omega_{X/S}$ is an ample sheaf on X by Corollary 3.13 and Remark 5.3.25. In particular, f is projective. We can even show that $\omega_{X/S}^{\otimes n}$ is very ample for every $n \geq 3$ ([26], Corollary of Theorem 1.2).

Definition 3.17. Let C be a semi-stable connected projective curve over a field k, with split ordinary double points, and irreducible components $\Gamma_1, \ldots, \Gamma_n$. The *dual graph of* C is the graph G whose vertices are the irreducible components of C; and each ordinary double point x defines an edge whose end vertices correspond to the irreducible components containing x (the two end vertices are equal if x only belongs to one component). For example, Figure 47 represents the dual graphs of the curves of Figure 45.

Figure 47. Dual graphs of semi-stable curves.

The following lemma is analogous to Proposition 1.51(b).

Lemma 3.18. *Let C, G be as above. Let $\beta(G)$ be the Betti number of G. Let Γ_i' denote the normalization of Γ_i. Then we have*

$$p_a(C) = \beta(G) + \sum_{1 \leq i \leq n} p_a(\Gamma_i').$$

Proof Let $\pi : C' \to C$ be the normalization morphism. Then we have an exact sequence

$$0 \to \mathcal{O}_C \to \pi_* \mathcal{O}_{C'} \to \mathcal{F} \to 0,$$

where $\mathcal{F}$ is a skyscraper sheaf with support equal to the set I of singular points of C, and we have $\mathcal{F}_x = k(x) = k$ for every $x \in I$. Consequently,

$$1 - p_a(C) = \chi_k(\mathcal{O}_C) = \chi_k(\mathcal{O}_{C'}') - \mathrm{Card}\, I = n - \sum_{1 \leq i \leq n} p_a(\Gamma_i') - \mathrm{Card}\, I,$$

whence the desired equality since $\beta(G) = \mathrm{Card}\, I - n + 1$. ☐

Remark 3.19. If k is algebraically closed, then $\beta(G)$ is equal to the toric rank of $\mathrm{Pic}^0(C)$ (Lemma 7.5.18).

10.3.2   Stable reduction

Let us return to curves over a Dedekind domain. We will begin by studying the formal structure of semi-stable curves over a Noetherian local ring.

Lemma 3.20. *Let $\rho : (A, \mathfrak{m}_A) \to (B, \mathfrak{m}_B)$ be a homomorphism of Noetherian local rings that induces an isomorphism $A/\mathfrak{m}_A \simeq B/\mathfrak{m}_B$, and $u, v \in \mathfrak{m}_B$ elements such that $uv \in \mathfrak{m}_A B$, and $\mathfrak{m}_B = uB + vB + \mathfrak{m}_A B$.*

(a) *There exist sequences $(u_n)_{n \geq 0}$, $(v_n)_{n \geq 0}$, $(\varepsilon_n)_{n \geq 0}$ with $u_0 = u$, $v_0 = v$, $u_n, v_n \in \mathfrak{m}_B$, $\varepsilon_n \in \mathfrak{m}_A$, such that*

$$u_{n+1} - u_n \in \mathfrak{m}_A^{n+1} B, \quad v_{n+1} - v_n \in \mathfrak{m}_A^{n+1} B, \quad \varepsilon_{n+1} - \varepsilon_n \in \mathfrak{m}_A^{n+1},$$

and $u_n v_n - \rho(\varepsilon_n) \in \mathfrak{m}_A^{n+1} B$.

(b) *There exist an $\alpha \in \mathfrak{m}_A$ and a surjective homomorphism*

$$\psi : \widehat{A}[[x, y]]/(xy - \alpha) \to \widehat{B}$$

such that $\psi(x) - u, \psi(y) - v \in \mathfrak{m}_A \widehat{B}$.

(c) *Let us suppose B is flat over A and that $(B/\mathfrak{m}_A B)^{\widehat{}} \simeq (A/\mathfrak{m}_A)[[w, z]]/(wz)$. Then ψ is an isomorphism.*

Proof   (a) For simplicity, we also denote the image $\rho(a)$ of an element $a \in A$ by a. We are going to construct the sequences by induction on n. We therefore take $u_0 = u$, $v_0 = v$, and $\varepsilon_0 = 0$. Let us suppose that we have constructed u_n, v_n, ε_n. We have $uB + vB \subseteq u_n B + v_n B + \mathfrak{m}_A B$. Hence

$$\mathfrak{m}_A^{n+1} B \subseteq \rho(\mathfrak{m}_A^{n+1}) + u_n \mathfrak{m}_A^{n+1} B + v_n \mathfrak{m}_A^{n+1} B + \mathfrak{m}_A^{n+2} B.$$

Let us write $(u_n v_n - \varepsilon_n) = \delta_{n+1} + u_n b_n + v_n c_n + d_n$ as in this decomposition. Then

$$(u_n - c_n)(v_n - b_n) = (\varepsilon_n + \delta_{n+1}) + (b_n c_n + d_n).$$

Let us set $u_{n+1} = u_n - c_n$, $v_{n+1} = v_n - b_n$, and $\varepsilon_{n+1} = \varepsilon_n + \delta_{n+1}$. It can then immediately be seen that $u_{n+1}, v_{n+1}, \varepsilon_{n+1}$ verify the required properties.

(b) For any $n \geq 0$, let us consider the homomorphism

$$\varphi_n : A[x, y]/(\mathfrak{m}_A, x, y)^n \to B/\mathfrak{m}_B^n$$

that sends x, y respectively to the images of u_n, v_n in $B/\mathfrak{m}_B^n$. Let I_n be the ideal $(\mathfrak{m}_A, x, y)^n$ of $A[x, y]$. This induces a homomorphism of projective systems $(A[x, y]/I_n)_n \to (B/\mathfrak{m}_B^n)_n$. As

$$\mathfrak{m}_B = \mathfrak{m}_A B + u_n B + v_n B = \mathfrak{m}_A + u_n A + v_n A + \mathfrak{m}_B^2,$$

it can immediately be seen that the φ_n are surjective and that $\varphi_{n+1}(I_n) = \mathfrak{m}_B^n/\mathfrak{m}_B^{n+1}$. This implies that the exact sequence of projective systems

$$0 \to (\operatorname{Ker} \varphi_n)_n \to (A[x, y]/I_n)_n \to (B/\mathfrak{m}_B^n)_n \to 0$$

verifies the conditions of Lemma 1.3.1, whence a surjective homomorphism $\varphi : \widehat{A}[[x, y]] \to \widehat{B}$. Let $\alpha \in \widehat{A}$ be the element induced by the sequence $(\varepsilon_n)_n$. Then

$\alpha \in \varprojlim_{n} (\mathfrak{m}_A/\mathfrak{m}_A^n) = \mathfrak{m}_A \widehat{A}$ (Theorem 1.3.16(a)). We have $\varphi(x)\varphi(y) = \varphi(\alpha)$. Hence φ induces a surjective homomorphism $\psi : \widehat{A}[[x,y]]/(xy - \alpha) \to \widehat{B}$. By construction, the image of $\psi(x) - u$ in $B/\mathfrak{m}_B^n$ coincides with that of $u_n - u_0$, and hence belongs to $(\mathfrak{m}_A B + \mathfrak{m}_B^n)/\mathfrak{m}_B^n$. Now $\varprojlim_{n} ((\mathfrak{m}_A B + \mathfrak{m}_B^n)/\mathfrak{m}_B^n) = \mathfrak{m}_A \widehat{B}$ (Corollary 1.3.14), whence $\psi(x) - u \in \mathfrak{m}_A \widehat{B}$. Likewise for $\psi(y) - v$. Finally, by multiplying α by an element of the form $1 + \delta$, $\delta \in \mathfrak{m}_A \widehat{A}$, and by dividing y by this same element, we reduce to $\alpha \in \mathfrak{m}_A$.

(c) By tensoring the exact sequence of $\widehat{A}$-modules

$$0 \to \operatorname{Ker} \psi \to \widehat{A}[[x,y]]/(xy - \alpha) \to \widehat{B} \to 0$$

by $\widehat{A}/\mathfrak{m}_A \widehat{A}$, and because $\widehat{B}$ is flat over $\widehat{A}$, we obtain an exact sequence

$$0 \to \operatorname{Ker} \psi/(\mathfrak{m}_A \operatorname{Ker} \psi) \to (A/\mathfrak{m}_A)[[x,y]]/(xy) \to (B/\mathfrak{m}_A B)\widehat{\ } \to 0$$

(Proposition 1.2.6). Now the homomorphism on the right is not necessarily an isomorphism, and hence $\operatorname{Ker} \psi = \mathfrak{m}_A \operatorname{Ker} \psi$, which implies that $\operatorname{Ker} \psi = 0$.   □

Lemma 3.21. *Let X be an integral flat curve over a discrete valuation ring $\mathcal{O}_K$ with residue field k. Let $x \in X_s$ be a closed point such that $X \setminus \{x\}$ is regular. We suppose that there exists a non-zero $c \in \mathcal{O}_K$ with valuation $e = \nu(c) \geq 1$ such that $\widehat{\mathcal{O}}_{X,x} \simeq \widehat{\mathcal{O}}_K[[u,v]]/(uv - c)$. Then we have a sequence of proper birational morphisms*

$$X_n \to \cdots \to X_1 \to X_0 = X$$

where X_i is normal, with a unique closed singular point x_i, and where $X_{i+1} \to X_i$ is the blowing-up of X_i with (reduced) center x_i. The sequence stops at $n = [e/2]$, and the fiber of $X_n \to X$ above x is a chain of $e - 1$ projective lines over k, of multiplicity 1 in $(X_n)_s$, that meet transversally at rational points. In particular, e depends only on $\mathcal{O}_{X,x}$ and not on the choice of u, v.

Proof   Let $Y = \operatorname{Spec} \mathcal{O}_{X,x}$ and $Z = \operatorname{Spec} \widehat{\mathcal{O}}_{X,x}$. As the desingularization is local on X, we can replace X by Y, and x by the closed point y of Y. Let us first compare the desingularization of Y to that of Z, without hypothesis on the structure of $\mathcal{O}_{X,x}$.

Let $f : Z \to Y$ be the canonical morphism, $\rho : Y_1 \to Y$ the blowing-up of Y with center y. Then the base change $\widehat{\rho} : Z_1 = Y_1 \times_Y Z \to Z$ is the blowing-up of Z with center its closed point z (Proposition 8.1.12(c)), and f induces a canonical morphism $f_1 : Z_1 \to Y_1$. Moreover, f_1 induces, by restriction, a bijection $\widehat{\rho}^{-1}(z) \to \rho^{-1}(y)$. For any closed point $z_1 \in \widehat{\rho}^{-1}(z)$, the canonical homomorphism $\widehat{\mathcal{O}}_{Y_1, f(z_1)} \to \widehat{\mathcal{O}}_{Z_1, z_1}$ is an isomorphism (proof analogous to that of Lemma 8.3.49(b)). If

$$Y_n \to \cdots \to Y_1 \to Y$$

is a sequence of blowing-ups of closed points in the special fibers, then we construct, as above, a sequence

$$Z_n \to \cdots \to Z_1 \to Z \tag{3.25}$$

of blowing-ups of closed points in the special fibers, the two sequences of morphisms having the same fibers and Y_i having the same formal completions at the closed points as Z_i. Hence Y_n is regular if and only if Z is regular. Conversely, such a sequence on Z is always constructed in this way, from a sequence on Y (proof analogous to that of Lemma 8.3.48).

By applying the result above with $X = \operatorname{Spec} \mathcal{O}_{S,s}[u,v]/(uv-c)$ and x the point $(0,0)$ of X_s, we obtain a sequence of blowing-ups (3.25) with $n = [e/2]$, described by Example 8.3.53. Getting back to our original X (without changing Z), and applying, once more, the comparison above, we obtain a desingularization $Y_n \to Y$ of Y whose fiber above y is a chain of $e-1$ projective lines represented by Figure 11, Section 8.3.   □

Corollary 3.22. *Let $X \to S$ be a semi-stable curve over a Dedekind scheme S of dimension 1. Let $s \in S$, $x \in X_s$ be a singular point of X_s.*

(a) *There exists a Dedekind scheme S', étale over S, such that any point $x' \in X' := X \times_S S'$ lying above x, belonging to a fiber $X'_{s'}$, is a split ordinary double point of $X'_{s'} \to \operatorname{Spec} k(s')$.*

(b) *Under the conditions of (a), we have an isomorphism*

$$\widehat{\mathcal{O}}_{X',x'} \simeq \widehat{\mathcal{O}}_{S',s'}[[u,v]]/(uv-c) \tag{3.26}$$

for some $c \in \mathfrak{m}_{s'}\mathcal{O}_{S',s'}$. If X_η is smooth, then $c \neq 0$.

(c) *Let e_x be the valuation of c for the normalized valuation of $\mathcal{O}_{S',s'}$. Then e_x is independent of the choice of S', s', and of x'.*

Proof (a) By virtue of Proposition 3.7(c), there exists a finite separable extension k' of $k(s)$ such that the singular points of $X_s \times_{\operatorname{Spec} k(s)} \operatorname{Spec} k'$ are split. As in the proof of Corollary 5.3.17, there exists a discrete valuation ring $\mathcal{O}_{K'}$ that is étale over $\mathcal{O}_{S,s}$, with residue field k'. It can immediately be seen that $S' = \operatorname{Spec} \mathcal{O}_{K'}$ verifies the desired properties.

(b) Let t' be a uniformizing parameter for $\mathcal{O}_{S',s'}$. By Exercise 1.3.8 and Proposition 3.7(c), we have isomorphisms

$$\widehat{\mathcal{O}}_{X',x'}/(t') \simeq \widehat{\mathcal{O}}_{X'_{s'},x'} \simeq \widehat{\mathcal{O}}_{S',s'}[[u,v]]/(uv,t) = k(s)[[u,v]]/(uv).$$

Lemma 3.20 then implies isomorphism (3.26). Let us suppose X_η is smooth; then X' also has a smooth generic fiber. When passing to the completion of $\mathcal{O}_{S',s'}$, X' remains semi-stable with smooth generic fiber, and we do not change $\widehat{\mathcal{O}}_{X',x'}$. We can therefore suppose $\mathcal{O}_{S',s'}$ is complete, and hence excellent (Theorem 8.2.39). As X' is normal, it follows that $\widehat{\mathcal{O}}_{X',x'}$ is normal (Proposition 8.2.41), and hence integral, which implies that $c \neq 0$.

(c) It is clear that e_x remains unchanged if we replace S' by an étale scheme $S'' \to S'$ and x' by a point $x'' \in X' \times_{S'} S''$ lying above x'. By Exercise 9.3.5, and by restricting S if necessary, we can then suppose that $S' \to S$ is finite Galois. In that case, the Galois group $G := \operatorname{Gal}(K(S')/K(S))$ acts on X' and $X'/G \simeq X$ (because $X' \to S$ is projective). Moreover, G acts transitively on the set of points $x' \in X'$ lying above x (Exercise 2.3.20(a)), and therefore induces isomorphisms between the rings $\mathcal{O}_{X',x'}$, which immediately implies that e_x is independent of the choice of the point x'.   □

Definition 3.23. Let $X \to S$ be a semi-stable curve over a Dedekind scheme S of dimension 1, with smooth generic fiber. Let x be a singular point of a closed fiber X_s. The integer $e_x \geq 1$ defined in the corollary above is called the *thickness of x in X*.

Example 3.24. Let E be an elliptic curve over a discrete valuation field K, with multiplicative reduction. Hence the reduction of E is of type I_n or $I_{n,2}$ with $n \geq 1$. Let W be the minimal Weierstrass model of E over $\operatorname{Spec} \mathcal{O}_K$. Then the thickness of the singular point of W_s is equal to n. Indeed, by definition we can make an étale extension to reduce the computation of the case $I_{n,2}$ (non-split multiplicative reduction) to the case I_n (split multiplicative reduction). Now in the split case, the thickness can be computed with the minimal desingularization as in the corollary above.

Corollary 3.25. *Let $X \to S$ be a semi-stable projective curve over a Dedekind scheme of dimension 1, with smooth generic fiber X_η. Let $\pi : X' \to X$ be the minimal desingularization, $x \in X_s$ a split ordinary double point, of thickness e in X. Then $\pi^{-1}(x)$ is made up of a chain of $e - 1$ projective lines over $k(s)$ that meet transversally at rational points. These lines are of multiplicity 1 in X'_s, and have self-intersection -2 in X'. See Figure 48.*

Figure 48. Minimal desingularization of an ordinary double point.

Proof By Lemma 3.21, we have a desingularization $X' \to X$ whose fiber above x is a chain of $e - 1$ projective lines over $k(s)$ represented by Figure 48. As these lines have self-intersection -2 (Proposition 9.1.21(b)), $X' \to X$ is indeed a minimal desingularization (Proposition 9.3.32). □

Remark 3.26. As desingularization is a local process on X, the $X \to S$ projective hypothesis is not necessary at all. We have added this hypothesis to be able to talk about the self-intersection and to use Castelnuovo's criterion. However, the result stays valid without $X \to S$ being projective. We just need to define the self-intersection of a divisor Γ in $\pi^{-1}(x)$ as being the number $\Gamma^2 := \deg_{k(s)} \mathcal{O}_{X'}(\Gamma)|_\Gamma$.

Definition 3.27. Let S be a Dedekind scheme of dimension 1. Let C be a smooth projective curve over $K(S)$. We say that C has *semi-stable reduction* (resp. *stable reduction*) at $s \in S$ if there exists a model $\mathcal{C}$ of C over $\operatorname{Spec} \mathcal{O}_{S,s}$ that is semi-stable (resp. stable) over $\operatorname{Spec} \mathcal{O}_{S,s}$. The special fiber $\mathcal{C}_s$ of a stable model over $\operatorname{Spec} \mathcal{O}_{S,s}$ is called the *stable reduction of C at s*. We will see below (Theorem 3.34) that the stable model (and therefore the stable reduction) is unique.

We say that C has *semi-stable* (resp. *stable*) *reduction over S* if the property is true for every $s \in S$. A model $\mathcal{C}$ of C over S is called a *stable model* if $\mathcal{C} \to S$ is a stable curve.

Example 3.28. If C has good reduction at s, then it has stable (and a fortiori semi-stable) reduction at s.

Example 3.29. Let C be a hyperelliptic curve over a discrete valuation field K, defined by an affine equation $y^2 = P(x)$, and of genus $g \geq 2$. Let us moreover suppose that the residue field k of $\mathcal{O}_K$ is of characteristic different from 2, that $P(x)$ is monic, with coefficients in $\mathcal{O}_K$, and that its image $p(x) \in k[x]$ only has roots (in $\bar{k}$) of order at most 2. Then C has stable reduction, and its stable reduction is the curve D of Example 3.5. Indeed, the stable model of C over $\mathcal{O}_K$ is the union of the affine open subschemes $\operatorname{Spec} \mathcal{O}_K[x, y]/(y^2 - P(x))$ and $\operatorname{Spec} \mathcal{O}_K[t, z]/(z^2 - P(1/w)w^r)$, where r is defined in Example 3.5.

Lemma 3.30. *Let S be a Dedekind scheme of dimension 1, C a smooth projective curve over $K(S)$, $\mathcal{C} \to S$ a model of C. Let S' be a Dedekind scheme of dimension 1 endowed with a surjective morphism $S' \to S$.*

(a) *Let us suppose that $\mathcal{C} \times_S S' \to S'$ is semi-stable (resp. stable). Then $\mathcal{C} \to S$ is semi-stable (resp. stable).*

(b) *Let $\mathcal{D}$ be a model of C over S such that there exists an isomorphism $\mathcal{D} \times_S S' \simeq \mathcal{C} \times_S S'$. Then the isomorphism comes from an isomorphism $\mathcal{D} \simeq \mathcal{C}$ over S.*

Proof (a) If we have $\mathcal{C} \times_S S'$ semi-stable over S', as $S' \to S$ is surjective, this clearly implies that $\mathcal{C} \to S$ is semi-stable. If $\mathcal{C} \times_S S'$ is stable, then $g \geq 2$. The same holds for $C \to \operatorname{Spec} K(S)$. We easily deduce from this that $\mathcal{C}$ is stable.

(b) Let Γ be the graph of the birational map $\mathcal{D} \dashrightarrow \mathcal{C}$. Then the projection morphism $p : \Gamma \to \mathcal{C}$ becomes an isomorphism over S'. Hence p is birational, projective, and quasi-finite, and as $\mathcal{C}$ is normal, p is an isomorphism (Corollary 4.4.6). We deduce from this that $\mathcal{D} \dashrightarrow \mathcal{C}$ is defined everywhere. By symmetry, its inverse is also defined everywhere. Consequently, $\mathcal{D} \to \mathcal{C}$ is an isomorphism. $\qquad \square$

Lemma 3.31. *Let C be as in Lemma 3.30 and $\mathcal{C} \to S$ a semi-stable model of C over S. Let s be a closed point of S. Let us suppose that $k(s)$ is separably closed. Let $\Delta_1, \ldots, \Delta_m$ be irreducible components of $\mathcal{C}_s$ verifying the following conditions:*

(i) $\Delta_i \simeq \mathbb{P}^1_{k(s)}$;

(ii) $\Delta_i \cap \Delta_j = \emptyset$ *if $|i - j| \geq 2$ and $\Delta_i \cap \Delta_{i+1}$ is reduced to a point;*

(iii) *the union $\cup_i \Delta_i$ meets the other irreducible components of $\mathcal{C}_s$ at at most two points, and these points belong to $\Delta_1 \cup \Delta_m$.*

In other words, the Δ_i form a chain of the $\mathbb{P}^1_{k(s)}$ that meet the other irreducible components at at most two points of the outer components Δ_1, Δ_m. Let us suppose that there exists a contraction morphism $f : \mathcal{C} \to \mathcal{D}$ of the components $\Delta_1, \ldots, \Delta_m$. Then $\mathcal{D} \to S$ is semi-stable.

Proof We can suppose S is local. Let us note that the intersection points are all rational over $k(s)$ (Proposition 3.7(b)). Let $r \in \{1, 2\}$ be the number of intersection points of $\cup_i \Delta_i$ with the other irreducible components of $\mathcal{C}_s$. Let

$\pi : \mathcal{E}_s \to \mathcal{D}_s$ be the finite birational morphism that consists of replacing the point $f(\cup_i \Delta_i)$ by a point that is regular if $r = 1$, and that is an ordinary double point if $r = 2$ (see the construction before Lemma 7.5.12). Then $\mathcal{E}_s$ is semi-stable, and its irreducible components are birational to the irreducible components of C_s that are not contracted by f. We are going to show that π is an isomorphism, which will prove the lemma.

The dual graph of $\mathcal{E}_s$ (Definition 3.17) is obtained by suitably deleting m vertices and m edges in the dual graph of C_s. See Figure 49 for the different possibilities (the solid vertices correspond to the components Δ_i). Looking at this figure, it becomes apparent that the dual graph of $\mathcal{E}_s$ has the same number of connected components and the same Betti number as the dual graph of C_s. By applying Lemma 3.18 to the connected components of these two curves, we obtain $p_a(\mathcal{E}_s) = p_a(C_s)$. It follows from Proposition 5.3.28 that

$$p_a(\mathcal{E}_s) = p_a(C_s) = p_a(C) = p_a(\mathcal{D}_s).$$

By Lemma 7.5.18(a), π is an isomorphism.   □

To study schemes over a discrete valuation ring, we sometimes need to do base changes that are not morphisms of finite type. In the two following lemmas, we are interested in this type of situation.

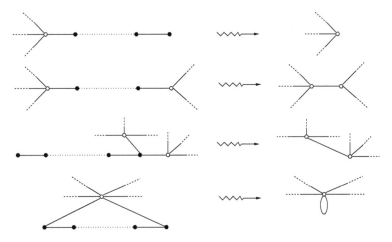

Figure 49. Dual graphs of C_s and of $\mathcal{E}_s$.

Lemma 3.32. *Let $\mathcal{O}_K$ be a discrete valuation ring, with field of fractions K and residue field k, and with uniformizing parameter t. Let k' be a (not necessarily finite) algebraic extension of k. Then there exists a discrete valuation ring $\mathcal{O}_L$ that dominates $\mathcal{O}_K$, with residue field k', ramification index $e_{\mathcal{O}_L/\mathcal{O}_K} = 1$, and such that L is separable algebraic over K. (See also [41], III.0.10.3.1, for the case of arbitrary Noetherian local rings).*

Proof Let K^s be the separable closure of K. Let us consider the set $\mathcal{F}$ of discrete valuation rings $\mathcal{O}_F$ dominating $\mathcal{O}_K$ such that $F \subseteq K^s$, $e_{\mathcal{O}_F/\mathcal{O}_K} = 1$, and that the residue field of $\mathcal{O}_F$ is contained in k'. This set is non-empty and stable under reunion of ascending sequences. By Zorn's lemma, it therefore admits a maximal element $\mathcal{O}_L$. It remains to show that the residue field l of $\mathcal{O}_L$ is equal to k'. Let us suppose the contrary. Then there exists an $\alpha \in k' \setminus l$. Let $P(T) \in \mathcal{O}_L[T]$ be a monic polynomial whose image in $l[T]$ is the minimal polynomial of α over l. Then $P(T)$ is irreducible in $K[T]$ and $A := \mathcal{O}_L[T]/(P(T))$ is a finite normal $\mathcal{O}_L$-algebra (Lemma 4.1.18) that is local because $A \otimes_{\mathcal{O}_L} l$ is a field. It is therefore a discrete valuation ring, with residue field $k'[\alpha]$, and we have $A \in \mathcal{F}$, which contradicts the maximality hypothesis on $\mathcal{O}_L$. □

Lemma 3.33. *Let $\mathcal{O}_K$ be a discrete valuation ring with residue field k, and let $\mathcal{O}_L$ be a discrete valuation ring that dominates $\mathcal{O}_K$, with field of fractions L algebraic over K. Then the following properties are true.*

(a) *For any projective scheme $X \to \operatorname{Spec}\mathcal{O}_L$, there exist a discrete valuation subring $\mathcal{O}_{K'}$ of $\mathcal{O}_L$ with K' finite over K, and a projective scheme $X' \to \operatorname{Spec}\mathcal{O}_{K'}$ such that $X \simeq X' \times_{\operatorname{Spec}\mathcal{O}_{K'}} \operatorname{Spec}\mathcal{O}_L$.*

(b) *Let us moreover suppose that L is separable over K, that $e_{\mathcal{O}_L/\mathcal{O}_K} = 1$, and that its residue field is separable algebraic over k. Let C be a smooth projective curve over $\mathcal{O}_K$. Then the formation of the minimal regular model and of the canonical model of C over $\mathcal{O}_K$ (if they exist) commutes with the base change $\operatorname{Spec}\mathcal{O}_L \to \operatorname{Spec}\mathcal{O}_K$.*

Proof (a) The scheme X is defined by a homogeneous $\mathcal{O}_L$-algebra

$$\mathcal{O}_L[T_0,\ldots,T_n]/(F_1,\ldots,F_m).$$

Let A be the sub-$\mathcal{O}_K$-algebra of $\mathcal{O}_L$ generated by the coefficients of the polynomials F_i; then the F_i define a projective scheme Y over A such that $Y_{\mathcal{O}_L} \simeq X$. Let $K' = \operatorname{Frac}(A)$, $\mathcal{O}_{K'} := \mathcal{O}_L \cap K' \supseteq A$, and $X' = Y \times_{\operatorname{Spec}A} \operatorname{Spec}\mathcal{O}_{K'}$. Then K' is finite over K, $\mathcal{O}_{K'}$ is a discrete valuation ring, and $X \simeq X' \times_{\operatorname{Spec}\mathcal{O}_{K'}} \operatorname{Spec}\mathcal{O}_L$, which shows (a).

(b) Under the hypotheses on $\mathcal{O}_L$, any $\mathcal{O}_{K'}$ as in (a) is étale over $\mathcal{O}_K$. By (a), any model of C_L over $\mathcal{O}_L$ is defined over an $\mathcal{O}_{K'}$ that is étale over $\mathcal{O}_K$. The lemma then results from Proposition 1.17. □

Theorem 3.34. *Let S be an affine Dedekind scheme of dimension 1. Let C be a smooth projective curve over $K = K(S)$, of genus $g \geq 1$. Let us suppose that C has semi-stable reduction over S.*

(a) *The minimal regular model $\mathcal{C}_{\min}$ of C over S is semi-stable over S.*

(b) *Let us suppose $g \geq 2$ and that C is geometrically connected over K. Then the canonical model $\mathcal{C}_{\operatorname{can}}$ of C over S is a stable curve over S, and it is the unique stable model of C over S.*

(c) *The curve C admits a semi-stable model over S. If C has stable reduction over S, then it admits a stable model over S.*

Proof (c) is an immediate consequence of (a), (b), and of the existence of the models $\mathcal{C}_{\min}, \mathcal{C}_{\operatorname{can}}$ (Propositions 1.8 and 9.4.20). For the proof of (a) and (b), we can suppose S is local. Hence C admits a semi-stable model $\mathcal{C} \to S$ (we will suppose $\mathcal{C}$ is stable in the proof of uniqueness in (b)). By Lemmas 3.30 and 3.33, we can suppose that $k(s)$ is separably closed. This implies that for any semi-stable curve over $k(s)$, the singular points are split and the irreducible components are geometrically integral. In particular, any irreducible component isomorphic to a line $\mathbb{P}^1_{k'}$ is isomorphic to $\mathbb{P}^1_{k(s)}$.

(a) Let $\mathcal{C}' \to \mathcal{C}$ be the minimal desingularization of $\mathcal{C}$. By Corollary 3.25, $\mathcal{C}'$ is semi-stable. We can therefore suppose that $\mathcal{C}$ itself is already regular. By definition, there exists a morphism of models $\pi : \mathcal{C} \to \mathcal{C}_{\min}$. Let $x \in \mathcal{C}_{\min}$ be a point such that $\dim \pi^{-1}(x) = 1$. Using the characterization of the exceptional divisors (Castelnuovo's criterion) and Proposition 9.1.21(b), we see that $\pi^{-1}(x)$ is a chain of the $\mathbb{P}^1_{k(s)}$ that meet the other irreducible components at a point on an outer component of the chain. It follows from Lemma 3.31 that $\mathcal{C}_{\min}$ is semi-stable.

(b) By (a), $\mathcal{C}_{\min}$ is semi-stable. Let $\rho : \mathcal{C}_{\min} \to \mathcal{C}_{\can}$ be the canonical morphism. By Corollary 8.3.6, $(\mathcal{C}_{\min})_s$ is geometrically connected over $k(s)$. Hence so is $(\mathcal{C}_{\can})_s$. Let us show that $\mathcal{C}_{\can}$ is semi-stable. Let $x \in \mathcal{C}_{\can}$ be a point such that $\dim \rho^{-1}(x) = 1$. By the definition of ρ (9.4.21) and Proposition 9.4.8, the irreducible components of $\rho^{-1}(x)$ are the $\mathbb{P}^1_{k(s)}$ with self-intersection -2. Using Proposition 9.1.21(b), we see that $\rho^{-1}(x)$ is a chain of projective lines that meet the other components of $(\mathcal{C}_{\min})_s$ at two points distributed between the outer components. It follows from Lemma 3.31 that $\mathcal{C}_{\can}$ is semi-stable.

Let us now show that $\mathcal{C}_{\can} \to S$ is stable. This comes down to saying that $Z := (\mathcal{C}_{\can})_s$ is stable over $k(s)$. Let Δ be an irreducible component of Z that is isomorphic to $\mathbb{P}^1_{k(s)}$. Let us suppose that Δ meets the other components at one or two points, a corollary 3.25 shows that the strict transform $\widetilde{\Delta}$ of Δ in $\mathcal{C}_{\min}$ is a projective line that meets the other components at at most two points (in fact exactly two, because otherwise it would be an exceptional divisor); hence $\widetilde{\Delta}^2 = -2$ and $\widetilde{\Delta}$ is contracted to one point by ρ, a contradiction. As the singular points of Z are split, it is clear that $Z_{\overline{k(s)}}$ is stable, whence Z is stable.

It remains to show that any stable model $\mathcal{C}$ of C over S is isomorphic to $\mathcal{C}_{\can}$. Let $\mathcal{C}' \to \mathcal{C}$ be the minimal desingularization. Using the description given in Corollary 3.25, we immediately see that $\mathcal{C}' \simeq \mathcal{C}_{\min}$, and that the morphism $\mathcal{C}_{\min} \to \mathcal{C}$ consists of the contraction of the $\mathbb{P}^1_{k(s)}$ with self-intersection -2 in $\mathcal{C}_{\min}$, which proves that $\mathcal{C} \simeq \mathcal{C}_{\can}$. $\qquad\square$

Example 3.35. Let E be an elliptic curve over $K(S)$. It follows from Theorem 3.34 that E has semi-stable reduction at a point $s \in S$ if and only if it has good reduction or multiplicative reduction at s (in the sense of Definition 2.2).

Corollary 3.36. *Let S be a Dedekind scheme of dimension 1, C a smooth projective curve over $K(S)$ with $p_a(C) \geq 1$, and S' a Dedekind scheme of dimension 1 that dominates S. Let $K' = K(S')$.*

(a) If C has semi-stable (resp. stable) reduction over S, then $C_{K'}$ has semi-stable (resp. stable) reduction over S'. If $\mathcal{C}$ is a semi-stable (resp. stable) model of C over S, then $\mathcal{C} \times_S S'$ is a semi-stable (resp. stable) model of $C_{K'}$ over S'.

(b) Let $\mathcal{C}_s$ be the stable reduction (of which we suppose the existence) of C at a point $s \in S$. Let $s' \in S'$ lie above s. Then the stable reduction of $C_{K'}$ at s' is isomorphic to $\mathcal{C}_s \times_{\operatorname{Spec} k(s)} \operatorname{Spec} k(s')$.

(c) Let us moreover suppose either that $S' \to S$ is étale surjective, or that $S = \operatorname{Spec} \mathcal{O}_K$ is local and $S' = \operatorname{Spec} \widehat{\mathcal{O}}_K$ or $S' = \operatorname{Spec} \mathcal{O}_L$ as in Lemma 3.33(b). If $C_{K'}$ has semi-stable (resp. stable) reduction over S', then C has semi-stable (resp. stable) reduction over S.

Proof   (a) is an immediate consequence of Proposition 3.15(a), and (b) results from (a). Let us show (c). We know that $\mathcal{C}_{\min} \times_S S'$ is the minimal regular model of $C_{K(S')}$ over S' (Proposition 1.17, Lemma 3.33), and it is semi-stable by Theorem 3.34(a). Hence $\mathcal{C}_{\min}$ is semi-stable (Lemma 3.30(a)).   □

Corollary 3.37. *Let S be a Dedekind scheme of dimension 1. Let C be a smooth projective curve over $K = K(S)$, admitting a stable model $\mathcal{C}$ over S. Then any K-automorphism of C extends uniquely to an S-automorphism of $\mathcal{C}$.*

Proof   As in Proposition 1.16, the existence of the extension comes from the uniqueness of the stable model. But let us explain this in more detail. By definition, a stable model $\mathcal{C}$ comes with an isomorphism $f : \mathcal{C}_K \simeq C$. Let $\sigma : C \to C$ be a K-automorphism. Then we have at our disposal two stable models $(\mathcal{C}, f)$ and $(\mathcal{C}, \sigma^{-1} f)$ of C over S. The uniqueness of the stable model can now be worded as saying that the birational map of models $f^{-1} \sigma f : (\mathcal{C}, f) \dashrightarrow (\mathcal{C}, \sigma^{-1} f)$ is an automorphism $\tau : \mathcal{C} \to \mathcal{C}$. This automorphism extends σ in the sense that τ_K makes the following diagram commutative

$$
\begin{array}{ccc}
\mathcal{C}_K & \xrightarrow{\ \tau_K\ } & \mathcal{C}_K \\
\downarrow{f} & & \downarrow{f} \\
C & \xrightarrow{\ \sigma\ } & C
\end{array}
$$

The extension is unique because $\mathcal{C}$ is separated (Proposition 3.3.11).   □

Proposition 3.38. *Let $X \to S$ be a stable curve over a Dedekind scheme S, with smooth generic fiber. Then for any $s \in S$, the canonical map $\operatorname{Aut}(X_\eta) \to \operatorname{Aut}(X_s)$ is injective.*

Proof   We can suppose S is local. Let S' be the spectrum of a discrete valuation ring $\mathcal{O}_{K'}$ with algebraically closed residue field that dominates $\mathcal{O}_{S,s}$ (Lemma 3.32). It suffices to show the proposition for $X_{S'} \to S'$ (Exercise 5.2.13). We can therefore suppose that $k(s)$ is already algebraically closed.

Let $\sigma \in \operatorname{Aut}(X_\eta)$. By Corollary 3.37, it acts on X, and it has finite order (see the references in Exercise 7.4.6(c)). Let us suppose that it acts trivially on X_s. Let us set $W = X/\langle \sigma \rangle$. As σ acts transitively on the fibers of $X \to W$, it follows that $X_s \to W_s$ is a bijective map. As $k(s)$ is algebraically closed, this implies that $X_s \to W_s$ is purely inseparable (Exercise 5.3.7(b)). Let Γ be an irreducible component of X_s and Δ its image in W_s. Then the morphism between the normalizations $\Gamma' \to \Delta'$ is purely inseparable, and hence bijective. Consequently, the image of an ordinary double point of X_s is a double point $w \in W_s$ (i.e., there exist two points lying above w in the normalization of W'_s). Let Y be the semi-stable curve over $k(s)$ associated to W_s (Definition 7.5.13). Then $p_a(X_s) = p_a(Y)$ and $p_a(Y) \le p_a(W_s)$, which implies that $2 \le p_a(X_\eta) \le p_a(W_\eta)$. By applying Hurwitz's formula to the finite separable morphism $X_\eta \to W_\eta$ that is of degree the order of σ, we obtain that $\sigma = 1$. □

Remark 3.39. Let $X \to S$ be a stable curve over a scheme S. We can show that there exists a finite unramified S-scheme $\mathcal{A}$ such that we have a bijection

$$\mathrm{Aut}_T(X_T) \simeq \mathrm{Mor}_S(T, \mathcal{A})$$

for every S-scheme T, compatible with morphisms $T' \to T$ of S-schemes in the natural sense. See [26], Theorem 1.11. We say that the functor $T \mapsto \mathrm{Aut}_T(X_T)$ is representable by the scheme $\mathcal{A}$. It is easy to show that the fact that $\mathcal{A}$ is finite and unramified implies Proposition 3.38.

10.3.3 Some sufficient conditions for the existence of the stable model

Corollary 7.5.24 characterizes curves with ordinary multiple singularities by the vanishing of the unipotent rank. We are going to apply this result to obtain the stable reduction.

Lemma 3.40. *Let $f : X \to Y$ be a birational morphism of arithmetic surfaces over a discrete valuation ring $\mathcal{O}_K$ with algebraically closed residue field k. Then X_s and Y_s have the same Abelian (resp. toric) rank. Moreover, if the gcd of the multiplicities of the irreducible components of X_s is equal to 1, then the unipotent rank is also the same for X_s and Y_s.*

Proof We can suppose that f is the blowing-up of a closed point $y \in Y_s$ (Theorem 9.2.2). Let $E \subset X_s$ be the exceptional divisor. As this curve is rational, we immediately deduce from this that the Abelian rank is invariant. Let m_y be the number of points lying above y in the normalization of $(Y_s)_{\mathrm{red}}$, and $x_1, \ldots, x_r$ be the points of $f^{-1}(y)$ belonging to the strict transform of Y_s. Taking into account the fact that E is normal, it is easy to establish that $m_y = \sum_i (m_{x_i} - 1)$. It follows from Lemma 7.5.18 that $\mathrm{Pic}^0(X_s)$ and $\mathrm{Pic}^0(Y_s)$ have the same toric rank. Finally, if the hypothesis on the gcd of the multiplicities is verified for X_s, then it is also verified for Y_s (Exercise 9.2.9(c)). It follows that

$$\dim_k H^1(X_s, \mathcal{O}_{X_s}) = p_a(X_\eta) = p_a(Y_\eta) = \dim_k H^1(Y_s, \mathcal{O}_{Y_s}),$$

whence the invariance of the unipotent rank since the sum of the Abelian, toric, and unipotent ranks of a projective curve over k is equal to $\dim_k H^1(C, \mathcal{O}_C)$. □

Remark 3.41. The hypothesis on the gcd of the multiplicities is not necessary. Indeed, using the Leray spectral sequence, we can show that there exists an isomorphism $H^1(Y, \mathcal{O}_Y) \simeq H^1(X, \mathcal{O}_X)$ (see the proof of [15], Theorem 9.7.1), which implies an isomorphism $H^1(Y_s, \mathcal{O}_{Y_s}) \simeq H^1(X_s, \mathcal{O}_{X_s})$ (Exercise 5.3.13(b)).

Proposition 3.42. *Let S be the spectrum of a discrete valuation ring with algebraically closed residue field k. Let $X \to S$ be an arithmetic surface such that X_s is connected, that the gcd of the multiplicities of the irreducible components of X_s is equal to 1, and that X_s has unipotent rank zero. Then the following properties are true.*

(a) *The curve $(X_s)_{\mathrm{red}}$ is semi-stable.*

(b) *Let us suppose $p_a(X_\eta) \geq 1$. Then the minimal regular model of X is semi-stable.*

Proof Let $Z_0 = (X_s)_{\mathrm{red}}$. If C is a projective curve over k, we will write $h^1(\mathcal{O}_C) = \dim_k H^1(C, \mathcal{O}_C)$.

(a) Let Y be the reduced projective curve with ordinary multiple singularities associated to Z_0 (Definition 7.5.13). Then the unipotent rank u of $\mathrm{Pic}^0(X_s)$ is the sum of two terms that are positive or zero

$$u = (h^1(\mathcal{O}_{X_s}) - h^1(\mathcal{O}_{Z_0})) + (h^1(\mathcal{O}_{Z_0}) - h^1(\mathcal{O}_Y)) \qquad (3.27)$$

(see formula (5.23) of the proof of Theorem 7.5.19). Let $\pi : Y \to Z_0$ denote the birational morphism. We have $h^1(\mathcal{O}_{Z_0}) = h^1(\mathcal{O}_Y)$. It follows from Lemma 7.5.18 that π is an isomorphism, and therefore that Z_0 has ordinary multiple singularities. As X is regular of dimension 2, we have $\dim_{k(x)} T_{Z_0,x} \leq \dim_{k(x)} T_{X,x} \leq 2$ for any $x \in X_s$. Therefore the singular points of Z_0 are ordinary double points (Proposition 7.5.15).

(b) Let us suppose that X_s is not reduced. Let us show that there exists an exceptional divisor E on X that is of multiplicity ≥ 2. With Lemma 3.40, this will imply that the minimal model is semi-stable. The proof of Corollary 9.1.24 gives a chain of divisors $Z_0 < Z_1 < \cdots < Z_{r+1} = X_s$ such that $Z_{i+1} - Z_i$ is a prime divisor Δ_i, that $\Delta_i \cdot Z_i > 0$, and that $H^0(Z_i, \mathcal{O}_{Z_i}) = H^0(Z_0, \mathcal{O}_{Z_0}) = k$ (see also Exercise 9.1.14). As $h^1(\mathcal{O}_{Z_{i+1}}) \geq h^1(\mathcal{O}_{Z_i})$, equality (3.27) implies that $h^1(\mathcal{O}_{X_s}) = h^1(\mathcal{O}_{Z_r})$. The exact sequence

$$0 \to \mathcal{O}_X(-Z_r)|_{\Delta_r} \to \mathcal{O}_{X_s} \to \mathcal{O}_{Z_r} \to 0 \qquad (3.28)$$

(see exact sequence (1.2) of Section 9.1) and equality (3.27) imply that

$$0 = \chi(\mathcal{O}_{X_s}) - \chi(\mathcal{O}_{Z_r}) = \chi(\mathcal{O}_X(-Z_r)|_{\Delta_r}) = -Z_r \cdot \Delta_r + 1 - p_a(\Delta_r).$$

Consequently, $p_a(\Delta_r) = 0$ and $\Delta_r^2 = -\Delta_r \cdot Z_r = -1$. Hence Δ_r is an exceptional divisor. Moreover, $2\Delta_r \leq Z_r + \Delta_r \leq X_s$; hence Δ_r has multiplicity ≥ 2 in X_s. $\square$

Remark 3.43. Using Remark 3.41, the method above allows us to show Proposition 3.42 without hypothesis on the gcd of the multiplicities.

When we want to compute the stable reduction of a curve given by explicit equations, it can occur that the computation is easier to do with affine models. We are going to give here a generalization of Corollary 1.25. The following proposition says that if we have the semi-stable reduction of certain pieces of C, and if these pieces are sufficiently large, then they glue to give a semi-stable reduction of C.

For any reduced curve U over a field k, we let $\hat{U}$ denote the projective curve obtained by adding regular points to U, as in Exercise 4.1.17. We will temporarily say that a semi-stable (not necessarily projective) curve U over k is *quasi-stable* if $U_{\overline{k}}$ does not contain any rational irreducible component (i.e., isomorphic to an open subscheme of $\mathbb{P}^1_{\overline{k}}$) that meets the other components in at most one point, nor any rational irreducible projective component that meets the other components at exactly two points. This implies that the connected components of $\hat{U}_{\overline{k}}$ have arithmetic genus $p_a \geq 1$.

Proposition 3.44. *Let $\mathcal{O}_K$ be a discrete valuation ring with perfect residue field k. Let C be a smooth, projective, geometrically connected curve over K, of genus $g(C) \geq 2$. Let $W_1^0, \ldots, W_m^0$ be flat quasi-projective schemes over $S = \operatorname{Spec} \mathcal{O}_K$, such that $W_{i,\eta}^0$ is isomorphic to an open subscheme of C. We make the following hypotheses:*

(i) *For every $i \leq m$, the curve $U_i := W_{i,s}^0$ is semi-stable over k with*

$$\sum_{1 \leq i \leq m} p_a(\hat{U}_i) \geq p_a(C).$$

(ii) *The valuations of $K(C)$ induced by the generic points of the U_i are pairwise non-equivalent.*

Then C has stable reduction over S. Moreover, if the U_i are quasi-stable, then the disjoint union of the U_i is isomorphic to an open subscheme of the stable reduction $(C_{\text{can}})_s$, and $\hat{U}_i$ is isomorphic to the Zariski closure of U_i in $(C_{\text{can}})_s$.

Proof The proof is done in several steps.

(1) *Some simplifications.* Using Lemmas 3.32–33, we reduce to the case when k is algebraically closed. With Corollary 3.36, we reduce to $\mathcal{O}_K$ complete. Without changing the condition on $p_a(\hat{U}_i)$, we can remove the rational irreducible components of U_i that meet the other irreducible components of U_i at at most one point. Each W_i^0 is an open subscheme of a model W_i of C over S. By contraction of the irreducible components of $W_{i,s}$ that do not meet U_i (Theorem 8.3.36), and of the irreducible components of U_i that are isomorphic to $\mathbb{P}^1_k$ and meet the other irreducible components of U_i at two points (the proof of Lemma 3.31 says that these components are contracted to ordinary double points), we can suppose that U_i is dense in $W_{i,s}$ and that U_i is quasi-stable. Finally, adding (a finite number of) semi-stable points to U_i if necessary (which increases $p_a(\hat{U}_i)$), and contracting the $\mathbb{P}^1_k$ in the new U_i, we can suppose that every semi-stable point of $W_{i,s}$ belongs to U_i.

(2) *Construction of a model that contains the disjoint union of the* U_i. Let W_0 be the closure of the diagonal of C^m in $W_1 \times_S W_2 \times_S \cdots \times_S W_m$. Let us consider the normalization $W \to W_0$ of W_0. Then W is the smallest model of C over S that dominates the W_i (this is a consequence of Theorem 8.3.20). Let $p_i : W \to W_i$ denote the canonical birational morphism and $\Delta_i \subset W_s$ the strict transform of $W_{i,s}$. By the minimality of W, we have $W_s = \cup_i \Delta_i$. In particular, the irreducible components of W_s have multiplicity 1, and hence W_s is reduced (Exercise 8.3.3). We are going to show that p_i is an isomorphism above U_i. The condition that $\sum_i p_a(\hat{U}_i) \geq p_a(C)$ will not be used.

We can suppose that $m = 2$ (reason by induction with U_1 and $\cup_{i \geq 2} U_i$ in the general case). Let $x \in U_1$ be a point above which p_1 is not an isomorphism. Then $p_1^{-1}(x)$ is a reduced connected curve (Proposition 8.3.22) contained in Δ_2, and therefore equal to a connected component Δ_{20} of Δ_2. By Lemma 3.45 applied to $X = W$ and $E = \Delta_{20}$, we have $p_a(\Delta_{20}) = 0$, and $F := \Delta_{20} \cap \Delta_1$ contains at most two points. The first condition implies that Δ_{20} has ordinary multiple singularities (Corollary 7.5.24) and is a union of projective lines. The morphism p_2 induces an isomorphism

$$p_2 : W_s \setminus \Delta_1 \to W_{2,s} \setminus p_2(\Delta_1)$$

because it is finite birational. Let $V_2 = p_2(\Delta_{20} \setminus F) \simeq \Delta_{20} \setminus F$. As $\Delta_{20} \setminus F$ is open in $W_s \setminus \Delta_1$, V_2 is open in $W_{2,s} \setminus p_2(F)$, and hence open in $W_{2,s}$. If V_2 has at least three irreducible components, then it is easy to see that the condition $p_a(\Delta_{20}) = 0$ implies that there exists an irreducible component Γ of V_2 that meets the other components of $W_{2,s}$ at at most one point. As $\Gamma \cap U_2$ is an irreducible rational component of U_2, this contradicts the U_2 quasi-stable hypothesis. Hence V_2 has at most two irreducible components, it is therefore semi-stable, and consequently contained in U_2. As $p_2(\Delta_{20})$ meets the other irreducible components of $W_{2,s}$ at $p_2(F)$, and hence at at most two points, we easily see that this again contradicts the U_2 quasi-stable hypothesis. We have therefore shown that $p_i^{-1}(U_i) \to U_i$ is an isomorphism for every $i \leq m$. As $p_i(U_j)$ is finite if $j \neq i$, we have $p_i^{-1}(U_i) \cap p_j^{-1}(U_j) = \emptyset$ if $i \neq j$.

(3) *Existence of the stable reduction.* Let $p : \mathcal{C} \to W$ be the minimal desingularization with normal crossings (Corollary 8.3.51 and Proposition 9.3.36). Let Z_i be the Zariski closure of $p^{-1}(p_i^{-1}(U_i))$ in $\mathcal{C}$, endowed with the reduced structure. For any double point $x \in p_i^{-1}(U_i)$, $p^{-1}(x)$ is a point or a chain of projective lines that meet the other irreducible components of $\mathcal{C}_s$ at two points (Corollary 3.25). We then easily see that $p_a(Z_i) = p_a(\hat{U}_i)$. As Z_i is semi-stable, we have the relation $p_a(Z_i) = a(Z_i) + t(Z_i)$ with the Abelian and toric ranks of Z_i. By Exercise 7.5.10, we have

$$a(\mathcal{C}_s) + t(\mathcal{C}_s) \geq \sum_{1 \leq i \leq m} \big(a(Z_i) + t(Z_i)\big) \geq p_a(C) = \dim_k H^1(\mathcal{C}_s, \mathcal{O}_{\mathcal{C}_s})$$

(the last equality comes from Corollary 9.1.24 and from the fact that $\mathcal{C}_s$ has irreducible components of multiplicity 1). It follows that the unipotent rank of $\mathcal{C}_s$ is zero (Definition 7.5.21), and therefore that C has stable reduction, by Proposition 3.42 and Theorem 3.34(b).

(4) *End of the proof.* We can show that $\cup_i p_i^{-1}(U_i) \subseteq W_s$ is an open subscheme of the stable reduction of C. Since it is semi-stable, it suffices to show that W is dominated by the canonical (hence stable, by Theorem 3.34) model $\mathcal{C}_{\text{can}}$ of C over S. For any regular model $\mathcal{C}$ that dominates W, the hypothesis that the U_i are quasi-stable implies that no irreducible component of the strict transform of Δ_i in $\mathcal{C}$ is a projective line with self-intersection -1 or -2. Hence the minimal regular model $\mathcal{C}_{\min}$, as well as the canonical model $\mathcal{C}_{\text{can}}$, dominate W.

Let Δ_i' be the strict transform of Δ_i in $\mathcal{C}_{\text{can}}$. As $p_i^{-1}(U_i)$ is isomorphic to a dense open subscheme of Δ_i', we have a birational morphism $\hat{U}_i \to \Delta_i'$. We then have $p_a(\Delta_i') \geq p_a(\hat{U}_i)$. The same reasoning as in (3) implies that $p_a(\Delta_i') = p_a(\hat{U}_i)$, and therefore that $\hat{U}_i \to \Delta_i'$ is an isomorphism. $\qquad\square$

Lemma 3.45. *Let $\mathcal{O}_K$ be a discrete valuation ring with algebraically closed residue field k. Let $X \to \operatorname{Spec} \mathcal{O}_K$ be a normal fibered surface, with X_s reduced and connected, $f : X \to Y$ a contraction morphism of a reduced connected curve $E \subseteq X_s$, and $y = f(E)$. Let us suppose that Y_s is semi-stable at y. Let Y_s' denote the strict transform of Y_s in X and r the cardinal of $E \cap Y_s'$. Then we have $r \leq 2$, and $p_a(E) = 0$.*

Proof Let us consider the birational morphism $\pi : Y_s' \to Y_s$. It is an isomorphism outside of y. We have $E \cap Y_s' = E \cap f^{-1}(y) = \pi^{-1}(y)$. Above y, π is either an isomorphism, or the normalization of y. Hence $r \leq 2$. As $p_a(Y_s) = p_a(Y_\eta) = p_a(X_\eta)$, this immediately implies the equality

$$p_a(Y_s') = p_a(X_\eta) - r + 1. \tag{3.29}$$

We have an exact sequence

$$0 \to \mathcal{O}_{X_s} \to \mathcal{O}_E \oplus \mathcal{O}_{Y_s'} \to \mathcal{F} \to 0$$

where $\mathcal{F}$ is a skyscraper sheaf whose support is equal to $E \cap Y_s'$. It follows that

$$p_a(X_s) = p_a(E) + p_a(Y_s') + \dim_k H^0(X_s, \mathcal{F}) - 1 \geq p_a(E) + p_a(X_\eta).$$

Now $p_a(X_s) = p_a(X_\eta)$, and hence $p_a(E) = 0$. $\qquad\square$

Example 3.46. Let us once more take the hyperelliptic curve C of Example 1.30, defined by a polynomial $P(x) = (x^3 + ax + b)(x^{2g-1} + tcx + td)$ with $(4a^3 + 27b^2)b \in \mathcal{O}_K^*$. Let us moreover suppose that k is perfect, of characteristic $\operatorname{char}(k)$ prime to $g - 1$, and that $c = t^{4g-5}c_2$, $d = t^{4g-3}d_2$ with $c_2 \in \mathcal{O}_K^*$ and $d_2 \in \mathcal{O}_K$. We have an affine scheme $W_1 := \operatorname{Spec} \mathcal{O}_K[x, y, 1/x]/(y^2 - P(x))$ whose special fiber U_1 is smooth, with $p_a(\hat{U}_1) = 1$. We have another affine scheme

$$W_2^0 = \operatorname{Spec} \mathcal{O}_K[x_2, y_2]/(y_2^2 - (t^6x_2^3 + at^2x_2 + b)(x_2^{2g-1} + c_2x_2 + d_2)),$$

with $x_2 = t^{-2}x$, $y_2 = t^{-(2g-1)}y$. Its generic fiber is an open subscheme of C, and its special fiber

$$U_2 = \operatorname{Spec} k[x_2, y_2]/(y_2^2 - (x_2^{2g-1} + \tilde{c}_2x_2 + \tilde{d}_2)),$$

where $\tilde{c}_2$ and $\tilde{d}_2$ are the images of c_2, d_2 in k, is semi-stable because $X^{2g-1} + c_2X + d_2$ has at most double roots in $\bar{k}$. We have $p_a(\hat{U}_2) = g - 1$. It follows from

Proposition 3.44 that C has stable reduction over $\mathcal{O}_K$, and that the special fiber of the stable model is the union of $\hat{U}_1$ and of $\hat{U}_2$. These two components meet at the unique point $\{q\} = \hat{U}_2 \setminus U_2$. Finally, $\hat{U}_2$ is smooth if $\operatorname{disc}(X^{2g-1} + c_2 X + d_2) \in \mathcal{O}_K^*$, otherwise it has a unique ordinary double point that is rational over k.

Example 3.47. Let $\mathcal{O}_K$ be a discrete valuation ring, with residue field k of characteristic $\operatorname{char}(k) \neq 3$ and containing the third roots of unity. Let C be the normal projective curve over K with field of functions defined by the equation

$$y^3 = (x^3 + at^{3n})(x^3 + 1), \quad n \geq 1, a \in \mathcal{O}_K^*. \tag{3.30}$$

We verify that C is geometrically connected (Corollary 3.2.14). The curve C is the union of the affine curves defined respectively by equation (3.30) and the equation $y_1^3 = (1 + at^{3n} x_1^3)(1 + x_1^3)$, with $x_1 = 1/x$ and $y_1 = y/x^2$. With the Jacobian criterion, we see that C is smooth over K. The inclusion $K(x) \subset K(x, y) = K(C)$ induces a morphism $f : C \to \mathbb{P}^1_K$ of degree 3. Over $\overline{K}$, f is ramified at six points, corresponding to the roots of $(x^3 + t^{3n} a)(x^3 + 1)$, with ramification index 3 in each of these points. By Hurwitz's formula 7.4.16, we deduce from this that $g(C) = 4$. Let us consider the affine scheme

$$W^0 = \operatorname{Spec} \mathcal{O}_K[x, u, v]/(ux - t^n, v^3 - (1 + au^3)(1 + x^3)).$$

Then W^0_η is isomorphic to an open subscheme of C (identifying v with y/x). Its special fiber is the union of two elliptic curves meeting at three points $x = u = 0$, $v^3 = 1$. Let Z be the curve $\hat{W}^0_s$. Then $a(Z) = 2$ and $t(Z) \geq 2$. As $p_a(Z) \leq g(C) = 4$, we deduce from this that $t(Z) = 2$, and therefore that $u(Z) = 0$. It follows that Z is semi-stable. By Proposition 3.44, Z is the stable reduction of C over $\mathcal{O}_K$. Each point q_i corresponds to $x = u = 0$ and v a third root of unity.

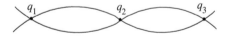

Proposition 3.48.   ([82], Appendice) *Let S be the spectrum of a discrete valuation ring $\mathcal{O}_K$. Let X be a semi-stable quasi-projective curve over S, endowed with the action of a finite group G. Then the quotient scheme $Y = X/G$ is semi-stable. More precisely, let $x \in X_s$ be a closed point, y its image in Y_s. Then we have the following properties:*

(a) *if X is smooth at x, then Y is smooth at y;*

(b) *if x is an ordinary double point of X_s, then y is a smooth or ordinary double point;*

(c) *let I' be the image of the inertia group I at x in $\operatorname{Aut}_{\mathcal{O}_K}(\mathcal{O}_{X,x})$. If x is split of thickness m, and if y is a double point, then y is split of thickness mn, where n is the order of I'.*

Proof We decompose $f : X \to Y$ into $X \to X/I \to Y$. Then $X/I \to Y$ is étale at the image of x in X/I (Exercise 4.3.19(d)). It therefore suffices to study $X \to X/I$. In other words, we can suppose that $G = I$. Then x is the unique point of $f^{-1}(y)$. Let $A = \mathcal{O}_{Y,y}$ and $B = \mathcal{O}_{X,x}$. Then $B = (f_*\mathcal{O}_X)_y$ is finite over A. Moreover, replacing I by I' if necessary, we can suppose that $I \subseteq \mathrm{Aut}_{\mathcal{O}_K}(B) \subseteq \mathrm{Aut}_{\hat{\mathcal{O}}_K}(\hat{B})$. Finally, as $X \to X/G$ commutes with flat base change (Exercise 4.3.18), and as the semi-stability of Y over $\mathcal{O}_K$ can be tested on a discrete valuation ring dominating $\mathcal{O}_K$ (Lemma 3.30(a)), we can replace $\mathcal{O}_K$ by a suitable discrete valuation ring, and suppose that $\mathcal{O}_K$ is complete, x is rational over $k(s)$, and that x is split if it is an ordinary double point in X_s. As $\hat{B} = B \otimes_A \hat{A}$ (Exercise 4.3.17), we have

$$\mathrm{Frac}(\hat{B}) = B \otimes_A \mathrm{Frac}(\hat{A}) = K(X) \otimes_{K(Y)} \mathrm{Frac}(\hat{A}).$$

For simplicity, we will suppose that X_η is smooth. In particular, $\hat{B}$ is integral (Corollary 3.22(b)). It follows that

$$[\mathrm{Frac}(\hat{B}) : \mathrm{Frac}(\hat{A})] = [K(X) : K(Y)] = n.$$

For the remainder of the proof, we let t denote a uniformizing parameter for $\mathcal{O}_K$, and k the residue field of $\mathcal{O}_K$, and we denote the class of $b \in B$ modulo t by $\tilde{b}$.

(a) If X is smooth at x, then $\hat{B} = \mathcal{O}_K[[u]]$ for some $u \in \hat{B}$. Let $w \in \hat{A}$ be the norm of $u \in \hat{B}$. Then we have the inclusions $C := \mathcal{O}_K[[w]] \subseteq \hat{A} \subseteq \hat{B}$, which are finite homomorphisms. We have $\mathrm{Frac}(\hat{B}) = \hat{B} \otimes_C \mathrm{Frac}(C)$ by the finiteness. By Exercise 5.1.15(a), we have

$$[\mathrm{Frac}(\hat{B}) : \mathrm{Frac}(C)] \le \dim_k(\hat{B} \otimes_C C/(t,w)) = \dim_k k[[\tilde{u}]]/(\tilde{w}).$$

As $\sigma(\tilde{u}) \in \tilde{u}k[[\tilde{u}]]^*$ for every $\sigma \in I$, we have $\tilde{w} \in \tilde{u}^n k[[\tilde{u}]]^*$. Hence the member on the right-hand side of the inequality above is equal to $n = [\mathrm{Frac}(\hat{B}) : \mathrm{Frac}(\hat{A})]$. Consequently, $\mathcal{O}_K[[w]] \to \hat{A}$ is finite and birational; it is therefore an equality since $\mathcal{O}_K[[w]]$ is normal, which implies that Y is smooth at y.

(b) Let us suppose that x is a double point of X_s. We have $\hat{B} = \mathcal{O}_K[[u,v]]$ with $uv = t^m$, $m \ge 1$. The group I acts on the set of minimal prime ideals of $\hat{B}/(t)$, which are $(\tilde{u})$ and $(\tilde{v})$. Let I_0 be the subgroup

$$I_0 = \{\sigma \in I \mid \sigma(\tilde{u}) \in \tilde{u}(\hat{B}/(t))\}$$

of elements that leave each of these minimal prime ideals globally invariant. Let us first suppose that $I = I_0$. Let u_1, v_1 be the respective norms of u, v in $\hat{A}$. Then $u_1 v_1 = t^{nm}$. Let us set $C = \mathcal{O}_K[[u_1, v_1]] \subseteq \hat{A}$ and let us estimate the degree $[\mathrm{Frac}(\hat{B}) : \mathrm{Frac}(C)]$. Let $\mathfrak{p} = (t, u_1) \in \mathrm{Spec}\, C$, $T = k[[\tilde{v}_1]] \setminus \{0\}$. As $\tilde{u}_1 \in \tilde{u}^n(\hat{B}/(t))^*$, we have

$$\hat{B} \otimes_C k(\mathfrak{p}) = T^{-1}(\hat{B} \otimes_C C/\mathfrak{p}) = T^{-1}k[[\tilde{u}, \tilde{v}]]/(\tilde{u}^n, \tilde{u}\tilde{v}) = T^{-1}k[[\tilde{v}]] = k((\tilde{v})).$$

Now $k((\tilde{v}))$ is of dimension n over $k((\tilde{v}_1)) = k(\mathfrak{p})$. We therefore have

$$[\mathrm{Frac}(\hat{B}) : \mathrm{Frac}(C)] \le \dim_{k(\mathfrak{p})}(\hat{B} \otimes_C k(\mathfrak{p})) = n.$$

As in (a), this implies that $C = \hat{A}$. Consequently, y is a split ordinary double point of Y_s, of thickness nm.

There remains the case when $I_0 \neq I$. It is then a subgroup of I of index 2. From the above, X/I_0 is semi-stable and has a split ordinary double point in the image of x in X/I_0. The quotient X/I is the quotient of X/I_0 by the group I/I_0 of order 2 that permutes the minimal prime ideal of $\hat{B}(t)$. We thus reduce to the case when $I_0 = 1$ and I is of order 2, generated by a permutation τ. Let $w = u + \tau(u)$, $C = \mathcal{O}_K[[w]]$. Then $\hat{B} = \mathcal{O}_K[[u, \tau(u)]]$, and $v \in (\tau(u), t)$. We have

$$\hat{B} \otimes_C C/(t, w) = \hat{B}/(t, w) = k[[\tilde{u}]]/(\tilde{u}\tau(\tilde{u})) = k[[\tilde{u}]]/(\tilde{u}^2).$$

A reasoning similar to that above shows that $[\mathrm{Frac}(\hat{B}) : \mathrm{Frac}(C)] \leq 2 = n$ and therefore that $\hat{A} = C = \mathcal{O}_K[[w]]$, which implies that Y is smooth at y. □

Remark 3.49. Let us suppose that X is integral (e.g., with smooth connected generic fiber), and that $G \to \mathrm{Aut}_S(X)$ is injective. Let D be the decomposition group at x. Then the canonical map $D \to \mathrm{Aut}_{\mathcal{O}_K}(\mathcal{O}_{X,x})$ is injective. Hence, the inertia group I at x can be identified with its image I'. In Examples 3.46 and 3.47, the curves C over K are Galois coverings of $\mathbb{P}^1_K$, with group G respectively equal to $\mathbb{Z}/2\mathbb{Z}$ and $\mathbb{Z}/3\mathbb{Z}$. In the first case, the quotient of the stable model of C by G is a semi-stable model having a unique ordinary double point of thickness 2 (this comes from the relation $x = t^2 x_2$); hence q has thickness 1. In the case of Example 3.47, the inertia groups at the double points q_1, q_2, q_3 are trivial. As the thickness of the image of the q_i in the quotient is n (this comes from the relation $xu = t^n$), the q_i have thickness n.

Remark 3.50. Proposition 3.48 is true in the following more general situation: let $f : X \to Y$ be a finite dominant morphism with X semi-stable over S, and Y normal. Then Y is semi-stable. Moreover, if X is smooth, then so is Y. See [82], Appendix, Proposition 5. See Exercise 3.19 for the case when f is purely inseparable. In the Galois case, we can replace S by a normal, excellent, locally Noetherian scheme. The proof is the same as the one above. See [50], Proposition 4.2.

Remark 3.51. By admitting the stable reduction theorem (4.3), we can show the following assertion: let $f : C \to E$ be a finite morphism of smooth, projective, geometrically connected curves over K, where K is the function field of an affine Dedekind scheme S. Let us suppose that C admits a stable (resp. smooth) model $\mathcal{C}$ over S and that $g(E) \geq 2$ (resp. $g(E) \geq 1$). Then E admits a stable (resp. smooth) model $\mathcal{E}$ over S, and f extends to a dominant (resp. finite) morphism $\mathcal{C} \to \mathcal{E}$. See [60], Corollaries 4.7 and 4.10. The particular case when $C \to E$ is Galois results immediately from Proposition 3.48.

Exercises

3.1. Let C be a stable curve over a field k. Show that $p_a(C) = \dim_k H^1(C, \mathcal{O}_C)$.

3.2. Let C be a semi-stable, projective, geometrically connected curve over a field k. Show that if $\omega_{C/k}$ is ample, then $p_a(C) \geq 2$ and C is stable. This is the converse of Corollary 3.13.

3.3. Let C be a semi-stable curve over a field k. Let $\pi : C' \to C$ be the normalization morphism. Show that we have canonical homomorphisms

$$\Omega^1_{C/k} \to \pi_* \pi^* \Omega^1_{C/k} \to \pi_* \Omega^1_{C'/k} \to \pi_* \omega_{C'/k}$$

whose composition is an isomorphism.

3.4. Let C be a semi-stable curve over a field k, and let $\pi : C' \to C$ be the normalization morphism. Show that for any closed point $x \in C$, we have

$$\dim_{k(x)} (\pi_* \mathcal{O}_{C'})_x \otimes_{\mathcal{O}_{C,x}} k(x) \le 2.$$

In particular, $\pi^{-1}(x)$ contains at most two points.

3.5. Let C be a reduced curve over a field k. Let $C' \to C$ be the normalization morphism. Show that C is semi-stable if and only if for every non-smooth closed point $x \in C$, we have $k(x)$ separable over k, $C' \times_C \operatorname{Spec} k(x)$ reduced, and $\dim_{k(x)} \mathcal{O}'_{C,x}/\mathcal{O}_{C,x} = 1$, where $\mathcal{O}'_{C,x}$ is the integral closure of $\mathcal{O}_{C,x}$ in its total ring of fractions.

3.6. Let C be a semi-stable curve over a field k, with split singular points.

(a) Let $\pi : C' \to C$ be the normalization morphism. Show that $\operatorname{Aut}_k(C)$ can be identified with the elements $\sigma \in \operatorname{Aut}_k(C')$ such that $\sigma(\pi^{-1}(x)) = \pi^{-1}(x)$ for every $x \in C$ (use Proposition 3.7(c)).

(b) Let us suppose C is geometrically connected over k and k infinite. Admitting the fact that the group of automorphisms of a smooth projective curve of genus ≥ 2 is finite (see the references in Exercise 7.4.6(c)), show that C is stable if and only if $\operatorname{Aut}_k(C)$ is finite.

3.7. Let C be a stable curve over a field k. Using Lemma 3.12(a), show that $\omega_{C/k}$ is generated by its global sections if and only if C is geometrically integral.

3.8. Let C be a smooth conic over a discrete valuation field K.

(a) Show that if C has semi-stable reduction over $\mathcal{O}_K$, then there exists an étale extension $\mathcal{O}_L/\mathcal{O}_K$ such that $C(L) \ne \emptyset$ (use Exercise 6.2.7(b)).

(b) Let $\mathcal{O}_L$ be a discrete valuation ring that is étale over $\mathcal{O}_K$. Let us suppose that $C(L) \ne \emptyset$. We want to show that C has semi-stable reduction over $\mathcal{O}_K$.

    (1) Show that there exists a ring R, finite, étale, Galois over $\mathcal{O}_K$, such that $L \subseteq \operatorname{Frac}(R)$.

    (2) Let $\mathcal{C}$ be a relatively minimal regular model of C over $\mathcal{O}_K$. Show that $\operatorname{Gal}(\operatorname{Frac}(R)/K)$ acts transitively on the irreducible components of the special fiber of $\mathcal{C}_R$ (Exercises 3.3.23 and 9.3.1(a)), and that $\mathcal{C}$ is semi-stable.

3.9. Let X, Y be schemes of finite type over a locally Noetherian scheme S, and let $f : X \to Y$ be a surjective étale morphism. Show that $X \to S$ is semi-stable if and only if $Y \to S$ is semi-stable.

3.10. Let $p > 2$ be a prime and $r \geq 1$. Let C be the smooth plane curve

$$C = \text{Proj } \mathbb{Q}_p[x, y, z]/((x^2 - 2y^2 + z^2)(x^2 - z^2) + p^r y^3 z).$$

(a) Show that the same equation defines a stable scheme $\mathcal{C}$ over $\mathbb{Z}_p$, whose special fiber is the union of two lines and a smooth conic.

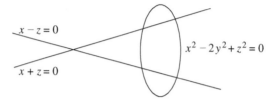

(b) Show that the thickness of the intersection point of the two lines is equal to $2r$, and that the other double points have thickness r.

3.11. Let $X \to S$ be a semi-stable projective curve over a Dedekind scheme of dimension 1, with smooth generic fiber. Show that there exists a finite flat morphism $S' \to S$ of degree 2 with S' Dedekind such that the minimal desingularization of $X \times_S S'$ is an arithmetic surface over S' with normal crossings.

3.12. Let S be a Dedekind scheme of dimension 1, C a smooth, projective, geometrically connected curve of genus $g \geq 2$ over $K(S)$. Show that any semi-stable model of C over S (if it exists) dominates the stable model of C over S.

3.13. Let S be a Dedekind scheme of dimension 1, $T \to S$ a smooth surjective morphism with T integral. Let C be a smooth projective curve over $K(S)$. Let us suppose that there exists a semi-stable curve $X \to T$ with generic fiber isomorphic to $C_{K(T)}$. Show that X has semi-stable reduction over S (use Exercise 6.2.7(b)).

3.14. Let C be a semi-stable curve over a field k, and $x \in C$ a split ordinary double point. Let A_0 be the localization of $k[u, v]/(uv)$ at the maximal ideal (u, v). Hence the completions of A_0 and of $A_1 := \mathcal{O}_{C,x}$ are isomorphic. We are going to show that there exists a local ring R that is étale over A_0 and over A_1.

(a) Show that we can take $R = A_1$ if A_1 is not integral. Let us suppose, in what follows, that A_1 is integral.

(b) Let A_1' be the integral closure of A_1. Then $\text{Spec } A_1'$ contains two closed points y_1, y_2. Show that there exists an $f \in A_1'$ such that $f(y_1) = 1$ and $f(y_2) = 0$. Show that $a := f^2 - f \in A_1$.

(c) Let us consider $D := A_1[w]/(w^2 - w - a)$. Show that D is étale over A_1. Let us write $f = a_1/a_2$ with $a_i \in A_1$. Show that we have the relation

$$(a_2 w - a_1)(a_2 w - a_2 + a_1) = 0$$

in D. Let $\mathfrak{m}$ be a maximal ideal of D. Show that $D_\mathfrak{m}$ is the local ring of a semi-stable curve at a split and non-integral ordinary double point. Using (a), show that $R = D_\mathfrak{m}$ has the desired properties.

3.15. Let X be an integral semi-stable curve over a discrete valuation ring $\mathcal{O}_K$, and let $x \in X_s$ be a split ordinary double point, of thickness $n \geq 1$. Let B_0 be the localization of $\mathcal{O}_K[u,v]/(uv - t^n)$ at the maximal ideal (u,v,t), so that the completions of B_0 and of $B_1 := \mathcal{O}_{X,x}$ are isomorphic. As in Exercise 3.14, we are going to show that there exists a ring R that is étale over B_0 and over B_1.

(a) Let us first suppose that $\mathcal{O}_{X_s,x}$ is not integral. Let $\tilde{u}, \tilde{v}$ be generators of $\mathfrak{m}_x \mathcal{O}_{X_s,x}$ such that $\tilde{u}\tilde{v} = 0$. Show that they lift to elements $u, v \in B_1$ such that $uv = t^n$. Show that we can take $R = B_1$.

(b) Let k be the residue field of $\mathcal{O}_K$. Show that there exists a finite étale homomorphism $B_1 \to D := B_1[w]/(w^2 - w - b)$ for some $b \in B_1$ such that $D \otimes_{\mathcal{O}_K} k$ is not integral. Conclude as in Exercise 3.14(c).

Remark. More generally, the *Artin approximation theorem* implies the following statement ([10], Corollary 2.6): let A_1, A_2 be two Noetherian local rings, localizations of finitely generated algebras over an excellent Dedekind ring $\mathcal{O}_K$. Let us suppose that their completions $\hat{A}_0, \hat{A}_1$ are $\mathcal{O}_K$-isomorphic. Then there exists a local ring R that is étale over A_0 and over A_1.

3.16. Let us replace the hypothesis of Proposition 3.44(i) by $U_i := (W_i^0)_{s,\mathrm{red}}$ smooth while keeping the same inequality. Show that C has semi-stable reduction and that the dual graph of any semi-stable reduction of C is a tree.

3.17. Let $\mathcal{O}_K$ be a discrete valuation ring with algebraically closed residue field k. Let Y be a normal fibered surface over $\mathcal{O}_K$ with Y_s connected. Let us recall that for a connected projective curve Z over k, $a(Z)$ and $t(Z)$ denote, respectively, the Abelian rank and the toric rank of Z (Definition 7.5.21).

(a) Let $f : X \to Y$ be a birational morphism with X normal. Let us suppose that the exceptional locus E of f is connected. Directly using the definition (Lemma 7.5.18(b)), show that

$$t(X_s) = t(Y_s) + t(E), \quad a(X_s) = a(Y_s) + a(E).$$

(b) Show that if Y_s is a reduced curve with ordinary multiple singularities, and if X is cohomologically flat over $\mathcal{O}_K$, then X_s has unipotent rank zero.

(c) Let Y be as in (b). Show that Y_K is smooth over K, and that the minimal desingularization of Y is a semi-stable curve over $\mathcal{O}_K$.

3.18. Let X be an integral semi-stable quasi-projective curve over a Dedekind scheme S. Let G be a finite subgroup of $\mathrm{Aut}_S(X)$ and $Y = X/G$. Let us fix a closed point $s \in S$.

(a) Let $\xi \in Y$ be a generic point of Y_s and $\xi_1, \ldots, \xi_r$ the points of X_s lying above ξ. Show that

$$\mathrm{Card}\, G = \sum_{1 \le i \le r} [k(\xi_i) : k(\xi)] = r[k(\xi_1) : k(\xi)].$$

(b) Let D (resp. I) be the decomposition (resp. inertia) group at ξ_1. Show that $\mathrm{Card}\, G = r\, \mathrm{Card}\, D$, and that the separable closure of $k(\xi)$ in $k(\xi_1)$ is Galois with Galois group D/I.

(c) Show that the morphism $X_s/G \to Y_s$ is purely inseparable. Deduce from this that if the order of G is prime to $\mathrm{char}(k(s))$, then $X_s/G \to Y_s$ is birational and is in fact an isomorphism. Moreover, with the notation of (b), we will have $D \to \mathrm{Aut}_{k(s)}(\overline{\{\xi_i\}})$ injective.

3.19. Let C be a smooth curve over a discrete valuation field K, with characteristic $\mathrm{char}(K) = p > 0$. We suppose that C admits a semi-stable model $\mathcal{C}$ over $\mathcal{O}_K$ (i.e., a semi-stable curve over $\mathcal{O}_K$ such that $\mathcal{C}_\eta \simeq C$).

(a) Let $F_{\mathcal{C}/\mathcal{O}_K} : \mathcal{C} \to \mathcal{C}^{(p)}$ be the relative Frobenius. Show that $\mathcal{C}^{(p)}$ is semi-stable over $\mathcal{O}_K$ (use Exercise 3.2.20(b)).

(b) Let D be a smooth curve over K and $f : C \to D$ a finite purely inseparable morphism. Show that f decomposes into a sequence of Frobenius morphisms (use Exercise 4.1.19). Deduce from this that D admits a semi-stable model $\mathcal{D}$ over $\mathcal{O}_K$ and that f extends to a finite purely inseparable morphism $\mathcal{C} \to \mathcal{D}$, with $\mathcal{D}$ stable (resp. smooth) if $\mathcal{C}$ is stable (resp. smooth).

3.20. Let K be a discrete valuation field with perfect residue field k and $\mathrm{char}(K) = p > 0$. Let $f : C \to D$ be a finite purely inseparable morphism of smooth, projective, geometrically connected curves of genus $g \ge 2$ over K. Let us suppose that D has stable reduction over $\mathcal{O}_K$ and that $C(K) \ne \emptyset$. We want to show that C also has stable reduction over $\mathcal{O}_K$. We can suppose K is complete and k algebraically closed.

(a) Let $\mathcal{D}$ be the stable model of D over $\mathcal{O}_K$. Let $\mathcal{C} \to \mathcal{D}$ be the normalization of $\mathcal{D}$ in $K(C)$. Show that it is a finite purely inseparable morphism (Theorem 8.2.39(a) and Exercise 5.3.9).

(b) Show that $a(\mathcal{C}_s) = a(\mathcal{D}_s)$ and $t(\mathcal{C}_s) = t(\mathcal{D}_s)$. Let $\widetilde{\mathcal{C}} \to \mathcal{C}$ be a desingularization of $\mathcal{C}$. Show that $\widetilde{\mathcal{C}}_s$ has unipotent rank zero (use Exercise 3.17(a)), and that C has stable reduction over $\mathcal{O}_K$.

10.4   Deligne–Mumford theorem

Let C be a smooth projective curve over a discrete valuation field K. It does not always have semi-stable reduction over $\mathcal{O}_K$. For example, by Theorem 3.34(a), the curves of Examples 1.12, 1.14 do not have semi-stable reduction.

Definition 4.1. Let C be a smooth, projective, geometrically connected curve over $K(S)$, where S is a Dedekind scheme of dimension 1. Let L be a finite extension of $K(S)$ and S' the normalization of S in L. This is a Dedekind scheme of dimension 1 (Proposition 4.1.31). We will say that C has *semi-stable* (resp. *stable*) *reduction* over S' if C_L has semi-stable (resp. stable) reduction over S'.

Remark 4.2. Let us suppose $g(C) = 0$. Let L/K be a finite separable extension such that $C(L) \neq \emptyset$ (Proposition 3.2.20). Hence $C_L \simeq \mathbb{P}_L^1$, which, of course, has good reduction over $\mathcal{O}_L$. Hence C has semi-stable reduction over $\operatorname{Spec} \mathcal{O}_L$. If C is of genus 1, then after a finite separable extension, we can suppose that $C(K) \neq \emptyset$, and hence that C is an elliptic curve. We have seen (Proposition 2.33, Example 3.35) that there exists a finite extension L/K such that C_L has semi-stable reduction over $\operatorname{Spec} \mathcal{O}_L$. We can even choose L/K separable (Exercise 2.3 or Exercise 4.2).

The aim of this section is in the first place to show the following theorem, which is the central result of the theory of reduction of algebraic curves:

Theorem 4.3 (Deligne–Mumford, [26]). *Let S be a Dedekind scheme of dimension 1, C a smooth, projective, geometrically connected curve of genus $g \geq 2$ over $K(S)$. Then there exists a Dedekind scheme S' that is finite flat over S such that $C_{K(S')}$ has a unique stable model over S'. Moreover, we can take $K(S')$ separable over $K(S)$.*

Remark 4.4. This theorem was first proven by Deligne and Mumford in [26] using the theorem of Grothendieck on the semi-Abelian reduction of Abelian varieties, and a theorem of Raynaud that links the reduction of a regular model of C over S to that of the Néron model of $\operatorname{Jac}(C)$ over S. See Remarks 4.24–26. Mumford had previously given a proof in characteristic $\neq 2$ using the theta function. Afterwards, there were different proofs: Artin–Winters [11], which we are going to present in this section; Bosch–Lütkebohmert [14] and van der Put [79] using rigid analytic geometry; Saito [84] with the theory of vanishing cycles. When S is local, this last work gives, moreover, a characterization of the case when $S' \to S$ is *wildly ramified*, i.e., not tamely ramified (see Theorem 4.47).

In the last subsection, we will give examples of computations of the stable reduction and will indicate some results making it possible to know better the extension $S' \to S$ needed to realize the stable reduction.

10.4.1 Simplifications on the base scheme

For the proof of Theorem 4.3, we can first carry out simplifications at the level of S. Let us note that to give S' is equivalent to giving the finite separable extension $K' = K(S')$ of $K(S)$ because S' is simply the normalization of S in $K(S')$ (see Definition 4.1.24 and Proposition 4.1.25). Note that the uniqueness of the stable model has already been proven (Theorem 3.34(b)).

Lemma 4.5. *Theorem 4.3 is true if it is true for every S that is the spectrum of a complete discrete valuation ring with algebraically closed residue field.*

Proof We will say that a Dedekind scheme S verifies property (P) if Theorem 4.3 is true for every curve C over $K(S)$ as in the statement of the theorem. Let us therefore suppose that S verifies (P) if it is the spectrum of a complete discrete valuation ring with algebraically closed residue field. If L/F is an algebraic extension and if $\mathcal{O}_F$ is a discrete valuation ring, then $\mathcal{O}_L$ will denote the integral closure of $\mathcal{O}_F$ in L.

(1) If S is the spectrum of a discrete valuation ring $\mathcal{O}_K$ with algebraically closed residue field, then S verifies (P). Let $\widehat{\mathcal{O}}_K$ be the completion of $\mathcal{O}_K$. Its residue field is that of $\mathcal{O}_K$, and hence algebraically closed. Therefore, there exists a finite separable extension $L/\widehat{K}$ such that C has stable reduction over $\operatorname{Spec}\mathcal{O}_L$. Let K'/K be a finite separable extension such that $L \simeq K' \otimes_K \widehat{K}$, and such that $\mathcal{O}_{K'}$ verifies $\widehat{\mathcal{O}}_{K'} \simeq \widehat{\mathcal{O}}_K \otimes_{\mathcal{O}_K} \mathcal{O}_{K'} = \mathcal{O}_L$ (Exercise 4.3). It follows from Corollary 3.36(c) that C has stable reduction over $\mathcal{O}_{K'}$.

(2) If S is the spectrum of a discrete valuation ring $\mathcal{O}_K$, then S verifies (P). The proof above shows that we can suppose $\mathcal{O}_K$ is complete. Let $\mathcal{O}$ be a discrete valuation ring dominating $\mathcal{O}_K$, with residue field equal to the algebraic closure of k, and with field of fractions L separable algebraic over K (Lemma 3.32). As $\mathcal{O}_K$ is complete, and hence Henselian (Example 8.3.34), $\mathcal{O}$ is equal to the integral closure $\mathcal{O}_L$ of $\mathcal{O}_K$ in L. By step (1) above, there exists a finite separable extension L'/L such that C has stable reduction over $\mathcal{O}_{L'}$. The stable model of $C_{L'}$ is defined over a finitely generated sub-$\mathcal{O}_K$-algebra B of $\mathcal{O}_{L'}$. It follows that C has stable reduction over $\mathcal{O}_{\operatorname{Frac}(B)}$.

(3) The general case. The curve C has good reduction at every $s \in S$ except for a finite number of points $s_1, \ldots, s_n$ (Proposition 1.21(a)). By step (2) above, there exist finite separable extensions L_i of K such that C has stable reduction over $\mathcal{O}_{L_i}$, the integral closure of $\mathcal{O}_{S,s_i}$ in L_i. Let L be the compositum of the L_i in a separable closure of K and S' the normalization of S in L. Let $s' \in S'$ lie above a point $s \in S$. If C has good reduction at s, then C_L has good reduction at s'. Otherwise, $s = s_i$ for some $i \leq n$. As $L_i \subseteq L$, the scheme $S' \times_S \operatorname{Spec}\mathcal{O}_{S,s}$ dominates $\operatorname{Spec}\mathcal{O}_{L_i}$. It follows that C_L has stable reduction over $S' \times_S \operatorname{Spec}\mathcal{O}_{S,s}$ (Corollary 3.36(a)); and therefore, it has stable reduction at s', which is a closed point of $S' \times_S \operatorname{Spec}\mathcal{O}_{S,s}$. $\square$

The case of residue characteristic 0

The proof of Theorem 4.3 is relatively simple when $\operatorname{char}(k) = 0$. It results from Theorem 9.2.26 (with $D = X_s$), from the following proposition, and from Theorem 3.34(b).

Proposition 4.6. Let $\mathcal{O}_K$ be a discrete valuation ring with algebraically closed residue field k. Let C be a smooth projective curve over K, $\mathcal{C}$ a regular model of C over S with normal crossings, and $d_1, \ldots, d_n$ the multiplicities of the irreducible components of $\mathcal{C}_s$. Let us suppose that $\operatorname{char}(k)$ is prime to the d_i. Then C has semi-stable reduction over $\mathcal{O}_L$ for any discrete valuation ring $\mathcal{O}_L$ dominating $\mathcal{O}_K$, with ramification index $e = e_{L/K}$ divisible by $d_1, \ldots, d_n$. More precisely, the normalization $\mathcal{C}'$ of $\mathcal{C} \times_{\operatorname{Spec}\mathcal{O}_K} \operatorname{Spec}\mathcal{O}_L$ is semi-stable over $\mathcal{O}_L$.

Proof Let $x \in C_s$ be a closed point. We are going to determine the integral closure of $\mathcal{O}_{C,x} \otimes_{\mathcal{O}_K} \mathcal{O}_L$. Let t be a uniformizing parameter for $\mathcal{O}_K$. By Proposition 9.2.34, in a neighborhood of x, C is a closed subscheme of a scheme Z that is smooth over $\mathcal{O}_K$, and there exists a system of parameters (t, U, V) for Z at x such that $\mathcal{O}_{C,x} = \mathcal{O}_{Z,x}/(F)$, where $F = U^{d_1} - tb$ or $F = U^{d_1}V^{d_2} - tb$, with $b \in \mathcal{O}_{Z,x}^*$ and where the d_i are the multiplicities of the irreducible components passing through x. Let τ be a uniformizing parameter for $\mathcal{O}_L$. Then $t = \tau^e \lambda$ with $\lambda \in \mathcal{O}_L^*$. Let

$$ B := \mathcal{O}_{Z,x} \otimes_{\mathcal{O}_K} \mathcal{O}_L, \quad A := \mathcal{O}_{C,x} \otimes_{\mathcal{O}_K} \mathcal{O}_L = B/(F). $$

Then B is the local ring at a closed point of a scheme that is smooth over $\mathcal{O}_L$; its maximal ideal is generated by (U, V, τ). Hence $\widehat{B} = \widehat{\mathcal{O}}_L[[U, V]]$. We let u, v, a denote the respective images of U, V, b in A, and A' the integral closure of A.

In preparation for the remainder of the proof, let us note that for any integer $q \geq 1$ prime to $\mathrm{char}(k)$, there exists a primitive qth root of unity $\xi_q \in \widehat{\mathcal{O}}_K$ (there exists one in k, and we apply Corollary 6.2.13 to the smooth scheme $\mathrm{Spec}\,\widehat{\mathcal{O}}_K[T]/(T^q - 1)$). Moreover, as k is algebraically closed, we have $\lambda b \in (B^*)^q(1 + \mathfrak{m}_B)$. There therefore exists a $c \in \widehat{B}^*$ such that $\lambda b = c^q$ (Exercise 1.3.9).

Let us first consider the case $F = U^{d_1} - tb$. Let $r = e/d_1$ and $w = u\tau^{-r} \in \mathrm{Frac}(A)$. Then $w^{d_1} = \lambda a$. Let us consider

$$ D := B[W]/(W^{d_1} - \lambda b, U - \tau^r W) = A[W]/(W^{d_1} - \lambda a, u - \tau^r W). $$

This is a finite algebra over A and $D \otimes_{\mathcal{O}_L} L \simeq A \otimes_{\mathcal{O}_L} L$. Let $c \in \widehat{B}^*$ be such that $c^{d_1} = \lambda b$. We have

$$ \begin{aligned} D \otimes_B \widehat{B} &= \widehat{\mathcal{O}}_L[[U, V]][W]/(W^{d_1} - c^{d_1}, U - \tau^r W) \\ &= \oplus_{1 \leq i \leq d_1} \widehat{\mathcal{O}}_L[[U, V]]/(U - \tau^r \xi_{d_1}^i c). \end{aligned} \tag{4.31} $$

Let $\mathfrak{m}$ be a maximal ideal of D. It follows from Exercise 4.3.17 that there exists an $i \leq d_1$ such that

$$ \widehat{D}_\mathfrak{m} \simeq \widehat{\mathcal{O}}_L[[U, V]]/(U - \tau^r \xi_{d_1}^i c) = \widehat{\mathcal{O}}_L[[V]]. \tag{4.32} $$

Consequently, D is regular and therefore normal; it therefore coincides with A'.

Let us now suppose $F = U^{d_1}V^{d_2} - tb$. Let $d = \gcd\{d_1, d_2\}$, $q_i = d_i/d$, $r = e/(q_1 q_2 d)$. Let us set $w = (u^{q_1} v^{q_2} \tau^{-q_1 q_2 r}) \in \mathrm{Frac}(A)$. Then $w^d = \lambda a$. Let $\alpha_1, \beta_1 \geq 1$ be such that $\alpha_1 q_1 = \beta_1 q_2 - 1$. Let us set $u_1 = u^{\beta_1} v^{\alpha_1} \tau^{-\alpha_1 q_1 r}$. Then $u_1^{q_2} = uw^{\alpha_1}$. In a similar way, we let $\alpha_2 = mq_1 - \beta_1, \beta_2 = mq_2 - \alpha_1 > 0$ for some integer $m > 0$. Then $\alpha_2 q_2 = \beta_2 q_1 - 1$, and we define $v_1 = v^{\beta_2} u^{\alpha_2} \tau^{-\alpha_2 q_2 r}$. We have $v_1^{q_1} = vw^{\alpha_2}$ and

$$ u_1 v_1 = u^{\beta_1 + \alpha_2} v^{\alpha_1 + \beta_2} \tau^{-(\alpha_1 q_1 + \alpha_2 q_2)r} = u^{mq_1} v^{mq_2} \tau^{-(\alpha_1 q_1 + (mq_1 - \beta_1)q_2)r} = w^m \tau^r. $$

In light of these relations, let us consider

$$ D := B[U_1, V_1, W]/I, $$

where I is the ideal generated by

$$U_1^{q_2} - UW^{\alpha_1}, \ V_1^{q_1} - VW^{\alpha_2}, \ U_1V_1 - \tau^r W^m, \ W^d - \lambda b,$$

and $U^{q_1}V^{q_2} - \tau^{q_1q_2r}W$ (the latter in fact belongs to the ideal generated by the first elements). We also have $F \in I$, and hence D is an A-algebra. It is easy to see that $D \otimes_{\mathcal{O}_L} L \simeq A \otimes_{\mathcal{O}_L} L$, $\mathrm{Frac}(D) = \mathrm{Frac}(A)$, and that D is finite over A. Let $c \in \widehat{B}^*$ be such that $\lambda b = c^d$. Then $\widehat{B} = \widehat{\mathcal{O}}_L[[Uc^{\alpha_1}, Vc^{\alpha_2}]]$. We easily see that

$$D \otimes_B \widehat{B} = \oplus_{1 \le j \le d} \widehat{\mathcal{O}}_L[[U_1, V_1]]/(U_1V_1 - \tau^r \xi_d^{mj} c^m). \tag{4.33}$$

For any maximal ideal $\mathfrak{m}$ of D, there exists a $j \le d$ such that

$$\widehat{D}_{\mathfrak{m}} \simeq \widehat{\mathcal{O}}_L[[U_1, V_1]]/(U_1V_1 - \tau^r \xi_d^{mj} c^m). \tag{4.34}$$

This is a normal ring (Lemma 4.1.18). Hence so is $D_{\mathfrak{m}}$ (Exercise 4.1.6). Consequently, D is normal and therefore coincides with A'.

Isomorphisms (4.32) and (4.34) imply that for any closed point $x' \in C'$, $\widehat{\mathcal{O}}_{C'_s, x'}$ is either isomorphic to $k[[V]]$, or isomorphic to $k[[U_1, V_1]]/(U_1V_1)$. Hence C' is semi-stable over $\mathrm{Spec}\,\mathcal{O}_L$.   □

Corollary 4.7. *Let $\mathcal{O}_K$ be a discrete valuation ring with perfect residue field k and residue characteristic $p \ge 0$. Let C be a smooth projective curve over K. Let us suppose that C admits a regular model with normal crossings $\mathcal{C}$ over $\mathcal{O}_K$, such that the multiplicities d_i of the irreducible components of $\mathcal{C}_s$ are all prime to p. Then $C_{K'}$ has semi-stable reduction over $\mathcal{O}_{K'}$ for any discrete valuation ring $\mathcal{O}_{K'}$ that dominates $\mathcal{O}_K$, with ramification index divisible by all of the d_i.*

Proof This is a standard descent from Proposition 4.6. Let $\mathcal{O}_L$ be a discrete valuation ring that dominates $\mathcal{O}_K$, with residue field $\overline{k}$ and with ramification index 1 over $\mathcal{O}_K$ (Lemma 3.32). Then $\mathcal{O}_L$ is a direct limit of discrete valuation rings which are finite and with ramification index 1 (hence étale) over $\mathcal{O}_K$. It follows that $\mathcal{C}_{\mathcal{O}_L}$ is a regular model with normal crossings of C_L. Let L' be the compositum of K' and L in an algebraic closure of K, $\mathcal{O}_{L'}$ a discrete valuation ring dominating $\mathcal{O}_{K'}$ and with field of fractions L' ($\mathcal{O}_{L'}$ is a localization of the integral closure of $\mathcal{O}_{K'}$ in $\mathcal{O}_L$). By Proposition 4.6, $C_{L'}$ has semi-stable reduction over $\mathcal{O}_{L'}$. Now $\mathcal{O}_{L'}$ has ramification index 1 over $\mathcal{O}_{K'}$ and residue field separable over that of $\mathcal{O}_{K'}$. It follows from Corollary 3.36(c) that $C_{K'}$ has semi-stable reduction over $\mathcal{O}_{K'}$.   □

Remark 4.8. Let us keep the hypotheses of Proposition 4.6. Let Γ_0 be an irreducible component of $\mathcal{C}_s$, of multiplicity d_0. Let Δ be an irreducible component of $\mathcal{C}'_s$ lying above Γ_0. Let us study the morphism $\Delta \to \Gamma_0$. Let $x_1, \dots, x_r$ be the intersection points of Γ_0 with the other components of $\mathcal{C}_s$, with x_i belonging to a component of multiplicity d_i ($i \ge 1$). Then isomorphisms (4.32) and (4.34) show that the morphism $\Delta \to \Gamma_0$ is étale outside of the points $x_1, \dots, x_r$, and that it is ramified with ramification index $e_i := d_0/(d_0, d_i)$ above x_i. Moreover, by (4.33), there are exactly (d_0, d_i) points in $\mathcal{C}'$ lying above x_i. Let $\Delta_1, \dots, \Delta_\ell$ be the irreducible components of $\mathcal{C}'_s$ lying above Γ_0. If $x \in \Gamma_0$ is not an intersection point, then (4.31) shows that there are exactly d_0 points of $\mathcal{C}'$ lying above x. It follows that

$$d_0 = \sum_{1 \le j \le \ell} [K(\Delta_i) : K(\Gamma_0)].$$

Let us suppose $\mathcal{O}_L$ is Galois over $\mathcal{O}_K$ (this is automatic if K is complete and e prime to $\mathrm{char}(k)$). We then have $d_0 = \ell[K(\Delta) : K(\Gamma_0)]$. Hence $\Delta \to \Gamma_0$ is of degree d_0/ℓ, and there are $(d_0, d_i)/\ell$ points of Δ lying above x_i. Hurwitz's formula (Theorem 7.4.16) implies that

$$2g(\Delta) - 2 = \frac{d_0}{\ell}(2g(\Gamma_0) - 2) + \sum_{1 \le i \le r} \frac{(d_0, d_i)}{\ell}\left(\frac{d_0}{(d_0, d_i)} - 1\right). \qquad (4.35)$$

On the other hand, there is no formula for the number of components ℓ (there is, however, a constraint: ℓ divides (d_0, d_i) as soon as Γ_i meets Γ_0). Once this number is known by an ad hoc examination, the preceding computations allow us to determine the stable reduction of C_L.

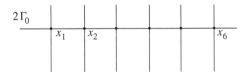

Example 4.9. Let us keep the notation of Proposition 4.6. Let us suppose that $\mathrm{char}(k) \ne 2$ and that $\mathcal{C}_s$ is as in the figure above. Then the number of irreducible components ℓ of $\mathcal{C}'_s$ lying above Γ_0 is 1 because it divides $(d_0, d_1) = 1$. We deduce from the above that $\mathcal{C}'_s$ is made up of a curve of genus 2 that meets six projective lines transversally at six points lying respectively over $x_1, \ldots, x_6$. Consequently, C_L has good reduction over every extension L of K of even ramification index.

Remark 4.10. When the residue characteristic p is zero or large enough with respect to g (if $p > g + 1$), we can also show Theorem 4.3 using coverings. See Remark 4.31.

10.4.2   Proof of Artin–Winters

We are going to give the principles of the proof of Theorem 4.3 by Artin–Winters [11]. Let C be a smooth projective curve over a discrete valuation field K, and let $\mathcal{C}$ be a regular model of C over $\mathcal{O}_K$. The principal ingredients of the proof are Proposition 4.17 which compares the torsion points of the groups $\mathrm{Pic}(\mathcal{C}_s)$ and $\mathrm{Pic}(C)$, and Theorem 4.19 which gives information on the 'component group' $\Phi(X)$.

Picard groups and component group

Let $\mathcal{O}_K$ be a discrete valuation ring with *algebraically closed residue field* k. Let $X \to S$ be an arithmetic surface over $\mathcal{O}_K$, with X_s connected. We are going to associate a finite commutative group $\Phi(X)$ to it. Let $\Gamma_1,\ldots,\Gamma_n$ be the irreducible components of X_s, of respective multiplicities $d_1,\ldots,d_n$. Let $D = Z^0(X_s)$ be the free commutative group generated by the Γ_i, and let $\Gamma_1^*,\ldots,\Gamma_n^*$ be a dual basis of the basis $\Gamma_1,\ldots,\Gamma_n$ of D. Let α be the homomorphism

$$\alpha:\ D \to D^\vee,\quad Z \mapsto \sum_{1\le i\le n}(Z\cdot\Gamma_i)\Gamma_i^*,$$

where $D^\vee$ is the dual of D.The kernel of α is generated by $d^{-1}X_s$, where d is the gcd of the d_i (Theorem 9.1.23). Let G be the group defined by the exact sequence

$$0 \to d^{-1}X_s\mathbb{Z} \to D \xrightarrow{\alpha} D^\vee \to G \to 0. \tag{4.36}$$

This is a finitely generated group. We can decompose G as follows. Let $\beta: D^\vee \to \mathbb{Z}$ be the homomorphism that sends $\sum_i a_i\Gamma_i^*$ to $\sum_i a_id_i$; then $\operatorname{Im}\alpha \subseteq \operatorname{Ker}\beta$.

Definition 4.11. We set $\Phi(X) = \operatorname{Ker}\beta/\operatorname{Im}\alpha$.

We have an exact sequence

$$0 \to \Phi(X) \to G \to d\mathbb{Z} \to 0.$$

Exact sequence (4.36) implies that $G\otimes_\mathbb{Z}\mathbb{Q}$ has dimension 1 as a vector space over $\mathbb{Q}$. Hence $\Phi(X)$ is a torsion group, and hence finite. It is also the subgroup of torsion points G_{tors} of G.

Remark 4.12. Let us suppose X_η is smooth and geometrically connected. Let $\mathcal{N}$ be the Néron model of $\operatorname{Jac}(X_\eta)$ over S and Φ its group of components (Definition 2.20). By a theorem of Raynaud ([15], Theorem 9.6.1), the group $\Phi(k)$ is isomorphic to $\Phi(X)$. This implies, in particular, that $\Phi(X)$ is a birational invariant in the class of regular models of X_η over $\mathcal{O}_K$. We leave it to the reader to verify this fact (Exercise 4.7) without using Raynaud's theorem.

We are going to link the Picard group $\operatorname{Pic}(X_\eta)$ to the group $\Phi(X)$ (Proposition 4.17). For this, we need two preliminary results. Let t be a uniformizing parameter for $\mathcal{O}_K$. For any scheme Y over S, let us set $Y_N = Y\times_S\operatorname{Spec}(\mathcal{O}_K/t^{N+1}\mathcal{O}_K)$ for any $N\ge 0$. We have $Y_0 = Y_s$.

Definition 4.13. Let G be a commutative group, and ℓ an integer. We say that G is ℓ-divisible if the multiplication by ℓ homomorphism $\ell_G: G \to G$ is surjective. We say that G is uniquely ℓ-divisible if ℓ_G is an isomorphism. Let us recall that $G[\ell]$ denotes the subgroup $\operatorname{Ker}\ell_G$.

Lemma 4.14. Let S be the spectrum of a discrete valuation ring $\mathcal{O}_K$ with residue field k. Let $Y \to S$ be a projective scheme.

(a) Let us suppose k is algebraically closed. Then for any $N\ge 0$ and for any integer $\ell\ge 1$ prime to $\operatorname{char}(k)$, the kernel of the canonical homomorphism $\operatorname{Pic}(Y_{N+1}) \to \operatorname{Pic}(Y_N)$ is uniquely ℓ-divisible. Moreover, $\operatorname{Pic}(Y_{N+1})[\ell] \to \operatorname{Pic}(Y_N)[\ell]$ is bijective.

(b) *The canonical homomorphism* $\mathrm{Pic}(Y) \to \varprojlim_N \mathrm{Pic}(Y_N)$ *is injective.*

Proof (a) As in the proof of Lemma 7.5.11, we have an exact sequence

$$0 \to V \to \mathrm{Pic}(Y_{N+1}) \to \mathrm{Pic}(Y_N) \to H^2(Y, t^{N+1}\mathcal{O}_{Y_{N+1}})$$

where V is isomorphic to $\mathrm{Ker}(H^1(Y, \mathcal{O}_{Y_{N+1}}) \to H^1(Y, \mathcal{O}_{Y_N}))$. Here we need the k algebraically closed hypothesis. As V is a k-vector space, and ℓ is prime to $\mathrm{char}(k)$, it is uniquely ℓ-divisible. The bijectivity of the homomorphism $\mathrm{Pic}(Y_{N+1})[\ell] \to \mathrm{Pic}(Y_N)[\ell]$ results from this property and from the fact that the multiplication by ℓ is injective in its cokernel.

(b) Let $\mathcal{L}$ be an invertible sheaf on Y such that $\mathcal{L}|_{Y_N}$ is free over $\mathcal{O}_{Y_N}$ for every $N \geq 0$. Let A_N be the set of bases ($\subset H^0(Y, \mathcal{L}|_{Y_N})$) of $\mathcal{L}|_{Y_N}$. By Nakayama, if $\pi_N : H^0(Y, \mathcal{L}|_{Y_{N+1}}) \to H^0(Y, \mathcal{L}|_{Y_N})$ denotes the canonical homomorphism, then we have $A_{N+1} = \pi_N^{-1}(A_N)$. By Lemma 5.3.10(b), the inverse system $(H^0(Y, \mathcal{L}|_{Y_N}))_N$ verifies the Mittag–Leffler condition (Exercise 1.3.15); the same therefore holds for $(A_N)_N$. Consequently, the inverse limit of sets $\varprojlim_N A_N$ is non–empty (Exercise 1.3.15(b)). Let

$$e \in \varprojlim_N A_N \subset \varprojlim_N H^0(Y, \mathcal{L}|_{Y_N}) \simeq H^0(Y, \mathcal{L}) \otimes_{\mathcal{O}_K} \hat{\mathcal{O}}_K$$

(the isomorphism comes from Corollary 5.3.11). By multiplying e by a suitable element of $\hat{\mathcal{O}}_K^*$, we can suppose that $e \in H^0(Y, \mathcal{L})$. Let $\mathcal{F} = \mathcal{L}/e\mathcal{O}_Y$. Then $\mathcal{F}|_{Y_0} = 0$. By Nakayama's lemma, $\mathcal{F}_y = 0$ for every $y \in Y_s$. Hence the support $\mathrm{Supp}\,\mathcal{F}$ is a closed subset of Y that does not meet Y_s. As Y is proper over S, this implies that $\mathrm{Supp}\,\mathcal{F} = \emptyset$ and $\mathcal{F} = 0$. Hence $\mathcal{L} = e\mathcal{O}_Y$. □

Remark 4.15. We can show that the homomorphism of Lemma 4.14(b) is surjective if $\mathcal{O}_K$ is complete. This is a consequence of the algebraization theorem of coherent sheaves of Grothendieck. See Exercise 4.4.

Lemma 4.16. *Let X be a regular, Noetherian, connected scheme, and U be an open subset of X. Then the canonical homomorphism $\mathrm{Pic}(X) \to \mathrm{Pic}(U)$ is surjective.*

Proof By identifying invertible sheaves with Weil divisors ((Propositions 7.2.14 and 7.2.16), it suffices to show that every prime divisor Z on U is the restriction to U of a Weil divisor Z' on X. For Z', we can take the Zariski closure of Z in X. Note that the homomorphism is injective if $\mathrm{codim}(X \setminus U, X) \geq 2$ (Lemma 9.2.17). □

Let us return to the situation before Definition 4.11. We have a homomorphism $\lambda : D \to \mathrm{Pic}(X)$ that to each divisor $Z \in D$ associates the isomorphism class of the sheaf $\mathcal{O}_X(Z)$, and a homomorphism $\varphi : \mathrm{Pic}(X) \to D^\vee$ defined by $\varphi(\mathcal{L}) = \sum_{1 \leq i \leq n} (\deg \mathcal{L}|_{\Gamma_i}) \Gamma_i^*$. We have $\varphi \circ \lambda = \alpha$.

Proposition 4.17. *Let S be the spectrum of a discrete valuation ring with algebraically closed residue field k. Let $X \to S$ be an arithmetic surface with X_s*

connected. Let d be the gcd of the multiplicities of the irreducible components of X_s. Then for any integer ℓ prime to $\mathrm{char}(k)$ and to d, we have an exact sequence

$$0 \to \mathrm{Pic}(X)[\ell] \to \mathrm{Pic}(X_\eta)[\ell] \to \Phi(X)[\ell], \qquad (4.37)$$

and $\mathrm{Pic}(X)[\ell]$ is isomorphic to a subgroup of $\mathrm{Pic}(X_s)[\ell]$.

Proof The open immersion $X_\eta \to X$ canonically induces a homomorphism $\mathrm{Pic}(X) \to \mathrm{Pic}(X_\eta)$ that is surjective by Lemma 4.16. Its kernel can clearly be identified with the divisors with support in X_s, that is to say $\mathrm{Im}\,\lambda$. We have a commutative diagram of exact sequences:

$$
\begin{array}{ccccccccc}
0 & \longrightarrow & \mathrm{Ker}\,\lambda & \longrightarrow & D & \overset{\lambda}{\longrightarrow} & \mathrm{Pic}(X) & \longrightarrow & \mathrm{Pic}(X_\eta) & \longrightarrow & 0 \\
 & & & & \| & & \downarrow{\scriptstyle\varphi} & & \downarrow{\scriptstyle\psi} & & \\
0 & \longrightarrow & d^{-1}X_s\mathbb{Z} & \longrightarrow & D & \overset{\alpha}{\longrightarrow} & D^\vee & \longrightarrow & G & \longrightarrow & 0
\end{array}
$$

There therefore exists a homomorphism $\psi : \mathrm{Pic}(X_\eta) \to G$ that completes the commutative diagram. This induces an exact sequence

$$0 \to \mathrm{Ker}\,\psi[\ell] \to \mathrm{Pic}(X_\eta)[\ell] \to G[\ell] = \Phi(X)[\ell].$$

The surjective homomorphism $\mathrm{Pic}(X) \to \mathrm{Pic}(X_\eta)$ induces a surjective homomorphism $\mathrm{Ker}\,\varphi \to \mathrm{Ker}\,\psi$, whose kernel is d-torsion because $d\,\mathrm{Ker}\,\alpha \subseteq \mathrm{Ker}\,\lambda$. As $D^\vee$ is free over $\mathbb{Z}$ and ℓ is prime to d, we have $\mathrm{Pic}(X)[\ell] = \mathrm{Ker}\,\varphi[\ell] \simeq \mathrm{Ker}\,\psi[\ell]$. It remains to show that $\mathrm{Pic}(X)[\ell]$ injects into $\mathrm{Pic}(X_s)[\ell]$. But that is just Lemma 4.14. $\qquad\square$

Remark 4.18. Let us suppose $\mathcal{O}_K$ is complete. With the help of the algebraization theorem (Remark 4.15), we can show ([11], Proposition 2.5, or [29], Corollaire 3.3) that we have an exact sequence

$$0 \to \mathrm{Pic}(X_s)[\ell] \to \mathrm{Pic}(X_\eta)[\ell] \to \Phi(X)[\ell] \to 0.$$

Generators of the component group

Let $T = (d_1,\ldots,d_n,k_1,\ldots,k_n,M)$ be a type (Definition 1.55). We can associate a group G to it in the following manner. Let $L = \mathbb{Z}^n$ with a basis $\varepsilon_1,\ldots,\varepsilon_n$, let $\alpha : L \to L^\vee$ be the linear map given by $\varepsilon_i \mapsto \sum_j m_{ij}\varepsilon_j^*$, where the m_{ij} are the coefficients of the matrix M, and where the ε_j^* form a dual basis of the ε_j. Set $G = L^\vee/\alpha(L)$. If T is the type associated to an arithmetic surface X (Example 1.56), then G coincides with the group defined by exact sequence (4.36). We let $\beta(T)$ denote the first Betti number of the graph associated to T (see Definition 1.46). The following result is a technical, but crucial, ingredient in the proof of Theorem 4.3.

Theorem 4.19 ([11], Theorem 1.16). *Let $g \geq 1$. Then there exists an integer $c \geq 1$, depending only on g, such that for any type T of genus g, there exists a subgroup H of G, of index $[G : H]$ dividing c, and generated by at most $\beta(T) + 1$ elements.*

The proof is very technical, and we will not reproduce it here. See [11], Section 1, and [29], Section 4. A very succinct idea of the proof is the following. We first show that we can suppose T is minimal (i.e., $k_i \geq 0$ for every i). Let $\mathcal{G}$ be the graph of T. By Proposition 1.57, $\mathcal{G}$ is made up of a subgraph $\mathcal{G}_0$ with a bounded (uniquely as a function of g) number of vertices and of edges, and of a bounded (in the same manner) number of paths $\sum_1, \ldots, \sum_r$ of the form $_dA_N$ with $N \geq 3$. We then delete an edge e in the middle of a path $\sum_i$. By suitably modifying the coefficients m_{ij} and k_i of the end vertices of the edge e, we once again obtain a type T' (possibly not minimal) of genus $\leq g$, or the 'disjoint union' (in the sense of graphs) of two types T_1, T_2 of genus $\leq g$. After 'contraction' of the exceptional curves (i.e., the vertices with $k_i < 0$) in the new types, and by repeating this operations for all of the paths $_dA_N$, we arrive at a graph bounded like $\mathcal{G}_0$. For such a graph, we can take $H = \{0\}$ as a subgroup. The real difficulty lies in the comparison of the group G when we pass from T to T' or $T_1 \coprod T_2$.

Corollary 4.20. *Let $\mathcal{O}_K$ be a discrete valuation ring with algebraically closed residue field. Let $g \geq 1$. Then there exists an integer $c \geq 1$, depending only on g, such that for any arithmetic surface X over $\mathcal{O}_K$, with smooth geometrically connected generic fiber of genus g, the group $\Phi(X)$ admits a subgroup of index dividing c and generated by at most $t(X_s)$ elements, where $t(X_s)$ is the toric rank of X_s. In particular, for any prime number ℓ not dividing c, we have $\dim_{\mathbb{F}_\ell} \Phi(X)[\ell] \leq t(X_s)$.*

Proof We know that $t(X_s)$ and $\Phi(X)$ are independent of the choice of a regular model of X_η over $\mathcal{O}_K$ (Lemma 3.40 and Exercise 4.7(c)). We can therefore suppose that X has normal crossings. Let T be the type associated to X_s, and $t = t(X_s)$. Then T is of genus g, and we have $\beta(T) \leq t$ (Proposition 1.51(c)). Let c be the constant given by Theorem 4.19. Let H be a subgroup of G of index dividing c and generated by $t+1$ elements. Then H_{tors} is a subgroup of $\Phi(X)$ of index dividing c. Let p be an arbitrary prime number. Then $\dim_{\mathbb{F}_p} H/pH \leq t+1$. As $H \otimes_{\mathbb{Z}} \mathbb{Q} \simeq G \otimes_{\mathbb{Z}} \mathbb{Q} \simeq \mathbb{Q}$ (see exact sequence (4.36)), we have $\dim_{\mathbb{F}_p} H_{\mathrm{tors}}/(p) \leq t$. We immediately deduce from this that H_{tors} is generated by t elements. Finally, if ℓ is a prime number not dividing c, then $\Phi(X)[\ell] = H_{\mathrm{tors}}[\ell]$. The latter is clearly an $\mathbb{F}_\ell$-vector space of dimension $\leq t$. $\square$

Remark 4.21. Using the description by Grothendieck of $\Phi(X)$ via the Tate module of $\mathrm{Jac}(X_\eta)$, and Lorenzini's filtration of this module [61], we can give an explicit bound for c, to wit: there exists a functorial subgroup $\Phi^3(X)$ of $\Phi(X)$, generated by $t(X_s)$ elements, and such that

$$[\Phi(X) : \Phi^3(X)] \leq 2^{2g-2a-2t}. \tag{4.38}$$

Indeed, if $\mathrm{char}(k) = 0$, this is an easy consequence of [61], Corollary 1.7 (see also [31], Corollary 3.4, for an optimal result for the prime-to-p part of $\Phi(X)$, where p is the residue characteristic of K). In the general case, we note that $\Phi(X)$ only depends on combinatorial data of the graph of X_s. Now this graph is the graph of the special fiber of an arithmetic surface over a discrete valuation ring of equal characteristic 0 ([95], Theorem 4.3), whence inequality (4.38) in the general case. Let us, however, note that the proof above depends on the theorem on the semi-Abelian reduction of Abelian varieties.

Proof of Theorem 4.3. By Lemma 4.5, we can suppose that S is the spectrum of a complete discrete valuation ring with algebraically closed residue field. Let C be a smooth, projective, geometrically connected curve of genus $g \geq 2$ over $K = K(S)$. Let ℓ be a prime number different from char(k) and strictly greater than the constant c given in Corollary 4.20. After a finite separable extension of K, we can suppose that $C(K) \neq \emptyset$ (Proposition 3.2.20), and that the points of $\text{Pic}(C)[\ell] = \text{Jac}(C)[\ell]$ are all rational over K (Theorems 7.4.38(a) and 7.4.39). Let X be a regular model of C over S, and a, t, u the Abelian, toric, and unipotent ranks of X_s respectively. Then X is cohomologically flat over S because X_s has at least one irreducible component of multiplicity 1 (Corollary 9.1.24), and therefore $g = a + t + u$. Exact sequence (4.37) of Proposition 4.17 implies that

$$2g = \dim_{\mathbb{F}_\ell} \text{Pic}(C)[\ell] \leq \dim_{\mathbb{F}_\ell} \Phi(X)[\ell] + \dim_{\mathbb{F}_\ell} \text{Pic}(X_s)[\ell].$$

By Corollaries 4.20 and 7.5.23, we then have $2g \leq t + (2a + t) = 2g - 2u$. Hence $u = 0$. By Proposition 3.42, this implies that C has stable reduction.   □

Remark 4.22. We have seen in the proof above that C has stable reduction as soon as $C(K) \neq \emptyset$ and that the points of $\text{Jac}(C)[\ell]$ are rational over K for an integer ℓ prime to char(k) and sufficiently large. In fact, we can replace this 'sufficiently large' by $\ell \geq 3$. See [1], Proposition 5.10, or [29], Théorème 5.15.

Definition 4.23. Let S be a Dedekind scheme of dimension 1 with function field K, and let A be an Abelian variety over K. Let $\mathcal{A}^0$ denote the identity component of the Néron model of A over S (Definition 2.7 and Exercise 2.6). We say that A has *semi-Abelian reduction at s* if $\mathcal{A}_s^0$ is a semi-Abelian variety. That is, there exists an exact sequence of algebraic groups

$$0 \to T \to \mathcal{A}_s^0 \to B \to 0$$

where T is a torus (that is to say that over $\overline{k(s)}$, T is isomorphic to a power of the multiplicative group $\mathbb{G}_m$) and where B is an Abelian variety over $k(s)$.

Remark 4.24. Let A be the Jacobian of a smooth projective curve C, and let $\mathcal{C}$ be a regular model of C over S. Let us suppose $k(s)$ is algebraically closed and that the gcd of the multiplicities of the irreducible components of $\mathcal{C}_s$ is equal to 1 (e.g., if $C(K) \neq \emptyset$). A theorem of Raynaud ([15], Theorem 9.5/4) then stipulates that we have an isomorphism of algebraic groups $\mathcal{A}_s^0 \simeq \text{Pic}^0_{\mathcal{C}_s/k(s)}$. Unfortunately, we have not defined the right-hand term. Let us say that in terms of abstract groups, we have, in particular, an isomorphism $\mathcal{A}_s^0(k(s)) \simeq \text{Pic}^0(\mathcal{C}_s)$.

Remark 4.25 (*Semi-Abelian reduction theorem of Grothendieck*, [42]). For any Abelian variety A over K, there exists a Dedekind scheme S', finite flat over S, such that $A_{K(S')}$ has semi-Abelian reduction at every point of S'.

Remark 4.26 ([26], Theorem 2.4). Let C be a smooth, projective, geometrically connected curve of genus $g \geq 2$ over K. Then C has stable reduction at s if and only if $\mathrm{Jac}(C)$ has semi-Abelian reduction at s. When $C(K) \neq \emptyset$, this is essentially Remark 4.24 and Proposition 3.42. But the proof is more difficult without this hypothesis. This theorem, together with the semi-Abelian reduction theorem, implies Theorem 4.3. This is the approach of Deligne–Mumford, and also that of Artin–Winters, who prove directly that $\mathrm{Jac}(C)$ has potential semi-Abelian reduction. Let us note that, conversely, one can show that Theorem 4.3 implies the semi-Abelian reduction theorem for every Abelian variety.

10.4.3 Examples of computations of the potential stable reduction

Theorem 4.3 guarantees that any smooth, projective, geometrically connected curve C of genus $g \geq 2$ over a discrete valuation field K obtains stable reduction after a suitable finite extension of K. However, there is no general method to compute this stable reduction when the residue field is of positive characteristic. We will state without proof Proposition 4.30 which describes a method to determine the potential stable reduction (definition below) of a curve that is a tame Galois covering of a curve of which we know the stable reduction. Then, we give a method that allows us to find 'pieces' of the potential stable reduction (Proposition 4.37). Some 'numerical' examples will be treated explicitly. Finally, we will give a quick word on the minimal extension that realizes the stable reduction.

Definition 4.27. Let C be a smooth, projective, geometrically connected curve of genus $g \geq 2$ over a discrete valuation field K. Let L be a finite separable extension of K such that C admits a stable model $\mathcal{C}$ over $\mathcal{O}_L$ (integral closure of $\mathcal{O}_K$ in L). Let $s' \in \mathrm{Spec}\,\mathcal{O}_L$ be a closed point. The curve $\mathcal{C} \times_{\mathrm{Spec}\,\mathcal{O}_L} \mathrm{Spec}\,\overline{k(s')}$ is called the *potential stable reduction* of C.

Lemma 4.28. *The potential stable reduction is unique up to isomorphism.*

Proof Let F/K be the Galois closure of L/K. Let $s', s'' \in \mathrm{Spec}\,\mathcal{O}_L$ be two closed points. Then they are the respective images of two closed points $t', t'' \in \mathrm{Spec}\,\mathcal{O}_F$. The latter are conjugated under the action of $\mathrm{Gal}(F/K)$. It follows that

$$\mathcal{C}_{\overline{s}'} = \mathcal{C}_{\overline{t}'} \simeq \mathcal{C}_{\overline{t}''} = \mathcal{C}_{\overline{s}''}$$

(where $\bar{x}$ means the algebraic closure of the residue field $k(x)$). Hence $\mathcal{C}_{\overline{s}'}$ does not depend on the choice of the point s'. A similar argument implies that it does not depend either on the choice of L: if we have another extension L', it suffices to consider the Galois closure of their compositum in a separable closure of K. $\qquad\square$

Remark 4.29. Let C be a smooth, projective, geometrically connected curve of genus 2 over a discrete valuation field K. Then it is a hyperelliptic curve (Proposition 7.4.9). It therefore possesses an affine equation of the form

$$y^2 + Q(x)y = P(x), \quad 5 \le \max\{2 \deg Q, \deg P\} \le 6$$

(Proposition 7.4.24). The potential stable reduction of C can then be computed explicitly as a function of the valuations of certain Igusa invariants associated to this equation. See [59], Théorème 1. Such an algorithm is not known in higher genus.

Proposition 4.30. *Let $\mathcal{O}_K$ be a discrete valuation ring with residue field k. Let $f : C \to D$ be a finite morphism (also called covering) of smooth, projective, geometrically connected curves over K. Let us suppose that f is Galois with group G of order prime to $\operatorname{char}(k)$, and that D admits a semi-stable model $\mathcal{D}_0$ over $\mathcal{O}_K$. Then the potential stable reduction of C can be obtained by following the steps below:*

(1) *Let $B \subset D$ be the branch locus of f. We take a finite separable extension M/K to make the points of B rational over M. We replace $\mathcal{D}_0$ by $\mathcal{D}_0 \times_{\operatorname{Spec} \mathcal{O}_K} \operatorname{Spec} \mathcal{O}_M$, where $\mathcal{O}_M$ is a discrete valuation ring that dominates $\mathcal{O}_K$ and has field of fractions M.*

(2) *By composing successive blowing-ups starting at $\mathcal{D}_0$, we obtain a birational morphism $\mathcal{D} \to \mathcal{D}_0$ with $\mathcal{D}$ semi-stable and such that the closure of B in $\mathcal{D}$ is a disjoint union of sections contained in the smooth locus of $\mathcal{D}$.*

(3) *Let $\mathcal{C}_0 \to \mathcal{D}$ be the normalization of $\mathcal{D}$ in $K(C_M)$. Let $\mathcal{F}$ be the set of irreducible components Δ of $\mathcal{D}_s$ such that either $p_a(\Delta) \ge 1$, or Δ contains at least three points of $\mathcal{B} \cup (\mathcal{D}_s)_{\mathrm{sing}}$. Let e_Δ denote the ramification index $e_{\Gamma/\Delta}$ for an irreducible component Γ of $(\mathcal{C}_0)_s$ lying above Δ (this integer is independent of the choice of Γ). Let us set $e = \mathrm{lcm}\{e_\Delta \mid \Delta \in \mathcal{F}\}$, and $e = 1$ if $\mathcal{F}$ is empty.*

Then for any extension of discrete valuation rings $\mathcal{O}_L/\mathcal{O}_M$ of ramification index divisible by e, the normalization $\mathcal{C}$ of $\mathcal{D}_{\mathcal{O}_L}$ in $K(C_L)$ is a semi-stable model of C_L.

Proof This result is classical. See, for example, [60], Theorem 2.3. The extension L/K obtained in this manner is almost the smallest possible ([60], Theorem 3.9). This proposition is convenient for computing the reduction of cyclic coverings of $\mathbb{P}^1_K$ of order prime to $\operatorname{char}(k)$. $\qquad\qquad\qquad\qquad\qquad\qquad\qquad\qquad\square$

Remark 4.31. Let $\mathcal{O}_K$ be a discrete valuation ring with $\operatorname{char}(K) \ne 2$ and residue characteristic $p \ge 0$. Let C be a smooth, projective, geometrically connected curve of genus $g \ge 2$ over K. Let us suppose that $p = 0$ or $p > g + 1$. Then we can show that C has stable reduction over a finite extension of K, independently of Theorem 4.3. Indeed, by [39], Proposition 8.1, there exists a finite extension M/K such that C_M admits a finite morphism $f : C_M \to \mathbb{P}^1_M$ of degree $g + 1$. Let N be the Galois closure of $K(\mathbb{P}^1_M)$ in $K(C_M)$ and $\widetilde{C}$ the normalization of C_M in N. Then $\widetilde{C} \to \mathbb{P}^1_M$ is a Galois morphism, of degree prime to p.

Proposition 4.30 implies that $\widetilde{C}$ has semi-stable reduction over a finite extension L of M, and Proposition 3.48 implies that C also has semi-stable (hence stable) reduction over L.

A particular case of Proposition 4.30 is the following theorem of Grothendieck:

Corollary 4.32 (Grothendieck). *Let $f : C \to D$ be a Galois covering with group G as in Proposition 4.30. Let us moreover suppose that f is étale, and that D admits a smooth model $\mathcal{D}$ over $\mathcal{O}_K$. Then there exists a discrete valuation ring $\mathcal{O}_L$ that is tamely ramified over $\mathcal{O}_K$, of degree $[L : K]$ dividing $\mathrm{Card}\, G$, such that C_L admits a smooth model $\mathcal{C}$ over $\mathcal{O}_L$ and that f extends to a finite étale Galois morphism $\mathcal{C} \to \mathcal{D}_{\mathcal{O}_L}$ with group G.*

Proof As f is étale, we do not need steps (1)–(2) of Proposition 4.30, and the integer e defined in step (3) is the ramification index of $\mathcal{D}_s$ in an irreducible component Γ of $(\mathcal{C}_0)_s$, where $\mathcal{C}_0$ is the normalization of $\mathcal{D}$ in $K(C)$. It therefore divides the order of G. Let $\mathcal{O}_L/\mathcal{O}_K$ be a totally ramified extension (i.e., there is no extension of the residue fields) of degree e. Let $\mathcal{C}$ be the normalization of $\mathcal{D}_{\mathcal{O}_L}$ in $K(C_L)$. Then $\mathcal{C} \to \mathcal{D}_{\mathcal{O}_L}$ is not ramified at the points of codimension 1, which implies that the branch locus is empty (Exercise 8.2.15). In other words, $\mathcal{C} \to \mathcal{D}_{\mathcal{O}_L}$ is étale. In particular, $\mathcal{C}$ is smooth over $\mathrm{Spec}\, \mathcal{O}_L$. This morphism factors through $\mathcal{C}/G \to \mathcal{D}_{\mathcal{O}_L}$, which is finite birational, and is therefore an isomorphism. $\square$

Remark 4.33. Let $f : C \to D$ be a Galois covering as in Proposition 4.30 but with group G of order divisible by $p = \mathrm{char}(k)$, then the situation is completely different. Few general results are known. Let us, however, cite a theorem of Raynaud [82]: let us suppose that f is étale, that G is a p-group, and that D has good reduction. Then the potential stable reduction $\mathcal{C}_{\bar{s}}$ of C has toric rank zero. In other words, the irreducible components of $\mathcal{C}_{\bar{s}}$ are smooth, and its dual graph is a tree. But in general, $\mathcal{C}_{\bar{s}}$ itself is not smooth. See Example 4.38. Moreover, contrarily to the tame case, it is in general difficult to produce explicitly an extension of K over which C has stable reduction, except by passing through the torsion points of $\mathrm{Jac}(C)$ (see, for instance, [29], Corollaire 5.18).

Lemma 4.34 (Going-down theorem). *Let $f : X \to Y$ be a surjective integral morphism of integral schemes with Y normal and $[K(X) : K(Y)]$ finite. Let $y \in Y$ and let Y_0 be an irreducible subset of Y containing y. Then for any $x \in f^{-1}(y)$, there exists an irreducible subset X_0 of X such that $x \in X_0$ and $f(X_0) = Y_0$.*

Proof Let Z be the normalization of Y in $K(X)$. Then Z is integral and surjective over Y, and it dominates X. It therefore suffices to show the assertion for $Z \to Y$. In other words, we can suppose that X is normal. Let L be the separable closure of $K(Y)$ in $K(X)$ and X' the normalization of Y in L. Then $X \to X'$ is a homeomorphism (Exercise 5.3.9). We can therefore replace X by X' and suppose $K(X)$ separable over $K(Y)$. The same reasoning as at the beginning shows that we can even suppose $K(X)/K(Y)$ Galois, with group G.

This then implies that $Y = X/G$, and that G acts transitively on the fibers of $X \to Y$ (Exercise 2.3.21). Let ξ_0 be the generic point of Y_0 and $\xi_0' \in f^{-1}(\xi_0)$. Then $f(\overline{\{\xi_0'\}}) = \overline{\{\xi_0\}}$. Let $x_0 \in \overline{\{\xi_0'\}} \cap f^{-1}(y)$. There exists a $\sigma \in G$ such that $\sigma(x_0) = x$. Then $X_0 = \sigma(\overline{\{\xi_0'\}})$ has the desired properties. See also [65], (5.E), Theorem 5(v). $\qquad\square$

Lemma 4.35. *Let Z be a curve over a field and $z \in Z$ a closed point. Let $m_{Z,z}$ be the number of points of the normalization Z' of Z that lie above z. Then $m_{Z,z}$ is equal to the number of irreducible components of $\widehat{\mathcal{O}}_{Z,z}$.*

Proof This is Proposition 8.2.41(c). But the proof is elementary in the case of curves. Let $\mathfrak{q}_1, \dots, \mathfrak{q}_n$ be the minimal prime ideals of $\widehat{\mathcal{O}}_{Z,z}$. Then

$$(\widehat{\mathcal{O}}_{Z,z})' = \oplus_{1 \le j \le n}(\widehat{\mathcal{O}}_{Z,z}/\mathfrak{q}_j)',$$

where the exponent $'$ means the integral closure in the total ring of fractions. Let $m = m_{Z,z}$ and let $z_1, \dots, z_m$ be the points of Z' lying above z. Then

$$\mathcal{O}_{Z,z}' \otimes_{\mathcal{O}_{Z,z}} \widehat{\mathcal{O}}_{Z,z} \simeq \oplus_{1 \le i \le m}\widehat{\mathcal{O}}_{Z',z_i}$$

(Exercise 4.3.17). The second member is regular (Lemma 4.2.26) since $\mathcal{O}_{Z',z_i}$ is regular. It follows that $(\widehat{\mathcal{O}}_{Z,z})' = \mathcal{O}_{Z,z}' \otimes_{\mathcal{O}_{Z,z}} \widehat{\mathcal{O}}_{Z,z}$. By comparing the number of idempotent elements in the identities above, we find that $n = m$. $\qquad\square$

Corollary 4.36. *Let $f : X \to Y$ be a finite surjective morphism of integral fibered surfaces over an excellent Dedekind scheme S. Let us suppose, moreover, that Y is normal. Let $y \in Y_s$ be a closed point and $x \in f^{-1}(y)$.*

(a) *Let c_y be the number of irreducible components of Y_s passing through y, and let c_x be defined in a similar manner. Then $c_x \ge c_y$.*

(b) *We have $m_{X_s,x} \ge m_{Y_s,y}$.*

Proof (a) Let $\Gamma_1, \Gamma_2, \dots, \Gamma_c$ be the irreducible components of Y_s passing through y. By Lemma 4.34, they each lift to irreducible components of X_s passing through x. Hence $c_x \ge c_y$.

(b) Let $A = \widehat{\mathcal{O}}_{Y,y}$. This is a normal ring (Proposition 8.2.41(b)). Let us consider the finite surjective morphism $g : Z = X \times_Y \operatorname{Spec} A \to \operatorname{Spec} A$. Then $g^{-1}(y)$ can be identified with X_y and we have $\widehat{\mathcal{O}}_{Z,x} = \widehat{\mathcal{O}}_{X,x}$. By Lemma 4.35 and (a), at least $m_{Y_s,y}$ irreducible components of Z_s pass through x. This is a fortiori true when we replace Z_s by $\operatorname{Spec} \widehat{\mathcal{O}}_{Z_s,x}$, whence $m_{X_s,x} = m_{Z_s,x} \ge m_{Y_s,y}$. $\qquad\square$

Let us recall that for a reduced curve U over a field k, the projective completion $\widehat{U}$ of U is the projective curve over k obtained by completing U by regular points, as in Exercise 4.1.17. The following proposition says that if we have a piece of reduction of C that is nearly stable, then this piece will stay in the potential stable reduction of C.

Proposition 4.37. *Let S be the spectrum of a discrete valuation ring $\mathcal{O}_K$, with residue field k. Let C be a smooth, projective, geometrically connected curve of genus $g \geq 2$ over K. Let W^0 be a normal quasi-projective scheme over S, with generic fiber isomorphic to an open subscheme of C. We suppose that the projective completion $\hat{U}$ of $U := (W^0_s)_{\mathrm{red}}$ is stable, or connected and smooth of genus 1. Then there exists a finite surjective morphism from a closed subscheme of the potential stable reduction of C onto $\hat{U}_{\overline{k}}$. It is an isomorphism if W^0_s is reduced.*

Proof   In the proof of Lemma 8.3.49(d), we saw that the normalization of $W^0 \times_S \operatorname{Spec} \widehat{\mathcal{O}}_K$ comes from the normalization of W^0 by base change. As W^0 is normal, $W^0 \times_S \operatorname{Spec} \widehat{\mathcal{O}}_K$ is also normal. We can therefore suppose $\mathcal{O}_K$ is complete. After a finite étale extension of $\mathcal{O}_K$, we can suppose that the singular points of $\hat{U}$ are split. Let W be a normal projective scheme over S such that W^0 is an open subscheme of W. After contracting some irreducible components of W_s (Theorem 8.3.36), we can suppose that U is dense in W_s. Let L be a finite extension of K such that C_L has stable reduction over $\mathcal{O}_L$, the integral closure of $\mathcal{O}_K$ in L. Let us set $S' = \operatorname{Spec} \mathcal{O}_L$.

Let us consider $f : W' \to W$, the finite morphism composed of the normalization $W' \to W \times_S S'$ and of the projection onto W. Let Γ be an irreducible component of W_s and Δ an irreducible component of $W'_{s'}$ lying above Γ. Let $y \in U \cap \Gamma$ be a singular point of U and $x \in f^{-1}(y)$. By Corollary 4.36, if y is an intersection point in W_s, then x is an intersection point in $W'_{s'}$; if U is irreducible at y, then $m_{\Delta,x} \geq m_{U,y} = 2$ because W is excellent since $\mathcal{O}_K$ is supposed complete.

Let $\pi : Z \to W'$ be the minimal desingularization. Let us show that $Z \to \mathcal{C}_{\min}$ is an isomorphism. This comes down to showing that the strict transform $\widetilde{\Delta} \subseteq Z_s$ of an arbitrary irreducible component Δ of $W'_{s'}$ is not an exceptional divisor. Let us suppose the contrary. As $\mathcal{C}$ is semi-stable and $Z \to \mathcal{C}_{\min}$ is a sequence of blowing-ups of closed points, it is easy to see that every exceptional divisor of Z is a projective line (over an extension of k) that meets the other irreducible components at at most two points. Hence $\widetilde{\Delta}$ is regular, and it is therefore the normalization of Δ. Let $x \in \Delta$ be a point lying above a singular point of U, and let $z \in \pi^{-1}(x)$. Then $\pi^{-1}(x)$ is either reduced to z, in which case π is an isomorphism in a neighborhood of z (Corollary 4.4.6), or a connected curve (Corollary 5.3.16). In both cases, z is an intersection point in $Z_{s'}$ if $f(x)$ is an intersection point in U. If U is irreducible at $f(x)$, then $m_{\Delta,x} \geq 2$, and $\pi^{-1}(x)$ is a connected curve that meets $\widetilde{\Delta}$ at $m_{\Delta,x}$ points. The $\hat{U}$ stable hypothesis then implies that $\widetilde{\Delta}$ contains at least three intersection points, a contradiction.

The properties of the $\widetilde{\Delta}$ that we have just shown imply that any irreducible component of $Z_{s'}$ that is contracted to a point of the stable model $\mathcal{C}_{\mathrm{can}}$ (Proposition 9.4.8) cannot be a $\widetilde{\Delta}$ and is therefore contracted to a point in W'. It follows that $Z \to W'$ factors through a birational morphism $\rho : \mathcal{C}_{\mathrm{can}} \to W'$. Let $\widetilde{W'_{s'}}$ be the strict transform of $W'_{s'}$ in $\mathcal{C}_{\mathrm{can}}$. This is a semi-stable curve; hence the finite surjective morphism $\widetilde{W'_{s'}} \to (W_s)_{\mathrm{red}}$ induces a finite surjective morphism $\widetilde{W'_{s'}} \to \hat{U}$. If W^0_s is reduced, and hence geometrically reduced since U is semi-stable, the same holds for W_s. Hence $W' = W \times_S S'$ (Lemma 4.1.18), and $\widetilde{W'_{s'}} \to \hat{U}$ is consequently finite birational. As these are semi-stable curves, we easily deduce from this that it is an isomorphism.

Note that the proof of the proposition follows immediately if no irreducible component of $\hat{U}$ is rational. □

Example 4.38. Let $K = \mathbb{Q}_2$. Let us consider the smooth, projective, hyperelliptic curve C over K defined by an affine equation

$$y^2 = x^8 - 5x^6 + 10x^4 - 10x^2 + 5.$$

Let σ be the automorphism of $K(C)$ defined by $\sigma(x) = -x$ and $\sigma(y) = -y$, and G the group of order 2 generated by σ. Then $C \to D := C/G$ is a finite étale morphism. Indeed, $K(C)^G = K(u,v)$, where $u = x^2 - 1$ and $v = xy$, and D is described by a hyperelliptic equation

$$v^2 = u^5 + 1.$$

Hurwitz's formula then implies that $C \to D$ is étale. We are going to see that D has potential good reduction, and that this is not the case for C.

Let us set $v = 2v_1+1$, $u = 2^{2/5}u_1$. Then $v_1^2+v_1 = u_1^5$. This is the equation of a hyperelliptic curve of genus $2 = g(D)$ over $\mathbb{F}_2$. Hence D has good reduction over $K[2^{2/5}]$ (Corollary 1.25). Let us replace K by $K[2^{2/5}]$. Let us set $x = 1+2^{1/5}x_1$. Then we have

$$v_1^2 + v_1 = (x_1^2 + 2^{4/5}x_1)^5.$$

As $K(v_1,x_1) = K(C)$, the equation above therefore defines an affine scheme W^0 over $\mathcal{O}_K$, whose generic fiber is an open subscheme of C and whose special fiber is smooth, defined by the equation $v_2^2 + v_2 = x_1^5$ (we translate v_1 by x_1^5 on $\mathbb{F}_2$) of a smooth hyperelliptic curve of genus 2. Hence C does not have potential good reduction, by Proposition 4.37. If we go further with the computation or if we admit Remark 4.33, then we find that the potential stable reduction of C consists of the union of a smooth curve of genus 2 (that we have just found) and a smooth curve of genus 1, the two irreducible components meeting transversally at one point.

Example 4.39. Let C be the plane curve $x^4 + y^4 + z^4 = 0$ that we have already come across in Example 1.12. We are going to show that its potential stable reduction at 2 consists of three elliptic curves that meet a projective line at three points. See Figure 50. We consider C as a curve over $K := \mathbb{Q}_2$. We have $K(C) = K(u,v)$, where $u = x/z$ and $v = y/z$, with the relation $u^4 + v^4 + 1 = 0$. Let E be the smooth projective curve of genus 1, with affine equation $w^2 + v^4 + 1 = 0$. Then we have a morphism $C \to E$ of degree 2 corresponding to the injection $K(E) \to K(C)$ that sends w to u^2 and v to v. In

Example 1.28 we have already studied the reduction of a curve with an equation close to that of E. Let $\xi_8 \in \overline{K}$ be a primitive eighth root of unity. Let us set

$$\xi_8 v = (1 + v_1 + \alpha)/(v_1 + \alpha), \quad w = (2w_1 + \beta v_1 + \gamma)/(v_1 + \alpha)^2$$

where $\alpha, \beta, \gamma \in \overline{K}$ are chosen so that $|\alpha| = |2|^{-1/4}$, $|\beta| = |2|^{1/2}$, $|\gamma| = 1$ and that

$$w_1^2 + (\beta v_1 + \gamma)w_1 = v_1^3.$$

See Example 1.28 for the computations. Let L/K be a finite extension containing ξ_8, α, β, and γ. Let Z^0 be the smooth scheme $\operatorname{Spec} \mathcal{O}_L[w_1, v_1]$. Its generic fiber is an open subscheme of E_L.

Let $W^0 \to Z^0$ be the normalization of Z^0 in $K(C_L)$. Then $W_s^0 \to Z_s^0$ is surjective and finite. Restricting W^0 if necessary, the curve $U = (W_s^0)_{\mathrm{red}}$ is then smooth with $g(\hat{U}) \geq 1$. By Proposition 4.37, the potential stable reduction $C_{\overline{s}}$ of C contains a curve Γ of geometric genus $p_a(\Gamma') \geq 1$ (where Γ' is the normalization of Γ). Let $\tau : C \to C$ be the automorphism of order 3 defined by $(x, y, z) \mapsto (y, z, x)$. It acts on $C_{\overline{s}}$. Let us show that $\tau(\Gamma) \neq \Gamma$. Let η be the generic point of U, and let ν_η be the valuation of $K(C_L)$ associated to the ring $\mathcal{O}_{W^0, \eta}$. We have

$$\nu_\eta(v) = \nu_\eta(1 + v_1 + \alpha) - \nu_\eta(v_1 + \alpha) = \nu_\eta(\alpha) - \nu_\eta(\alpha) = 0,$$

which means that the image of v in $k(\eta) \subseteq k(\Gamma)$ is non-zero. We have $\tau(v) = 1/u$. As

$$2\nu_\eta(u) = \nu_\eta(w) = \nu_\eta(2w_1 + \beta v_1 + \gamma) - 2\nu_\eta(v_1 + \alpha) = 0 - 2\nu_\eta(\alpha) > 0,$$

$\nu_\eta(\tau(v)) = \nu_\eta(1/u) < 0$ and $\tau(\Gamma) \neq \Gamma$. Consequently, $C_{\overline{s}}$ contains three irreducible curves of geometric genus ≥ 1. As it is of genus $g(C) = 3$, these curves are smooth of genus 1. Taking into account the symmetry imposed by the automorphism τ, we see that the only possibility for $C_{\overline{s}}$ is as in Figure 50.

Let σ be the automorphism of order 2 of C defined by $u \mapsto -u$ and $v \mapsto v$. Then σ acts on Γ and leaves invariant the intersection point q of Γ with the projective line, and $\Gamma/\langle\sigma\rangle$ is isomorphic to the good reduction Z_s of E. As $\Gamma \to Z_s$ is ramified at q (Exercise 4.3.19), Hurwitz's formula implies that $\Gamma \to Z_s$ is purely inseparable. In other words, σ acts trivially on Γ. The same type of reasoning and Proposition 3.38 imply that σ must permute the two other elliptic components of $C_{\overline{s}}$.

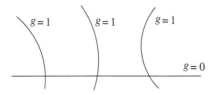

Figure 50. Potential stable reduction of the curve $x^4 + y^4 + z^4 = 0$.

Example 4.40. Let $K = \mathbb{Q}_p$ with $p > 2$. Let a, b be integers such that $1 \le a, b \le p - 1$ and that $a + b$ is prime to p. Let us consider the normal projective curve $F_{a,b}$ over K whose function field is

$$K(C) = K(x, y), \quad y^p = x^a(x - 1)^b.$$

This is a smooth projective curve (because $\mathrm{char}(K) = 0$), geometrically connected because $K(C) \cap \overline{K} = K$. The inclusion $K(x) \subset K(x, y)$ induces a finite morphism $C \to \mathbb{P}^1_K$ that is ramified at the points $x = 0, 1, \infty$. Hurwitz's formula implies that $g(C) = (p - 1)/2$. We are going to show that the potential stable reduction of $F_{a,b}$ is smooth and hyperelliptic.

Let $F(x) = x^a(x - 1)^b$ and $\alpha = a/(a + b) \in \mathcal{O}_K^*$. Then $F'(\alpha) = 0$ and $F(\alpha), F''(\alpha) \in \mathcal{O}_K^*$. Let ξ_p be a primitive pth root of unity and $\lambda = \xi_p - 1$. Then $|\lambda| = |p|^{1/(p-1)}$. Let L be a finite extension of K containing λ and $F(\alpha)^{1/p}$. Let us set $x = \lambda^{p/2} u + \alpha$ and $y = \lambda v + F(\alpha)^{1/p}$. Then

$$F(x) = F(\alpha) + \lambda^p \left((F''(\alpha)/2)u^2 + \sum_{i \ge 3} a_i \lambda^{(i-2)p/2} u^i \right), \quad a_i = F^{(i)}(\alpha)/i! \in \mathbb{Z}[\alpha]$$

and a simple computation shows that

$$v^p + (p\lambda^{-(p-1)})F(\alpha)^{1/p} v = (F''(\alpha)/2)u^2 + \lambda P(u) + \lambda Q(v)$$

with $P(u), Q(v)$ polynomials with coefficients in $\mathcal{O}_L$. This equation defines an affine scheme W^0 whose special fiber is given by

$$v^p + \beta v = \gamma u^2 \tag{4.39}$$

with $\beta, \gamma \in \mathbb{F}_p^*$. Restricting W^0 if necessary, we have W^0 smooth over $\mathcal{O}_L$, and its special fiber is a smooth curve that is an open subscheme of a smooth projective curve of genus $(p - 1)/2 = g(C)$. Hence C_L has good reduction over $\mathcal{O}_L$ (Corollary 1.25), and the reduction is none other than the hyperelliptic curve with equation (4.39). After a change of variables with coefficients in $\overline{\mathbb{F}}_p$, equation (4.39) becomes $v^p - v = u^2$.

Remark 4.41. Let $m \ge 3$, and let $a, b, c \in \mathbb{Z}$ be such that $a + b + c = 0$. Let $F_{a,b,c}^m$ be the smooth projective curve over $K = \mathbb{Q}_p$ whose function field is

$$K(F_{a,b,c}^m) = K(x, y), \quad y^m = (-1)^c x^a (1 - x)^b.$$

Let us suppose that $p \ge 5$, or that $p = 3$ and $(3, m/(m, abc), abc) = 1$. Then Coleman and McCallum ([22], Theorem 3.4) have computed a semi-stable (and stable if $g(F_{a,b,c}^m) \ge 2$) model of $F_{a,b,c}^m$ over a finite extension of $\mathcal{O}_K$. This allowed them to compute certain Jacobi sums.

Example 4.42. Let C be the Klein curve $x^3y + y^3z + z^3x = 0$ over $\mathbb{Q}$. It has good reduction outside of $p = 7$ (Exercise 1.4). Let ξ_7 be a primitive seventh root of unity. It is well known that the group of automorphisms of $C_{\overline{\mathbb{Q}}}$ is of order 168 and contains an element τ of order 7 defined over $K = \mathbb{Q}(\xi_7)$ ([19], Section 232). Hurwitz's formula implies that $C/\langle\tau\rangle$ is of genus 0, and hence isomorphic to $\mathbb{P}_K^1$ because $C(K) \neq \emptyset$, and that the morphism $C \to C/\langle\tau\rangle$ is ramified at three points of $\mathbb{P}^1(\overline{K})$. This implies that over a finite extension L of K, we have $K(C) = L(x,y)$ with $y^7 = x^a(x-1)^b$, $1 \le a, b \le 6$, and $a + b$ prime to 7. Example 4.40 implies that C has potential good reduction, and that the latter is the hyperelliptic curve $y^2 = x^7 - x$.

Remark 4.43. Given a curve C, using an explicit equation of C to determine its potential stable reduction is not always possible nor always desirable. On the other hand, in interesting cases, the potential stable reduction can be found by exploiting the intrinsic properties of the curve. For example, the potential stable reduction of the modular curves $X_0(N)$ at p is known if at most a power 2 of p divides N ([27] if p^2 does not divide N, and [32] if p^2 divides N exactly and $p \ge 5$).

To conclude, we mention some results concerning the field extensions necessary to obtain the stable reduction. From here to the end of the section, we fix a Henselian (e.g., complete) discrete valuation ring $\mathcal{O}_K$ with *algebraically closed residue field* k of characteristic $p = \text{char}(k) \ge 0$. Further, C will be a smooth, projective, geometrically connected curve of genus $g \ge 2$ over K. We will say that an extension L/K *realizes the stable reduction of* C if C_L has stable reduction over $\mathcal{O}_L$.

Theorem 4.44. *There exists a finite extension L of K with the following properties:*

(a) *The extension L realizes the stable reduction of C, and it is minimal in the sense that any extension that realizes the stable reduction of C contains L; moreover, L/K is Galois.*

(b) *Let $C_{\overline{s}}$ be the potential stable reduction of C; then we have an injective canonical homomorphism $\text{Gal}(L/K) \hookrightarrow \text{Aut}_k(C_{\overline{s}})$.*

Proof (a) We use the analogous statement for the semi-Abelian reduction of Abelian varieties over K ([29], théorème 5.15), and the equivalence between the stable reduction of C and the semi-Abelian reduction of $\text{Jac}(C)$ ([26], Theorem 2.4). (b) Use [29], lemme 5.16. $\quad\square$

Proposition 4.45. *If $p = 0$ or if $p > 2g + 1$, then C has stable reduction over a tamely ramified extension of K.*

Proof Indeed, this is true for the semi-Abelian reduction of Abelian varieties (see the proof of [88], Section 2, Corollary 2(a)). We can also show that for any stable curve Z of genus g over k, the prime factors of $\text{Card}(\text{Aut}_k(Z))$ are at most equal to $2g + 1$, and then apply the theorem above. $\quad\square$

Proposition 4.46. *Let C be as above and let $\mathcal{C}$ be the minimal regular model with normal crossings of C over $\mathcal{O}_K$. Let us suppose that the multiplicities of the irreducible components of $\mathcal{C}_s$ are all prime to p, so that C has stable reduction over a tamely ramified extension of K (Corollary 4.7). Let d be the lcm of the multiplicities of the irreducible components Γ such that $p_a(\Gamma) \geq 1$ or that Γ meets the other components at at least three points. Then the minimal extension L/K is the unique extension of K with ramification index d.*

Proof   This result is relatively easy to show. See, for example, [96], Proposition 1.   ☐

The following theorem, due to T. Saito, characterizes the L/K tamely ramified condition by numerical properties of the special fiber of the minimal regular model with normal crossings.

Theorem 4.47. *Let $\mathcal{C}$ be as above and let us suppose $p > 0$. Let $\mathcal{C}$ be a minimal regular model with normal crossings of C. Let L/K be the minimal extension that realizes the stable reduction of C. Then L/K is tamely ramified if and only if the following condition is verified:*

(*) *Every irreducible component of $\mathcal{C}_s$, of multiplicity divisible by p, is isomorphic to $\mathbb{P}_k^1$, meets the other irreducible components at exactly two points, and these components are of multiplicity prime to p.*

Proof   See [84], Theorem 3.11. The proof is also presented in [1], Sections 2–3. In [84], the condition of the normal crossings is weaker than that which we have defined. But condition (*) is equivalent in either definition.   ☐

Exercises

4.1. Let $\mathcal{O}_K$ be a discrete valuation ring with residue field k of characteristic $\neq 2, 3, 5$. Compute the model $\mathcal{C}'$ of Proposition 4.6 explicitly when C is an elliptic curve, using Remark 4.8.

4.2. Let C be as above and let C be a smooth, projective, geometrically connected curve of genus $g \geq 2$ over a discrete valuation field K. We suppose that $p = \mathrm{char}(K) > 0$, that the residue field k of $\mathcal{O}_K$ is perfect, and that $C(K) \neq \emptyset$. We want to show that if there exists a purely inseparable extension K'/K such that $C_{K'}$ has stable reduction over $\mathcal{O}_{K'}$, then C has stable reduction over $\mathcal{O}_K$. We can suppose K is complete, k algebraically closed, and that $[K' : K] = p$.

    (a) Let $\mathcal{C}'$ be the stable model of $C_{K'}$ over $\mathcal{O}_{K'}$ and $\mathcal{C}' \to \mathcal{C}'^{(p)}$ the relative Frobenius over $\mathcal{O}_{K'}$. Show that there exists an $\mathcal{O}_K$-scheme $\mathcal{D}$ such that $\mathcal{C}'^{(p)} = \mathcal{D} \times_{\mathrm{Spec}\, \mathcal{O}_K} \mathrm{Spec}\, \mathcal{O}_{K'}$ (use Lemma 3.2.25).

    (b) Show that $\mathcal{D}$ is a stable curve over $\mathcal{O}_K$, with generic fiber isomorphic to $C^{(p)}$. Deduce from this that C has stable reduction over $\mathcal{O}_K$ (Exercise 3.20).

4.3. Let $\mathcal{O}_K$ be a discrete valuation ring with field of fractions K, and let L be a finite separable extension of the completion $\widehat{K}$ of K, defined by a monic polynomial $P(X) \in \widehat{\mathcal{O}}_K[X]$. We want to show that L comes from a finite separable extension of K.

(a) Let π denote the image of X in L and let t be a uniformizing parameter for $\mathcal{O}_K$. Let $c = P'(\pi)$. Let $Q(X) \in \mathcal{O}_K[X]$ be a separable monic polynomial of the same degree as $P(X)$ and such that $Q(X) - P(X) \in tc^2\widehat{\mathcal{O}}_K[X]$. Show that for any $\varepsilon \in tc\mathcal{O}_L$, $Q'(\pi + \varepsilon)$ has the same valuation as c.

(b) Applying the approximation method used in the proof of Proposition 6.2.15, show that there exists a $y \in \mathcal{O}_L$ such that $Q(y) = 0$ and that $y - \pi \in \mathfrak{m}_L$.

(c) Let $K' = K[Y]/(Q(Y))$. Show that K' is a finite separable extension of K and that $K' \otimes_K \widehat{K} \simeq L$.

(d) Let $\mathcal{O}_{K'}$ (resp. $\mathcal{O}_L$) be the integral closure of $\mathcal{O}_K$ in K' (resp. of $\widehat{\mathcal{O}}_K$ in L). Show that we have $\widehat{\mathcal{O}}_K \otimes_{\mathcal{O}_K} \mathcal{O}_{K'} \simeq \mathcal{O}_L$.

4.4. (*Theorem of algebraization of coherent sheaves*). Let X be a projective scheme over a complete discrete valuation ring $\mathcal{O}_K$. Let t be a uniformizing parameter for $\mathcal{O}_K$. Let us recall that $X_N = X \times_{\operatorname{Spec} \mathcal{O}_K} \operatorname{Spec} \mathcal{O}_K/(t^{N+1}\mathcal{O}_K)$ for every $N \geq 0$. Let $(\mathcal{F}_N)_{N \geq 0}$ be an inverse system of coherent sheaves: that is to say that $\mathcal{F}_N$ is a coherent sheaf on X_N and that

$$\mathcal{F}_{N+1} \otimes_{\mathcal{O}_{X_{N+1}}} \mathcal{O}_{X_N} = \mathcal{F}_N.$$

We will say that $(\mathcal{F}_N)_N$ is algebraizable if there exists a coherent sheaf $\mathcal{F}$ on X such that $\mathcal{F} \otimes_{\mathcal{O}_X} \mathcal{O}_{X_N} = \mathcal{F}_N$ for every $N \geq 0$. We are going to show that this is always the case. See [41], III.5.1.6, for the case when the base is an arbitrary complete Noetherian local ring.

(a) Let us consider the $\mathcal{F}_N$ as coherent sheaves on X killed by t^{N+1}. Show that for any $N, n \geq 0$, we have an exact sequence

$$0 \to t^{N+1}\mathcal{F}_{N+n} \to \mathcal{F}_{N+n} \to \mathcal{F}_N \to 0$$

and natural surjective homomorphisms

$$\rho_{N,n} : t^n \mathcal{F}_n \to t^{N+n}\mathcal{F}_{N+n}$$

of $\mathcal{O}_{X_0}$-modules. Considering the ascending sequence of the $\operatorname{Ker} \rho_{N,0} \subseteq \mathcal{F}_0$, show that there exist a coherent sheaf $\mathcal{G}$ on X_0, quotient of $\mathcal{F}_0$, and $N_0 \geq 0$ such that $\rho_{N,0} : \mathcal{F}_0 \to t^N\mathcal{F}_N$ induces an isomorphism $\mathcal{G} \simeq t^N\mathcal{F}_N$ for every $N \geq N_0$.

(b) Let $\mathcal{O}_X(1)$ be an ample sheaf on X. Let $d_1 \geq 1$ be such that $\mathcal{F}_{N_0}(d_1) := \mathcal{F}_{N_0} \otimes \mathcal{O}_X(d_1)$ is generated by its global sections and that $H^1(X, \mathcal{G}(d_1)) = 0$. Show that for any $N \geq N_0$, the canonical homomorphism $H^0(X, \mathcal{F}_{N+1}(d_1)) \to H^0(X, \mathcal{F}_N(d_1))$ is surjective, and that

$\mathcal{F}_N(d_1)$ is generated by its global sections. Deduce from this that there exist a coherent sheaf $\mathcal{L}_1 = \mathcal{O}_X(-d_1)^r$ on X and a homomorphism of inverse systems $(\mathcal{L}_1 \otimes \mathcal{O}_{X_N})_N \to (\mathcal{F}_N)_N$ that is surjective for every $N \geq N_0$, and therefore for every $N \geq 0$.

(c) Show that there exist a coherent sheaf $\mathcal{L}_2$ on X and an exact sequence of inverse systems

$$(\mathcal{L}_2 \otimes \mathcal{O}_{X_N})_N \xrightarrow{\alpha} (\mathcal{L}_1 \otimes \mathcal{O}_{X_N})_N \to (\mathcal{F}_N)_N \to 0.$$

Applying Corollary 5.3.11 to the sheaf $\mathcal{H}om_{\mathcal{O}_X}(\mathcal{L}_1, \mathcal{L}_2)$, show that α comes from a homomorphism of coherent sheaves $f : \mathcal{L}_1 \to \mathcal{L}_2$. Let $\mathcal{F} = \mathrm{Coker}(f)$. Show that $\mathcal{F} \otimes \mathcal{O}_{X_N} \simeq \mathcal{F}_N$ for every $N \geq 0$.

(d) Let C be as above and let us suppose that $\mathcal{F}_N$ is invertible on X_N for every N. Show that $\mathcal{F}$ is invertible. Deduce from this that $\mathrm{Pic}(X) \to \varprojlim_N \mathrm{Pic}(X_N)$ is surjective.

4.5. Let $\mathcal{O}_K$ be a discrete valuation ring with algebraically closed residue field. Let $X \to \mathrm{Spec}\, \mathcal{O}_K$ be an arithmetic surface with X_s connected. Let d be the gcd of the multiplicities of the irreducible components. Let us suppose that $\mathrm{Pic}^0(X_s)$ is unipotent. Show, using Lemma 4.14, that d is a power of $\mathrm{char}(k)$.

4.6. Let $f : X \to Y$ be a dominant projective morphism of regular locally Noetherian schemes, with geometrically integral fibers. Let us suppose that Y is irreducible with generic point ξ.

(a) Show that $\mathcal{O}_Y \to f_*\mathcal{O}_X$ is an isomorphism, and therefore that $\mathrm{Pic}(Y) \to \mathrm{Pic}(X)$ is injective (use the projection formula, Proposition 5.2.32).

(b) Let D be a prime divisor on X whose support does not meet X_ξ. Show that there exists a prime divisor E on Y such that $D = X \times_Y E$. Deduce from this that the canonical complex

$$1 \to \mathrm{Pic}(Y) \to \mathrm{Pic}(X) \to \mathrm{Pic}(X_\xi) \to 1$$

is exact.

4.7. Let X be an arithmetic surface over a discrete valuation ring. Let $\Gamma_1, \ldots, \Gamma_n$ be irreducible components of X_s and $M = (\Gamma_i \cdot \Gamma_j)_{i,j}$ the intersection matrix of X_s.

(a) Show that there exist two invertible matrices $P, Q \in M_{n \times n}(\mathbb{Z})$ such that PMQ is a diagonal matrix $\mathrm{Diag}(a_1, \ldots, a_{n-1}, 0)$ with $a_i \in \mathbb{N}$. The product with P, Q comes down to repeatedly replacing a line V (resp. a column) of M by $\pm V$ plus a linear combination of the other lines (resp. columns). We say that M and PMQ are *equivalent* matrices. Show that

$$\Phi(X) \simeq \prod_{1 \leq i \leq n-1} \mathbb{Z}/a_i\mathbb{Z}.$$

(b) Compute $\Phi(X)$ for the minimal regular model of an elliptic curve of type I_n^*.

(c) Let $\tilde{X} \to X$ be the blowing-up of a closed point $x \in X_s$ and $\widetilde{M}$ the intersection matrix of $\tilde{X}_s$. Let $E \subset \tilde{X}_s$ denote the exceptional divisor and $\tilde{\Gamma}_i$ the strict transform of Γ_i in $\tilde{X}$. Show that for any $i \leq n$, we have

$$\tilde{\Gamma}_i \cdot \tilde{\Gamma}_j + \mu_x(\Gamma_i)(E \cdot \tilde{\Gamma}_j) = \Gamma_i \cdot \Gamma_j$$

(Exercise 9.2.9(d)). Deduce from this that $\widetilde{M}$ is equivalent to the matrix

$$\begin{pmatrix} M & 0 \\ 0 & 1 \end{pmatrix}$$

and that $\Phi(Y) = \Phi(X)$ for any regular model Y of X_η over $\mathcal{O}_K$.

4.8. Let $\mathcal{O}_K$ be a discrete valuation ring with algebraically closed residue field k. Let C be a smooth, projective, geometrically connected curve of genus $g \geq 2$ over K. Let us fix a minimal regular model with normal crossings $\mathcal{C}$ of C over $\mathcal{O}_K$. Let $\mathcal{O}_L/\mathcal{O}_K$ be an extension of discrete valuation rings such that C_L has stable reduction over $\mathcal{O}_L$. Let $e_{L/K}$ denote the ramification index of this extension. Let Γ be an irreducible component of $\mathcal{C}'_s$ of multiplicity d.

(a) Let ξ be the generic point of Γ. Let $\mathcal{C}'$ be the stable model of C_L over $\operatorname{Spec} \mathcal{O}_L$ and let us suppose that there exists a generic point ξ' of $\mathcal{C}'_s$ such that $\mathcal{O}_{\mathcal{C}',\xi'}$ dominates $\mathcal{O}_{\mathcal{C},\xi}$. Show that $e_{L/K}/d$ is equal to the ramification index $e_{\mathcal{O}_{\mathcal{C}',\xi'}/\mathcal{O}_{\mathcal{C},\xi}}$. In particular, d divides $e_{L/K}$.

(b) Let $\Gamma_1,\dots,\Gamma_n$ be irreducible components of $\mathcal{C}_s$ such that $\cup_i \Gamma_i$ is a stable curve. Let $d_1,\dots,d_n$ be their respective multiplicities in $\mathcal{C}_s$. Show that d_i divides $e_{L/K}$ (use Proposition 4.37).

(c) Show that if $p_a(\Gamma) \geq 1$, then the multiplicity of Γ divides $e_{L/K}$.

4.9. Let C be a smooth, projective, geometrically connected curve over a discrete valuation field K, with algebraically closed residue field. Let us define the Abelian rank $a(C)$ as being the maximum of the Abelian ranks $a(\mathcal{C}_s)$ of the normal models $\mathcal{C}$ of C over $\operatorname{Spec} \mathcal{O}_K$. Likewise for the toric rank $t(C)$.

(a) Show that $a(C)$ and $t(C)$ are finite (use Exercise 3.17(a) to reduce to the regular models and show that the ranks are then invariant).

(b) Let f be a map $G_1 \to G_2$ between two connected graphs. Let us suppose that f is surjective at the level of the vertices, and that for any edge e_2 of G_2, with end vertices v_2, v'_2, and for any vertices v_1, v'_1 in G_1 lying respectively above v_2, v'_2, there exists an edge e_1 lying above e_2 that joins v_1 to v'_1. We want to show that $\beta(G_1) \geq \beta(G_2)$. We can suppose $\beta(G_2) \geq 1$.

(1) By deleting the terminal vertices (i.e., those with only one adjacent edge) and the adjacent edges in G_2, show that we can reduce to the case when G_2 has no terminal vertices (Exercise 1.18).

(2) Let $E(v)$ denote the set of edges adjacent to a vertex v. Show that $2(\beta(G_i) - 1) = \sum_v(\text{Card } E(v) - 2)$, the sum being taken over the set of vertices of G_i.

(3) Show that $\beta(G_1) \geq \beta(G_2)$.

(c) Let $\mathcal{O}_L$ be a discrete valuation ring that dominates $\mathcal{O}_K$ and such that L is finite over K. Show that

$$a(C_L) \geq a(C), \quad t(C_L) \geq t(C)$$

(use the going-down theorem as in the proof of Proposition 4.37).

Bibliography

[1] A. Abbes, *Réduction semi-stable des courbes d'après Artin, Deligne, Grothendieck, Mumford, Saito, Winters, ...*, in Courbes semi-stables et groupe fondamental en géométrie algébrique (Luminy, 1998), Prog. Math., **187**, Birkhäuser, Basle, 2000, 59–110.

[2] S. Abhyankar, *Resolution of singularities of arithmetical surfaces*, in Arithmetical Algebraic Geometry (Proc. Conf. Purdue Univ., 1963), ed. O.F.G. Schilling. Harper & Row, New York (1965), 111–152.

[3] _____, *On the valuations centered in a local domain*, Am. J. Math. **78** (1956), 321–348.

[4] D. Abramovich, K. Karu, K. Matsuki, J. Wlodarczyk, *Torification and factorization of birational maps*, J. Amer. Math. Soc. **15** (2002), 531–572.

[5] A. Altman, S. Kleiman, *Introduction to Grothendieck Duality Theory*, Lect. Notes Math., **146**, Springer, New York–Heidelberg–Berlin, 1970.

[6] M. Artin, *Lipman's proof of resolution of singularities for surfaces*, in *Arithmetic Geometry* [25], 267–287.

[7] _____, *Néron Models*, in *Arithmetic Geometry* [25], 213–230.

[8] _____, *Some numerical criteria for contractibility of curves on algebraic surfaces*, Am. J. Math. **84** (1962), 485–496.

[9] _____, *On isolated rational singularities of surfaces*, Am. J. Math. **88** (1966), 129–136.

[10] _____, *Algebraic approximation of structures over complete local rings*, Publ. Math. IHES **36** (1969), 23–58.

[11] _____, G. Winters, *Degenerated fibres and stable reduction of curves*, Topology **10** (1971), 373–383.

[12] E. Bombieri, *Canonical models of surfaces of general type*, Publ. Math. IHES **42** (1973), 171–219.

[13] S. Bosch, U. Güntzer, R. Remmert, *Non-Archimedean analysis; A systematic approach to rigid analytic geometry*, Grund. Math. Wiss. **261**, Springer, New York–Heidelberg–Berlin, 1984.

[14] _____W. Lütkebohmert, *Stable reduction and uniformization of abelian varieties,* I, Math. Ann. **270** (1985), 349–379.

[15] S. Bosch, W. Lütkebohmert, M. Raynaud, *Néron Models*, Ergeb. Math. Grenz., **21**, Springer, New York–Heidelberg–Berlin, 1990.

[16] N. Bourbaki, *Algèbre*, Diffusion CCLS, Paris, 1970.

[17] _____, *Algèbre Commutative*, Masson, Paris, 1985.

[18] _____, *Topologie générale*, Hermann, Paris, 1971.

[19] W. Burnside, *Theory of groups of finite order*, 2nd edn, Dover, New York, 1955.

[20] T. Chinburg, *Minimal models for curves over Dedekind rings*, in *Arithmetic Geometry* [25], 309–326.

[21] _____, R. Rumely, *The capacity pairing*, J. Reine Angew. Math. **434** (1993), 1–44.

[22] R. Coleman, W. McCallum, *Stable reduction of Fermat curves and Jacobi sum Hecke characters*, J. Reine Angew. Math. **385** (1988), 41–101.

[23] J.-L. Colliot-Thélène, S. Saito, *Zéro-cycles sur les variétés p-adiques et groupe de Brauer*, Int. Math. Res. Not., **4** (1996), 151–160.

[24] B. Conrad, *Grothendieck duality and base change*, Lect. Notes Math., **1750**, Springer, New York–Heidelberg–Berlin, 2000.

[25] G. Cornell, J. Silverman (editors), *Arithmetic Geometry*, Storrs Conference, Springer, New York–Heidelberg–Berlin, 1986.

[26] P. Deligne, D. Mumford, *The irreducibility of the space of curves of given genus*, Publ. Math. IHES **36** (1969), 75–110.

[27] _____, M. Rapoport, *Les schémas de modules des courbes elliptiques*, in Modular Functions of One Variable II, Lect. Notes Math. **349** (1973).

[28] M. Demazure, P. Gabriel, *Introduction to algebraic geometry and algebraic groups*, North-Holland Math. Stud., **39**, North-Holland, Amsterdam, 1980.

[29] M. Deschamps, *Réduction semi-stable*, in Séminaire sur les pinceaux de courbes de genre au moins deux, ed. L. Szpiro, Astérisque **86** (1981), 1–34.

[30] P. Du Val, *On isolated singularities of surfaces which do not affect the condition of adjunction* (Parts I, II, III), Proc. Cambridge Math. Soc. **30** (1934), 453–459, 460–465, 483–491.

[31] B. Edixhoven, *On the prime-to-p part of the groups of connected components of Néron models*. Special issue in honour of F. Oort, Compositio Math. **97** (1995), 29–49.

[32] _____, *Minimal resolution and stable reduction of $X_0(N)$*, Ann. Inst. Fourier **40** (1990), 31–67.

[33] D. Eisenbud, *Commutative algebra with a view toward algebraic geometry*, Grad. Texts Math., **150**, Springer, New York–Heidelberg–Berlin, 1994.

[34] H. Farkas, I. Kra, *Riemann surfaces*, Grad. Texts Math., **71**, 2nd edition, Springer, New York–Heidelberg–Berlin, 1992.

[35] J.-M. Fontaine, *Il n'y a pas de variété abélienne sur* $\mathbb{Z}$, Inv. Math. **81** (1986), 515–538.

[36] J. Fresnel, M. van der Put, *Rigid Geometry and Applications*, Prog. Math, Birkhäuser, Basle, 2002.

[37] W. Fulton, *Intersection Theory*, Ergeb. Math. Grenz., **2**, Springer, New York–Heidelberg–Berlin, 1984.

[38] ———, *Algebraic curves, an introduction to algebraic geometry*, Math. Lect. Note Ser., Benjamin, New York, 1969.

[39] ———, *Hurwitz schemes and irreducibility of moduli of algebraic curves*, Ann. Math. **90** (1969), 542–575.

[40] R. Godement, *Topologie Algébrique et Théorie des Faisceaux*, Hermann, Paris, 1958.

[41] A. Grothendieck, J. Dieudonné, *Éléments de Géométrie Algébrique* (EGA), Publ. Math. IHES, **4, 8, 11, 17, 20, 24, 28, 32**, 1960–1967.

[42] ———, *Modèle de Néron et monodromie*, SGA 7, exposé IX, Lect. Notes Math. **288**, Springer, New York–Heidelberg–Berlin (1972), 313–523.

[43] R. Hartshorne, *Algebraic Geometry*, Grad. Texts Math., **52**, Springer, New York–Heidelberg–Berlin, 1977.

[44] ———, *Residues and duality*, Lect. Notes Math., **20**, Springer, New York–Heidelberg–Berlin, 1966.

[45] H. Hauser, J. Lipman, F. Oort, A. Quirós (editors), *Resolution of singularities: A research textbook in tribute to Oscar Zariski*, Prog. Math., **181**, Birkhäuser, Basle, 2000.

[46] H. Hironaka, *Resolution of singularities of an algebraic variety over a field of characteristic zero. I, II*, Ann. Math. **79** (1964), 109–203; 205–326.

[47] J. Humphreys, *Reflection groups and Coxeter groups*, Cambridge Stud. Adv. Math., **29**, Cambridge Univ. Press, Cambridge, 1992.

[48] J. I. Igusa, *Arithmetic variety of moduli for genus two*, Ann. Math. **72** (1960), 612–649.

[49] A. J. de Jong, *Smoothness, semi-stability and alterations*, Publ. Math. IHES **83** (1996), 51–93.

[50] ———, *Families of curves and alterations*, Ann. Inst. Fourier **47** (1997), 599–621.

[51] S. Kleiman, *Relative duality for quasi-coherent sheaves*, Compositio Math. **40** (1980), 39–60.

[52] K. Kodaira, *On compact analytic surfaces, II*, Ann. Math. **77** (1963), 563–626.

[53] E. Kunz, *Introduction to Commutative Algebra and Algebraic Geometry*, Birkhäuser, Basle, 1985.

[54] S. Lang, *Introduction to Arakelov Theory*, Springer, New York–Heidelberg–Berlin, 1988.

[55] S. Lang, *Algebra*, 2nd edition, Addison-Wesley, Reading, MA, 1984.

[56] S. Lichtenbaum, *Curves over discrete valuation rings*, Am. J. Math. **90** (1968), 380–405.

[57] J. Lipman, *Rational singularities, with applications to algebraic surfaces and unique factorization*, Publ. Math. IHES **36** (1969), 195–279.

[58] ———, *Desingularization of two-dimensional schemes*, Ann. Math. **107** (1978), 151–207.

[59] Q. Liu, *Courbes stables de genre 2 et leur schéma de modules*, Math. Ann. **295** (1993), 201–222.

[60] ———, D. Lorenzini, *Models of curves and finite covers*, Compositio Math. **118** (1999), 61–102.

[61] D. Lorenzini, *On the group of components of a Néron model*, J. Reine Angew. Math. **445** (1993), 109–160.

[62] W. Lütkebohmert, *On compactification of schemes*, Manuscripta Math. **80** (1993), 95–111.

[63] W. McCallum, *The degenerate fibre of the Fermat curve*, in Number theory related to Fermat's last theorem, Prog. Math., **26**, Birkhäuser, Basle, 1982, 57–70.

[64] M. Matignon, J. Ohm, *A structure theorem for simple transcendental extensions of valued fields*, Proc. Am. Math. Soc. **104** (1988), 392–402.

[65] H. Matsumura, *Commutative Algebra,* 2nd edn, Benjamin, New York, 1980.

[66] J. S. Milne, *Étale cohomology*, Princeton Univ. Press, Princeton, New Jersey, 1980.

[67] ———, *Abelian Varieties*, Chapter V in *Arithmetic Geometry* [25], 103–150.

[68] ———, *Jacobian Varieties*, Chapter VII in *Arithmetic Geometry* [25], 167–212.

[69] D. Mumford, *The Red Book of Varieties and Schemes*, Lect. Notes Math., **1358**, Springer, New York–Heidelberg–Berlin, 1988.

[70] ———, *Abelian Varieties*, 2nd edition, Oxford Univ. Press, Oxford, 1974.

[71] ———, *The topology of normal singularities of an algebraic surface and a criterion for simplicity*, Publ. Math. IHES **9** (1961), 5–22.

[72] ———, J. Fogarty, F. Kirwan, *Geometric Invariant Theory*, Ergeb. Math. Grenz. **34**, 3rd enlarged edition, Springer, New York–Heidelberg–Berlin, 1994.

[73] M. Nagata, *Imbedding of an abstract variety in a complete variety*, J. Math. Kyoto Univ. **2** (1962), 1–10.

[74] Y. Namikawa, K. Ueno, *The complete classification of fibers in pencils of curves of genus two*, Manuscripta Math. **9** (1973), 143–186.

[75] A. Néron, *Modèles minimaux des variétés abéliennes sur les corps locaux et globaux.*, Publ. Math. IHES **21** (1964), 1–128.

[76] A. P. Ogg, *On pencils of curves of genus two*, Topology **5** (1966), 355–362.

[77] C. Peskine, *Une généralisation du "Main Theorem" de Zariski*, Bull. Sci. Math. **90** (1966), 119–127.

[78] M. Polzin, *Prolongement de la valeur absolue de Gauss et problème de Skolem*, Bull. SMF **116** (1988), 103–132.

[79] M. van der Put, *Stable reductions of algebraic curves*, Indag. Math. **46** (1984), 461–478.

[80] M. Raynaud, *Anneaux locaux henséliens*, Lect. Notes Math. **169**, Springer, New York–Heidelberg–Berlin, 1970.

[81] _____, *Spécialisation du foncteur de Picard*, Publ. Math. IHES **38** (1970), 27–76.

[82] _____, *p-groupes et réduction semi-stable des courbes*, in The Grothendieck Festschrift, Vol. **III**, Prog. Math., **88**, Birkhäuser, 1990, 179–197.

[83] M. Rosenlicht, *Automorphisms of function fields*, Trans. AMS **79** (1955), 1–11.

[84] T. Saito, *Vanishing cycles and geometry of curves over a discrete valuation ring*, Am. J. Math. **109** (1987), 1043–1085.

[85] J.-P. Serre, *Faisceaux algébriques cohérents*, Ann. Math. **61** (1955), 197–278.

[86] _____, *Groupes algébriques et corps de classes*, Hermann, Paris, 1959.

[87] _____, *Corps locaux*, Hermann, Paris, 1968.

[88] _____, J. Tate, *Good reduction of abelian varities*, Ann. Math. **88** (1968), 492–517.

[89] I. R. Shafarevich, *Basic Algebraic Geometry*, Vol 1–2, Springer, New York–Heidelberg–Berlin, 1994.

[90] _____, *Lectures on minimal models and birational transformations of two dimensional schemes*, Tata Inst. Fundam. Res., **37**, Bombay, 1966.

[91] J. Silverman, *The Arithmetic of Elliptic Curves*, Grad. Texts Math., **106**, Springer, New York–Heidelberg–Berlin, 1986.

[92] _____, *Advanced Topics in the Arithmetic of Elliptic Curves*, Grad. Texts Math., **151**, Springer, New York–Heidelberg–Berlin, 1994.

[93] J. Tate, *Algorithm for determining the type of a singular fiber in an elliptic pencil*, Lect. Notes Math. **476** (1975), 33–52.

[94] W. Waterhouse, *Introduction to affine group schemes*, Grad. Texts Math., **66**, Springer, New York–Heidelberg–Berlin, 1979.

[95] G. Winters, *On the existence of certain families of curves*, Am. J. Math. **96** (1974), 215–228.

[96] G. Xiao, *On the stable reduction of pencils of curves*, Math. Z. **203** (1990), 279–389.

[97] O. Zariski, P. Samuel, *Commutative Algebra*, Van Nostrand, Princeton, New Jersey, 1960.

Index

A

Abelian rank of a curve, 314, 316,
    521, 531, 555
Abelian variety, 298
    good reduction, 502
    models, 456
    potential good reduction, 502
    semi-Abelian reduction,
        see reduction
    torsion points, 299
Abhyankar's theorem on the
        desingularization of
        surfaces, 362
Abhyankar's theorem on the
        exceptional locus of
        birational morphisms, 411
absolute value, 467
Adjunction formula, 239
adjunction formula, 390
affine line, 27, 43
affine morphism, 172, 191
    and base change, 193
affine open subset, 44
    complement has codimension
        1, 124
    of a projective space, 76
    of an affine space, 76
affine scheme, 43
    and quasi-coherent
        sheaves, 160
    cohomology of sheaves, 186
    is separated, 100
    Serre's criterion, 187
affine space, 47
    open subset, 192
affine variety, 55
algebra, 5
    finite, 29
    finitely generated, 20, 29
    flat, 6
    graded, 20
    homogeneous, 53
    simple, 227
algebraic group, 307, 314;
        see also group scheme

algebraic set, 30
    and rational points, 49
algebraic variety, 55
    complete, 107
    Jacobian criterion of
        smoothness, 130
    proper, 105, 109, 112
    smooth, 141–142, 219, 220;
        see also smooth morphism
algebraization of coherent
        sheaves, 553
alteration, 361, 374, 407
ample divisor, 266; *see also* ample
        sheaf
    on a curve, 315
    the support is connected, 266
ample sheaf, 169–178, 194, 197, 198
    on a curve, 305
    on a projective scheme over a
        Dedekind scheme, 205
    on a proper scheme, 196
annihilator, 13, 173
approximation theorem, 379, 531
arcwise connectedness, 331
arithmetic surface, 353
    minimal, 418, 423–424
    relatively minimal, 418
    smooth, 368–370, 388, 418,
        526
    with normal crossings, 404,
        406–408, 425, 426, 530
Artin–Rees lemma, 21
associated point, 254
associated prime ideal, 253
automorphisms
    of a curve, 300
    of a minimal arithmetic
        surface, 418
    of a projective space, 176
    of a stable curve, 520, 529
    of models, 460

B

base change, 81
base point, 176
Betti number, 472

Z

Zariski closure, 56
Zariski tangent space, *see* tangent
    space
Zariski topology, 27

Zariski's connectedness
    principle, 200
Zariski's Main Theorem, 152
Zariski's purity theorem, 347
zero divisor, 32, 253
zeros divisor, 269

Milton Keynes UK
Ingram Content Group UK Ltd.
UKHW020013160924
448397UK00001B/1